Biofiber Reinforcement in Composite Materials

Related titles:

Residual stresses in composite materials
(ISBN 978-0-85709-270-0)

Natural fibre composites
(ISBN 978-0-85709-514-4)

Rehabilitation of metallic civil infrastructure using fiber reinforced polymer (FRP) composites
(ISBN 978-0-85709-653-1)

Woodhead Publishing Series in Composites Science and Engineering: Number 51

Biofiber Reinforcement in Composite Materials

Edited by
Omar Faruk and Mohini Sain

ELSEVIER

AMSTERDAM • BOSTON • CAMBRIDGE • HEIDELBERG • LONDON
NEW YORK • OXFORD • PARIS • SAN DIEGO
SAN FRANCISCO • SINGAPORE • SYDNEY • TOKYO
Woodhead Publishing is an imprint of Elsevier

WP WOODHEAD PUBLISHING

Woodhead Publishing is an imprint of Elsevier
80 High Street, Sawston, Cambridge, CB22 3HJ, UK
225 Wyman Street, Waltham, MA 02451, USA
Langford Lane, Kidlington, OX5 1GB, UK

British Library Cataloguing-in-Publication Data
A catalogue record for this book is available from the British Library

Library of Congress Control Number: 2014940871

ISBN 978-1-78242-122-1 (print)
ISBN 978-1-78242-127-6 (online)

For information on all Woodhead Publishing publications visit our website at http://store.elsevier.com/

Typeset by Toppan Best-set Premedia Limited

Printed and bound in the United Kingdom

Contents

Contributor contact details

(* = main contact)

Editors

O. Faruk and M. Sain
Centre for Biocomposites and Biomaterials Processing
University of Toronto
33 Willcocks Street
Toronto, Ontario, M5S 3B3, Canada

E-mail: o.faruk@utoronto.ca; m.sain@utoronto.ca

Chapter 1

J. A. Khan
Narsingdi Government College
National University of Bangladesh
Narsingdi – 1602, Bangladesh

M. A. Khan*
Institute of Radiation and Polymer Technology
Bangladesh Atomic Energy Commission
PO Box 3787
Dhaka – 1000, Bangladesh

E-mail: makhan.inst@gmail.com

Chapter 2

J. Müssig* and K. Haag
Hochschule Bremen – University of Applied Sciences
Biomimetics/The Biological Materials Group
Neustadtswall 30
D – 28199 Bremen, Germany

E-mail: joerg.muessig@hs-bremen.de

Chapter 3

H. N. Dhakal* and Z. Zhang
Advanced Polymer and Composites Research Group
School of Engineering
University of Portsmouth
Anglesea Road
Anglesea Building
Portsmouth, PO1 3DJ, UK

E-mail: hom.dhakal@port.ac.uk

Chapter 4

Y. Du and N. Yan*
Faculty of Forestry
University of Toronto
33 Willcocks Street
Toronto, Ontario, M5S 3B3, Canada

E-mail: yicheng.du@utoronto.ca; ning.yan@utoronto.ca

M. T. Kortschot
Department of Chemical Engineering and Applied Chemistry
University of Toronto
200 College Street
Toronto, Ontario, M5S 3E5, Canada

Chapter 5

H. M. Akil*, M. H. Zamri and M. R. Osman
School of Materials and Mineral Resources Engineering
Cluster for Polymer Composites (CPC)
Science and Engineering Research Centre (SERC)
Universiti Sains Malaysia (USM)
Engineering Campus
14300 Nibong Tebal
Pulau Pinang, Malaysia

E-mail: hazizan@eng.usm.my

Chapter 6

Y. Li* and Y. O. Shen
School of Aerospace Engineering and Applied Mechanics
Tongji University
1239 Siping Road
Shanghai, 200092, China

E-mail: liyan@tongji.edu.cn

Chapter 7

A. L. Leão*
Department of Rural Engineering
São Paulo State University (UNESP)
Botucatu 18610-307
São Paulo, Brazil

E-mail: alcideslopesleao@gmail.com

B. M. Cherian and S. Narine
Departments of Physics and Astronomy and Chemistry
Trent University
1600 West Bank Drive
Peterborough, Ontario, K9J 7B8, Canada

S. F. Souza and M. Sain
Centre for Biomaterials and Biocomposite Processing
33 Willcocks Street
University of Toronto
Toronto, Ontario, M5S 3B3, Canada

Chapters 8

A. A. Mamun* and H. P. Heim
Polymer Engineering
Institute of Material Engineering
University of Kassel
Mönchebergstrasse-3
34125 Kassel, Germany

E-mail: mamun@uni-kassel.de; mithu05@gmail.com

O. Faruk
Centre for Biocomposites and Biomaterials Processing
University of Toronto
33 Willcocks Street
Toronto, Ontario, M5S 3B3, Canada

A. K. Bledzki
Institute of Materials Science and Engineering
West Pomeranian University of Technology
19 Piastow Avenue
70310 Szczecin, Poland

Chapter 9

D. Kocak* and S. I. Mistik
Department of Textile Engineering
Faculty of Technology
Marmara University
34722 Istanbul, Turkey

E-mail: dkocak@marmara.edu.tr; imistik@marmara.edu.tr

Chapter 10

D. Verma*
Department of Mechanical Engineering
Indian Institute of Technology
BHU (Banaras Hindu University)
Varanasi – 221002, Uttar Pradesh, India

E-mail: dverma.mech@gmail.com

P. C. Gope
Department of Mechanical Engineering
College of Technology
Pantnagar, Uttarakhand – 263445, India

Chapter 11

S. K. Bajpai* and G. Mary
Department of Chemistry
Government Model Science College (Autonomous)
Jabalpur, Madhya Pradash – 482001, India

E-mail: mnlbpi@rediffmail.com; gracemary9@gmail.com

N. Chand
Advanced Materials and Processes Research Institute (AMPRI) (CSIR)
Habibganj Naka
Bhopal – 462026, India

E-mail: navinchaud15@yahoo.co.in

Chapter 12

M. D. H. Beg*, M. F. Mina, R. M. Yunus and A. K. M. Moshiul Alam
Faculty of Chemical and Natural Resources Engineering
Universiti Malaysia Pahang
Gambang 26300, Kuantan, Malaysia

E-mail: mdhbeg@ump.edu.my

Chapter 13

M. Bassyouni*
Department of Chemical and Materials Engineering
King Abdulaziz University
Rabigh 21911, Saudi Arabia

and

Department of Chemical Engineering
Higher Technological Institute
Zip code 11111, Tenth of Ramadan City, Egypt

E-mail: migb2000@hotmail.com

S. Waheed ul Hasan
Department of Chemical and Materials Engineering
King Abdulaziz University
Rabigh 21911, Saudi Arabia

Chapter 14

S. Panthapulakkal* and M. Sain
Department of Chemical Engineering and Applied Chemistry and Centre for Biocomposites and Biomaterials Processing
Faculty of Forestry
33 Willcocks Street
University of Toronto
Toronto, Ontario, M5S 3B3, Canada

E-mail: s.panthapulakkal@utoronto.ca

Chapter 15

A. A. Mamun* and H. P. Heim
Polymer Engineering
Institute of Material Engineering
University of Kassel
Mönchebergstrasse-3
34125 Kassel, Germany

E-mail: mamun@uni-kassel.de; mithu05@gmail.com

A. K. Bledzki
Institute of Materials Science and Engineering
West Pomeranian University of Technology
19 Piastow Avenue
70310 Szczecin, Poland

Chapter 16

H. P. S. Abdul Khalil*, M. S. Alwani and Y. M. H'ng
School of Industrial Technology
Universiti Sains Malaysia
11800 Penang, Malaysia

E-mail: akhalil@usm.my; akhalilhps@gmail.com

M. N. Islam
School of Industrial Technology
Universiti Sains Malaysia
11800 Penang, Malaysia

and

School of Life Science
Khulna University
Khulna – 9208, Bangladesh

S. S. Suhaily
School of Industrial Technology and Product Design Department
School of the Arts
Universiti Sains Malaysia
11800 Penang, Malaysia

R. Dungani
School of Life Sciences and Technology
Institut Teknologi Bandung
Gedung Labtex XI
Jalan Ganesha 10
Bandung 40132, West Java, Indonesia

and

School of Industrial Technology
Universiti Sains Malaysia
11800 Penang, Malaysia

M. Jawaid
Laboratory of Biocomposite Technology
Institute of Tropical Forestry and Forest Products (INTROP)
Universiti Putra Malaysia
43400 UPM Serdang
Selangor, Malaysia

Chapter 17

H. Hajiha* and M. Sain
Department of Chemical Engineering and Applied Chemistry and Centre for Biocomposites and Biomaterials Processing
Faculty of Forestry
33 Willcocks Street
University of Toronto
Toronto, Ontario, M5S 3B3, Canada

E-mail: hamideh.hajiha@mail.utoronto.ca

Chapter 18

R. A. Shanks
School of Applied Sciences
RMIT University
GPO Box 2476
Melbourne, VIC 3001, Australia

E-mail: robert.shanks@rmit.edu.au

Chapter 19

S. Bandyopadhyay-Ghosh* and S. B. Ghosh
Centre for Biocomposites and Biomaterials Processing
Faculty of Forestry
33 Willcocks Street
University of Toronto
Toronto, Ontario, M5S 3B3, Canada

and

Birla Institute of Technology and Science, Pilani
Pilani Campus
Rajasthan 333031, India

E-mail: sanchita.bandyopadhyayghosh@utoronto.ca

M. Sain
Centre for Biocomposites and Biomaterials Processing
Faculty of Forestry
33 Willcocks Street
University of Toronto
Toronto, Ontario, M5S 3B3, Canada

Chapter 20

L. M. Matuana*
School of Packaging
448 Wilson Road
Packaging Building
Michigan State University
East Lansing, MI 48824-1223, USA

E-mail: matuana@msu.edu

N. M. Stark
US Department of Agriculture
Forest Service
Forest Products Laboratory
One Gifford Pinchot Drive
Madison, WI 53705-2398, USA

E-mail: nstark@fs.fed.us

Chapter 21

D. Kocak*, S. I. Mistik and M. Akalin
Department of Textile Engineering
Faculty of Technology
Marmara University
34722 Istanbul, Turkey

E-mail: dkocak@marmara.edu.tr; imistik@marmara.edu.tr

N. Merdan
Department of Textile Engineering
Faculty of Engineering and Design
Istanbul Commerce University
Istanbul, Turkey

Chapter 22

S. F. Souza* and M. Ferreira
CCNH – Center of Natural and Human Science
Universidade Federal do ABC – UFABC
Av. dos Estados, 5001
Santo André – SP – Brazil, CEP 09210-580

E-mail: sivoneyfds@gmail.com

M. Sain
Centre for Biocomposites and Biomaterials Processing
Faculty of Forestry
33 Willcocks Street
University of Toronto
Toronto, Ontario, M5S 3B3, Canada

M. Z. Ferreira, H. F. Pupo, B. M. Cherian and A. L. Leão
Department of Rural Engineering
São Paulo State University (UNESP)
Botucatu 18610-307
São Paulo, Brazil

Editor biographies

Dr Omar Faruk

Dr Omar Faruk completed his B.S. and M.S. in Chemistry at the University of Chittagong, Bangladesh. With a DAAD (German Academic Exchange Service) scholarship, he joined at University of Kassel, Germany. He achieved his PhD in Mechanical Engineering at 2005. He worked at the Department of Forestry, Michigan State University, USA as a Visiting Research Associate from 2006 to 2009. Since 2010, he is working at the Centre for Biocomposites and Biomaterials Processing, University of Toronto, Canada. He has more than 70 publications (including a book) to his credit which have been published in different international journals and conferences. He is also an invited reviewer of 48 international reputed journals.

Professor Mohini Sain

Professor Mohini Sain is Dean and professor at Faculty of Forestry, University of Toronto. He specializes in advanced nanocellulose technology, biocomposites and bio-nanocomposites. He is cross-appointed to the Department of Chemical Engineering and Applied Chemistry. He is a fellow of Royal Society of Chemistry, UK. Besides, he is also an adjunct professor of the Chemical Engineering Departments at the University of New Brunswick, Canada; King Abdulaziz University, Jeddah Saudi Arabia; University of Guelph, Canada, University of Lulea, Sweden, Honorary Professor at Slovak Technical University and Institute of Environmental Science at the University of Toronto, and collaborates with American and European research institutes and universities. Prof. Sain holds several awards; the most recent one is the Plastic Innovation Award and KALEV PUGI Award for his innovation and contribution to Industry. Author of more than 300 papers and hi-cited researcher Professor Sain hugely contributed to society at large by translating research to commercialization which led to three new companies making products ranging from packaging to automotive to building construction.

Woodhead Publishing Series in Composites Science and Engineering

1 **Thermoplastic aromatic polymer composites**
F. N. Cogswell

2 **Design and manufacture of composite structures**
G. C. Eckold

3 **Handbook of polymer composites for engineers**
Edited by L. C. Hollaway

4 **Optimisation of composite structures design**
A. Miravete

5 **Short-fibre polymer composites**
Edited by S. K. De and J. R. White

6 **Flow-induced alignment in composite materials**
Edited by T. D. Papthanasiou and D. C. Guell

7 **Thermoset resins for composites**
Compiled by Technolex

8 **Microstructural characterisation of fibre-reinforced composites**
Edited by J. Summerscales

9 **Composite materials**
F. L. Matthews and R. D. Rawlings

10 **3-D textile reinforcements in composite materials**
Edited by A. Miravete

11 **Pultrusion for engineers**
Edited by T. Starr

12 **Impact behaviour of fibre-reinforced composite materials and structures**
Edited by S. R. Reid and G. Zhou

13 **Finite element modelling of composite materials and structures**
F. L. Matthews, G. A. O. Davies, D. Hitchings and C. Soutis

14 **Mechanical testing of advanced fibre composites**
Edited by G. M. Hodgkinson

15 **Integrated design and manufacture using fibre-reinforced polymeric composites**
Edited by M. J. Owen and I. A. Jones

16 **Fatigue in composites**
Edited by B. Harris

17 **Green composites**
Edited by C. Baillie

18 **Multi-scale modelling of composite material systems**
Edited by C. Soutis and P. W. R. Beaumont

19 **Lightweight ballistic composites**
Edited by A. Bhatnagar

20 **Polymer nanocomposites**
Y-W. Mai and Z-Z. Yu

21 **Properties and performance of natural-fibre composite**
Edited by K. Pickering

22 **Ageing of composites**
Edited by R. Martin

23 **Tribology of natural fiber polymer composites**
N. Chand and M. Fahim

24 **Wood-polymer composites**
Edited by K. O. Niska and M. Sain

25 **Delamination behaviour of composites**
Edited by S. Sridharan

26 **Science and engineering of short fibre reinforced polymer composites**
S-Y. Fu, B. Lauke and Y-M. Mai

27 **Failure analysis and fractography of polymer composites**
E. S. Greenhalgh

28 **Management, recycling and reuse of waste composites**
Edited by V. Goodship

29 **Materials, design and manufacturing for lightweight vehicles**
Edited by P. K. Mallick

30 **Fatigue life prediction of composites and composite structures**
Edited by A. P. Vassilopoulos

31 **Physical properties and applications of polymer nanocomposites**
Edited by S. C. Tjong and Y-W. Mai

32 **Creep and fatigue in polymer matrix composites**
Edited by R. M. Guedes

33 **Interface engineering of natural fibre composites for maximum performance**
Edited by N. E. Zafeiropoulos

34 **Polymer-carbon nanotube composites**
Edited by T. McNally and P. Pötschke

35 **Non-crimp fabric composites: Manufacturing, properties and applications**
Edited by S. V. Lomov

36 **Composite reinforcements for optimum performance**
Edited by P. Boisse

37 **Polymer matrix composites and technology**
R. Wang, S. Zeng and Y. Zeng

38 **Composite joints and connections**
Edited by P. Camanho and L. Tong

39 **Machining technology for composite materials**
Edited by H. Hocheng

40 **Failure mechanisms in polymer matrix composites**
Edited by P. Robinson, E. S. Greenhalgh and S. Pinho

41 **Advances in polymer nanocomposites: Types and applications**
Edited by F. Gao

42 **Manufacturing techniques for polymer matrix composites (PMCs)**
Edited by S. Advani and K-T. Hsiao

43 **Non-destructive evaluation (NDE) of polymer matrix composites: Techniques and applications**
Edited by V. M. Karbhari

44 **Environmentally friendly polymer nanocomposites: Types, processing and properties**
S. S. Ray

45 **Advances in ceramic matrix composites**
Edited by I. M. Low

46 **Ceramic nanocomposites**
Edited by R. Banerjee and I. Manna

47 **Natural fibre composites: Materials, processes and properties**
Edited by A. Hodzic and R. Shanks

48 **Residual stresses in composite materials**
Edited by M. Shokrieh

49 **Health and environmental safety of nanomaterials: Polymer nanocomposites and other materials containing nanoparticles**
Edited by J. Njuguna, K. Pielichowski and H. Zhu

50 **Polymer composites in the aerospace industry**
Edited by P. E. Irving and C. Soutis

51 **Biofiber reinforcement in composite materials**
Edited by O. Faruk and M. Sain

52 **Fatigue and fracture of adhesively-bonded composite joints: Behaviour, simulation and modelling**
Edited by A. P. Vassilopoulos

Preface

Recent trends such as increasing oil prices, depletion of fossil resources and increasing greenhouse gas emissions have encouraged the development of new biodegradable materials produced from renewable resources. In this respect natural fiber-reinforced polymer composites have been developed to replace synthetic composites. There are more than 1000 species of cellulose plants available in fiber form and a number of them are being investigated as composite reinforcement materials. This is part of an increasing interest in investigating new biofibers from a range of sources (such as petiole bark, rachis, rachilla, spatha, palmyrah, talipot, *Sansevieria cylindrica*, sea grass, coconut tree leaf sheath, vakka, okra, elephant grass, abaca leaf fiber, *Sansevieria rifasciata*, *Phormium tenax*, alfa, piassava, isora, *Sansevieria ehrenbergii*, sunflower stalk flour and *Opuntia ficus indica*). Composites with biofibers as reinforcements have potential applications as low-cost building materials, automobile components and other biomedical applications.

There has been research in biocomposites for well over a decade which has demonstrated such advantages of cellulosic fibers as excellent stiffness and strength. However, this has not led to the hoped-for range of applications because of their drawbacks. One problem is variability in fiber quality due to factors such as variations in plant growth, harvesting and extraction. Problems of interfacial adhesion between biofiber and polymer matrix, moisture absorption and long-term durability (affected by ultraviolet light, temperature, and humidity) are also important issues which needed to be resolved.

In recent years, there have been a number of books published on biofiber-reinforced composites covering general processing, properties, performance criteria and applications. This book focuses specifically on biofibers as reinforcements in composite materials. The main biofibers are sub-categorized based on their origin (Part I Bast fibers, Part II Leaf fibers, Part

III Seed fibers, Part IV Grass, reed and cane fibers, and Part V Wood, cellulosic and other fibers including cellulosic nanofibers). Chapters on a specific biofiber review their sources and cultivation, production, fiber properties and modification, integration into matrices, performance and current applications. The book will be helpful to researchers, engineers, chemists, technologists and professionals who would like to know more about the development and potential of natural fiber-reinforced composites.

Omar Faruk and Mohini Sain

Part I

Bast fibers

1 The use of jute fibers as reinforcements in composites

J. A. KHAN, National University of Bangladesh, Bangladesh and M. A. KHAN, Bangladesh Atomic Energy Commission, Bangladesh

DOI: 10.1533/9781782421276.1.3

Abstract: This chapter discusses the physico-mechanical properties of jute-reinforced polymer composites and the applications of jute composites in different areas. The chapter describes the various types of surface modifications such as grafting, mercerization, oxidation, ultraviolet and gamma radiation, etc., which are commonly used to improve the interfacial adhesion between jute fibers and polymer matrices. Jute hybrid composites, interfacial bonding characteristics of grafted jute fibers and fabrication of jute composites are also included.

Key words: jute-reinforced polymer composites, physico-mechanical properties of jute composites, applications of jute composites, surface modifications of jute fibers, jute hybrid composites, interfacial adhesion, fabrication of jute composites.

1.1 Introduction

Jute is extracted from the stem of the jute plant which belongs to the genus *Corchorus*, family Tiliaceae. Only two species of *Corchorus*, *C. capsularis* L. and *C. olitorius* L., are grown commercially. *Corchorus capsularis* is known as white jute whilst *Corchorus olitorius* is known as tossa jute. Olitorius and capsularis jute have very similar chemical compositions [1–3]. Jute plants grow to about 2.5 to 3.5 meters in height. The fiber runs along the length of the stem in the form of a lacework sheath. The fiber is removed from the stem by a process of biological retting.

Jute is grown in South Asia. The major jute-producing countries are Bangladesh, India, China and Thailand. Bangladesh provides over 90% of the world's raw jute and allied fiber exports. Jute is the second most important vegetable fiber after cotton, in terms of usage, global consumption, production and availability. It is one of the cheapest and the strongest of all natural fibers [2, 4]. Jute fiber has traditionally been used for the manufacture of sacks, hessian cloth, carpet and twines, ropes and cords. Jute fiber is used as a reinforcing material in the automotive, construction and packaging industries [5–8].

Table 1.1 Energy inputs and greenhouse gas (GHG) outputs for PP plastic resin and jute hessian [10]

Material	Energy (GJ/1000 kg)	GHG (tonnes CO_2 eq.)
Polypropylene (PP)	63	1340
Jute hessian	2	0.15

Jute fiber is biodegradable and eco-friendly. Jute products compare well with other fibers in terms of energy use, greenhouse gas emissions, eutrophication and acidification. It has been reported that one hectare of jute plants absorbs 15 tons of CO_2 from the atmosphere and adds 11 tons of O_2 during their lifespan of 120 days. Moreover, the decomposed leaves and roots of jute plants increase the fertility of the soil, reducing fertilizer costs. It was reported that the manufacture of 1 kg of fabric of jute shopping bags saves 80 MJ of energy in comparison to 1 kg of polyhydroxyalkanoid (PHA) [9]. Jute hessian cloth consumes lesser amounts of energy and emits negligible amounts of greenhouse gas (GHG) in comparison to thermoplastic polypropylene resin [10] (Table 1.1). Jute fiber is renewable and cheap.

1.2 Composition and properties of jute fibers

The chemical composition of jute fiber is as follows [11]:

- Cellulose (61–71%)
- Hemicelluloses (13.6–20.4%)
- Lignin (12–13%)
- Ash (0.5–2%)
- Pectin (~0.2%)
- Wax (~0.5%)
- Moisture (~12.6%).

The different structural components of a fiber play an important role in the properties of the fiber. High cellulose content and low microfibril angle are the desirable properties of a fiber to be used as reinforcement in a polymer matrix. It is notable that jute fiber has high cellulose content and low microfibril angle (~8.0°) [11]. The crystalline portion of cellulose is resistant to strong alkali (17.5 wt%) but is easily hydrolyzed by acid. Cellulose is relatively resistant to oxidizing agents. The cellulose fibrils provide rigidity and high tensile and flexural strength. The fibrils are composed of microfibrils which form a microfibrillar angle with respect to the fiber axis. The microfibrillar angle determines the stiffness of the fibers which in turn governs the mechanical properties of the composite. Low

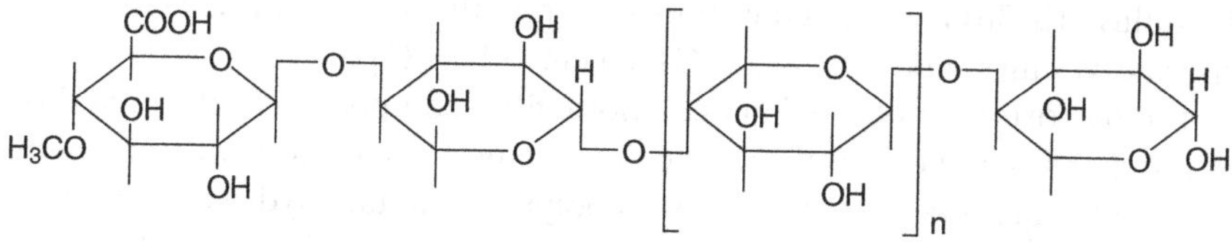

1.1 Structure of β-D-xylopyranose with terminal α-D-4-*o*-methylglucuronic acid, a main constituent of jute hemicelluloses.

microfibril angle makes the fiber more rigid, inflexible and mechanically more strong. The value of the microfibrillar angle varies from one fiber to another [11–13].

The large number of hydroxyl groups in cellulose gives a hydrophilic character to jute fiber which is responsible for its poor compatibility with hydrophobic polymer matrices and also for its dimensional instability. However, these hydroxyl groups make the fibers more reactive toward the different surface modifiers. The reinforcing efficiency of jute fiber is related to the nature of cellulose and its crystallinity.

The hemicellulose fraction of plant fibers consists of several different sugar units. The principal constituent of jute hemicellulose is composed of the backbone of β-D-xylopyranose units with every seventh unit carrying a terminal α-D-4-*o*-methylglucuronic acid residue (Fig. 1.1). Hemicellulose is very hydrophilic, soluble in alkali and easily hydrolyzed in acids [14–16].

Lignin is an amorphous, highly complex substance, consisting mainly of aromatic phenylpropane units. It is very susceptible to oxidation and readily undergoes condensation reactions [17]. Jute lignin contains predominantly synapyl alcohol (syringyl propane). A small portion of jute lignin dissolves in dilute alkali at room temperature [1]. Lignin is considered to be responsible for ultraviolet (UV) degradation and is also known to resist microbial degradation [15, 16].

The properties of jute fiber, as with other plant fibers, depend mainly on the nature of the plant, the locality in which it is grown, the age of the plant, and the extraction methods used. Along with the individual fiber properties, the fiber aspect ratio, the volume fraction of fibers and the orientation of fibers must be considered in the design of a fiber-reinforced composite [18–20]. Fibers like jute with a high aspect ratio have a large surface to volume ratio. The large bonding surface area provides a greater opportunity for contact and bonding with the matrix material, greater stress transfer to the fiber component and, ultimately, greater composite strength.

Jute fiber has low density and is light compared to glass, carbon and aramid fibers. Key properties include high specific strength and stiffness. The various mechanical parameters of jute fiber are as follows: density 1.3–1.46 $g.cm^{-3}$), elongation 1.5–1.8%, tensile strength 393–800 MPa, Young's modulus 10–30 GPa, specific tensile strength 302–547 MPa/g.cm^{-3} and specific Young's modulus 8–20.5 GPa/g.cm^{-3}) [21–23].

The performance of any lignocellulosic fiber composite is determined by the properties of the fiber itself. Jute fiber changes dimensions with changing moisture content because the cell wall polymers contain hydroxyl and other oxygen-containing groups that attract moisture through hydrogen bonding.

Jute fiber is degraded biologically by organisms. Biodegradation of the high-molecular-weight cellulose weakens the fiber cell wall. Strength is lost as the cellulose polymer undergoes degradation through oxidation, hydrolysis, and dehydration reactions. Jute fiber which is exposed outdoors undergoes photochemical degradation caused by ultraviolet light. Jute fiber composites are also affected by environmental conditions. Exposure to UV can cause changes in the surface chemistry of the composite, which may lead to discoloration, making the products aesthetically unappealing. Prolonged UV exposure may ultimately lead to loss in mechanical integrity [24–26].

There are therefore several disadvantages associated with jute fibers as a reinforcement in polymer matrices. Due to the presence of hydroxyl and other oxygen-containing groups in the fiber, jute fibers are polar and hydrophilic. Polymer matrices are mostly non-polar thermoplastics. This results in poor dispersion and interfacial adhesion between the fiber and matrix phases. This is a major disadvantage of jute fiber-reinforced composites. High moisture absorption is another drawback of jute fibers which results in poor mechanical properties and reduces dimensional stability of the composites. Jute fiber is vulnerable to degradation at higher processing temperature (above 250°C) which restricts the choice of matrices. The variations in properties within the same fiber also create problems in producing composites with uniform properties.

The strong interfacial adhesion between the matrix and reinforcement phases, essential for the transfer of load from the former to the latter, requires surface modification of jute fibers. Surface characteristics, such as wetting, adhesion, surface tension and porosity of the fibers, can be improved by the modification of the jute surface. The irregularities of the fiber surface play an important role in the mechanical interlocking at the interface. The interfacial properties can be improved by making appropriate modifications to the components, which gives rise to changes in the physical and chemical interactions at the interface. The different surface modifications of jute fiber have achieved various levels of success in improving fiber strength and fiber/matrix adhesion in jute fiber composites. Different types of surface modification have been carried out to improve the mechanical properties of jute fiber, mainly grafting and mercerization. Other chemicals have also been successfully employed for the development of the mechanical properties of jute fiber.

1.3 Processing and properties of grafted jute fibers

Jute yarns have been modified by photo-grafting and photo-curing with different types of acrylic monomers such as the following:

- Acrylamide (AA)
- 2-Hydroxyethylmethacrylate (HEMA)
- Methylacrylate (MA)
- Ethylacrylate (EA)
- 2-Hydroxyethylacrylate (HEA)

The effects of monomer concentration and irradiation time on polymer loading (PL), tensile factor (T_f) and elongation factor (E_f) of jute yarn have been studied extensively. The tensile factor is the ratio of the tensile strength of the treated and untreated jute yarn. The elongation factor is calculated in the same way. Monomer solutions were prepared in methanol together with photoinitiators with different formulations. Jute yarns were soaked in these solutions and irradiated with different levels of UV radiation.

Grafting of AA onto jute yarn has been carried out by UV radiation together with the photoinitiator in order to improve the physico-mechanical properties of jute yarn [27–29]. Jute yarns were treated with different concentrated solutions of AA in methanol (5–50% w/v) and for different irradiation times (15–300 minutes). The highest tensile factor T_f (2.00) and elongation factor E_f (2.56) are obtained after 60 minutes of irradiation and at 30% AA. The highest tensile factor T_f was obtained with the samples with a PL value of 22%. At the highest PL value T_f was 1.78. The decrease of T_f with the increase of the PL value may be due to the formation of excess three-dimensional crosslinked structures in the grafting zone of the cellulose, which form a brittle polymer product [30, 31].

A small amount (1%) of additives such as 3-(trimethoxysilyl) propylmethacrylate (silane), urea (U), poly *N*-vinylpyrrolidone (PNVP), urethane acrylate (UA) and urethane diacrylate (UDA) were used in the AA (30%) solution. Among the additives used, silane and urea significantly influenced the PL, tensile strength (TS) and elongation at break (E_b) values of AA-treated jute yarns [32]. The tensile properties of jute yarn are found to increase with the addition of urea and PNVP in AA. In the case of UA and UDA, the T_f values fall with the increase of grafting, which may be caused by the more brittle character of the yarns at higher grafting. The E_b factor of all additive-treated samples except silane remains above unity, indicating that the yarns are stretchable even at high grafting [33].

The jute yarns were treated with 3–20% HEMA solutions and with irradiation periods of 5–120 min [34]. The highest PL (10.9%) is obtained with 30 min of irradiation, whereas the highest T_f (1.6) is achieved with 20 min of irradiation at 3% HEMA. At the highest T_f, the PL value is about 5%. Some of the samples, particularly the sample containing 20% HEMA irradiated beyond 40 min, attain a tenacity that is even lower than that of the virgin jute sample. Towards the maximum PL values, the jute samples become brittle and break very easily. This may be one of the reasons why the highest TS values are not obtained with the samples containing the highest PL values. The maximum E_b values (1.6–1.8) are obtained mostly after 30 min of irradiation at any concentration of the monomer solution. The highest E_b is attained by the sample treated with 5% HEMA and the lowest was observed with the 3% HEMA sample.

The elongation of the yarn may be enhanced with a small (1%) addition of urethane acrylate, and simultaneous irradiation and grafting, as well as preirradiation grafting of jute fibers under UV radiation with 1-hydroxycyclohexyl-phenylketone as the photoinitiator, produced significant improvement in the mechanical properties of jute fibers [35]. The latter method produced up to 76% graft weight compared to the 45% otherwise obtained. Jute samples grafted with poly(hydroxyethyl methacrylate) at a level of 12% graft weight exhibited up to a 20% increase in hydrophilicity.

Jute yarns were also grafted with different types of vinyl monomers with different functionalities such as methylacrylate (MA), ethylacrylate (EA) and 2-hydroxyethylacrylate (HEA) [36–38]. MA, EA and HEA produced enhanced tensile strengths of 87, 78 and 85%, respectively. The monomers MA, EA and HEA showed improved elongations at break of 118, 91 and 76%, respectively.

1.4 Processing and properties of alkali-treated jute fibers

In a study by Hasan *et al.* [39], jute yarns were pretreated with alkali (5% NaOH) and were grafted with two types of monomer such as 3-(trimethoxysilyl)propylmethacrylate (silane) and acrylamide (AA) under UV radiation of different intensities. The alkali-treated silane- and AA-grafted jute yarn produced enhanced TS and E_b than that of the virgin fiber.

Jute yarns were further pretreated by alkali and either UV or gamma radiation at different intensities, then grafted with silane and AA. The jute yarns that were pretreated with alkali and UV radiation and grafted with silane showed the best improvement in properties. Alkali treatment increases the amorphous region via the dissolution and leaching out of fatty acids and some other lignin components from jute yarns. The jute yarns pretreated under UV conditions exhibited better tensile properties than

Table 1.2 Mechanical properties of acrylic monomer grafted jute fiber

Type of monomer	Concentration of monomer (%)	Irradiation time (min)	T_f	E_f	PL (%)[a]
AA	30	60	2.00	2.56	22.0
HEMA	3	20	1.60	–	5.0
Silane	30	30	2.60	3.40	26.2
Alkali + silane	30	30	3.65	4.50	27.0
Alkali + AA	30	60	3.10	3.70	23.0
HDDA	5	5	1.60	–	11.0
Alkali + UV + HDDA	5	5	2.03	1.46	12.0

[a] PL values are at highest T_f.

Table 1.3 Percent loss of weight and shrinkage in dimension of jute fabrics during mercerization

NaOH (%)	Loss of weight (%)			Shrinkage in length (%)			Shrinkage in width (%)		
	30 min	60 min	90 min	30 min	60 min	90 min	30 min	60 min	90 min
5	13.7	15.1	16.3	6.3	6.9	7.1	8.1	12.5	12.8
10	17.4	18.6	19.8	13.6	13.9	14.1	14.8	16.8	17.1
20	17.6	19.3	20.4	23.9	25.4	26.0	21.8	24.2	24.5

those pretreated with gamma radiation. This is because, in the latter case, the jute material loses strength due to the high degree of polymer loading resulting in brittleness.

Jute yarns have also been pre-soaked in hexanedioldiacrylate (HDDA) prior to treatment with an alkali (5% NaOH) solution and then irradiated with either UV or gamma radiation at varying intensities. These treatments were able to significantly improve the mechanical properties in comparison to virgin jute yarn [40, 41]. and the best values in terms of mechanical properties (T_f and E_f) of various acrylic monomer grafted jute fibers under optimum conditions of monomer concentration and irradiation time are shown in Table 1.2. The PL values are shown at the maximum T_f.

Mercerization is an important method for the surface modification of jute fiber. Mercerization has a significant effect on the crystallinity, fineness, textile properties and dimensions of the fiber, the magnitude of which depends on the strength of the alkali solution, treatment time and temperature. The effect of mercerization on the weight and dimension of jute fabrics was studied by Khan *et al.* [42]. It is observed that loss of weight and shrinkage increases with increasing soaking time and strength of the NaOH solution (Table 1.3). The final properties of the fabric are also dependant on the process temperature, as shown in (Table 1.4).

Table 1.4 Percent reduction of weight and dimension of jute fabrics during mercerization at different temperatures

Temperature (°C)	Reduction (%)		
	Weight	Length	Width
0	19.5	28.6	31.3
30	19.3	25.4	24.2
70	18.1	27.0	25.0

Ray *et al.* [43–45] studied the effect of mercerization on the properties of jute fiber for different treatment times (2–8 h). The desorption of moisture and the degradation of hemicelluloses were observed during mercerization at 5% NaOH. The fibers treated with alkali for 6 and 8 h exhibited an increase in crystallinity which reduces loss of moisture [43]. A milder alkali concentration was observed to increase the linear density, strength and tenacity of the fibers as higher concentrations lower the fibre crystallinity.

1.5 Characterization of jute fibers

Jute fibers both untreated and treated (silane and acrylic monomers) were characterized by X-ray photoelectron spectrometry (XPS) and Fourier-Transform Infrared Spectroscopy (FTIR). It was observed that both silane [46] and acrylic monomers such as HEMA [47] might be reacted with or deposited on the jute surface.

Untreated jute yarn (JY) and acrylic monomer-treated (AMJY) and silane-treated (SJY) jute yarn samples were characterized by FTIR in order to understand the chemical reaction between the monomers and the cellulose of the jute fibers. The characteristic absorption peak of SJY, observed at around 766 cm^{-1}, could be attributed to the Si–C stretching bond. A weak peak found at 847 cm^{-1} also corresponds to an Si–C bond, and a broad peak, apparent in the range 925–1105 cm^{-1}, could possibly be due to asymmetric stretching of an Si–O–Si or Si–O–C (1014–1090 cm^{-1}) bond. Such an absorption band for Si–O–Si is an indication of the presence of polysiloxane deposition on the jute fibers. A distinct absorption peak is also observed at around 1200 cm^{-1}, which corresponds to the Si–O–C bond.

The characteristic absorption band of the carbonyl group (>C=O) of lignocellulose fiber at 1730 cm^{-1} is observed in both virgin jute and acrylic monomer (AM) treated jute yarns. The intensity of the OH group of the AM treated jute is lower, and the intensity of the carbonyl group is found to be higher, by comparison to virgin jute. A sharp band observed at

Table 1.5 XPS analysis of surface composition of treated and untreated jute fibers

Element	Concentration (%)		
	JY	AMJY	SJY
C_{1s}	55.04	36.82	33.98
O_{1s}	39.38	50.68	51.55
Ca_{2p}	0.87	0.48	0.53
N_{1s}	4.71	2.02	3.22
Si_{2p}	None	None	10.72

1539 cm^{-1} may be attributed to C=C stretching of the AM. Another band at 1396 cm^{-1} corresponds to CH_3 deformation of the AM. The above peaks are indicative of the existence of AM deposition onto the cellulose backbone of the jute fiber.

The chemical environments of untreated (JY) and grafted jute (AMJY and SJY) yarns were analyzed by X-ray photoelectron spectroscopy and the atomic concentrations of carbon, oxygen, nitrogen and Si obtained for both treated and untreated fibers are presented in Table 1.5. The carbon concentration in the virgin jute yarn is higher and the oxygen concentration is lower than that of the grafted jute yarns. The carbon to oxygen ratios of treated and untreated samples are 0.73 and 1.40, respectively. The nitrogen concentrations in the treated jute fibers are lower compared to untreated fibers. From these findings it can be ascertained that AM deposition on the jute surface occurred, or that a chemical reaction took place with the cellulose backbone of the jute fibers.

1.6 Manufacture of jute fiber composites

The most common methods for fabrication of jute fiber composites are hand lay-up, compression molding and injection molding, all of which involve placing of the uncured composite into or onto a mold so that the material can be shaped into the final part. Different methods are suitable for the manufacture of thermoset and thermoplastic matrix composites.

1.6.1 Hand lay-up

In this method, jute fabric is laid into or onto a mold and the resin is sprayed, brushed or poured over the mat. Al-Kafi *et al.* [48] prepared jute fabric (hessian cloth) and E-glass fiber (mat) hybrid composites using the hand lay-up technique with unsaturated polyester (USP) resin. The working surfaces of the mold were first treated with waxes to facilitate easy removal of the samples from the mold surfaces. Cobalt naphthenate (catalyst) and methyl ethyl ketone peroxide, MEKP (curing agent) were then mixed with the USP with various formulations. At the beginning of fabrication, a gel coat with 2% MEKP was uniformly brushed onto the male and female parts of the mold, and after 1 h, when curing of the gel coat was complete, each layer of the fiber was pre-impregnated with formulations of USP and the samples were then placed one over another. This sandwich was placed into the mold, both parts of which were tightened and given 3 h for curing.

1.6.2 Compression molding

This is a conventional and simple method for the manufacture of jute fiber-reinforced composites and is also known as 'hot-pressed' molding. Use of molds allows for the production of composites with simple shapes and curved surfaces. In this method, the sandwich is prepared by placing jute fabric or yarn between the polymer sheets. This is then laid inside the mold and hot-pressed at pressure. The fabrication temperature in the hot-press machine depends on the nature of the polymer used. For example, the fabrication temperature of jute–PP composites is maintained at 180–190°C, jute–polycarbonate composites at 200°C, and jute fiber–polyester amide composites at 135°C. The sandwich is kept under a pressure of 4–5 MPa (in some cases 20 MPa) for about 5 min during which time the resin melts into the reinforcement phase and is cured.

1.6.3 Injection molding

This is one of the most versatile and widely used methods for producing relatively complex shapes with excellent accuracy and at high volume production. Before feeding into the injection mold, the jute and polymer are extruded [49–51] or poltruded [52, 53] into pellets. The pellets are then injected into the cavity of a closed metallic die at high pressure.

1.7 Preparation and properties of irradiated jute composites

The effects of gamma radiation on the mechanical properties of jute fabric-reinforced polypropylene (PP) composites were studied by Khan *et al.* [54] and Zaman *et al.* [55]. These authors investigated the composite properties obtained for a variety of radiation doses and for different combinations of irradiation of the matrix and/or reinforcement phases.

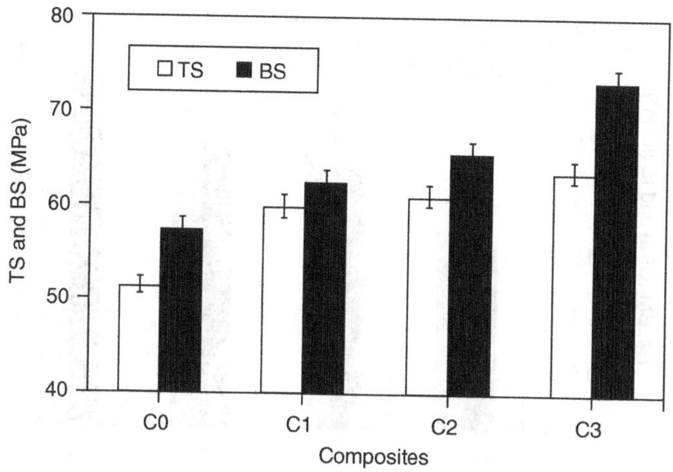

1.2 Tensile and bending strength of various types of composites at 500 krad of gamma dose. Key: TS = tensile strength; BS = bending strength; C0 = non-irradiated jute fabric/non-irradiated PP; C1 = non-irradiated jute fabric/pre-irradiated PP; C2 = pre-irradiated jute fabric/non-irradiated PP; C3 = pre-irradiated jute fabric/pre-irradiated PP.

The results obtained are summarized in Figs 1.2 and 1.3. For doses between 250 and 1000 krad, 500 krad appears to be the optimal level of irradiation.

The ionizing gamma radiation results in the generation of three reactive species: ionic, radical and peroxide. Peroxides occur when the polymers are irradiated in the presence of oxygen. High-energy free radicals will be formed which may react with the polymer in a manner that affects its structure and properties. Such mechanisms are shown in Figs 1.4, 1.5 and 1.6 and indicate that bonding between the reinforcement and matrix phase is improved by pre-irradiation.

For UV and gamma treatments, 100 passes of UV radiation and 500 krad of gamma dose produced the best mechanical properties in the composites. The UV-treated jute fabric–PP composites showed higher values of tensile strength (TS), tensile modulus (TM), bending strength (BS), bending modulus (BM) and impact strength (IS) when compared to the gamma-treated jute fabric–PP composites. UV radiation simultaneously causes photo crosslinking and photodegradation in the polymeric materials and crosslinking between the neighboring radical species may be responsible for the enhanced mechanical properties of these composites. However, photodegradation of cellulose molecules causes the opposite phenomenon. At lower intensities of UV radiation, free radicals produced from the

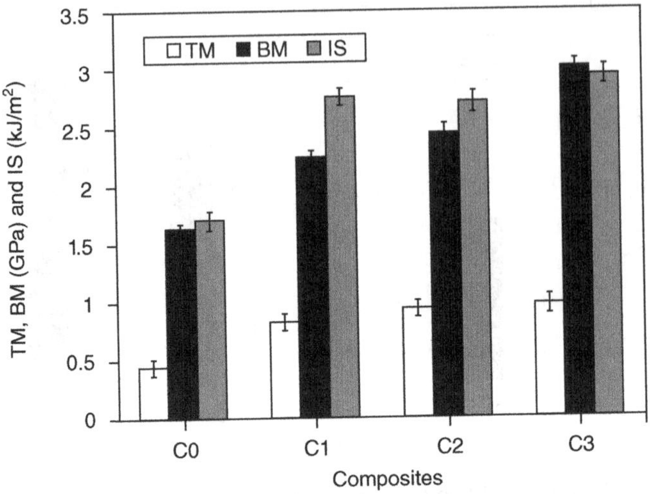

1.3 Tensile and bending modulus and impact strength of various types of composites at 500 krad of gamma dose. Key: TM = tensile modulus; BM = bending modulus; IS = impact strength; C0 = non-irradiated jute fabric/non-irradiated PP; C1 = non-irradiated jute fabric/pre-irradiated PP; C2 = pre-irradiated jute fabric/non-irradiated PP; C3 = pre-irradiated jute fabric/pre-irradiated PP.

Initiation:

$$RH \xrightarrow{\gamma\text{-radiation}} R^{\bullet} \quad \cdots\cdots \quad (1)$$

Propagation:

$$R^{\bullet} + O_2 \longrightarrow ROO^{\bullet} \quad \cdots\cdots \quad (2)$$

$$RH + ROO^{\bullet} \longrightarrow ROOH + R^{\bullet} \quad \cdots\cdots \quad (3)$$

$$ROOH \longrightarrow RO^{\bullet} + {}^{\bullet}OH \quad \cdots\cdots \quad (4)$$

$$2ROOH \longrightarrow ROO^{\bullet} + RO^{\bullet} + H_2O \quad \cdots\cdots \quad (5)$$

$$RO^{\bullet} + RH \longrightarrow ROH + R^{\bullet} \quad \cdots\cdots \quad (6)$$

$$HO^{\bullet} + RH \longrightarrow H^{\bullet} + H_2O \quad \cdots\cdots \quad (7)$$

Termination:

$$R^{\bullet} + R^{\bullet} \longrightarrow R_2 \quad \cdots\cdots \quad (8)$$

$$R^{\bullet} + ROO^{\bullet} \longrightarrow ROOR \quad \cdots\cdots \quad (9)$$

$$ROO^{\bullet} + ROO^{\bullet} \longrightarrow ROOR + O_2 \quad \cdots\cdots \quad (10)$$

RH = Jute cellulose

$R^{\bullet}$ = Possible free radicals formed by the abstraction of H and OH from jute cellulose, cleavage of C2–C3 bonds and chain scission of cellulose backbone

1.4 Possible free radical mechanism of jute cellulose in presence of O_2 and gamma radiation.

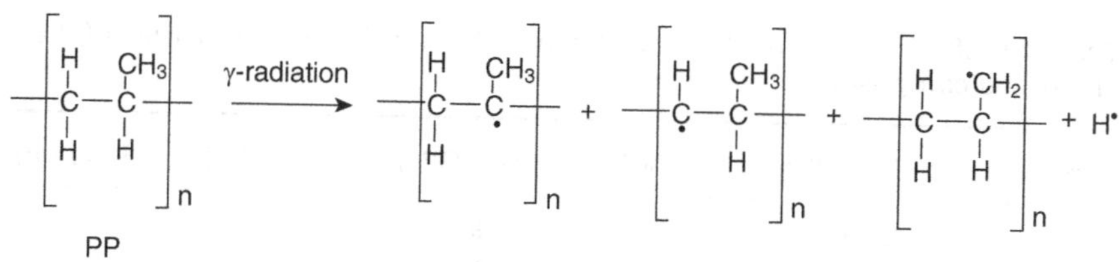

1.5 Free radical formation from polypropylene in presence of gamma radiation.

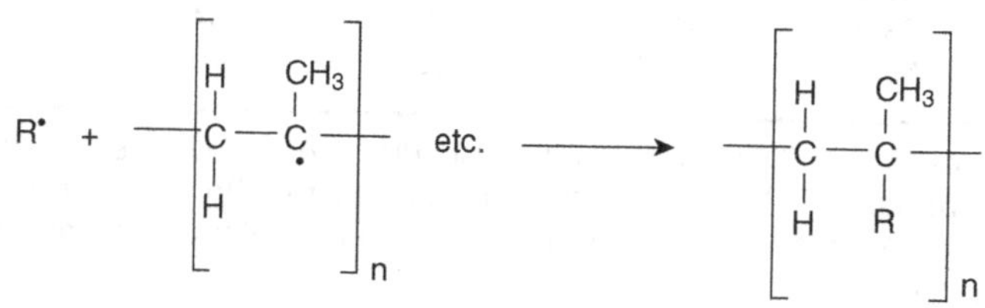

1.6 Reactions between jute free radicals and PP free radicals.

polymer molecules are stabilized by a combination reaction and, as a result, photo crosslinking occurs. The higher the number of active sites generated on the polymeric substrate, the greater the grafting efficiency. But at higher radiation intensities, the main chain may be broken down and the polymer may degrade, together with the properties of the composite.

Soaking with starch may be used to enhanced the properties of both gamma and UV irradiated composites. Stress transfer from the fiber to the matrix is improved in the presence of starch, although at higher starch concentrations (10% w/v) the fibers become sticky and adhesion to the matrix becomes worse. Table 1.6 shows properties of starch-treated composites, from which it can be seen that the optimal properties were achieved with the starch and UV-treated composite. Similar results were seen by Khan *et al.* [56] who, however, employed only gamma-treated materials. These results are summarized in Fig. 1.7.

Acrylate monomers in the presence of a photoinitiator and possibly also plasticizers may also be used to enhanced the properties of irradiated composite materials, and further information on this subject may be found in References 57–60 as well as in Table 1.7 and 1.8.

1.8 Preparation and properties of oxidized jute composites

Oxidation is an effective method to bring about chemical as well as physical changes in cellulosic materials. The chemical and physical properties of the oxidized products depend on the nature of the oxidizing agents. Surface

Table 1.6 Effect of gamma, UV and starch on the mechanical properties of the jute-based composites

Composite[a]	TS (MPa)	BS (MPa)	TM (GPa)	BM (GPa)	IS (kJ/m^2)
PP	25.9	34.1	0.49	0.54	5.1
GPP	32.1	40.3	0.55	0.63	10.2
UVPP	33.9	44.4	0.63	0.65	10.4
UC	48.0	51.2	0.95	1.12	17.1
GC	57.9	62.4	1.24	1.58	24.2
UVC	59.6	65.3	1.32	1.62	26.1
SGC	63.2	72.2	1.35	1.64	31.5
SUVC	66.2	72.7	1.52	1.60	28.2

[a] PP = polypropylene; GPP = gamma-treated PP; UVPP = UV-treated PP; UC = untreated composite; GC = gamma-treated composite; UVC = UV-treated composite; SGC = starch and gamma-treated composite; SUVC = starch and UV-treated composite.

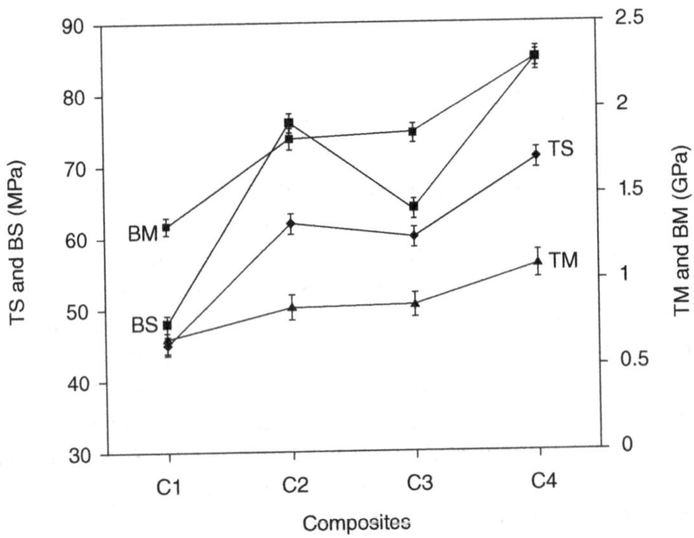

1.7 Effect of starch on the mechanical properties of gamma-treated jute–PP composites. Key: TS = tensile strength; BS = bending strength; TM = tensile modulus; BM = bending modulus; C1 = untreated jute/untreated PP; C2 = starch treated jute/untreated PP; C3 = irradiated jute fabric/irradiated PP; C4 = irradiated + starch treated jute fabric/irradiated PP.

modification of jute fabrics with oxidizing agents has been investigated by different authors [49, 61–68] and the properties of the resulting composites have been described. Khan *et al.* [61–66] worked on potassium permanganate and potassium dichromate treated jute reinforced-PP composites. Jute fabric was treated with different concentrations of potassium permanganate

Table 1.7 Effect of acrylic monomers on the mechanical properties of the jute-based composites

Composite[a]	TS (MPa)	BS (MPa)	TM (GPa)	BM (GPa)
Control	70	75	1.16	2.35
BDDA	90	95	2.20	3.20
EHA	70	83	2.00	2.70
TPGDA	70	75	2.00	2.35

[a]BDDA = 1,4-butanediodiacrylate; EHA = ethylexylacrylate; TPGDA = trypropyleneglycoldiacrylate.

Table 1.8 Effect of thermal and photoinitiators on the mechanical properties of acrylic monomer treated jute-based composites

Sample	TS (MPa)	BS (MPa)	TM (GPa)	BM (GPa)
Untreated composite	45	54	0.643	1.8
Treated composite + photoinitiator	54	62	0.738	2.03
Treated composite + thermal initiator	54	63	0.998	2.15

in aqueous, acid (sulfuric acid and oxalic acid) and alkaline media (potassium hydroxide) [61, 62], and of potassium dichromate in aqueous and acid (sulfuric acid and oxalic acid) media [63–65]. For each type of oxidizing treatment, an optimal concentration was determined. The optimum concentrations of sulfuric acid, oxalic acid and alkaline solution were 0.5, 5 and 5% respectively.

The effects of the various types of oxidizing agent at optimal concentration on the mechanical properties of the composites are shown in Fig. 1.8. It is clear that oxidation has had a positive impact on the mechanical properties of the composites, potassium dichromate and potassium permanganate in oxalic acid medium yielding the best results.

On oxidation, the surface of cellulose fibers becomes rough and the enhanced effective surface area promotes interpenetration between the fibers and the PP matrix, improving the strength of the interfacial adhesion and thereby the mechanical properties of the composite. If the concentration of the oxidizing agent is too high, however, it can degrade the fibrous material by penetrating into and attacking the amorphous regions of cellulose and the surfaces of the crystallites [69–71]. This can cause agglomeration of the fibers in the matrix or inhomogeneous stress transfer when load is applied. The investigation also showed that the thermal stability of the PP is higher than that of both untreated and treated jute fabrics in nitrogen atmosphere and that treatment of jute fabrics with oxidizing

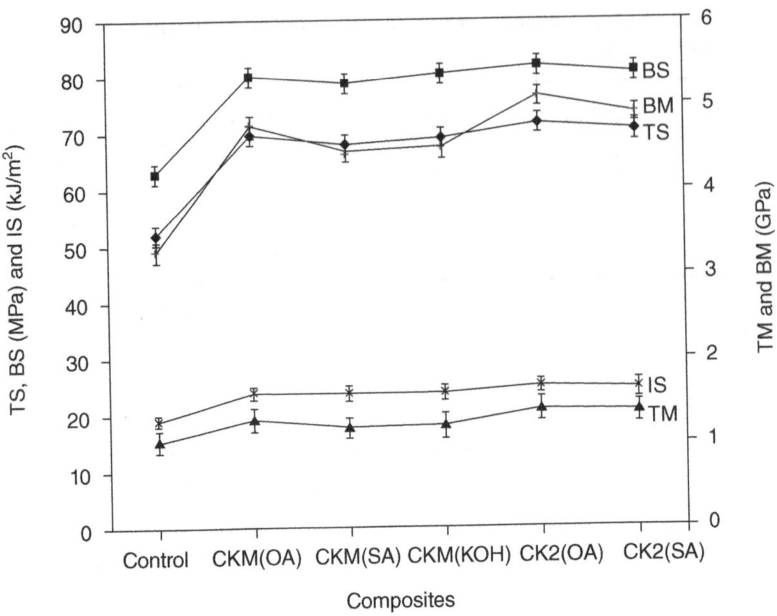

1.8 Effect of oxidizing agents on the mechanical properties of jute–PP composites. Key: TS = tensile strength; BS = bending strength; IS = impact strength; TM = tensile modulus; BM = bending modulus; CKM(OA), CKM(SA), CKM(KOH) = jute fabric-PP composites treated with $KMnO_4$ in oxalic acid (OA), sulphuric acid (SA) and potassium hydroxide (KOH) media, respectively; CK2(OA) and CK2(SA) = jute fabric-PP composites treated with $K_2Cr_2O_7$ in oxalic acid (OA) and sulphuric acid (SA) media, respectively.

agents increases the thermal degradation temperature of the composites (Tables 1.9 and 1.10).

1.9 Preparation and properties of mercerized jute composites

Alkaline treatment, or mercerization, is one of the most common chemical treatments applied to natural fibers when they are used to reinforce thermoplastics and thermosets. Alkaline treatment disrupts the hydrogen bonding in the network structure, thereby increasing the surface roughness. This treatment removes a certain amount of hemicelluloses, lignin, wax and oils covering the external surface of the fiber cell wall, and exposes the short-length crystallites [72]. During alkaline treatment, the fibers are immersed in NaOH solution of different concentrations for a given period of time and at a given temperature. The effects of NaOH concentrations, soaking time and temperatures on the thermo-mechanical and degradation

Table 1.9 Comparative thermal stability of PP, untreated, and $KMnO_4$ and $K_2Cr_2O_7$ treated jute fabrics in different media

Sample[a]	Degradation temperature (°C)		
	Onset	50%	Maximum
PP	371.6	398.0	413.8
Raw jute	327.7	352.1	365.0
$J_{KM(OA)}$	326.8	349.7	362.8
$J_{KM(SA)}$	320.3	347.2	361.6
$J_{KM(KOH)}$	324.2	350.7	362.5
$J_{K2(OA)}$	330.4	353.4	365.2
$J_{K2(SA)}$	325.5	352.2	362.6

[a] $J_{KM(OA)}$, $J_{KM(SA)}$ and $J_{KM(KOH)}$ stand for jute fabrics treated with $KMnO_4$ in oxalic acid (OA), sulphuric acid (SA) and potassium hydroxide (KOH) media, respectively. $J_{K2(OA)}$ and $J_{K2(SA)}$ stand for jute fabrics treated with $K_2Cr_2O_7$ in oxalic acid (OA) and sulfuric acid (SA) media, respectively.

Table 1.10 Comparative thermal stability of PP, control composite, and $KMnO_4$ and $K_2Cr_2O_7$ treated jute fabric–PP composites in different media

Sample[a]	MDT (°C)	Increase from neat PP (°C)
PP	413.8	–
Control composite	426.7	12.9
$C_{KM(OA)}$	455.8	42.0
$C_{KM(SA)}$	454.2	40.4
$C_{KM(KOH)}$	454.3	40.5
$C_{K2(OA)}$	465.4	51.6
$C_{K2(SA)}$	456.6	42.8

[a] $C_{KM(OA)}$, $C_{KM(SA)}$, $C_{KM(KOH)}$ stand for jute fabric-PP composites treated with $KMnO_4$ in oxalic acid (OA), sulphuric acid (SA) and potassium hydroxide (KOH) media, respectively. $C_{K2(OA)}$ and $C_{K2(SA)}$ stand for jute fabric-PP composites treated with $K_2Cr_2O_7$ in oxalic acid (OA) and sulphuric acid (SA) media, respectively.

characteristics of jute–PP composites were extensively studied by Khan *et al.* [42]. Jute fabrics were soaked with different concentrated solutions of NaOH (5, 10 and 20%) in aqueous medium for periods of 30, 60 and 90 min at room temperature (30°C). The results of this treatment are shown in Table 1.11. The mechanical properties of the composites increase with increasing NaOH concentration and also with increasing soaking time except at 20% NaOH concentration, where the mechanical properties of the composites were found to decrease after 60 min soaking.

During mercerization, fibrillation occurs in jute fabrics, increasing the effective fiber surface available for wetting by the matrix. The removal of cementing materials and an increase in crystallinity also contribute in

Table 1.11 Effect of temperature, soaking time and concentration of NaOH solution on the mechanical properties of jute fabric–PP composites

Conc. of NaOH (%)	Soaking time	Temp. (°C)	TS (MPa)	BS (MPa)	TM (GPa)	BM (GPa)	% E_b	IS (kJ/m^2)
5	30	30	56.8	66.9	1.10	3.41	17.90	18.0
	60	30	59.8	68.8	1.14	3.52	15.75	17.1
	90	30	61.7	72.4	1.17	3.61	14.30	16.2
10	30	30	58.5	68.0	1.16	3.49	16.20	16.9
	60	30	61.8	70.2	1.20	3.60	14.50	15.2
	90	30	62.5	73.2	1.19	3.70	10.60	13.2
20	30	30	61.4	72.0	1.20	3.85	14.00	13.7
	60	30	64.7	75.6	1.25	4.06	11.40	11.6
	90	30	57.2	70.4	1.09	3.30	8.30	9.5
20	60	0	68.5	80.0	1.30	4.25	10.00	14.2
	60	70	54.0	63.4	1.12	3.40	7.80	9.5
Control composite	–	30	52.0	63.0	1.03	3.27	10.72	19.1

improving the mechanical properties. However, with 20% NaOH treatment and after 60 min of soaking the fibers became somewhat brittle owing to excessive fibrillation. On application of stress, these fibers suffered breakage as a result and could not effectively transfer stress at the interface.

The effects of temperature on the mechanical properties of 20% NaOH-treated jute composite at 60 min of soaking are shown in Table 1.11. It is observed that the mechanical properties of the composites increase when mercerization is carried out at lower temperature (0°C). The study also shows that decomposition of the hemicelluloses and α-cellulose of mercerized jute fabrics occur at 294.2°C and 363.6°C respectively. NaOH treated composites exhibit higher thermal stability compared to control composites and PP by 12.9°C and 39.8°C respectively. The mercerized jute–PP composite shows a lower water uptake tendency due to better fiber–matrix adhesion, reduction of polar groups and removal of hemicelluloses from the fibers during mercerization. The mercerized jute composite is less degradable in soil and water and less sensitive to weather conditions.

1.10 Preparation and properties of jute composites modified by other processes

Jute fibers were cyanoethylated and vinyl (acrylonitrile, AN and methyl methacrylate, MMA) grafted in order to improve the mechanical properties of jute composites. Cyanoethylene and vinyl monomers were coupled with jute fibers through their functional groups (Fig. 1.9). Jute fabrics have also been modified with 2-hydroxyethylmethylacrylate (HEMA) and 2-ethylhexylacrylate (EHA) by soaking in solutions at various concentrations of monomer (2–5%) in methanol for 5 min [73]. In these studies, dicumylperoxide was used as the thermal initiator. The TS and BS of the treated composites were found to be highest at 3.0% monomer. At low concentrations of vinyl monomer, the increased crystallinity of the short vinyl polymeric units on the fiber surface results in enhanced fiber matrix adhesion. At higher concentrations of monomer, the monomer–monomer recombination reaction increases rather than the cellulose–monomer reaction. The tensile and bending strengths of virgin PP increase from 25 to 32 MPa (28% increase) and from 31 to 54 MPa (74% increase) respectively as a result of 20% jute reinforcement in PP.

The superior mechanical properties obtained for EHA and HEMA treated jute fabric-based composites are due to the fact that monomers improve the adhesive properties of the fiber and produce a rough surface which offers a better fiber–matrix interaction. Vinyl monomers react with –OH groups of cellulose through a graft copolymerization reaction and with PP through a free radical reaction. The bending E-modulus values of EHA and HEMA treated jute composites were found to be higher than those of

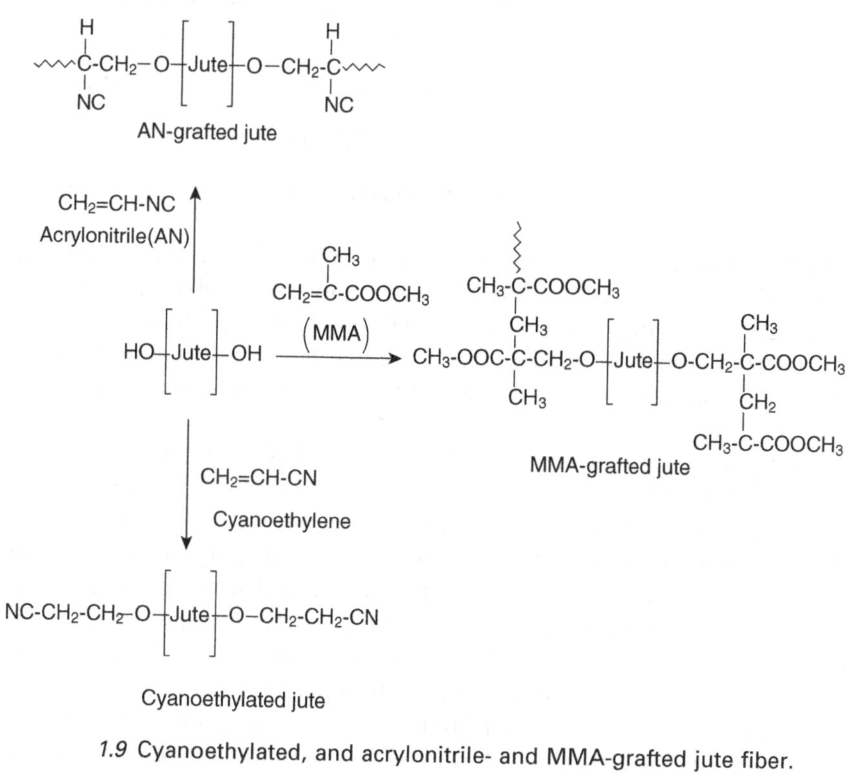

1.9 Cyanoethylated, and acrylonitrile- and MMA-grafted jute fiber.

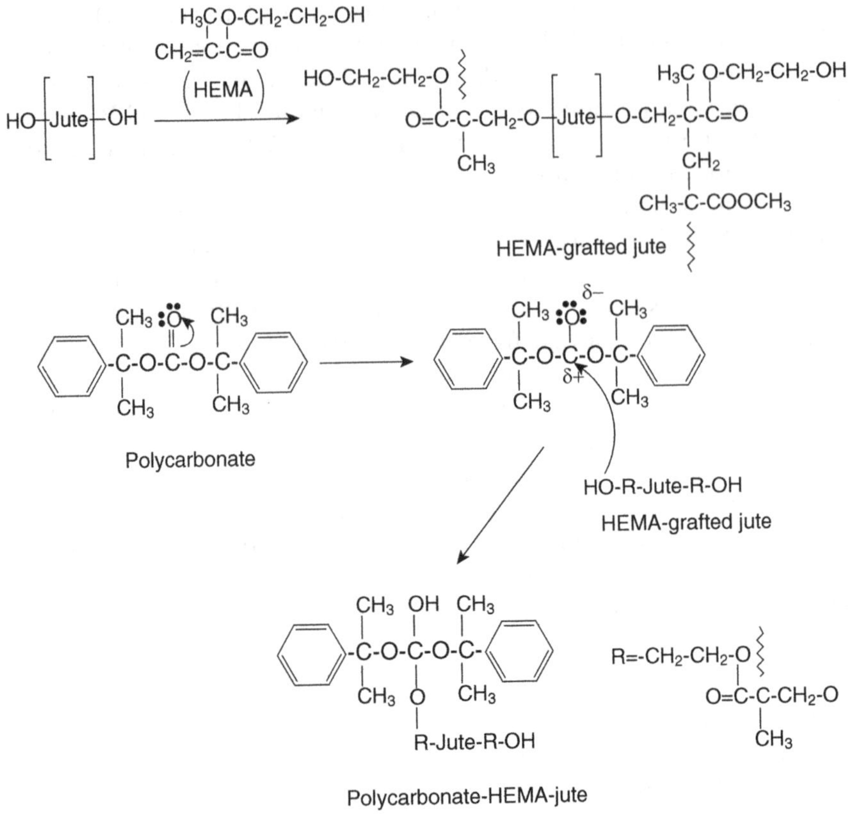

1.10 **Reaction between polycarbonate and HEMA-grafted jute fiber.**

pure PP (three times) and untreated jute composite (1.6 times). In another study [47], it is observed that the hydroxyl groups of HEMA react with the carbonate groups of polycarbonate (PC) through nucleophilic addition (Fig. 1.10). Better dispersion of the HEMA treated fibers in PC along with the increased interfacial adhesion between jute and PC were also reported in this study (Table 1.12).

The thermal, dynamic mechanical and aging behavior of injection molded short jute fiber–PP composites were investigated both with and without the matrix modifier maleic anhydride grafted polypropylene, MAPP [51]. The thermal gravimetric behavior of jute fiber and polypropylene resin composites was determined under both nitrogen and air purge gas and was found to be significantly different in the two cases. This was due to oxidative degradation of α-cellulose which occurs at a temperature of 322°C in an air atmosphere and at 353°C in a nitrogen atmosphere.

The composite modified with MAPP, at fixed fiber content, is found to be more thermally stable (by 10°C) than that of the non-modified composites. The authors suggest that this effect is due to the stronger interaction

Table 1.12 Effect of HEMA on the mechanical properties of jute–polycarbonate composite

Composite[a]	TS (MPa)	BS (MPa)	BM (GPa)	Shear strength (MPa)	Shear modulus (GPa)
PCJ	63.5	87	4.1	5.26	1.3
PCJH	70.6	93.3	5.5	6.56	1.6

[a]PCJ = untreated polycarbonate jute composite; PCJH = HEMA treated jute composites.

between the fiber and matrix caused by the formation of the covalent bond at the interface. The overall thermal resistance of the modified composites decreased with increasing fiber content in a nitrogen atmosphere, which is in agreement with the higher thermal resistance of PP compared to jute fiber in nitrogen. Conversely, overall thermal resistance of the composites increases with fiber content due to the lower thermal stability of PP compared to jute fiber in an air atmosphere.

Moisture absorption also increases with increasing fiber content and the modified PP composites absorb less moisture than the unmodified ones. The hydrophilic moieties on the fiber surface act as passageways for water entry and the MAPP reduces the number of hydrophilic fiber surface moieties, thus reducing water uptake.

A study of fire behavior and mechanical properties for PP, PP/jute and PP/rayon (Cordenka®) composites in presented in Ref. 52. The matrix material was a polypropylene/ethylene block copolymer, and maleic acid anhydride grafted PP (MAPP) was used as a coupling agent. Three types of fire retardants (15wt%) such as Mgnifin (magnesium hydroxide), Exolit AP-750 (ammonium polyphosphate) and expandable graphite and their mixture (1:1 by weight) were used in the composites and the pristine PP. It was observed that expandable graphite led to the lowest burning speed and the mixture of Mgnifin and Exolit gave the highest burning speeds. All of the fire retardants and their mixtures had a negative effect on the mechanical properties of the composites with graphite having the largest impact in this respect. Among the fire retardants and their mixtures, the combination of expandable graphite and Mgnifin for jute-based composites and the mixture of expandable graphite with Exolit for the Cordenka composites yielded the best compromise between mechanical performance and burning behavior.

Gassan and Bledzki [74] showed that the flexural strength of composites treated with MAPP was higher than that of unmodified fibers, and increased with fiber loading. The cyclic-dynamic values at an increasing load indicated that the coupling agent hinders the progression of damage. Dynamic strength (dynamic failure stress at load increasing test) of the MAPP-modified composites was raised by approximately 40%. The improved properties were attributed to improved fiber–matrix adhesion.

The dynamic mechanical response and the short term creep–recovery behavior of composites made from bi-directional jute fabrics and polypropylene were studied by Acha *et al.* [75]. The effect of coupling agents and the chemical modification of the fibers on the properties of the composites were compared. In the first case, two commercial maleated polypropylenes and lignin, a natural polymer, were used. In the second approach, the fibers were esterified using a commercial alkenyl succinic anhydride. The maleated polypropylenes acted as compatibilizers since they were able to join the fibers to the neat PP, locating themselves in the interphase region. A clear separation between fibers and matrix could be observed when lignin was used as the compatibilizing agent and when the chemically modified fibers were used to prepare the composite. The creep deformation could be directly related to the interfacial properties. The jute/MAPP composites exhibited less fiber pullout, smoother fiber surface, and higher tensile and impact strength than the uncompatibilized one [76].

1.11 Types and properties of hybrid jute composites

The effect of gamma radiation and starch on the mechanical properties of jute yarn/coir fiber-based hybrid composites was studied by Zaman *et al.* [77, 78]. It was found that 20% coir and 80% jute reinforced PP composites gave the best results in terms of mechanical properties. In a hybrid composite, the mechanical properties are mainly dependent on the moduli of the individual reinforcing fibers. The enhanced mechanical properties through the addition of jute fiber to coir are due to the higher modulus of the jute fiber. All of the materials (coir, jute and PP) were irradiated with gamma radiation of 400 to 1000 krad dose. The irradiated composites (20% coir and 80% jute) showed the best mechanical properties at 600 krad of total gamma dose. The irradiated yarns were further treated with starch with different concentrations (2–10%) and different soaking times. The maximum starch loading (SL) value was found to be 30% at 5% starch for 5 min soaking time and the results are shown in Figs 1.11 and 1.12.

Zaman *et al.* [79,80] prepared jute-reinforced composites with polyethylene (PE), polypropylene (PP) and a mixture of PP and PE. The effects of green dye and UV radiation [79] and gamma radiation [80] on the mechanical properties of the hybrid composites were studied. The investigation showed that composites with mixtures of PP and PE showed improved mechanical properties. This was attributed to the fact that during fabrication at higher temperature, the high melt fluidity of PE impregnated into the fiber and the PP, causing better fiber–matrix adhesion.

Al-Kafi *et al.* [48] studied unsaturated polyester (USP) resin-based jute fabric (hessian cloth) and E-glass fiber (mat) hybrid composites, together with the effects of UV radiation on their mechanical properties. Jute fiber

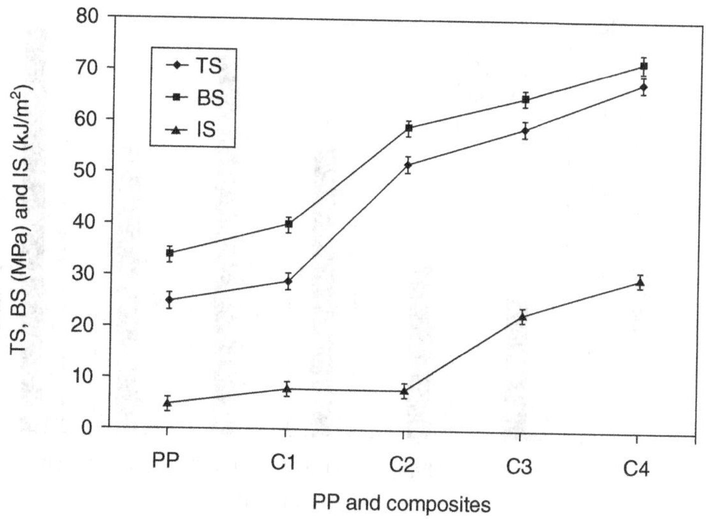

1.11 Effect of starch and gamma radiation on tensile, bending and impact strength of jute–coir hybrid composites. Key: TS = tensile strength; BS = bending strength; IS = impact strength; PP = polypropylene sheet; C1 = 20% coir/PP, C2 = 20% coir + 80% jute/PP; C3 = 600 krad gamma treated 20% coir + 80% jute/PP; C4 = 5% starch and 600 krad gamma treated 20% coir + 80% jute/PP.

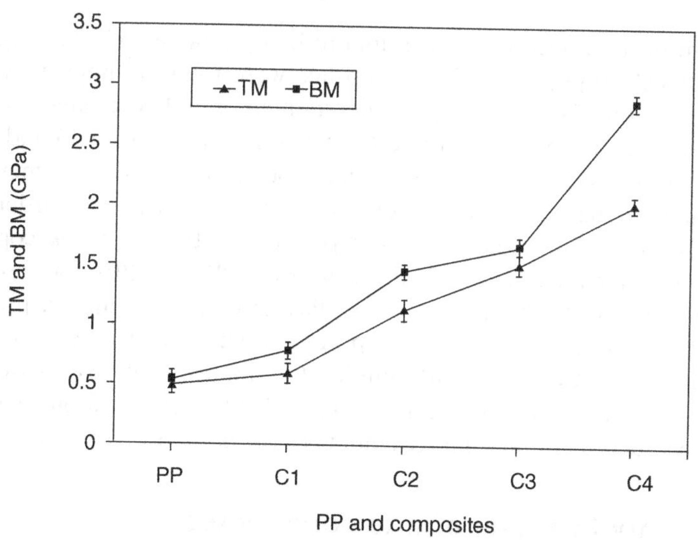

1.12 Effect of starch and gamma radiation on tensile and bending modulus of jute–coir hybrid composites. Key: TM = tensile modulus; BM = bending modulus; PP = polypropylene sheet; C1 = 20% coir/PP, C2 = 20% coir + 80% jute/PP; C3 = 600 krad gamma treated 20% coir + 80% jute/PP; C4 = 5% starch and 600 krad gamma treated 20% coir + 80% jute/PP.

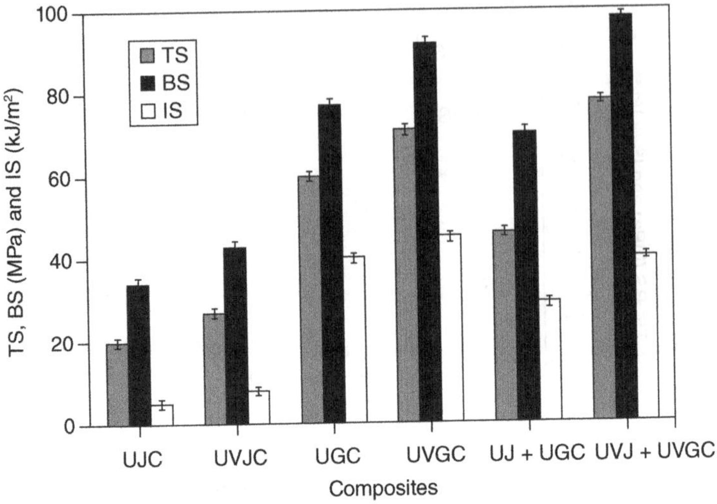

1.13 Effect of UV radiation on tensile, bending and impact strength of jute–glass hybrid composites. Key: TS = tensile strength; BS = bending strength; IS = impact strength; UJC = untreated jute composite; UVJC = UV-treated jute composite; UGC = untreated glass composite; UVGC = UV-treated glass composite; UJ + UGC = untreated jute/glass hybrid composite; UVJ + UVGC = UV-treated jute/glass hybrid composite.

content in the composites was found to be optimal at 25% by weight. The mechanical properties of the composites were found to improve with the incorporation of glass fiber up to a certain amount. UV irradiation further improved the properties of these hybrid composites (Figs 1.13 and 1.14).

A number of hybrid composites have been made with jute, mercerized jute, and the high-tenacity synthetic cellulose tyre cord yarn Cordenka [53, 81]. The matrix material was a polypropylene/ethylene block copolymer (PP), and a maleic acid anhydride grafted PP (MAPP) was used as a coupling agent. High strength (>70 MPa) and excellent impact properties (>80 kJ/m^2) were achieved with pure Cordenka reinforcement. Partial substitution of Cordenka with jute led to enhanced stiffness as well as an increased heat distortion temperature (HDT). Mercerization of the jute fibers yielded moderate improvements in the composite properties.

1.12 Applications of jute composites

Jute has been in use mainly as a packaging material, and for roping and household purposes. Jute fiber composites have also been used for producing door panels, roofing and sanitary products. Jute-reinforced thermoplastic laminates and composites represent a promising substitution for thermoplastic and synthetic fiber-reinforced composites with good physical

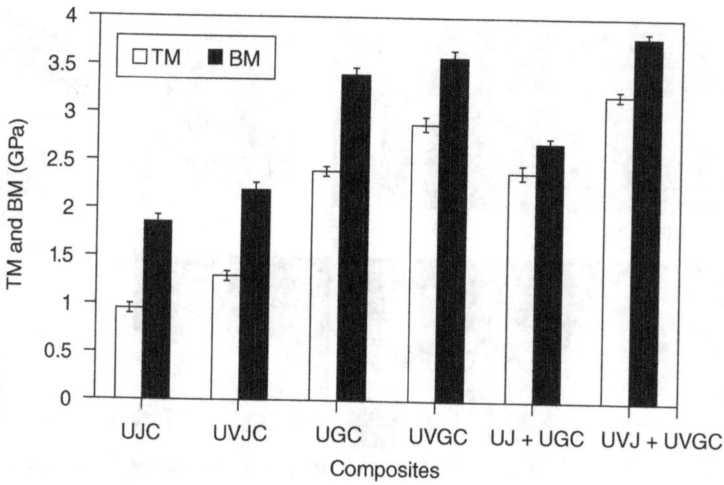

1.14 Effect of UV radiation on tensile and bending modulus of jute–glass hybrid composites. Key: TM = tensile modulus; BM = bending modulus; UJC = untreated jute composite; UVJC = UV-treated jute composite; UGC = untreated glass composite; UVGC = UV-treated glass composite; UJ + UGC = untreated jute/glass hybrid composite; UVJ + UVGC = UV-treated jute/glass hybrid composite.

properties and excellent performance at low weight. Potential future applications of jute fiber composites include the automobile industry, the footwear industry, building/construction, home/garden furniture, and the toy sectors.

The Bangladesh Atomic Energy Commission has successfully made a wide range of products from jute-reinforced polymer composite (Fig. 1.15). The products include:

- Jute/polymer corrugated sheet (Jutin) [82]
- False ceilings, roof tiles, kitchen sinks
- Durable chairs, tables, etc.
- Sanitary latrine accessories such as the slab, ring, etc.
- Decorative materials
- Helmets, chest guards, leg guards, etc.

The main applications of Jutin (jute + tin) include:

- Covering of industrial buildings
- Roofing of residential houses
- Settings of bungalows
- Commercial buildings
- Prefabricated sheds, etc.

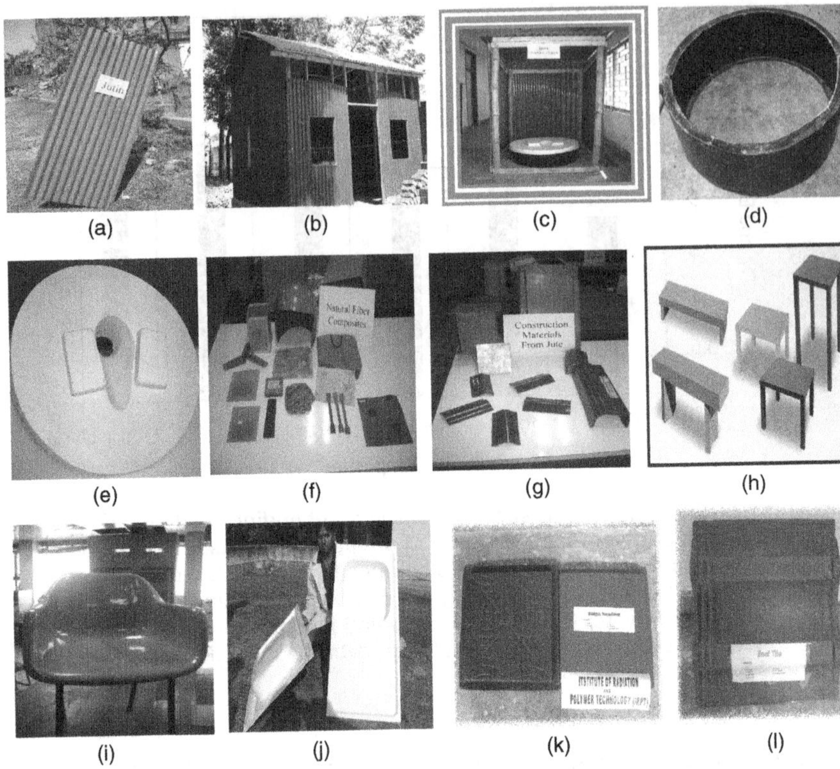

1.15 Various types of products from jute composites: (a) jutin sheet, (b) model hut, (c) model latrine, (d) and (e) ring and pan of latrine, (f) and (g) helmet, fan, bag, construction materials, etc., (h) benches and tables, (i) chair, (j) lightweight and durable kitchen sink, (k) false ceiling, (l) roof tile.

Table 1.13 Mechanical properties of Jutin

	TS (MPa)	BS (MPa)	TM (GPa)	BM (MPa)	IS (kJ/m^2)
Jutin	117	96	4	4.5	44
UV-cured Jutin	150	140	6.5	5.2	68

These sheets can also be used for hut making in coastal areas because of their saline resistant properties, and in earthquake regions for their light weight. Jutin may last as long as 50 years under normal conditions. Jutin has outstanding mechanical (Table 1.13), thermal (Table 1.14), and aging (Table 1.15) properties which make it suitable for roofing and wall cladding.

Table 1.14 Thermal conductivity of different structural materials

	Jutin	Brick	Concrete	Iron	Aluminum	Natural fiber	USP
Thermal conductivity (W/mK)	0.05–0.07	0.86	1.51	58.15	250	0.13–0.17	0.17

Table 1.15 Effect of thermal aging on the tensile strength of Jutin

Sample	Tensile strength (MPa)				
	Room temp.	0°C	4°C	50°C	70°C
Jutin	96	116	99.8	93	78

1.13 Conclusion

Traditionally jute fiber has been used for making rope, hessian cloth, carpet, wall-mat, bags, etc., although the use of jute fiber as the reinforcement phase in polymer matrix composites has opened up new possibilities for applications in the area of structural materials. Jute as a single or hybrid fiber has proved promising as a reinforcement material for sustainable and eco-friendly applications, although further research is required to fully exploit the potential of such composite materials.

1.14 References

1. P. Ghose and P. K. Ganguly, Jute, in: *Polymeric Materials Encyclopedia*, ed. J. C. Salamone, CRC Press, Boca Raton, FL, 1996, **5**: 3504–3513.
2. B. C. Kundu, K. C. Basak and P. B. Sarkar, *Jute in India*, Indian Central Jute Committee, Calcutta, India, 1959.
3. A. Majumder, S. Samajpati, P. K. Ganguly, D. Sardar and P. C. Das Gupta, Swelling of jute: Heterogeneity of crimp formation, *Textile Research Journal*, 1980, **50**: 575–578.
4. A road map for jute, International Jute Study Group (IJSG) Dhaka, Bangladesh and International Trade Centre UNCTAD/WTO (ITC) Geneva, Switzerland.
5. P. Wambua, U. Ivens and I. Verpoest, Natural fibers: Can they replace glass in fiber-reinforced plastics? *Composites Science and Technology*, 2003, **63**: 1259–1264.
6. D. N. Saheb and J. P. Jog, *Advances in Polymer Technology*, 1999, **18**: 351–363.
7. A. K. Mohanty, M. Misra and L. T. Drzal, *Journal of Polymers and the Environment*, 2002, **10**: 19–26.
8. D. Gon, K. Das, P. Paul and S. Maity, Jute composites as wood substitute, *International Journal of Textile Science*, 2012, **1**(6): 84–93.

9. C. R. A. Chavez, S. Edwards, R. M. Eraso and K. Geiser, Sustainability of bio-based plastics: General comparative analysis and recommendations for improvement, *Journal of Cleaner Production*, 2012, **23**: 47–56.
10. C. K. Saha and S. Sagorika, Carbon credit of jute and sustainable environment, *Jute Matters*, 2013, **1**: 1–4.
11. R. M. Rowell, J. S. Han and J. S. Rowell, Characterization and factors affecting fiber properties, in: *Natural Polymers and Agrofibers Based Composites*, eds E. Frollini, A. L. Leão and L. H. C. Mattoso, EAI, São Carlos, Brazil, 2000, 115–134.
12. T. P. Nevell and S. H. Zeronian, *Cellulose Chemistry and its Application*, Wiley, New York, 1985.
13. K. Goda, H. Takagi and A. N. Netravali, Fully biodegradable green composites reinforced with natural fibers, in: *Natural Fiber Reinforced Polymer Composites*, eds S. Thomas and L. A. Pothan, Old City Publishing, Philadelphia, PA, 2009, 329–360.
14. R. M. Rowell and H. P. Stout, Jute and Kenaf, in: *Handbook of Fiber Chemistry*, ed. M. Lewin, Taylor & Francis, Boca Raton, FL, 2007.
15. C. Heitner, Light induced yellowing of wood containing papers, in *Photochemistry of Lignocellulostic Materials*, ACS Symposium Series 531, *Journal of the American Chemical Society*, 1993, 3–25.
16. D. N. S. Hon, Weathering and photochemistry of wood, in: *Wood and Cellulosic Chemistry*, 2nd edn, Marcel Dekker, New York, 2000, 512–546.
17. K. Joseph, L. H. C. Mattoso, R. D. Toledo, S. Thomas, L. H. de Carvalho, L. Pothan, S. Kala and B. James, Natural fiber reinforced thermoplastic composites, in: *Natural Polymers and Agrofibers Based Composites*, eds E. Frollini, A. L. Leão and L. H. C. Mattoso, EAI, São Carlos, Brazil, 2000, 159–201.
18. C. Gupta and A. P. Gupta, *Polymer Composite*, New Age International (P) Ltd., New Delhi, India, 2005.
19. A. N. Netraveli and S. Chabba, Composites get greener, *Materials Today*, 2003, **6**: 22–29.
20. M. Jacob, R. D. Anandjiwala and S. Thomas, Lignocellulosic fiber reinforced rubber composites, in: *Natural Fiber Reinforced Polymer Composites*, eds S. Thomas and L. A. Pothan, Old City Publishing, Philadelphia, PA, 2009, 256–257.
21. A. K. Bledzki and J. Gassan, Composites reinforced with cellulose based fibres, *Progress in Polymer Science*, 1999, **24**: 221–274.
22. A. Paul, K. Joseph and S. Thomas, Environmental effects on the degradation behaviour of sisal fibre reinforced polypropylene composites, *Composites Science and Technology*, 1997, **57**(1): 67–79.
23. T.W. Frederick and W. Norman, *Natural Fibers Plastics and Composites*, Kluwer Academic Publishers, New York, 2004.
24. N. M. Stark and L. M. Matuana, Ultraviolet weathering of photostabilized wood-flour filled high-density polyethylene, *Journal of Applied Polymer Science*, 2003, **90**(10): 2609–2617.
25. L. M. Matuana, D. P. Kamdem and J. Zhang, Photoaging and stabilization of rigid PVC/wood-fibre composites, *Journal of Applied Polymer Science*, 2001, **80**(11): 1943–1950.

26. L. M. Matuana and D. P. Kamdem, Accelerated ultraviolet weathering of PVC/wood fibre composites, *Polymer Engineering Science*, 2002, **42**(8): 1657–1666.
27. M. M. Hasan, M. R. Islam, M. A. Sawpan and M. A. Khan, Effect of silane on mechanical and degradable properties of photo-grafted jute yarn with acrylamide, *Journal of Applied Polymer Science*, 2003, **89**: 3530–3538.
28. M. M. Hasan, M. R. Islam and M. A. Khan, Role of amino acid on mechanical properties of photo-grafted jute yarn with acrylamide, *Polymer-Plastics Technology and Engineering*, 2003, **42**(5): 779–793.
29. M. M. Hasan, M. A. Khan and M. R. Islam, Influence of mercerization and UV radiation treatment on the improvement of mechanical properties of photo-grafted jute yarn with silane and acrylamide, *Polymer-Plastics Technology and Engineering,* 2003, **42**(4): 515–531.
30. S. K. Datta, T. K Chaki and A. K. Bhowmick, Structure-property relationship of ethylene vinyl acetate copolymer grafted with triallyl cyanurate in presence of electron beam irradiation, *Radiation Physics and Chemistry*, 1996, **47**: 913.
31. R. Clough, *Encyclopedia of Polymer Science and Technology*, Wiley, New York, 1986, 666.
32. M. M. Hasan, M. A. Khan and M. R. Islam, Effect of additives on the improvement of mechanical and degradable properties of photo-grafted jute yarn with acrylamide, *Journal of Polymers and the Environment,* 2002, **10**(4): 139–145.
33. M. A. Khan and K. M. I. Ali, Swelling and thermal properties of wood and wood plastic composite (WPC), *Polymer-Plastics Technology and Engineering*, 1997, **36**(2): 179–187.
34. K. M. I. Ali, M. A. Khan and K. S. Akhunzada, *In situ* jute yarn composite with HEMA via UV radiation, *Journal of Applied Polymer Science*, 1999, **71**: 841–846.
35. F. Khan and S. R. Ahmad, Graft copolymerization and characterization of 2-hydroxyethyl methacrylate onto jute fiber by photoirradiation, *Journal of Applied Polymer Science*, 2006, **101**: 2898–2910.
36. M. Montazer and A. Salehi, Novel jute yarns grafted with methyl methacrylate, *Journal of Applied Polymer Science*, 2008, **107**: 2067–2073.
37. M. I. H. Mondal and M. M. U. Haque, Effect of grafting methacrylate monomers onto jute constituents with a potassium persulfate initiator catalyzed by Fe(II), *Journal of Applied Polymer Science*, 2007, **103**: 2369–2375.
38. M. A. Sawpan, M. A. Khan and M. Z. Abedin, Surface modification of jute yarn by photografting of low-glass transition temperature monomers, *Journal of Applied Polymer Science*, 2003, **87**: 993–1000.
39. M. M. Hasan, M. R. Islam, S. Shehrzade and M. A. Khan, Influence of mercerization along with ultraviolet (UV) and gamma radiation treatment on physical and mechanical properties of jute yarn by grafting with 3-(trimethoxysilyl)-propylmethacrylate (silane) and acrylamide under UV radiation, *Polymer-Plastics Technology and Engineering*, 2003, **42**: 515–531.
40. M. A. Khan, S. Shehrzade and M. M. Hassan, Effect of alkali and ultraviolet (UV) radiation pretreatment on physical and mechanical properties of 1,6-hexanediol diacrylate-grafted jute yarn by UV radiation, *Journal of Applied Polymer Science*, 2004, **92**: 18–24.
41. M. A. Khan, S. Shehrzade, A. M. S. Chowdhory and M. M. Rahman, Effect of pre-treatment with UV-radiation on physical and mechanical properties

of photocured jute yarn with 1,6,hexanediol diacrylate (HDDA), *Journal of Polymers and the Environment*, 2002, **9**(5): 115–124.
42. J. A. Khan, M. A. Khan and R. Islam, Effect of mercerization on mechanical, thermal and degradation characteristics of jute fabric-reinforced polypropylene composites, *Fibers and Polymers*, 2012, **13**(10): 1300–1309.
43. D. Ray, B. K. Sarkar, R. K. Basak and A. K. Rana, Study of the thermal behavior of alkali-treated jute fibers, *Journal of Applied Polymer Science*, 2002, **85**: 2594–2599.
44. D. Ray and B. K. Sarkar, Characterization of alkali-treated jute fibers for physical and mechanical properties, *Journal of Applied Polymer Science*, 2001, **80**: 1013–1020.
45. D. Ray, M. Das and D. Mitra, A comparative study of the stress-relaxation behavior of untreated and alkali-treated jute fibers, *Journal of Applied Polymer Science*, 2012, **123**: 1348–1358.
46. M. M. Hasan, M. A. Khan and M. R. Islam, Improvement of physico-mechanical properties of jute yarn by photo-grafting with 3-(trimethoxysilyl)-propyl methacrylate, *Journal of Adhesion Science and Technology*, 2003, **17**(5): 737–750.
47. M. A. Khan, M. M. Hassan and L. T. Drzal, Effect of 2-hydroxyethylmethacrylate on the mechanical and thermal properties of jute–polycarbonate composite, *Composites: Part A*, 2005, **36:** 71–81.
48. A. Al-Kafi, M. Z. Abedin, M. D. H. Beg, K. L. Pickering and M. A. Khan, Study on the mechanical properties of jute/glass fiber-reinforced unsaturated polyester hybrid composites: Effect of surface modification by ultraviolet radiation, *Journal of Reinforced Plastics and Composites*, 2006, **25**: 575–588.
49. S. Sultana, M. M. Huque and M. M. Helali, Studies on the physico-mechanical properties of sodium periodate oxidized jute reinforced polypropylene composites, *Polymer-Plastics Technology and Engineering*, 2007, **46**: 385–391.
50. M. A. Kabir, M. M. Huque, M. R. Islam and A. K. Bledzki, Mechanical properties of jute fiber reinforced polypropylene composite: Effect of chemical treatment by benzenediazonium salt in alkaline medium, *BioResources*, 2010, **5**(3): 1618–1625.
51. T. T. L. Doan, H. Brodowsky and E. Mäder, Jute fiber/polypropylene composites II. Thermal, hydrothermal and dynamic mechanical behaviour, *Composites Science and Technology*, 2007, **67**: 2707–2714.
52. M. A. Khan, J. Ganster, R. Rihm and H. P. Fink, *Sixth Global Wood and Natural Fiber Composites Symposium*, 5–6 April 2006, Kassel, Germany.
53. M. A. Khan, J. Ganster and H. P. Fink, Hybrid composites of jute and man-made cellulose fibers with polypropylene by injection molding, *Composites: Part A*, 2009, **40**: 846–851.
54. M. A. Khan, R. A. Khan, H. U. Zaman, A. Hossain and A. H. Khan, Effect of gamma radiation on the physico-mechanical and electrical properties of jute fiber-reinforced polypropylene composites, *Journal of Reinforced Plastics and Composites*, 2009, **28**: 1651–1660.
55. H. U. Zaman, R. A. Khan, M. A. Khan, A. H. Khan and M. A. Hossain, Effect of gamma radiation on the performance of jute fabrics–reinforced polypropylene composites, *Radiation Physics and Chemistry*, 2009, **78**: 986–993.
56. M. A. Khan, R. A. Khan, H. U. Zaman, S Ghoshal, M. N. A. Siddiky and M. Saha, Study on the physico-mechanical properties of starch-treated jute yarn-

reinforced polypropylene composites: Effect of gamma radiation, *Polymer-Plastics Technology and Engineering*, 2009, **48**: 542–548.

57. S. M. M. A. Rahman, A. I. Mustafa and M. A. Khan, Jute reinforced polypropylene composite: Effect of surface pre-treatment by photocuring with acrylic monomers, *Journal of Reinforced Plastics and Composites*, 2009, **28**: 1733–1745.
58. M. K. Uddin, M. A. Khan and K. M. I. Ali, Degradable jute plastic composites, *Polymer Degradation and Stability*, 1997, **55**: 1–7.
59. M. A. Khan, R. A. Khan, H. U. Zaman, M. N. A. Alam and M. A. Hoque, Effect of surface modification of jute with acrylic monomers on the performance of polypropylene composites, *Journal of Reinforced Plastics and Composites*, 2010, **29**: 1195–1205.
60. H. U. Zaman, M. A. Khan, N. Akter, S. Ghoshal and R. A. Khan, Role of gamma radiation and EGDMA on the physic-mechanical properties of jute fabrics/polypropylene composites, *Polymer Composites*, 2011, **32**: 1888–1894.
61. J. A. Khan, M. A. Khan and R. Islam, Mechanical, thermal and degradation properties of jute fabric-reinforced polypropylene composites: Effect of potassium permanganate as oxidizing agent, *Polymer Composites*, 2013, **34**(5): 671–680.
62. J. A. Khan, M. A. Khan and R. Islam, Effect of potassium permanganate on mechanical, thermal and degradation characteristics of jute fabric-reinforced polypropylene composite, *Journal of Reinforced Plastics and Composites*, 2012, **31**: 1725–1736.
63. J. A. Khan, M. A. Khan and N. Rahman, Effect of oxidizing agents on thermo-mechanical behavior of jute fabric-reinforced polypropylene composites, *Advanced Materials Research*, 2010, **123–125**: 1127–1130.
64. J. A. Khan, M. A. Khan, R. Islam and A. Gafur, Mechanical, thermal and interfacial properties of jute fabric-reinforced polypropylene composites: Effect of potassium dichromate, *Materials Sciences and Applications*, 2010, **1**: 350–357.
65. J. A. Khan, Jute fiber reinforced polymer composites: effect of surface modification, PhD thesis, Department of Chemistry, Jahangirnagar University, Bangladesh, 2010.
66. J. A. Khan, M. A. Khan and R. Islam, Studies on mechanical and thermal behavior of *N*, *N*-dimethylaniline and potassium dichromate treated jute fabric-reinforced polypropylene composites, *Eighth Global WPC and Natural Fiber Composites Congress and Exhibition*, 22–23 June 2010, Stuttgart, Germany.
67. T. P. Nevell, *Methods in Carbohydrate Chemistry*, Vol. III, Academic Press, New York, 1963.
68. L. F. McBurney, *Cellulose and Cellulose Derivatives*, Interscience Publishers, New York, 1954.
69. J. T. Marsh and F. C. Wood, *An Introduction to the Chemistry of Cellulose*, 2nd edn, Chapman & Hall, London, 1942.
70. M. R. Rahman, M. M. Huque, M. N. Islam and M. Hasan, Improvement of physico-mechanical properties of jute fiber reinforced polypropylene composites by post-*treatment*, *Composites: Part A*, 2008, **39**: 1739–1747.
71. S. S. Tripathy, G. Levita and L. D. Landro, Interfacial adhesion in jute—polyolefin composites. *Polymer Composites*, 2001, **22**: 815–822.
72. M. S. Sreekala, M. G. Kumaran, S. Joseph, M. Jacob and S. Thomas, Oil palm fibre reinforced phenol formaldehyde composites: Influence of fiber surface

modifications on the mechanical performance, *Applied Composite Materials*, 2000, **7**: 295–329.

73. M. A. Khan, G. Hinrichsen and L. T. Drzal, Influence of novel coupling agents on mechanical properties of jute reinforced polypropylene composite, *Journal of Materials Science Letters*, 2001, **20**, 1711–1713.
74. J. Gassan and A. K. Bledzki, The influence of fiber-surface treatment on the mechanical properties of jute–polypropylene composites, *Composites: Part A*, 1997, **28**: 1001–1005.
75. B. A. Acha, M. M. Reboredo and N. E. Marcovich, Creep and dynamic mechanical behavior of PP–jute composites: Effect of the interfacial adhesion, *Composites: Part A*, 2007, **38**: 1507–1516.
76. B. A. Acha, M. M. Reboredo and N. E. Marcovich, Effect of coupling agents on the thermal and mechanical properties of polypropylene–jute fabric composites, *Polymer International*, 2006, **55**: 1104–1113.
77. H. U. Zaman, A. H. Khan, M. A. Hossain, M. A. Khan and R. A. Khan, Mechanical properties of the coir fiber-reinforced polypropylene composites: Effect of the incorporation of jute fiber, *Journal of Composite Materials*, 2010, **44**(4): 401–416.
78. H. U. Zaman, M. A. Khan and R. A. Khan, Comparative experimental measurements of jute fiber/polypropylene and coir fiber/polypropylene composites as ionizing radiation, *Polymer Composites*, 2012, **33**: 1077–1084.
79. H. U. Zaman, M. A. Khan and R. A. Khan, Improvement of mechanical properties of jute fibers–polyethylene/polypropylene composites: Effect of green dye and UV radiation, *Polymer-Plastics Technology and Engineering*, 2009, **48**: 1130–1138.
80. H. U. Zaman, A. H. Khan, M. A. Hossain, M. A. Khan and R. A. Khan, Mechanical and electrical properties of jute fabrics reinforced polyethylene/polypropylene composites: Role of gamma radiation, *Polymer-Plastics Technology and Engineering*, 2009, **48**: 760–766.
81. M. A. Khan, J. Ganster and H. P. Fink, Natural and man-made cellulose fiber reinforced hybrid polypropylene composites, *Fifth Global Wood and Natural Fiber Composites Symposium*, Kassel, Germany, 2004.
82. M. A. Khan and J. A. Khan, Jute reinforced polymer corrugated sheet (JUTIN) and its opportunities, *Eighth Global WPC and Natural Fiber Composites Congress and Exhibition*, 22–23 June 2010, Stuttgart, Germany.

2 The use of flax fibres as reinforcements in composites

J. MÜSSIG and K. HAAG, Hochschule Bremen – University of Applied Sciences, Germany

DOI: 10.1533/9781782421276.1.35

Abstract: The flax fibre composite production chain consists of four main segments, from up- to downstream: (i) agricultural production of flax plants, (ii) the stem-to-fibre converting process, (iii) the converting process in which short and long flax are formed into semi-finished and finished textiles products and (iv) the industrial application of flax in composites. Within this production chain the quality management is key to introducing flax fibres into the composite market. Starting with the plant anatomy the reader will learn more about the hierarchical structure of the flax plant (*Linum usitatissimum* L.) and properties of the fibres. Finally a broad overview of already realized composite products is given in the application section and future trends are discussed.

Key words: flax, *Linum usitatissimum* L., flax fibre properties, cultivation and processing, textile and composites production applications.

2.1 Introduction

When talking about flax fibre one has to differentiate between fibre bundles and fibres (single plant cells) (Fig. 2.1). The properties of the plant material depend strongly on the size of the sample. The single plant cell is approximately 10 to 25 microns in cross-section and has a length of 10 to 25 mm (Bos and Donald, 1999). A fibre bundle consists of two to 20 elementary fibres (Charlet and Béakou, 2011) held together by the pectic substances of the middle lamella (Fig. 2.1) (Bos and Donald, 1999).

Flax fibre is a bast fibre of the flax plant (*Linum usitatissimum* L.) that, from a botanical point of view, belongs to the phloem tissue (Eder and Burgert, 2010). Figure 2.2 illustrates the location of the fibre bundles in the flax stem and shows its strongly hierarchical structure. The fibre bundles are located in the outer parts. The flax fibres forming the fibre bundle consist of several cell wall layers. One of the most important constituents is cellulose in the microfibrils in the cell wall which are several nanometres wide and many micrometres long (Astley and Donald, 2001). More precisely, the microfibrils are an agglomeration of micelle units that consist of several

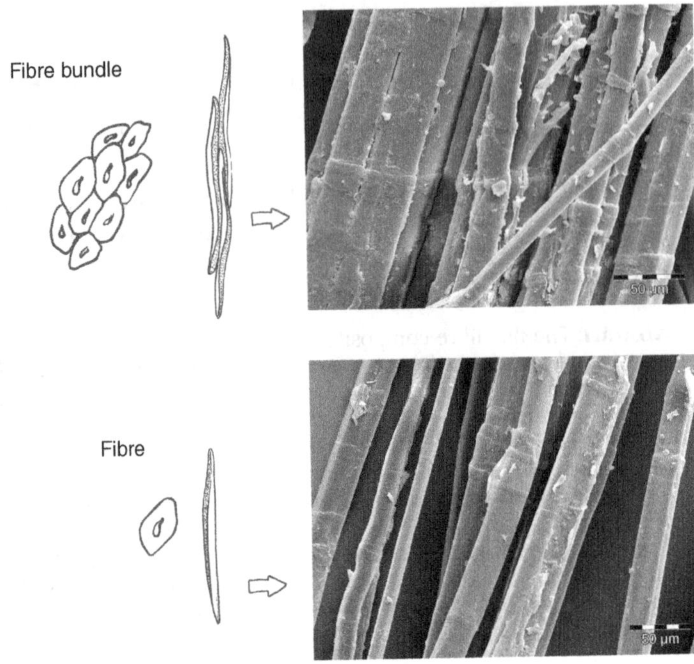

2.1 Schematic and SEM images of flax fibre bundles (upper part) and single flax fibres (lower part). Adapted from Müssig and Hughes (2012), pp. 46 and 48, by permission of JEC.

cellulose molecule chains, highly crystalline β-1,4-linked glucose units with a degree of polymerization of 1200 to 1500 (Bledzki *et al.*, 2008) (Fig. 2.2).

The cell wall of the flax fibre as for most plant fibres consists of different layers surrounding the lumen: an approximately 0.2 µm thin primary layer (Bos and Donald, 1999) and a strongly developed secondary layer subdivided in three layers S1 to S3, of which the S2 layer has the biggest dimensions (Eder and Burgert, 2010). The cellulose microfibrils and their arrangement in the layers are responsible for the mechanical strength and stiffness of the fibre. Their orientation is highly parallel within the layers. The angle of the microfibril orientation with respect to the main fibre axis is the so-called microfibril angle (MFA) (Fig. 2.3, Eder and Burgert, 2010).

An overview of the relationship between the microfibril angle and the Young's modulus for different plant fibres is given in Fig. 2.4. It can be seen that small microfibril angles correlate with high stiffness values. This can be explained by the high orientation of the load-carrying cellulose structures in the load direction. In plant fibres with high MFA values like coir (*Cocos nucifera*, mesocarp fibre with MFA of 45°, E = 3.3 to 5 GPa (Eder and Burgert, 2010)) the cellulose microfibrils are tilted at a 45° angle to the cell

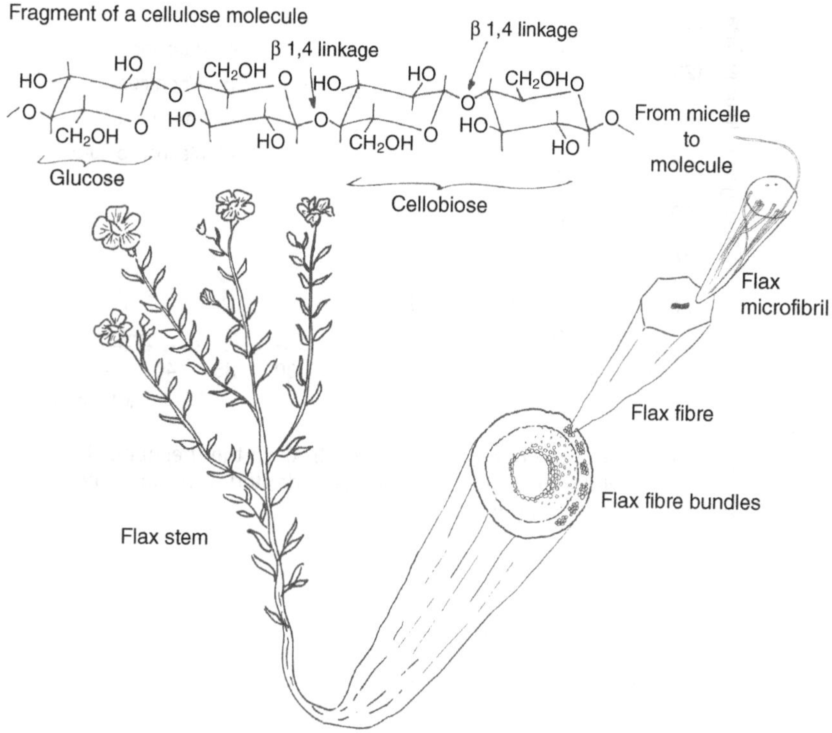

2.2 Hierarchical structure of the flax plant (*Linum usitatissimum* L.). The technically used fibre bundles are located in the outer part of the flax stem and consist of several flax fibres (Müssig and Hughes, 2012, p. 43, by permission of JEC).

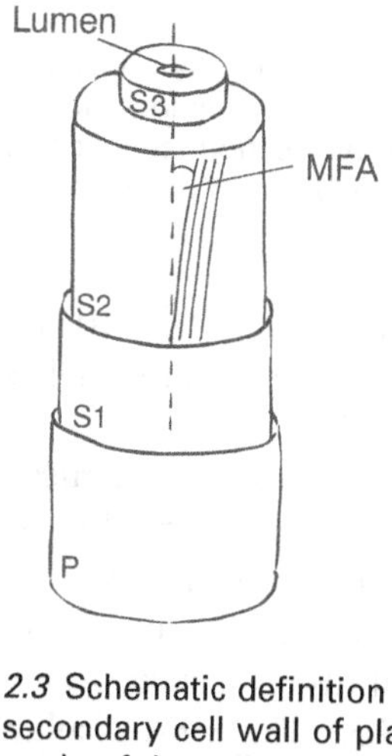

2.3 Schematic definition of the microfibril angle (MFA) in the secondary cell wall of plant fibres. It describes the main orientation angle of the cellulose fibrils in the S2 layer of the secondary cell wall in respect to the cell axis. Adapted from Müssig and Hughes (2012), p. 49, by permission of JEC.

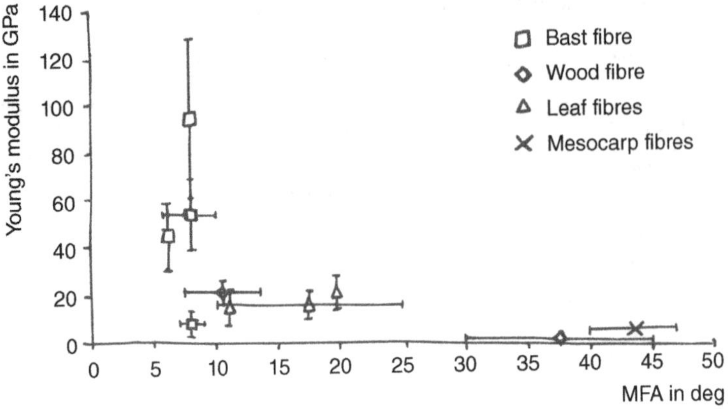

2.4 Influence of the microfibril angle (MFA) on the mechanical stiffness of different plant fibres. Values adapted from Eder and Burgert (2010), pp. 30 ff.

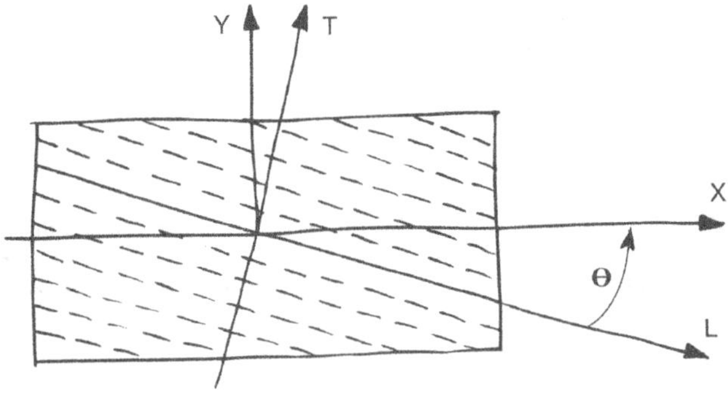

2.5 Model of the plant cell wall as a unidirectionally reinforced composite material. The reinforcing elements (cellulose fibrils) are tilted with the MFA θ against the loading direction (cell axis X). Adapted from Baley (2002), with permission from Elsevier.

axis, resulting in low stiffness values. With a MFA of 6 to 10° (Eder and Burgert, 2010) flax fibres are among the plant fibres with the highest tensile stiffness with a Young's modulus of up to 70 GPa.

The strong correlation between fibre orientation and mechanical stiffness shows a strong similarity to theories developed for fibre-reinforced composites such as glass fibre-reinforced composites. If these theories are applied to the flax fibre, the cell wall can be seen as a composite consisting of unidirectionally oriented cellulose microfibrils in a matrix of surrounding substances (mainly hemicelluloses and lignins). The MFA displays the tilt angle between fibre orientation and load direction (Baley, 2002, Fig. 2.5).

Baley (2002) calculated theoretical stiffness values for the flax fibre using the Halpin–Tsai model. Therefore a number of assumptions were made:

1. The S2 layer of the flax cell wall corresponds to a short fibre-reinforced composite.
2. Its composition is homogeneous.
3. The adhesion between fibre and matrix is perfect.
4. Fibres are free of defects.

Taking the following adaptations into account the elastic properties of a unidirectional composite can be estimated:

1. The Young's modulus varies with deformation.
2. The Young's modulus depends on the MFA and the cellulose content of the cell wall.
3. The cellulose microfibrils reorient themselves in the direction of the applied load during tensile loading.

The results of the calculation can be seen in Fig. 2.6. The microfibril angle was assumed to be 10° at the start of the experiment and decreased to 0° during deformation. As the cellulose content reported in the literature varies between 64% (solid line) and 74% (dashed line), the calculation was performed for both values and compared to experimental values (diamonds) (Fig. 2.6). Except for the first region of low displacement the calculated

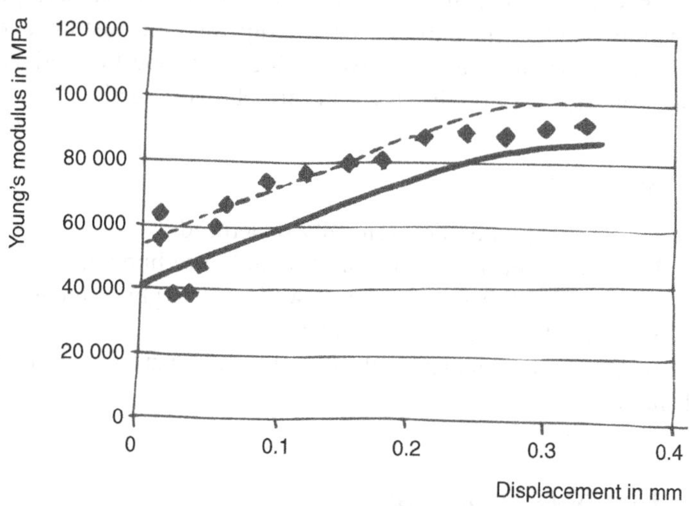

2.6 Comparison of experimental (black diamonds) and calculated (lines) Young's modulus of a flax fibre. For the calculation a cellulose content of 64% (solid line) and 74% (dashed line) was approximated referring to the literature. Adapted from Baley (2002), with permission from Elsevier.

Table 2.1 Factors influencing the fibre properties and calculation models

Factor	Influence on fibre properties	Model used	References
Cellulose content	Strength and Young's modulus increase with increasing cellulose content	Rule of mixtures	Used by many authors as a basic calculation model
Cellulose content	Strength and Young's modulus increase with increasing cellulose content	Halpin–Tsai Model	Baley (2002)
MFA	Young's modulus increases with decreasing MFA	Halpin–Tsai Model	Baley (2002)

values show a good agreement with the experimental data. The deviation at short displacements can be explained by the extension of kink bands and effects of plasticity associated with the reorientation processes of the fibrils (Baley, 2002).

An overview of different calculation models and the influence of different factors on the fibre properties is given in Table 2.1. If theoretically calculated values for the mechanical properties of flax fibres are compared to experimental results, the experimental setup and the testing conditions should be noted. Their influence as well as the key properties and variations of those properties will be discussed in the following section.

2.1.1 Summary

- The flax stem is highly hierarchically structured.
- One has to differentiate between fibre and fibre bundle.
- The structure of the flax fibre cell wall is strongly influenced by the microfibril angle of the cellulose fibrils.
- The mechanical properties of the cell wall can be described with calculation models.

2.2 Key fibre properties

As already discussed above, it is important to distinguish between a single cell (fibre) and a group of fibres forming a bundle in the stem (Fig. 2.7). This is not only important from a terminology point of view but may be very useful for selecting appropriate methods to test fibre properties and

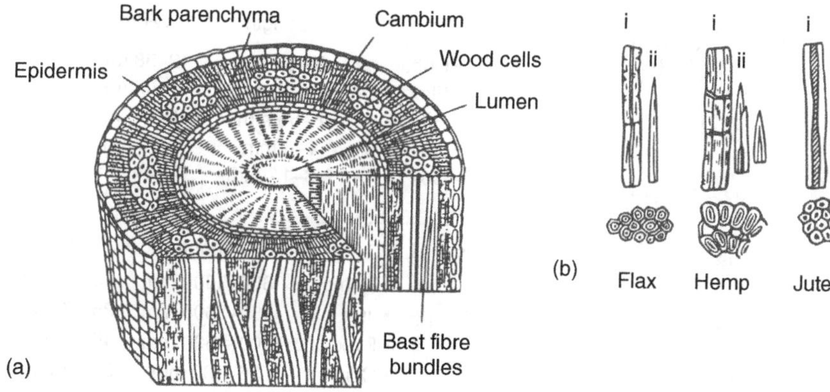

2.7 (a) Schematic cross-section of a flax stem; (b) fibre bundles of flax, hemp and jute in cross-sectional view with (i) the middle part of the fibre and (ii) the final part of the fibre. Adapted from Schönfeld (1955), p. 67.

may be essential for the correct interpretation of mechanical and morphological test results and literature values.

As can be seen in Fig. 2.7(b), the fibre and the fibre bundle morphology often vary considerably between different bast fibre plants. Herzog (1930) described the variation of cell morphology, cell lumen and cell wall thickness as early as the beginning of the 1920s for flax. In Fig. 2.8(a) flax fibres arranged in bundles with large variations in cell wall thickness are illustrated from flax fibre field experiments in Sorau (today Żary, Poland) from the early 1920s (Herzog, 1930). Figure 2.8(b) illustrates the vast possible variations in cell morphology in one bast fibre genus. Hence, the identification of bast fibres based on a cross-sectional surface investigation can sometimes be unclear or ambiguous.

As the morphology of the fibres differs, so do their dimensions. In Fig. 2.8(b) the length and width values of single fibres (i) and fibre bundles (ii) from selected fibre plants are summarized from literature data.

As well as the difficulties in the interpretation of morphological data from literature if the authors mix up the terms for fibre and fibre bundle, problems can occur during testing and evaluating the mechanical properties of bast fibres. Both can lead to inaccurate or questionable results. Fig. 2.9 shows possible compositions of the samples used to investigate the tensile properties of flax.

Briefly summarized, the essential points of the different compositions of the samples, as shown in Fig. 2.9, are:

- *Single-element tests versus collective tests.* Single-element tests provide more detailed data on tensile properties. Single-element tests show

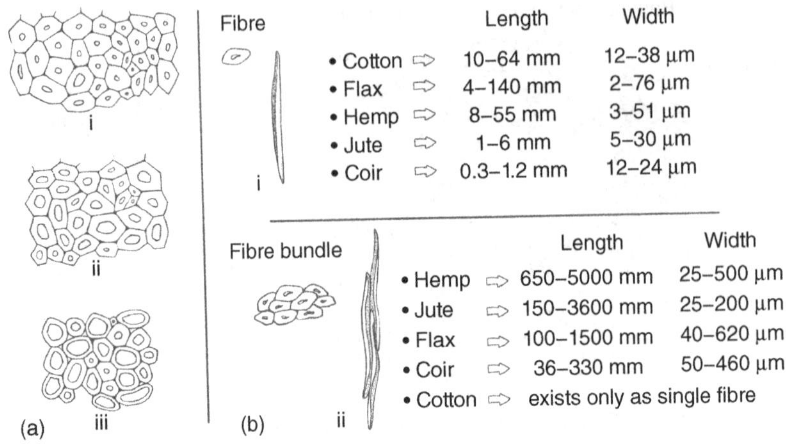

2.8 (a) Flax fibres arranged in bundles with large variations in cell wall thickness; three flax fibre lines (Sorau (today Żary, Poland), 2000 seeds/m^2, standard mineral fertilization, plant height: 90 cm, cross-section out of the middle of the stem), (i) favourable dense type, (ii) unfavourable type and (iii) very unfavourable low density type. Adapted from Herzog (1930), p. 147. (b) Length and width values of single fibres (i) and of fibre bundles (ii) (values taken from a literature survey from Müssig *et al.*, 2010).

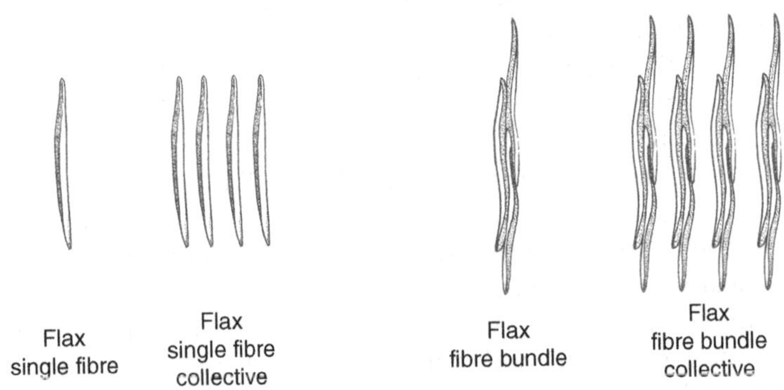

2.9 Various forms of flax fibres/fibre bundles. The illustrations show the difference between a single element (single fibre or single fibre bundle) and a collective (collective of single fibres or collective of single fibre bundles). Adapted from Müssig (2001).

higher scattering, and compared to collective tests more experiments are necessary to get reliable values.

- *Collective tests versus single-element tests.* In general, collective tests show lower strength values. The reduction in collective strength efficiency within a collective is realized by (1) the slack in the collective and (2) the variation in single-element breaking elongation (Suh *et al.*, 1994).

- *Fibre versus fibre bundle.* Fibres tested in a single-element test tend to show higher strength values compared to tested bundles. This effect increases with increasing gauge length, especially when the gauge length of the fibre bundle gets larger compared to the length of the single fibre (Bos *et al.*, 2002).

As described by Müssig and Hughes (2012), it should be noted that several important factors strongly influence the tensile properties of bast fibres like flax. These include (1) the testing speed, (2) the cross sectional area, (3) the gauge length, (4) the number of specimens tested and (5) the entity (i.e. fibre or fibre bundle) being tested.

The influence of changing the gauge length is shown in Fig. 2.10. For a gauge length shorter than the length of the single fibre cell, the fibre is clamped at both ends and the properties of the cell wall are measured. Using slightly longer gauge lengths of approximately 10–20 mm, some fibres will still be fixed at one end but others will not. Thus, the properties will vary between those of the cell wall and those achieved by fibre bundle testing at higher gauge lengths. In the case of gauge lengths longer than the length of the fibre cell the shear forces within the fibre bundle are measured (Charlet and Béakou, 2011).

Table 2.2 gives an overview of the physical and mechanical properties of selected natural fibres. Taking the previous discussion into account, these values should be seen as representative guiding values only.

Besides the above, the chemical composition of a flax fibre is an important factor in determining the performance properties of a natural fibre product. Next to the physical properties of natural fibres listed in Table 2.2, the

Table 2.2 Mechanical properties of selected natural fibres

Property	Flax	Hemp	Jute	Coir	Cotton
Tensile strength (MPa)	700	800	500	200	450
Young's modulus (GPa)	70	65	30	5	8
Elongation at break (%)	3	3	1.8	30	8
MFA (°)	5–10	2–6	7–10	30–49	20–30
Density (g/cm^3)	1.4–1.52	1.4–1.6	1.3–1.5	1.15–1.5	1.5–1.6

Note: Naturally the measured properties show a range. To give the reader an overview the most frequently published data are shown (mean values taken from a literature review by Müssig *et al.*, 2010).

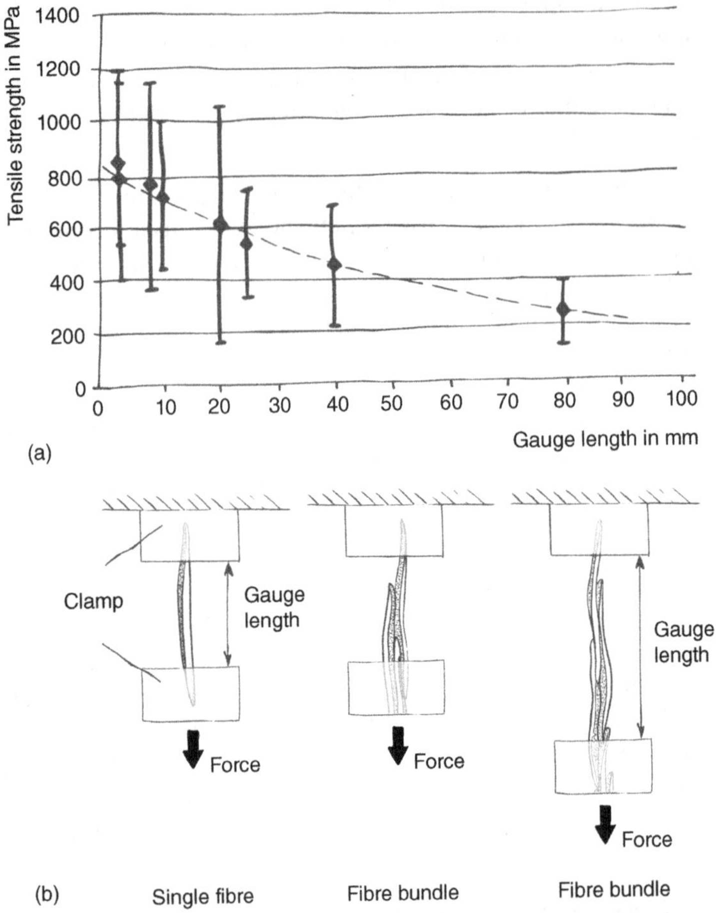

2.10 Effect of gauge length on tensile strength of flax fibre bundles. (a) Dependency of tensile strength from gauge length. Adapted from Charlet and Béakou (2011), with permission from Elsevier. (b) Effects of gauge length explaining the dependency. Adapted from Müssig and Hughes (2012), p. 50, by permission of JEC.

chemical composition of the fibre has to be taken into account, if natural fibres are used as reinforcement in a composite structure. The most important aspects to consider are (1) fibre/matrix adhesion, (2) thermal stability during processing and use, (3) chemical stability during processing and use, (4) hygroscopic behaviour and (5) resistance towards microorganisms. Table 2.3 gives an overview of the chemical composition of flax compared to some other natural fibres. As these values depend on the degree of treatment during processing, the data are provided for comparison purposes only. In conclusion, flax shows a high cellulose content that increases with

Table 2.3 Chemical composition (%) of selected natural fibres

Substance	Flax	Hemp	Jute	Coir	Cotton
Cellulose	70	70	65	40	90
Hemi-cellulose	17	16	15	0.2	4
Lignin	2.5	6	10	43	0.7
Pectin	2	1	1.5	3	4
Fat/wax	1.5	0.7	0.5		0.6
Ash	1.5	1.5	0.4		1.4
Water solubles	6	1	1	4.5	0.7

Note: As a result of the considerable naturally occurring variation in these values, the data shown represent typical values most frequently published (values taken from a literature review by Müssig *et al.*, 2010).

higher retting degrees[1] (compare Section 2.3, 'Cultivation and quality issues') as well as after enzymatic or chemical separation, because water solubles, waxes and pectins can be removed during retting and separation. The lignification is lower compared to other bast fibres, which makes the separation and textiles processing gentler.

As discussed before, the chemical composition of the flax fibre is an important factor to consider during many different steps of composite manufacturing and the usage phase. Thermal stability is one of the most important factors during the processing of composite materials, as thermal damage of the fibre strongly decreases the composite properties. Van de Velde and Baetens (2001) tested different long flax samples for their thermal properties. The thermal stability was, *inter alia*, evaluated by thermogravimetric analysis (TGA). Mass changes versus temperature for scutched long flax (longitudinal flax) varying in retting degree were measured. Figure 2.11 shows the TG and DTG curves of the fibre bundles measured in nitrogen and air atmosphere at increasing temperature.

As can be seen in Fig. 2.11, generally three peaks can be identified in the DTG curve (Van de Velde and Baetens, 2001):

- The first peak is related to moisture release.
- The second peak corresponds to degradation of hemicellulose and cellulose.
- The third peak is related to degradation of non-cellulosic components.
- The remaining mass after the third peak at stabilization represents the percentage of ash; this corresponds, according to Table 2.3, to a value of 1.5% for flax.

[1] 'Flax [...] undergoes a process called retting which is usually microbial in nature, to loosen and separate the bast fibre bundles from the non-fibre fractions of the flax stem.' (Akin, 2010).

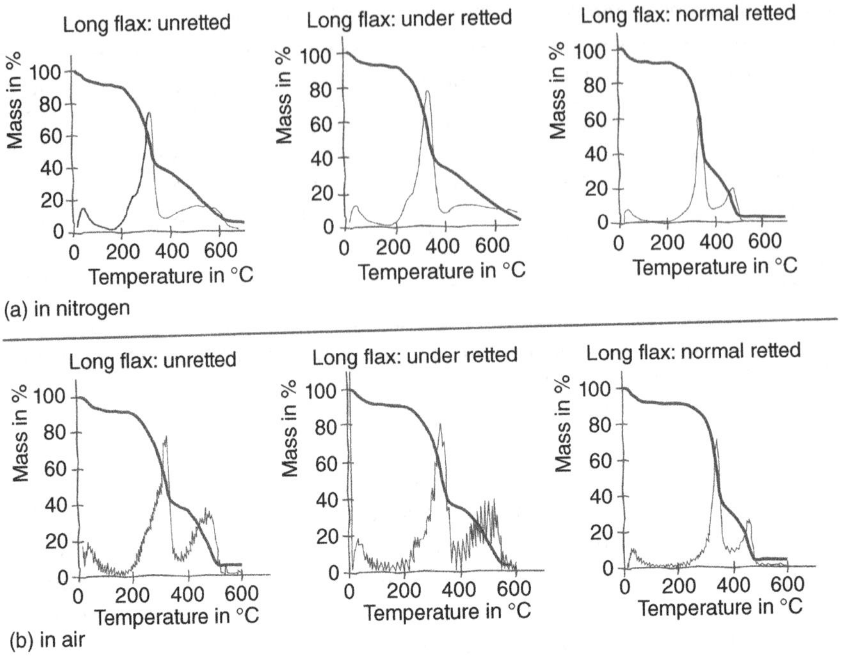

2.11 (a) TG curves of scutched long flax with varying retting degrees measured in nitrogen atmosphere at increasing temperature; (b) TG curves of scutched long flax with varying retting degrees measured in air atmosphere at increasing temperature. The mass losses are marked by thick lines; in addition to the TG curve the differential curve (DTG – thin lines) is given as change in sample mass as a function of time (d*m*(T)/d*t*). Adapted from Van de Velde and Baetens (2001), © 2001 Wiley-VCH Verlag GmbH, Weinheim, Germany.

Clear differences in thermal properties concerning the varying retting degree can be observed (see Fig. 2.11) (Van de Velde and Baetens, 2001):

- Exposure of flax fibres to a temperature of 120°C results in removal of moisture and degradation of waxes. As retting leads to removal of the waxes, the degradation effect decreases with increasing retting degree.
- A higher moisture content for fibres of lower retting degree can be found for measurements conducted under a nitrogen atmosphere.
- Mass changes at the second peak correspond to cellulosic components and increase with a higher retting degree (under air and nitrogen atmosphere).
- The shoulder at the second peak is suppressed or disappears at a higher retting degree. Van de Velde and Baetens (2001) suppose a change in pectin content and other components which are degraded and extracted during retting.

Based on their thermal and mechanical evaluation, Van de Velde and Baetens (2001) proposed the following: 'During production of natural fibre-reinforced plastics, short periods of exposure to high temperatures are allowed. Composite production temperatures higher than 180°C have to be avoided, unless short periods are respected to minimise the damage of the fibre.'

Besides the fact that temperature has to be controlled during composites production, the moisture content of the fibres is an often discussed issue in the field of natural fibre-reinforced composite production. According to Bourmaud *et al.* (2010) and Van de Velde and Baetens (2001) flax fibres contain approximately 7 to 10 mass% of moisture measured by TGA dynamic heating. A drying step before composite production will lead to a degradation of the cell wall components depending on the temperature (at 120°C for wax, above 180°C for pectin and above 230°C for cellulose). Stamboulis *et al.* (2001) and Baley *et al.* (2005) have already shown that exposure of flax fibres to high temperature leads to a decrease in strength and to an increase of defects inside the cell wall. Temperature treatment influences the sorption and swelling properties of lignocellulosic materials, too. It is well known that thermally treated wood shows changes in the void structure and chemisorption compared to untreated wood. Additionally, crosslinking between cellulose molecules takes place at higher temperatures. Depending on the humidity, water can be attached in the lignocellulosic cell wall by chemisorption, physisorption or capillary condensation (Popper *et al.*, 2005).

Water in flax fibres is mainly bound to pectins (Baley *et al.*, 2012) as well as inside the polymeric matrix between the cellulose mesofibrils[2] (Morvan *et al.*, 2003). Depending on the kind of sorption, the energy to separate the water from the cell wall may vary markedly. Baley *et al.* (2012) have investigated the change in moisture content in flax by TGA dynamic heating. Figure 2.12 shows the TG curves of dew-retted, hackled long flax held at 105°C for 14 hours and at 150°C for 5 hours.

The observations made by Baley *et al.* (2012) based on the TGA dynamic heating can be summarized as follows: heating of the flax at 105°C for 14 hours will not lead to a complete extraction of the strongly linked water located in the polysaccharide matrix. Increasing the temperature to 150°C enables a complete release of water. The observation shows that the final mass loss after 5 hours at 150°C was 7.1% and that the bound water needs high temperatures to be extracted.

The exact percentages of different cell wall constituents depend on the variety as well as on cultivation and processing, as addressed in the following section.

[2] The mesofibril has a diameter of 0.5 μm and is a substructure of the cell wall between the cell (single fibre) and the microfibril (see Fig. 2.2).

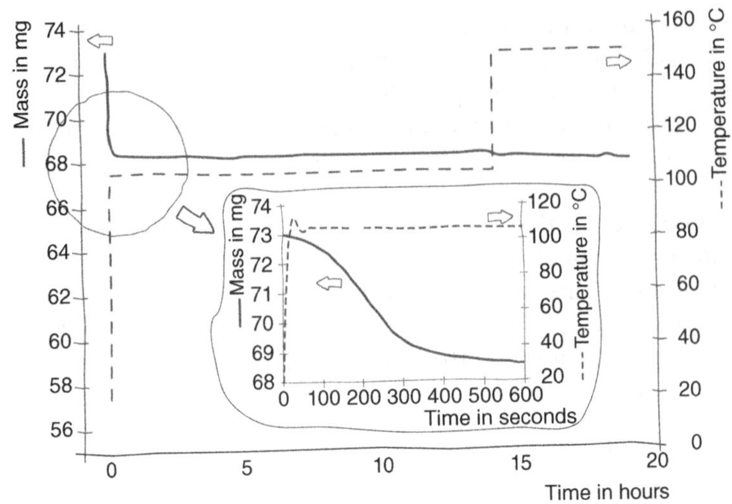

2.12 TG curves of dew-retted, hackled long flax (variety 'Ariane') held at 105 °C for 14 hours and at 150 °C for 5 hours. The insert shows the initial part of the TG curve at higher magnification. Adapted from Baley *et al.* (2012), with permission from Elsevier.

2.2.1 Summary

- For a single flax fibre a mean fibre length of 25 mm and a mean fibre width of 20 μm can be used as values for guidance purposes.
- Because of a very low MFA (5–10°) flax fibres show high strength (more than 1000 MPa) and Young's modulus values up to 70 GPa.
- Flax shows a high cellulose content and a relatively low lignification which makes the fibres much better processable compared to higher-lignified bast fibres like jute.
- The water content of flax fibre after storage in a standard climate shows values between 7 and 10 mass%. Drying at high temperature will lead to a decrease in strength, to an increase of defects inside the cell wall and to a possible change of the cell wall chemistry.
- Composite production temperatures should be significantly below 200°C to minimize the risk of an eventual degradation of the fibres.

2.3 Cultivation and quality issues

According to Schilling (1930), flax is amongst the oldest cultivated crops in the world. The flax plant as a resource for oil and fibres was already being used in ancient Egypt. Unger (1866) found parts of flax yarn and capsules of flax in bricks used for the Dahshur brick pyramids of the fourth dynasty.

2.13 An old representation of the preparation of flax – cultivation, harvesting, retting, drying, breaking, scutching and hackling. Cited in Müller (1930), p. 216.

Flax-based products were used in Thebes as a sacrifice found in the necropolis of Dra' Abu el-Naga' (2400–2200 BC) (Schweinfurth, 1884). The work of Vogelsang-Eastwood provides a comprehensive representation of ancient Egyptian materials and technology in the field of textiles, pointing out the importance of flax as an agricultural crop. The majority of ancient Egyptian textiles were made of flax fibres (Vogelsang-Eastwood, 2000). Since that time, flax has been an important fibre resource for mankind. The steps for processing the plant have not undergone significant changes regarding the main processing concepts illustrated in Fig. 2.13.

Flax plants have been cultivated in Europe for centuries in order to use their fibres for textile applications. An overview of the traditional flax value-added chain is given in Fig. 2.14.

Based on the classification by Kulpa and Danert proposed in 1962, Mansfeld's Encyclopedia of Agricultural and Horticultural Crops divides *Linum usitatissimum* L. into four convarieties (Mansfeld, 2001):

- *Linum usitatissimum* convar. *crepitans* (taxon with shattering capsules; not commercially cultivated)
- *Linum usitatissimum* convar. *elongatum* (taxon with small seeds and long stems grown for fibre production; see Fig. 2.15(a) (i))

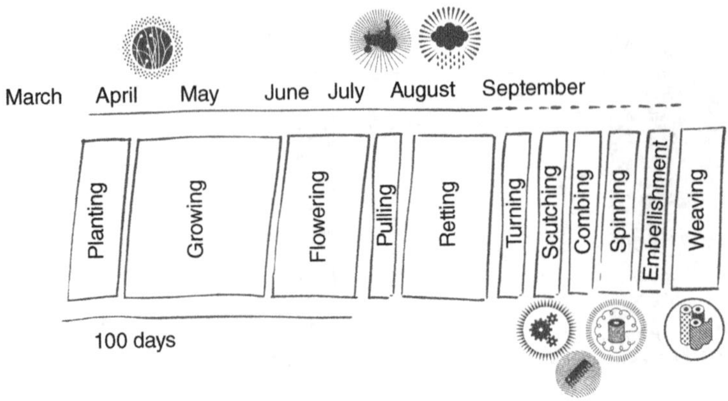

2.14 The traditional value-added chain from flax plants to textile products (Müssig and Hughes, 2012, by permission of JEC).

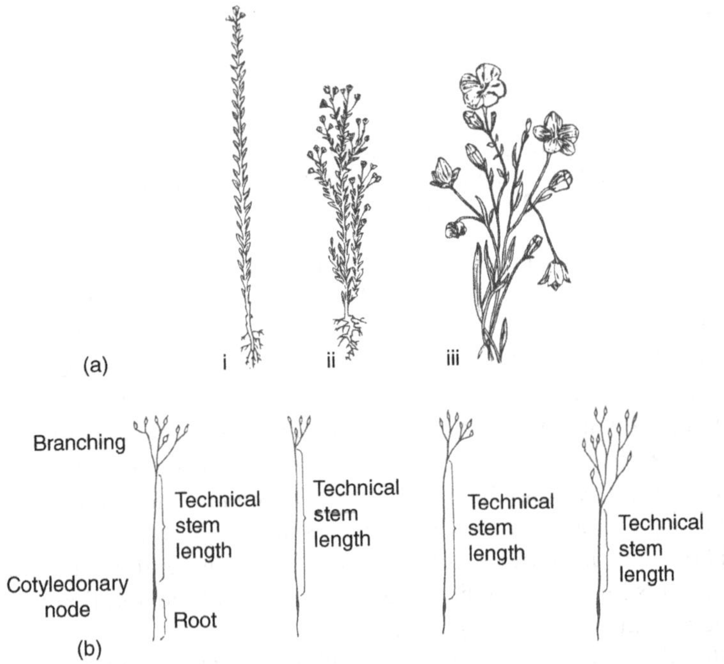

2.15 (a): (i) Fibre flax plant (unbranched), (ii) oil flax plant (branched) and (iii) flax flower; (b): Flax stems with wide variation in branching morphology resulting in different technical stem lengths. Adapted from Schönfeld (1955), p. 2.

- *Linum usitatissimum* convar. *mediterraneum* (taxon with short stems and large capsules and seeds for obtaining linseed oil preferably in warmer regions; see Fig. 2.15(a) (ii))
- *Linum usitatissimum* convar. *usitatissimum* (taxon with elongated stems and small capsules and seeds often basal-branched; intermediate forms have been selected which produce fibre and good yield of seeds (namely combination flax)).

Within the classical concept of fibre production (compare Fig. 2.14) unbranched stems at the highest possible length are favoured. The term 'technical stem length' (compare Fig. 2.15) is used to quantify the processability of the flax stems to get the longest possible fibre bundles.

Currently, breeding efforts are being made to develop new flax varieties optimized for use in composites. The main objectives are high yields and good mechanical properties, easier decortication of the fibre bundles and an optimized interface between the flax fibre bundles and polymeric matrix systems in the composite (Plant 2030, 2013; Haag and Müssig, 2013).

The process of separating the fibres or fibre bundles from the flax stems generally starts in the field. During harvesting, the stems are pulled and spread on the ground. Flax then traditionally undergoes a process known as retting to loosen and separate the bast fibre bundles from the remaining stem tissue. During retting, microorganisms, in particular fungi, colonize the stem and the enzymes secreted by these organisms degrade the pectins that are largely responsible for binding the bast fibres to the stem tissue (Müssig and Hughes, 2012).

After the retting process, the stems are collected and transported to the decortication and separation facilities. In the decortication process the woody core material of the stem, known as the 'shive', is broken away and mechanically separated from the fibre bundles. Further separation processes may be carried out to further refine the fibre bundles. As can be seen in Fig. 2.16 there are different possible separation processes. Currently, only the mechanical separation and to some extent the chemical and physical/chemical separation processes are used industrially. Enzymatic separation is currently in a transition state from research to commercial reality (Akin, 2010; Fischer and Müssig, 2010).

In the traditional long flax processing the orientation of the stems and the fibre bundles is maintained during the entire process from harvest to the final product to prevent entanglement. In so-called 'total fibre lines', however, the orientation is not controlled in the same manner, resulting in the fibres being more or less randomly oriented. The following definitions help to distinguish between the processing techniques: (1) longitudinal flax, in which the fibre bundles are oriented in only one direction, and (2) disordered flax, in which the fibre bundles have no preferred orientation.

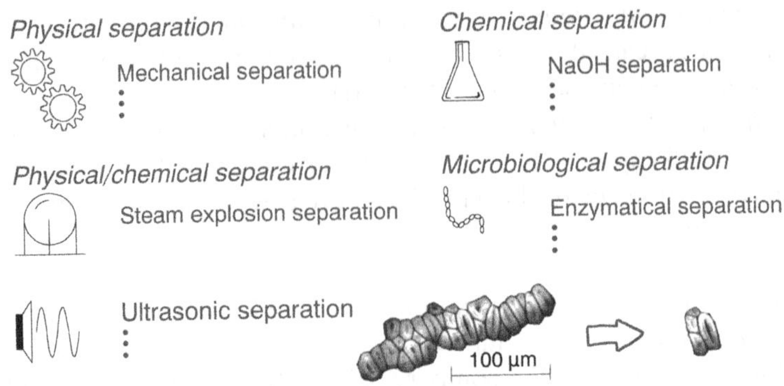

2.16 Separation techniques to separate and refine fibre bundles from flax or hemp. Adapted from Müssig (2001).

Figure 2.17 displays the process definition for longitudinal and disordered flax as a flow diagram.

Figures 2.17 (right-hand side) and 2.18 provide an overview of the traditional processing of long flax which begins with dew or water retted flax being broken to remove the woody core material of the stem, known as the 'shive', and fibre apart. This is followed by scutching and hackling, where the fibre bundles are further cleaned to remove any remaining shives. The orientation of the stems and the fibre bundles is maintained during the entire process from harvest to the final product so as to prevent entanglement and to facilitate easier spinning. During scutching and hackling short fibre bundles are separated in the form of tows.

The harvested area of *Linum usitatissimum* used for seed and oil production (Fig. 2.19) is much larger compared to the cultivation of fibre flax (Fig. 2.20). Next to the latest data from 2011 the summarized data from 2001 to 2011 are presented.

According to the data in Fig. 2.19, Canada, India, China, the USA and Ethiopia played a dominant role in worldwide flax oil production between 2001 and 2011. In 2011 the harvested area was 2 million ha, dominated by the Russian Federation, China, India and Canada (FAO, 2013).

Flax for fibre production was cultivated between 2001 and 2011 on about 390,000 ha per year (see Fig. 2.20). In this period the five main producing countries were China, the Russian Federation, France, Belarus and Belgium. In 2011 the worldwide cultivated area was around 250,000 ha dominated by France and Belarus (FAO, 2013).

According to Reux and Verpoest (2012b) the average yield of fibre flax straw has been around 7 t/ha over the last 19 years in France. A 13% variability from year to year and an upward annual increase of 5% were

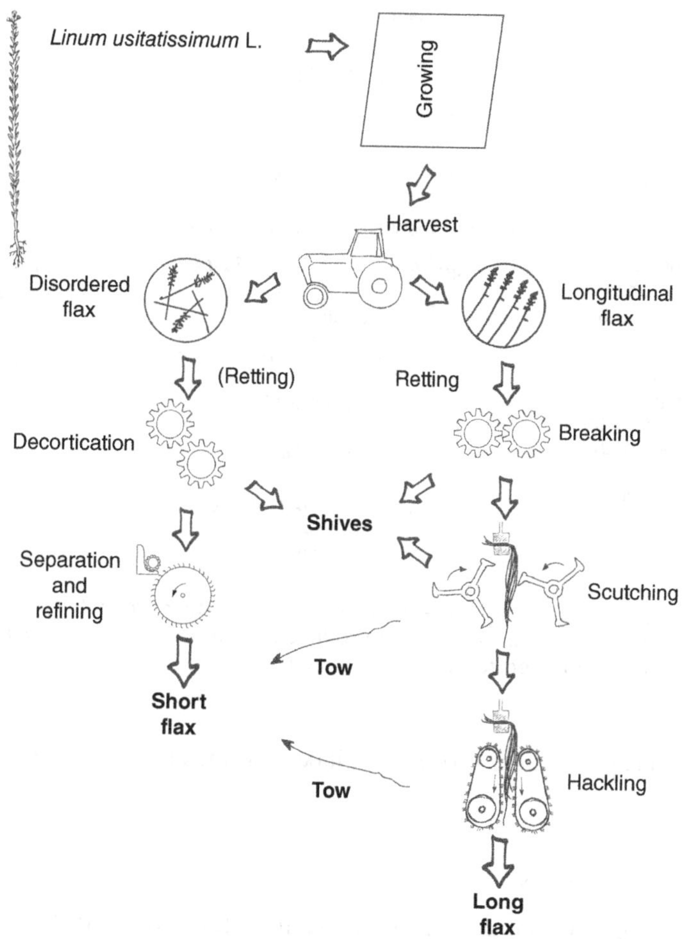

2.17 Longitudinal flax processing resulting in long flax versus disordered flax processing producing short flax.

observed for that period. Based on a study of Fibres Recherche Développement – FRD, Troyes, France, the straw yield of fibre flax was 6.9 t/ha in 2011. For example, in France, the harvested fibre flax is dew- or field-retted and further processed to long flax (see Fig. 2.18). The amount of long flax, according to the study of FRD, is 15–25% (1.0–1.7 t/ha) whereas the amount of short flax (tow) is 10–15% (0.7–1.0 t/ha). The shive content is 45–50%, resulting in a value of 3.1–3.5 tonnes of shives per hectare. The accumulated mass of dust and losses is about 0.7 t/ha and comprises 10% of the total harvested mass (Reux and Verpoest, 2012b).

After the first converting process short and long flax can be further processed to semi-finished textiles products for composite applications. The

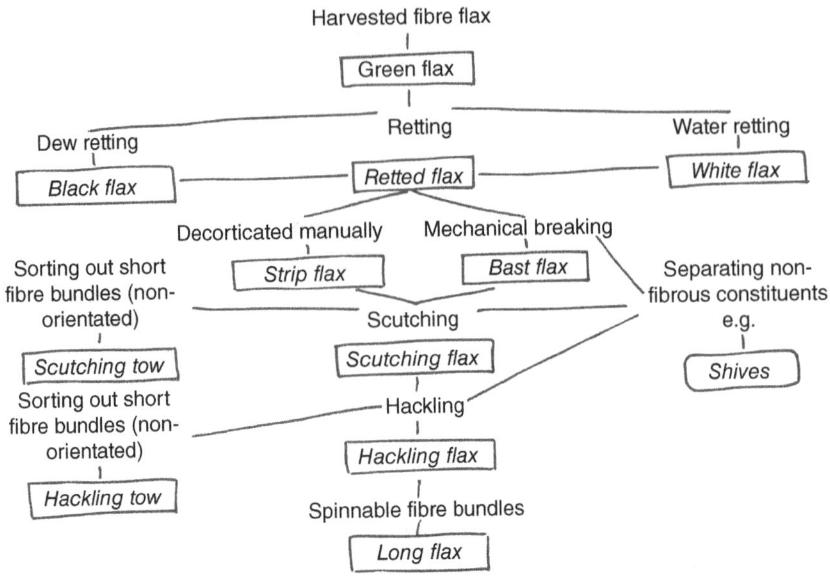

2.18 A schematic overview of the systematic nomenclature used in traditional flax processing products. From hemp (Schnegelsberg, 1999) adapted to flax (Müssig, 2001).

main concepts of these processes will be described in detail in the following section.

2.3.1 Summary

- The flax fibre industry is a mature industry and the different process steps are well adapted to the production of high-quality fibres.
- Long flax as well as short flax is suitable for different kinds of composite applications.
- The harvested area of oil flax displays a considerable quantity of short oil flax fibres. The processing technique for oil flax fibre use needs to be adapted and optimized to develop alternative fibre supply sources for the composite industry.

2.4 Processing as a fibre reinforcement for composites

As already mentioned in the previous section, the flax fibre composite production chain consists of four main segments. Following the agricultural production of flax plants and the stem-to-fibre conversion, the downstream

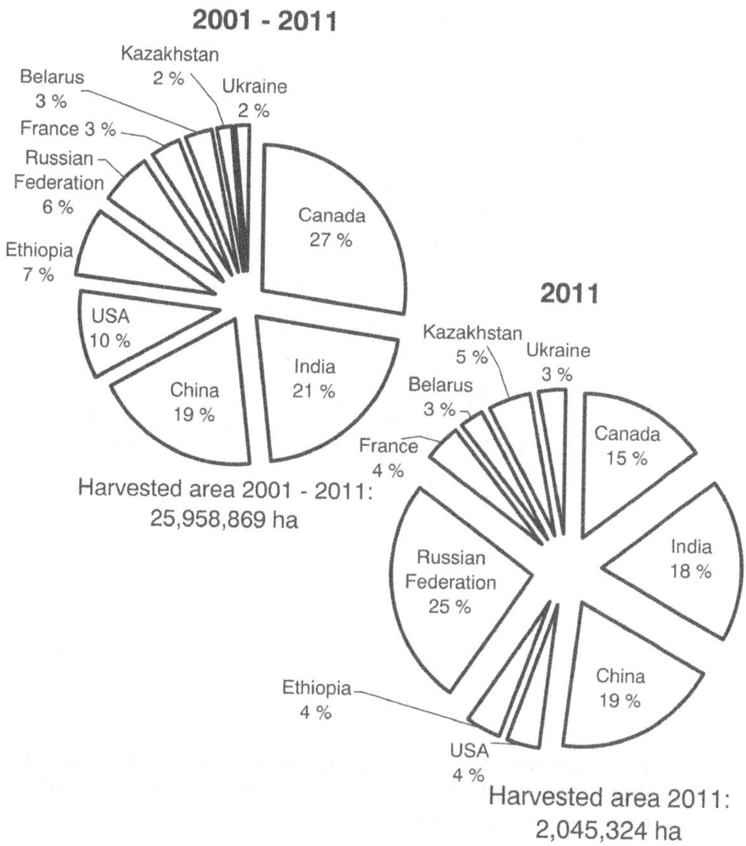

2.19 **Left: harvested area of oil flax from 2001 to 2011. Right: harvested area in 2011; the 10 most important countries are mentioned (FAO, 2013).**

process starts with the converting process during which short and long flax are transformed into semi-finished and finished textile products.

Figure 2.21 gives an overview of the different flax textiles which are used not only for technical textiles and apparel fabrics but also for composite manufacturing. Composites are produced by the addition of a polymer matrix to raw fibres or to textile products in the form of special granules or fibre pellets, sliver, yarn, weaves, knitted goods, fleeces or needle felts. Sliver is a textile ribbon-shaped entity (morphologically). Normally it is an assemblage of untwisted parallelized staple fibres or fibre bundles produced by a carding or combing machine and ready for drawing, roving or spinning (Schnegelsberg, 1971; Von Acker, 2012).

Ropes, weaves, knitted goods and knotted nets can be produced from yarns. Wadding fleeces and felts, for example, are produced by aerodynamic or mechanical laying of short flax.

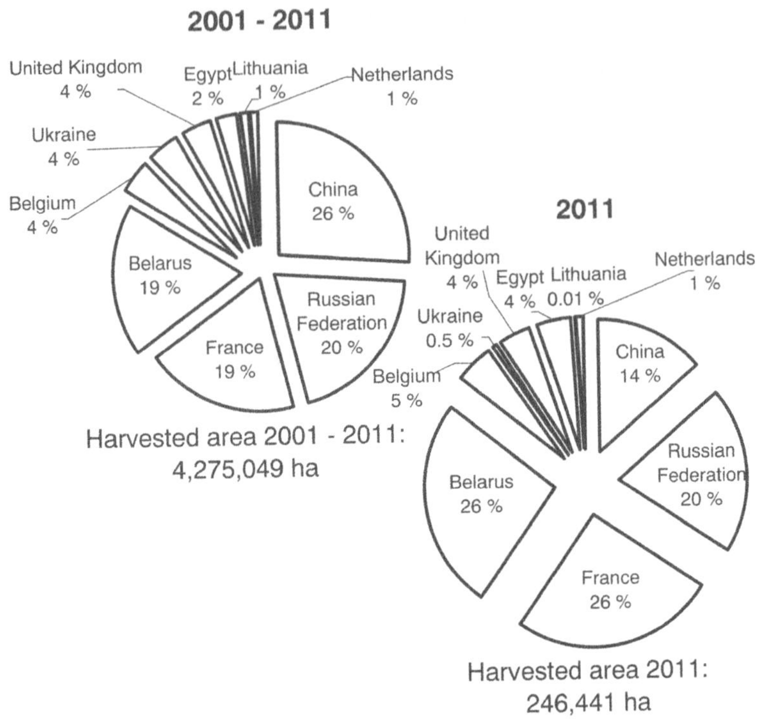

2.20 Left: harvested area of flax fibre and tow from 2001 to 2011. Right: harvested area in 2011; the 10 most important countries are mentioned (FAO, 2013).

The textile terms used in this chapter have been prepared in accordance with the *systematic of textiles* provided by Schnegelsberg (1971, 1999). The definition of the terms for the latest developments in the field of new flax textiles for composite applications (e.g. low twist yarn, low crimp weave or comingled roving) are taken from Von Acker (2012) and are also based on the *systematic of textiles* provided by Schnegelsberg (1971, 1999).

The various types of planar arranged flax fibre textiles (often referred to as 'non-woven') are shown in Fig. 2.22. This can be done by aerodynamic, mechanical or hydraulic laying systems. For mechanical laying, carding machines (card or roller card) are used in most cases. This process converts the flock (a tangled mass of fibres/fibre bundles) to a planar web while cleaning and mixing the flax. According to Fig. 2.22 three different types of planar arranged flax fibre textiles can be distinguished, depending on the nature of bonding. While in a wadding the fibres/fibre bundles are held together by friction, a fleece is, morphologically, a planar arranged entity which is bonded by a binder or other vinculat (Schnegelsberg, 1971). The fibres for a felt can be oriented by a roller carding machine or an aerodynamic

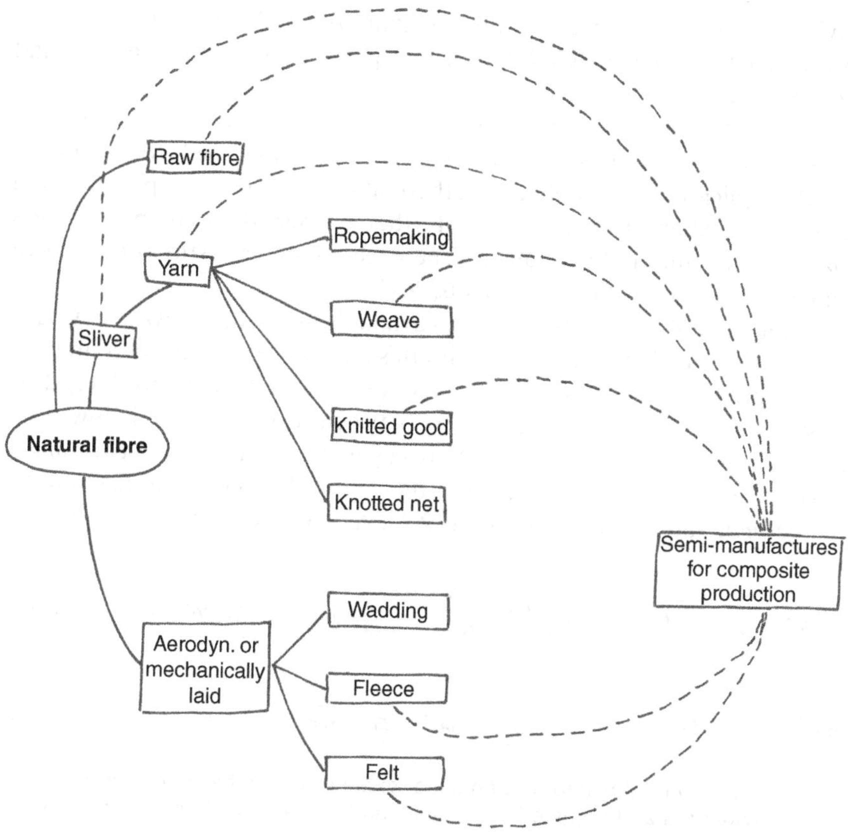

2.21 Different uses of flax in textile and composite applications. Adapted and adjusted from Graupner and Müssig (2010).

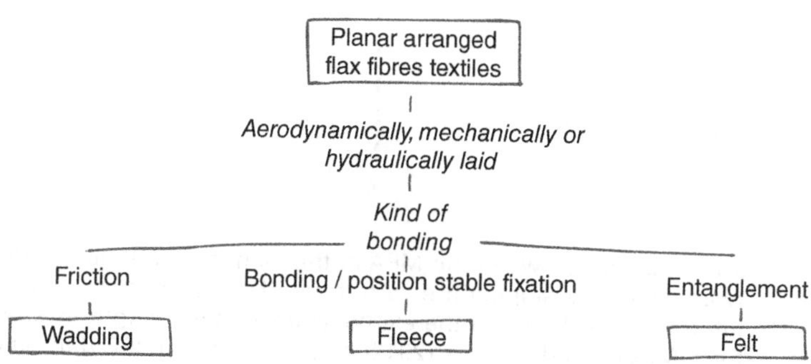

2.22 Planar arranged flax fibre textiles.

laying machine and processed into a planar arranged entity. They are then processed to a needle felt by a needle process using fibre tangles and adjustable needle density.

Most of the planar arranged flax fibre textiles in the form of wadding, fleece and felt (mainly based on short flax) are used for technical textile applications like upholstery, thermal insulation, sound proofing and geotextiles. Alternatively, they are further processed to composite parts mainly in the automotive industry. The traditional flax textile process with long flax is shown in Fig. 2.23 (Baets, 2012).

As the natural fibre bundles are not available in an endless way, they have to be transformed into yarns and with this intermediate step a certain twist of the fibre bundles is introduced to the semi-finished textile products. This twist angle present in the yarn can vary between very low values in low-twist yarns and 45° in high-twist yarns. A strong analogy between the twist angle in yarns and the microfibril angle in the cell wall described in Section 2.1 can be seen and is illustrated in Fig. 2.24. As observed for different MFAs

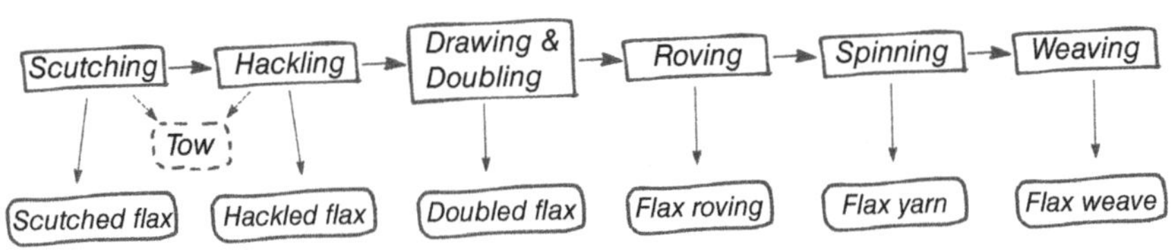

2.23 Semi-finished and finished textile products based on long flax. Adapted and adjusted from Baets and Pariset (2012), by permission of JEC.

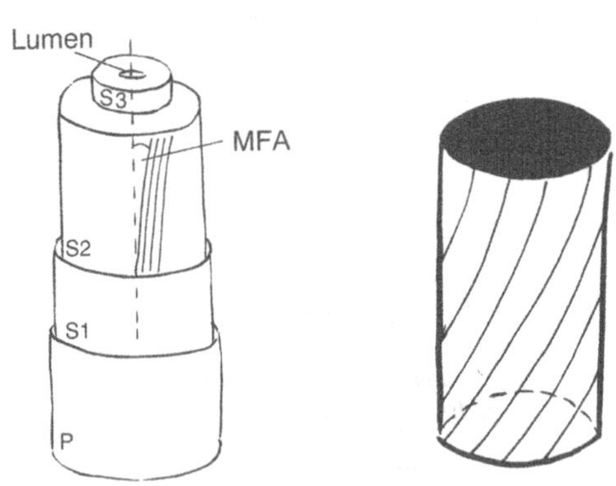

2.24 Analogy between the MFA in the plant cell wall (left) and the twist angle in a staple fibre yarn (right). Both angles are defined as a tilt in the orientation of the elements against the length axis. Adapted from Müssig and Hughes (2012) by permission of JEC (left) and from Shah *et al.* (2012), © 2012 by Sage, reprinted by permission of Sage Publications (right).

(Fig. 2.4) a higher twist angle leads to reduced tensile properties also for yarns (Fig. 2.25) (Shah *et al.*, 2012). The maximum tensile strength at a surface twist angle of 10–15° can be explained by two overlapping effects illustrated in Fig. 2.26. With increasing twist angle the resistance of the fibre bundles against fibre slippage increases due to increased fibre–fibre cohesion but the probability of fibre breakage is lowest at low twist angles and increases with twist angle, leading to a reduction of yarn strength. The combination of these two described effects results in an optimum twist angle of 10–15° (Shah *et al.*, 2012). This orientation angle is in the same order as the MFA of a flax fibre (10–11° according to Eder and Burgert, 2010).

Based on the evaluation of Baets *et al.* (2011) and Shah *et al.* (2012), the effect of yarn twist on the tensile strength of unidirectional plant fibre yarn composites can be better understood. Recently, this has led to the development of new flax textiles optimized for the composites industry within the European flax industry. These developments are presented in Fig. 2.27 (Baets, 2012).

To better understand the latest developments in the field of flax textiles, certain definitions need to be clarified:

- Roving: commonly used for glass. A textile ribbon-shaped structure of endless glass fibres. Adapted to flax: A number of slivers, pre-yarns or

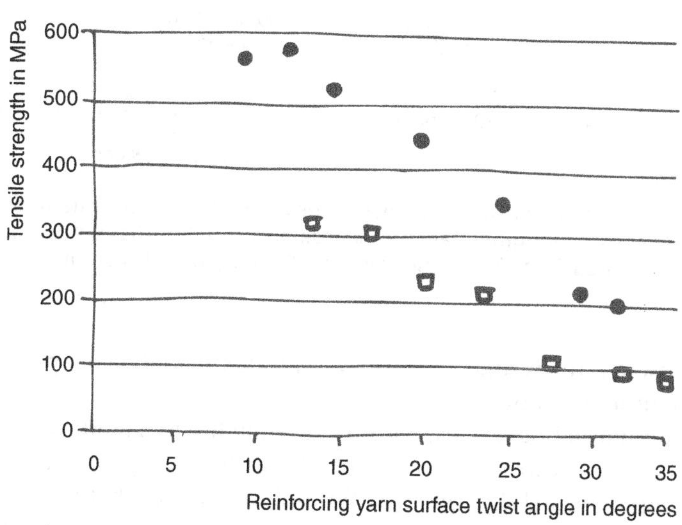

2.25 Influence of yarn twist angle on the tensile strength of long (solid plots) and short (open plots) flax infiltrated with an epoxy resin. Adapted from Shah *et al.* (2012), © 2012 by Sage, reprinted by permission of Sage Publications.

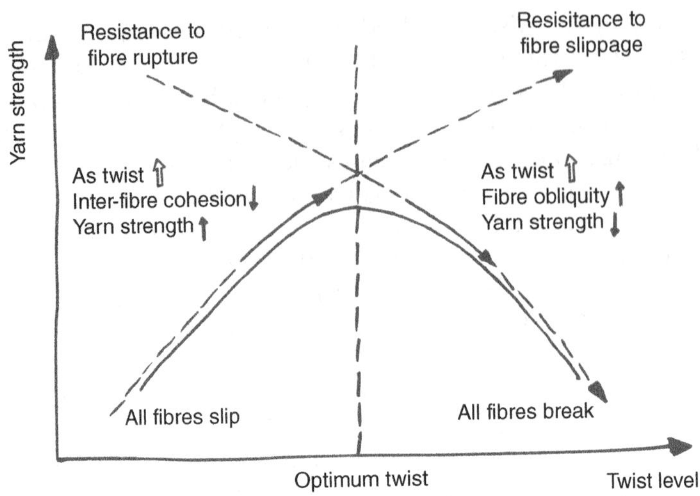

2.26 Explanation of overlapping effects influencing the correlation between yarn strength and twist angle. Adapted from Shah *et al.* (2012), © 2012 by Sage, reprinted by permission of Sage Publications.

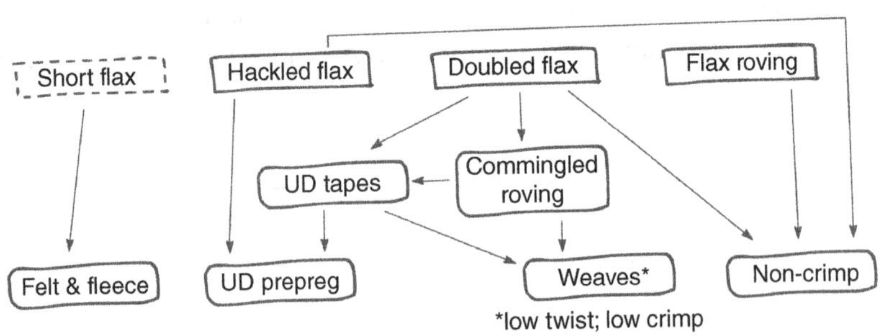

2.27 Textile products especially developed for composite application based on longitudinal and disordered flax (short flax). Adapted and adjusted from Baets and Pariset (2012), by permission of JEC.

low-twist threads collected into a parallel textile ribbon-shaped structure with little or no twist.

- Non-crimp fabric: consists of different unidirectional fibre layers, each layer with its own direction, stitched together using very fine threads.
- Unidirectional (UD): refers to fibres/fibre bundles that are oriented in the same direction, such as unidirectional textile, tape or laminate (Von Acker, 2012).
- Commingled roving and yarn: flax fibre bundles and thermoplastic synthetic fibres are combined in one hybrid textile structure.

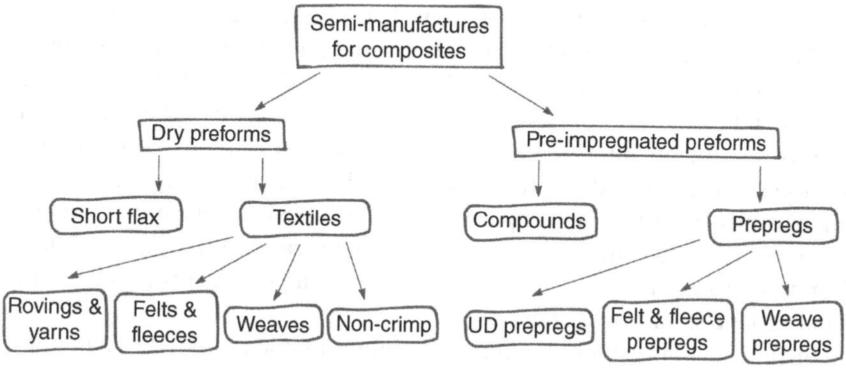

2.28 Dry and pre-impregnated flax preforms for composite applications. Adapted and adjusted from Baets and Pariset (2012), by permission of JEC.

Based on more traditional flax textiles as well as newly developed textiles, a broad range of dry and pre-impregnated flax preforms for composite applications are available (Baets, 2012). Figure 2.28 gives an overview of the different kinds of preforms. The term prepreg refers to textiles and fibres/fibre bundles which are already pre-impregnated with the matrix. The impregnation of the wet prepreg is completed during manufacturing when the matrix is consolidated for thermoset and thermoplastic systems respectively (Von Acker, 2012).

After the conversion process of short and long flax to semi-finished textile products these preforms can be further processed in composite applications. The main concepts of these processes will be described in detail in the following section.

2.4.1 Summary

- The flax textile industry is a mature sector which has long experience in producing high-quality textiles.
- In recent years the European flax industry has developed specially designed flax preforms for composite manufacturing to exploit the full potential of flax.

2.5 Integration into the matrix

The integration of the flax fibre into the polymeric matrix is one of the main challenges within the natural fibre composites production chain. For composites with good properties the fibres (1) have to be well impregnated with the used resin/polymeric system, (2) have to be evenly distributed within the component, and (3) may not be damaged during processing.

The different semi-finished products described in Section 2.4 demand different techniques for the integration into the matrix. For short flax reinforcement an intermediate processing step known as compounding is necessary to introduce the fibre into the matrix. Compounding can significantly influence the properties of the composite as it can lead to fibre breakage, change the fibre morphology and cause separation of fibre bundles into single fibres and therefore also influence the properties of the final product (Barkoula *et al.*, 2010).

In Table 2.4 three compounding techniques that are applied to flax fibres are compared. Using the techniques of kinetic mixing and twin screw extrusion, which are more adapted to the industrial scale, leads to two major problems:

- Dosage: homogeneous feeding and distribution of the fibres to the compounding system is difficult and inhomogeneous compounds can cause problems in the feeding process within the following injection steps (Barkoula *et al.*, 2010).
- Damage: the fibre bundles usually are significantly shortened during processing, which reduces the reinforcing effect in the composites due to a decrease in aspect ratio.[3]

These two problems could be significantly reduced by using natural fibre pellets. Haag *et al.* (2013) showed that for hemp fibres, both fibre length reduction and homogeneity of extruded profiles were significantly improved with the use of this technique. The pellets allow gravimetric dosage of the natural fibres and cause only very limited fibre damage.

The third compounding technique shown in Table 2.4, the LFT technique specially developed for the manufacture of natural fibre composites by Fraunhofer ICT, Pfinztal, Germany, leads to optimal compound properties. However, this technique is not preferable for granule production from an economic point of view as expensive hybrid yarns/slivers have to be used.

The compound as a semi-finished product can be processed into a component via injection moulding. A schematic view of the process can be seen in Fig. 2.29. Before introducing the compound into the injection moulding machine, a drying step depending on the polymer and the used process can be useful due to the hygroscopic nature of the fibres (Huber *et al.*, 2010). In the injection moulding machine the compound is heated and transported by a screw. During this so-called plastification process the viscosity of the compound within the screw is reduced and the molten

[3] The aspect ratio is defined as the ratio of fibre length to its diameter and plays an important role in the performance of composites (Eder and Burgert, 2010).

Table 2.4 Overview of common compounding techniques applied for flax

Technique	Kinetic mixing	Twin screw extrusion	LFT technique
Scheme	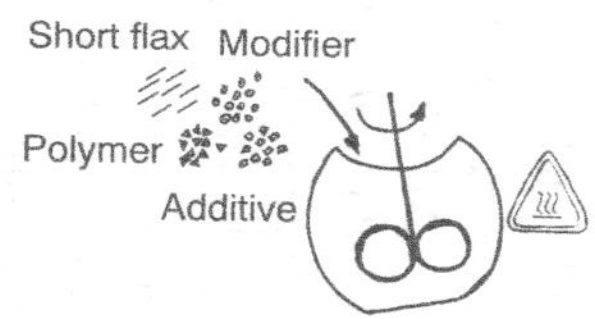	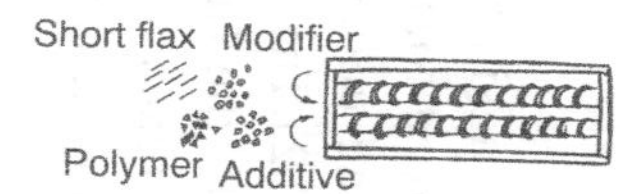	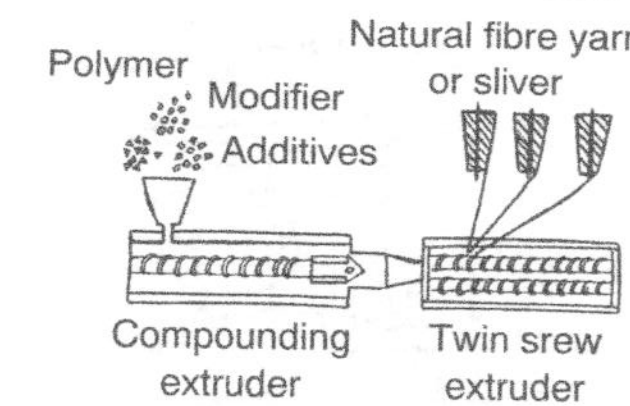
Raw material used	Short fibre or fibre pellets	Short fibre or fibre pellets	Sliver or yarn
Polymers used	PP, PE, PLA, etc.	PE, PP, PLA, etc.	PE, PP, PLA, etc.
Advantages	• Costs	• Continuous process • Costs	• Long fibre granules • Homogeneous properties • Expensive yarns necessary
Disadvantages	• Discontinuous process • May lead to fibre damage and inhomogeneity • Additional granulation process is necessary	• May lead to fibre damage and inhomogeneity	

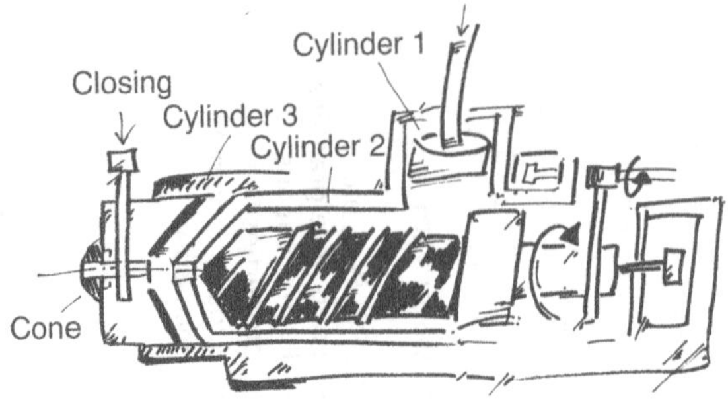

2.29 Schematic view of the injection moulding process (Huber *et al.*, 2010).

mixture of polymer and fibre (and additives) is again homogenized. The processing temperatures for natural fibre-reinforced composites usually lie between 175 and 190°C as fibre degradation and undesired odours may appear at higher temperatures (compare Section 2.2). The melt is transported to the tip of the screw and then transferred into the (usually preheated) mould. The pressures within the mould may reach 500–2000 bar. The pressure is kept during cooling until the component is solidified and can then be released (Huber *et al.*, 2010).

The injection moulding process is a suitable technique for high-quantity items with a high degree of complexity and without back tapers using short cycle times (Barkoula *et al.*, 2010). Most often, polypropylene is used as thermoplastic matrix (Huber *et al.*, 2010) as it has a low melting temperature below the thermal decomposition temperature of the fibres and is used in many technical applications, e.g. automotive. Additionally, polypropylene can be easily and ecologically recycled (Barkoula *et al.*, 2010). Natural fibres are hydrophilic and polyolefins such as polypropylene are hydrophobic. However, this phase incompatibility is reduced by the use of compatabilizers such as maleic anhydride (Bos, 2004; Barkoula *et al.*, 2010).

The fibre orientation within the component depends on the direction of the melt flow, which can be influenced by the number and location of the injection points within the mould. Nonetheless, additional research is necessary to reliably predict the fibre orientation in the component. The behaviour of the fibre bundles within the polymer melt is not yet fully understood and variation in fibre content and fibre length in different regions of one component may occur.

Composites made of randomly oriented short fibres in thermoplastic matrices show a lower performance compared to oriented fibres and the good mechanical properties of the fibres are not fully exploited (Lehtiniemi *et al.*, 2011).

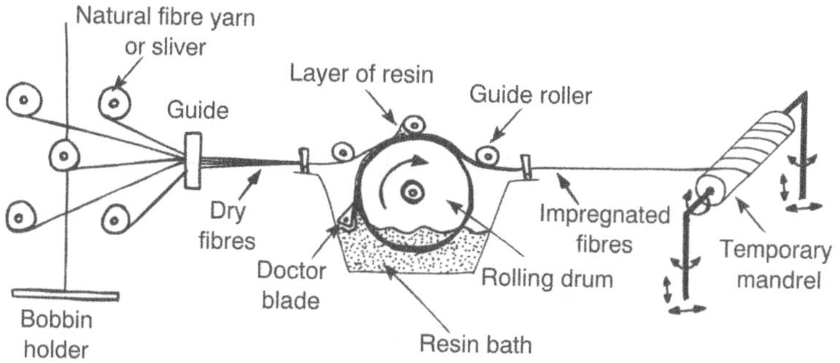

2.30 Schematic view of the winding technique. The yarns or slivers are led from a bobbin holder (left) over an impregnation unit and afterwards are wound with a specific lay-up that defines the fibre orientation.

Other processing techniques that allow control of the fibre orientation are winding and pultrusion techniques. For these, the fibre has to be available in a quasi-endless way in the form of yarn, sliver or roving. As schematically shown in Fig. 2.30, the fibres are taken from a bobbin holder and guided through an impregnation unit where the resin is transferred to the fibre.

Using filament winding techniques, rotationally symmetric composite products can be produced in a commercially important way. Great accuracy must be applied for the fibre lay-up. The technique was developed and is applied to glass and carbon fibre, but it can also be used for flax rovings. Fibres are usually oriented unidirectionally and the fibre content is rather high. For structural composites (in which the fibre has to carry load) long fibres/fibre bundles are required. Among other factors, the fibre tension during the winding process has a strong influence on the fibre and void content of the laminates (Lehtiniemi *et al.*, 2011).

Pultrusion is another method for the production of composites using flax yarns, sliver or rovings. With this technique, mainly applied for thermoset matrices, continuous profiles with a constant cross-section can be produced (Angelov *et al.*, 2007). A schematic view of the pultrusion technique is given in Fig. 2.31. The pulling speed and the length of the tool have to be adapted to the used resin. The cross-section of the profile is defined by the tool used and the composites are cut to any desired length behind the pulling unit. The orientation of the fibre yarn, sliver or roving used is unidirectional within the profile.

To use the advantages of the pultrusion process such as high fibre orientation, reproducible properties and endless profiles with thermoplastic polymers some adaptations of the processing techniques have to be made (Fig. 2.32). Hybrid semi-finished products such as yarns (e.g. commingling

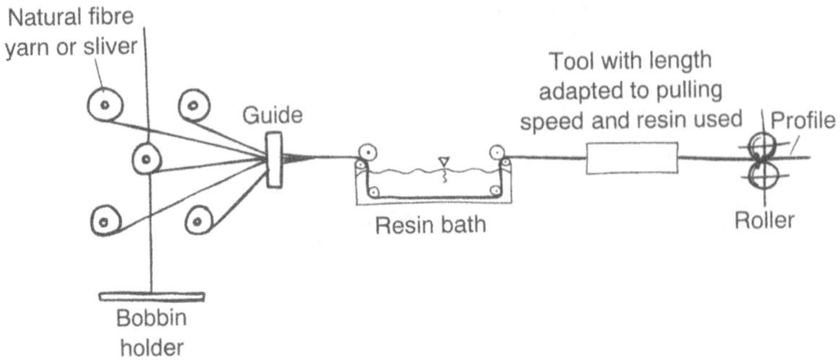

2.31 Schematic view of a pultrusion line. The flax fibre yarns or rovings are stored on a bobbin holder, pulled through a resin bath and a die by a pulling unit arranged after the die. The length of the die and the pulling speed have to be adapted to the specific pultrusion profile and material combination (resin and fibre/resin ratio).

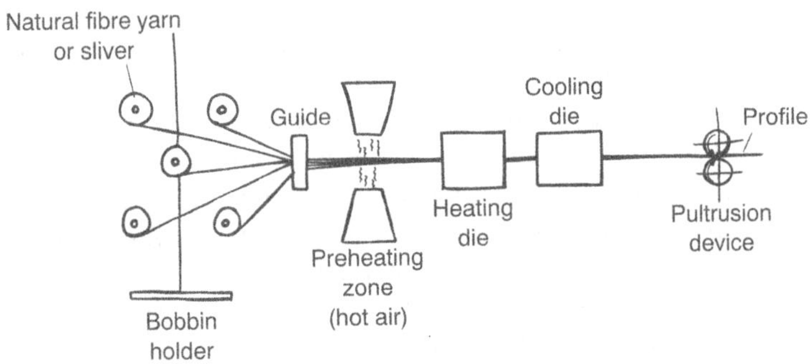

2.32 Schematic view of a thermoplast pultrusion line. Hybrid flax fibre thermoplast yarns are used. Compared to the traditional thermoset pultrusion (Fig. 2.31) the matrix is already within the yarn, so no resin bath is needed. The temperature profile of the line has to be adapted to the thermoplastic polymer with a preheating zone in the beginning and heated and cooling dies (Huber *et al.*, 2010). Adapted from Friedrich *et al.* (2007).

yarns) are used. The yarn is led through a preheating zone where the temperature is close to the melting temperature of the polymer. The formable strand is consolidated in a heating die where the polymer is melted, infiltrates the fibres and is processed into the final profile shape which is consolidated in a cooled die (Friedrich *et al.*, 2007).

If three-dimensional components with a specific fibre orientation are required, none of the above described techniques can be used. However, the prepreg technology allows combination of a matrix with the fibres and design of a specific laminate lay-up. The thermoplastic prepreg process is schematically shown in Fig. 2.33. Natural fibre yarns or slivers are oriented parallel and combined with thermoplastic films on both sides. The viscosity

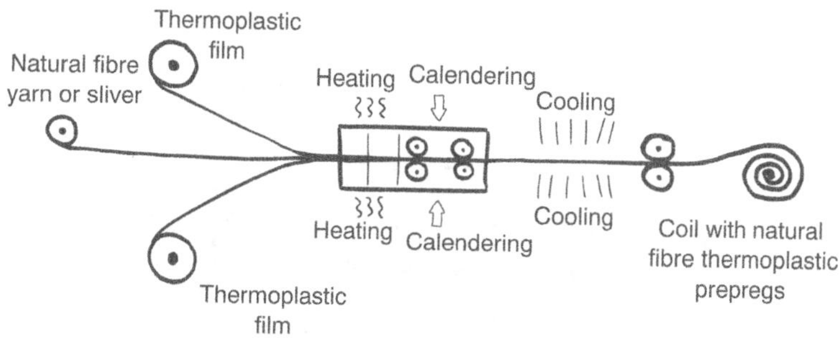

2.33 Schematic view of an UD-prepreg production line. A thin and homogeneous layer of flax sliver is combined with polymer films from both sides under the influence of elevated temperature. The semi-finished products are coiled and can be stored till further application.

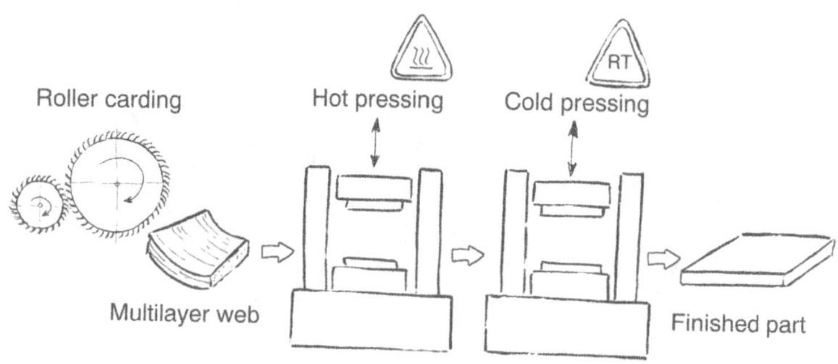

2.34 Schematic view of the compression moulding technique. The semi-finished product, e.g. hybrid needle felts or prepregs, is put into the mould and consolidated with a specific temperature and pressure profile.

of the polymer decreases in a heating zone and the following calendering impregnates the fibres. After cooling, the prepreg can be coiled and stored. Using prepregs an ideal laminate can be laid-up and adapted optimally to the specific requirements of the component (Baets, 2012).

The most common processing technique for natural fibre-reinforced composites in automotive applications is compression moulding (Prömper, 2010). In most cases semi-finished products with thermoplastic matrices are used. In special cases prepregs can be used, but it is more common to use needle felts made of natural fibres and polymer fibres as described in Section 2.4. The fibre content can vary from 35–90%. The components made with this technique show a two- or three-dimensional geometry limited by the impression depth of up to 10–20 cm (Huber *et al.*, 2010). The semi-finished product undergoes a hot and a cold pressing step as shown in Fig. 2.34.

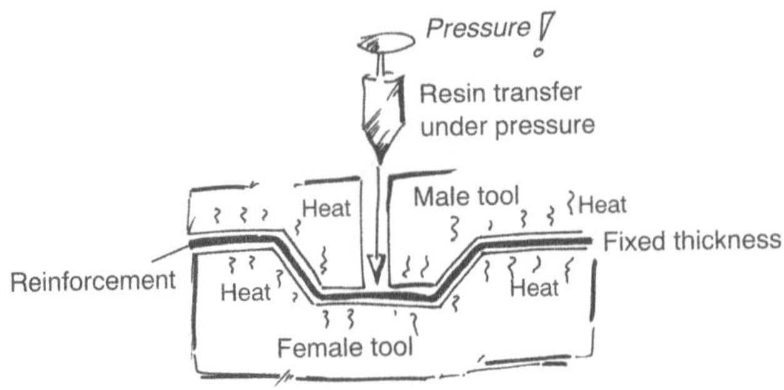

2.35 Schematic view of the RTM process. The resin is transferred into the mould at high pressure and infiltrates the fibres (Huber *et al.*, 2010).

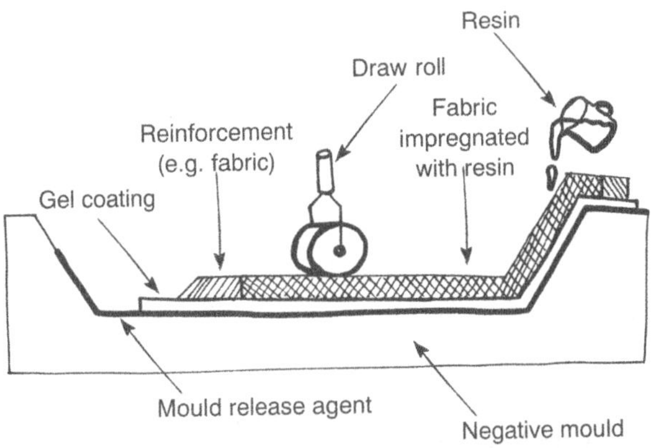

2.36 Schematic view of the hand lay-up process. Flax preforms and matrix are manually added layer by layer.

Another possibility for the use of felts or fabrics is so-called resin transfer moulding (RTM). The dry semi-finished product is put into a closed cavity and is most often infiltrated with a thermoset resin such as epoxy or polyester (Fig. 2.35). RTM uses relatively low pressures of up to 5 bar that are sufficiently high to fill the whole cavity and displace air from the cavity. Often a vacuum trap is used to make the resin transfer easier (vacuum assisted RTM, VARTM). The process needs more time compared to injection or compression moulding but high-quality components with specific fibre orientation can be achieved (Huber *et al.*, 2010).

Another technique for composites manufacturing with flax preforms is the manual hand lay-up process. It is suitable for prototypes and low quantities and is often chosen for design applications (compare Section 2.7, Applications). As shown in Fig. 2.36, the mould is treated with a mould release agent and then, layer by layer, the laminate is set up: the flax preforms, e.g. a weave, are put into the mould and impregnated with the resin using a laminate roller. Depending on the resin system, curing takes place at either room or elevated temperature for a certain time after which the part can then be released from the mould. Caution has to be taken that the resin is homogeneously distributed and that the fabric is in the right position and orientation. The part quality strongly depends on operator skill and experience.

The performance of the different composites produced with the above techniques will be further discussed in the following section and their applications will be described in Section 2.7.

2.5.1 Summary

- Different semi-finished products demand different techniques for further processing.
- Various techniques known from plastics and composites processing have been adapted for the use of flax fibres.
- A broad range of techniques is used for the integration of flax into the matrix, resulting in a wide range of applications.
- The thermal and chemical properties (e.g. hydrophilicity) have to be considered during processing.

2.6 Assessing the performance of the composites

The performance of the composites depends strongly on the properties of the flax fibre and the polymeric matrix used as well as on their interface. A very general picture of the influence of the variation in flax fibre properties is given in Table 2.5.

The selection of materials to reduce the mass of a given component is strongly connected with the density of the material (Verpoest, 2012). In Fig. 2.37 the fibre volume fraction dependent density values – calculated by the rule of mixtures – of composites (carbon fibre/polypropylene (PP), glass fibre/PP and flax fibre/PP) are compared to metals.

Flax fibre composites show excellent mechanical properties with a high ratio of stiffness to density. In Fig. 2.38 the calculated fibre volume fraction-dependent longitudinal specific stiffness (E/ρ) values of composites (carbon fibre/polypropylene (PP), glass fibre/PP and flax fibre/PP) are compared to metals (aluminium and steel) for the 'rod under tension' load case (Verpoest, 2012). This kind of material selection procedure (tie; minimum mass; stiffness prescribed) is based on Ashby's concept for 'materials selection in mechanical design' (Ashby, 2005). Selecting an optimal material for an application with pure tension forces even at a low fibre load of 20%, the

Table 2.5 Influence of fibre properties and characteristics on composite properties

Fibre property	Composite properties		
	Strength	Young's modulus	Impact
Length ⇑	+	≈	++
Diameter ⇓	+	+	+
Strength ⇑	++	++*	+
Young's modulus ⇑	++*	++	– –
Elongation ⇑	–**	–**	++
Interphase ⇑	++	+	–
Dislocations ⇑	–	–	– –

– decrease, – – strong decrease, + increase, ++ strong increase, ≈ low influence
* In general fibres with a high Young's modulus show high strength values and vice versa.
** In general fibres with high elongation values show lower strength and Young's modulus values.
Source: Reprinted from Müssig and Hughes, 2012, by permission of JEC.

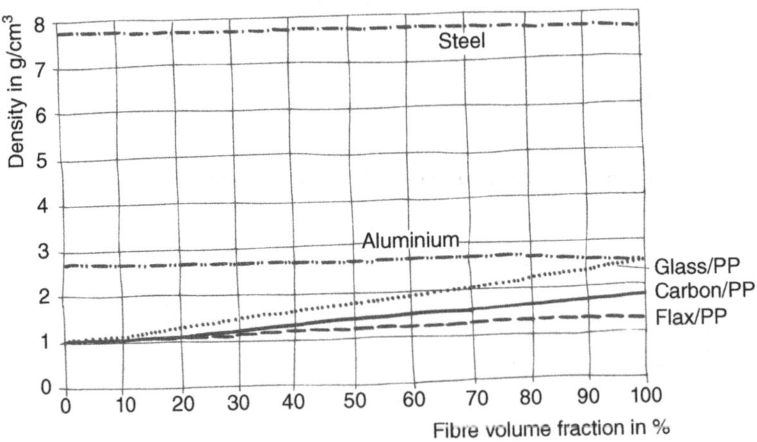

2.37 Fibre volume fraction-dependent density values of composites compared to metals. Adapted from Verpoest and Baets (2012), by permission of JEC.

specific stiffness for flax fibre/PP (fibre load higher than approximately 45%) is theoretically higher than for glass fibre-reinforced plastics and metals like steel and aluminium (Fig. 2.38).

In Fig. 2.39 the calculated fibre volume fraction-dependent longitudinal specific stiffness ($E^{1/3}/\rho$) values of composites (carbon fibre/PP, glass fibre/PP and flax fibre/PP) are compared to metals (aluminium and steel) for the 'plate under bending (area load)' load case (Verpoest, 2012). Selecting a

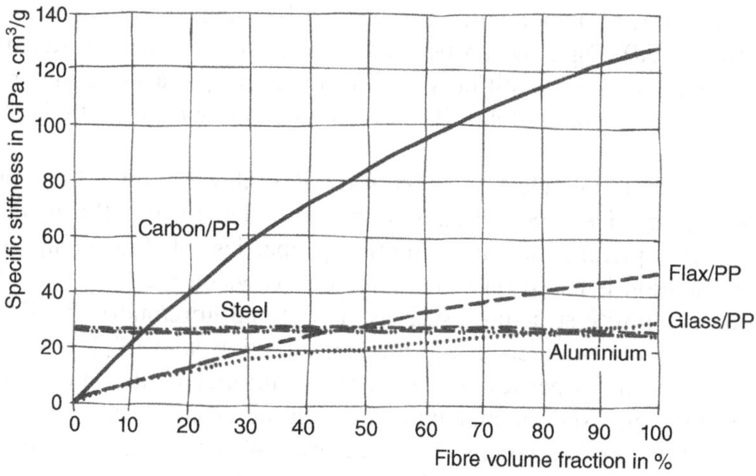

2.38 Fibre volume fraction-dependent longitudinal specific stiffness (E/ρ) values – calculated by the rule of mixtures – of composites compared to metals for the 'rod under tension' load case. Adapted from Verpoest and Baets (2012), by permission of JEC.

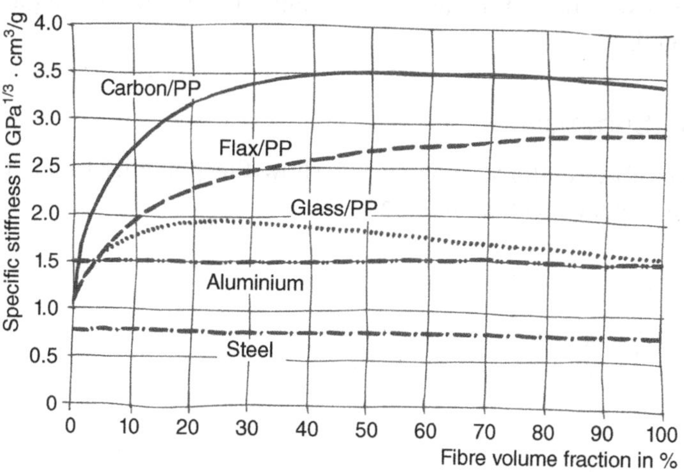

2.39 Fibre volume fraction-dependent longitudinal specific stiffness ($E^{1/3}/\rho$) values – calculated by the rule of mixtures – of composites compared to metals for the 'plate under bending (area load)' load case. Adapted from Verpoest and Baets (2012), by permission of JEC.

material for a panel under area load in bending (minimum mass; stiffness prescribed), Fig. 2.39 can be used for a material selection. Regarding the theoretical specific stiffness for the load case of a plate under bending ($E^{1/3}/\rho$) the benefit of flax fibre/PP compared to glass fibre/PP is even more obvious.

The material selection process for flax fibre-reinforced polypropylene based on Figs 2.38 and 2.39 is only intended to give the reader a very general picture of the potential properties of flax composites. The calculation is based on some assumptions which are hard to find in real-life composite structures, such as (1) optimal fibre/matrix bonding, (2) all fibres being oriented in the loading direction, and (3) low scattering of mechanical properties of the fibre. Additional to this more theoretical evaluation, experimental data are helpful in conducting a coordinated selection process (Baets, 2012). Figure 2.40 gives experimental data for flax composites based on the different composite processing techniques described in Section 2.5.

Based on the given data, flax shows a high potential for a very broad range of possible applications from injection-moulded mass markets up to structural components, for example in sports applications. In the following section a selection of already implemented flax composite applications is presented.

2.6.1 Summary

- Flax fibre-reinforced composites show the potential to compete with classical engineering materials.

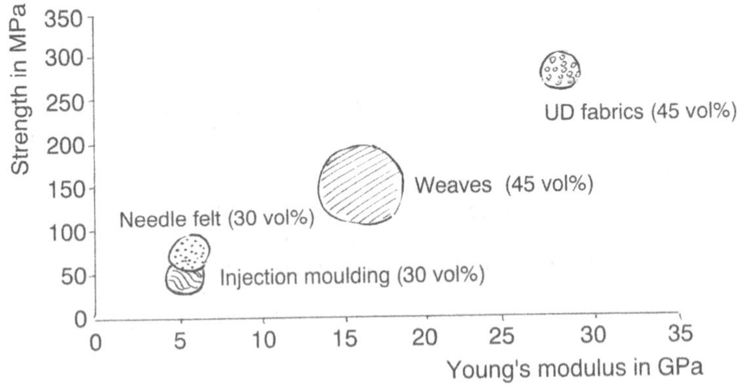

2.40 The performance playground for flax composites. Adapted from Verpoest and Baets (2012), by permission of JEC.

- The performance playground for flax composites allows the use of flax fibre-reinforced injection-moulded components for mass production with moderate properties as well as structural composites by using specially designed flax textiles like UD prepregs.

2.7 Applications

An overview of already used applications of flax fibre-reinforced composites is given in Table 2.6. It can be seen that depending on the demands of the application and the processing conditions different preforms and techniques are used.

As can be seen in the tabular overview, most applications are in the automotive and sports sectors. The automotive industry can be seen as one of the leading sectors as it has to face the strong demand of mass reduction and sustainable alternatives to traditional materials requested by consumers and policy makers. However, in the automotive industry the materials costs are also very important and only cost-efficient applications with a high quantity output are realized using felts or fleeces and short flax. As a result, the reinforcement potential of flax fibres is not fully utilized (compare Fig. 2.40). Other techniques with a higher degree of efficiency of the fibre are not competitive and as well not ready for series production. Research on optimization and industrialization has to be done to access market share in structural applications, too.

In sports applications the performance and to a lower extent the green image of the flax fibres compensates the higher costs. Thus, more expensive and less automated techniques are used.

As very strict certifications have to be fulfilled for an application in aeronautics, flax fibre is not yet used in this very promising sector even if the mass reduction and eco-friendly effects of the flax fibre are very interesting for this industry. According to Martin (2012), one of the main limitations of flax use in cabin components is the fire safety, especially the heat release rate in case of fire. If this factor could be controlled, flax could even have a safety advantage compared to glass fibre-reinforced plastics as the risk of cutting is lower (Gomina, 2012) and benefit can be taken from the high specific stiffness. For the aeronautics sector as well as for the whole transportation sector including railways, trucks and packaging, the advantages in bending stiffness (compare Fig. 2.39) compared to other construction materials will be very interesting.

2.8 Summary: strengths and weaknesses

In the previous section it could be shown that flax-reinforced composites already have a very broad range of applications in nearly all sectors of

Table 2.6 Overview of applications of flax-reinforced composite

Application	Example	Preform used	Process	Reason for using flax
Automotive	 Wheel housing, rear window shelf, dashboard, rear shelf, front seat, door insert, etc. (© PSA)	Fleece and felts (fabrics)	Compression moulding (infusion, hand lay-up)	Mass production possible; more complex shapes; possible to recycle; possible to use recycled PP
	 Rear-view mirror support frame (© AFT Plasturgie), car with mirror (© PSA) and coolant degas tank cap frame (© AFT Plasturgie)	Compound	Injection moulding, LFT	Recycling possible with limited decrease in properties; stiffness increase compared to unreinforced polymer with only small increase in density; mass production; complex shapes

Building	 Decking (© CELC)	Compound with short fibre bundles and shives	Extrusion	Mass production; complex 2D shapes; costs
	 Window profiles (© Innobat)	Roving	Pultrusion	Mechanical properties; damping behaviour; thermal insulation

Continued

Table 2.6 Continued

Application	Example	Preform used	Process	Reason for using flax
Transportation	 Body parts of three-wheeled scooter (© Dehondt Technologies)	Non-twist roving (50% per mass)	Infusion/RTM	Mechanical properties; reduced mass; design
Consumer goods	 Lipstick case (© Uruku Lip Pigment)	Short fibre bundles/shives compound	Injection moulding	Design; costs; renewability; complex structures
	 Smartphone case (© Fibreshell)	Hybrid weave with 50% flax, 50% thermoplastic	Compression moulding	Design; mechanical properties

	 Suitcase (© Delsey)	Hybrid weave	Compression moulding	Design; damping behaviour; mechanical properties
Sports	 Surfboards (© NoTox)	Weaves	Hand lay-up	Stiffness; performance; ecological aspects
	 Helmet (© Urge Bike)	Weaves	Hand lay-up	Stiffness; design; mechanical properties

Continued

Table 2.6 Continued

Application	Example	Preform used	Process	Reason for using flax
	 Tennis and padel racquets (© Artengo)	UD prepregs (thermoset)	Autoclave	Mass reduction; damping behaviour leads to fewer injuries and less risk of tennis elbow
	 Fishing rod (© Caperlan)	UD prepregs (thermoset)	Autoclave	Mass reduction; vibration; damping behaviour
	 Bike (© Museeuw Bikes)	UD prepregs (thermoset)	Autoclave	Damping
	 Racing sail boat (© CELC)	Prepregs (thermoset)	Hand lay-up	Mass reduction; mechanical properties; ecological aspects

Design	 Chair (© Saint Luc)	Non crimp textile	Hand lay-up	Design; mechanical properties; ecological aspects
	 Chair (© Magis)	Needle felt	Compression moulding	Mass production; complex shapes; possible to use recycled PP
	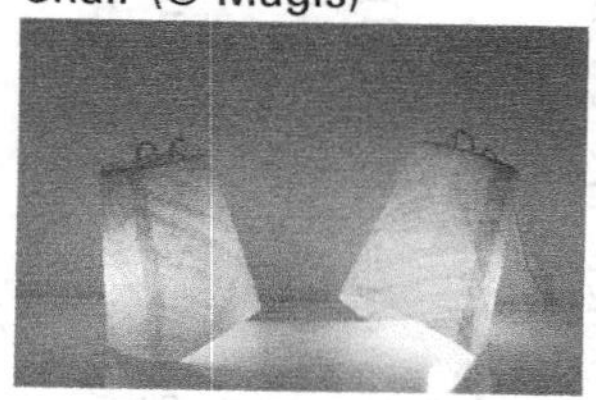 Lamps (© AZ&MUT)	Noncrimp textile	Infusion/RTM	Design; mass reduction

Source: Baets and Pariset, 2012; Gomina, 2012.
Copyright of illustrating pictures is given in the captions.

everyday life. This is possible due to the process chain of flax that has been developed over centuries and consists of four main segments, from up- to downstream:

1. Agricultural production of flax plants.
2. The stem-to-fibre conversion process.
3. The conversion process in which short and long flax are formed into semi-finished and finished textile products.
4. The industrial application of flax in composites.

This process chain demands expertise and experience during all steps. Especially, the retting step, during which the fibre bundles are loosened and separated from the stem tissue, is crucial to the properties of the resulting components.

The hydrophilic nature of the flax fibre and its limited thermal stability are a challenge for process and polymer engineers. The use of specialized polymeric systems and the use of coupling agents necessary for a better bonding of fibre and matrix are critical in terms of economic aspects.

The viscoelastic behaviour of the composites is not fully understood yet and may also be a limiting factor in some applications.

As the flax fibres/fibre bundles are of finite length their reinforcement potential is not fully used yet. Adaptations to the textile processes have been made and are currently in progress to develop semi-finished textiles that are better adapted to composites applications.

Flax fibre composites show excellent mechanical properties with a high ratio of stiffness to density. Even at a low fibre load of 20% the specific stiffness is higher than for glass fibre-reinforced plastics and at a fibre load higher than approximately 45% it is higher than for metals like steel and aluminium. Regarding the specific stiffness for the load case of a plate under bending ($E^{1/3}/\rho$) the benefit is even more obvious.

2.9 Future trends

As fibrous material from oil (linseed) flax and shives are available in large quantities their use in composites will be further developed in the coming years.

In the field of high-quality applications the potential of fibre flax has to be used to a higher degree. Therefore several parts of the process chain can be optimized:

1. Gentle compounding techniques for integration of the flax fibres into the matrix without thermal degradation and critical reduction of the fibre bundles' aspect ratio have to be developed.
2. The interface between the flax fibres/fibre bundles and the polymer matrix has to be improved and engineered.

3. New varieties specially bred for composites application will release new potential for the whole process chain.

For a broader acceptance in the industry the establishment of worldwide testing standards and databases for flax fibres and flax fibre compounds is very important. The development and realization of simulation models for process simulation and structural calculation of natural fibre composites has to be further developed.

We hope that while reading this chapter the reader has experienced some of the fascination of flax and agrees with us that flax has a high potential as reinforcement for composite applications.

2.10 Sources of further information and advice

In the following we have compiled some addresses, standards and books which will hopefully provide the reader with some additional interesting information.

Agriculture and statistics

- Food and Agriculture Organization of the United Nations: <www.fao.org>
- FAOSTAT database of the United Nations: <http://faostat.fao.org>

Flax agro-industrial organizations

- The European Confederation of Linen and Hemp (CELC): <http://www.mastersoflinen.com>
- The Flax Council of Canada: <http://www.flaxcouncil.ca>

The latest book on flax fibre-reinforced composites

- **Reux, F. and Verpoest, I. (editors) 2012:** *Flax and Hemp fibres: a natural solution for the composite industry*, first edition. Paris, JEC Composites, prepared for JEC by the European Scientific Committee of the CELC, ISBN 978-2-9526276-1-0.

ISO flax standard

- ISO 2370 Standard Test Method: Textiles – Determination of fineness of flax fibres – Permeametric methods, 1980.

ASTM flax standards to date under subcommittee D13.17 of ASTM International

- Standard Terminology Relating to Flax and Linen, D-6798-02, 2002.
- Standard Test Method for Color Measurement of Flax Fiber, D-6961-03, 2003.

- Standard Test Method for Assessing Clean Flax Fiber Fineness, D-7025-04a, 2004.
- Standard Test Method for the Measurement of Shives in Retted Flax, D-7076-05, 2005.

French flax standards (AFNOR) – flax in composite applications

- Reinforcement fibres – Flax fibres for plastic composites – Part 1: Terminology and characterisation of flax fibres, XP T25-501-1, July 2010; Experimental standard.
- Reinforcement fibres – Flax fibres for plastics composites – Part 2: Determination of tensile properties of elementary fibres, XP T25-501-2, October 2009; Experimental standard.
- Reinforcement fibres – Flax fibres for plastic composites – Part 3: Determination of tensile properties of technical fibres, XP T25-501-3, July 2010; Experimental standard.

2.11 Acknowledgements

Using mainly handwritten graphics, we have tried a new way of illustration in this chapter. Our special thanks to Ms Kathina Müssig for her creative work. We wish to thank Dr Tim Huber, University of Canterbury, Christchurch, New Zealand for revision of the manuscript and for his critical comments. We would also like to thank Ms Julie Pariset from CELC, Paris, France for providing us with the illustrations pictures of different applications of flax-reinforced composites.

2.12 References

Akin, D. 2010: Flax – ASTM standardisation and harmonisation. In: Müssig, 2010, pp. 371–379.

Angelov, I., Wiedmer, S., Evstatiev, M., Friedrich, K. and Mennig, G. 2007: Pultrusion of a flax/polypropylene yarn. *Composites Part A* 38, 1431–1438.

Ashby, M.F. 2005: *Materials Selection in Mechanical Design*. Oxford, UK, Elsevier Butterworth-Heinemann, 603 pp.

Astley, O.M. and Donald, A.M. 2001: A small-angle X-ray scattering study of the effect of hydration on the microstructure of flax fibers. *Biomacromolecules* 2, 672–680.

Baets, J. 2012: V – Semi-products with flax and hemp fibres. In: Reux and Verpoest, 2012a, pp. 87–98.

Baets, J. & Pariset, J. 2012: The range of marketed semi-finished products/preforms, how they are industrialised, and their multisector applications. In: JEC composites conference in partnership with CELC (organizers): *Flax and Hemp Fibres: A Natural Solution for the Composite Industry – High-performance operational green-chemistry solutions* (Leuven, Belgium, 4 December 2012), Paris: JEC and CELC, CD-ROM, Conference Proceedings, 24 pp.

Baets, J., Plastria, D., Ivens, J. and Verpoest, I. 2011: Determination of the optimal flax fibre preparation for use in UD flax epoxy composites. In: Editor(s): Whysall, Claire, NetComposites, Chesterfield, UK, *Proceedings of the 4th international conference on sustainable materials, polymers and composites*. Studio Venue (Birmingham), UK, 6–7 July 2011.

Baley, C. 2002: Analysis of the flax fibres tensile behaviour and analysis of the tensile stiffness increase. *Composites Part A* 33, 939–948.

Baley, C., Morvan, C. and Grohens, Y. 2005: Influence of the absorbed water on the tensile strength of flax fibres. *Macromolecular Symposia* 222(1), 195–202.

Baley, C., Le Duigou, A., Bourmaud, A. and Davies, P. 2012: Influence of drying on the mechanical behaviour of flax fibres and their unidirectional composites. *Composites Part A: Applied Science and Manufacturing* 43, 1226–1233.

Barkoula, N.M., Garkhail, S.K. and Pejs, T. 2010: Effect of compounding and injection molding on the mechanical properties of flax fiber polypropylene composites. *Journal of Reinforced Plastics and Composites* 29(9), 1366–1385.

Bledzki, A.K., Mamun, A.A., Lucka-Gabort, M. and Gutowski, V.S. 2008: The effects of acetylation on properties of flax fibre and its polypropylene composites. *eXPRESS Polymer Letters* 2(6), 413–422. DOI: 10.3144/expresspolymlett.2008.50.

Bos, H.L. 2004: The potential of flax fibres as reinforcement for composite materials. Dissertation, Department of Chemical Engineering, Technische Universiteit Eindhoven, The Netherlands.

Bos, H. and Donald, A.M. 1999: In situ ESEM study of the deformation of elementary flax fibres. *Journal of Materials Science* 34, 3029–3034.

Bos, H.L., van den Oever, M.J.A. and Peters, O.C.J.J. 2002: Tensile and compressive properties of flax fibres for natural fibre reinforced composites, *Journal of Materials Science* 37(8), 1683–1692.

Bourmaud, A., Morvan, C. and Baley, C. 2010: Importance of fibre preparation to optimize the surface and mechanical properties of unitary flax fibre. *Industrial Crops and Products* 32(3), 662–667.

Charlet, K. and Béakou, A. 2011: Mechanical properties of interfaces within a flax bundle – Part I: Experimental analysis. *International Journal of Adhesion & Adhesives* 31, 875–881.

Eder, M. and Burgert, I. 2010: Natural fibres – function in nature. In: Müssig, 2010, pp. 23–40.

Fakirov, X. and Bhattacharyya, D. (editors) 2007: *Handbook of Engineering Biopolymers: Homopolymers, Blends and Composites*, Munich, Germany, Hanser Verlag.

FAO 2013: Production of flax fibre and linseed oil. <http://faostat.fao.org/> (2013-04-22).

Fischer, H. and Müssig, J. 2010: Bast fibre processing and uses. In: Singh, 2010, pp. 326–348.

Friedrich, K., Estatiev, M., Angelov, I. and Mennig, G. 2007: Pultrusion of flax–polypropylene composite profiles. In: Fakirov and Bhattacharyya, 2007, pp. 223–235.

Gomina, M. 2012: VIII – Flax and hemp composite applications. In: Reux and Verpoest, 2012a, pp. 141–162.

Graupner, N. and Müssig, J. 2010: Technical applications of natural fibres: an overview. In: Müssig, 2010, pp. 63–71.

Haag, K. and Müssig, J. 2013: FIBRAGEN – Flax fibre for improved biomaterials through applied genomics. Poster presentation, PLANT 2030 Status Seminar, 6–8 March 2013, Kongresshotel am Templiner See, Potsdam, Germany.

Haag, K., Scholz, S., Schirp, A., Carus, M. and Müssig, J. 2013: Pellets für die Extrusion. *Kunststoffe* 103(4), pp. 67–69, in German.

Herzog, R.O. (editor) 1930: *Der Flachs – erste Abteilung – Botanik, Kultur, Aufbereitung, Bleicherei, und Wirtschaft des Flachses*. Bd. V.1,1. Berlin, Verlag von Julius Springer (Technologie der Textilfasern), in German.

Huber, T., Graupner, N. and Müssig, J. 2010: Natural fibre composite processing: a technical overview. In: Müssig, 2010, pp. 407–421.

Kulpa, W. and Danert, S. 1962: Zur Systematik von *Linum usitatissumum* L. *Kulturpflanze*, 3, 341–388, in German.

Lehtiniemi, P., Dufva, K., Berg, T., Skrifvars, M. and Järvelä, P. 2011: Natural fiber-based reinforcements in epoxy composites processed by filament winding. *Journal of Reinforced Plastics and Composites* 30(23), 1947–1955.

Mansfeld, 2001: Taxonomy and nomenclature of *Linum usitatissimum* L. (April 2001) <http://mansfeld.ipk-gatersleben.de> (2013-05-25). In Hanelt, P. and Institute of Plant Genetics and Crop Plant Research 2001: *Mansfeld's Encyclopedia of Agricultural and Horticultural Crops*, published by Springer, electronic version: IPK Gatersleben.

Martin, P.P. 2012: New generation of materials for the aerospace industry. In: JEC composites conference in partnership with CELC, (Organizers): *Flax and Hemp Fibres: A Natural Solution for the Composite Industry – High-performance operational green-chemistry solutions* (Leuven, Belgium, 2012-12-04) Paris, France: JEC & CELC, 2012.

Morvan, C., Andème-Onzighi, C., Girault, R.; Himmelsbach, D.S., Driouich, A. and Akin, D.E. 2003: Building flax fibres: more than one brick in the walls. *Plant Physiology and Biochemistry* 41(11–12), 935–944.

Müller, W. 1930: Die Aufbereitung des Flachses. In: Herzog, 1930, pp. 213–280, in German.

Müssig, J. 2001: *Untersuchung der Eignung heimischer Pflanzenfasern für die Herstellung von naturfaserverstärkten Duroplasten – vom Anbau zum Verbundwerkstoff*, VDI Verlag GmbH, Düsseldorf, Germany (Fortschritt-Bericht VDI, Series 5, no. 630), ISBN 3-18-363005-2, 214 pp, in German.

Müssig, J. (editor) 2010: *Industrial Applications of Natural Fibres: Structure, Properties and Technical Applications*. Chichester, UK, John Wiley & Sons, 538 pp.

Müssig, J. and Hughes, M. 2012: II – Flax reinforcements: fibres. In: Reux and Verpoest, 2012a, pp. 39–60.

Müssig, J., Fischer, H., Graupner, N. and Drieling, A. 2010: Testing methods for measuring physical and mechanical fibre properties (plant and animal fibres). In: Müssig, 2010, pp. 269–309.

Nicholson, P.T. and Shaw, I. (editors) 2000: *Ancient Egyptian Materials and Technology*. Cambridge, UK, Cambridge University Press, 2000.

Plant 2030. 2013: Flax fibre for improved biomaterials through applied genomics. <http://www.pflanzenforschung.de>, search: FIBRAGEN, (2013-05-28).

Popper, R., Niemz, P. and Eberle, G. 2005: Untersuchungen zum Sorptions- und Quellungsverhalten von thermisch behandeltem Holz. *Holz als Roh- und Werkstoff*, 63, 135–148, in German.

Prömper, E. 2010: Natural fibre-reinforced polymers in automotive interior applications. In: Müssig, 2010, pp. 423–436.

Reux, F. and Verpoest, I. (editors) 2012a: *Flax and Hemp Fibres: a natural solution for the composite industry*, first edition. Paris, JEC Composites, prepared for JEC by the European Scientific Committee of the CELC, ISBN 978-2-9526276-1-0.

Reux, F. and Verpoest, I. 2012b: Availability and accessibility of flax and hemp for use as materials. In: Reux and Verpoest, 2012a, pp. 181–188.

Schilling, E. 1930: Botanik und Kultur des Flaches. In: Herzog, 1930, pp. 49–212, in German.

Schnegelsberg, G. 1971: *Systematik der Textilien*. Verbesserter, fotomechanisch vervielfältigter Nachdruck der 1. Auflage. München, Wilhelm Goldmann Verlag (Das Wissenschaftliche Taschenbuch, Abteilung Technik), in German.

Schnegelsberg, G. 1999: *Handbuch der Faser – Theorie und Systematik der Faser*. Frankfurt am Main, Germany, Deutscher Fachverlag, in German.

Schönfeld, H. 1955: *Bastfasern*. Leipzig, VEB Fachbuchverlag, in German.

Schweinfurth, G. 1884: Über Pflanzenreste aus altägyptischen Gräbern. *Ber. Dtsch.* Botan. Ges. 2, 354–371, in German.

Shah, D.U., Schubel, P.J. and Clifford, M.J. 2012: Modelling the effect of yarn twist on the tensile strength of unidirectional plant fibre yarn composites. *Journal of Composite Materials* published online 13 March 2012.

Singh, B.P. (editor) 2010: *Industrial Crops and Uses*. Cambridge, MA, USA, CABI.

Stamboulis, A., Baillie, C.A. and Peijs, T. 2001: Effects of environmental conditions on mechanical and physical properties of flax fibres. *Composites Part A: Applied Science and Manufacturing* 32(8), 1105–1115.

Suh, M.W., Cui, X. and Sasser, P.E. 1994: Understanding on HVI tensile data based on mantis single fiber test results. *Proceedings of the Beltwide Cotton Conferences*, 5–8 January, San Diego, CA, Vol. 3 of 3, pp. 1400–1403.

Unger, F. 1866: Botanische Streifzüge auf dem Gebiete der Culturgeschichte – Ein Ziegel der Dashurpyramide in Ägypten nach seinem Inhalte an organischen Einschlüssen. *Sitzungsberichte der Kaiserlichen Akademie der Wissenschaften in Wien (SAWW): mathematisch-naturwissenschaftliche Klasse* 54, 33–62, in German.

Van de Velde, K. and Baetens, E. 2001: Thermal and mechanical properties of flax fibres as potential composite reinforcement. *Macromolecular Materials and Engineering* 286, 342–349.

Verpoest, I. 2012: I – A general introduction to composites, highlighting the advantages of flax and hemp composites. In: Reux and Verpoest, 2012a, pp. 15–37.

Verpoest, I. and Baets, J. 2012: Flax and hemp fibre composites. In: JEC composites conference in partnership with CELC (organizers): *Flax and Hemp Fibres: A Natural Solution for the Composite Industry – High-performance operational green-chemistry solutions* (Leuven, Belgium, 4 December 2012), Paris: JEC and CELC, CD-ROM, Conference Proceedings, 17 pp.

Vogelsang-Eastwood, G. 2000: Textiles. In: Nicholson and Shaw, 2000, pp. 268–298.

Von Acker, J. 2012: Keywords on natural fibre composites. In: Reux and Verpoest, 2012a, pp. 189–200.

3

The use of hemp fibres as reinforcements in composites

H. N. DHAKAL and Z. ZHANG, University of Portsmouth, UK

DOI: 10.1533/9781782421276.1.86

Abstract: The use of hemp fibres as reinforcement in composite materials has received increasing attention by both academia and industry because of their unlimited availability, lower density and much higher specific strength than conventional fibres such as carbon and glass, and the fact that they are renewable. This chapter is concerned with the evaluation of the key fibre properties, cultivation and quality issues, surface modification and its effect on the properties as well as the applications of hemp fibre-reinforced composite materials.

Key words: hemp, chemical composition, properties, cultivation, processing, modification, interaction, applications.

3.1 Introduction

It is well known that the overall performance of composites involves both compatibility between fibre and matrix as well as physical and chemical properties of reinforcements. The use of hemp fibres as reinforcements in polymer matrix composites (PMCs) has generated much interest in recent years due to environmental impact considerations, the need to redirect agricultural production from solely the food industry to other applications, and the urgency to find alternative, more energy-saving materials [1–4]. The use of natural fibres and in particular hemp, also called *Cannabis sativa* L., fibre bundles as reinforcing agents in composites offers many advantages over synthetic fibres (glass, carbon or aramid) such as acceptable specific strengths and modulus, unlimited and sustainable availability, low density, reduced tool wear, enhanced energy recovery, reduced dermal and respiratory irritation, good thermal and acoustic insulating properties and good biodegradability. Hemp fibre-reinforced composites have been successfully used for lightweight and low-cost applications in recent years, but significant barriers for structural applications of these composites still exist. These barriers include lack of confidence in the use and performance of natural plant fibres and their composites limited understanding of diffusion behaviour and poor resistance to moisture [5–7]. Synthetic fibres like glass and carbon can be produced with a definite range of reproducible characteristics, whereas the properties of natural fibres vary depending on the fibre diameter structure, climatic conditions, the age of the fibres and the digestion process. This chapter aims at highlighting the important parameters that govern the property of hemp fibres as reinforcement in polymeric composites.

3.2 Hemp fibre

Hemp fibre has been used by mankind for over a thousand years. It is a tall (2–5 m), robust, annual herbaceous plant, which is sown in spring and harvested in autumn. Originally native to Central Asia, it has since spread to every inhabited continent, region and country and is widely cultivated in Europe [8].

Before the 21st century hemp plants were common agricultural crops grown in temperate climates for the production of ropes and shipping sails, illustrating their outstanding strength. Hemp is one of the commercial sources of long natural fibres grown in the UK [9]. These fibres are finding increasing use as reinforcement in composite materials due to their high specific strength and stiffness and their ecological advantages over glass fibres, as they are renewable and can be incinerated. The fibre bundles extend continuously from bottom to top of the hemp plant; however, the single fibres are smaller units with lengths in the range of 5–55 mm [10].

Hemp fibres require very little maintenance to grow and substantially less energy to produce than the more traditional glass fibres, for example. Due to their abrasive nature, glass fibres have increased manufacturing costs as damage to the tools occurs during construction and repair, as well as adverse effects on the operator as they can cause discomfort and irritation if they come in contact with skin or are inhaled. Those problems are greatly reduced when working with hemp fibre [11].

Hemp fibre is a multi-celled structure bast fibre plant. Figure 3.1 shows non-woven hemp mat and a scanning electron micrograph (SEM) of hemp and glass fibres.

The chemical composition and structural parameters of hemp fibres are listed in Table 3.1. The chemical composition of hemp varies with the species of plant and with the part of the plant that is used. Hemp mainly consists of lignin, cellulose and hemicellulose, the least of which is similar to cellulose but does not form a polymer of chain-like structure, giving it little strength. The bast fibres in hemp, which are generally seen as the most valuable part of the plant, are made up of approximately 74% cellulose, 18% hemicellulose, 4% lignin, 0.9% pectins and 0.8% wax [14].

Figure 3.2 shows the typical structure of hemp fibres. The cell wall of the fibre is made up of a number of layers: the so-called primary wall (the first layer deposited during cell development) and the secondary wall (L), which

3.1 Non-woven hemp mat and SEM of hemp and glass fibres [12, 13]. Reproduced from Dhakal *et al.* [12, 13] by permission from Elsevier Science Ltd, UK.

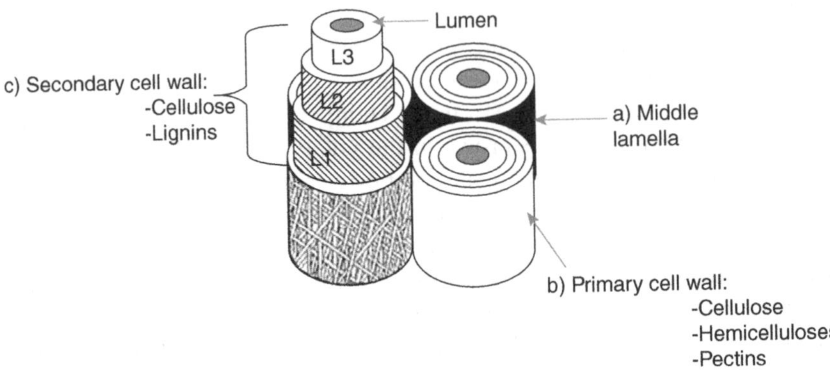

3.2 Structure of hemp fibres [15]. Reproduced from Troedec *et al.* [15] by permission from Sage Publications.

Table 3.1 Chemical composition and structural parameters of hemp fibre [14]

Cellulose (%)	Hemicellulose (%)	Lignin (%)	Pectins (%)	Wax (%)	Cell length (mm)	Spiral angle (deg)	Moisture content (%)
74.4	17.9	3.7	0.9	0.8	23.00	6.2	10.8

Source: Reprinted from Bledzki *et al.* (1996) [14] by permission from Elsevier Science Ltd, UK.

again is made up of three layers (L_1, L_2 and L_3). As in all lignocellulosic fibres, these layers mainly contain cellulose, hemicellulose and lignin in varying amounts. The individual fibre is bonded together by a lignin-rich region known as the middle lamella. Cellulose attains its highest concentration in the L_2 layer (about 50%) and lignin is most concentrated in the middle lamella (about 90%). The L_2 layer is by far the thickest layer (32–150 laminas) and dominates the properties of the fibres.

3.3 Key fibre properties

Industrial hemp fibre is one of the strongest and stiffest available natural fibres and therefore has a great potential for use as reinforcement in composite materials. As briefly mentioned in the above section, some of the key reasons why hemp fibre is becoming increasingly attractive for use in the composites industry as reinforcement are its high specific strength to low density, its low cost of production and its ecofriendliness and biodegradability properties. This last feature has become an increasingly appealing key selling point for products as companies aim for their products to be ecofriendly and easily recyclable.

Comparative values of physical and mechanical properties of hemp and glass fibres are presented in Table 3.2. The results demonstrate that glass fibres are superior to hemp fibres, and moreover the ultimate tensile strength in particular is higher for glass fibre than for hemp fibre. However, if one considers the specific modulus of hemp fibres (modulus/density) then one obtains an average value of 35.5 GPa. A similar calculation for glass fibres using a modulus of 70 GPa gives a specific modulus of 28 GPa. This indicates that in certain respects the mechanical properties of hemp fibres are approaching the properties of glass fibres. Because the density of hemp fibre is lower than that of glass fibre, the reinforcement of hemp fibre to the polymeric matrix reduces the density of the composite as a whole. These properties make hemp fibres attractive as environmentally friendly reinforcing materials.

Unlike conventional glass fibres, the cross-sectional shapes of hemp fibres show wide variation (Fig. 3.1). As a result, compared to composites from synthetic reinforcements, natural plant fibre-reinforced composites have high variation on properties and dimensions, sensitivity to moisture and wet environments, and incompatibility between the hydrophilic natural fibres and generally hydrophobic thermoplastic and thermoset matrices, requiring appropriate use of physical and chemical treatments to enhance the adhesion

Table 3.2 Comparative values of physical and mechanical properties of hemp fibres (bundles) with E-glass [16, 17]

Fibre type	Density (g/cm^3)	Moisture absorption (%)	Elongation at break (%)	Tensile strength (MPa)	Young's modulus (GPa)	Specific modulus (E/ρ) (GPa)
Hemp	1.4	8	1.6	690	30–70	21–50
E-glass	2.5	–	2.5	2000–3500	70.0	28

Source: Reprinted from Bledzki *et al.* (2001) [16] and Bledzki and Gassan (1999) [17] by permission from Elsevier Science Ltd, UK.

between the fibre and the matrix or the fibre–matrix interface. These limitations can be disadvantageous in structural applications.

3.4 Cultivation and quality issues

Hemp cultivation is the process that the plants undergo to become bast fibres, which are found in the stems of the plant. Hemp fibres exhibit wide variation in plant height, stem diameter and plant density. Hemp has a rigid woody stem with height ranging from 1.2 to 5 m. The fibre of commercial interest is derived from the stem. The factors affecting the quality of hemp fibres are genotype, sex, precipitation, time of harvest, stem part, length and diameter. Commercial fibres of hemp consist of single fibre cells which are joined together and form fibre bundles. Fibres such as hemp have several quality issues, which include variations in fibre quality and processing limitations. Fibre length, diameter and linear density vary depending on many different parameters. Some of the parameters which create variance in fibre properties are the time of harvest, the type of fibre processing and production, and the type of soil in which the plant is cultivated [18].

Variations in different parameters contribute variations in morphological structure and chemical composition which in turn affect the physical properties of fibres. The more densely they are sown will also determine the stem diameter, and thus the fibre properties as well. Tightly packed fields will produce thinner and taller stems, which in turn will produce thinner and longer fibres. The closer they are sown will also help to increase the yield and produce regular straight fibres. Hemp when cultivated can grow up to 5 m in height and around 2 cm in diameter. Due to the plant being tightly bunched or having dense foliage, weeds and other unwanted plants are prevented from growing, which in turn prevents the need for pesticides and herbicides. Hemp also requires very little or no irrigation due to its deep root system. All these factors mean that it is a very economical crop as once the plants have been sown they need little care until they are harvested.

There is a clear relationship between fibre properties and harvest time. It is generally accepted that spring-harvested fibres exhibit poor strength and elongation compared to those harvested in the autumn. Sowing density and harvest time are important parameters which influence the quality of fibre.

Furthermore, the qualities of hemp fibres can be linked to the processing temperature at which they are processed. Synthetic fibres can be processed at a much higher temperature, while natural fibres such as hemp can only be processed at a maximum of several hundred degrees Celsius. It is expected that when the processing temperature is higher than 200°C, the irregular fracture morphology of hemp contributes to a decrease in average diameter which results in decomposition of hemicelluloses and lignin. This can affect the properties of the composite after further processing as they are significantly determined by the fibre properties and the fibre geometry [19].

3.5 Processing of hemp as fibre reinforcement for composites

3.5.1 Fibre retting

Once the hemp plant has been harvested it must undergo a process called retting. Retting is controlled degradation of plant stems (a microbial process) to allow the fibre to be separated from the woody core. The traditional method to separate the fibres from the plant is to cut and leave the hemp stems on the field, where they are soaked during the night by the dew, allowing natural bacterial degradation to take place. Under these conditions, microorganisms grow and produce enzymes, which degrade pectic substances, and the cortex fibres are progressively disassociated into fibre bundles and sub-bundles. This method is known as dew retting and is current in practice, but the quality of the fibres may vary due to variations in climatic conditions. Alternatively, the plant stems are retted in water tanks to make the retting process more controlled. Other approaches include drying of stems that are artificially subjected to multiple passes in a system of rollers. This can be an expensive process and can cause too much fibre damage. Two main processes are commonly used to separate the fibres in this way: field and water retting.

In field retting, the harvested stems of the hemp are left to rot in the field. The moisture and heat generated in the process breaks the fibres apart. The advantages of using field retting are that it does not need excess watering and it is relatively inexpensive. When the straw is not sufficiently retted (under-retted), the fibre resulting from mechanical processing will be coarse, with high content of shives and with residues of surrounding tissues still adhering to the fibre. Over-retting can lead to reduction of fibre strength and reduction of long fibre efficiency. Therefore, field retting needs constant monitoring to ensure that the hurd and bast fibres separate.

It is believed that water retting produces a much higher quality, more uniform fibre. In this process, fibres are extracted by a controlled warm-water retting procedure which is an effective method that avoids extreme weather situations that may come from field retting. However, it is very labour intensive compared to field retting and requires large amounts of fresh water, which ultimately increases the production costs and produces high volumes of fermentation waste. If water retting is conducted in a controlled environment, microorganisms or enzymes can be added to the water to aid the retting process to produce very high-quality fibres that retain their high mechanical properties. It is plausible to note that the mechanical actions on the fibres, such as scutching and hackling, will have an important effect on the reinforcing capability of these fibres in a matrix material. Even more critical is the effect of the retting process, during which the chemical composition of the fibres is changed.

3.5.2 Fibre extraction

After the stems of the hemp plants have been retted and dried they are ready to be further processed, separating the different parts of the plant. There are two common methods, mechanical separation and steam explosion.

Steam explosion

Steam explosion is the chemical removal of the cellulose and other glue-like substances which hold the fibres together. The traditional way to separate the hemp fibres from the plant is to cut and leave the hemp stems on the field, where the stems are soaked during the night by the dew and dried by the sun during the day. Under such conditions microorganisms grow and produce enzymes which degrade pectic substances, and the cortex fibres are progressively dissociated into fibre bundles and sub-bundles.

The work carried out by Vignon *et al.* [20] on hemp fibre using steam explosion treatment suggests that the tensile strength and modulus of hemp/PP composite were improved as a result of improved fibre matrix adhesion following treatment. However, there is an argument that these methods separate the material into its component fibre cells and hence destroy the structure of fibre bundles that the plant has provided.

Mechanical separation

Mechanical separation involves the stems of the plant being passed through fluted rollers that crush the hurds into short pieces, separating the majority of the hurds from the outer bast fibres (hackling and decortications). In this process, the required time is shortened and better quality can be achieved compared to the traditional retting process.

3.6 Surface modifications of hemp fibre and their effects on properties

3.6.1 Physical modifications

Physical methods in hemp fibre processing are used mainly for two reasons: to separate fibre bundles into individual filaments and to modify the surface structure of the fibres.

Physical methods include, amongst others, corona and cold plasma treatment. The work carried out by Prasad and Sain [21] reported that thermally treated hemp fibres showed less moisture uptake than untreated hemp fibres. Similarly, the work of Jovancic *et al.* [22] suggested that the wettability and dyeability of hemp fibres are significantly enhanced after plasma treatment. As plasma treatment promotes delignification, if it is left for a longer period of time, the percentage of delignification increases which in turn reduces the strength of the fibres. Therefore, the time duration of plasma treatment is important. Ragoubi *et al.* [23] used a corona treatment to modify hemp fibres, and their results suggest that the tensile strength and Young's modulus were increased by 32% and 30% respectively compared to untreated fibres. The improvement in mechanical properties can be related to improved fibre matrix adhesion as a result of physical methods and can be explained by the increased surface roughness after modification, resulting in better mechanical interlocking between fibre and matrix, and better wettability of the fibre by the resin due to more compatible surface energies of both components.

3.6.2 Chemical modifications

The chemical modification methods can be generally divided into two main classes: 'compatibilisation' methods that improve the efficiency of wetting by the matrix polymer, mostly carried out in the case of thermoplastic composites, and 'coupling' methods, which utilise a chemical reagent to enable a chemical reaction between the fibre surface and the matrix polymer.

Dhakal *et al.* [24] carried out an experimental study on the effect of 10% sodium hydroxide (NaOH) treatment on the thermal and surface energy characteristics of hemp fibre-reinforced unsaturated polyester composites. Their work suggests that NaOH treatment significantly improved both the thermal and surface energy characteristics. From the thermogravimetry (TGA) and derivative thermogravimetry (DTG) curves, the untreated composite prepared with hemp/UP showed a higher decomposition rate with lower char yield compared to NaOH-treated hemp/UP samples. The lowering of the decomposition rate and an increase in the yield of residue char of treated hemp composite can be ascribed to its superior thermal stability compared to the untreated composite and is attributed to the improved fibre–matrix interaction as a result of the chemical treatment of the fibre surface. The morphological change that occurred after the treatment is illustrated in Fig. 3.3. The treated fibre shows reduced surface impurities compared with the untreated one.

In the same study, the wetting behaviour of treated and untreated hemp fibre-reinforced UP composites was investigated using a contact angle measurement technique and it was found that the treated hemp composites

(a) (b)

3.3 SEM micrographs of surface morphology of (a) untreated and (b) treated hemp fibre surface [24]. Reproduced from Dhakal *et al.* [24] by permission from Elsevier Science Ltd, UK.

exhibited lower contact angle and increased surface energy, indicating better wettability than the untreated one.

Similar observations were reported by Mwaikambo and Ansel [25] who investigated the effect of mercerisation, also known as alkali treatment, on the mechanical properties of hemp fibre composites and reported that the tensile strength and Young's modulus of hemp fibre increased with 6% and 4% NaOH treatment.

It has been reported that chemically treated hemp fibre can show a considerable decrease in mechanical properties and this decrease is attributed to a substantial delignification and degradation of cellulosic chains during chemical treatment. It is therefore important to consider the concentration of chemical, the time of soaking and the temperature which all affect the final properties of composites [26].

Grafting

The work carried out on modified hemp fibres by Pracella *et al.* [27] using melt grafting reactions reported that the tensile strength and modulus of hemp fibre-based composites were improved significantly compared to the hemp composites without treatment. The improvement is attributed to improved fibre–matrix interactions due to grafting.

All the fibre modification methods aim at ameliorating the composite quality by affecting the interface characteristics, which results in better dimensional stability, lower water uptake and swelling and improved mechanical performance. The results of adhesion are essential in the production process, as the mechanical properties of the composites such as fracture toughness are transformed through these processes, hence altering their use for structural applications.

3.7 Fibre–matrix interaction

In a hemp fibre-reinforced polymer composite, the hemp fibres serve as reinforcement whereas the task of the matrix is to hold the fibres together, to transmit the forces, and to work as a coating. Therefore, the mechanical properties of the hemp fibres determine the stiffness and strength of the composites decisively. Generally, very thin fibres showing a large surface to volume ratio are used for good adhesion of the fibres and the matrix. The matrix in the composite serves three major functions:

- To support and transfer the stresses to the fibres
- To protect the fibres against physical damage and the environment
- To reduce propagation of cracks in the composite.

There are several reported works on hemp fibre-reinforced composites in which the importance of good adhesion between fibre and matrix are highlighted (Fig. 3.4). The work carried out by Dhakal *et al.* [28] on the creep behaviour of hemp/UP composite suggests that as a result of good fibre–matrix interaction, the creep properties of hemp composites exhibited higher creep resistance behaviour at various stress and temperature levels. The work suggests that the creep strain decreases with increase in hemp fibre volume fraction up to its threshold value. The reason for this decrease in creep properties above the fibre threshold value can be related to the resin being able to 'wet' the fibres, as a result of which the fibre–matrix interaction gets reduced.

To optimise the performance of hemp fibre-reinforced composites, effective utilisation of the fibre properties is essential. The improvement of

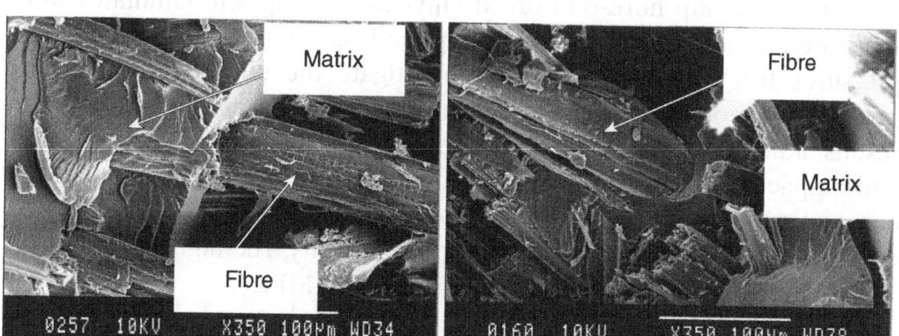

3.4 SEM micrographs of fractured surface of untreated hemp UPE composites showing fibre–matrix interaction.

adhesion between hemp fibre and matrix increases the tensile strength of the composites. The value of tensile strength, for example, increases with the increasing interface shear strength for which the fibre length and the fibre diameters are critical characteristics.

In order to evaluate the effect of various concentrations of hemp fibre on the creep behaviour of hemp fibre-reinforced unsaturated polyester composites, Dhakal and co-workers [28] further reported that the creep behaviour of a three-layered hemp/UPE composites sample exhibited comparable creep strain similar to the same fibre volume fraction of E-glass/UPE composites. The results further showed that the percentage of creep strain of hemp composites fell significantly at higher temperature due to the degradation of the fibre–matrix interface as a result of moisture uptake.

A similar observation was made by Dhakal *et al.* on the effect of water absorption on the mechanical properties of hemp fibre-reinforced unsaturated polyester composites [29]. They studied the various fibre volume fractions of hemp composites, immersing them in room temperature and at elevated temperature. The results showed that the tensile and flexural properties of hemp composites fell significantly due to the degradation of the fibre–matrix interface as a result of moisture uptake.

The effect of water absorption on the nano-indentation behaviour of hemp composite was investigated at different temperatures, and the results showed that nanomechanical properties such as hardness and elastic modulus for water-immersed samples at elevated temperature were lowest compared to dry and water-immersed samples at room temperature [30]. The reduction in nanohardness properties for water-immersed specimens can be related to the formation of a weaker fibre–matrix interface. When water uptake reaches its saturation level, the bound water and free water stays as a reservoir in the composite and this leads to softening of the hemp fibres and weakens the fibre–matrix adhesion, resulting in reduced material properties.

Untreated hemp fibre-reinforced vinyl ester composite laminates were fabricated by a compression moulding method using varying compaction pressures from 1.5 to 4.5 MPa to investigate the effect of compaction pressure on moisture absorption, flexural and impact properties [31]. The results indicated that the water repellence behaviour (Fig. 3.5) and flexural and impact properties (Table 3.3) were improved by increasing the compaction pressure from 1.5 MPa to 3.5 MPa as a result of increased composite density and reduced composite porosity. The highest properties were achieved at the compaction pressure of 3.5 MPa.

The work carried out by Dhakal *et al.* [31] further suggests that the strengths of composites depend on processing conditions such as temperature, compaction pressure and fibre volume fraction amongst others. Below the optimum value, the properties will be reduced to varying

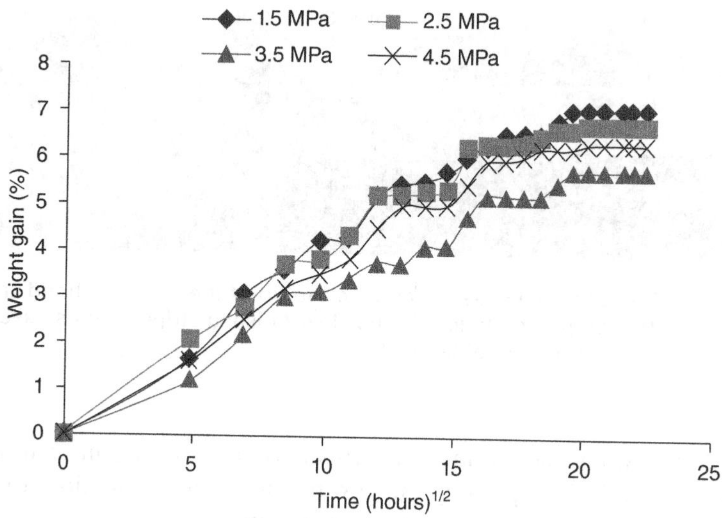

3.5 Comparison of water absorption behaviour of different samples [31].

Table 3.3 Impact test results of hemp/vinyl ester composites [31]

Sample type	Peak load (N)	Energy absorbed (J)
Hemp/VE 1.5 MPa	2636	20
Hemp/VE 2.5 MPa	3892	25
Hemp/VE 3.5 MPa	4456	27
Hemp/VE 4.5 MPa	2774	21

degrees depending on the various processing parameters. The results demonstrated that the right processing parameters, such as compaction pressure, are crucial in achieving optimum mechanical properties.

3.8 Current applications of hemp fibres

Hemp fibres are predominantly used in the automotive industries mainly for reinforcement of door panels, passenger rear decks, pillars and boot linings. Hemp fibres are also used in the pulp and paper industry. In addition, insulation is another important area of hemp fibre application. Non-woven mat as a final product provides excellent properties such as being non-irritant as an insulating material. The use of hemp for seeds and oil is another area of hemp application. Recently, an attempt has been

3.6 Lotus Eco Elise car and a close-up of the Eco Elise hood showing the hemp fibre pattern in the clear-finished stripe. Photos used with kind permission of Lotus, UK.

made to explore hemp fibre as a cobalt(II) removal agent from aqueous solutions [32]. This opens up the new possibility for hemp fibre to be used as an alternative in Co(II) wastewater treatment.

3.8.1 Automotive applications

Short hemp fibres are used by the automotive industry as reinforcement in moulded composites with thermoplastic and thermoset matrices which account for large percentages in Europe [33]. The applied fibre semi-products are raw fibres and non-woven mats, and the composites therefore possess moderate mechanical properties which make them well qualified for non-structural parts (Fig. 3.6).

Due to the high sensitivity of hemp fibres to moisture absorption, interior parts are mostly made of natural fibre-reinforced composites. These fibres are currently used in the interior of passenger cars and truck cabins. Besides their use in trim parts such as door panels and cabin linings, hemp fibres are used extensively for thermo-acoustic insulation.

3.8.2 Ecological thermal insulation materials

It is believed that one of the most important markets for short hemp fibres is their use in ecological thermal insulation materials. In many countries this market is growing faster than the total market for insulation materials. The hemp fibre contains 10–40 single elementary fibres which consist of layers and lumen inside the cell. The porous structure of hemp fibres makes them ideal for thermal insulation. Due to their outstanding characteristics, hemp fibres are used to strengthen cement in many construction applications [34, 35]. Hemp hurds combined with lime to form concrete blocks termed 'Hemcrete' (Fig. 3.7) have been developed to provide combined strength and acoustic properties [36].

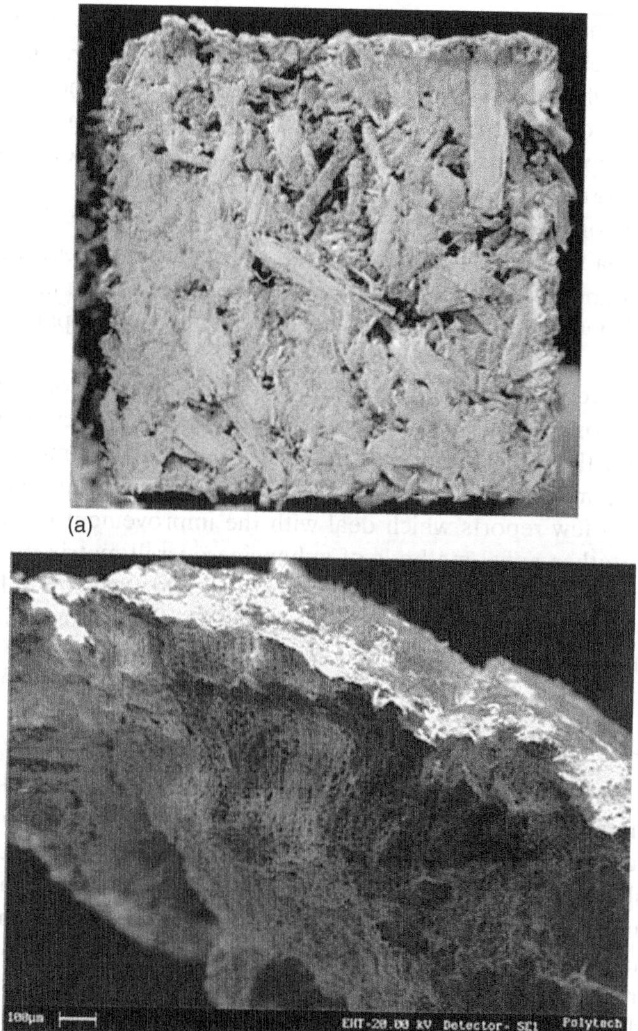

3.7 (a) Typical micrograph of projected lime and hemp concrete; (b) SEM image of hemp shive [36]. Reproduced from Elfordy *et al.* [36] by permission from Elsevier Science Ltd, UK.

3.9 Future trends

Composites and bio-composites suitable for the automotive industry reinforced with fibres such as hemp will see an increased market trend. A similar trend of hemp fibre use can be seen in building materials, hemp shives and lime applications. The enhancement of the fibre–matrix interface is crucial and more effort is required in this area. The development of composite materials based on reinforcement of two or more fibres in a single matrix, which leads to the development of hybrid composites with a great diversity of material properties, can be one way of increasing the mechanical properties as well as decreasing the water absorption behaviour of hemp-reinforced composites.

Processing of hemp fibres to produce bio-composites is another important area of future research. So far, it seems that press moulding is the only established method. Some new methods need to be investigated so that a higher production rate as well as better-quality parts can be achieved.

Most of the reported studies on hemp fibre composites involve study of mechanical properties as a function of fibre content, effect of various treatments on the mechanical properties, and prediction of modulus and strength using models and comparison with experimental data. However, there are very few reports which deal with the improvement of fibre and matrix using other novel methods of enhancing overall matrix properties. It is recommended that future research work should focus on enhancing the properties of fibre and matrix by developing homogeneously dispersed nanoparticles into the composite system.

3.10 Summary

The use of renewable crop materials, such as hemp, in various applications offers economic and social benefits not only to the automotive, pulp and paper and construction industries, but also to rural economies through new agricultural markets for farmers and associated industries. With rising prices of petroleum-based products, strong consumer pressure for environmentally friendly materials and government legislation on eco-friendly products, natural fibres such as hemp have tremendous future growth potential. Due to its good strength-to-weight ratio, hemp fibre is a desirable composite reinforcement from the technical, ecological and economic points of view. However, applications have so far been confined to non-structural components using short fibres. It is evident that there is some gap in the research on hemp fibre-reinforced composites towards using this reinforcement in structural load-bearing applications using long fibres. The automotive industry is the main segment using hemp fibre composites and it is important for future research to employ more advanced fibre processing and manufacturing techniques and the use of bio-resin so that these materials are kept light and affordable. This in turn will lead to improved performance which can cater for more applications not just in the automotive, pulp and paper, insulation and food sectors but also in the aerospace and construction industries.

3.11 References

1. Faruk, O., Bledzki, A.K., Fink, H.P. and Sain, M. (2012). Biocomposites reinforced with natural fibres: 2000–2010. *Progress in Polymer Science*, 37: 1552–1596.
2. Eichhorn, S.J., Baillie, C.A., Zafeiropoulos, N., Maikambo, L.Y., Ansell, M.P., Dufresne, A., Entwistle, K.M., Herrera-Franco, P.J., Escamilla, G.C., Groom, L., Hughes, M., Hill, C., Rials, T.G. and Wild, P.M. (2001). Review. Current international research into cellulosic fibres and composites. *Journal of Materials Science*, 36: 2107–2131.
3. Wambua, P.W., Ivens, J. and Verpoest, I. (2003). Natural fibres: can they replace glass in fibre reinforced plastics? *Composites Science and Technology*, 63: 1259–1264.
4. Bledzki, A.K., Sperber, V.E. and Faruk, O. (2002). *Natural and wood fibre reinforcement in polymer*. Rapra Review Reports, Rapra Shrewsbury, UK.
5. Dhakal, H.N., Zhang, Z.Y. and Richardson, M.O. (2007). Effect of water absorption on the mechanical properties of hemp fibre reinforced unsaturated polyester composites. *Composites Science and Technology*, 67(7–8): 1674–1683.
6. Eichhorn, S.J., Sirichaist, R.J. and Young, R.J. (2001). Deformation mechanisms in cellulose fibres, paper and wood. *Journal of Materials Science*, 36: 3129–3135.
7. Evans, W.J., Isaac, D.H., Suddel, B.C. and Crosky, A. (2002). Natural fibres and their composites: A global perspective. In *Proceedings of the 23rd Risø International Symposium on Materials Science. Sustainable Natural and Polymeric Composites*, ed. Lilholt, H. *et al.* Risø National Laboratory, Roskilde, Denmark, pp. 1–14.
8. Richardson, M.O.W., Santana, M.T.J. and Hague, J. (1998). Natural fibre composites–the potential for the Asian markets. *Progress in Rubber and Plastics Technology*, 14: 174–188.
9. Hepworth, D.G., Hobson, R.N., Bruce, D.M. and Farrent, J.W. (2000). The use of hemp fibre in composite manufacture. *Composites Part A: Applied Science and Manufacturing*, 31: 1279–1283.
10. Bledzki, A.K. and Gassan, J. (1999). Composites reinforced with cellulose base fibres. *Progress in Polymer Science*, 24: 221–274.
11. Bolton, J. (1995). The potential of plant fibres as crops for industrial use. *Outlook on Agriculture*, 24: 85–89.
12. Dhakal, H.N., Zhang, Z.Y., Richardson, M.O.W. and Errajhi, O.A.Z. (2007). The low velocity impact response of non-woven hemp reinforced unsaturated polyester composites. *Composite Structures*, 81: 559–567.
13. Dhakal, H.N., Zhang, Z.Y., Bennett, N. and Reis, P.N.B. (2012). Low-velocity impact response of non-woven hemp fibre reinforced unsaturated polyester composites: Influence of impactor geometry and impact velocity. *Composite Structures*, 94: 2756–2763.
14. Bledzki, A.K., Reihmane, S. and Gassan, J. (1996). Properties and modification methods for vegetable fibre composite. *Journal of Applied Polymer Science*, 59: 1329–1336.
15. Troedec, M.L., Dalmay, P., Patapy, C., Peyratout, C., Smith, A. and Chotard, T. (2011). Mechanical properties of hemp–lime reinforced mortars: influence of the chemical treatment of fibres. *Journal of Composite Materials*, 45: 2347–2357.

16. Bledzki, A.K., Zhang, W. and Chate, A. (2001). Natural fibre reinforced polyurethane microfoams. *Composites Science and Technology*, 61: 2405–2411.
17. Bledzki, A.K. and Gassan, J. (1999). Composites reinforced with cellulose base fibres. *Progress in Polymer Science*, 24: 221–274.
18. Hobson, R.N., Hepworth, D.G. and Bruce, D. (2001). Quality of fibre separated from unretted hemp stems by decortification. *Journal of Agriculture Engineering Research*, 78: 153–158.
19. Summerscales, J., Dissanayake, N.P.J., Virk, A.S. and Hall, W. (2010). A review of bast fibres and their composites. Part 1 – Fibres as reinforcements. *Composites Part A*, 41: 1329–1335.
20. Vignon, M.R., Dupeyre, D. and Jaldon, C.G. (1996). Morphological characterisation of stem-exploded hemp fibres and their utilization in polypropylene-based composites. *BioResource Technology*, 58: 203–215.
21. Prasad, B.M. and Sain, M.M. (2003). Mechanical properties of thermally treated hemp fibres in inert atmosphere for potential composite reinforcement. *Materials Research Innovations*, 7: 231–238.
22. Jovancic, P., Jocic, D., Radetic, M., Topalovic, T. and Petrovic, Z.L. (2005). The influence of surface modification on related functional properties of wool and hemp. *Materials Science Forum*, 494: 283–290.
23. Ragoubi, M., Bienaime, D., Molina, S., George, B. and Merlin, A. (2010). Impact of corona treated hemp fibres on to mechanical properties of polypropylene composites. *Industrial Crops and Products*, 31: 344–349.
24. Dhakal, H.N., Zhang, Z.Y and Bennett, N. (2012). Influence of fibre treatment and glass fibre hybridisation on thermal degradation and surface energy characteristics of hemp/unsaturated polyester composites. *Composites Part B*, 43: 2757–2761.
25. Mwaikambo, L.Y. and Ansel, M.P. (2002). Chemical modification of hemp, sisal, jute and kapok fibers by alkalization. *Journal of Applied Polymer Science*, 84: 2222–2234.
26. Aziz, S.H. and Ansell, M.P. (2004). The effect of alkalization and fibre alignment on the mechanical and thermal properties of kenaf and hemp bast fibre composites. Part 2–Cashew nut shell liquid matrix. *Composites Science and Technology*, 64: 1231–1238.
27. Pracella, M., Chionna, D., Anguillesi, I., Kulinski, Z. and Piorkowska, E. (2006). Functionalization, compatibilization and properties of polypropylene composites with hemp fibres. *Composites Science and Technology,* 66: 2218–2230.
28. Dhakal, H.N., Zhang, Z.Y. and Richardson, M.O.W. (2009). Creep behaviour of hemp fibre reinforced unsaturated polyester composites. *Journal of Biobased Materials and Bioenergy*, 3: 232–237.
29. Dhakal, H.N., Zhang, Z.Y. and Richardson, M.O.W. (2007). Effect of water absorption on the mechanical properties of hemp fibre reinforced unsaturated polyester composites. *Composites Science and Technology*, 67(7–8): 1674–1683.
30. Dhakal, H.N., Zhang, Z.Y., Reis, P.N.B., Surip, S.N. and Bennett, N. (2012). Evaluation of water absorption damage in hemp fibre reinforced unsaturated polyester composites by the nanoindentation testing. *Journal of Biobased Materials and Bioenergy*, 6(4): 493–499.

31. Dhakal, H.N., Zhang, Z.Y. and Ghasemnejad, H. (2013). Effect of compaction pressure on water absorption behaviour and mechanical properties of hemp fibre reinforced vinyl ester composites. *Proceedings of the American Society for Composites, the 28th Technical Conference*, ed. Bakis, C.E., 8–11 September, Pennsylvania State University, USA.
32. Tofan, L., Teodosiu, C., Carmen, P. and Wenkert, R. (2013). Cobalt(II) removal from aqueous solutions by natural fibres: Batch and fixed-bed column studies. *Applied Surface Science Part A*, 285: 33–39.
33. Karus, M. and Kaup, M. (2002). Natural fibres in the European automotive industry. *Journal of Industrial Hemp*, 7: 117–129.
34. Sedan, D., Pagnoux, C., Smith, A. and Chortard, T. (2008). Mechanical properties of hemp fibre reinforced cement: Influence of the fibre/matrix interaction. *Journal of the European Ceramic Society*, 28: 183–192.
35. Pickering, K.L., Beckermann, G.W., Alam, S.N. and Foreman, N.J. (2007). Optimising industrial hemp fibre for composites. *Composites Part A – Applied Science and Manufacturing*, 38: 461–468.
36. Elfordy, S., Lucas, F., Tancret, F., Scudeller, Y. and Goudet, L. (2008). Mechanical and thermal properties of lime and hemp concrete (hempcrete) manufactured by projection process. *Construction and Building Materials*, 22: 2116–2123.

4

The use of ramie fibers as reinforcements in composites

Y. DU, N. YAN, and M. T. KORTSCHOT,
University of Toronto, Canada

DOI: 10.1533/9781782421276.1.104

Abstract: This chapter provides an overview of research in the area of ramie fiber for fiber-reinforced polymer composite applications based on a literature review in this subject since the mid-1980s. Research directions introduced in this chapter include fundamental properties of ramie fibers, ramie fiber/polymer matrix interfacial properties, ramie fiber/polymer composite manufacturing processes, composite mechanical properties, processability, biodegradability, and properties related to specific applications.

Key words: ramie fiber; properties, bonding, polymer composites, applications.

4.1 Introduction

Ramie, *Boehmeria nivea*, is a perennial plant originating in China, and is also known as China grass (Fig. 4.1). Ramie was recorded in AD 1300 as one of the plant fibers for cloth making in China prior to the introduction of cotton [1]. There are other statements of its occurrence in Egypt during 5000–3000 BC [2] or origin in Indonesia [3]. However, there are no authentic records of the utilization of ramie in Egypt [1].

Ramie canes may grow up to 3–10 feet in height with few branches [1–3], and usually are less than half an inch in diameter [1]. Ramie is planted from roots instead of seed. It is started with root cuttings planted three inches in the ground, in good loose soil [3]. If the ramie canes are cut during the growing season to obtain the fiber, a new crop of canes begins to grow, and three or four crops may be obtained in one season. Ramie is a semitropical plant [1]. It is best cultivated in warm and humid regions with an annual rainfall of at least 1000 mm [1, 2]. It has been reported that ramie grows better where the winters are cool enough to induce a resting period. However, severe freezing and repeated freezing and thawing are likely to kill the roots [1].

Ramie is cultivated primarily for its fiber. The ancient Chinese prepared the fiber by hand. The outside skin of the stalk was stripped off, then the ribbon layer was peeled off. The ribbon layer contains normally 20–30%,

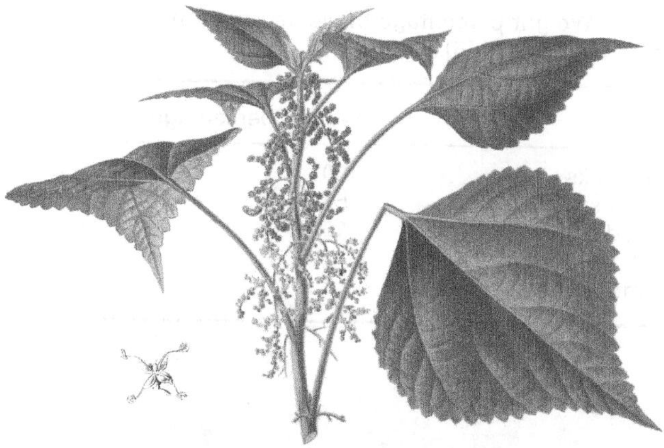

4.1 Ramie plant [4].

(a) (b)

4.2 Ramie fiber: (a) raw ramie fiber; (b) degummed ramie fiber [2].

by weight of gum [5]. After the gum was removed, soft fibers were made ready for spinning [1]. Typically, fibers extracted from ramie are white (as shown in Fig. 4.2), fine, long, and of a silk luster [3]. Table 4.1 summarizes the weight percentage of the ramie stalks [5]. The overall degummed fiber constitutes only about 1–1.5% by weight of the whole stalk (with leaves) [1,5]. Global production of ramie in 2011 was 128,000 tonnes [6]. Over 96% of the total global ramie production was from China in 2011 (Fig. 4.3).

Ramie fibers are used for a wide variety of industrial applications. Typical applications are summarized below [7]:

- Industrial sewing thread, packing materials, fishing nets, handkerchiefs, parachute fabrics, woven fire hoses, narrow weaving canvas, filter cloth, etc.

Table 4.1 Weight percentage of yielded dry ramie fiber to green stalks [5]

Material	Weight percentage
Green stalks	100
Stripped stalks	52
Air-dried stalks	10.4
Decorticated fiber	2.1
Degummed fiber	1.2

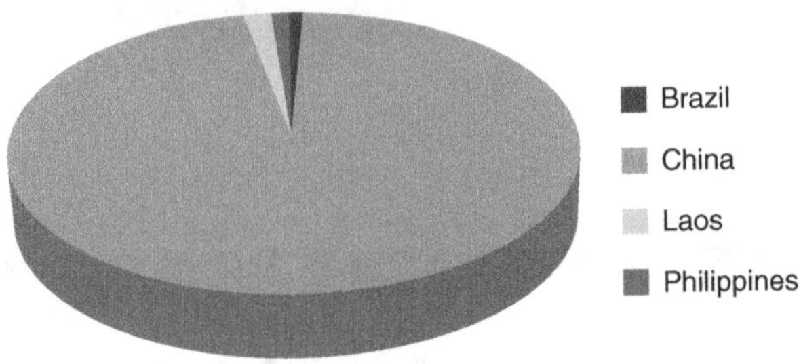

4.3 Global ramie production quantity by countries in 2011 [6].

- Fabrics for household furnishings, e.g. upholstery and canvas, and clothing, frequently in blends with other textile fibers (for instance when used in admixture with wool, shrinkage is reported to be greatly reduced when compared with pure wool), curtains, draperies, upholstery, bedspreads, table linens, sheets, dish towels.
- High-quality papers, e.g. bank notes and cigarette papers.
- Apparel, e.g. dresses, suits, skirts, jackets, pants, blouses, shirts, children's wear, mixed with cotton in knitted sweaters.

Ramie fiber has comparable specific modulus and strength to traditional synthetic fibers for composite production. As one renewable and biodegradable fiber resource, ramie fiber has numerous environmental advantages over synthetic fibers, e.g. occupational health advantages and low carbon footprint. With the increasing awareness of global environmental concerns, natural fiber-based eco-friendly materials are finding more industrial applications. This chapter focuses mainly on the research results related to the utilization efforts of ramie fiber for fiber-reinforced composite material applications dating from the earliest 1980s.

4.2 Ramie fiber properties

Ramie has exceptionally long fiber cells ranging from 120 to 150 mm [7]. This is nearly six times more than cotton, 10 times more than flax, and eight

Table 4.2 Chemical composition and microfibril angle of natural fibers [8]

Fiber	Cellulose (wt%)	Lignin (wt%)	Hemicellulose (wt%)	Pectin (wt%)	Wax (wt%)	Microfibril angle (deg)
Jute	61–71.5	12–13	13.6–20.4	0.2	0.5	8.0
Flax	71	2.2	18.6–20.6	2.3	1.7	10.0
Hemp	70.2–74.4	3.7–5.7	17.9–22.4	0.9	0.8	6.2
Ramie	68.6–76.2	0.6–0.7	13.1–16.7	1.9	0.3	7.5
Kenaf	31–39	15–19	21.5	–	–	–
Sisal	67–78	8.0–11.0	10.0–14.2	10.0	2.0	20.0
Palm	70–82	5–12	–	–	–	14.0
Henequen	77.6	13.1	4–8	–	–	–
Cotton	82.7	–	5.7	–	0.6	–
Coir	36–43	41–45	0.15–0.25	3–4	–	41–45

times more than silk. Ramie fiber is extremely white in color and the color does not change with exposure to sunlight, etc. Ramie fiber also has a higher resistance to chemicals than other fibers and deters growth of bacteria and fungus, including mildew.

4.2.1 Chemical properties

Chemical compositions of commonly used natural fibers for composite production are summarized in Table 4.2 [8]. Ramie fiber has a generally similar chemical composition, except for a distinctively lower lignin content, to other natural fibers, which explains why it is extremely white and soft.

4.2.2 Mechanical properties

A comprehensive chart of Young's modulus and tensile strength of natural materials (Ashby plot) is shown in Fig. 4.4 [9]. Ramie fiber has one of the highest values of Young's modulus and tensile strength among the wide variety of natural fibers, second only to flax and hemp and a little less stiff than kenaf. In terms of specific Young's modulus (Fig. 4.5), ramie fiber is even superior to steel wire, and close to E-glass fibers [9].

Table 4.3 summarizes the data for tensile strength, Young's modulus, elongation, and corresponding fiber diameter of ramie fibers reported in the literature. The mechanical properties of ramie fibers vary in the literature. The differences could be due to fiber production methods, fiber diameters, and/or differences in testing methods.

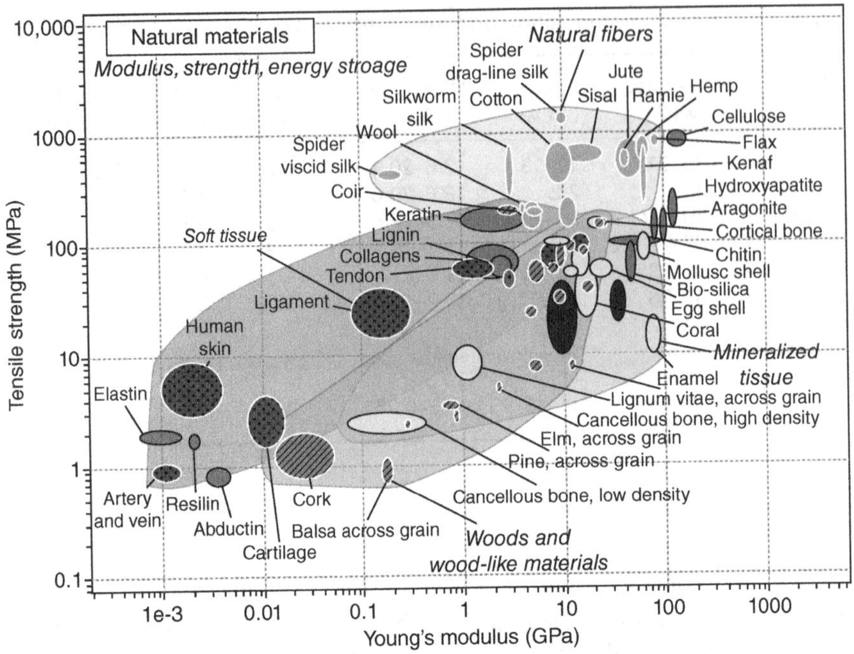

4.4 Young's modulus and tensile strength of selected natural materials [9].

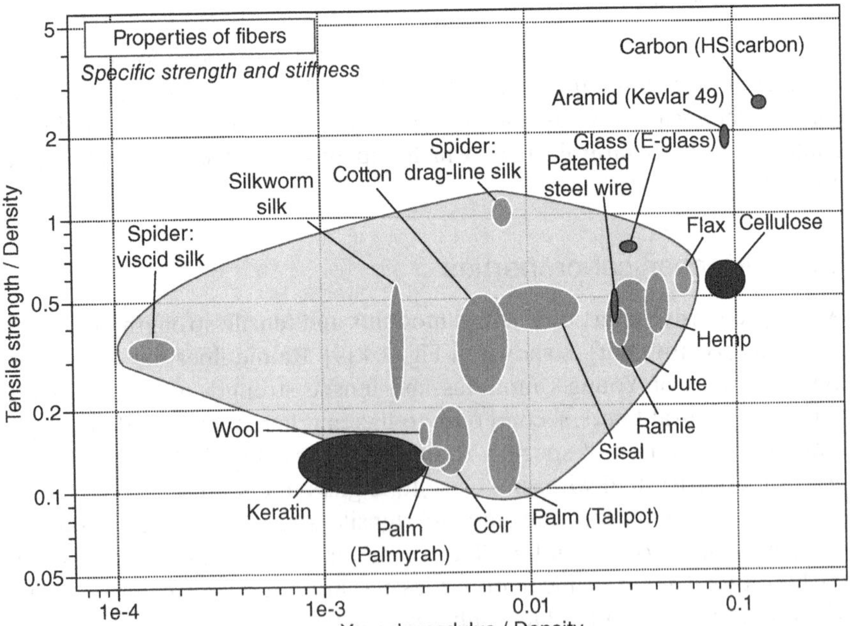

4.5 Specific Young's modulus and tensile strength of selected natural fibers compared to synthetic fibers [9].

Table 4.3 Mechanical properties of ramie fibers from literature

Tensile strength (MPa)	Young's modulus (GPa)	Elongation (%)	Diameter (μm)	Reference
800–1000	65	1.7–2.3	40–60	[10]
621	47.5	1.9	48.5	[11]
402	30–40	–	36	[12]
730	42	–	30	[13]
526	15.5	3.5	120	[14]
560	–	2.5	34	[15]
–	48.6	2.4	–	[16]
627	31.8	2.7	120	[17]
1250	36.0	3.8	20.3	[18]
768	42	2.6	–	[19]
494	24	2.57	33.6	[20]
333	24.7	1.63	25–30	[21]
936	25	–	15–40	[22]

4.2.3 Factors affecting mechanical properties of ramie fibers

Fiber diameter and gauge length

Weibull developed a statistical approach for material strength that assumes a distribution of weak spots (for example cracks or dislocations) that result in material failure [10]. According to this approach, a material is only as strong as the volume element containing the weakest flaw, and larger volumes of material are more likely to have more flaws, so the strength of a Weibull material is strongly dependent on its volume. Tensile strengths of ramie fibers were found to follow this prediction. Bevitori *et al.* [11] and Monteiro *et al.* [12] reported that tensile strengths of ramie fibers displayed a hyperbolic type of inverse relationship with the fiber diameter. Angelini *et al.* found that strengths of ramie and three other types of fibers (either natural or synthetic) increased with a decrease in the gauge length (Fig. 4.6) [13].

Thermal treatment

There is an upper temperature limit at which degradation of natural fibers occurs. Processing at elevated temperatures with either a thermoset or a thermoplastic polymer might affect mechanical properties of the fibers. Nam and Netravali investigated the effect of short-term heating on tensile properties of ramie fibers [14]. Individual ramie fiber specimens were heated at 100, 120, 140, 160 and 200°C in an air circulating oven for two

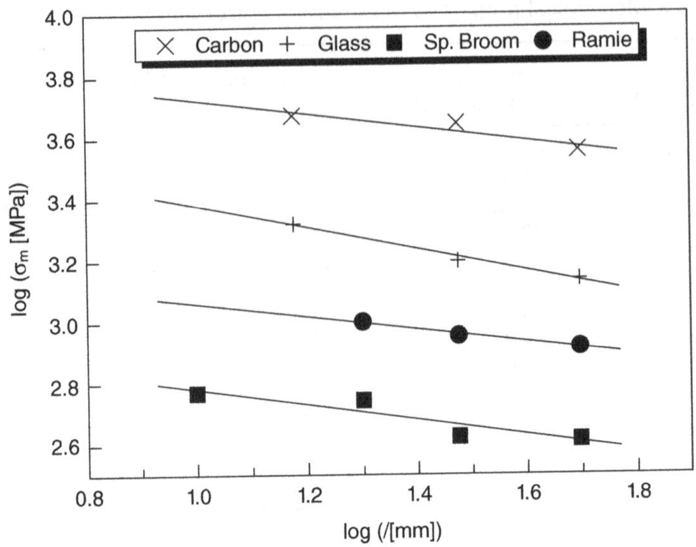

4.6 Influence of gauge length on strength of Spanish broom, ramie, carbon and glass fibers (reprinted from [13], © 2000, with permission from Elsevier).

time periods of 0.5 and 2 hours. The heat treatment did not influence the tensile strength and modulus of ramie fibers up to 160°C. At a temperature of 200°C, the fibers lost their strength significantly after two hours of heat treatment. The heat treatment affected the fiber tensile strength more than Young's modulus.

Chemical treatment

Choi and Lee compared mechanical properties of ramie fibers before and after treatments by alkaline, silane, and peroxide [15]. The tensile strength of ramie fibers did not change after alkaline treatment. Further treatment by silane and peroxide slightly abated fiber tensile strengths. Alkaline, silane and peroxide all reduced the tensile modulus of the ramie fibers. Fourier transform infrared (FTIR) spectrum analysis results showed the removal of hemicellulose and lignin by the alkalization. Goda *et al.* evaluated the effect of mercerization on ramie fiber tensile properties [16]. Ramie fibers were treated with 25% NaOH solution for two hours with or without loads applied. As compared to the untreated ramie fibers, NaOH treatment improved the elongation of the fiber almost twofold, but no improvement in tensile strength was observed. If a load was applied on the fiber during mercerization, tensile strength was improved for up to 101 MPa. However, all treatments caused a significant reduction in the fiber Young's modulus.

Table 4.4 Interfacial shear strength between ramie fibers and polymers

Matrix polymer	Shear strength (MPa)	Testing method	Reference
Epoxy	79	Fragmentation	[10]
Soy protein isolate	30	Micro droplet	[11]
Stearic acid modified soy protein isolate	13.1	Micro droplet	[14]
Polypropylene (PP)	5	Micro droplet	[27]
Epoxy	16	Micro droplet	[28]
Phenolic	9.6	Fiber pullout	[29]
Soy flour	9.5	Micro droplet	[21]
PP (2% MAPP)	24.9	Fragmentation	[30]
PP	16.1	Micro droplet	[22]
Acrylated epoxidized soybean oil	3.7	Micro droplet	[31]

4.3 Improving fiber/matrix interfacial bonding

The interfacial shear strengths between ramie fiber and various matrices reported in the literature are summarized in Table 4.4. Three methods were commonly used for fiber/matrix interfacial shear strength characterization: fiber fragmentation, micro droplet, and fiber pullout. There was a large variation in these reported results.

Like most natural fibers, the polar and hydrophilic nature of ramie fibres has resulted in their poor compatibility with hydrophobic polymer matrices. A significant amount of research has been done in the area of surface modification of ramie fibers for improving the fiber/matrix interfacial bonding. Methods that have been used to improve fiber/matrix interfacial bonding, either by chemical treatment or by addition of a coupling agent, are summarized below.

4.3.1 Alkali treatment or mercerization

Alkali treatment has been the most commonly used method for fiber modification. It can be used either exclusively or as a pre-treatment combined with other treatments. Goda *et al.* evaluated the effect of alkali treatment on tensile properties of ramie fiber-reinforced biodegradable polymer composites [17]. The results showed that 5% NaOH treatment of ramie fibers caused no change in the composite tensile strength, while the composite Young's modulus was 29% higher than that of the untreated fiber composite. He *et al.* treated ramie fibers with 5%, 10%, and 15% NaOH solutions for 24 hours [18]. Composites were made using the untreated and treated ramie fibers and PP using extrusion followed by an injection molding

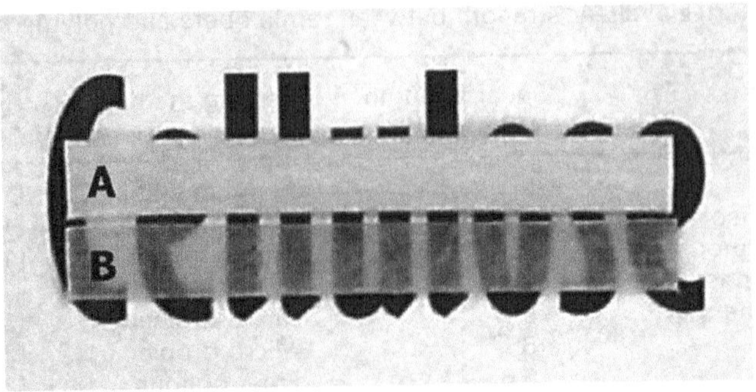

4.7 Optical photograph of the ramie–cellulose composites (prepared with 4% cellulose solution): (A) unmercerized and (B) mercerized (reprinted from [20], © 2008, with permission from Elsevier).

process. The highest composite strength was obtained with the 15% NaOH treatment. Tensile, flexural, and compression strengths were improved. However, no improvement in impact strength was observed. Yu *et al.* evaluated the effect of NaOH treatment on ramie fiber-reinforced PLA composites [19]. Significant improvements in the tensile strength, elongation, flexural strength, impact strength, and storage modulus were obtained. The NaOH treatment was shown to be more effective than the silane treatment. Qin *et al.* prepared all-cellulose composites with ramie fibers in cellulose solutions at concentrations of 1–7% [20]. The composites were treated with a 9% NaOH solution for one hour at room temperature. After the NaOH solution treatment, the composite tensile strength increased significantly from 440 MPa (for the untreated composites) to 540 MPa. The mercerized composites showed a better optical transparency (Fig. 4.7) indicating improved fiber/matrix interfacial bonding since less light is scattered at the fiber/matrix interface. Suizu *et al.* compared the effect of applied load during mercerization on ramie yarn reinforced biodegradable polymer (polycaprolactone and cornstarch) composites [21]. Ramie yarns were treated in the 15 wt% NaOH solution for two hours with or without load application. The alkali treatment improved the composite elongation and impact resistance. However, tensile strength and modulus both decreased. The application of load on the ramie yarns during mercerization diminishes the reduction in composite tensile strength and Young's modulus.

4.3.2 Polyester amide polyol as the interfacial agent

Unidirectional degummed ramie/unsaturated polyester composites were made using a matched-die molding process [22]. Polyester amide polyol was used as an interfacial agent. Flexural strength and modulus of the control composite samples and boiling water-treated composite samples were

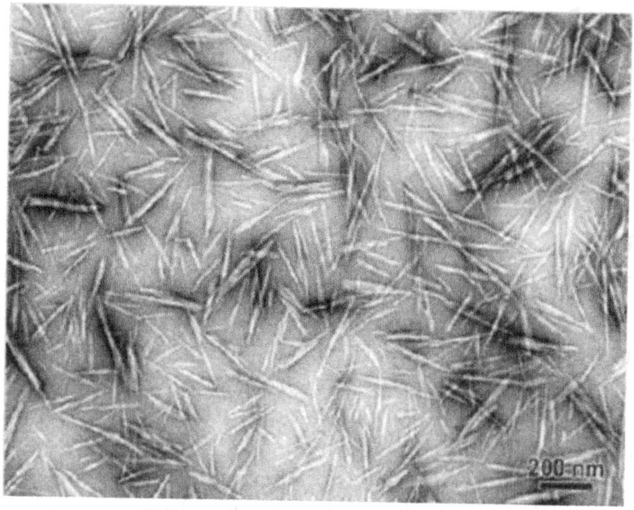

4.8 A transmission electron microscope (TEM) image of a ramie whisker suspension (reprinted from [26], with kind permission from Springer Science and Business Media).

tested. The polyester amide polyol improved composite flexural strength from 197.6 to 223.0 MPa. No significant change was observed in composite flexural modulus. The incorporation of the interfacial agent also increased the retention of composite flexural strength after the boiling water treatment and reduced composite water uptake amounts.

4.3.3 Acid hydrolysis

Acid hydrolysis treatment of ramie fibers is mainly used to remove amorphous domains in cellulose and leave mostly crystalline domains, such as cellulose whiskers or cellulose nanocrystals. The process is typically done by first treating ramie fibers with NaOH in order to remove residual additives. The purified ramie fibers are then hydrolyzed in 65 wt% H_2SO_4 solution. The acid hydrolysis leads to aqueous suspensions containing cellulose nanocrystals (Fig. 4.8). The suspensions are finally neutralized using NaOH solutions [23–28]. Cellulose nanocrystals with high aspect ratios can be produced by the acid hydrolysis process (Table 4.5).

4.3.4 Solvent treatment

Marsyahyo *et al.* treated ramie fiber with acetone, methyl ethyl ketone, and alcohol, separately [29]. Ramie fiber was immersed in each solvent at 110°C for one hour. Solvent immersion changed the fiber surface at the molecular

Table 4.5 Diameter and length of cellulose nanocrystals by acid hydrolysis of ramie fiber after acid hydrolosis

Diameter (nm)	Length (nm)	Reference
85	539	[36]
30	250	[38]
6–8	200	[39]

level as indicated by FTIR results. The specific surface areas of the untreated and treated ramie fibers were measured by the Brunauer–Emmet–Teller (BET) method. The BET and scanning electron microscopy (SEM) observations indicated that the porosity and surface area of the ramie fiber increased after the solvent treatment. All the treated fiber surfaces showed topography with macropores. It was thought the porous surface of the fiber might promote more mechanical interlocking between the fiber and the matrix, resulting in a stronger fiber/matrix interfacial bonding.

4.3.5 Steam treatment

Munawar *et al.* used steam treatment of the fiber to improve interfacial bonding between the ramie fiber and phenolformaldehyde (PF) resin [30]. The fibers were steamed above boiling water in a tank covered with screen-plaited bamboo at a steam pressure of about 0.1 MPa for two hours. The steam-treated fibers were air-dried to a moisture content of 6–8%. The steam treatment was found to have improved the interfacial adhesion between the ramie fiber and PF resin (the PF resin contact angle decreased from 64.7° to 47.3°). The steam treatment also significantly improved internal bonding strength and flexural properties of the ramie/PF composites.

4.3.6 Silane treatment

Silane has been extensively used as a coupling agent for fiberglass. Recently, it has been proved to be also effective in treating natural fiber [31]. After silane treatment, the surface free energy of the untreated ramie fiber (63.4 mJ/m^2) was reduced to 17.4 mJ/m^2 for the treated fiber. The water contact angle on the ramie fiber increased from 30° to 104° before and after the treatment. Yu *et al.* found that silane treatment also improved tensile strength, flexural strength, impact strength, and Vicat softening temperature of ramie/PLA composites [19]. Silane treatment increased the interfacial shear strengths (IFSS) between the ramie fiber and the phenolic resin from 9.6 MPa to 15.6 MPa [32].

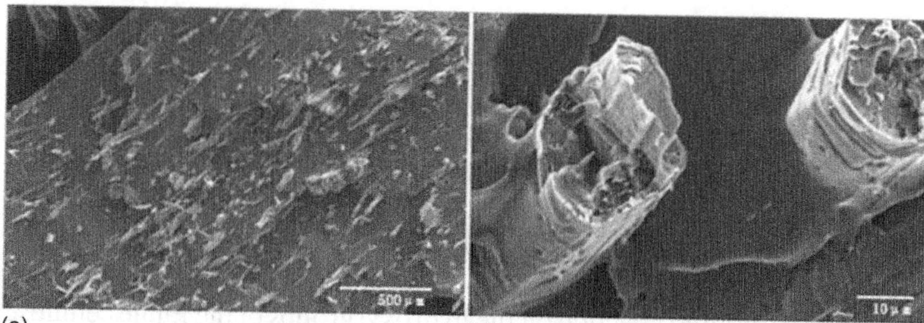

4.9 The impact fracture surfaces of silane-treated ramie fiber-reinforced PLLA-PCL composites at (a) low and (b) high magnification levels (reprinted from [33], with kind permission from Springer Science and Business Media).

Xu *et al.* developed ramie fiber reinforced poly(L-lactic acid)-poly(ε-caprolactone) (PLLA-PCL) composites by an in-situ polymerization process [33]. Ramie fiber was first treated by silane. The treated ramie fiber was grafted with $-NH_2$ functional groups which then reacted with NCO end groups of PCL to form acylamino. Then composites were prepared by in-situ polymerization of PLLA oligomer with NCO-terminated PCL prepolymer. Interfacial adhesion between PLLA-PCL matrix and ramie fiber was improved when the ramie was treated by silane (Fig. 4.9). The silane treatment of the ramie fiber caused an increase of 93% in tensile strength, 196% in impact strength, and 84% in elastic modulus.

4.3.7 Silicone oil treatment

A laboratory-made silicone oil emulsion that was composed of a mixture of epoxy-silicone oil, alkylphenols polyoxyethylene, and cetyl alcohol was used to treat ramie fibers [34]. To prepare the emulsion, the three components were mixed in a flask with distilled water, then stirred and ultrasonic-dispersed for three hours at 50°C. The ramie fiber was treated with the laboratory-made silicone oil emulsion in a reactor at 160°C for three hours under an argon atmosphere. The treatment improved tensile strength, impact strength, and elongation of the ramie fiber/polypropylene composite in the studied fiber loading range from 10 to 30 wt%. Thermal gravimetric analysis (TGA) results indicated that the modification also enhanced the heat resistance of the composites.

4.3.8 Polymer grafting

Chen *et al.* converted the hydroxyl groups of the ramie fibers to 2-dithiobenzoyl isobutyrate as a reversible addition–fragmentation chain transfer (RAFT) agent [35]. The hydrophobic polymers, poly (methyl methacrylate) (PMMA) and poly (methyl acrylate) (PMA), were then directly grafted on the surface of the ramie fibers via the RAFT method. About 33% of the hydroxyl groups in the raw ramie fiber were substituted by 2-bromoisobutyryl bromide and the grafted ramie fibers had hydrophobic surface properties (Fig. 4.10). The water contact angles were 136° and 133°, respectively, after PMMA and PMA were grafted on the fiber.

Goffin *et al.* grafted L-lactide on the hydroxyl groups of the ramie cellulose nanowhisker surface by ring-opening polymerization to yield cellulose nanowhisker-g-PLA nanohybrids [27]. The PLA graft cellulose nanowhisker was extruded and injection-molded with PLA polymer. The PLA graft cellulose nanowhisker was found to induce a limited reinforcing effect on the composite nanohybrids below composite T_g due to the plasticization effect of the grafted PLA chains on the polymer matrix. Above T_g, the presence of nanowhiskers increased the stiffness of the composite. The surface-grafted cellulose nanowhiskers acted as nucleation sites and promoted the degree of crystallinity in the nanocomposites.

4.3.9 Acid chloride

Junior de Menezes *et al.* modified ramie cellulose whiskers with three organic acid chlorides: hexanoyl chloride, lauroyl chloride, and stearoyl chloride [28]. The modification changed the hydrophilic fiber surface to hydrophobic (the water equilibrium contact angle increased from 35° to a range of 72–101°). Functionalized nanofibers were compounded with low-density polyethylene by extrusion to prepare nanocomposite materials. The homogeneity of the developed nanocomposites was found to increase with the length of the grafted chains. Composite tensile strengths and elongation were improved significantly. The longer the polymer chains grafted on the surface of the nanoparticles, the higher tensile strength and elongation at break were observed.

4.3.10 Acetylation

Lee *et al.* reported that the surface free energy of ramie fiber decreased from 63.4 mJ/m^2 for untreated ramie fiber to 41.7 mJ/m^2 after acetic anhydride treatment [31]. The water contact angle increased from 30° to 66°. The experimentally measured IFSS between the treated ramie fiber and the acrylated epoxidized soybean oil resin increased to 16.2 MPa as compared with 3.7 MPa between the control ramie fiber and the resin (data taken from the original figure).

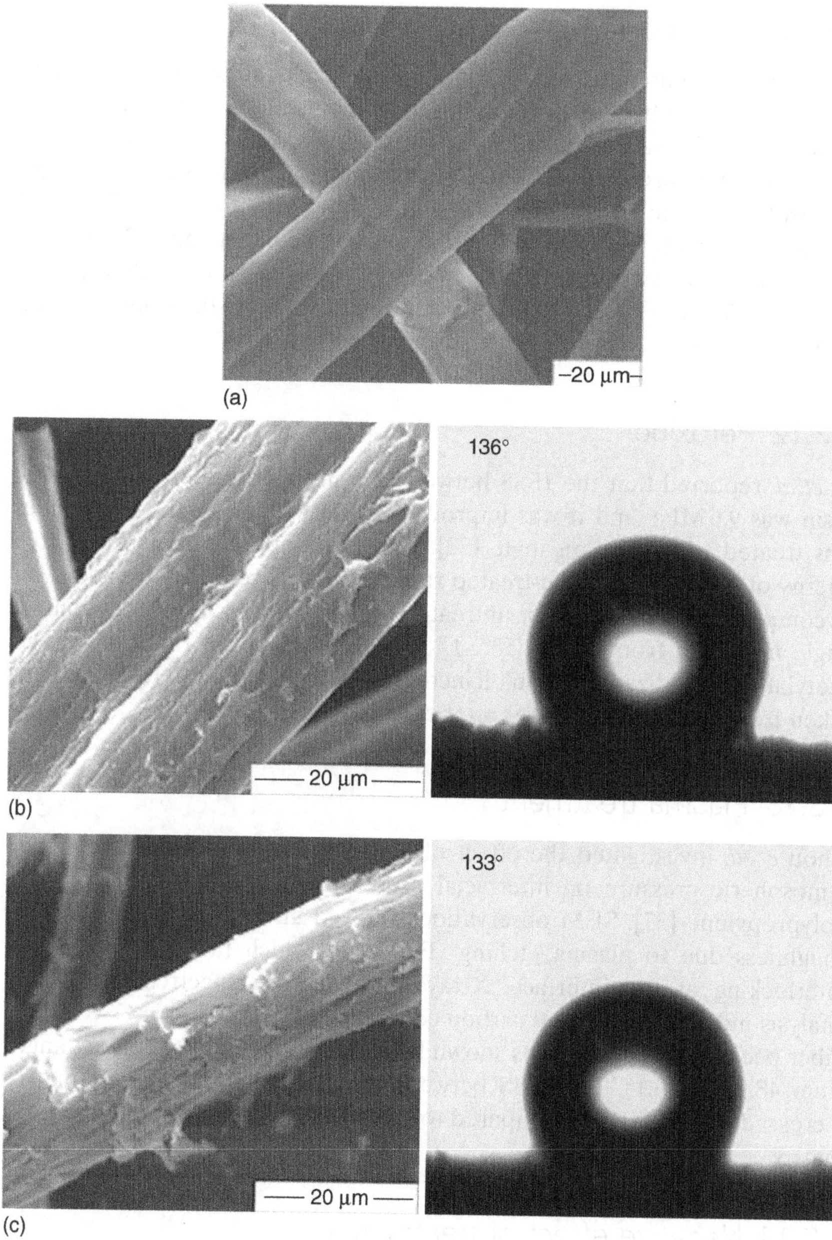

4.10 SEM images of (a) original ramie cellulose, (b) PMMA-grafted ramie fiber and its corresponding water contact angle, and (c) PMA-grafted ramie fiber and its corresponding water contact angle (reprinted from [35], with kind permission from Springer Science and Business Media).

4.3.11 Arylation

Kumar and Zhang improved fiber/matrix interfacial adhesion by arylation of the ramie fiber/soy protein isolate composites [36]. Significant improvement in water resistance, tensile modulus, and thermal stability of the composites was observed. The 26-hour water absorptions of ramie fiber/soy protein isolate composites with and without arylation were 20% and 113%, respectively. The equilibrium water droplet contact angle (30 seconds later) on these two composites were 70° and 33°, respectively. The hydrophobic property of the composites was attributed to the water-induced biphasic system.

4.3.12 Peroxide

Li *et al.* reported that the IFSS between the untreated fiber and phenolic resin was 9.6 MPa, and it was improved to 14.9 MPa after the ramie fiber was treated with permanganate [32]. Lee *et al.* reported the surface free energy of dicumyl peroxide-treated ramie fiber was reduced to 48.1 mJ/m^2 as compared to 63.4 mJ/m^2 for untreated ramie fiber [31]. The water contact angle increased from 30° to 53°. The IFSS between the ramie fiber and acetylated epoxidized soybean oil increased from 3.7 MPa to 11.8 MPa (data taken from the original figure) after the fiber was treated.

4.3.13 Plasma treatment

Zhou *et al.* investigated the effect of helium plasma jet treatment under atmospheric pressure on interfacial properties between ramie fiber and polypropylene [37]. SEM observations showed an increased fiber surface roughness due to plasma etching (Fig. 4.11), which favored mechanical interlocking at the interface. X-ray photoelectron spectroscopy (XPS) analysis indicated increased carbon contents and hydrophobic C–C bonds. Fiber became hydrophobic as shown by an increase in water contact angle from 48.9° to 73.1°. The IFSS between the treated fiber and the matrix increased to 23.5 MPa as compared to 16.1 MPa for the untreated fiber and matrix.

4.3.14 Negative effect of treatments

The above research studies showed positive effects of fiber treatment on composite mechanical properties and enhanced fiber–matrix interfacial bonding. However, negative effects of treatments have been reported too. For example, Choi and Lee treated ramie fibers with alkaline, silane, and peroxide in sequence. The tensile modulus of the treated fiber was lower than that of the untreated fiber. Tensile moduli and strengths of treated

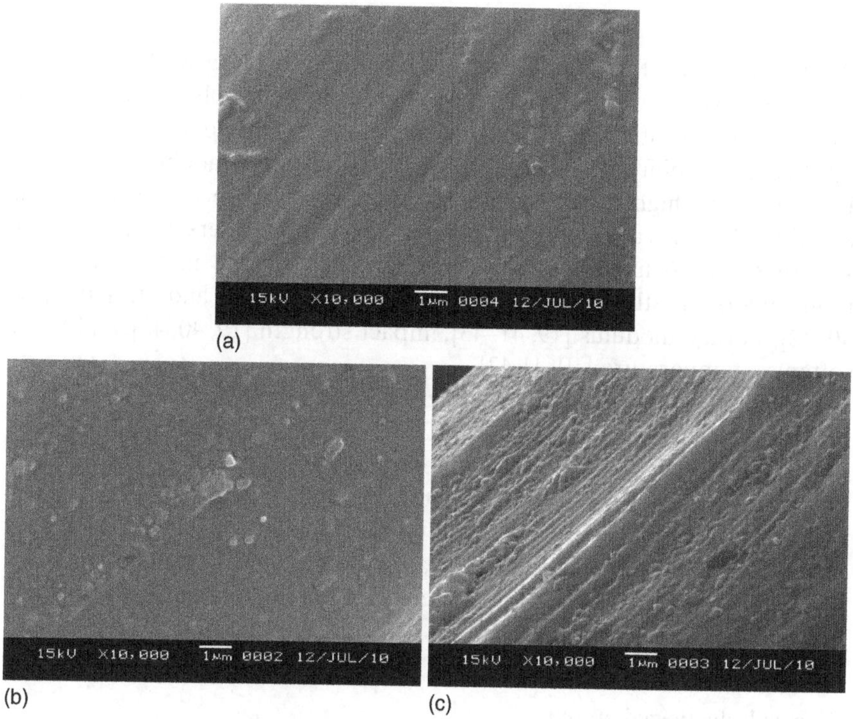

4.11 SEM photographs of control and treated ramie fibers: (a) control; (b) plasma treated for 16 s with ethanol pretreatment; (c) plasma treated for 24 s with ethanol pretreatment (reprinted from [37], © 2011, with permission from Elsevier).

ramie fibers/PLA composites were all lower than those of the untreated ramie fiber/PLA reinforced composite [15]. The fiber treatment should be carried out with caution with optimized conditions in terms of treatment solution concentration, treatment time, etc.

4.4 Ramie fiber-reinforced polymer composites

4.4.1 Polymer matrix

Thermoplastic polymers

Polyester, polysaccharide, starch blends

Wollerdorfer and Bader evaluated the reinforcement effect of ramie fibers to six polymer matrices: one polyester, two polysaccharides, and three starch blends [38]. Incorporation of the ramie fiber increased the composite tensile strength of three of these six polymers due to the chemical compatibility between the three thermoplastics and the ramie fibers.

Poly(lactic acid)

Poly(lactic acid) (PLA) is a biodegradable polyester derived completely from annually renewable resources. For fiber-reinforced polymer composites, PLA has an advantage over the conventional petroleum-derived polyolefin due to the possibility of producing fully green and biodegradable composites. PLA also has higher stiffness and strength than those of polyolefin. In addition, PLA has a better compatibility with natural fibers due to its polar structure [39]. Numerous studies have shown that the incorporation of ramie fibers greatly improved the composite stiffness and strength [19, 40–42], storage modulus [19, 41–43], impact strength [19, 40, 42], and Vicat softening temperature [19, 41,42].

The ramie fiber was shown to affect crystallinity of PLA (Fig. 4.12) [43]. Differential scanning calorimetry (DSC) results (in the heating cycle) showed that the cold crystallization temperature decreased gradually from 92.7°C for a neat PLA to 88.4°C after 30 wt% ramie fibers were added, indicating that ramie fiber was an efficient nucleating agent for PLA. Meanwhile, the crystallization time was shortened during the cooling cycle. The incorporation of ramie fibers increased polymer crystallinity and the glass transition temperature.

Glycerol plasticized starch

Lu *et al.* prepared environmentally friendly glycerol plasticized starch biocomposites with 0–40 wt% ramie cellulose nanocrystals [23]. After ramie

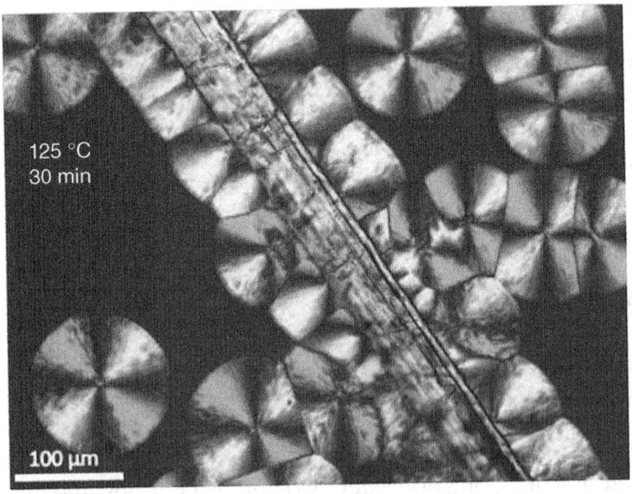

4.12 Transcrystallization of PLA in the vicinity of ramie fiber (reprinted from [43], © 2012 Wiley Periodicals, Inc.).

cellulose nanocrystal was blended in, the tensile strength of the composite increased to 6.9 MPa as compared to 2.8 MPa for the neat glycerol plasticized starch film, and Young's modulus increased significantly from about 55.9 MPa to 479.8 MPa. The ramie cellulose nanocrystals improved the composites' glass transition temperature and water resistance properties as well. However, the elongation at break of the composites decreased from 94.2% to 13.6% after addition of ramie cellulose nanocrystal.

Polyolefin

Polyolefin is the most extensively used matrix polymer for natural fiber composites. Literature concerning ramie fiber-reinforced polyolefin composites was mostly focused on interfacial bonding enhancement [28, 34] and process improvement [44, 45]. Chen *et al.* prepared low-density ramie/PP composites by blending ramie and PP fibers into non-woven mats involving carding and needle-punching, followed by a compression molding process [44]. Mechanical, thermal, water absorption and acoustic absorption properties were tested. Junior de Menezes *et al.* modified hydrophilic ramie cellulose nanowhiskers using acid chlorides [28]. The modified cellulose nano whiskers were compounded with LDPE by extrusion. Most studies on cellulose nanowhisker-reinforced composites involved using a casting/evaporation processing technique. The research of Junior de Menezes *et al.* demonstrated that industrial techniques, like extrusion, can also be used. Feng *et al.* fabricated short ramie fiber-reinforced PP composites by extrusion and injection molding processes (with 3% MAPP) [45]. The tensile strength and flexural strength were as high as 67 and 80 MPa after incorporation of 30 wt% ramie fibers, which the authors believed to be one of the best results for short natural fiber-reinforced PP composites. The average fiber length of the ramie fiber before compounding was 6.0 mm. It was reduced to 2.0 mm after processing, which was still much higher than what was reported by others. So, the high tensile strength in this study could be attributed mainly to the long fiber length retained in the matrix.

Ecoflex

Ecoflex® is a commercial biodegradable resin developed by BASF (1,4-benzenedicarboxylic acid polymer with 1,4-butanediol and hexanedioic acid). Nakamura *et al.* studied tensile properties at high temperatures (100, 115 and 130°C) and deep drawing properties of ramie woven fabric-reinforced Ecoflex® composites [46]. Young's modulus and tensile strength of the composites were 1.67 GPa and 75.9 MPa, respectively, at room temperature. Tensile strengths at higher temperatures were significantly lower than those tested at room temperature. The same polymer was used by Goda *et al.* in determining the optimum forming pressure [17] and by Krasowska *et al.* in evaluating composite degradation [47].

Polycaprolactone and cornstarch

Polycaprolactone and cornstarch blend is a hydrophilic emulsion resin. Laminated green composites were made with this resin combined with unidirectional ramie yarn sheets [21]. As compared to untreated fiber-reinforced composites, the composites with the mercerized ramie fiber had a slightly lower tensile strength, lower tensile modulus, and higher elongation. Mercerized fiber-reinforced composites absorbed almost twice as much impact energy as the composites made with the untreated fiber. Impact energies of the laminated green composites using mercerized ramie woven fabrics increased with the increasing water content.

Poly(oxyethylene)

Alloin *et al.* made nanocomposite polymer electrolytes using poly (oxyethylene) with ramie nanowhiskers and four other natural fiber-based nanowhiskers or microfibrils [24]. The addition of 6% cellulose fiber led to a higher storage modulus at high temperatures as compared to the unfilled PEO-lithium bis (trifluoromethylsulfonyl)imide (LiTFSI) sample.

Thermoset polymers

Soy protein

Soybeans, obtained from an annual plant, typically contain 20% oil and up to 50% protein [48]. This protein consists of polypeptide chains of various lengths of which about 62% are polar and reactive amino acid residues. Some researchers have investigated ramie fiber-reinforced soy protein composites.

Lodha and Netravali measured the tensile properties of ramie fiber soy protein isolate composites and the fiber/matrix IFSS [48]. The IFSS was 29.8 MPa, indicating a good interfacial bonding between the ramie fiber and soy protein resin. The highest tensile properties (modulus 1.7 GPa; strength 33 MPa) were achieved at a ramie fiber loading of 30 wt% and a fiber length of 15 mm. A similar study was carried out by Nam and Netravali on ramie fiber and soy protein concentrate composites [49]. Lodha and Netravali modified soy protein isolate with stearic acid [50]. The stearic acid performed better than glycerol which was used previously. The ramie/stearic acid-modified soy protein isolate composites had adequate tensile properties for use in various commercial applications requiring moderate strength. Kim and Netravali fabricated ramie fiber soy flour composites [51]. The soy flour was modified by filtering out soluble sugars to increase protein content. The modification improved IFSS and composite Young's modulus and tensile strength.

Kumar and Zhang made ramie fiber-reinforced arylated soy protein composites [36]. Ramie/arylated soy protein composites had a better water resistance, thermal stability, and tensile modulus than their counterparts with nonarylated soy protein. The same authors made composites with the same matrix and cellulose nanowhisker hydrolyzed from the ramie fiber and found similar results [25]. Composite tensile strengths were also improved by the incorporation of ramie nanowhiskers.

Unsaturated polyester

Paiva Júnior *et al.* evaluated effects of fiber volume fraction and orientation on tensile strengths of hybrid ramie–cotton fabric reinforced unsaturated polyester composites [52]. The composite tensile strength was improved up to 118 MPa at a fiber loading of 60 vol%, 338% higher than that of the neat polymer (24.2 MPa). Thermal conductivity, thermal diffusivity and specific heat of the unsaturated polyester and its composites with ramie, jute, cotton, and sisal fibers were studied by Alsina *et al.* [53]. The effect of hybrid glass and ramie fibers on the unsaturated polyester composites' flexural and impact properties was evaluated by Romanzini *et al.* [54].

Epoxy

Müssig evaluated the tensile strength, modulus, and impact properties of two natural fiber-reinforced epoxy composites (ramie and reclaimed cotton) for automotive applications [55]. Müssig *et al.* also compared experimentally measured tensile moduli with micromechanical model predictions for three epoxy composites reinforced by cotton, ramie and lyocell fiber, respectively [56]. Marsyahyo and Rochardjo made ramie fiber woven fabric-reinforced epoxy composites and evaluated whether the composite can be used for bulletproof panel applications [57]. Yang and Li made composites with epoxy and ramie fiber, glass fiber, and carbon fiber, respectively, and compared their acoustic properties [58].

Wood-based epoxy

A wood-based epoxy resin was made from resorcinol liquefied wood [59]. Wood was depolymerized and liquefied by reacting with resorcinol. The phenolic-OH groups of the liquefied wood then reacted with epichlorohydrine under an alkali condition. By glycidyl etherification, epoxy functionality was introduced to the liquefied wood. Composites were made with ramie fibers and the wood-based epoxy resin and compared with the composites using a petroleum-based bisphenol-A type epoxy in terms of flexural strength and modulus. The flexural modulus and strength of the ramie/wood based epoxy composite were similar to those of ramie/bisphenol-A type epoxy composites. The T_g of the resorcinol-liquefied wood-based epoxy resin itself was lower than that of the bisphenol-A type epoxy resin. Wood-based epoxy resin had a broad T_g region, indicating a diversity of crosslinked structures in the cured resin.

Cellulose

Nishino *et al.* made an all-cellulose composite with ramie fiber and cellulose by drying the cellulose solution-impregnated ramie fiber under a reduced pressure [60]. The tensile modulus and strength of the unidirectional all-cellulose composite at room temperature were 45 GPa and 480 MPa, respectively, which were comparable with or even higher than those of conventional fiberglass. The dynamic storage modulus was as high as 20 GPa at 300°C. This is significantly higher than most polymer-based composites due to the good thermal stability of cellulose to conventional polymer matrices. Qin *et al.* made all-cellulose composites with ramie fiber (unidirectional) and 1–7% cellulose solution [20]. The highest composite tensile strength, 440 MPa, was reached for the composite prepared in cellulose solution with a concentration of 4% (ramie fiber volume fraction 85%). Mercerization of ramie fiber further increased composite tensile strength to 540 MPa. Yang *et al.* made all-cellulose composite films with short ramie fiber and regenerated cellulose [61]. The resultant composite had a tensile strength and a tensile modulus of up to 125 MPa and 5.9 GPa, respectively. The biocomposites also had good optical transmittance, thermal stability, and biodegradability.

Soybean oil

Resin was made of acrylated epoxidized soybean oil with styrene monomer in a 2:1 weight ratio [31]. It was cured by 1 wt% (with respect to the total resin weight) free radical initiator benzoyl peroxide (BPO). Ramie fiber was treated by acetic anhydride, silane, or dicumyl peroxide. The ramie fiber/matrix had good interfacial shear strengths, ranging from to 9.9 to 18.5 MPa, after chemical treatment.

4.4.2 Processing

Processing with thermoplastic polymers

Various conventional processing techniques have been employed to make ramie fiber-reinforced thermoplastic polymer composites. These process

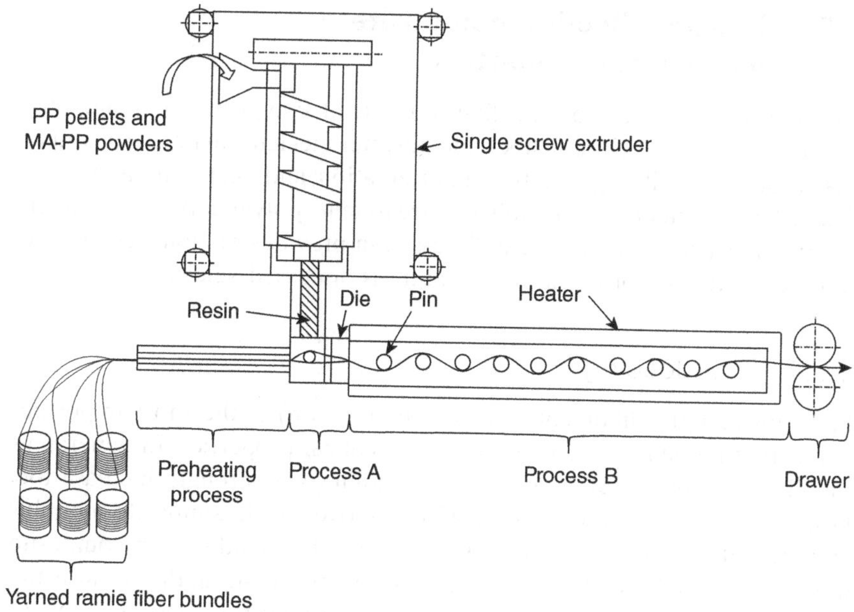

4.13 Schematic view of the continuously compounding process for ramie yarn with PP [63].

techniques included extrusion with injection molding [18, 34, 38, 43, 45], brabender compounding plus compression molding [19, 42, 62], ramie fiber/polymer fiber mixing followed by thermo-forming [15, 44], and fabric layup followed by compression molding [47].

Kim *et al.* developed a process to continuously compound ramie yarn with PP (MAPP) into continuous strands (Fig. 4.13) [63]. The continuous composite strands were then chopped into pellets and injection molded into composites. As there was no need for an extrusion process, fibers experienced less mechanical abrasion. The authors' results showed that the mean fiber length after injection molding was 1.56 mm, 0.44 mm less than the original fiber length prior to injection molding. It kept the fiber length to be still above the critical fiber length after processing.

Processing with thermoset polymers

The reported process techniques used for making ramie fiber-reinforced thermoset polymer composites were all conventional methods: compression molding [17, 40, 53, 55], hot pressing [22, 50], filament winding [64], prepreg plus hot pressing or molding [59], and resin transfer molding [54]. Low-volume processes were also used for making laboratory-scale composite samples, including solvent evaporation [24, 33, 60], and suspension cast [23, 49].

4.5 Factors affecting composite mechanical properties

Mechanical properties of the fiber-reinforced composites depend on a number of factors, including fiber and polymer mechanical properties, fiber/matrix adhesion, fiber geometry and orientation, and fiber volume fraction. The effects of fiber and/or polymer treatment, polymer type, and fiber–matrix interfacial bonding were discussed in previous sections. The impact of other factors reported in the literature is discussed below.

4.5.1 Fiber loading

Fiber loading (i.e. fiber content or fiber fraction) is the most important factor that influences the composite mechanical properties. In general, as the fiber loading increases, composite modulus and strength increase too; however, there are critical fiber loadings above which composite modulus and strength start to decrease. The critical fiber loading for composite modulus is usually higher than that for composite strength, because the modulus is less sensitive to local defects, such as unwetted fiber, that appear at high fiber loadings. An increase in fiber loading generally results in lower composite elongation at failure. Table 4.6 summarizes the observations of the optimum fiber loadings for mechanical performance according to the literature.

4.5.2 Fiber length

It is well known from composite micromechanics that higher fiber length (aspect ratio) is desirable for composite mechanical property enhancement. He *et al.* evaluated the strengths of ramie fiber/PP composites with fibers with lengths of 3 mm, 5 mm, 8 mm, and 10 mm, respectively [18]. Tensile, flexural, and compression strengths of ramie/PP composites all increased as the fiber length increased. Lodha and Netravali found that short ramie fiber/soy protein composites showed increased strength and modulus as the fiber length increased [48]. Xu *et al.* evaluated the effect of the ramie fiber length (1–2 mm, 5–6 mm, and 10–12 mm) on ramie fiber/PLA/PCL composite mechanical properties [33]. They found 10–12 mm was the best for tensile strength and modulus, while 5–6 mm was the best for impact properties.

From the perspective of processing, higher fiber length could cause poor fiber distribution in polymers if compounding was not handled properly. So, higher fiber length could lead to lower mechanical properties if the material was not properly processed. Romanzini *et al.* compared flexural and impact strength of ramie fiber/unsaturated polyester composites at fiber lengths of 25, 35, 45, and 55 mm, respectively [54]. When the fiber length increased

Table 4.6 Optimum fiber loadings for composite mechanical performance from literature

Material	Fiber loading range studied	Optimum fiber loading			Reference
		Modulus	Strength	Impact	
Ramie–cotton woven fabric/unsaturated polyester	45.3–60.9 vol%	–	58.0 vol%	–	[58]
Ramie/epoxy	30–60 wt%	60 wt%	–	–	[16]
Ramie/PP	5–20 wt%	–	20 wt% [a]	5 wt%	[32]
Ramie/epoxy	30–60 wt%	60 wt%	–	60 wt%	[18]
Ramie/PTP® (plant-oil resin)	20–50 wt%	50 wt%	–	50 wt%	[18]
Ramie/PLA-PCL	15–60 wt%	45 wt%	–	45 wt%	[44]
Ramie nanowhisker/LDPE	3–15 wt%	15 wt% [b]	5 wt%	–	[41]
Ramie/PP	10–30 wt%	30 wt%	30 wt%	30 wt%	[55]
Ramie/PLA	10–50 wt%	30 wt%	–	30 wt%	[52]
Ramie/cellulose	0–25 wt%	25 wt%	15 wt%	–	[64]
Ramie/PP	10–30 wt%	–	30 wt%	10 wt%	[45]
Ramie/PLA	10–30 wt%	30 wt%	30 wt%	20 wt%	[53]
Ramie/PP	10–50 wt%	50 wt%	40 wt%	–	[66]
Ramie cellulose Nanocrystallites/starch	5–40 wt%	40 wt%	40 wt%	–	[36]
Ramie whisker/soy protein	5–25 wt%	15 wt%	15 wt% [c]	–	[38]

[a] 20 wt% for tensile and flexural strength, 15 wt% for compressive strength.
[b] 5 wt% for one of three acid chloride treatments.
[c] 25 wt% for non-arylated soy protein.

from 45 mm to 55 mm, the flexural and impact strengths of the composites both decreased.

4.5.3 Processing

Alloin *et al.* investigated the effect of two processing techniques, extrusion and casting evaporation, on ramie cellulose nanowhisker/PEO composites [26]. The viscoelastic and creep measurements showed the cast/evaporated film had a solid-like behavior. It exhibited a spectacular reinforcement even above polymer melting temperatures. This high mechanical performance for the casting/evaporation process was ascribed to the formation of a rigid cellulosic network. The extruded composites had a liquid-like behavior indicating weak reinforcement by the ramie nanowhisker, as the extrusion process prevented the formation of a strong whisker network.

4.5.4 Hybrid with glass fiber

The effect of hybridization of ramie fiber with glass fiber on composite flexural and impact strength was evaluated by Romanzini *et al.* [54]. Glass fiber and ramie fiber were mixed at four volume ratios: 0:100, 25:75, 50:50, and 75:25. The hybrid with glass fiber increased the tensile strength only when the glass fiber percentage reached 50 vol% and above. Impact strength of the composite increased with the increase of glass fiber percentage from 25 to 75 vol%.

4.6 Other studies of ramie fiber-reinforced composites

4.6.1 Simulation

Müssig *et al.* found that there was good consistency between theoretical composite Young's moduli predicted by the rule of mixtures and experimental values [56]. The model didn't work well for cotton and Lyocell fibers. They attributed this observation to the fact that the tensile elongation behavior of ramie fiber was more like that of glass fiber than the other two fibers. Nam and Netravali calculated the Young's modulus and tensile strength of unidirectional ramie fiber-reinforced soy protein concentrate composites [49]. However, the experimental measured values for tensile strength and modulus, in the longitudinal direction of the composite, were much lower than the theoretical values. They believed this could be due to fiber misalignment, voids, and/or resin shrinkage.

4.6.2 Thermal conductivity

Alsina *et al.* studied thermal properties of ramie–cotton hybrid fabric-reinforced polyester matrix composites [53]. The thermal conductivity, thermal diffusivity and specific heat of the ramie–cotton composites were higher than those of the resin matrix itself. The thermal diffusivity was shown to be anisotropic. The highest values were obtained when the heat flux was parallel to the plane of the fabrics. This information can be very useful for composite manufacturing. These properties are also important for some applications where the material's thermal insulation properties and thermal capacity need to be considered.

4.6.3 Deformability

Lankford values (an indicator of the formability) of woven ramie fabric-reinforced Ecoflex® polymer composites were tested by Nakamura *et al.*

[46]. Deep drawability of the composite laminates was tested at 115°C using cylindrical and square punches. When alkali-treated ramie woven fabrics were used as the reinforcement, a deep drawing depth of 20 mm was achieved. Processing the green composites using deep drawing technology is promising.

4.6.4 Sound absorption

Chen *et al.* measured sound absorption properties of ramie fiber/PLLA composites [62]. They found that the composites with short ramie fibers had better sound absorption properties than the composites with ramie fabric. SEM studies revealed morphologically that there were microphase separations in the composites and pores within the individual ramie fiber bundles and in between short ramie fibers or ramie fabrics in the PLLA composites. These features are the main reason for better acoustic absorptivity. Yang and Li compared the sound absorption property of three natural fiber-reinforced epoxy composites with glass or carbon fiber-reinforced composites [58]. Ramie fiber/epoxy composites possessed better acoustic absorption behavior than glass and carbon fiber-reinforced composites, especially at high frequencies, which could be desirable in aeronautical applications.

4.6.5 Degradation

Environmental degradability of two ramie fiber-reinforced biocomposites, ramie fiber/Ecoflex® and ramie fiber/cellulose nanofiber-reinforced cornstarch composites, were evaluated both in Baltic Sea water and in compost with activated sludge [47]. The degradation processes of the two biocomposites were monitored by observing changes in polymer surfaces and composite weight, and by measuring tensile strength decrease during incubation under natural conditions (Fig. 4.14). The biocomposites were degraded in compost faster than in seawater. The ramie/Ecoflex® biocomposite was degraded faster than ramie/cellulose nanofiber-reinforced cornstarch biocomposite in both natural environments.

4.6.6 Fiber moisture effect on PLA degradation

The effects of water content in the ramie fiber on PLA degradation and composite mechanical performance were evaluated by van den Oever *et al.* [65]. It is concluded that relatively large amounts of moisture present in the fibres did not cause serious PLA degradation during 13 minutes of compounding. The PLA degradation is mainly attributed to a difference in fiber–PLA contact surface area, which is related to fiber or fiber bundle

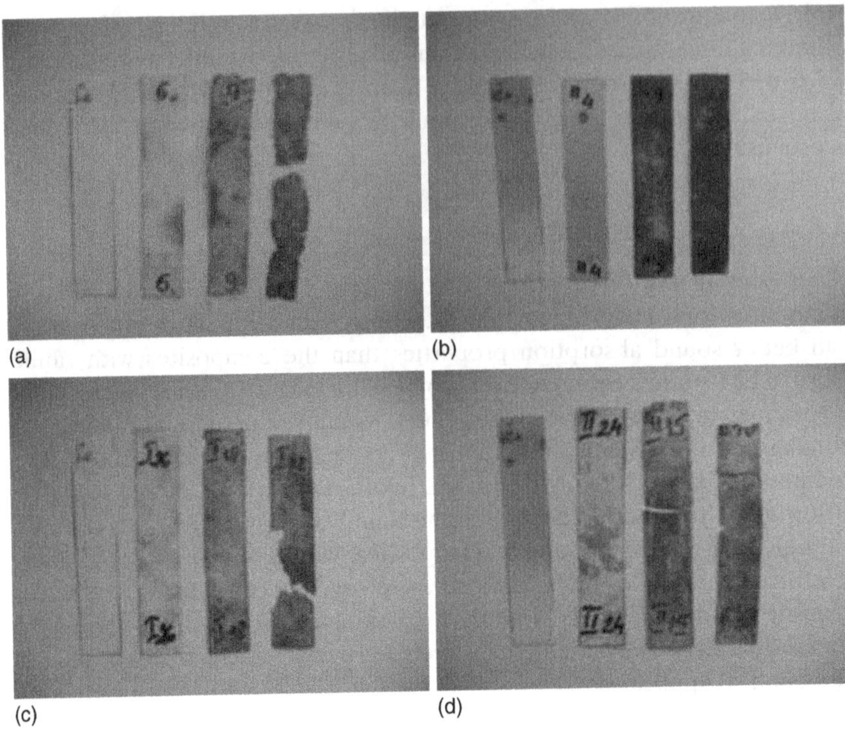

4.14 (a) Ramie/Ecoflex® and (b) ramie/cellulose/starch in seawater for 0, 4, 9, and 12 months; (c) ramie/Ecoflex® and (d) ramie/cellulose/starch in compost for 0, 4, 8, and 12 months [47].

diameter. Difference in fiber moisture content had little effect on composite morphological and mechanical properties.

4.6.7 Reactive hydroxyl group determination

He *et al.* proposed a simple and practical methodology to measure the amount of reactive hydroxyl groups in ramie fibers with toluene-2,4-diisocyanate [66]. The results indicated that the number of reactive hydroxyl groups involved in the modification reaction was about 150 mg KOH/g, which was much lower than that of the theoretical total hydroxyl value (1037 mg KOH/g). This suggested that some of the hydroxyl groups could not react with the chemical modifier.

4.6.8 UV irradiation hydrothermal aging

The reduction in interfacial adhesion because of water permeation was the main reason for the decline in mechanical properties of ramie fiber/PLLA composites during aging [41,67]. Fiber surface treatment with permanganate and silane improved fiber/PLA interfacial adhesion. However, the chemical treatment caused a higher degree of water absorption than in the untreated ramie fibers which in turn caused an accelerated decline in mechanical properties of PLLA biocomposites after UV irradiation hydrothermal aging [67].

4.7 Applications

The majority of studies on ramie fiber-reinforced polymer composites have been either fundamental or targeted mainly at applications as engineering materials. A few studies focused on specialty applications are summarized below.

4.7.1 Bulletproof armor

Ramie woven fabric-reinforced epoxy composite bulletproof panel prototypes were designed by Marsyahyo and Rochardjo [57]. The prototypes were lighter and less expensive than conventional bulletproof panels made from ceramic plate, Kevlar/aramid composites, or steel-based material. The ballistic testing results showed that ramie fiber-reinforced composites panels could resist penetration of level II high-impact projectiles. In another study, ramie fiber was used to reduce the amount of Kevlar used and the production cost for economic armor production [68]. The hybrid Kevlar–ramie polyester composite armor (Fig. 4.15) was capable of meeting the third level of protective ballistic limits in the National Institute of Justice

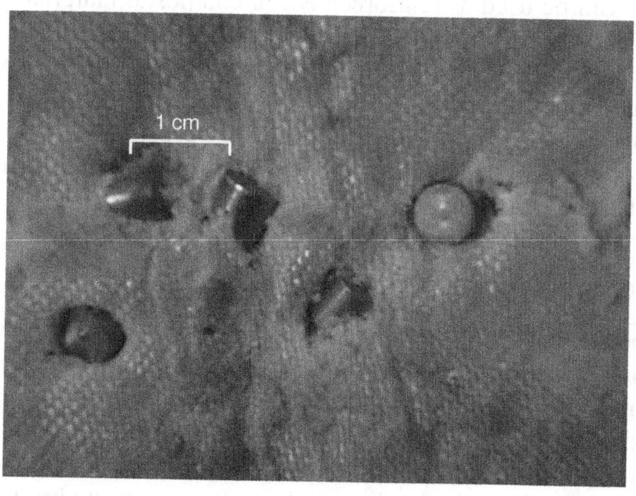

4.15 Kevlar–ramie hybrid composite after the multi-shots test [68].

(NIJ) standards. A Kevlar 29-ramie fiber-reinforced polyester laminate hybrid composite was developed for hard body armor [69]. Ballistic limit test data indicated that maximum energy absorptions of 1362 J at 623.97 m/s and 3185 J at 837 m/s were reached for composites with thicknesses of 15 mm and 25 mm, respectively. The composite targets were improved in terms of the impact resistance with increasing relative humidity in the range of 50 ± 20%.

4.7.2 Socket prosthesis

Irawan *et al.* developed a socket prosthesis with ramie fiber-reinforced epoxy composites [64]. The tensile strength, elastic modulus, and flexural strength of the composite material were 86 MPa, 9.56 GPa, and 103 MPa, respectively, which were higher than those of ramie/polyester and fiberglass/polyester composite materials. The ramie fiber/epoxy composite has the potential to be further developed as an alternative material for socket prosthesis, replacing fiberglass polyester composites. Campbell *et al.* investigated the feasibility of using a ramie fiber-reinforced renewable plant oil-based polycarbonate–polyurethane copolymer composite, instead of conventional materials, to improve safety and accessibility of prosthetic limb manufacture [70]. The ramie fiber/plant resin composite socket had a similar failure load to the conventional composite sockets, exceeding the ISO 10328 standard. The ramie fiber/plant resin composite socket showed potential to replace conventional sockets.

4.7.3 Civil applications

Ramie fiber can be used as reinforcement in composite materials for civil applications. Industrial aluminum sheets were sandwiched with ramie fiber/epoxy prepreg to make laminate composites [71]. The tensile strength of the laminate composite was higher than that of the aluminum. The specific properties of the ramie fiber/aluminum laminate were greatly improved due to the lower density of the ramie/epoxy core. The cost of ramie fiber/aluminum laminate was only two-thirds of that of the aluminum.

4.7.4 Electrolyte

Nanocomposite polymer electrolyte was made with ramie nanowhiskers and polyoxyethylene [24]. The high-performance ramie nanowhisker/polyoxyethylene nanocomposite electrolytes had a higher storage modulus at elevated temperatures than the unfilled PEO–LiTFSI sample. Ionic conductivity measurements showed that the presence of nanowhiskers induced a weak decrease of polymer electrolyte conductivity. The small decrease in ionic conduction was largely compensated by enhancement in mechanical properties, especially at high temperatures, obtained with the use of natural, biodegradable, and low-density nanowhiskers.

4.8 Conclusions

Ramie fiber has excellent specific strength and modulus, and has been used as a papermaking and fabric fiber for centuries. There has been an increasing interest in the manufacture and use of ecologically friendly natural fiber composite materials over the past two decades, so it is no surprise that ramie fibers and composites made from these fibers have been extensively studied. Research has shown that it is possible to manufacture and surface-treat the ramie fibers to make effective reinforcements in a wide variety of thermoplastic and thermosetting resins. In addition, when combined with biologically sourced resins such as PLA, fully green composites with good mechanical properties can be manufactured. With the continuous emergence of new technologies and processes, ramie fiber composite materials may be suitable for use in a wide variety of commercial applications in the near future.

4.9 References

1. Robinson BB (1940) Ramie Fiber Production. Circular, United States Department of Agriculture 585.
2. Swicofil (2008) http://www.swicofil.com/products/007ramie.html#Properties. Accessed 2suly 2013.
3. Lefranc EB (1869) *The Ramie. Its Origin, Its Value, Its Advantages, Its Culture, Its Extracting Process*. Imprimerie de la Renaissance louisianaise, Nouvelle Orléans (New Orleans), USA.
4. Blanco FM (1880) *Flora de Filipinas*, vol. 2013. Establecimiento tipográfico de Plana y C, Manila, Philippines.
5. Lewin M (2010) *Handbook of Fiber Chemistry*. third Edition. CRC Press, Boca Raton, FL.
6. FAOSTAT – Food and Agriculture Organization of the United Nations (2012). http://faostat.fao.org/site/567/DesktopDefault.aspx?PageID=567#ancor. Accessed July 2013.
7. Sen T, Jagannatha Reddy H (2011) Various industrial applications of hemp, kinaf, flax and ramie natural fibres. *International Journal of Innovation, Management and Technology* 2 (3): 192–198.
8. Mohanty A, Misra M, Hinrichsen G (2000) Biofibres, biodegradable polymers and biocomposites: an overview. *Macromolecular Materials and Engineering* 276 (1): 1–24.
9. Ashby M (2008) The CES EduPack database of natural and man-made materials. Granta Material Inspiration – Bioengineering, Cambridge, UK.
10. Weibull W (1939) *A Statistical Theory of the Strength of Materials*, vol. 151. Ingeniörsvetenskapsakademiens Handlingar, Stockholm.

11. Bevitori AB, Silva ILAD, Carreiro RS, Margem FM, Monteiro SN (2012) Elastic modulus variation with diameter for ramie fibers. Paper presented at the *Characterization of Minerals, Metals, and Materials – TMS 2012 Annual Meeting and Exhibition*, Orlando, FL.
12. Monteiro SN, Satyanarayana KG, Ferreira AS, Nascimento DCO, Lopes FPD, Silva ILA, Bevitori AB, Inácio WP, Bravo Neto J, Portela TG (2010) Selection of high strength natural fibers. *Revista Materia* 15 (4): 488–505.
13. Angelini LG, Lazzeri A, Levita G, Fontanelli D, Bozzi C (2000) Ramie (*Boehmeria nivea* (L.) Gaud.) and Spanish Broom (*Spartium junceum* L.) fibres for composite materials: Agronomical aspects, morphology and mechanical properties. *Industrial Crops and Products* 11 (2–3): 145–161.
14. Nam S, Netravali AN (2006) Green composites. I. Physical properties of ramie fibers for environment-friendly green composites. *Fibers and Polymers* 7 (4): 372–379.
15. Choi HY, Lee JS (2012) Effects of surface treatment of ramie fibers in a ramie/poly(lactic acid) composite. *Fibers and Polymers* 13 (2): 217–223.
16. Goda K, Sreekala MS, Gomes A, Kaji T, Ohgi J (2006) Improvement of plant based natural fibers for toughening green composites – Effect of load application during mercerization of ramie fibers. *Composites Part A: Applied Science and Manufacturing* 37 (12): 2213–2220.
17. Goda K, Asai T, Yamane T (2003) Development of ramie fiber reinforced biodegradable resin matrix composites by press forming and effect of chemical treatments. *Zairyo/Journal of the Society of Materials Science, Japan* 52 (10): 1245–1252
18. He LP, Tian Y, Wang LL (2008) Study on ramie fiber reinforced polypropylene composites (RF-PP) and its mechanical properties. *Advanced Materials Research* 41–42: 313–316
19. Yu T, Ren J, Li S, Yuan H, Li Y (2010) Effect of fiber surface-treatments on the properties of poly(lactic acid)/ramie composites. *Composites Part A: Applied Science and Manufacturing* 41 (4): 499–505.
20. Qin C, Soykeabkaew N, Xiuyuan N, Peijs T (2008) The effect of fibre volume fraction and mercerization on the properties of all-cellulose composites. *Carbohydrate Polymers* 71 (3): 458–467.
21. Suizu N, Uno T, Goda K, Ohgi J (2009) Tensile and impact properties of fully green composites reinforced with mercerized ramie fibers. *Journal of Materials Science* 44 (10): 2477–2482.
22. Pal SK, Mukhopadhyay D, Sanyal SK, Mukherjea RN (1988) Studies on process variables for natural fiber composites – Effect of polyesteramide polyol as interfacial agent. *Journal of Applied Polymer Science* 35 (4): 973–985.
23. Lu Y, Weng L, Cao X (2006) Morphological, thermal and mechanical properties of ramie crystallites – Reinforced plasticized starch biocomposites. *Carbohydrate Polymers* 63 (2): 198–204.
24. Alloin F, D'Aprea A, Kissi NE, Dufresne A, Bossard F (2010) Nanocomposite polymer electrolyte based on whisker or microfibrils polyoxyethylene nanocomposites. *Electrochimica Acta* 55 (18): 5186–5194.
25. Kumar R, Zhang L (2010) Investigation into ramie whisker reinforced arylated soy protein composites. *Frontiers of Chemistry in China* 5 (1): 104–108.
26. Alloin F, D'Aprea A, Dufresne A, Kissi NE, Bossard F (2011) Poly(oxyethylene) and ramie whiskers based nanocomposites: Influence of processing: Extrusion and casting/evaporation. *Cellulose* 18 (4): 957–973.

27. Goffin AL, Raquez JM, Duquesne E, Siqueira G, Habibi Y, Dufresne A, Dubois P (2011) From interfacial ring-opening polymerization to melt processing of cellulose nanowhisker-filled polylactide-based nanocomposites. *Biomacromolecules* 12 (7): 2456–2465.
28. Junior de Menezes A, Siqueira G, Curvelo AAS, Dufresne A (2009) Extrusion and characterization of functionalized cellulose whiskers reinforced polyethylene nanocomposites. *Polymer* 50 (19): 4552–4563
29. Marsyahyo E, Soekrisno, Rochardjo HSB, Jamasri (2008) Identification of ramie single fiber surface topography influenced by solvent-based treatment. *Journal of Industrial Textiles* 38 (2): 127–137.
30. Munawar SS, Umemura K, Kawai S (2008) Manufacture of oriented board using mild steam treatment of plant fiber bundles. *Journal of Wood Science* 54 (5): 369–376.
31. Lee TS, Choi HY, Choi HN, Lee KY, Kim SH, Lee SG, Yong DK (2013) Effect of surface treatment of ramie fiber on the interfacial adhesion of ramie/acetylated epoxidized soybean oil (AESO) green composite. *Journal of Adhesion Science and Technology* 27 (12): 1335–1347.
32. Li Y, Lv F, Deng H, Ronald K, Zeng S (2009) Effects of fiber surface treatments on the moisture absorption and interfacial properties of natural fibers and their composites. *Materials Science Forum* 610–613: 728–733.
33. Xu H, Wang L, Teng C, Yu M (2008) Biodegradable composites: Ramie fibre reinforced PLLA-PCL composite prepared by *in situ* polymerization process. *Polymer Bulletin* 61 (5): 663–670.
34. Li X, He L, Zhou H, Li W, Zha W (2012) Influence of silicone oil modification on properties of ramie fiber reinforced polypropylene composites. *Carbohydrate Polymers* 87 (3): 2000–2004.
35. Chen J, Yi J, Sun P, Liu ZT, Liu ZW (2009) Grafting from ramie fiber with poly(MMA) or poly(MA) via reversible addition–fragmentation chain transfer polymerization. *Cellulose* 16 (6): 1133–1145.
36. Kumar R, Zhang L (2009) Aligned ramie fiber reinforced arylated soy protein composites with improved properties. *Composites Science and Technology* 69 (5): 555–560.
37. Zhou Z, Liu X, Hu B, Wang J, Xin D, Wang Z, Qiu Y (2011) Hydrophobic surface modification of ramie fibers with ethanol pretreatment and atmospheric pressure plasma treatment. *Surface and Coatings Technology* 205 (17–18): 4205–4210.
38. Wollerdorfer M, Bader H (1998) Influence of natural fibres on the mechanical properties of biodegradable polymers. *Industrial Crops and Products* 8 (2): 105–112.
39. Shanks R, Hodzic A, Wong S (2004) Thermoplastic biopolyester natural fiber composites. *Journal of Applied Polymer Science* 91 (4): 2114–2121
40. Kimura T, Kurata M, Matsuo T, Matsubara H, Sakobe T (2004) Compression molding and mechanical properties of green-composite based on ramie/PLA non-twisted commingled yarn. *Zairyo/Journal of the Society of Materials Science, Japan* 53 (7): 776–781.
41. Chen D, Li J, Ren J (2012) Biocomposites based on ramie fibers and poly(L-lactic acid) (PLLA): Morphology and properties. *Polymers for Advanced Technologies* 23 (2): 198–207
42. Yu T, Li Y, Ren J (2009) Preparation and properties of short natural fiber reinforced poly(lactic acid) composites. *Transactions of Nonferrous Metals Society of China* (English Edition) 19 (Suppl. 3): 651–655.

43. Xu H, Liu CY, Chen C, Hsiao BS, Zhong GJ, Li ZM (2012) Easy alignment and effective nucleation activity of ramie fibers in injection-molded poly(lactic acid) biocomposites. *Biopolymers* 97 (10): 825–839.
44. Chen Y, Sun L, Negulescu I, Wu Q, Henderson G (2007) Comparative study of hemp fiber for nonwoven composites. *Journal of Industrial Hemp* 12 (1): 27–45.
45. Feng Y, Hu Y, Zhao G, Yin J, Jiang W (2011) Preparation and mechanical properties of high-performance short ramie fiber-reinforced polypropylene composites. *Journal of Applied Polymer Science* 122 (3): 1564–1571.
46. Nakamura R, Goda K, Noda J, Ohgi J (2009) High temperature tensile properties and deep drawing of fully green composites. *Express Polymer Letters* 3 (1): 19–24.
47. Krasowska K, Brzeska J, Rutkowska M, Janik H, Sreekala MS, Goda K, Sabu T (2010) Environmental degradation of ramie fibre reinforced biocomposites. *Polish Journal of Environmental Studies* 19 (5): 937–945.
48. Lodha P, Netravali AN (2002) Characterization of interfacial and mechanical properties of 'green' composites with soy protein isolate and ramie fiber. *Journal of Materials Science* 37 (17): 3657–3665.
49. Nam S, Netravali AN (2006) Green composites. II. Environment-friendly, biodegradable composites using ramie fibers and soy protein concentrate (SPC) resin. *Fibers and Polymers* 7 (4): 380–388.
50. Lodha P, Netravali AN (2005) Characterization of stearic acid modified soy protein isolate resin and ramie fiber reinforced 'green' composites. *Composites Science and Technology* 65 (7–8): 1211–1225.
51. Kim JT, Netravali AN (2010) Mechanical, thermal, and interfacial properties of green composites with ramie fiber and soy resins. *Journal of Agricultural and Food Chemistry* 58 (9): 5400–5407.
52. Paiva Júnior CZ, De Carvalho LH, Fonseca VM, Monteiro SN, D'Almeida JRM (2004) Analysis of the tensile strength of polyester/hybrid ramie-cotton fabric composites. *Polymer Testing* 23 (2): 131–135.
53. Alsina OLS, De Carvalho LH, Ramos Filho FG, D'Almeida JRM (2005) Thermal properties of hybrid lignocellulosic fabric-reinforced polyester matrix composites. *Polymer Testing* 24 (1): 81–85.
54. Romanzini D, Junior HLO, Amico SC, Zattera AJ (2012) Preparation and characterization of ramie-glass fiber reinforced polymer matrix hybrid composites. *Materials Research* 15 (3): 415–420.
55. Müssig J (2008) Cotton fibre-reinforced thermosets versus ramie composites: A comparative study using petrochemical- and agro-based resins. *Journal of Polymers and the Environment* 16 (2): 94–102.
56. Müssig J, Herrmann AS, Rau S (2006) Influence of fineness, stiffness and load–displacement characteristics of natural fibres on the properties of natural fibre-reinforced polymers. *Journal of Natural Fibers* 3 (1): 59–80.
57. Marsyahyo E, Rochardjo HSB (2009) Preliminary investigation on bulletproof panels made from ramie fiber reinforced composites for NIJ Level II, IIA, and IV. *Journal of Industrial Textiles* 39 (1): 13–26
58. Yang WD, Li Y (2012) Sound absorption performance of natural fibers and their composites. *Science China Technological Sciences* 55 (8): 2278–2283.
59. Kishi H, Fujita A (2008) Wood-based epoxy resins and the ramie fiber reinforced composites. *Environmental Engineering and Management Journal* 7 (5): 517–523.

60. Nishino T, Matsuda I, Hirao K (2004) All-cellulose composite. *Macromolecules* 37 (20): 7683–7687.
61. Yang Q, Lue A, Zhang L (2010) Reinforcement of ramie fibers on regenerated cellulose films. *Composites Science and Technology* 70 (16): 2319–2324.
62. Chen D, Li J, Ren J (2010) Study on sound absorption property of ramie fiber reinforced poly(L-lactic acid) composites: Morphology and properties. *Composites Part A: Applied Science and Manufacturing* 41 (8): 1012-1018.
63. Kim HB, Goda K, Noda J, Aoki K (2013) Developing simple production of continuous ramie single yarn reinforced composite strands. *Advances in Mechanical Engineering* 2013: article 496274.
64. Irawan AP, Soemardi TP, Widjajalaksmi K, Reksoprodjo AHS (2011) Tensile and flexural strength of ramie fiber reinforced epoxy composites for socket prosthesis application. *International Journal of Mechanical and Materials Engineering* 6 (1): 46–50.
65. Van den Oever MJA, Beck B, Müssig J (2010) Agrofibre reinforced poly(lactic acid) composites: Effect of moisture on degradation and mechanical properties. *Composites Part A: Applied Science and Manufacturing* 41 (11): 1628–1635.
66. He L, Li X, Li W, Yuan J, Zhou H (2012) A method for determining reactive hydroxyl groups in natural fibers: Application to ramie fiber and its modification. *Carbohydrate Research* 348: 95–98.
67. Chen D, Li J, Ren J (2011) Influence of fiber surface-treatment on interfacial property of poly(L-lactic acid)/ramie fabric biocomposites under UV-irradiation hydrothermal aging. *Materials Chemistry and Physics* 126 (3): 524–531.
68. Radif ZS, Alil A, Abdan K (2011) Development of a green combat armour from ramie–Kevlar–polyester composite. *Pertanika Journal of Science and Technology* 19 (2): 339–348.
69. Ali A, Shaker ZR, Khalina A, Sapuan SM (2011) Development of anti-ballistic board from ramie fiber. *Polymer – Plastics Technology and Engineering* 50 (6): 622–634.
70. Campbell AI, Sexton S, Schaschke CJ, Kinsman H, McLaughlin B, Boyle M (2012) Prosthetic limb sockets from plant-based composite materials. *Prosthetics and Orthotics International* 36 (2): 181–189.
71. Shihong L, Benlian Z, Qiyun Z, Xianrong B (1994) A new kind of super-hybrid composite material for civil use – ramie fibre/Al. *Composites* 25 (3): 225–228.

5

The use of kenaf fibers as reinforcements in composites

H. AKIL, M. H. ZAMRI and M. R. OSMAN, University of Sains, Malaysia

DOI: 10.1533/9781782421276.1.138

Abstract: Kenaf fiber (*Hibiscus cannabinus* L.) is a type of natural fiber offering many advantages and high potential as reinforcement in composite materials, especially polymer composites. Conventionally, synthetic fibers such as carbon, glass and aramid are commonly used in the production of polymer composites, but kenaf fibers have comparable specific properties and relatively low processing costs favoring their substitution for conventional synthetic fibers. Kenaf fibers may be modified either physically or chemically and used for either thermoset or thermoplastic polymer systems. Fields of use for kenaf fiber-reinforced composites include construction, automotive, corrosion resistance, electrical, marine and transportation.

Key words: kenaf fiber, processing, properties, fabrication, fiber modification, polymer system, application.

5.1 Introduction

5.1.1 The kenaf plant

The kenaf plant (Fig. 5.1) is composed of multiple useful components (for example, stalks, leaves and seeds) and within each of these plant components there are various usable portions (for example, fibers and fiber strands, proteins and oils) [1]. Kenaf plants usually can be harvested several times a year depending on the origin of the plants. Kenaf is the name given to fiber from the bast of stems of plants of the genus *Hibiscus*, family Malvaceae and species *H. cannabinus* L. Kenaf plants are usually grown from seeds, but some are grown as perennials in the tropics and subtropics. Kenaf can grow to 2.5–4.0 m in height at optimum temperature between 22 and 30°C with a minimum moisture requirement of 150 mm in a soil pH between 6.0 and 6.8, as shown in Table 5.1 [2–4].

The yield and composition of these plant components can be affected by many factors including cultivar, planting date, photosensitivity, length of growing season, plant populations and plant maturity [5]. Kenaf filaments consist of discrete individual fibers, generally 2–6 mm long [6]. Filaments and individual fiber properties can vary depending on the sources, age,

(a) (b)

5.1 Kenaf plants [11].

Table 5.1 Climatic requirements for growing kenaf [4]

Plant	Optimum temperature (°C)	Minimum moisture (mm)	Optimum soil (pH)	Growing cycle (days)	Fiber yield (kg/ha)
Kenaf	22–30	120	6.0–6.8	150–180	1700

separation technique and history of the fibers. The stem is unbranched and straight and is composed of an outer layer (bark) and a core [7]. It is easy to separate the stem into bark and core by chemical and/or enzymatic retting. The bark constitutes 30–40% of the dry weight of the stem and shows a rather dense structure. On the other hand the core is wood-like and makes up the remaining 60–70% of the stem [8]. The core reveals an isotropic and almost amorphous pattern, while the bark shows an orientated, highly crystalline fiber pattern. The macrofibril size and chemical content of kenaf stem is shown in Table 5.2. On average, kenaf fibers contain 60–80% cellulose, 5–20% lignin (pectin) and up to 20% moisture [9, 10].

5.1.2 Properties of kenaf fibers

Although the mechanical properties of natural fibers, kenaf included, are lower than those of conventionally used fibers such as glass, carbon and Kevlar, because of their low density the properties of these natural fibers are comparable to those of the conventional synthetic fibers. A single fiber

Table 5.2 Macrofibril size and chemical content of kenaf stem [1]

	Bark	Core
Fibril length, *L* (mm)	2.22	0.75
Fibril width, *W* (mm)	17.34	19.23
L/W	128.00	39.00
Lumen diameter (mm)	7.50	32.00
Cell wall thickness (mm)	3.60	1.50
Cellulose (%)	69.20	32.10
Lignin (%)	2.80	25.21
Hemicellulose (%)	27.20	41.00
Ash content (%)	0.80	1.80

Table 5.3 Tensile properties of kenaf fiber bundles test for standard kenaf fiber [3]

Kenaf fiber bundle	Tensile strength (MPa)		Failure strain (%)		Young's modulus (MPa)	
	Mean	Std. dev.	Mean	Std. dev.	Mean	Std. dev.
Standard	118.30	25.54	8.31	1.22	2416.50	460.14

of kenaf can have a tensile strength and modulus as high as 11.9 GPa and 60 GPa, respectively [12]. Edeerozey *et al.* (2007) in their study performed series of fiber bundle tensile tests to evaluate the effect of treatments on the fiber tensile strength. They also reported that alkalization treatment improved the mechanical properties of the kenaf fiber significantly as compared to untreated kenaf fiber [13]. Nosbi *et al.* (2011), in their study of the behavior of kenaf fibers after immersion in several water conditions reported the tensile properties of the kenaf fiber bundle [3]. The standard kenaf fiber tensile properties show remarkable results and thus prove that kenaf can be used as a reinforcement material in the composite systems. The results of the tensile properties test for standard kenaf fiber bundles are shown in Table 5.3.

5.1.3 Cultivation of kenaf plants

Rouison *et al.* (2004) indicated that kenaf has been actively cultivated in recent years for two main reasons [14]. One is that kenaf absorbs nitrogen and phosphorus that is present in the soil. The average absorption rate for kenaf is 0.81 g/m^2 per day for nitrogen and 0.11 g/m^2 per day for phosphorus; these rates are several times higher than those with a variety of stress. The other reason is that kenaf accumulates carbon dioxide at a significantly high rate. The photosynthesis rate of kenaf is much higher than photosynthesis rates of conventional trees [1, 15].

The evaluation of procedures for harvesting kenaf continues to be an important aspect of commercialization. The harvest method depends on the production location, the equipment availability, the processing method, and the final product use.

Over the last 6000 years, since its first domestication, kenaf has consistently been hand-harvested for use as a cordage crop (rope, twine, and sackcloth) [16]. The bast fiber strands, located in the kenaf bark, are the source for these cordage products. When hand-harvested, the tall, cylindrical-shaped stalks were cut at or near ground level with a curved blade or machete [16]. Usually plants were still actively growing, nearing or already flowering at the time of harvest.

The USDA, universities and private industry have developed a range of mechanical harvesters and post-harvest equipment to separate the bark from the core material, and the bast fibers from the core fibers. As result of the USDA's initial interest in kenaf as an alternative cordage source during World War II, a tractor-drawn harvester–ribboner was developed. This machine harvested green plants, removed the leafy, low fiber top portion of the plant, ribboned the bark, bundled the ribbons, and tied the ribbon bundles [16]. Ribboning is the process of removing the bark from the core material. The same process is also referred to as decorticating, the removal of the core from the bark. The original objective of the ribboners/decorticators was to harvest the bark for its valuable bast fiber and discard the unwanted core material.

Newer ribboners/decorticators have been developed specifically for the kenaf industry [17] or adapted from other fiber industries (hemp and jute). Unlike the older equipment, the newer ribboner/decorticator was built specifically for kenaf and actually intended to be an in-field harvester–separator. The objective is no longer to harvest only bark ribbons, but to separate and harvest the core material for other uses [17]. As with the earlier ribboners/decorticators, the newer equipment must also achieve a number of outcomes to produce positive economic advantages. These issues typically include integrating the equipment into a kenaf production, harvesting and processing system, increasing efficiency as a result of product quality and/or throughput, and improving equipment reliability (durability, safety, ease of use, and maintenance). The advantages of these newer ribboner/decorticator harvesters over other types of kenaf harvesters, such as sugarcane-type or forage-type harvesters, include the ability to produce a cleaner separation between the bark and core components, quicker drying of the separated components, and greater flexibility in determining the cutting length of the fiber strands [17].

Following the successful evaluation of kenaf's bast and core fibers as a combined cellulose fiber source for paper products (newsprint, bond paper, and corrugated liner board) in the 1950s and early 1960s, the development of harvesting equipment shifted away from in-field ribboners/decorticators to whole-stalk harvesters [16, 18]. The development of these whole-stalk harvesters has taken two major approaches; sugarcane-type harvesters and forage-type harvesters. In both approaches scientists and industry have concentrated on using or adapting existing equipment, rather than developing a totally unique kenaf harvester.

A dry kenaf stalk without leaves is a lightweight material with a low density, 0.31 g/cm^3. Chopped uncompressed kenaf fiber will have an even lower density of approximately 0.1 g/cm^3. The low bulk density of either the individual kenaf stalk or the chopped kenaf stalk will affect management decisions concerning the economic transportation and storage of the kenaf material. Industry and the USDA have cooperated to develop methods to increase the density of the kenaf material for increased transportation and storage efficiency.

Webber and Bledsoe (1993) [19] reported that pelletizing kenaf increased the kenaf material density by approximately 390%, to 1.21 g/cm^3. The same researchers successfully cubed bast fibers to a density of 0.89 g/cm^3, core fiber to a density of 1.22 g/cm^3, and whole stalk (bast and core) to a density of 0.93 g/cm^3. The kenaf cubes produced were 3×3 cm square with lengths ranging from 3 to 13 cm. Although these kenaf pellet and cube densities refer to average densities of the items produced rather than total bulk densities, the advantages of compressing the kenaf material would also translate into advantages in bulk transportation or storage of these materials compared to unprocessed kenaf stalks or non-compressed chopped kenaf.

5.1.4 Potential of kenaf as a reinforcement in composite materials

The use of kenaf fiber composites as reinforcements is currently one of the more interesting areas of research. Various sectors, especially from the research field involved in the use of natural materials as a reinforcement fiber, claim kenaf fiber, composite materials to be among the best available alternatives to replace synthetic fiber. Rowell *et al.* [20] studied the potential of kenaf as a reinforcing fiber in a polypropylene matrix and compared the mechanical properties with other commonly used composite systems as shown in Table 5.4. Kenaf core fibers are also used in product applications such as animal bedding, summer forage, and potting media [21].

Generally, the tensile and flexural properties of kenaf-reinforced composites vary depending on the type of fiber, its orientation (random or unidirectional), content and form (fiber or fabric), and the type of blending/

Table 5.4 Properties of reinforced polypropylene composites [12]

Filler/reinforcement in PP	Unit	None	Kenaf	Glass	Talc	Mica
Filler by weight	%	0	50	40	40	40
Filler by volume	%	0	39	19	18	18
Specific gravity	–	0.9	1.07	1.23	1.27	1.26
Tensile modulus	GPa	1.7	8.3	9	4	7.6
Specific tensile modulus	GPa	1.9	7.8	7.3	3.1	6.0
Tensile strength	MPa	33	65	110	35	39
Specific tensile strength	MPa	37	61	89	28	31
Flexural strength	MPa	41	98	131	63	62
Specific flexural strength	MPa	46	92	107	50	49
Flexural modulus	GPa	1.4	7.3	6.2	4.3	6.9
Specific flexural modulus	MPa	1.6	6.8	5.0	3.4	5.5
Elongation at break	%	>>10	2.2	2.5	–	2.3
Notched Izod impact	J/m	24	32	107	32	27
Water sorption, 24-h	%	0.02	1.05	0.006	0.02	0.03

plasticizer used. For example, Ochi [22] reported that the tensile and flexural strength of kenaf-reinforced PLA composites increase linearly with fiber contents up to 50%. Statically, the tensile and flexural strength of kenaf-reinforced composites were approximately 223 MPa and 254 MPa, respectively, in samples with a fiber fraction of 70%. According to the results obtained by this researcher, this proved that kenaf fiber exhibits higher strength values in terms of tensile and flexural properties, as compared to other natural fibers, when reinforcing PLA [1, 23, 24].

The favorable aspect ratios and high specific properties at low cost make kenaf fibers an ecologically friendly alternative to conventional reinforcing fibers in composite materials [25]. The ecological character, the biodegradability and the price of natural fibers are very important for their acceptance in large-volume engineering markets such as the construction industry [26]. Natural fibers are currently used in considerable quantities in various applications in the automotive industry only in the interior of passenger cars and truck cabins [27].

5.2 Processing of kenaf fibers

5.2.1 Kenaf fiber extraction

Kenaf, one of the bast fiber subdivisions, is well known as a cellulosic source with economic and ecological advantages: in three months after sowing the

Table 5.5 Chemical composition and moisture content of kenaf fibers [9]

Cellulose (%)	Hemicellulose (%)	Lignin (wt%)	Pectin (wt%)	Pentosan (%)	Ref.
45.0–70.6	12.3–21.5	8.0–13.0	3.0–5.0	–	[9]
68.1–87.4	–	5.3–7.3	–	28.9	[3]

seeds, it can grow under a wide range of weather conditions to a height of more than 3 m and a base diameter of 3 to 5 cm [1, 28]. This statement is supported by Lee and Eitenam (2001) who mention that the growing speed may reach 10 cm/day under optimum ambient conditions [8]. The price of kenaf was $400 per ton in 1995 and $278 to $302 per ton in 2000. From the viewpoint of energy consumption, it takes 15 MJ of energy to produce 1 kg of kenaf, whereas it takes 54 MJ to produce 1 kg of glass fibers [1].

5.2.2 Geometry and morphology of kenaf fibers

A better understanding of the chemical composition and surface adhesive bonding of natural fiber is necessary for developing natural fiber-reinforced composites. Natural fibers are composite materials designed by nature [29]. The fibers are basically rigid, crystalline cellulose, amorphous lignin and/or hemicelluloses. Most natural fibers, except for cotton, are composed of cellulose, hemicelluloses, lignin, waxes, and some water-soluble compounds, where cellulose, hemicelluloses, and lignin are the major constituents [30]. The composition of kenaf fiber is shown in Table 5.5 [9]. The composition may differ with the growing condition and test methods even for the same kind of fiber.

Edeerozey *et al.* (2007) and Nosbi *et al.* (2011) also reported some kenaf fiber microstructural images to check the kenaf fiber's surface morphology [3, 13]. The microscopic analysis of fiber surface morphology is of utmost importance in characterizing the structural changes that have occurred upon treatment. Edeerozey *et al.* presented surface morphology of untreated and treated kenaf fibers via the differences in terms of their level of smoothness and roughness surfaces. A Scanning Electron Microscopy (SEM) micrograph clearly shows the impurities observed on the surface of the untreated fiber as shown in Fig. 5.2(a). On the other hand Fig. 5.2(b) shows the SEM micrograph of 6% NaOH treated kenaf fiber. It can be observed that almost all impurities have been removed from the fiber surface. The fiber surface topography could provide vital information on the level of interfacial adhesion that would exist between

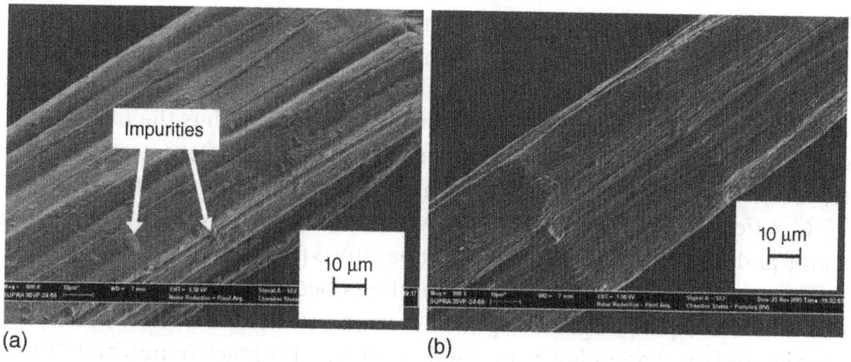

5.2 SEM micrograph of (a) an untreated kenaf fiber and (b) 6% NaOH treated kenaf fiber [12].

the fiber and the matrix later when used as reinforcement fiber with and without treatment [13].

5.2.3 Modification of natural fibers

A few major drawbacks of natural fibers as reinforcement in composites are poor compatibility between fiber and matrix and their relatively high moisture absorption. Thus, modifications of the natural fiber's surface are considered very crucial in order to improve the adhesion between fiber and matrices [31]. There are two types of fiber modification method that have been carried out by researchers and gave positive results: physical and chemical.

Physical method

Fiber surface modifications via a physical method consist of two types of treatment: corona treatment and plasma treatment. The main purpose of corona treatment is surface oxidation activation which changes the surface energy of the cellulose fibers. It was found that this treatment was able to improve the compatibility between the hydrophilic fibers and the hydrophobic matrix [32]. Similar to corona treatment, plasma treatment is another type of physical treatment that can be used to enhance the natural fiber's compatibilization with the matrices used in composites. This method exploits the plasma properties and induces changes on the surface of the material. Surface modification can be varied depending on the type and nature of the gases used.

Chemical method

A study was carried out by Edeerozey *et al.* [13] where kenaf fibers were immersed in NaOH solution at different concentrations (3%, 6% and 9%

NaOH) for 3 hours at room temperature. In the study, various concentrations of NaOH were used and the optimum concentration of NaOH to alkalize kenaf fibers was determined. NaOH at 6% concentration was found to be the optimum level of concentration for the treatment. Thus the 6% NaOH concentration treatment were performed under two different conditions: (i) immersion at room temperature and (ii) immersion in a water bath at 95°C. After the treatment, the fibers were thoroughly washed with running water and allowed to dry at room temperature for 48 hours.

The main factor that makes natural fibers become highly hydrophilic is the silane coupling agent; especially amino-functional silanes, such as epoxies and urethane silanes are used as primers for reactive polymers. One solution that may be effective in order to enable the use of silane is blending hydrophilic silanes with hydrophobic silanes such as phenyltrimethoxysilane. In addition, a mixture of siloxane primers has the ability to improve thermal stability which is typical for aromatic silicones.

Research has been carried out on the kenaf fiber surface to improve the adhesion between kenaf fiber and polystyrene matrix composites [32]. A reaction occurred between alkoxysilane and hydroxyl groups of kenaf cellulose to increase the fiber–matrix adhesion of the composites. Consequently, the storage modulus of kenaf/polystyrene composites is higher than that of untreated fibers and proves that a greater interaction between the matrix resin and the fiber present is due to the surface fiber treatment.

Some research has the aim of reducing the hydrophilicity property of natural fibers. Silane treatment is one of the modifications that can be made and proves able to reduce the hydrophilicity of natural fibers. The procedure is to mix or blend the hydrophilic silanes with the hydrophobic silanes such as phenyltrimethoxysilane. Apart from that, mixed siloxane primers can improve the thermal stability of the fibers.

Xu *et al.* (2009) modified the surface of kenaf fiber using silane coupling agent to promote adhesion with the polystyrene matrix. A condensation reaction between alkoxysilane and the hydroxyl group of kenaf cellulose increased the adhesion between fiber and matrix [32].

The most common method used in the modification of natural fiber surface is alkaline treatment. The treatment removes a certain amount of lignin, wax and oils covering the external surface of the fiber cell wall. A good adhesion between fiber and matrix is obtained by the alkaline treatment as the surface roughness of the fibers increases due to disruption of hydrogen bonding in the network structure.

Another method of modifying the surface of natural fibers is acetylation, which also makes the fibers more hydrophobic. This method introduces the acetyl functional group into an organic compound. OH groups of the fibers which contribute to their hydrophilic nature are coated with molecules that have a more hydrophobic nature during acetylation.

Maleated coupling, a treatment that uses maleic anhydride, is another method that is widely used nowadays to strengthen natural fiber-reinforced composites. It is a bit different from other chemical treatments as it modifies not only the fiber surface but also the polymeric matrix, resulting in better interfacial bonding between fiber and matrix, which improves the mechanical properties of the composites.

Finally, natural fiber modification by enzyme treatment is getting more popular in natural fiber processing. The main reason why this method is increasingly used nowadays is that the use of enzymes is environmentally friendly. The reaction catalyzed from this treatment is very specific and can focus on the particular performance required [33].

5.2.4 Fiber bundle tensile test

Fiber bundle tensile strength tests were carried out in order to characterize the effects of the NaOH treatment upon the tensile strength of the fiber, and the results are shown in Fig. 5.3. From the figure, the average unit break of the bundle of 3% NaOH treated kenaf fibers is higher than that of the untreated kenaf fiber bundle. Increment of NaOH concentration to 6% shows a further increment of the average unit break of the fibers. Kenaf fibers treated with 6% NaOH at high temperature show the highest average unit break among all. This is explained by the increase of uniformity that contributes to the increase in strength, due to the removal of the impurities. However, when NaOH concentration further increased up to 9%, the fiber bundle tensile strength recorded was suddenly decreased. The value recorded was even lower than that of the untreated fibers. This could be the result of the damage caused by high concentration of NaOH [13].

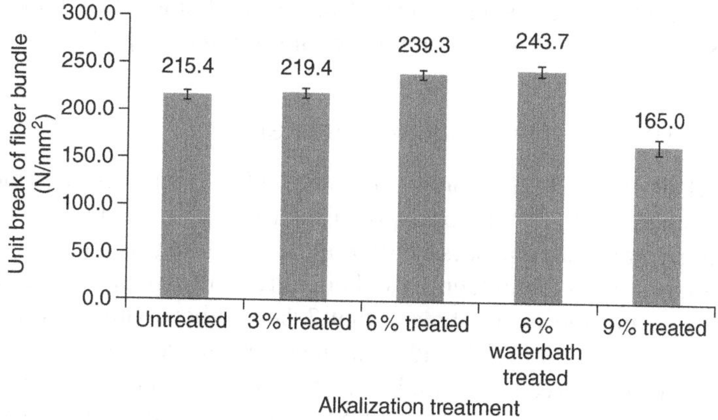

5.3 Effects of the NaOH treatment upon the tensile strength of the fiber bundle test.

5.3 Matrices for kenaf fiber-reinforced composites

The composite materials are composed of two phases; one is termed the matrix, which is continuous and surrounds the reinforcement materials or fibers. Kenaf fiber-reinforced composites lay under a polymer–matrix composites (PMCs) system. Polymer materials are used in the greatest diversity of kenaf fiber composite applications and can be divided into two main polymer systems: thermoset and thermoplastic polymer base composite materials. These two polymer composite systems have been extensively studied with the incorporation of kenaf fiber as reinforcing materials. Polyester resin and epoxy resin are commonly used for the thermoset matrices, whereas polypropylene (PP), polyurethane (TPU) and polylactic acid (PLA) are commonly used for the thermoplastic matrices.

5.3.1 Thermoset polymer systems

Unsaturated polyester resin was used to make pultruded kenaf samples by using a pultrusion technique. The flexural and compression strength of the pultruded kenaf samples increase with the increase of kenaf volume fraction during sample fabrication. Hybridization of glass with kenaf fiber in pultruded samples also improves the properties of composite samples [27, 34]. Salim *et al.* (2011) used epoxy resin to prepare a sample by a resin transfer molding (RTM) technique. Optimization of the stitching density of nonwoven kenaf fibre mat manages to increase the mechanical behavior of the kenaf composite sample [35]. Aziz and Ansell (2004) also used a kenaf–polyester composites system in their study of alkalization and fiber alignment on the mechanical and thermal properties for hot-pressed samples [36]. Besides, kenaf fiber–unsaturated polyester composites also were manufactured using a resin transfer molding (RTM) process by Rouison *et al.* (2004) in his study for curing simulation [14].

5.3.2 Thermoplastic polymer systems

El-Shekeil *et al.* (2012) prepared kenaf fiber-reinforced thermoplastic polyurethane (TPU) composites by a melt-mixing method followed by a compression molding process. Different fiber loadings of kenaf fiber were prepared to study the influence of fiber content on mechanical and thermal properties and the result showed that 30% of fiber loading exhibited the best tensile strength [37]. Besides, the mechanical properties of soil-buried kenaf fibre-reinforced thermoplastic polyurethane (TPU) composites are presented by Sapuan *et al.* (2013) [38]. Polypropylene (PP) is also used as matrix material in kenaf fiber-reinforced composites manufacture. Kenaf fibre-reinforced polypropylene composites were prepared by compression molding to study the effects of alkali-silane treatment on the tensile and flexural properties of short fiber. The results show that alkali treatment significantly improves the tensile and flexural properties of short fiber non-woven kenaf polypropylene composites [39]. Judawisastra *et al.* (2011) studied the influence of processing temperatures and fiber chemical treatments on the mechanical properties of polypropylene (PP) waste–kenaf fiber composites [40]. Moreover, natural fibre-reinforced polylactic acid (PLA) based biocomposites likewise have been widely investigated by materials researchers in the last decade to compete with non-renewable petroleum-based products. PLA is a very interesting material as it is a degradable thermoplastic polymer with excellent mechanical properties and is produced on a large scale from fermentation of cornstarch to lactic acid and subsequent chemical polymerization. Pure PLA can degrade to carbon dioxide, water and methane in the environment over a period of several months to two years, compared to other petroleum plastics needing a very longer period [41, 42]. Avella *et al.* (2008) reported the use of kenaf fiber in their biocomposites samples [43]. Ochi (2008) presented mechanical properties of kenaf fibers and kenaf/PLA composites. The unidirectional kenaf fiber-reinforced PLA composites showed tensile and flexural strengths of 223 MPa and 254 MPa, respectively. Furthermore, tensile and flexural strength and elastic moduli of the kenaf fiber-reinforced composites improved linearly up to a fiber content of 50% [22].

5.4 Fabrication of kenaf fiber-reinforced composites (KFRC)

Composites fabrication techniques range in nature from the low capital-intensive, high-labor content, to exact opposites of high capital and low-labor content technique. In total, there are around 20 well-established techniques, and with only limited restriction it can be feasible to select that process that best satisfied the quantity, economics, dimensional, shape complexity, and mechanical/physical property performance specifications demanded by the customer for a particular component or application.

In recent years the processing and production technologies for biocomposites have also advanced. To date, injection molding, extrusion, compression molding, sheet molding and resin transfer molding are the major manufacturing processes for natural fiber-reinforced plastic composites [31]. Recently, rapidly expanding usage of composite components in automotive, construction, sports and leisure and other mass-production industries has focused on continuous processing techniques [44, 45].

5.4.1 Fiber kenaf type

The kenaf fiber content is generally important for promoting the sustainability of a composite and will determine its fabrication method. Furthermore, the fiber length (short or long), aspect ratio (length/diameter) and chemical composition have a great influence on the processing, and therefore processing parameters and accessories should be developed. The compounding process significantly influences the shortening and fibrillation as well as the thermal deterioration of the fibers in the early stages; the final properties of the product are already determined at the beginning of the production process [31].

The technique for the manufacture of structural profiles from composites on a continuous basis is definitely pultrusion. The pultrusion method, a manufacturing process for producing continuous lengths of reinforced plastic structural shapes with constant cross-section, is particularly important in composites manufacture [45]. The raw materials are a liquid resin mixture that contains resin, fillers and specialized additives, and flexible textile reinforcing fibers. The process involves pulling these raw materials (rather than pushing, as in extrusion) through a heated steel-forming die using a continuous pulling device. The reinforcing materials are in continuous forms such as a roll of fiber mat and natural fibers. As the reinforcements are saturated with the resin mixture (wet-out) in the resin bath and pulled through the die, a rigid, cured profile is formed that corresponds to the shape of the die [34, 45, 46].

Sanadi *et al.* (1995), in their study on kenaf fibers and polypropylene (PP), used a blending thermokinetic mixer followed by an injection molding technique to get their composites samples [47]. Biodegradable kenaf composite specimens also were fabricated by hot pressing by Ochi (2008) in his study on mechanical properties of kenaf/PLA composites [22]. Besides that, biocomposites from kenaf fiber and based on soy were also fabricated by extrusion, followed by injection or compression molding techniques [48]. Since so many researchers are now involved with kenaf fibers, many processing techniques are being developed.

5.5 Performance of KFRC

5.5.1 Thermal performance

Thermal analysis studies are another essential characterization which needs to be considered to fully distinguish the overall behavior of kenaf fiber-reinforced composite. Three commonly used characterization methods are differential scanning calorimetry (DSC), thermogravimetric analysis (TGA), and dynamic mechanical analysis (DMA). From the DSC scan, several crucial parameters can be estimated, such as the glass transition temperature (T_g), melting temperature, crystallization level, and oxidation [49]. Meanwhile, from the TGA measurement, the mass of the sample as a function of the temperature can be measured. Changes of mass usually occur during sublimation, evaporation, decomposition and chemical reaction, and magnetic or electrical transformation of the material that is directly related to thermal stability [50].

Julkapli and Akil [50] performed both DSC and TGA on kenaf-filled chitosan biocomposites in order to evaluate their thermal characteristics. During the DSC analysis, the heating scan was performed twice and the researchers found that all samples showed a broad endothermic peak during the first heating scan, which was associated with the hydration of water. Meanwhile, the second heating scan showed a decreasing endothermic temperature caused by the addition of kenaf dust content in the chitosan film. The authors claimed that this trend was attributed to the formation of hydrogen bonding.

However, no clear evidence of changes in terms of enthalpy values (DH) was observed with the increasing kenaf fiber content. In the meantime, the TGA results show that the addition of kenaf dust in the chitosan film does not give any significant change in the thermal stability of chitosan film. On the other hand, DMA is another versatile thermal facility, which can be manipulated to study the correlation between the thermo-mechanical properties of a material as a function of temperature and deformation. The DMA measurement consists of observation of the time-dependent deformation behavior (t) of a sample under periodic, mostly sinusoidal deformation force with very small amplitudes $F(t)$.

Another interesting DMA result was achieved by Mazuki *et al.* [27] who studied the degradation of the dynamic mechanical properties of kenaf fiber-reinforced composites after immersion in various solutions. From their overall observations, the conclusion was that the thermo-mechanical properties of kenaf fiber-reinforced composite, in terms of tan d, storage modulus and loss modulus, were reduced after immersion in various solutions. Statistically, kenaf fiber-reinforced composite immersed in seawater (pH 8.9) showed the highest reduction, followed by acidic rainwater (pH 5.5) and distilled water (pH 7).

Theoretically, the reduction pattern recorded for tan d is attributed to the increase of ductility caused by the restriction of movement of polymer molecules, after the immersion process for all samples tested. Incorporation of reinforcing fibers restricts the mobility of the polymer molecules under deformation, and thus raises the storage modulus values and reduces the viscoelastic lag between the stress and the strain. Hence, the tan d values of composites are reduced [51, 52]. Meanwhile, the decreased value of the storage modulus for immersed kenaf fiber-reinforced composite was due to the damaged matrix, deteriorated interfacial adhesion, and bond strength between matrix and fiber. The behaviors of the chemical combination of chains, the Van der Waals bonding, and the hydrogen bonding in the molecular construction of the polymer, were responsible for the ability of the material to bear foreign stress [52].

When water molecules entered the polymer, the hydrolysis and plasticization of the matrix damaged the chemical combinations and bonding. In the case of experiencing stress, when a greater strain was induced, it would indirectly lead to a decrease of the storage modulus. Instead of normal practice, the DMA machine can also be manipulated to determine the heat deflection temperature of kenaf fiber-reinforced composites, as revealed by Liu *et al.* [48]. Generally, the heat deflection temperature (HDT) is defined as the temperature at which a material deflects by 0.25 mm under the application of a load (0.45 MPa).

According to Liu *et al.* [48] the heat deflection temperature of kenaf/soy-based plastic biocomposites showed a significant increase of approximately 36°C due to the processing method, where the modulus at room temperature is independent of the processing method. However, the rate at which the modulus decreases with temperature for compression-molded samples is lower than that observed for injection-molded samplings. This improvement will cause a decrease in the creep rate under a given load, resulting in higher HDT values. In addition, HDT values increased with increases in fiber content (as well as fiber length), reflecting the dependence of the modulus on these variables. Ultimately, the authors believed that improving a composite's HDT by optimizing fiber size, fiber content and processing method could reveal new uses for kenaf/soy-based plastic biocomposites.

5.5.2 Mechanical properties

A study by Zampaloni *et al.* [28] fabricated kenaf fiber-reinforced polypropylene sheets which can be further thermoformed to produce a variety of applications and products with comparable properties to those of existing synthetic composites. Compression molding was chosen to be an optimal technique, utilizing a layered sifting of a microfine polypropylene powder and shopped kenaf fibers. The results obtained showed that the flexural strength of 40% kenaf fiber/polypropylene composites performed significantly better compared to 30% kenaf/polypropylene composites. Compared with other natural fiber composites systems, the flexural strength of 40% kenaf/polypropylene composites was equivalent to that of flax/polypropylene composites, larger than the hemp/polypropylene composites and almost double that of coir/polypropylene and sisal/polypropylene composites. Meanwhile, the flexural strength of 30% kenaf/polypropylene composites showed results equivalent to those of the 40% hemp/

polypropylene composites, while also outperforming the coir/polypropylene and sisal/polypropylene composites [28].

Yousif *et al.* (2012), in their study of the properties of unidirectional long kenaf fiber-reinforced epoxy (KFRE) composites, reported that reinforcement of epoxy with treated kenaf fibers increased the flexural strength of the composite by about 36% while untreated fibers introduced 20% improvement. This was mainly due to the high improvement by the chemical treatment (NaOH) in the interfacial adhesion of the fibers and the porosity of the composites which prevented debonding, detachment or pull-out of fibers [53].

5.6 Applications of KFRC

Recently, the rapidly expanding use of composite components in automotive, construction, sports and leisure, and other mass production industries, has been focused on sustainable and renewable reinforced composites [54]. This interest encompasses a wide variety of shapes and materials ranging from synthetic to natural, in order to fulfill the demands of producing composites with desired properties. The incorporation of reinforcements such as fibers and fillers into composites affords a means of extending and improving the properties of the composites that meet the requirements. Improvements will be associated with economic advantages, such as low production costs and low resin consumption [36]. As a result, the demands for natural fiber-reinforced composites have increased dramatically over the past few years, for various commercial applications in the industrial sector of most engineering applications.

5.6.1 Constructions and building structures

In terms of volume, construction is the largest market that can with confidence be seen as offering the greatest opportunity to pultrusion well into the future (Fig. 5.4). However, it has always been seen as a difficult market to penetrate, particularly as Codes of Practice and Product Specifications are rarely written with the composites in mind. There is therefore a clear requirement for greater user education of track-record, case-history examples and better application as well as amplification of the existing pultruded profile design data [55].

5.6.2 Automotive parts

This wonderful versatility means that this material can be used in a huge variety of applications. Industrial designers are using it to make interior door panels, seat backs, armrests, consoles and other automotive parts for

(a)

(b)

5.4 Twenty-first century building structure [27].

5.5 Automobile components made of natural fiber-reinforced composites [56].

popular manufacturers such as Mercedes-Benz and Jeep. Because of the uncompressed state of kenaf mat, it can be used as an acoustic absorber and sound barrier as shown in Fig. 5.5.

5.6.3 Corrosion resistance

In terms of future growth this market is seen as an excellent 'number two', and one of the principal reasons is the fact that in many situations composite is the best material to employ. Consequently, given an additional structural requirement as in the case of walkways, fencing, stairs, ladders, staging and drain covers, pultruded profiles become very much the preferred option (Fig. 5.6). They have the additional advantages of lightweight and easy and inexpensive shipment and installation [55].

5.6.4 Electrical

Cable tray support members and ladders have always been significant markets for pultruded composites and no decline is foreseen (Fig. 5.7). However, to both must be increasingly added such items as transmission poles and towers, which find pultrusion of benefit for similar reasons to the

(a) (b)

5.6 (a) Walkways; (b) drain cover [27].

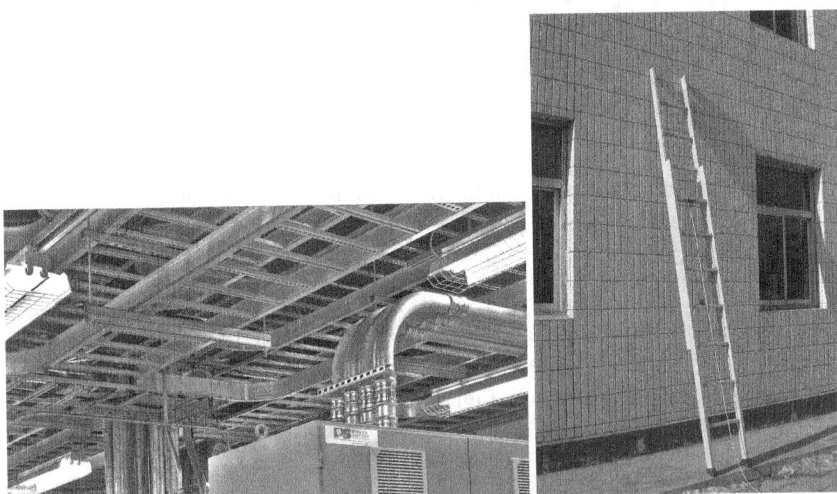

(a) (b)

5.7 (a) Cable tray support; (b) ladder [27].

corrosion-resistance applications, as well as their electrical insulating qualities [55].

5.6.5 Marine

Although seawater-, freshwater- and fungi/biological-resistant components required for such applications as marine and marina structures, piers, boat docks (Fig. 5.8) and quays could justifiably feature under the construction heading, this whole waterfront market is already receiving close attention through the installation of trial constructions [55].

5.8 Boat docks [27].

5.6.6 Transportation

Pultruded profiles have already found use, for example in bus luggage racks (Fig. 5.9) and exterior paneling, while composites generally have been increasingly and successfully considered over recent years for many transport applications, whether road, rail or sea. The latter examples are now legion, including window masks, structural elements, seats, partitions, cargo containers and even the filament winding of the whole envelope of rail rolling-stock. Given close collaboration, therefore, with vehicle builders and users and not excluding the transport infrastructure requirement, pultruded profiles can with confidence be expected to find a growing market acceptance [55].

5.7 Conclusion

Research on kenaf fiber-reinforced composites has caught many researchers' attention for reasons such as their ecological benefits and excellent specific properties. Apart from that, it can also help reduce the waste of composite

5.9 Bus luggage racks [27].

materials and contribute to a healthier environment. As discussed above alongside the results of research carried out over the years, kenaf fiber-reinforced composites have a great potential as alternative composite materials to offer an alternative material in addition to conventional materials or synthetic fibers as reinforcement in composites. In addition, kenaf fibers are relatively flexible for use in a variety of processing techniques, including pultrusion techniques. Therefore, the development of kenaf fiber-reinforced composites is expected to continue progressively with various other potential applications in the future.

5.8 References

1. Nishino, T., *et al.*, Kenaf reinforced biodegradable composite. *Composites Science and Technology*, 2003. **63**(9): 1281–1286.
2. Lewin, M., *Handbook of Fiber Chemistry*. 2010: Boca Raton, FL: CRC Press.
3. Nosbi, N., *et al.*, Behavior of kenaf fibers after immersion in several water conditions. *BioResources*, 2011. **6**(2): 950–960.
4. Chawla, K.K., *Fibrous Materials*. 2005: Cambridge: Cambridge University Press.
5. Coates, W., Kenaf performance in northwestern Argentina. *Industrial Crops and Products*, 1996. **5**(3): 223–228.
6. Cao, Y., S. Shibata, and I. Fukumoto, Mechanical properties of biodegradable composites reinforced with bagasse fibre before and after alkali treatments. *Composites Part A: Applied Science and Manufacturing*, 2006. **37**(3): 423–429.
7. Han, Y.H., *et al.*, Kenaf/polypropylene biocomposites: effects of electron beam irradiation and alkali treatment on kenaf natural fibers. *Composite Interfaces*, 2007. **14**(5–6): 559–578.

8. Lee, S.A. and M.A. Eiteman, Ground kenaf core as a filtration aid. *Industrial Crops and Products*, 2001. **13**(2): 155–161.
9. Saheb, D.N. and J. Jog, Natural fiber polymer composites: a review. *Advances in Polymer Technology*, 1999. **18**(4): 351–363.
10. Mohanty, A., M. Misra, and L. Drzal, Sustainable bio-composites from renewable resources: opportunities and challenges in the green materials world. *Journal of Polymers and the Environment*, 2002. **10**(1): 19–26.
11. Akil, H., *et al.*, Kenaf fiber reinforced composites: a review. *Materials and Design*, 2011. **32**(8): 4107–4121.
12. Karnani, R., M. Krishnan, and R. Narayan, Biofiber-reinforced polypropylene composites. *Polymer Engineering and Science*, 2004. **37**(2): 476–483.
13. Edeerozey, A., *et al.*, Chemical modification of kenaf fibers. *Materials Letters*, 2007. **61**(10): 2023–2025.
14. Rouison, D., M. Sain, and M. Couturier, Resin transfer molding of natural fiber reinforced composites: cure simulation. *Composites Science and Technology*, 2004. **64**(5): 629–644.
15. Baillie, C., *Green Composites: Polymer Composites and the Environment*. 2005: Boca Raton, FL: CRC Press.
16. Dempsey, J.M., *Fiber Crops*. 1975: Gainesville, FL: University of Florida Press.
17. Chen, L., *et al.* Kenaf bast and core separation. In *Proc. Int. Kenaf Assoc. Conf.*, Irving, TX, 1995.
18. White, G.A., *et al.*, *Cultural and Harvesting Methods for Kenaf: An Annual Crop Source of Pulp in the Southeast*. 1970: United States Department of Agriculture.
19. Webber III, C.L. and R.E. Bledsoe, Kenaf: Production, harvesting, processing, and products. In Janick, J. and J.E. Simon (eds), *New Crops*, Wiley, New York, 1993: 416–421.
20. Rowell, R.M., *et al.*, Properties of kenaf/polypropylene composites. *Kenaf Properties, Processing and Products*, 1999: 381–392.
21. Ramaswamy, G.N., *et al.*, Kenaf nonwovens as substrates for laminations. *Industrial Crops and Products*, 2003. **17**(1): 1–8.
22. Ochi, S., Mechanical properties of kenaf fibers and kenaf/PLA composites. *Mechanics of materials*, 2008. **40**(4–5): 446–452.
23. Oksman, K., M. Skrifvars, and J.F. Selin, Natural fibres as reinforcement in polylactic acid (PLA) composites. *Composites Science and Technology*, 2003. **63**(9): 1317–1324.
24. Lee, S.H. and S. Wang, *Biodegradable* polymers/bamboo fiber biocomposite with bio-based coupling agent. *Composites Part A: Applied Science and Manufacturing*, 2006. **37**(1): 80–91.
25. Ma, X., J. Yu, and J.F. Kennedy, Studies on the properties of natural fibers-reinforced thermoplastic starch composites. *Carbohydrate Polymers*, 2005. **62**(1): 19–24.
26. Puglia, D., J. Biagiotti, and J. Kenny, A review on natural fibre-based composites – Part II. *Journal of Natural Fibers*, 2005. **1**(3): 23–65.
27. Mazuki, A.A.M., *et al.*, Degradation of dynamic mechanical properties of pultruded kenaf fiber reinforced composites after immersion in various solutions. *Composites Part B: Engineering*, 2011. **42**(1): 71–76.
28. Zampaloni, M., *et al.*, Kenaf natural fiber reinforced polypropylene composites: A discussion on manufacturing problems and solutions. *Composites Part A: Applied Science and Manufacturing*, 2007. **38**(6): 1569–1580.

29. Puglia, D., J. Biagiotti, and J. Kenny, A review on natural fibre-based composites – part II: Application of natural reinforcements in composite materials for automotive industry. *Journal of Natural Fibers*, 2005. **1**(3): 23–65.
30. Pandey, J.K., *et al.*, Recent advances in biodegradable nanocomposites. *Journal of Nanoscience and Nanotechnology*, 2005. **5**(4): 497–526.
31. Faruk, O., *et al.*, Biocomposites reinforced with natural fibers: 2000–2010. *Progress in Polymer Science*, 2012. **37**(11): 1552–1596.
32. Xu, Y., *et al.*, Thermomechanical properties of the silanized-kenaf/polystyrene composites. *Express Polymer Letters*, 2009. **3**(10): 657–664.
33. Bledzki, A.K., *et al.*, Polypropylene composites with enzyme modified abaca fibre. *Composites Science and Technology*, 2010. **70**(5): 854–860.
34. Zamri, M.H., *et al.*, Effect of water absorption on pultruded jute/glass fiber-reinforced unsaturated polyester hybrid composites. *Journal of Composite Materials*, 2012. **46**(1): 51–61.
35. Salim, M.S., Z.A.M. Ishak, and S. Abdul Hamid, Effect of stitching density of nonwoven fiber mat towards mechanical properties of kenaf reinforced epoxy composites produced by Resin Transfer Moulding (RTM). *Key Engineering Materials*, 2011. **471**: 987–992.
36. Aziz, S.H. and M.P. Ansell, The effect of alkalization and fibre alignment on the mechanical and thermal properties of kenaf and hemp bast fibre composites: Part 1 – Polyester resin matrix. *Composites Science and Technology*, 2004. **64**(9): 1219–1230.
37. El-Shekeil, Y.A., *et al.*, Influence of fiber content on the mechanical and thermal properties of Kenaf fiber reinforced thermoplastic polyurethane composites. *Materials and Design*, 2012. **40**(0): 299–303.
38. Sapuan, S.M., Fei-ling Pua, Y.A. El-Shekeil, and Faris M. AL-Oqla, Mechanical properties of soil buried kenaf fibre reinforced thermoplastic polyurethane composites. *Materials and Design*, 2013. **50**: 467–470.
39. Asumani, O.M.L., R.G. Reid, and R. Paskaramoorthy, The effects of alkali-silane treatment on the tensile and flexural properties of short fibre non-woven kenaf reinforced polypropylene composites. *Composites Part A: Applied Science and Manufacturing*, 2012. **43**(9): 1431–1440.
40. Judawisastra, H., F. Falencia, and A.H.D. Abdullah, Effects of processing temperature and chemical treatment on tensile strength of kenaf fiber – Reinforced polypropylene waste composites. *Key Engineering Materials*, 2011. **471–472**: 963–968.
41. Shibata, M., *et al.*, Biodegradable polyester composites reinforced with short abaca fiber. *Journal of Applied Polymer Science*, 2002. **85**(1): 129–138.
42. Ghorpade, V.M., A. Gennadios, and M.A. Hanna, Laboratory composting of extruded poly(lactic acid) sheets. *Bioresource Technology*, 2001. **76**(1): p. 57–61.
43. Avella, M., et al., Poly(lactic acid)-based biocomposites reinforced with kenaf fibers. *Journal of Applied Polymer Science*, 2008. **108**(6): 3542–3551.
44. Van de Velde, K. and P. Kiekens, Thermoplastic pultrusion of natural fibre reinforced composites. *Composite Structures*, 2001. **54**(2): 355–360.
45. Starr, T.F. and J. Ketel, Composites and pultrusion. In Starr, T.F. (ed.), *Pultrusion for Engineers*, 2000: Cambridge: Woodhead Publishing, p. 1.
46. Joshi, S.V., *et al.*, Are natural fiber composites environmentally superior to glass fiber reinforced composites? *Composites Part A: Applied Science and Manufacturing*, 2004. **35**(3): 371–376.

47. Sanadi, A.R., *et al.*, Renewable agricultural fibers as reinforcing fillers in plastics: Mechanical properties of kenaf fiber–polypropylene composites. *Industrial and Engineering Chemistry Research*, 1995. **34**(5): 1889–1896.
48. Liu, W., *et al.*, Influence of processing methods and fiber length on physical properties of kenaf fiber reinforced soy based biocomposites. *Composites Part B: Engineering*, 2007. **38**(3): 352–359.
49. Qu, X., A. Wirsen, and A.-C. Albertsson, Effect of lactic/glycolic acid side chains on the thermal degradation kinetics of chitosan derivatives. *Polymer*, 2000. **41**(13): 4841–4847.
50. Julkapli, N.M. and H.M. Akil, Thermal properties of kenaf-filled chitosan biocomposites. *Polymer–Plastics Technology and Engineering*, 2010. **49**(2): 147–153.
51. Rana, A., *et al.*, Short jute fiber-reinforced polypropylene composites: effect of compatibilizer. *Journal of Applied Polymer Science*, 1998. **69**(2): 329–338.
52. Gu, H., Dynamic mechanical analysis of the seawater treated glass/polyester composites. *Materials and Design*, 2009. **30**(7): 2774–2777.
53. Yousif, B.F., *et al.*, Flexural properties of treated and untreated kenaf/epoxy composites. *Materials and Design*, 2012. **40**(0): 378–385.
54. Pothan, L.A., Z. Oommen, and S. Thomas, Dynamic mechanical analysis of banana fiber reinforced polyester composites. *Composites Science and Technology*, 2003. **63**(2): 283–293.
55. Starr, T.F., *Pultrusion for Engineers*. 2000: Cambridge: Woodhead Publishing.
56. Mazuki, A.A.M., Fabrication, characterization and properties of pultruded kenaf reinforced unsaturated polyester composites. MSc. thesis, University of Sains, Malaysia, 2010.

Part II

Leaf fibers

6

The use of sisal and henequen fibres as reinforcements in composites

Y. LI and Y. O. SHEN, Tongji University, China

DOI: 10.1533/9781782421276.2.165

Abstract: Firstly, a brief overview of sisal fibres, including global sources and availability, chemical compositions, key properties, extraction and quality control is given. Their multi-scale structures and dimensional parameters are revealed by micro-observation, and their mechanical properties are measured a by single fibre tensile test. Several theoretical models are proposed and the results are compared with experimental values. Secondly, manufacturing techniques of sisal fibre-reinforced thermoplastics and thermosets are introduced. Effects of fibre surface treatments and moisture absorption on the interfacial shear strength, interlaminar shear strength and interlaminar fracture toughness are investigated. Mechanical properties, moisture absorption behaviour, and effects of moisture absorption and UV light radiation are revealed. Finally, current applications and further research aspects are proposed.

Key words: sisal fibre, composites, mechanical properties, microstructure, interfacial properties, water absorption, durability.

6.1 Introduction

Sisal (*Agave sisalana*) is an herbaceous monocotyledonous plant from the Agavaceae family that consists of a rosette of sword-shaped leaves about 100–150cm tall and 13–15cm wide. A sisal plant has a 7–10 year life-span and produces about 200–250 leaves. Each leaf contains approximately 1000 fibre bundles. Sisal fibre is a hard fibre extracted from the leaves of the sisal plant. Nearly 4.5 million tons of sisal fibres are produced each year throughout the world.[1] Sisal is considered a plant of the tropics and subtropics, as it thrives in temperatures above 25°C and in sunshine. Brazil and Tanzania are currently the two main producing countries.

Sisal fibres can be extracted from the plants after two years' growth. Sisal leaves are ripe when they reach a length of 80–100cm, which provide high extraction rate and superior mechanical properties of fibres. During the extraction process, which is known as decortication, sisal leaves are crushed and beaten until only fibres remain. The sisal fibres are then dried, classified and baled after the pectin and impurities are washed off. The extracted fibres are generally ivory or cyan in colour with a length of no less than

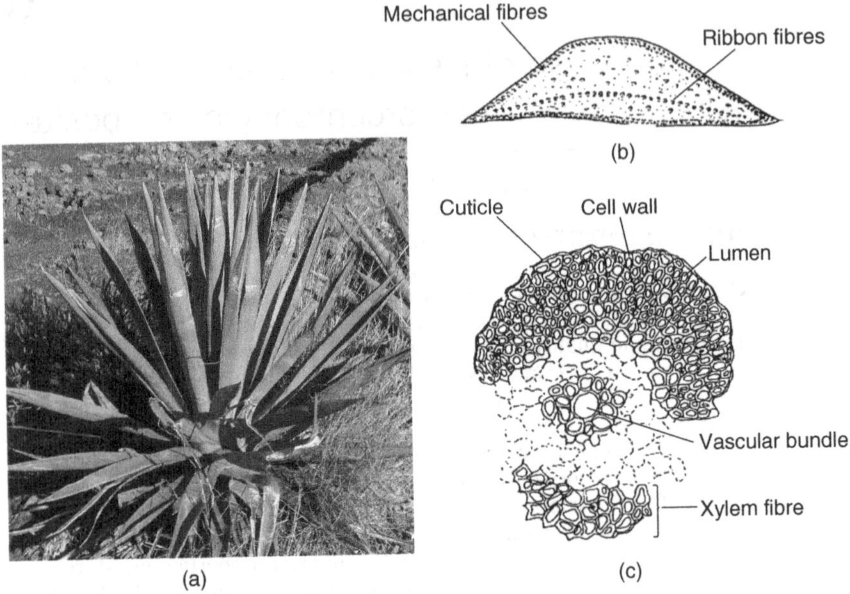

6.1 (a) Photograph of a sisal plant, (b) a sketch of the cross-section of a sisal leaf[9] and (c) cross-section of a ribbon-fibre bundle.[9] Reproduced with permission from Elsevier from Ref [9].

90 cm. The impurities and water content of the fibres should be less than 3% and 11.5%, respectively.[2]

Three types of fibres can be extracted from sisal leaves, including mechanical, ribbon and xylem fibres.[3] Figure 6.1(a) is a photograph of a typical sisal plant and Fig. 6.1(b) shows a sketch of the cross-section of a sisal leaf. The most commercially useful fibres (mechanical) are extracted from the periphery of the leaf. Ribbon fibres are the longest fibres that can be obtained from the median line of the leaf. The related conducting tissue structure of the ribbon fibres provides them with considerable mechanical strength. Xylem fibres are composed of thin-walled cells as shown in Fig. 6.1(c) and are therefore easily lost during the extraction process.

Sisal fibres account for only 4% of the total weight of a sisal leaf. In addition to 0.75% cuticle, 8% mesophyll and dregs, the remainder (87.25%) of the sisal leaf is composed of water.[4] The chemical compositions of sisal fibres have been reported by several researchers.[5–9] However, many influencing factors, such as growing and harvesting conditions, measurement methods, etc., cause a lot of variations. Generally, the sisal fibre consists of 67–78% cellulose, 10–14.2% hemicelluloses, 8–10% lignin, 2% waxes, 1% ash and 1% water.[9] The general physical and mechanical properties of sisal fibres and the other three kinds of natural fibres are compared with those

Table 6.1 General physical and mechanical properties of five types of fibres[4, 9]

Fibre	Density (g/cm^3)	Elongation at break	Young's modulus (GPa)	Specific modulus (GPa cm^3/g)	Production of energy (GJ/t)	Price (kg/¥)[a]	Water absorption
Sisal	1.33	2–3%	38	28.6	4	2.88	11%
Ramie	1.55	2%	44	28.4	4	7.2	11%
Jute	1.46	1.80%	10–30	7–21	4	2.64	12%
Flax	1.4	2–4%	60–80	43–57	4	5.53	12%
Glass	2.55	3%	73	28.6	30	26	–

[a] RMB = renminbi.

of glass fibres in Table 6.1.[4,9] Sisal fibre is superior to glass fibre in many respects, such as low density, high specific modulus and low energy consumption.

Traditionally, sisal fibre has been the leading material for the manufacture of ropes, twines, and general cordages due to its strength, durability, ability to stretch, affinity for certain dyestuffs, and resistance to deterioration in saltwater. The low-grade fibre with a high cellulose and hemicellulose content is utilised in the paper industry. The medium-grade fibre is used in the cordage industry for making ropes, baler and binder twine, which are widely employed for marine, agricultural and general industrial uses. The high-grade fibre after treatment is converted into yarns and used by the carpet industry. In recent years sisal fibre has been utilised as an environmentally friendly reinforcement to replace asbestos and fibreglass in making composite structures in various applications, such as the automobile industry.

6.2 The microstructures of sisal fibres

Microstructures of natural fibres are quite complicated and unique in that each kind of natural fibre has its own characteristics. Observing and studying the compositions and microstructures of sisal fibre will help to further understand the mechanisms in modified fibres and the related properties of its reinforcing composites.

The microstructures of sisal fibre are observed with the aid of a scanning electronic microscope (SEM) and a transmission electronic microscope (TEM). In Fig. 6.2(a), the cross-section of sisal fibres shows multi-scale structures. It can be seen from Fig. 6.2(b) that the sisal fibre itself is a composite structure in micro-scale. Its cell walls are reinforced with spirally oriented cellulose in a hemicellulose and lignin matrix.[10] It can be observed from Fig. 6.2(c) that the cell tissue structure of the sisal fibre is composed

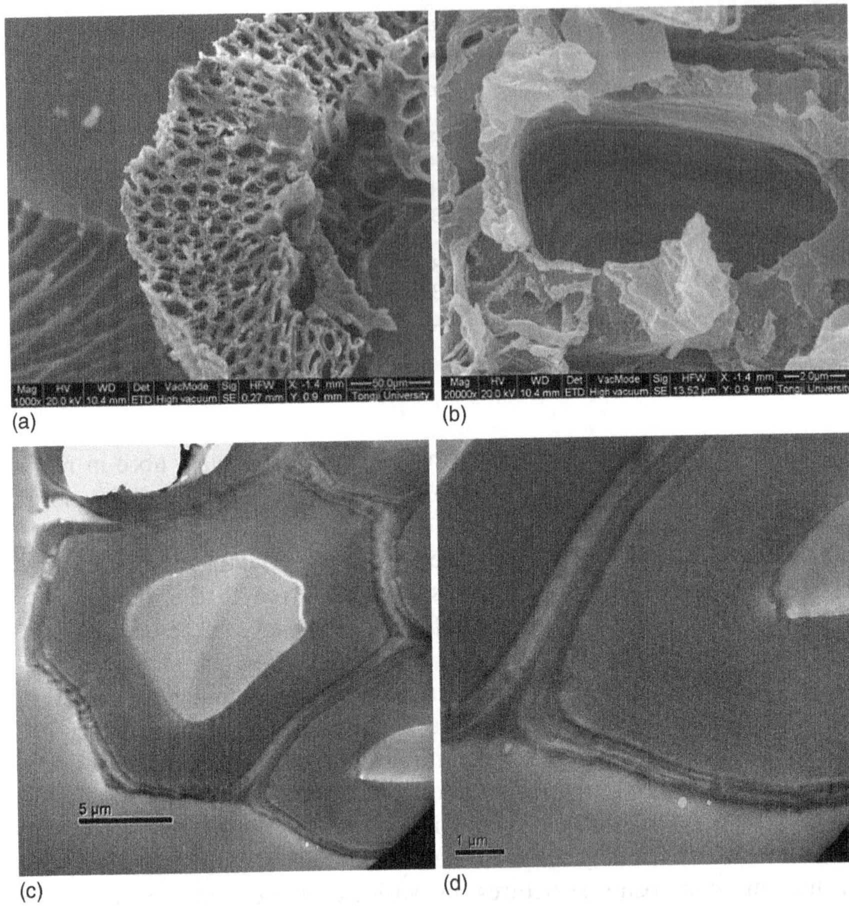

6.2 (a) Cross-section of a sisal fibre and (b) a single cell of sisal fibre observed by SEM, (c) cell tissue structure of sisal fibre and (d) textures of sisal cell observed by TEM.

of lumen, secondary wall and primary wall from interior to exterior. The textures of the secondary wall and primary wall are clearly revealed in Fig. 6.2(d).

The multi-scale structures of sisal fibre can then be simplified at two scales, as revealed in Fig. 6.3. At the first scale, several cells connect to each other through lignin, hemi-lignin and pectin to form a single natural fibre, or technical fibre. At the second scale, several micro-fibrils connect with each other through lignin, hemi-lignin and pectin to form the cell wall of the single cell surrounding a lumen, which is also called elementary fibre.

Figure 6.4 shows the cross-section of a single sisal fibre. Dimensional parameters of the microstructures of sisal fibres at the two above-mentioned

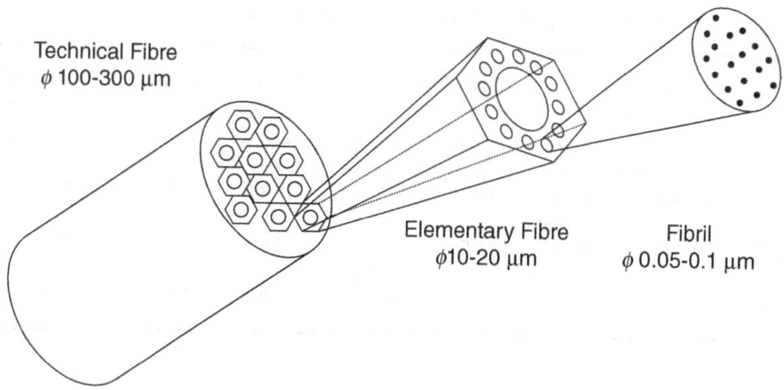

6.3 Sketch of multi-scale structures of a single sisal fibre.

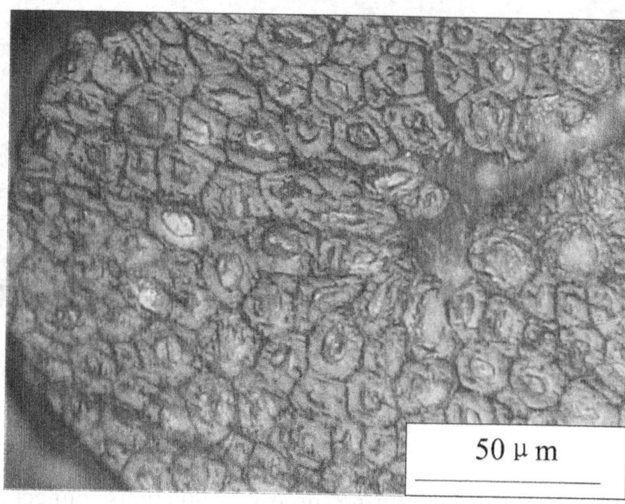

6.4 Microphotograph of cross-section of a single sisal fibre.

different scale levels are obtained with the aid of optical microscopy by the QuantLab-MG software and shown in Table 6.2, based on the measurements of 70 samples. It can be seen that the dimensions of the sisal lumen, cell and fibre are highly scattered due to natural growing characteristics. In this study, the average cell number of 70 sisal fibres is 134. Meanwhile, the lumen ratio is obtained by calculating the percentage content of the lumen in a cell, which is 17.5%.

It is found in Fig. 6.5(a) that the middle lamellae between the sisal cell walls have clear geometric characteristics. The middle lamellae form a three-way junction with 120° symmetry between cell walls of three

Table 6.2 Dimensional parameters of a single sisal fibre

	Area (μm²)		Perimeter (μm)		Major axis (μm)		Minor axis (μm)	
	Average	Range	Average	Range	Average	Range	Average	Range
Fibre	20,490	12,535–36,988	533.9	420–721	198.8	144–256	141	99.7–204.7
Cell	189.6	100.7–334.1	51.5	36.2–69.5	18.8	11.9–25.1	13.2	8.31–20.0
Lumen	–	–	–	–	10.4	3.5–17.5	4.18	1.2–9.54

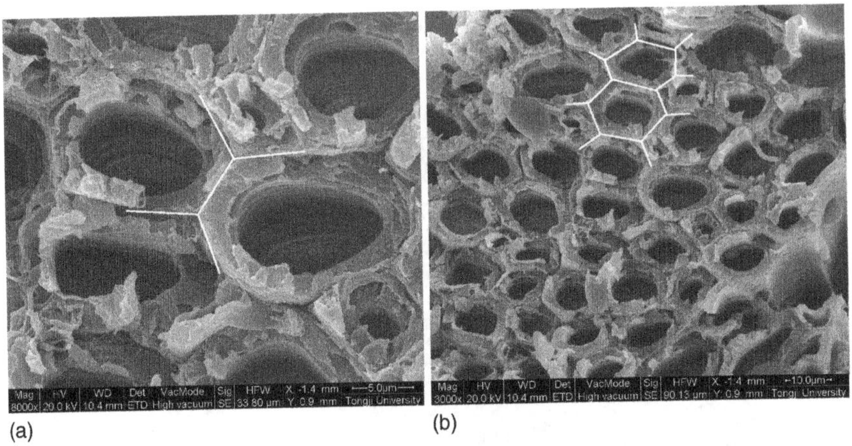

6.5 (a) Three-way junction with 120° symmetry and (b) the Steiner ring of sisal fibres formed by Steiner trees.

neighbouring cells. The neighbouring three-way junctions connect together to form an invariable topological network known as a Steiner tree with angular symmetry, which fits the geometrical conservation law created by Yin *et al.*[11–14] A network of Steiner rings can be generated with several Steiner trees connected as shown in Fig. 6.5(b). Idealised cell walls and the middle lamellae are dominated by the Steiner geometry. This geometry not only depicts the geometric symmetry, the topological invariability and minimal property of the middle lamellae, but importantly it controls the mechanics of sisal fibres.

6.3 The mechanical properties of sisal fibres

It has been well accepted that the properties of materials strongly depend on their microstructures. The unique structural characteristic of sisal fibre

Table 6.3 Tensile properties of sisal fibre before and after water taken

	Weibull shape parameter	Weibull scale parameter (MPa)	Average tensile strength (MPa)
Dry fibre	2.17	383.73	338.62
Fibre with water absorption	3.79	552.97	528.99

would definitely lead to special mechanical properties that differ from synthetic fibres. As the main chemical composition of sisal fibres is cellulose, which has a large amount of hydrophilic hydroxyl groups, sisal fibres are water-absorbent. However, most of the matrices are hydrophobic, which causes an obvious bonding problem between the fibre and matrix and leads to a strong detrimental effect on the application of sisal fibres in manufacturing composites.

A series of single fibre tensile tests were conducted on both dry and water immersed sisal fibres. At least 30 samples were randomly picked for each group. The moisture absorption ratios of sisal fibres were also measured by a gravimetric method to be approximately 11.5%.

The tensile test results showed a lot of scattering due to their natural growth characteristics, thus the results were processed with the aid of the two-parameter Weibull statistic method, and are given in Table 6.3. The Weibull shape parameter is a measure of the variability of strength, with larger values indicating smaller scattering of the data. The Weibull scale parameter is the characteristic strength (often assumed to be even, though not equal, to the average strength). It can be found that the tensile strength increases and the scatter of the results decreases after water absorption.

6.3.1 Predicting the tensile modulus of sisal fibres using the Halpin–Tsai model

The Halpin–Tsai model is introduced to obtain the elastic modulus of single sisal fibre, which includes the geometrical factor of reinforcement based on the rule of mixture. It gives:

$$\frac{M_c}{M_m} = \frac{1+\xi\eta V_f}{1-\eta V_f} \qquad [6.1]$$

$$\eta = \frac{\dfrac{M_f}{M_m} - 1}{\dfrac{M_f}{M_m} + \xi} \qquad [6.2]$$

Table 6.4 Theoretical modulus of single sisal fibres predicted by Halpin–Tsai model

Fibre	Fibre volume fraction (%)	Theoretical modulus (GPa)
Sisal	63.9	12.3–25.7

where ξ is the geometrical factor, M and V are the effective moduli and fibre volume fraction, respectively, and subscripts c, f and m present composite, fibre and matrix, respectively.

The factor ξ is dependent on the geometry of the fibres. It is given by:

$$\xi = 2l/d \qquad [6.3]$$

where l and d are the length and diameter of the fibre, respectively. In the case of continuous fibre, ξ approaches infinity, and the Halpin–Tsai model is changed into the rule of mixtures (ROM). The fibre volume fraction of sisal fibres (being regarded as a composite structure) is obtained using an optical microscope analysed by QuantLab-MG software. The measured content of pectin is 22.8%. Therefore, the fibre volume fraction of sisal fibres can be calculated as 100% – (22.8% + (1 – 17.5%) × 22.8%)), which is 63.9%. The modulus of pectin can be neglected due to its much smaller value compared to that of cellulose, thus resulting in the theoretical modulus as shown in Table 6.4.

6.3.2 Predicting the tensile strength of sisal fibres using the fibre bundle model and ROM

Sisal fibres possess multi-scale structures. A cell is considered as the smallest representative volume element (RVE). Each cell can be simplified as being made up of the cell wall with a lumen inside, as shown in Fig. 6.6. In order to determine the tensile strength of one cell for sisal fibres, the following formula is given as:

$$\sigma_f = (1 - V_{hole})\sigma_0 \qquad [6.4]$$

where σ_f is the strength of a cell. V_{hole} is the lumen ratio, and σ_0 is the strength of the cell wall without a hole inside. Since ramie fibre is made of only one cell, the tensile strength of ramie fibre is assumed to be σ_f. The tensile strength of one cell for sisal fibre can be calculated based on the strength of ramie fibre (σ_f^{ramie}). The results are shown in Table 6.5.

A fibre bundle model[15, 16] is introduced to determine the tensile strength of sisal fibres which possess a multi-cell structure. This model is established

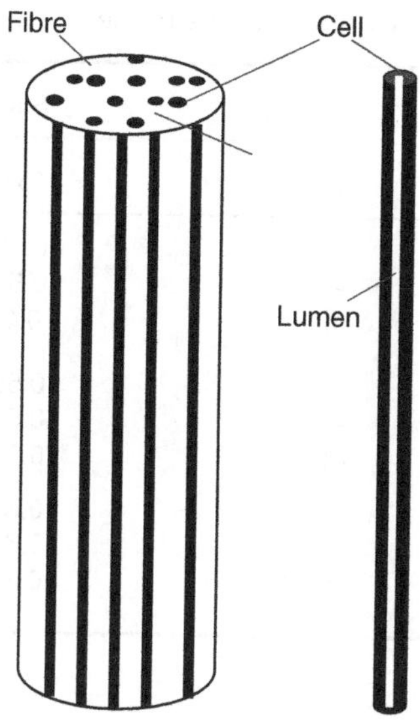

6.6 Microstructure model of a single sisal fibre.

Table 6.5 Tensile strength of ramie and sisal fibres

Fibre	Ramie	Sisal
Lumen ratio (%)	5.3	17.5
Strength	σ_f^{ramie}	$0.871\sigma_f^{ramie}$

on the basis of probabilistic fracture mechanics. Here, a modification of the tensile strength of sisal fibre from countless numbers to a limited number of cells is conducted.

By derivation, the formula for the tensile strength σ_b of a single sisal fibre is:

$$\overline{\sigma_b^N} = \alpha\gamma\sigma_f \qquad [6.5]$$

where,

$$\alpha = \left[\beta^{1/\beta}\exp\left(\frac{1}{\beta}\right)\Gamma\left(1+\frac{1}{\beta}\right)\right]^{-1} \qquad [6.6]$$

$$\gamma = 1 + 0.996N^{-2/3}e^{2/3\beta}\beta^{-1/3} \qquad [6.7]$$

σ_f is the tensile strength of a cell, β is the configuration parameter of the Weibull distribution for the strength of a single cell, and N is the number

Table 6.6 Strength of sisal fibres obtained by fibre bundle model

Number of cells		Ramie	Sisal
		1	134
$\beta = 2$	α	–	0.484
	γ	–	0.504
	Strength	σ_f^{ramie}	$0.213\sigma_f^{ramie}$
$\beta = 3$	α	–	0.556
	γ	–	0.574
	Strength	σ_f^{ramie}	$0.278\sigma_f^{ramie}$
$\beta = 4$	α	–	0.608
	γ	–	0.625
	Strength	σ_f^{ramie}	$0.331\sigma_f^{ramie}$

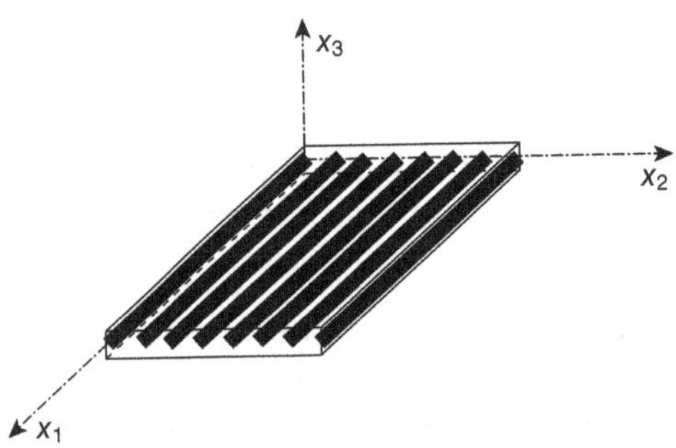

6.7 The model of transverse isotropy composite structure.

of cells within a single fibre. The value of β for brittle fibres usually changes from 2 to 4. Thus, for different values of β, the strength of sisal fibres can be calculated and the results are listed in Table 6.6.

The rule of mixtures (ROM) is widely used in determining the properties of composite materials. From the aforementioned observation, sisal fibres can be treated as fibre-cells reinforced lignin and hemicellulose composites Figure 6.7 is a sketch of the transverse isotropy composite structure model; the helical structure along the length direction is not considered in this model, and the tensile strength of the matrix is neglected due to its naturally weak strength and the fact that most of the colloid are removed during the extraction process. Cell walls of sisal fibres are considered as uniform mixtures of cellulose, and hemicellulose, etc.

Assuming equal strain within a single fibre, the tensile strength relationship between a fibre and one cell can be given by the ROM formula (equation

Table 6.7 Tensile strength of sisal fibre predicted by ROM

Fibre	Ramie	Sisal
Matrix volume (%)	0	22.8
Strength	σ_f^{ramie}	$0.673\sigma_f^{ramie}$

Table 6.8 Comparison of tensile strength of sisal fibre obtained from theoretical predictions and experimental results[16]

Tensile strength	Ramie	Sisal
Theoretical results (MPa)	938	200–310
Fibre bundle model ROM	938	631
Experimental data (MPa)	938	511–635

6.8). Substituting equation 6.4 into 6.8, the tensile strength can be obtained. The matrix volume means the content of pectin obtained by direct measurement using an optical microscope with QuantLab-MG software. The results are shown in Table 6.7.

$$\sigma_b = E_f\varepsilon = (1 - V_m)\sigma_f \qquad [6.8]$$

Where ε is the strain of fibhre during tensile test.

6.3.3 Comparison and discussion

Taking the measured data of ramie fibres in reference [16] as σ_f^{ramie}, and substituting the values in Tables 6.6 and 6.7, the theoretical tensile strength calculated from the fibre bundle model together with the ROM method can be obtained and is shown in Table 6.8. These results are also compared with the experimental results.[16] They indicate that ROM gives a much better prediction than the fibre bundle model in this case.

Since sisal fibres are extracted from the leaves of the plants while ramie fibres are produced from the stems, it is logical that the strength of leaf fibres is much lower than that of stem fibres. Our theoretical results, as well as the experimental data given in reference [16] prove this. The main function of leaves is photosynthesis, and a huge number of cells are necessary. However, the stem is used to support the whole plant and transport water and nourishment to the upper part of the plant. In this respect, it is necessary that the stem fibres must have enough strength as a supportive component. Normally the ramie plants are about 0.5 m in height.

Basically, nature dictates good plant design, and these differences are perfectly explained by nature. The theoretical prediction values from different models are aimed at giving reference to the experimental results, which are still lacking, and also providing a basis for the design of natural fibre composites.

Having reviewed the calculations, a conclusion can be drawn: the number of cells and the volume fraction of cells in one fibre are the critical parameters in the tensile strength of a single sisal fibre. This gives quite useful instructions for the design of new bio-mimic materials.

6.4 Manufacture of sisal fibre-reinforced composites

Different matrix systems possess different properties. The processing parameters of sisal fibre-reinforced composites are mainly determined by the properties of the matrix. In this section, several manufacturing processes of sisal fibre-reinforced thermoplastic/thermoset composites are introduced.

6.4.1 Sisal fibre-reinforced thermoplastics composites

Two conventional manufacturing techniques of fibre-reinforced thermoplastic composites are extrusion compounding and injection moulding. When applying them for the fabrication of short fibre-reinforced thermoplastic composites, fibre length may be reduced significantly through breakage in the intermediate steps, which results in a wide distribution of fibre lengths.[17] However, when employing other techniques, e.g. hot-pressing, fibres are already chopped before being compounded with resin and usually in a uniform length.

The choice of fibre length is a crucial factor for the final quality of the composites, as it affects the rotational freedom of fibres in the thermoplastics and the distribution of the fibres in the composites. Sisal fibre is one of the largest natural fibres and its diameter ranges widely from 50 to 300 μm, averaging about 150 μm (Fig. 6.8(a)). Considering the theory of 'maximum packing fraction' for randomly oriented short fibres, expressed in equation 6.9, the free volume fraction drops dramatically as the sisal fibre length increases from 1 to 10 mm (Fig. 6.8(b)). Apparently, when the fibre length exceeds 6 mm, the optimal fraction of fibres should be controlled to lower than 10%. However, raising the fibre content and ensuring good wetting can be achieved in the meantime, when methods like surface modification of sisal fibres and dilution of the matrix are employed. Figure 6.9 shows the increased mechanical properties of sisal fibre-reinforced PMMA composites with an increase in fibre length and modification of the fibre surfaces. Similar trends are found by other researchers.[18–21] However, if the fibre length exceeds a threshold, the mechanical properties show a decreasing

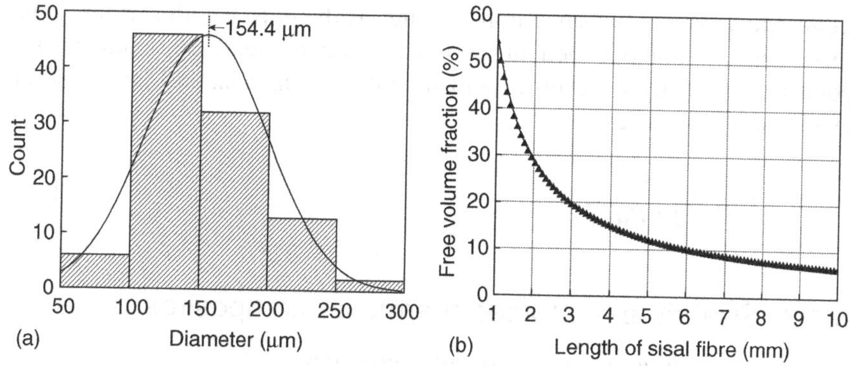

6.8 (a) Scattering of diameter of sisal fibres; (b) 'Maximum Packing Fraction' of short sisal fibres vs. their lengths.

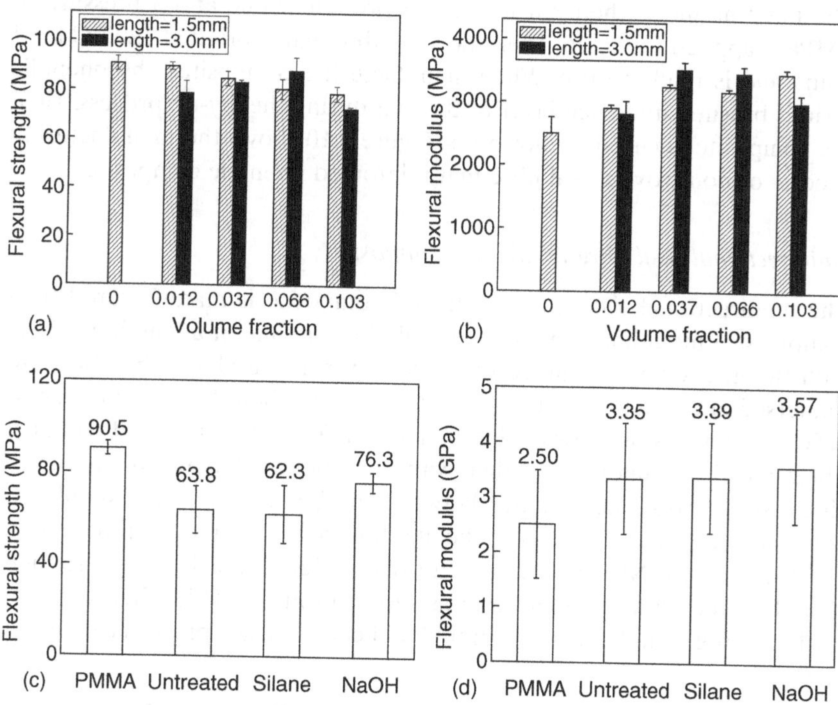

6.9 The (a) flexural strength and (b) flexural modulus of composites vs. fibre volume fraction; the effect of modification of fibre surfaces on the (c) flexural strength and (d) flexural modulus of sisal fibre reinforced PMMA composites.

trend. Moreover, at the same fibre loading, as the fibre length increases, the space distribution between fibres is expected to decrease, leading to a poorer dispersion of the fibres and the polymer matrix, and hence lower mechanical properties:

$$V_{\max}^{\mathrm{f}} = \frac{4}{1 + l/d} \qquad [6.9]$$

6.4.2 Sisal fibre-reinforced thermoset composites

Non-woven sisal fibre mat reinforced composites

Sisal fibre-reinforced composites are manufactured using non-woven short sisal fibre mat and phenolic resin by a hot press. First, the weighted sisal fibres are cut into proper lengths and spread evenly on the bottom mould. The filled mould is then put into a hot press machine and a pressure of 1 MPa is applied for 2 min. Then the sisal fibre mat is dried at 1200°C for 4 hours and is ready to use. When manufacturing composites, the phenolic resin is brushed into each layer of the mat during the lay-up process. Then the composite is cured by hot press. Figure 6.10 shows the manufacturing process of non-woven sisal fibre mat reinforced phenolic composite.

Unidirectional sisal fibre-reinforced composites

The unidirectional sisal fibre-reinforced composite is manufactured by a compressive moulding process. The sisal fibres are cut at a length of 35 cm, with the messy fibres combed off as shown in Fig. 6.11(a). The fibres are immersed and straightened in water for 2 hours, then dried in an oven at 60°C for 24 hours, as shown in Fig. 6.11(b). In Fig. 6.11(c), the unidirectional fibres are then bundled together with each bundle having five to seven single sisal fibres. The distance between the bundles of fibres should be controlled at 1.0–1.5 mm to make unidirectional sisal non-crimp fabric (NCF). The sisal NCF fabric is then dried at 120°C for 4 hours, and then hand-brushed with phenolic resin as shown in Fig. 6.11(d). Then the sisal NCF prepreg is laid inside the mould and cured by hot-press machine.

6.5 Mechanical properties of sisal fibre-reinforced composites: interfacial properties

The mechanical performance of fibre-reinforced polymeric composites depends primarily on three factors: (1) strength and modulus of the fibre, (2) strength and chemical stability of the resin, and (3) effectiveness of the bond between resin and fibre in transferring stress across the interface.[22,23]

(a) (b) (c) (d)

6.10 The manufacturing process of non-woven sisal fibre mats reinforced phenolic resin. (a) Evenly spread sisal fibres, (b) sisal fibre mat after being pressed at 1 MPa of pressure, (c) brush resin into the fibre mat and (d) the short sisal fibre mat reinforced composite.

The poor fibre/matrix interface can be improved by physical and chemical treatments of the fibres. The effects of surface modification on the mechanical properties of sisal fibre-reinforced composites are investigated here. In terms of physical treatment, dried sisal fibres are immersed in 0.055% potassium permanganate acetone solution for 2 min, then washed with acetone, and dried at 60°C for 4 hours. As for the chemical coupling treatment, dried sisal fibres are immersed in 6% silane acetone solution for 24 hours, washed with acetone to get rid of the residual silane, then dried at 60°C for 4 hours. In addition, the effect of moisture absorption on the mechanical properties of sisal fibre-reinforced composites is also investigated.

Sisal fibre is relatively coarse and inflexible. It has moderate specific strength and stiffness, durability, elasticity and resistance to corrosion.[24] The presence of hydroxyl and other polar groups in sisal fibres makes them exhibit highly hydrophilic properties, which leads to incompatibility and poor wettability in a hydrophobic polymer matrix, thus weak interfacial property[25] and consequently lower mechanical properties. Single fibre pullout tests, short beam shear tests, double cantilever beam (DCB) tests

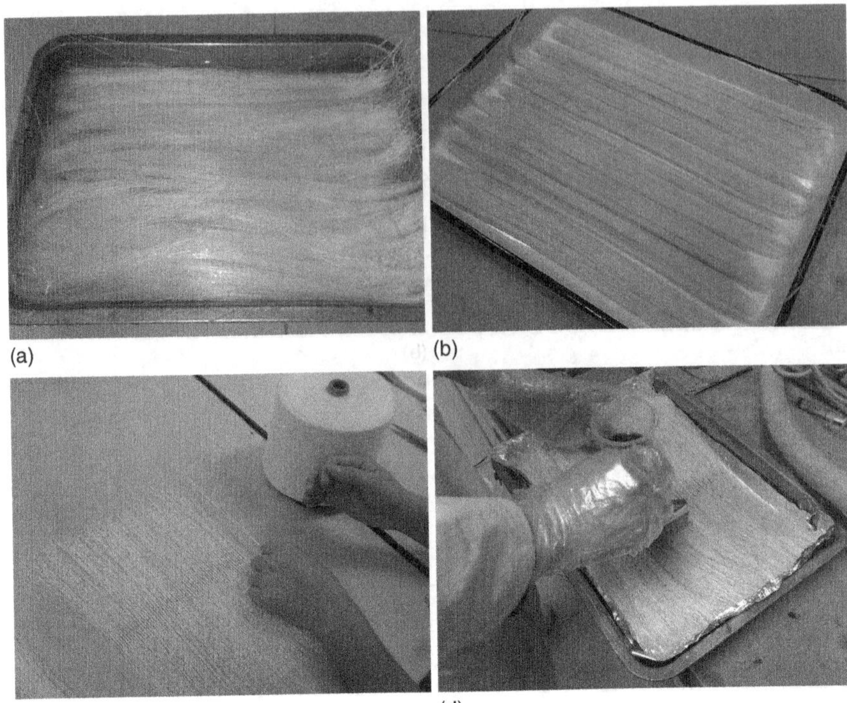

(a) (b) (c) (d)

6.11 The manufacturing process of unidirectional sisal fibre reinforced composite. (a) Sisal fibre with a length of 35 mm, (b) straight and dried sisal fibre, (c) a layer of bundled sisal fibres, and (d) brushing resin onto the bundled sisal fibres.

and end notch flexure (ENF) tests are performed to measure the interfacial shear strength (IFSS), interlaminar shear strength (ILSS), and mode I and mode II interlaminar fracture toughness of sisal fibre-reinforced composites.

The single fibre pull-out test is a widely used method to measure the IFSS between fibre and matrix. Equation 6.10 is used to calculate the results.

$$\tau = \frac{P}{\pi d l_e} \qquad [6.10]$$

where τ is the interfacial shear strength, P is the maximum pull-out force, d is the fibre diameter and l_e is the embedded fibre length.

According to the previous study, it is known that the diameter of sisal fibres is not a fixed value but varied along the fibre length. The cross-section of the sisal fibre is not circular. Therefore, it is not appropriate to use equation 6.10 to calculate the IFSS of sisal fibre reinforced composites and a modified equation (6.11) is proposed to try to get more accurate results by using perimeter rather than diameter:

$$\tau = \frac{P}{l_e C} = \frac{P}{l_e EX} \qquad [6.11]$$

where C is the perimeter of the fibre cross-section, and

$$EX = t_0 \Gamma\left(1 + \frac{1}{m}\right) \qquad [6.12]$$

where EX is the mean Weibull distribution of the perimeter, and t_0 and m are factors of the Weibull distribution which can be obtained by statistical sampling.

Figure 6.12 shows the typical pullout force against displacement curves of untreated, potassium permanganate-treated, silane-treated and water-immersed sisal fibre-reinforced composites. Two types of curves are observed: one is the stable pullout process for potassium permanganate-treated specimens, untreated specimens and with moisture absorption specimens due to the mainly mechanical bonding between the sisal fibre and the matrix. The other type of curve shows the unstable pullout process for silane-treated sisal fibre-reinforced composites, which clearly indicates the chemical bonding mechanism between silane-treated sisal fibre and the matrix. The IFSS are calculated according to equations 6.11 and 6.12, and the results are given in Fig. 6.13. From this investigation it is clearly evident that both permanganate and silane treatment can improve the interfacial adhesion between sisal fibre and polymeric matrices. The IFSS greatly decreases after moisture absorption.

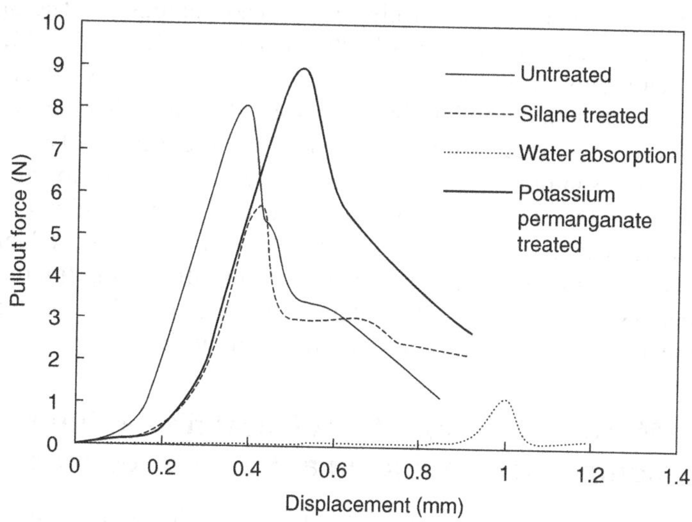

6.12 Pull-out force versus displacement during single fibre pull-out tests for untreated and treated sisal fibre-reinforced composites.

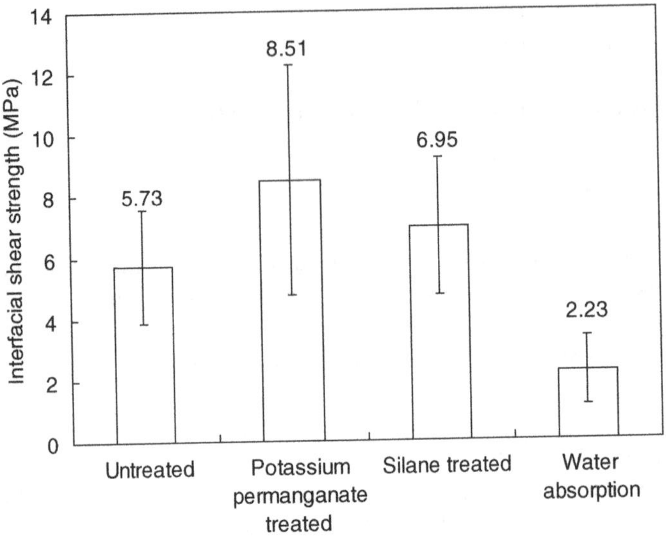

6.13 The IFSS obtained by single fibre pull-out test of sisal fibre-reinforced composites.

Short beam shear tests are conducted to compare the apparent interlaminar shear strength (ILSS) of treated and untreated sisal textile reinforced polymers. Figure 6.14 shows the effects of fibre surface treatments on the ILSS of sisal fibre-reinforced composites. They are very consistent with the results for interfacial shear strength. Both silane and permanganate can improve the ILSS of sisal fibre-reinforced composites. The apparent ILSS has been improved by nearly 40% from 3.65 MPa to 5.49 MPa after silane treatments. Silane treatment can improve the ILSS effectively by introducing chemical bonding between sisal fibre and phenolic resin. The ILSS of the composites after water absorption reduces to 70% of that of the untreated samples due to the higher water absorption capacity of sisal fibre compared to phenolic resin, which results in the generation of internal stress in the interlaminar area. It can be seen from Fig. 6.15 that delamination is observed for the water-absorption composites after being fractured by a short beam shear test, which illustrates that interlaminar damage dominates in the shear failure of sisal fibre reinforced composites.

6.6 Mechanical properties of sisal fibre-reinforced composites: interlaminar fracture toughness

For laminated composites, delamination is the most common failure mode due to weak interlaminar bonding. Thus, interlaminar fracture toughness is very important for a composite material.

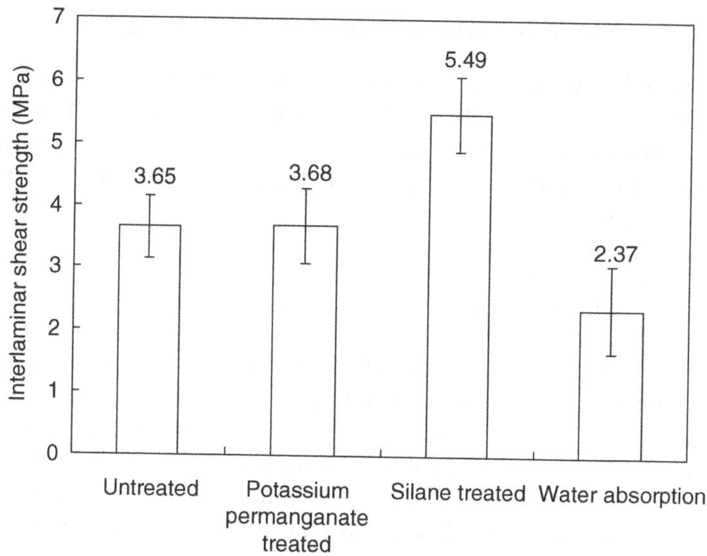

6.14 The ILSS of sisal fibre-reinforced composites with different surface treatments.

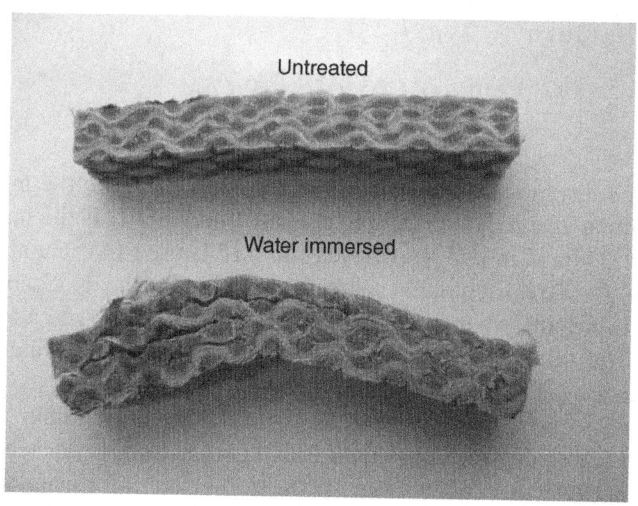

6.15 Photographs of shear failure of sisal fibre-reinforced composite laminates.

6.6.1 Mode I interlaminar fracture toughness

Treated and untreated sisal textile-reinforced vinyl ester composites are manufactured to investigate the mode I interlaminar fracture toughness, G_I. At least five specimens are tested for each group. In this study, the modified beam theory is used to interpret the experimental results as shown in equation 6.13.

$$G_I = \frac{3P\delta}{2b(a+|\Delta|)}\frac{F}{N} \quad [6.13]$$

where P is the load, δ is the load point displacement, b is the specimen width, and a is the delamination length. The absolute value of Δ is a correlation of the delamination length, which can be determined experimentally by generating a least-squares plot of the cube root of compliance, $C^{1/3}$, as a function of the delamination length. The compliance, C, is the ratio of the load point displacement to the applied load, δ/P. Sometimes, large displacement effects need to be corrected by the inclusion of a parameter, F, and a loading blocks parameter, N, in the calculation of G_I. The expressions for F and N are given as:

$$F = 1 - \frac{3}{10}\left(\frac{\delta}{a}\right)^2 - \frac{3}{2}\left(\frac{\delta t}{a^2}\right) \quad [6.14]$$

$$N = 1 - \left(\frac{L'}{a}\right)^3 - \frac{9}{8}\left[1 - \left(\frac{L'}{a}\right)^2\right]\left(\frac{\delta t}{a^2}\right) - \frac{9}{35}\left(\frac{\delta}{a}\right)^2 \quad [6.15]$$

where t is the distance from the centre of the loading pin to the mid-plane of the specimen arm to which the block is attached and L' is the half-width of the loading block. In this study, equation 6.13 is used to calculate the G_I values. Two fracture toughness values are defined. One is G_{IC} which is defined at the point when the pre-crack is visually observed to start to propagate. The other is G_{mc} which is defined as the plateau value of the crack resistance curve.

Figures 6.16 and 6.17 show the R-curves and interlaminar fracture toughness of treated and untreated sisal textile-reinforced vinyl ester composites. The composites that possess better interfacial bonding properties also have higher mode I interlaminar fracture toughness.

6.6.2 Mode II interlaminar fracture toughness

In this study, the direct beam theory is employed to calculate the mode II interlaminar fracture toughness by using the maximum load during the experiment. The direct beam theory is expressed as:

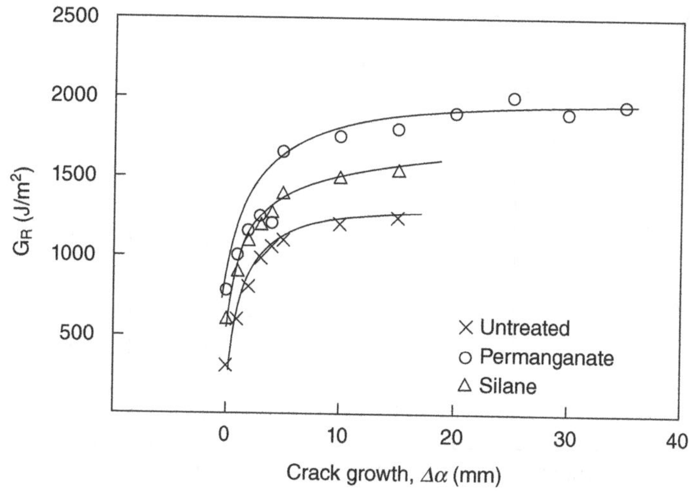

6.16 R-curves of sisal textile reinforced vinyl ester composites.

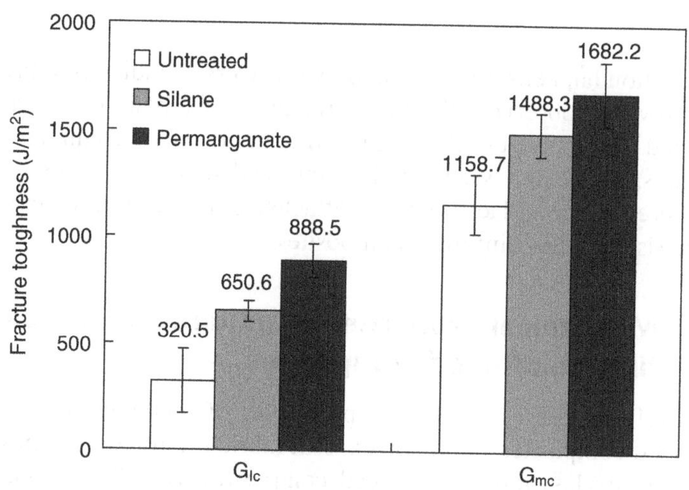

6.17 Mode I interlaminar fracture toughness of sisal textile reinforced vinyl ester composites.

$$G_{IIC} = \frac{9a^2 P\delta}{2B(2L^3 + 3a^3)} \qquad [6.16]$$

where a is the distance from one supporter to the tip of the crack, P is the maximum load, δ is the corresponding displacement, B is the width of the specimen and L is the span of the flexural tests.

Figure 6.18 shows the mode II interlaminar fracture toughness of treated and untreated sisal textile-reinforced vinyl ester composites. The

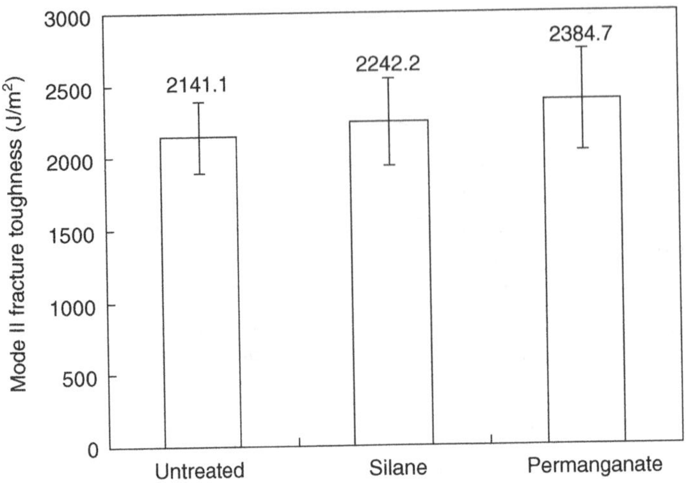

6.18 Mode II interlaminar fracture toughness of sisal textile-reinforced vinyl ester composites.

delamination happens first, followed by transverse bending fracture, and no crack growth is observed. The better the interfacial bond between the sisal fibre and the polymer, the higher the mode II interlaminar fracture toughness. Untreated sisal textile-reinforced composite shows a rather clean fibre surface and less matrix is attached onto the fibres compared with treated sisal textile-reinforced composites.

6.7 Mechanical properties of unidirectional sisal fibre-reinforced composites

Figure 6.19 compares the tensile properties of unidirectional sisal fibre-reinforced composites, plain woven sisal fabric-reinforced composites, and plain woven glass fabric-reinforced composites as well as unidirectional glass fibre-reinforced composites. The manufacture of unidirectional sisal fibre-reinforced composites has already been introduced in Section 6.2.2. It can be seen that the tensile strength of unidirectional sisal fibre-reinforced composite is about one-third that of unidirectional glass fibre-reinforced composite. The tensile moduli of these two composites are very close. However, both the tensile strength and tensile modulus of sisal woven fabric-reinforced composites are much lower than those of glass woven fabric-reinforced composites due to the twist structures of sisal fibre yarns, which dramatically decreases their mechanical properties.

The same tendency is found for the flexural properties of the four types of composites as shown in Fig. 6.20, which indicates that by reducing the

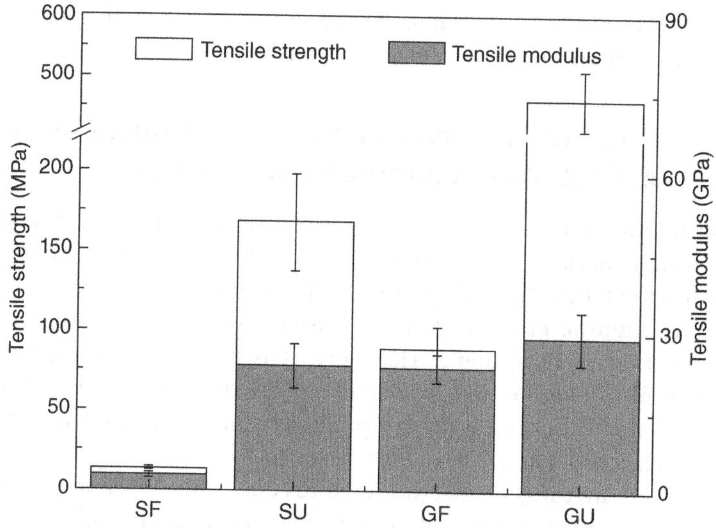

6.19 The comparison of tensile properties of fibre-reinforced composite. Unidirectional sisal fibre-reinforced composite (SU), sisal fabric-reinforced composite (SF), unidirectional glass fibre-reinforced composite (GU) and glass fabric-reinforced composite (GF). The tensile properties of glass fibre-reinforced composites for both GU and GF are adapted from Ref [26].

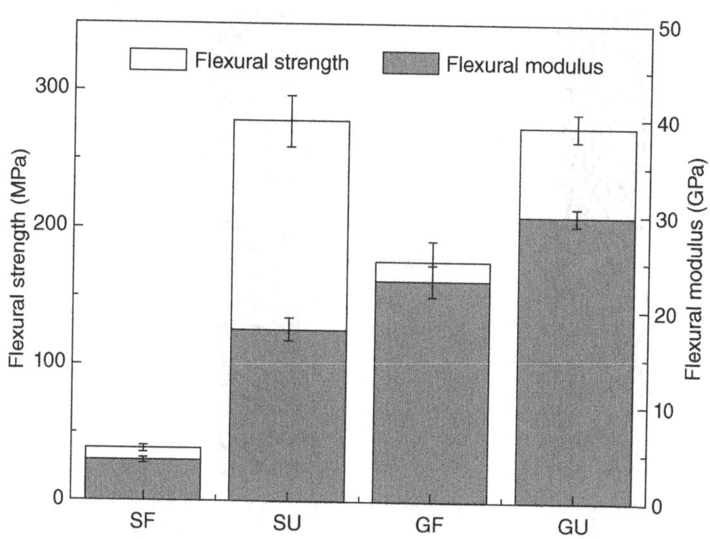

6.20 The comparison of flexural properties of fibre-reinforced composite. The flexural properties of glass fibre-reinforced composites for both GU and GF are adapted from Ref [26].

twist of sisal fibres, the mechanical properties of the sisal fibre-reinforced composites might be significantly improved.

6.8 Effect of fibre twist on the mechanical properties of sisal fibre-reinforced composites

As with other natural fibres, sisal fibres are always short and discontinuous due to their natural growth characteristics. The production of continuous sisal fibres can be achieved by twisting short fibres together as shown in Fig. 6.21.[27] The tensile properties of the twisted sisal yarns are often expressed as a function of twist angle. Therefore, it is necessary to understand the effect of fibre twist on the mechanical properties of sisal fibre-reinforced composites, in order to establish a predictive model and also to optimise its mechanical properties.

In order to investigate the effect of fibre twist on the mechanical properties of sisal fibre yarn-reinforced composites, non-twisted and two different levels of twisted sisal fibre yarns (30 turns/m and 50 turns/m respectively) are prepared and then combined with resin to make unidirectional composites using hot press. The fibre volume fractions of the three types of composites are approximately 66%.

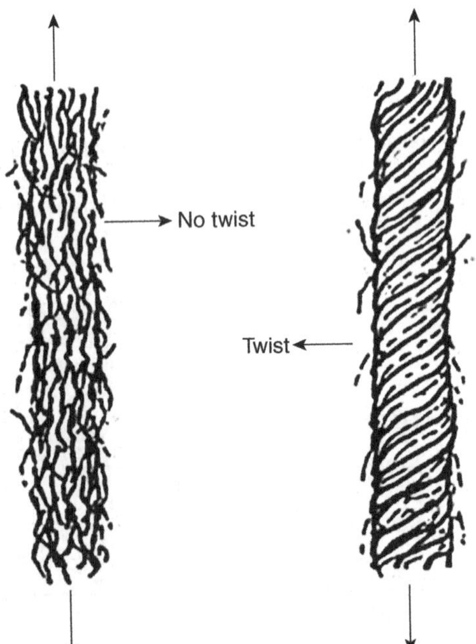

6.21 Non-twisted and twisted sisal fibre yarns. Reproduced with permission from Springer from Ref. [27].

The effects of different levels of fibre twist on the flexural modulus and strength are shown in Fig. 6.22. The results show that both the flexural strength and modulus of sisal fibre reinforced composites decrease with an increasing fibre twist level. The reduction of the flexural properties could be very high; a 44% decline of the flexural modulus from non-twist sisal fibre yarns to 50 turns/m twisted sisal fibre yarns is evident. With an increase

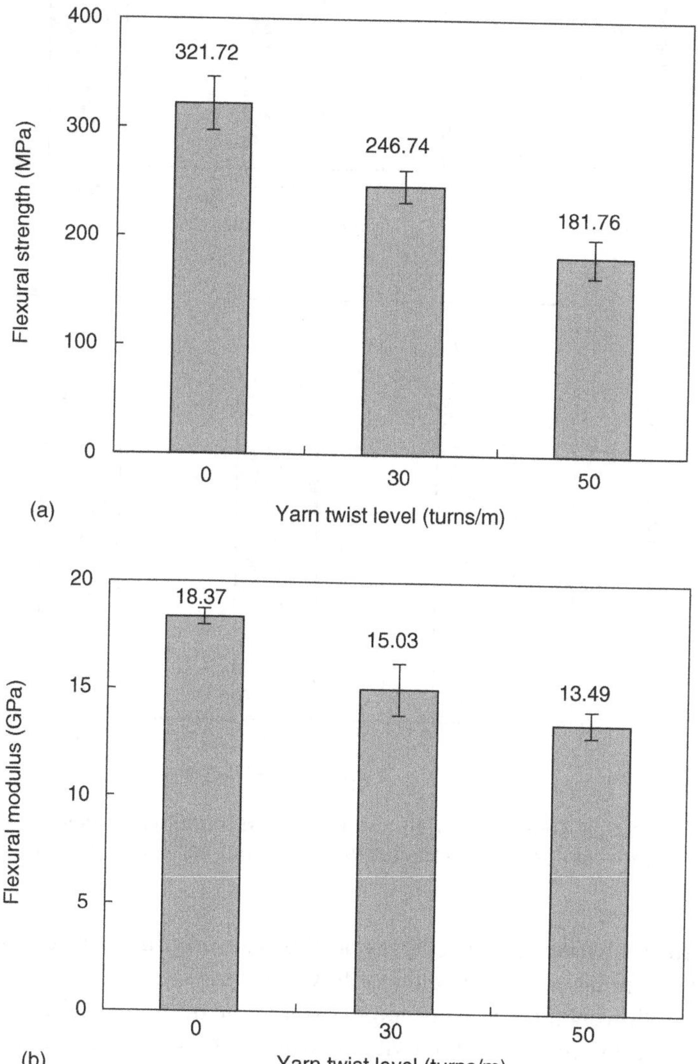

6.22 (a) Flexural strength and (b) flexural modulus of sisal fibre-reinforced composites with different yarn twist levels.

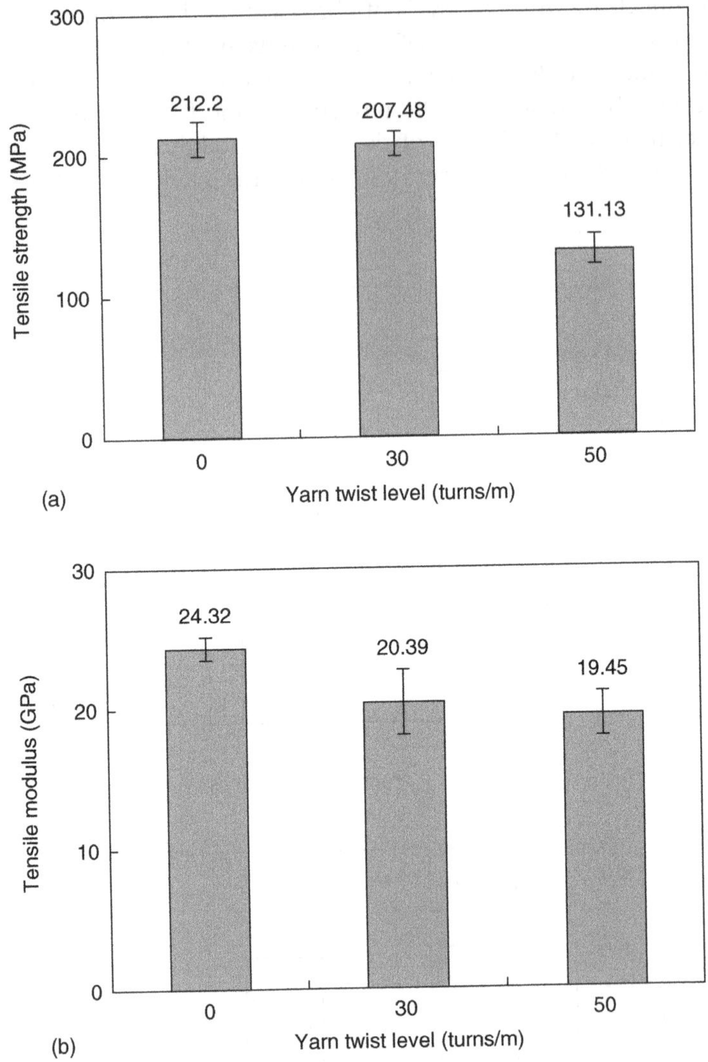

6.23 (a) Tensile strength and (b) tensile modulus of sisal fibre-reinforced composites with different yarn twist levels.

in yarn twist level, the tensile properties also drop as shown in Fig. 6.23, which is similar to the tendency of flexural results.

The following three reasons might explain the decrease of the mechanical properties. Firstly, during the twisting process, the yarns are pre-stressed on each component fibre. The pre-stress also increases with the increase in twisting level, which would lower the loading capacity of the reinforcing fibres in the composites. Moreover, the angle (twist angle) between fibre

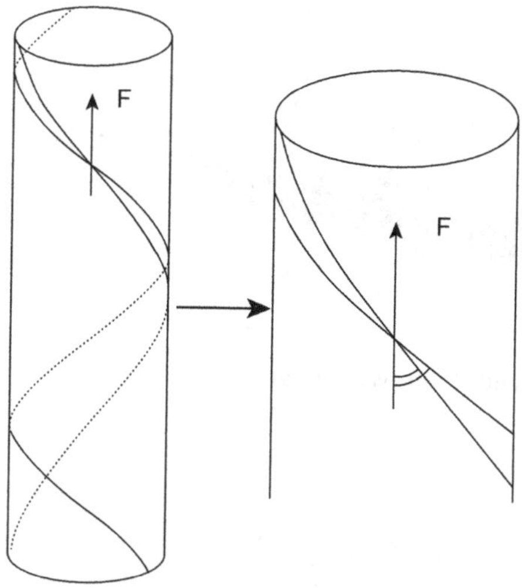

6.24 Model of the twisted fibre structure in a ring spun yarn.

Table 6.9 Relationship of twisting angles with fibre twist level

Twist level (turns/m)	Periphery twist angle (°)	Mean twist angle (°)
0	0	0
30	18.96	13.18
50	39.87	27.91

orientation and the direction of the applied load increases with the increase in the fibre twist level, as illustrated in Fig. 6.24. The twist angles of sisal fibre yarns with different twist levels are shown in Table 6.9. Thirdly, the resin impregnation would also get poorer due to the lower permeability of the twisted yarns caused by the high level of the twist (Fig. 6.25). More manufacturing defects could appear. The fracture modes of non-twisted and 50 turns/m twisted sisal fibre yarn-reinforced composites are compared in Fig. 6.26. Very different fracture modes can be observed. Fibre breakage is the major fracture mode for the non-twisted yarn-reinforced composites, while interfacial debonding is the major fracture mode for the twisted counterparts, which indicates the poor impregnation of resin in the twisted yarns.

It is referenced[28] that the fracture strength of yarns (σ_{fy}) can be given by equation 6.17:

$$\sigma_{fy} = \frac{P_y}{A_y} = \frac{N\sigma_f A_f \cos\theta_{mean}}{A_y} \quad [6.17]$$

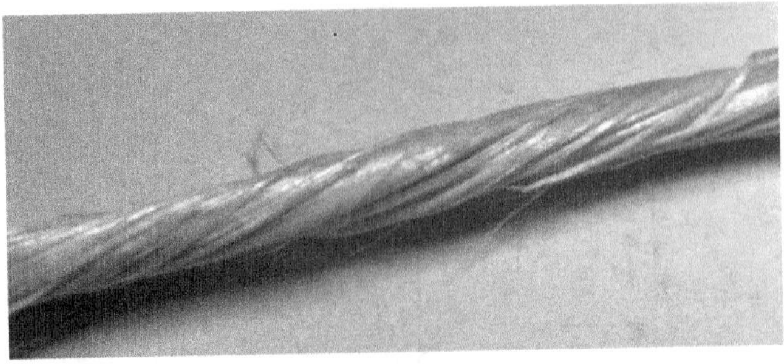

6.25 Sisal fibre yarn with high twist level.

(a)

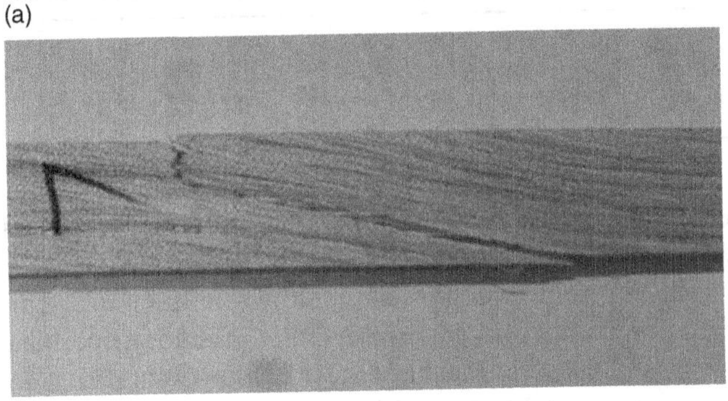

(b)

6.26 The fracture modes of (a) non-twist yarns and (b) 50 turns/m twist yarns reinforced composites.

where σ_f is the axial tensile strength of single sisal fibre. P_y is the loading capabilities of the sisal fibre yarns, A_f and A_y are the cross-sectional area of sisal fibres and sisal yarns, respectively, and θ_{mean} is the mean twist angle.

Then by substituting equation 6.17 into the rule of mixtures, the ultimate stress of sisal fibre-reinforced composites can be calculated by equation 6.18 and the results are shown in Table 6.10. For comparison, the experimental results are also listed in the table. It can be seen that the calculated result

Table 6.10 Comparison of theoretical and experimental strength of different levels of twist sisal yarn-reinforced composite

Twist level (turns/m)	Theoretical strength (MPa)	Experimental strength (MPa)
0	–	212.2
30	206.6	207.5
50	177.5	131.1

is very similar to the experimental value for the 30 turns/m sisal yarn-reinforced composites. However, for the 50 turns/m sisal yarn-reinforced composites, the calculated strength is higher than the experimental one. The greater number of manufacturing defects caused by poor permeability, as discussed earlier, lead to the variation.

$$\sigma_1 = V_f\sigma_f + V_m\sigma_m = \frac{V_f N\sigma_f A_f \cos\theta_{mean}}{A_y} + V_m\sigma_m \quad [6.18]$$

where σ_m is the tensile strength of the resin, V_m is its volume fraction, and V_f is the fibre volume fraction.

During the manufacturing process of sisal fibre-reinforced composite, the reinforcement needs to be in the form of stable fibre yarns. Due to the limited length of technical sisal fibres, twisting is unavoidable. Thus, fibre obliquity caused by fibre twisting might result in non-linear behaviour of sisal fibre-reinforced composites.

Figure 6.27 shows the stress–strain curves of unidirectional non-twist and twisted (50 turns/m) sisal fibre-reinforced phenolic composites following a tensile test with a crosshead speed of 1 mm/min. It can be seen from the figure that fibre twist clearly led to a nonlinear behaviour for the composite.

Figure 6.28(a) gives the nonlinear curve fitting for the twist sisal fibre-reinforced composite, which gives the functional relationship as shown in equation 6.19. The variation of tensile modulus should have the same tendency with the tangent of the stress–strain curve, which can then be expressed by the derivation of equation 6.19, shown in Fig. 6.28(b). The function of the curve is given by equation 6.20.

$$y = -1.07393 + 213.7887x - 243.4035x^2 + 148.68231x^3 \quad [6.19]$$

$$\frac{dy}{dx} = 213.7887 - 427.5774x + 446.04693x^2 \quad [6.20]$$

From the change in the slope of the fitted stress–strain curve of twisted sisal fibre-reinforced composites as shown in Fig. 6.28(b), it can be seen that

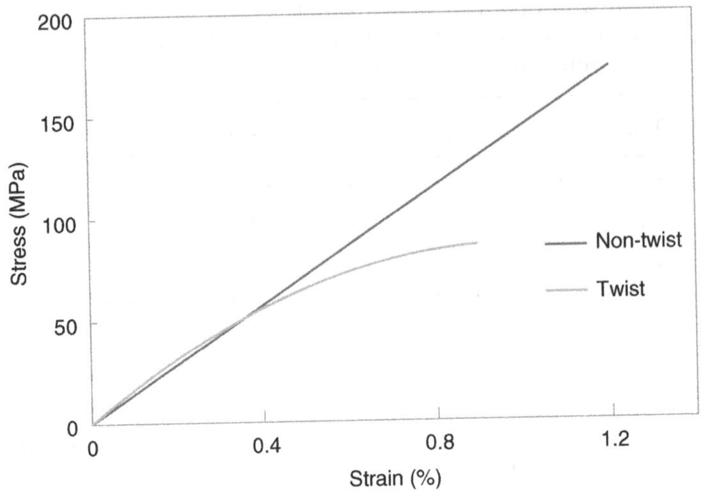

6.27 The stress-strain curves of the non-twist and twist sisal fibre composite.

the slope decreases first and then increases. A schematic drawing of twisted sisal fibre yarn-reinforced composite is given in Fig. 6.29. The black region in the sketch presents matrix and silver strips present in the twisted sisal fibres. At the beginning of the loading process, the twisted fibre would rotate and lead to interfacial cracks, as shown in Fig. 6.29(b). The restriction of the matrix to the sisal fibre yarn decreases. Therefore the deformation ability of the composite increases, which leads to a decreased tensile modulus of the composite.

This also explains why the twisted sisal fibre-reinforced composites show much lower elongation at break compared to the untwisted composites. With a further increase in the applied load, the matrix breaks and the debonding between the sisal fibre and the matrix is so great that the sisal fibres start to sustain most of the load by themselves. Therefore, the modulus of the composite can be obtained according to the rule of mixtures by equation 6.21:

$$E = V_m E_m + V_f E_f \qquad [6.21]$$

where

$$E_f = E_{fw} \cos^2 \theta \qquad [6.22]$$

where E_{fw} is the modulus of non-twisting fibre bundle and θ is the twist angle. With an increase in tension loads, the twisting fibre bundles would

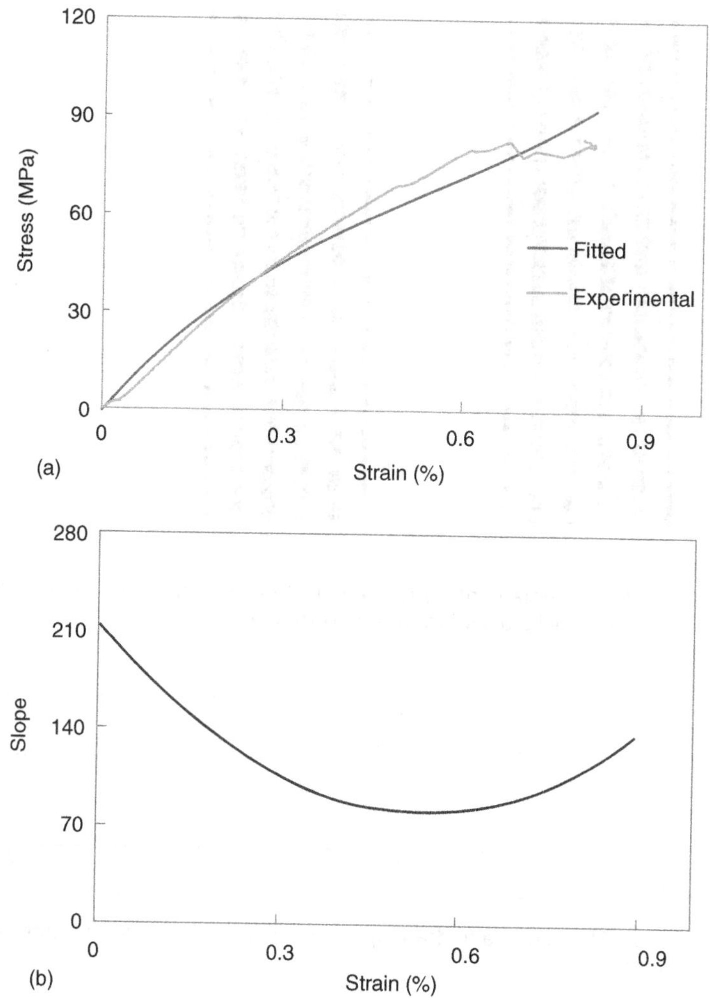

6.28 (a) The curve fitting of stress-strain relation of twisted sisal fibre-reinforced composites and (b) the fitting gradient of stress-strain curve of the twisted sisal fibre-reinforced composites.

experience an elongation as shown in Fig. 6.30. The decreased twist angle would lead to an increased modulus of the sisal fibre-reinforced composite.

6.9 Durability of sisal fibre-reinforced composites: effects of moisture absorption

The mechanical properties and fracture properties of sisal fibre-reinforced composites deteriorate when the composites are exposed to various environments that will affect their service life. Thus, one of the main

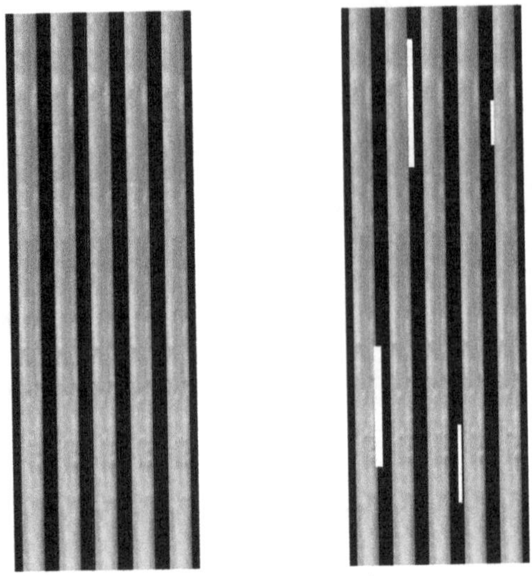

6.29 A schematic of (a) twisted sisal fibre yarn-reinforced composite and (b) cracks between fibre and matrix.

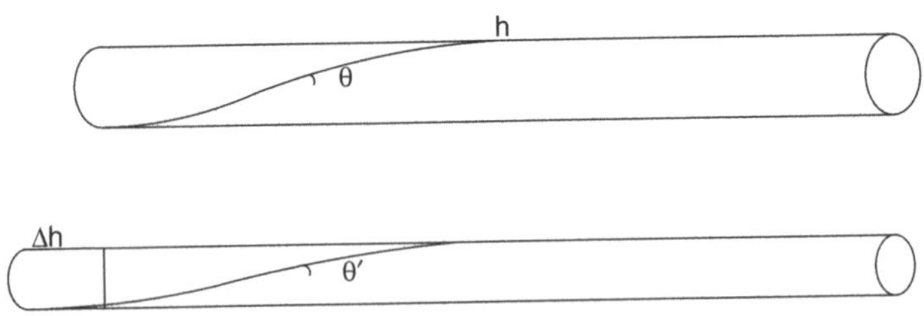

6.30 The schematic drawing of twisting angle variation under tension loads.

concerns for the application of sisal fibre-reinforced composite materials is their susceptibility to moisture, temperature and ultraviolet (UV) light, which may cause the degradation of their physical, mechanical and thermal properties.[29] Therefore, it is important to know the effect of those environmental factors on the properties of sisal fibre-reinforced composites. In this and the following section, the effects of moisture absorption and UV light on the mechanical properties of sisal fibre-reinforced composites are discussed.

Sisal fibres absorb water from the air and their surroundings. Absorbed moisture deforms the surface of the composites by swelling and creating voids. The absorption of moisture leads to the degradation of the fibre–matrix interface and creates poor stress transfer efficiencies, thus causing a reduction of mechanical properties.[30]

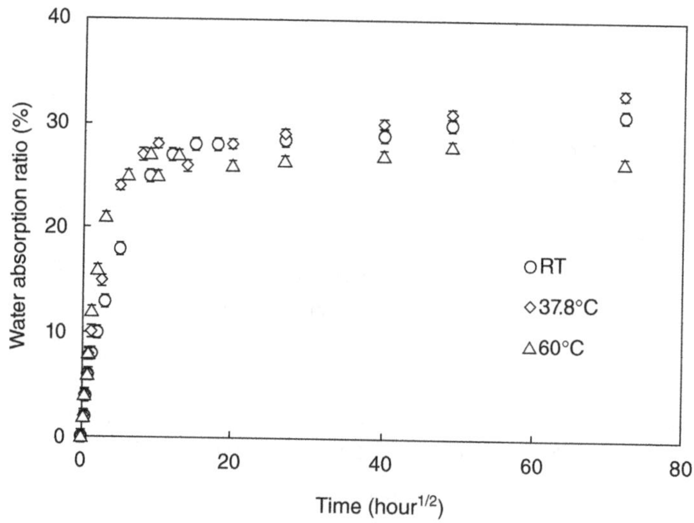

6.31 Water absorption ratio at different temperatures against time curves.

Figure 6.31 shows the percentage of weight gain as a function of the square root of time for sisal fibre-reinforced composites at different temperatures (RT, 37.8°C, 60°C). It can be seen from the curves that water absorption at all studied temperatures increases linearly at the beginning, which fits the Fickian behaviour in equation 6.23. Water molecules first enter into the free space, air bubbles or cracks and defects inside the composite.

$$\frac{M_t}{M_\infty} = kt^n \qquad [6.23]$$

where M_t is the water absorption at time t, M^∞ is the saturated water absorption of the material and k is the diffusion constant of water. When $n \leq 0.5$, the water absorption process is diffusion controlled.

At the middle and later stages, the composite absorbs water gradually, and is saturated after four days, which is much faster than that of glass or carbon fibre-reinforced composites. After reaching the saturated water absorption stage, more water starts to enter into the composite, and the curve no longer fits the Fickian diffusion law. Newly created cracks form caused by the different expansions of the fibre and matrix and/or the leaching of the degraded small molecules. Considering the unique microstructures of sisal fibres, after the water gets into the lumens and gets absorbed by the hydroxyl groups on the sisal fibres, the different expansion of the sisal fibre and the matrix would lead to the residual stress at the

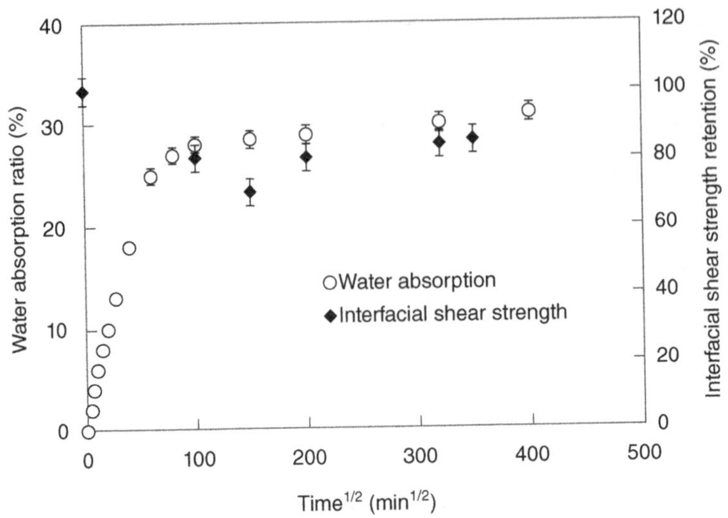

6.32 The moisture absorption curves and shear strength retention curves of sisal fibre-reinforced polymer.

interface, thus leading to the formation of a new crack. The leaching out of the degraded small molecules from both the sisal fibre and the matrix would also lead to new spaces for more water to get in. The effects of moisture absorption on the interlaminar shear strength and tensile properties of sisal fibre reinforced composites are discussed below.

6.9.1 Interlaminar shear strength

The variations of interlaminar shear strength with immersion time measured by the short beam shear test are shown in Fig. 6.32. In order to easily understand the effects of moisture absorption, the weight gain versus immersion time curve is also plotted in the same figure. It can be seen that with increased moisture absorption, the IFSS drops continuously. When the moisture absorption reaches saturation point, the IFSS starts to level off. The debonding between fibre and matrix gradually happens and the interface starts getting poorer and poorer. When the moisture absorption reaches saturation point, cracks between layers form and the IFSS drops to the lowest value.

6.9.2 Tensile properties

Figure 6.33 gives the variations in tensile strength and modulus of sisal fibre-reinforced phenolic composites with the increase in immersion time.

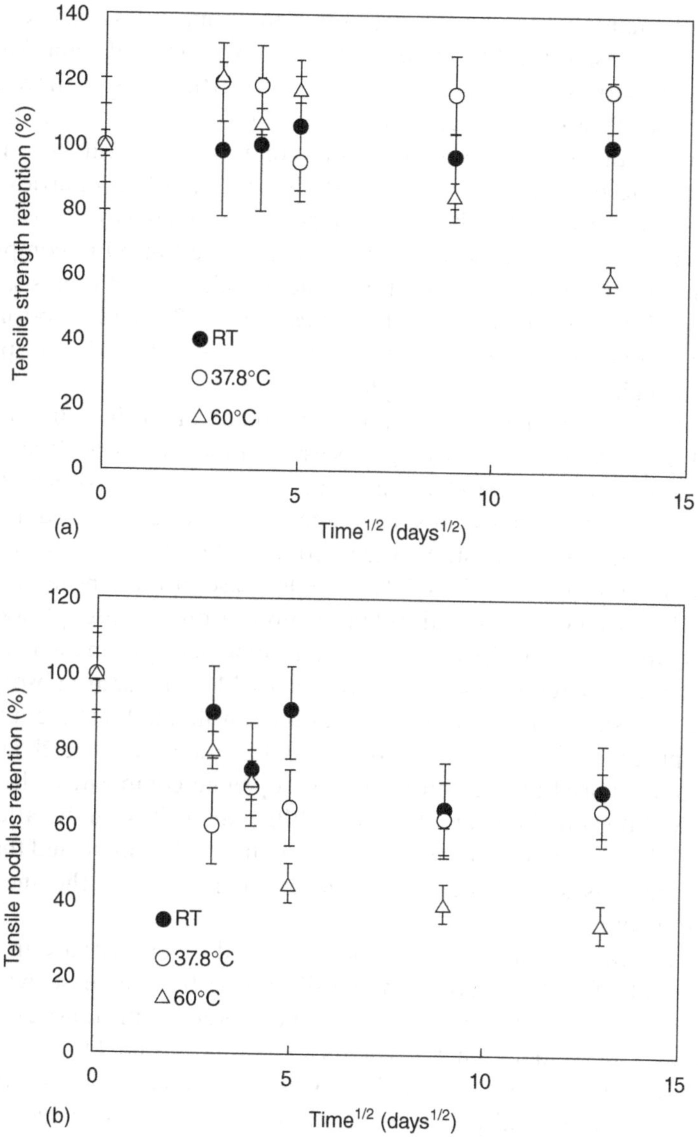

6.33 Variations of (a) tensile strength and (b) tensile modulus with the exposure time of sisal fibre-reinforced composites under hydrothermal ageing.

It can be found that the tensile strength of the composites does not change too much considering the large scatters of experimental data. Since the tensile properties are dominated by the properties of sisal fibres, if the properties of sisal fibres do not change with the increase in moisture absorption, the tensile strength of the composites should not show an obvious difference. From a previous study,[31] the tensile properties of sisal fibres do not decease but show a small increase instead. However, since the degradation of the composites is more accelerated at 60°C compared to that at the other two temperatures, the tensile strength of sisal fibre-reinforced composites reduces dramatically after 30 days' exposure. The severe debonding between sisal fibre and matrix, as discussed in the previous section, might contribute to this phenomenon.

The tensile modulus of sisal fibre-reinforced composites drops quickly from the beginning of the ageing process, and then starts to decease more slowly after a certain period of time, which varies for immersion at different temperatures. At higher exposure temperature, the tensile modulus of the composites drops the most. At a temperature of 60°C the reduction of the tensile modulus is more than 50%. The decrease of the tensile modulus is caused by the increased ductility of the composite due to water plasticisation. The failure tensile strain value for water-immersed specimens is found to be higher compared to that of dry specimens. More moisture absorption at higher exposure temperature decreases the tensile modulus the most.

Joseph *et al.*[32] studied the environmental effects on the degradation behaviour of sisal fibre-reinforced polypropylene composites. They found that the water uptake properties of sisal/PP composites can be affected by fibre loading, temperature and chemical treatment. It is indicated that water uptake increases with increase of fibre loading due to the increase in cellulose content.

The increase in temperature is accompanied by an increase in the rate and extent of water absorption. Tensile properties decrease with water uptake, time of immersion and fibre loading. Moreover, the tensile behaviour is strongly dependent on the chemical treatment and fibre orientation. Similar conclusions are obtained by Athijayamani *et al.*[33] on short roselle and sisal fibres hybrid polyester composites. They also conclude that the tensile and flexural strength show a high-level reduction with an increase in fibre length in the water absorption state compared to those in the dry state, as shown in Fig. 6.34.

6.10 Effects of ultraviolet (UV) light on the mechanical properties of sisal fibre-reinforced composites

The accelerated UV ageing test on sisal fibre-reinforced composite is conducted according to GB/T 16422.3. A UV-B313 artificial fluorescent UV

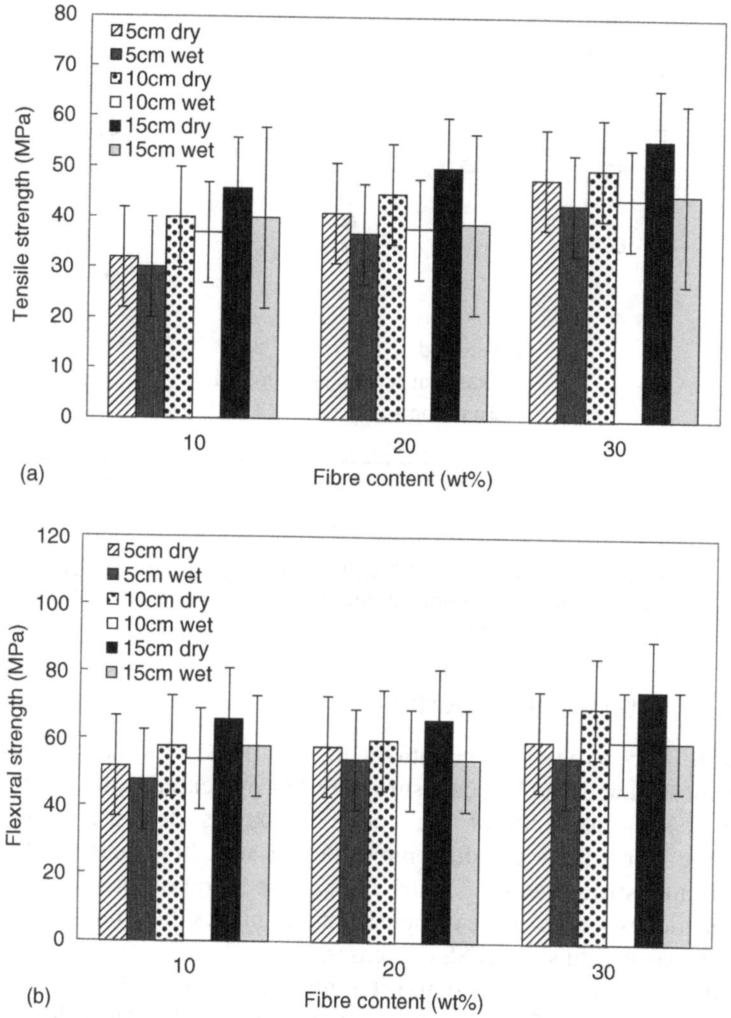

6.34 Effect of fibre content on (a) tensile strength and (b) flexural strength at dry and wet conditions. Reproduced with permission from Elsevier from Ref. [33].

lamp is used as the light source. The specimens are aged for 12 hours as a cycle for 28 days. In each cycle, the specimens are irradiated at 0.6 W/m^2 at 60 ± 1°C for 8 hours to simulate degradation from the sunlight, and then exposed without irradiation at 50 ± 1°C for 4 hours to imitate the dew from the atmospheric environment. The heating tube at the bottom of the chamber evaporates water. When steam touches the cold surface of the specimens, it condenses into water to simulate the dew. Effects of fibre surface treatments on the UV degradation performance are investigated.

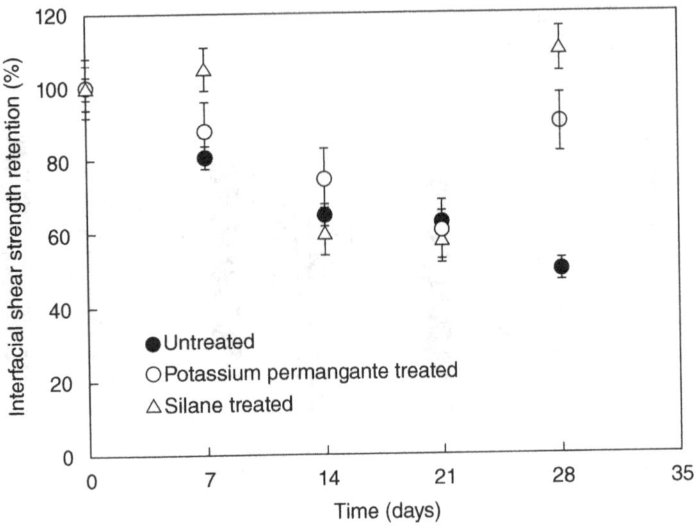

6.35 The changes of ILSS with the UV aging time of sisal fibre-reinforced phenolic composites before and after fibre surface treatments.

6.10.1 Interfacial properties

Figure 6.35 shows the variations in the ILSS of sisal fibre-reinforced composites with the UV exposure time. For untreated sisal fibre-reinforced composite, the ILSS of the composite keeps decreasing due to the degradation of the cellulose molecules of sisal fibre and the long chain molecular structure of the matrix. The degradation of these polymers leads to brittleness of the entire composite, which makes the interfacial crack easy to propagate, thus the IFSS decreases.

Morphologies of the composites after undergoing UV irradiation for different times (0, 7, 14 and 28 days) are compared in Fig. 6.36. The colour of the composites becomes darker with an increasing UV exposure time, which might be caused by the degradation of the chromophoric groups of the matrix through absorbing UV irradiation.

Figure 6.37 compares specimens of sisal fibre-reinforced composites being fractured by the short beam shear test after being exposed to UV for different times. More obvious crack propagation along the layers of the composites in the specimens exposed to UV for 28 days can be observed.

However, surface treatments can change the UV ageing performance of sisal fibre-reinforced composites as shown in Fig. 6.35. It can be seen that both silane treatment and permanganate treatment increase the IFSS of the composites after 28 days' exposure to UV following the initial decease, with silane treatment showing the highest IFSS. The main chemical

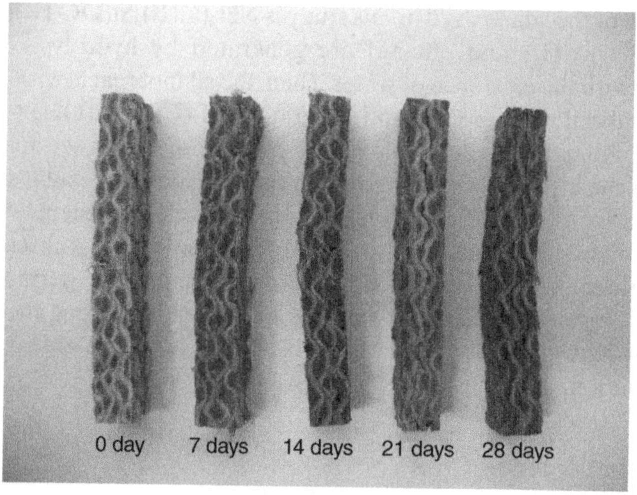

6.36 Specimens of sisal fibre-reinforced phenolic composites after UV exposure for 0, 7, 14, 21 and 28 days.

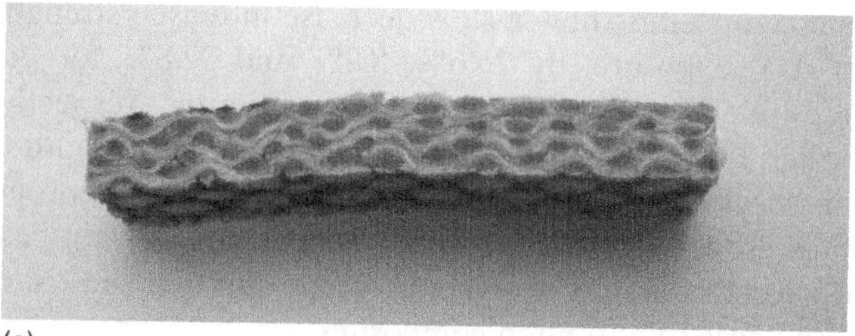

(a)

(b)

6.37 Fractured specimens by shear beam shear test for composites exposed to UV for (a) 0 day and (b) 28 days.

composition of the silane used in this study is $NH_2(CH_2)_3Si(OC_2H_5)_3$. Silanol ($(HO)_3Si(CH_2)_3NH_2$) and ethanol are generated by hydrolysis reaction from silane with the existence of water. Then, the silanol reacts with hydroxy from the sisal cell wall (G-OH) and generates $NH_2(CH_2)_3Si(OH)_2OG$. Thus chemical bonding between the sisal fibre and the matrix is set up. Though the UV can cause the degradation of both the sisal fibre molecular structure and the matrix molecular structure, there is no clear evidence that the chemical bonding between the sisal fibre and the matrix caused by the silane treatment can be degraded. In terms of potassium permanganate treatment, the rough surface of sisal fibre would lead to strong mechanical interlocking between sisal fibre and matrix which would also not be affected by UV radiation.

6.10.2 Tensile properties

Figure 6.38 shows the variations of tensile strength and modulus of sisal fibre-reinforced composites with increased UV exposure time. Effects of different fibre surface treatments are also shown. It can be seen that all three kinds of composites show a slow decrease in tensile strength during the 28 days' UV exposure, by 26.6%, 8.8% and 32.8% for untreated, permanganate-treated and silane-treated composites respectively. As discussed earlier, UV irradiation causes the degradation of both the fibre and matrix molecules. Therefore, the fibre-dominated tensile properties of the composites drop, but not too much. Since silane does not change the fibres, silane-treated sisal fibre-reinforced composites thus show similar UV ageing performances to untreated composites.

However, permanganate roughens the sisal fibre by getting rid of the lignin, which would degrade under UV irradiation. This might be the reason why the composites with permanganate treatment show the smallest decrease in tensile strength. The tensile modulus of sisal fibre-reinforced composites drops dramatically after 28 days' UV ageing. The maximum decrease is as high as 60%. This degradation of both sisal fibre and matrix causes a decrease of the degree of polymerisation and the relative molecular weight.

Joseph *et al.*[32] also obtained similar results, namely that the tensile properties of both untreated and chemically treated sisal fibre-reinforced PP composites decrease after exposure to UV radiation due to chain scission as a result of photo-oxidation. Chemically treated composites show a decrease in tensile strength with increased exposure time. SEM images show a badly damaged surface appearance of chemically treated sisal fibre-reinforced PP composites. The fibres lying near the surface of the specimens become exposed upon UV irradiation. It is observed that chemically treated composites undergo severe degradation compared to the untreated composites.

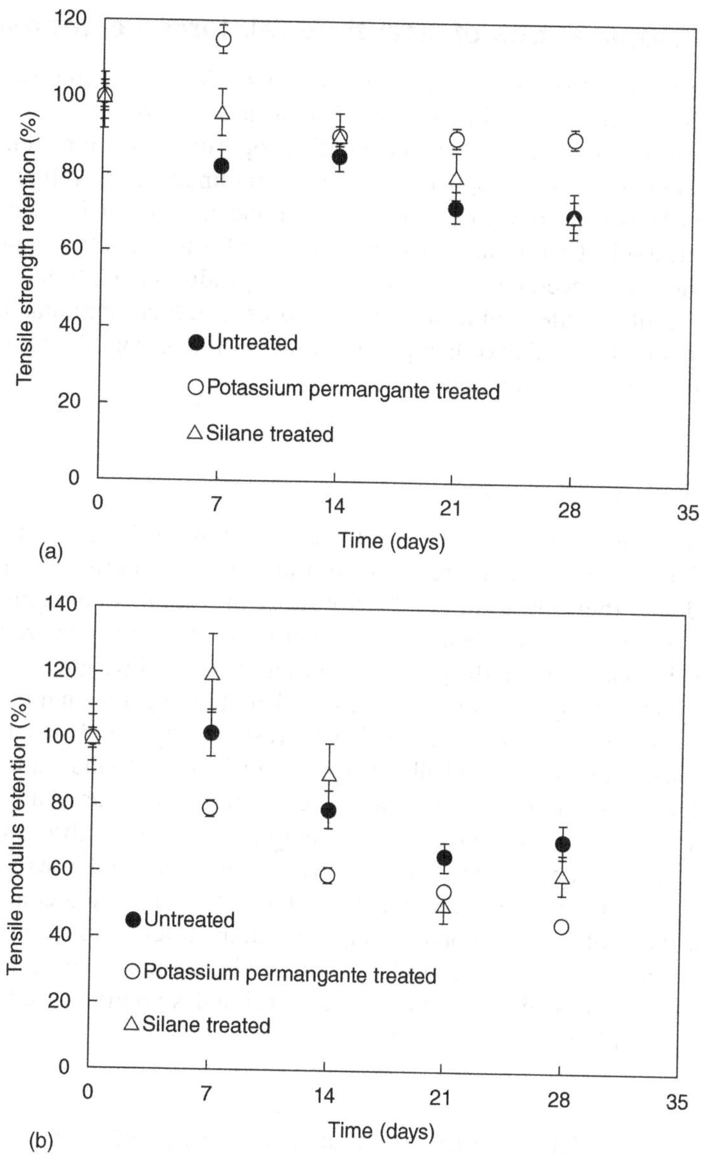

6.38 (a) Tensile strength and (b) tensile modulus of sisal fibre-reinforced composites under UV radiation.

6.11 Applications of sisal fibre-reinforced composites

From a series of studies on sisal fibre and its reinforced composites, it is clear they have potential as structural materials as a result of their good mechanical, environmental and economic properties. Sisal fibre-reinforced composites have been applied in the areas outlined in the following few sections. However, they might have wider applications in engineering in the future. The selection of the type of natural fibre for manufacturing products is strongly influenced by their availability. The products from India and Asia mainly contain jute, ramie and kenaf fibres. Products manufactured in Europe tend to use flax or hemp fibre, while products from South America tend to use sisal, curaua and ramie.

6.11.1 Automobile industry

In recent years, natural fibre-reinforced composites have attracted the attention of car manufacturers for both automobile interior and exterior parts due to their low cost, low density, superior sound and heat insulation, and environmentally friendly characteristics. Natural fibre-reinforced composites can achieve the goal of lowering the overall weight of a vehicle, thereby increasing fuel efficiency, as well as making the manufacturing process more sustainable. According to research by the Nova Institute, about 4000 tons of natural fibres were used in the German automotive industry in 1996. In Europe, many companies such as Audi, BMW, DaimlerChrysler, etc., have already employed natural fibre-reinforced composites in vehicles, and are still trying to expand their uses.[34–37] Since 1994, DaimlerChrysler, Audi and BMW have started using a combination of flax and sisal fibre-reinforced composites in the interior trim components for their automobiles. This results in extremely low mass per unit volume, and the products also exhibit high dimensional stability, including high strength and impact resistance.[38, 39]

6.11.2 Construction industry and civil engineering

Natural fibre has long been used in building and civil applications. In recent years, the mechanical properties, cost and quality of plant fibres have been significantly improved with the development of technology. The construction industry constitutes the second largest sector to employing natural materials in a range of products including light structural walls, insulation materials, floor and wall coverings, geotextiles and thatch roofing.[39] Sisal fibre-reinforced cement products such as roof tiles and building blocks are being produced in countries such as Tanzania and Brazil.

6.11.3 Other fields

A range of products can be manufactured due to sisal fibres being environmentally friendly, cheap and a renewable resource. Appliances such as computer casings can be manufactured with injection moulding techniques. Leisure products such as skis, golf clubs and butt stocks can also be manufactured using sisal fibre reinforced composites.

6.12 Conclusion and future trends

The predominant chemical make-up of sisal fibre, as a kind of natural fibre, is cellulose and hemicellulose. It has a multi-scaled structure, which makes it a composite material. Therefore, theoretical models can be set up to calculate the mechanical properties of sisal fibre-reinforced composites, which are quite consistent with experimental results.

The unique chemical structure and microstructure of sisal fibre lead to unique mechanical properties of sisal fibre-reinforced composites. Fibre twist is a must for sisal fibre-reinforced composites and is proved to have a great influence on the mechanical properties of the composites. Non-twist sisal fibre-reinforced composites show comparable mechanical properties to glass fibre-reinforced composites. Fibre twisting is also the reason for the nonlinear behaviour of sisal fibre-reinforced composites.

The interfacial and tensile properties of sisal fibre-reinforced composites decrease with hydrothermal and UV ageing. Fibre surface treatments could change the ageing behaviour of sisal fibre-reinforced composites.

Research into sisal fibre and its reinforced composites started about 25 years ago. Many research outcomes have been achieved, especially on the interfacial bonding properties of sisal fibre-reinforced composites. However, the real application for such composites is still quite limited due to an incomplete understanding of some mechanisms of both the mechanical and physical performances of sisal fibre and its composites. Some future research trends are proposed.

- The properties of natural fibres vary depending on the sources, environment, processing, etc. It is necessary to control the properties of natural fibres from growing and production so that the consistency of the properties can be guaranteed.
- Natural fibres are discontinuous. Twist is essential for forming the yarns and fabrics. Therefore, proper twist and non-twist natural fibre prepreg will be an important research direction to fully fulfil the mechanical properties of natural fibres.
- The thermal stability of natural fibres is poor. A suitable manufacturing technique for making natural fibre-reinforced composites is also a key point.

- Further improvement of the mechanical properties of sisal fibre-reinforced composites by the optimisation of the manufacturing and structural design is crucial for its application.
- Fully biodegradable sisal fibre-reinforced composites will be an important direction. However, the cost of the biodegradable polymers, repeatability and recyclable properties are key factors for this development.

6.13 Acknowledgements

The authors would like to thank the support of the National Basic Research Program ('973' program) (Grant No. 2010CB631105) and the National Natural Science Foundation (Project. No. 11172212).

6.14 References

1. Chand, N., Tiwary, R.K. and Rohatgi, P.K., Bibliography resource structure properties of natural cellulosic fibres: an annotated bibliography, *Journal of Materials Science*, 1998, 23, 381–387.
2. Pan, Y.R. and Lv, Q., *The Production and Processing of Sisal Fibre* (in Chinese), Guangxi Scientific and Technical Publishers, 2007, 80–88.
3. Murherjee, P.S. and Satyanarayana, K.G., Structure and properties of some vegetable fibres, part 1. Sisal fibre, *Journal of Materials Science*, 1984, 19, 3925–3934.
4. Bisanda, E.T.N. and Ansell, M.P., Properties of sisal-CNSL composites, *Journal of Materials Science*, 1992, 27, 1690–1700.
5. Wilson, P.I., *Sisal*, Vol. II. In *Hard Fibres Research Series*, No. 8, Rome: FAO, 1971.
6. Rowell, R.M., Schultz, T.P. and Narayan, R., Emerging technologies for materials and chemicals for biomass, *American Chemical Society*, 1991, 476, 469–481.
7. Joseph, K., Thomas, S. and Pavithran, C., Effect of chemical treatment on the tensile properties of short sisal fibre-reinforced polyethylene composites, *Polymer*, 1996, 37, 5139–5149.
8. del Rio, J.C., Martinez, A.T. and Gutierrez, A., Presence of 5-hydroxyguaiacyl units as native lignin constituents in plants as seen by Py-GC/MS, *Journal of Analytical and Applied Pyrolysis*, 2007, 79, 33–38.
9. Li, Y., Mai, Y.W. and Ye, L., Sisal fibre and its composites: a review of recent developments, *Composites Science and Technology*, 2000, 60, 2037–2055.
10. Li, H., Zadorecki, P. and Flodin, P., Cellulose fibre–polyester composites with reduced water sensitivity (1) – chemical treatment and mechanical properties, *Polymer Composite*, 1987, 8, 199–202.
11. Yin, Y. and Yin, J., Geometrical conservation law for cells or vesicles with membrane nanotubes or singular points, *Journal of Nanobiotechnology*, 2006, 4, 6–11.
12. Yin, Y., Symmetry in the mechanics and geometry of biomembrane nanotube networks (in Chinese), *Mechanical Engineering*, 2008, 30, 1–10.

13. Yin, Y., Chen, Y., Yin, J. and Huang, K.Z., Geometric conservation laws for perfect Y-branched carbon nanotubes, *Nanotechnology*, 2006, 17, 4941–4945.
14. Yin, Y., Zhang, T., Yang, F. and Qiu, X.M., Geometric conditions for fractal super carbon nanotubes with strict self-similarities, *Chaos, Solitons Fractals*, 2008, 37, 1257–1266.
15. Fan, F., *Micromechanics of Composite Materials* (in Chinese), Science Publications, China, 1998.
16. Lu, B., Zhang, L. and Zeng, J., *Natural Fibre Reinforced Composites* (in Chinese), Chemistry and Industry Publications, China, 2005.
17. Fu, S.Y., Lauke, B. and Mai, Y.W., *Science and Engineering of Short Fibre Reinforced Polymer Composites*, Woodhead Publishing Limited, Cambridge 2009.
18. Sreekumar, P.A., Joseph, K., Unnikrishnan, G. and Thomas, S.A., Comparative study on mechanical property of sisal leaf fibre reinforced polyester composites prepared by resin transfer and compression moulding techniques, *Composite Science Technology*, 2007, 67(3–4), 453–461.
19. Joseph, K., Varghese, S., Kalaprasad, G., Thomas, S., Prasannakumari, L. and Koshy, P., Influence of interfacial adhesion on the mechanical properties and fracture behaviour of short sisal fibre reinforced polymer composites. *European Polymer Journal*, 1996, 32(10), 1243–1250.
20. Sangthong, S., Pongprayoon, T. and Yanumet, N. Mechanical property improvement of unsaturated polyester composite reinforced with admicellar-treated sisal fibres. *Composites Part A: Applied Science and Manufacturing*, 2009, 40, 687–94.
21. Meddahi, A., AitTahar, K. and Bibi, M., Studies of sisal fibre-containing composites, *Journal of Natural Fibres*, 2008, 5, 36–46.
22. Zhou, L.M., Kim, J.K. and Mai, Y.W., On the single fibre pull-out problem: effect of loading method, *Composites Science and Technology*, 1992, 45, 153–160.
23. Zhou, L.M., Kim, J.K. and Mai, Y.W., Micromechanical characterization of fibre/matrix interfaces, *Composites Science and Technology*, 1993, 48, 227–236.
24. Joseph, P.V., Joseph, K. and Thomas, S., Effect of processing variables on the mechanical properties of sisal-fibre-reinforced polypropylene composites, *Composites Science and Technology*, 1999, 59, 1625–1640.
25. Li, X., Tabil, L.G. and Panigrahi, S., Chemical treatments of natural fibre for use in natural fibre reinforced composites: a review, *Journal of Polymers and the Environment*, 2007, 15(1), 25–33.
26. St John, N.A. and Brown, J.R., Flexural and interlaminar shear properties of glass-reinforced phenolic composites, *Composites Part A*, 1998, 29, 939–946.
27. Goutianos, S., Peijs, T., Nystrom, B. and Skrifvars, M., Development of flax fibre based textile reinforcements for composite applications, *Applied Composite Materials*, 2006, 1, 199–215.
28. Ma, H., Li, Y. and Luo, Y., The effect of fibre twist on the mechanical properties of natural fibre reinforced composites, *18th International Conference on Composite Materials*, Jeju island, Korea, August 2011.
29. Thwe, M.M. and Liao, K., Effects of environmental ageing on the mechanical properties of bamboo-glass fiber reinforced polymer matrix hybrid composites, *Composites Part A*, 2002, 33, 43–52.

30. Yang, G.C., Zeng, H.M., Li, J.J., Jian, N.B. and Zhang, W.B., Relation of modification and tensile properties of sisal fiber, *Acta Science National University Sunyatseni*, 1996, 35, 53–57.
31. Hu, Y.P., Microstructures and Mechanical Properties of Natural Fibres, Bachelor thesis, Tongji University, China, 2008.
32. Joseph, P.V., Rabello, M.S., Mattoso, L.H.C., Joseph, K. and Thomas, S., Environmental effects on the degradation behaviour of sisal fibre reinforced polypropylene composites, *Composites Science and Technology*, 2002, 62, 1357–1372.
33. Athijayamani, A., Thiruchitrambalam, M., Natarajan, U. and Pazhanivel, B. Effect of moisture absorption on the mechanical properties of randomly oriented natural fibres/polyester hybrid composite, *Materials Science and Engineering A*, 2009, 517, 344–353.
34. Faruk, O., Bledzki, A.K., Fink, H.P. and Sain, M., Biocomposites reinforced with natural fibres: 2000–2010, *Progress in Polymer Science*, 2012, 37, 1552–1596.
35. Gao, Y.C., Mai, Y.W. and Cotterell, B., Fracture of fibre reinforced materials, *Journal of Applied Mathematics and Physics*, 1988, 39, 550–572.
36. Holbery, J. and Houston, D., Natural-Fiber-Reinforced Polymer Composites in Automotive Applications, *JOM*, 2006, 11, 80–86.
37. Carus, M., Kaup, M. and Ortmann, S., Use of natural fibres in the German and Austrian automotive industry. Market survey 2002: Status, analysis and trends, Nova-Institute GmbH, 2002.
38. Carus, M., Kaup, M. and Lohmeyer, D., Study on markets and prices for natural fibres (Germany and EU), Nova-Institute GmbH, 2000.
39. Sudden, B.C., Industrial fibres: recent and current developments, *Proceedings of the Symposium on Natural Fibres*, Rome, 2008, 71–82.

7
The use of pineapple leaf fibers (PALFs) as reinforcements in composites

A. L. LEÃO, São Paulo State University (UNESP), Brazil,
B. M. CHERIAN and S. NARINE, Trent University, Canada,
S. F. SOUZA and M. SAIN, University of Toronto, Canada
and S. THOMAS, Mahatma Gandhi University, India

DOI: 10.1533/9781782421276.2.211

Abstract: Pineapple leaf fibers (PALFs) can be utilized as fabrics for textile materials and in the manufacture of yarns and handicrafts in many countries. The excellent mechanical properties and environmentally sustainable characteristics exhibited by PALFs have triggered the interest of researchers to use the material as a potential reinforcement in structural and non-structural applications. This chapter reviews the properties and applications of PALFs and their composites, and focuses on PALFs extraction, characterization, modification, fabrication and the properties of PALFs reinforced micro- and nanocomposites.

Key words: PALFs, reinforcement, composites, mechanical strength, chemical treatments.

7.1 Introduction

There is a global trend towards seeking alternative natural resources that may benefit mankind, creating new products, in addition to creating jobs and generating income through the development of new technologies. Brazil is no exception of this principle.

Bleached softwood pulp is required for the production of white paper and packaging that requires high mechanical strength. There is a significant deficit of bleached softwood pulp in the Brazilian market, with just one single producer of this type of fiber, operating at a capacity of 200 tonnes per day. Much of the national demand for bleached long fiber is supplied by imports from Argentina, Chile and Canada. Limited production of bleached softwood in Brazil comes from the limited availability of softwood (pine, *Araucaria*, bamboo, etc.) and its high production cost in comparison to, for example, bleached hardwood eucalyptus. Therefore, alternative sources of long fibers are encouraged.

One proposed alternative is the use of pineapple leaves as post-harvest waste, since this material has a significant amount of fiber, and is currently discarded. In Brazil, up to 40% of industrial products are wasted (Manica,

1999). The pineapple leaf is currently being wasted in most regions that grow this plant, with only the fruit itself being utilised. In recent decades, pineapple has become a crop of great importance for the north-east of Brazil, with respect to the regional economy (IBGE, 2004). However, after harvesting, the fruit's leaves (rich in fiber) are placed for pasture for cattle, or in most cases left to rot. This is due to lack of technology suitable for fiber utilization, as well as growers' ignorance of the possible uses of fibers, which can generate them extra income.

The leaves of the pineapple plant, when properly extracted and processed, provide long vegetable fibers, which can be used in the manufacture of yarn, woven fabrics, woven knitted, non-woven mats and handmade products. After extraction, the resulting mucilage can also generate numerous other products, including pharmaceuticals and animal feed.

The fibers derived from pineapple leaves are long, with one of the best fineness indexes among vegetal fibers, which make them suitable for many industrial applications. It is estimated by Leao *et al.* (2010) that production of pineapple fibers per hectare/year yields around 15 tonnes.

The fibers can be considered to produce a negative cost, as they are currently dumped for composting. If this material is transformed into something more valuable, this negative cost can be translated into income for growers. Therefore this raw material results in a very low cost of US$15 per tonne in the field, compared to softwood fibers at US$250 per tonne in the field, representing an excellent opportunity. Pineapple leaf fibers (PALFs) could be a viable solution to the deficiency of long fibers in Brazil, and in many other tropical countries.

Brazil is the world's third largest producer of pineapple plant residue, with Thailand and the Philippines as the first and second, respectively. The total planted area in Brazil is about 60,000 hectares, with each hectare resulting in 1.5 tonnes of dry fiber, according to data obtained from Leao *et al.* (2010) for the Hawaiian variety. Through this raw material it will be possible to identify the crop production chain and bring a new and important resource for growers.

Because there are new vegetable fibers being brought to market, there is a need to study in-depth the chemical and morphological characteristics of PALFs and evaluate in detail the optical, surface, and mechanical properties of these fibers. Through these characterizations, it will be possible to make a precise assessment of the market for special papers and composites applications, adding more value to these long fibers. The nanofibers and nanowhiskers also represent an emerging alternative, thanks to the composition of the pineapple leaf (low lignin content), its high fineness index, and high aspect ratio.

The pineapple is a tropical plant native to warm climates with dry or irregular rainfall. It originates in the region between 15° to 30° south latitude and 40° to 60° west longitude, which includes the central and southern areas of Brazil's coast, north-eastern Argentina and Paraguay. From Brazil and Paraguay, homeland of the Guarani natives, came the most primitive forms of pineapple. The Guarani natives spread this plant and its name in the successive inroads they made in Brazil, Guyana, Venezuela, Ecuador, Colombia, Panama and the West Indies, according to Cunha *et al.* (1994). The pineapple was found by Christopher Columbus on the island of Guadalupe in 1493 and then later in other islands of the West Indies, and was taken to Europe where it became greatly appreciated. In the New World, pineapple was offered to the European invaders as a gesture of hospitality and welcome. For its resemblance to the fruit of the pine in Europe, the fruit was called *Piña*, as it is still known today. In 1557, a priest used the name 'dwarf' for pineapple, derived from the Tupi Guarani Indian name 'nana'. In Portuguese the name 'abacaxi' (pineapple) comes from the union of two words from the Tupi language: iba (fruit) and cati (smell).

The spread of pineapples across the world was due to Hispanic–Portuguese maritime expansion during the sixteenth century, when navigators took it on board for their own use, and crowns left in ports naturally propagated (Margarido, 1991). The Spaniards introduced the pineapple into the Philippines, Hawaii and Guam in the sixteenth century. The Portuguese introduced it in India and the west coast of Africa. The plant arrived in China in 1594 and South Africa in 1655. The pineapple reached Europe in 1650, the Netherlands in 1686, and England and France in 1700. At the end of the seventeenth century the pineapple was already dispersed in most regions of the world where the climate was favorable to its development. Lutheran missionaries introduced the pineapple to Australia in 1838 and to Malaysia in 1888. From 1900 to 1919, the pineapple was marketed on an industrial scale in Hawaii, Taiwan, the Philippines, Malaysia, Australia, South Africa and the Caribbean. The pineapple trade industry took shape after 1924 and developed further after the Second World War, with commerce in many sea and land routes, according to Cunha *et al.* (1994).

7.2 The pineapple plant

The pineapple belongs to the order Bromeliales, family Bromeliaceae, subclass monocotyledons, genera *Ananas* and *Pseudoananas*, and species *Ananas comosus* (L.) Merrill which is now called *Ananas comosus* var. *comosus* and varieties such as Pearl Cayenne, Queen and Smooth Cayenne. For a better understanding, all of the world's pineapple cultivars are grouped in one of four main classes, according to Purdue University. The cultivars share some semblance of common lineage. Smooth Cayenne, Red Spanish,

7.1 Pineapple (*Ananas comosus*) mature plant.

Queen and Abacaxi are the four umbrella classes for all pineapples. In 1917, Merrill established the binomial *Ananas comosus*. Currently, the flora includes 46 genera, about 1700 species of herbaceous plants and 11 hybrids of bromeliads. A pineapple plant with a detailed view of the leaves is shown in Fig. 7.1.

The pineapple is an herbaceous perennial monocotyledonous plant. After producing fruit, its development continues through its base, growing for the following year, producing a new fruit. The plant needs 200 days or 6–9 months on average to bear fruit and is propagated vegetatively by cuttings grown from the plant itself. The pineapple consists of a short and thick stem (stalk), in which leaves grow in narrow, rigid, and axillary roots. The system fasciculated root is superficial and fibrous, usually found at a depth of 15 to 30 cm beneath the soil surface. How the leaf falls from the pineapple stem determines the type of this plant: erect, semi-erect and prone to horizontal. The mature plant reaches 1 to 1.2 meters in height and a width of 0.80 to 1.50 meters and consists of:

- Primary (seed embryo), secondary (lateral branches) and adventitious roots, developed from the stems (axillary type)
- Stem or stalk, 25 to 30 cm in length and 2.5 to 3.5 cm wide where there are inserted organs
- Leaves, which fully involve the stem and shoots. Adult plants hold up to 80 leaves which vary in shape, with lengths of up to 1.3 meters. They are rigid with waxes in the upper and lower surfaces, which reduces perspiration to a minimum. At the base of the stem, older leaves are located, with a spear format. The younger leaves are elongated, conserving a wider baseline.

The most metabolically active and mature leaves are those at an angle of 45° with the floor level and with the axis of the plant. Leaves develop from the stem in a spiral shape:

- The peduncle, prolongation of the stem (lower part of the stem), which holds the inflorescence and fruit
- Saplings and seedlings that develop from axillary buds located on the stem in peduncle below the fruit
- Flowers (about 100 to 200) arranged around the single axis
- Multiple fruit (the fruit itself is not a single one, but a composite fruit) resulting from the development of the inflorescence
- Crown foliaceus or small stem which is formed on top of the fruit.

The younger leaves are elongated, conserving a wider basal portion. Plants have leaves in ranks ranging in size from about 14 rows, each containing an average of five leaves. Among the varieties, the most widely cultivated is the Smooth Cayenne due to its industrial characteristics (Margarido, 1991).

7.3 Pineapple production

The major world producers are Thailand, the Philippines, China, Brazil, Indonesia and India, of which Brazil ranks fourth, producing over 700,000 tonnes of fruit annually. Although planting is widespread throughout the Brazilian territory, the cultivation of pineapple is primarily concentrated in the north, north-east and south-east regions. Currently three varieties are grown in Brazil: the Pearl, the Jupy and the Cayenne. The fruits of these varieties are used for 'natural' consumption, and are industrialized for the production of juice and fruit-shaped slices for canning. While maintaining a cultivation area much larger than the other producing countries, Brazil has not yet developed fully efficient cultivation techniques that allow the same high yields as are obtained in Asian pineapple plantations. Basically, the most cultivated varieties in Brazil are those of low acidity, the Pearl, marketed for raw consumption, and Smooth Cayenne or Hawaiian, which produces a larger, more acid-resistant fruit which, therefore, is normally destined for export and the jam and juice industries.

Pineapple in Brazil is well accepted in horticulture, not only by its quality but also by its profitability, being the fifth most harvested tropical fruit in the world and the third in cultivated areas in Brazil.

7.4 Pineapple culture in Brazil and worldwide

The pineapple industry is an important source of income for tropical countries. Since 1830 the commercial exploitation of pineapple has expanded and has gained in economic importance for the producer countries.

The economic potential of fibers from pineapple leaf should be evaluated as an extra source of income, as they can be used for industrial applications, mainly pulp and paper, handicraft products, textiles and composites, with an economic and social impact, generating jobs and income. Furthermore, PALFs can be used as reinforcement in polymeric composites, being a renewable source of natural raw material that could replace synthetic fibers (Alexandre *et al.*, 2004; Leão *et al.*, 2010).

7.5 Fiber extraction

Pineapple leaf fibers can be extracted either manually or mechanically. The first method involves combining layered fibers in water (about 18 days to become saturated) before they are manually scraped. After this process, the fibers are washed and naturally dried. The extraction process of the plant fibers is of great importance to the quality as well as the yield of the fibers. The fiber extraction of pineapple leaf in the Philippines is undertaken by the traditional method of scraping a small knife or a piece of broken porcelain throughout the leaf – a painstaking and skillful process.

The manual process begins with shredding through beating, scraping and husking the leaves, as shown in Fig. 7.2. After this process the fibers are soaked in water. Chemicals may be added to accelerate the activity of microorganisms that digest the unwanted materials and separate the fibers. Using this process, the wetting time is reduced from five days to 26 hours. The wetting material is washed and cleaned, dried in the sun and combed or carded. The mechanical process is carried out in a decorticator machine

7.2 Manual shredding of pineapple leaf (source: http://site.thegreenlifeonline.org/category/bright-green-ideas/).

with the same principles of sisal fiber extraction. On average, about 22 units of pineapple leaf weigh a kilogram. The reported fiber yield is about 2.7 to 3.5% of fibers (Alexandre, 2005; Leão *et al.*, 2010).

In India, both methods are used in the extraction of the fibers. In the Philippines, the manufacturers use a more efficient process utilizing microbiological attack, which results in better-looking fibers and good strength (Mukherjee and Satyanarayana, 1986).

Doraiswamy and Chellamani (1993) reported that, in India, the extraction of PALFs is made by machine or the so-called decorticator. The extraction of 100kg of leaves yielded 3.5 to 4kg of the PALF by the process of shredding. The refining process consists of scraping the leaf (removal of bark), after which it is manually entered on a rotor beater (coupled to a motor), with blind scraping blades, with an output of about 16kg/day of fibers, which are washed and transported to other centers. In these centers the fibers are beaten, brushed, sorted and baled. The values are similar to those obtained in LABTEX (Alexandre, 2005).

Among the various types of natural fibers, PALFs exhibits excellent mechanical properties. The superior mechanical properties of PALFs are associated with their high cellulose content and low microfibril angle. The fiber properties vary according to their dimensional appearance, length and thickness or diameter, planting conditions, age, type of leaves, and so on. The characteristics and properties of the fiber in the nature of PALFs are shown in Table 7.1. The properties of the cell structure, the unit fibers and the PALF strands are shown in Table 7.2.

Table 7.1 Characteristics and pineapple leaf fiber properties

Characteristics	Properties
Length	Technical fiber: 50 to 120cm Individual fiber: 2.5 to 4.5mm (ultimate fiber)
Diameter	105μm
Purity	Great
Color	White-cream
Brightness	Glossy surface
Fiber surface	Smooth
Touch	Soft (soft fiber)
Combustion test	Yellow flame, quick, smell of burned paper
Acid behavior	Very sensitive
Density	1.44g/cm^3
Hygroscopicity	11%
Dyeability	Great: light colors and bright

Source: Reprinted from Satyanarayana, K. G., Sukumaran, K., Mukherjee, P. S. and Pillai, S. (1986) Materials science of some lignocellulosic fibers, *Metallography*, 19, 389–400,

Table 7.2 Properties of the structure of the pineapple leaf fiber: cell, elemental fiber and basic fiber

Cell	Length *L* (mm)	3–9
	Width *B* (μm)	4–8
	Aspect ratio (*L/D*)	450
Elementary fiber	Linear density (denier)	14
	Tenacity (mN/m^2)	710
	Elongation at break (%)	2–6
	Torsional stiffness (mN/m^2)	360
	Bending stiffness (mN/ m^2)	3–8
	Transverse expansion in water (%)	18–20
Basic fiber	Tenacity (mN/m^2)	370
	True density (kg/m^3)	1480
	Apparent density (kg/m^3)	1350
	Porosity (%)	9.0
	Moisture retention (%) to 65% RH	11.8
	Moisture retention (%) of 100% RH	41.0

Source: With kind permission from Springer Science, Business Media and the original publisher (Mukherjee, P. S. and Satyanarayana, K. G. (1986) Structure and properties of some vegetable fibres. Part 2 pineapple fibre (*Anannus comosus*), *Journal of Material Science*, 21, 51–56, Original copyright notice.

PALFs are used in India for the manufacture of yarns and fabrics for textile and handicraft products, as shown in Fig. 7.3. In the Philippines, these fibers are used in linings for shorts, mats, rugs, bags, blankets, thermal and acoustic insulating material (Mukherjee and Satyanarayana, 1986).

7.6 Potential of fiber production plant

Brazil is a country of great natural resources, with an extensive arable area. However, there are economically deprived areas with socio-economic problems. In addition to under-utilization of available natural resources in the form of natural fibers, by-products are also wasted and misused, which could be used for the creation of new jobs and income, preventing rural exodus. Properly exploited and processed, vegetable fibers can provide long fibers with suitable characteristics and properties for application in textiles, the manufacture of yarns, woven, non-woven and knitted fabrics, handicraft products, and in the area of new materials such as reinforcement for composites, and use in the construction, mechanical, automotive, naval, aeronautical and space industries.

7.3 Textiles and handcraft made from pineapple leaf fiber.

PALFs are strong, white and silky, and have been extracted in the Philippines since the late 1500s. Certain varieties grown specifically for the production of fibers and their fruits are removed to give maximum plant vitality. The 'Perolera' is an ideal cultivar for fiber extraction because its leaves are long, broad and rigid. From the stem and the leaves of the pineapple can also be extracted the bromelain, used in pharmaceuticals, in compositions for animal feed and fiber production for human consumption.

Some species have only ornamental value, whilst others have fibers that produce excellent pulp for papermaking, the manufacture of ropes and rustic materials (sacks) for crafts, weaving fine fabric, etc.

A shirt made of sheer fabric from pineapple fiber, are hand-woven in Philippines. The fiber is extracted from the pineapple plant from Aklan province where the majority of the fibers are woven.

7.7 Fiber properties

Analyses of the physical and mechanical properties of natural fibers such as sisal, coir, banana, PALF and talipot palm were made by Satyanarayana *et al.* (1986), who suggested modifications or surface treatments to improve their mechanical strength properties, for use in composite materials, replacing synthetic fibers.

It is important to mention that when natural fibers are transformed into nanocellulose, all potential defects due to extraction are removed, resulting in fibers with very high specific modulus (Leão *et al.*, 2010). Using the manual process through a decortication machine, a yield of 2.5 to 3.5% and a fiber length of 900 to 1500 μm was obtained (Satyanarayana *et al.*, 1986).

Table 7.3 presents the main properties – tensile strength, density, diameter, moisture content and Young's modulus – to be considered during the design of new composites.

Washing of fibers can be carried out by placing them in water tanks or through the use of a large centrifugal washing machine. Afterwards, the fibers are dried, placed under the shade, in centrifugal drying chambers, or through heated steam. Centrifugal machines are used in Indonesia, with an output of 1.8 to 2.3 dry tonnes of fibers every 20–25 minutes at a speed of 800 to 900 rpm with about 50% moisture. Indonesia has developed a series of machines that match the decortication machine extractors and fiber washers and offer advantages that can process 200 to 300 leaves per minute (Doraiswamy and Chellamani, 1993).

An engineering company in Mumbai, India, has developed a semiautomatic decortication machine that can process sisal leaves, pineapple and the stalks of bananas. The material can be fed on one side and withdrawn on the other side. A 5 HP electric motor (1440 rpm) has a capacity of 150 kg of sisal fiber, 60 kg of banana fiber, and 45 kg of PALFs per eight hours of operation. The methods of wetting and removal of waste fibers can be processed in two ways:

- Natural biological process through bacteria (soaking in water) or yeast (wetting by dew) which are the active ingredients
- Chemistry with dilute acids or enzymes, which are processes based on active ingredients.

Table 7.3 Data from several authors for the main fiber properties

Cell length (mm)	Diameter (μm)	Density (kg/m³)	Elongation (%)	Micro-fibrillar angle (°)	Moisture content (%)	Tensile strength (MPa)	Young's modulus (GPa)	Reference
–	5.0–30.0	1.44	1.6	12	–	170	6.26	George *et al.* (1993)
–	–	1.36	3.37	–	4.45	–	13.21	Luo and Netravali (1999)
–	20–80	–	1.6	1.4	11.8	413–1627	34.5–82.5	Mohanty *et al.* (2000)

It is important to mention that when the natural fibers are transformed into nanocellulose, all of their potential defects due to extraction are removed, resulting in fibers with very high specific modulus (Leão *et al.*, 2010). Using the manual process through a decortication machine a yield of 2.5 to 3.5% and a fiber length of 900 to 1500 μm was obtained (Satyanarayana *et al.*, 1990).

The chemical compositions of PALFs studied by various researchers are shown in Table 7.4. The content of cellulose in the PALFs was in the range of 67.12–82%; hemicelluloses 9.45–18.80%, hollocellulose 80–87.56%, lignin 4.4–15.4%, pectin 1.2–3%, fat and wax 3.2–4.2%, and ash 0.9–2.7%. Differing composition may be attributed to several factors, including source of fibers, plant age, climatic conditions and the fiber extraction process.

According to Sharma (1981), PALFs contain 2.5% waxy matter, 13.1% lignin and 3.8% ortho-acetyl groups. Sharma isolated hemicellulose from PALFs under various conditions. A study of the properties of these hemicelluloses gave direct evidence of ester linkages between the hemicellulose and the lignin in the fibers. PALFs also contain other minor constituents such as fat, pectin, uronic anhydride, pentosan and certain inorganic substances.

7.8 Pineapple leaf fiber (PALF)-reinforced polymer composites

Palmario *et al.* (1976) studied the tensile strength of PALF. Their tests indicated that the extraction and removal of pectin from the fiber did not cause any significant change in the tensile strength of the fibers except in the case of overwetting.

Yu (2001) studied the properties and processing characteristics of PALF and compared them with similar natural fibers. The properties of PALF were similar to those of many other bast and leaf fibers such as ramie, flax and jute. The structural parameters of PALF determined by Yu (2001) were length 3–8mm; diameter ~7–18 μm, ultimate tensile elongation ~3.42%, initial tensile modulus ~10.2 CN/tex, and density 1.543 g/cm^3.

Mukherjee and Satyanarayana (1986) reported that, among various natural fibers, PALFs exhibit excellent mechanical properties. The chemical constituents of pineapple fiber include cellulose (70–82%), lignin (5–12%) and ash (1.1%). Ultimate tensile strength, initial modulus and elongation are in the range of 362–748 MN m^{-2}, 25–36 GN m^{-2} and 2.0–2.8% respectively, for fibers with diameters ranging from 45 to 205 μm.

Samal and Ray (1997) used chemical reagents to modify the surface characteristics of PALFs. Primary requisites such as hydrophobicity, mechanical strength and chemical inertness can be imparted to the fibers in such a way that may make them suitable for use as reinforcements

Table 7.4 Chemical composition of PALF

Cellulose	Hemicellulose	Hollocellulose	Lignin	Pectin	Fat and wax	Ash	Extractives	reference
			13.1		2.5			Sharma (1981)
70–82	–	–	5–12.7	–	–	–	–	Mohanty *et al.* (2000)
56–62	16–19		9–13.0	2.0–2.5	4.7	2.0–3.0	–	Yu (2001)
81.3	12.3	–	3.5	–	–	–	–	Cherian *et al.* (2010)

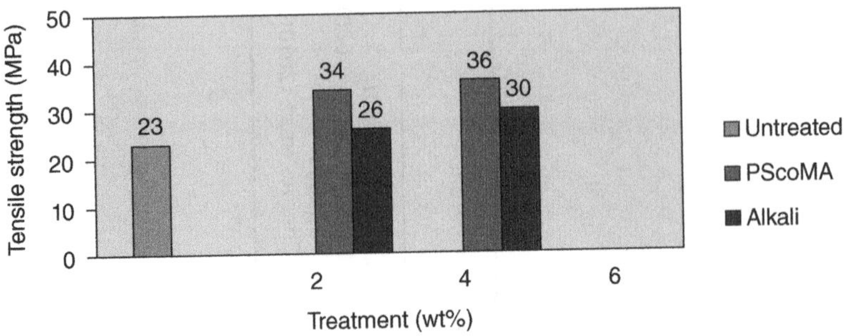

7.4 The effect of treatments on the tensile strength of PALF/HIPS composites (Siregar *et al.*, 2012).

in fiber-reinforced polymers. Mohanty *et al.* (2000) stated that graft copolymerization onto PALF is of considerable importance for physicochemical modification and improvement of the textile performance of fibers.

PALF treated with NaOH associated with the use of coupling agent, poly(styrene-*co*-maleic anhydride), presented better adhesion with the matrix of high impact polystyrene (HIPS), proved by the observed results of tensile strength for the composites as shown in Fig. 7.4 (Siregar *et al.*, 2012). The comparison was made between composites with the addition of 50% PALF fibers without treatment and without coupling agent – so-called 'untreated' in the graph. The composites called alkali were treated with 2% or 4% of NaOH but without compatibilizer, and the composites called PScoMA were treated with NaOH 2% and 4% respectively, and the same percentage of poly(styrene-*co*-maleic anhydride), 2% and 4%.

George *et al.* (1995) employed an injection-molding technique for making PALF composites. They reported that PALFs were initially separated from undesirable foreign matter such as pith. This is the normal practice when extracting fibers. PALFs were chopped to various lengths of 2.6 and 10 mm. Composites of PALF and low density polyethylene (LDPE) were prepared using two methods, namely melt mixing and solution mixing. Both mixtures were then extruded at a temperature of 120°C and the samples were produced using an injection-molding machine. The extruded samples were cut into 120 mm long rods. The oriented PALF composites were prepared by closely aligning the cylindrical extruded samples in a mold at 120°C. Randomly oriented PALF composites were prepared by direct extrusion of the mixture into the mold.

PALF was used in blends with ABS and fiber retardants, which could be a potential market for composites in the automotive sector. It reported blends with 10–20% by weight of PALF. Two different flame retardants, i.e.

bisphenol-A bis (diphenyl phosphate) (BDP) and 9,10-dihydro-9-oxa-10-phosphaphenanthrene-10-oxide (DOPO), were investigated. Plasticizers were added (diisononyl phthalate 1% w/w) at the different flame retardant concentrations. The compounder used was a co-rotating twin-screw extruder. In their conclusions, DOPO, a phosphorous-containing flame-retardant, enabled the natural fiber-enhanced composites, resulting in superior mechanical properties, in this case Young's modulus and tensile strength (Threepopnatkul *et al.*, 2013).

George *et al.* (1993) carried out research into the development of randomly oriented PALF composites that had been prepared using a ram-type injection molding machine. The orientated PALF composites were prepared by a combination of injection and compression molding techniques. They were initially extruded into 4 mm thick cylindrical rods in an injection-molding machine. Samples were then prepared by closely aligning the extruded rods in a mold under compression at 120°C.

Bhattacharya *et al.* (1986) studied the mechanical properties of short PALF reinforced rubber composites, which have attracted considerable interest because of their processing advantages and improved mechanical properties such as tensile strength, stiffness and damping. These researchers also found that bonding agents such as resorcinol and hexamethylene tetramine are essential for good fiber–rubber adhesion.

Luo and Netravali (1999) studied the tensile strength of the PALF composites and found that they increased in the longitudinal direction with an increase in fiber content. However, they found a decreasing trend in the transverse direction with an increase in fiber content. The flexural strength in the transverse direction decreased, and the flexural modulus in the transverse direction showed slight changes with increasing fiber content.

Pavithran *et al.* (1987) carried out a comparative study of the impact properties of unidirectionally aligned polyester composites reinforced with sisal, pineapple, banana and coir fibers and found that sisal fiber composites had the highest toughness followed by PALF composites. The toughness of the sisal fiber composites was 98.7 kJm^{-2} and PALF achieved 79.5 kJm^{-2}. The extent of fracture of these natural fiber composites was measured using an unnotched Charpy impact test.

Attharangsan and Saikrasun (2013) investigated thermal and thermo-oxidative decomposition behavior and decomposition kinetics of recycled PE/PALF composite systems in relation to PALF and compatibilizer loadings. The thermal stability of the polymer progressively decreases with PALF loading. From the calculated activation energy in nitrogen and in air, PALF showed much lower thermal and thermo-oxidative stability than recycled PE and its composites.

Cherian *et al.* (2010) developed a methodology employed for the successful extraction of cellulose nanofibrils from PALF for the first time applying acid coupled steam treatment. The steam explosion process resulted in the isolation of PALF nanofibers having a width in the range of 5–60 nm. The high-pressure defibrillation contributed a unique morphology of the interconnected web-like structure of nanofibers.

Cherian *et al.* (2011) investigated the reinforcing effect of PALF nano-sized cellulose fibrils in the polyurethane matrix, especially for medical implants. They reported that high-tensile-strength and high-strain-to-failure nanocomposites with strongly improved modulus were synthesized based on nanocellulose and polyurethane. The produced nanocellulose and its composites proved to be a highly versatile material, having a wide range of medical applications, including cardiovascular implants, scaffolds for tissue engineering, repair of articular cartilage, vascular grafts, urethral catheters, mammary prostheses, penile prostheses, adhesion barriers and artificial skin.

The development of PALF nanocellulose–polyurethane valve (Fig. 7.5(a))design with good biological durability, fatigue resistance and haemodynamics, and a new generation of biostable polyurethanes have proven themselves of superior biostability in a demanding six-month, strained, rat implant model. PALF-derived nanocellulose embedded polyurethane has also been utilized as an attractive and readily available range of materials for the fabrication of vascular prostheses. Nanocellulose–PU vascular grafts (Fig. 7.5(b)) with a wall thickness of 0.7–1.0 mm showed elongation at a break of 800–1200%, and withstood hydraulic pressures up to 300 kPa.

Costa *et al.* (2013) developed a bionanocomposite PVA/pineapple nanofibers/*Stryphnodendron adstringens* bark extract, especially for medical implants, by an electrospinning technique. Morphological analysis of the compounded materials showed changes in PVA nanofibers mats with the presence of 1% (w/w) of *Stryphnodendron adstringens* bark extract and nanocellulose from pineapple. DSC and TGA showed higher thermal properties and changed crystallinity of the developed bionanocomposite mainly due to the nanocellulose and *Stryphnodendron adstringens* bark extract.

7.9 Application of pineapple fibers and composites

Sena Neto *et al.* (2013) compared the PALFs extracted from six different cultivars, and characterized them in terms of mechanical properties, TGA, XRD, tensile, SEM and IR. It was described that, within the pineapple family, there are intrinsic variabilities indicating different potential applications for each variety. The differences observed were due to the fiber structure and morphology. All six cultivars showed onset temperature at around 245°C, indicating they are to be compounded with matrices that

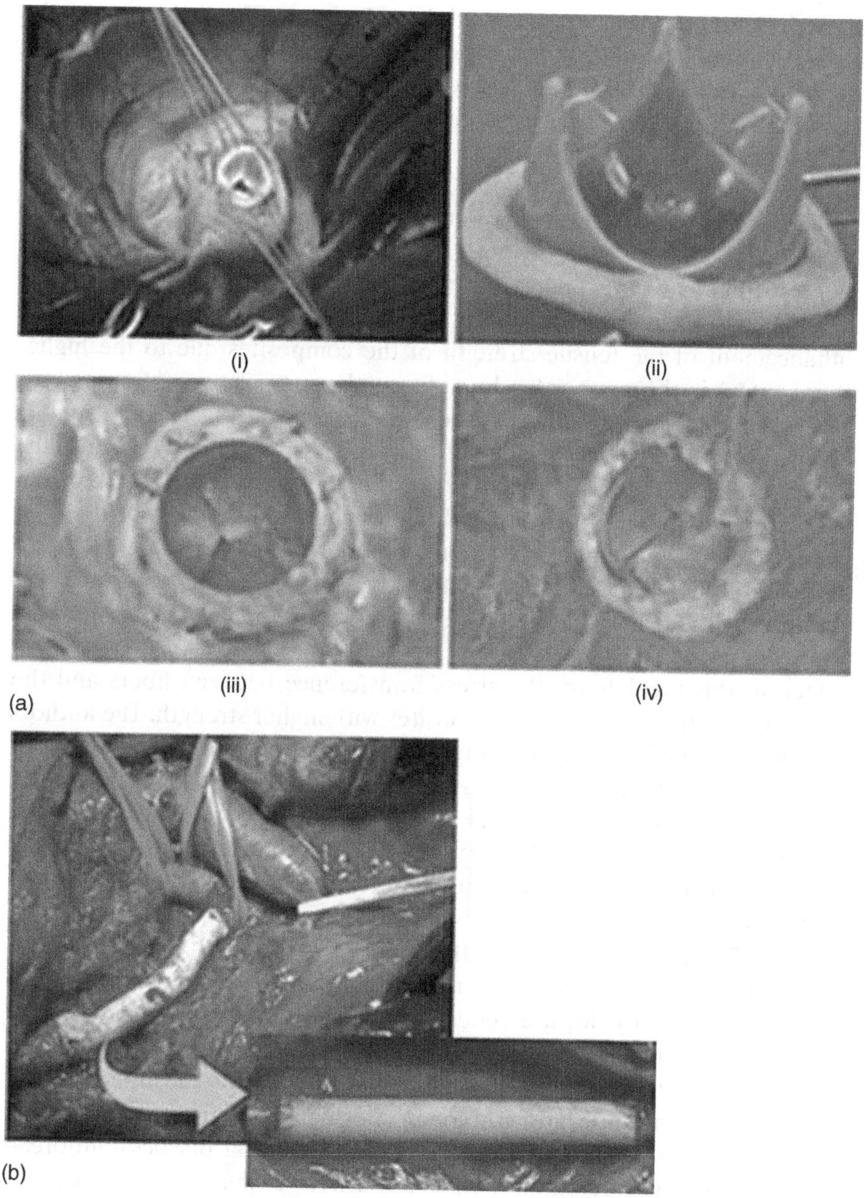

7.5 (a) Nanocellulose–polyurethane prosthetic heart valve: (i) valve implant, (ii) heart valve, (iii) viewed *in situ* immediately prior to explant (inflow surface), (iv) viewed *in situ* immediately prior to explant (outflow surface) and (b) vascular prostheses made of nanocellulose–polyurethane placed between the brachiocephalic trunk and the right common carotid artery in a 26-year-old male patient with multiple endocrine neoplasia 2B (MEN 2B) (Cherian *et al.*, 2011).

show low melting temperature, mainly polyolefins. The reduction in the cross-section of the fiber implies a lower presence of amorphous material, impurities, waxes, fats and defects, while maintaining the crystalline phase. For applications that require high mechanical strength and modulus, varieties with smaller dimensions and higher crystallinity indeces are recommended.

Pineapple leaf fiber is used as a reinforcement in polyolefins. Polypropylene (PP) and low-density polyethylene (LDPE) composites with different fiber lengths (long and short fibers) and fiber contents (0–25%) were prepared and characterized by Chollakup *et al.* (2010). The results showed enhancement of the tensile strength of the composites due to the higher fiber content in the composites. In addition, the composites containing long fiber PALF were stronger than the short fiber composites, by tensile strength. Aiming to better identify the compatibility and dispersion of PALF into the matrix, a SEM study was carried out on the tensile fractured surface, and the pictures confirmed a better and homogeneous dispersion of the long fibers in the polymer matrices when compared to the short fibers. Due to the unidirectional arrangement of the long fibers, a better interface was obtained with a good interfacial bonding between the PALF and the polymer matrix.

Due to this bondability, the stress transference between fibers and the matrix was better, resulting in composites with higher strength. The authors reported a reduction in crystallinity of the composites, which was observed by the XRD and DSC analysis, in addition to a better stress transference for the LDPE when compared to the PP. A possible explanation for why the higher-modulus fiber (PALF) had better load transfers is due to the fact that the LDPE matrix is much more ductile than the PP, which is more brittle. This is more evident with higher fiber loads, as can be seen in Figs. 7.6 and 7.7 (Chollakup *et al.*, 2010).

Luo and Netravali (1999) have reported the use of PALF composites to produce several parts for Mercedes-Benz commercial vehicles, particularly those produced in Brazil.

Devi *et al.* (2012) reported an important aspect for the utilization of PALF for high-end applications in composites, namely hybrid reinforcements (glass and natural fibers). The main issue for natural plastics reinforced composites is impact resistance, which is much lower than for neat polymers. Therefore, mainly for automotive applications, impact resistance must be improved.

The authors studied the tensile and impact performance of intimately mixed (IM) hybrid composites based on glass fiber (GF) and pineapple leaf fiber (PALF). The composite was fabricated at a constant volume fraction of fiber, $V_f = 0.3$ (fiber 0.3 and matrix 0.7). The volume fraction of the matrix was constant (0.7), and only the PALF/GF ratio was varied, from 0 to 1. It

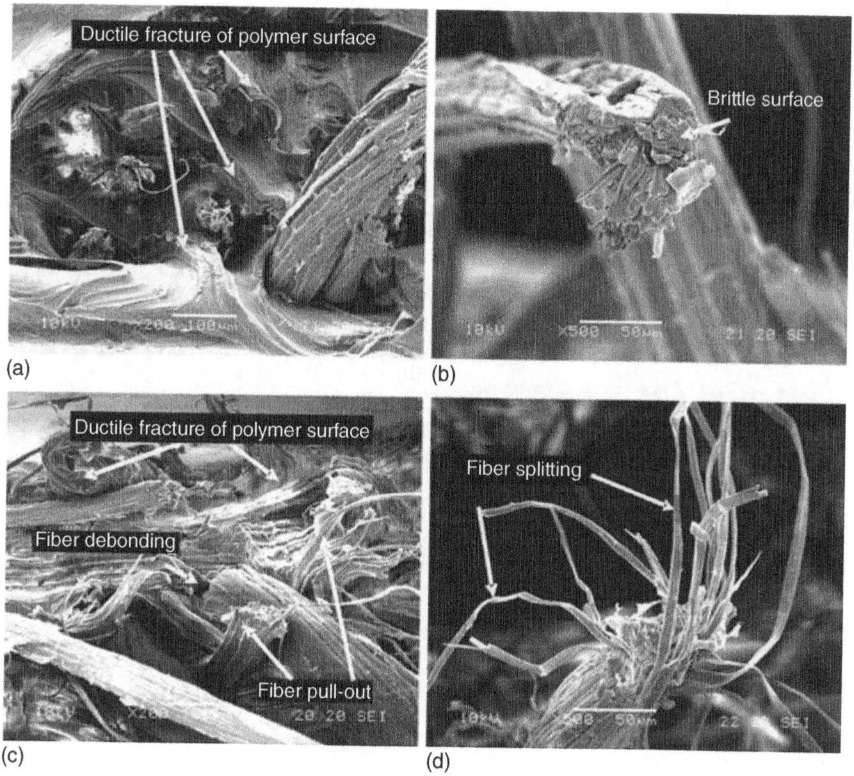

7.6 SEM photographs of tensile fracture of LDPE/PALF composites with 15% PALF. (a) LDPE/PALF-L at magnification of ×200; (b) LDPE/PALF-L at magnification of ×500; (c) LDPE/PALF-S at magnification of ×200; (d) LDPE/PALF-S at magnification of ×500 (Chollakup *et al.*, 2010).

was related that the incorporation of 0.1 volume fraction of GF increased the tensile strength of the hybrid composite by about 28%. However, the most important result was that intimately mixed hybrid composites exhibited higher impact strength than the individual fiber composites, where a 70:30 PALF/GF ratio showed maximum impact strength of 1203 J/m. Therefore, from these findings, the PALFs can be recommended to reinforce GF composites, which could help to reduce the cost and weight of the parts. The incorporation of both PALF and GF into the polyester matrix has resulted in an increase in tensile and impact strengths with higher fiber loading, and thus a composite of outstanding mechanical performance (Devi *et al.*, 2012).

An interesting approach was the production of composites with the association of two fibers – pineapple leaf and kenaf fiber – and high-density

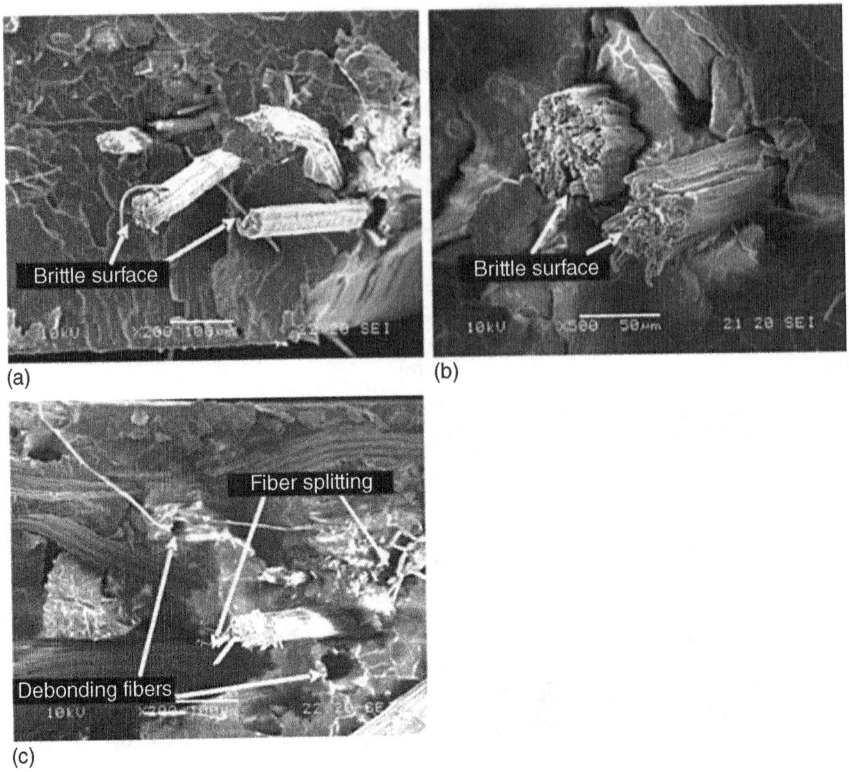

7.7 SEM photographs of tensile fracture of PP/PALF composites with 15% PALF. (a) PP/PALF-L at magnification of ×200; (b) PP/PALF-L at magnification of ×500; (c) PP/PALF-S at magnification of ×200 (Chollakup *et al.* 2010).

polyethylene (Aji *et al.*, 2013a). One important problem to be solved in the use of natural fibers in thermoplastics composites is hydrophilicity, which interferes in the final product. The composite with 70% PALF and 30% kenaf resulted in the least overall water uptake, probably due to better adhesion to the matrix (Fig. 7.8). The composite with 50% PALF and 50% kenaf obtained the best tensile and flexural strength; PALF improved the tensile and flexural strength, but kenaf increased the impact strength and reduction of water uptake, presenting synergy when utilized together. The compositions of fibers were kept at 40% in the total composites but just changing the proportions, where K7P3 represents 70% kenaf and 30% PALF, K6P4 60% kenaf and 40% PALF, K4P6 40% kenaf and 60% PALF, K3P7 30% kenaf and 70% PALF, and KP 50% kenaf and 50% PALF (Aji *et al.*, 2013a).

Electron beam radiation was used, avoiding the application of a coupling agent, where by increasing the radiation it was possible to achieve higher

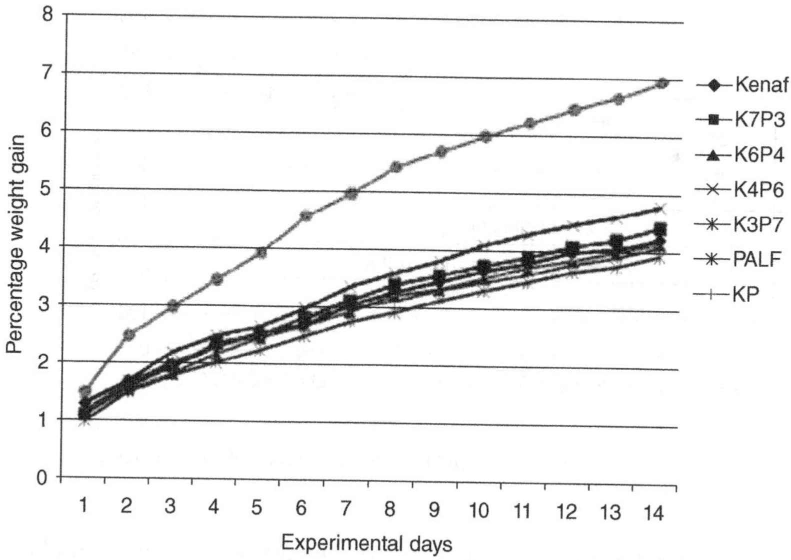

7.8 Water immersion behavior of hybridized composite (Aji *et al.*, 2013a). © 2013, reprinted by permission of Sage Publication.

tensile strength for hybrid composites of HDPE and 50% of PALF and kenaf fibers (ratio 1:1) using a mixer followed by a compression molding processes (Aji *et al.*, 2013b). As shown in Fig. 7.9, the tensile strength and modulus increased with a higher radiation dose, where it was possible to form the radicals and consequently improve the chemical bonding of the matrix and the fibers (Aji *et al.*, 2013b).

Natural fiber-reinforced composites are highly sensitive to thermal degradation, and therefore the process temperature must be kept as low as possible. The thermal degradation behavior of hybridized kenaf (bast fiber)/pineapple leaf fiber (PALF) reinforced high density polyethylene (HDPE) composites was analyzed by thermogravimetric and derivative thermogravimetric analyses (TG/DTG) with respect to the proportions of fiber in the composite, variation in fiber loading, and fiber length. It was reported that the thermal decomposition of all the samples occurred in the temperature range of 35–615°C, with an initial peak between 236.9 and 331°C, corresponding to a mass loss of between 23 and 26%. Decomposition temperature of the natural fiber reinforced composites was a little higher than the temperature for net HDPE. Kenaf and PALF composite have shown weaker thermal stability compared to neat HDPE. The introduction of higher levels of fiber into the matrix directly decreased the thermal stability of the hybridized composite (Aji *et al.*, 2012).

Kengkhetkit and Amornsakchai (2012) described the application of pineapple agricultural wastes for thermoplastics composites. The use of

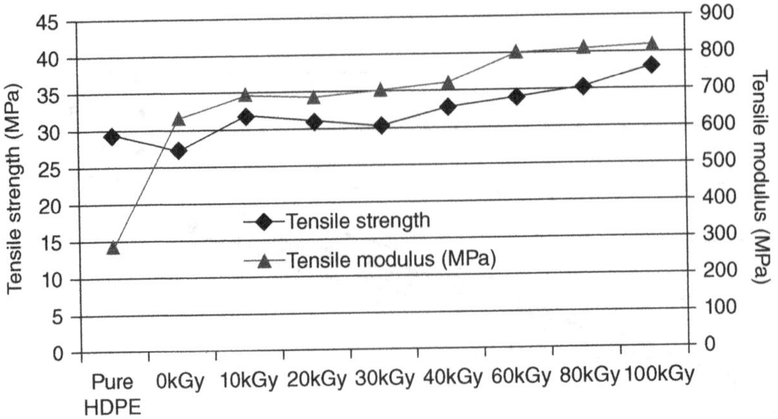

7.9 Variation in the tensile properties of hybridized composite (Aji *et al.* 2013b).

short fibers in composites can reduce the production costs of natural fiber-reinforced composites, since there is no need for long fibers in some processes, such as extrusion and injection molding. The mechanical extraction process, through ball mill and disc mill, resulted in higher fiber yield and smaller fibers, when compared to conventional extraction methods. The wet milling process, with a disc mill, was faster and with a higher proportion of elementary fibers, while the dry mill process resulted in very poor-quality fibers. Therefore, the use of a higher ratio of elementary fibers will result in a composite with better properties. These fibers could be used directly in the thermoplastics industry.

7.10 Conclusions

The utilisation of PALFs and their composites has triggered the interest of researchers due to their environmental advantages. Unlike some fibers, PALFs seldom cause problems at the end-of-product life stage. The high cellulose content, relatively low cost and abundant availability have made PALFs a potential reinforcement in polymer composites on both the micro- and nanoscales. The positive results show that PALF micro- and nanocomposites are suitable for automotive parts, building, non-structural and medical applications. PALFs bring together performance and sustainability, and are therefore recommended for value-added products, especially in the textile and composites sector. However, some constraints (such as constant supply, quality control, reliable and replicable standards) must still be overcome to establish PALFs in some key sectors, such as the paper and pulp industry. Future research should be focusing on PALFs for more sophisticated applications due to their unique properties.

7.11 References and further reading

Aji, I. S., Zainudin, E. S., Khalina, A., Sapuan, S. M. and Khairul, M. D. (2012) Thermal property determination of hybridized kenaf/PALF reinforced HDPE composite by thermogravimetric analysis. *J Therm Anal Calorim*, 109, 893–900.

Aji, I. S., Zainudin, E. S., Abdan, K., Sapuan, S. M. and Khairul, M. D. (2013a) Mechanical properties and water absorption behavior of hybridized kenaf/pineapple leaf fibre-reinforced high-density polyethylene composite. *J Compos Mater*, 47, 979–990.

Aji, I. S., Zinudin, E. S., Khairul, M. Z., Abdan, K. and Sapuan, S. M. (2013b) Induced tensile properties with EB cross linking of hybridized kenaf/PALF reinforced HDPE composite. *Pertanika J Sci & Technol*, 21, 135–140.

Alexandre, M. E. O. (2005) Matrix composites reinforced with polyester fiber pineapple leaf: mechanical property and water absorption. PhD. Thesis, UFRN, Natal, Brazil.

Alexandre, M. E. O., Ladchumananandasivam, R., Verissimo, S. A., Menezes, P. L., Batista, J. H. and Araujo, R. C. (2004) Aproveitamento da Fibra da Folha do abacaxi (FFA) como recurso economico para o Estado do Rio Grande do Norte. Anais do II Sientex, Natal, Brazil.

Attharangsan, S. and Saikrasun, S. (2013) Kinetic analysis of thermal and thermo-oxidative decomposition of recycled Pe/PALF bio-based composites. *Int J Plast Technol*, 17, 94–110.

Bhattacharya, T. B., Biswas, A. K., Chatterjee, J. and Pramnick, D. (1986) Short pineapple leaf fibre reinforced rubber composites. *Plast. Rubber Process Appl*, 6, 119–125.

Cherian, B. M., Leao, A. L., Souza, S. F., Thomas, S., Pothan, L. A. and Kottaisamy, M. (2010) Isolation of nanocellulose from pineapple leaf fibres by steam explosion. *Carbohyd Polym*, 81, 720–725.

Cherian, B. M., Leão, A. L., Souza, S. F., Costa, L. M. M., Olyveira, G. M., Kottaisamy, M., Nagarajan, E. R. and Thomas, S. (2011) Cellulose nanocomposites with nanofibres isolated from pineapple leaf fibers for medical applications. *Carbohyd Polym*, 86, 1790–1798.

Chollakup, R., Tantatherdtam, R., Ujjin, S. and Sriroth, K. (2010) Pineapple leaf fiber reinforced thermoplastic composites: Effects of fiber length and fiber content on their characteristics. *J Appl Polym Sci*, 119, 1952–1960.

Costa, L. M. M., Olyveira, G. M., Cherian, B. M., Leao, A. L., Souza, S. F. and Ferreira, M. (2013) Bionanocomposites from electrospun PVA/pineapple nanofibers/*Stryphnodendron adstringens* bark extract for medical applications. *Ind Crop Prod*, 41, 198–202.

Cunha, G. A. P., Matos, A. P., Cabral, J. R. S., Souza, L. F. S., Sanches, N. F. and Reinhardt, D. H. R. C. (1994) *Pineapple for export aspects – the technical production*. Ministry of Agriculture, SDR – FRUPEX, Embrapa – SPI, Brasilia.

Devi, L. U., Bhagawan, S. S. and Thomas, S. (2012) Polyester composites of short pineapple fiber and glass fiber: tensile and impact properties. *Polym Compos*, 33, 1064–1070.

Doraiswamy, I. and Chellamani, P. (1993) *Pineapple Leaf Fibre*. The Textile Institute, Manchester, UK, 24, 1–37.

George, J., Joseph, K., Bhagawan, S. S. and Thomas, S. (1993) Influence of short pineapple fiber on the viscoelastic properties of low-density polyethylene. *Mater Lett*, 18, 163–170.

George, J., Bhagawan, S., Prabhakaran, N. and Thomas, S. (1995) Short pineapple-leaf-fiber-reinforced low-density polyethylene composites. *J Appl Polym Sci*, 57, 843–854.

IBGE. Local agricultural production. Available at www.sidra.ibge.gov.br, accessed 4 October 2013.

Kengkhetkit, N. and Amornsakchai, T. (2012) Utilisation of pineapple leaf waste for plastic reinforcement: 1 A novel extraction method for short pineapple leaf fiber. *Ind Crop Prod*, 40, 55–61.

Leão, A. L., Souza, S. F., Cherian, B. M., Frollini, E., Thomas, S., Pothan, L. A. and Kottaisamy, M. (2010) Agro-based biocomposites for industrial applications. *Mol Cryst Liq Cryst*, 522, 18–27.

Luo, S. and Netravali, A. N. (1999) Mechanical and thermal properties of environment friendly green composites made from pineapple leaf fibers and poly(hydroxybutyrate-*co*-valerate) resin. *Polym Compos*, 20, 367–378.

Manica, I. (1999) *Tropical Fruit Crops 5. Pineapple*. Cinco Continentes, Porto Alegre, Brazil, 501, 1999.

Margarido, S. M. F. (1991) *Pineapple King of Fruits – Practical Methods for Growing*. Coleção Brasil Agricola, São Paulo.

Mohanty, A. K., Tripathy, Y. P. C., Misra, M., Parija, S. and Sahoo, S. (2000) Chemical modification of pineapple leaf fiber: Graft copolymerization of acrylonitrile onto defatted pineapple leaf fibers. *J Appl Polym Sci*, 77, 3035–3043.

Mukherjee, P. S. and Satyanarayana, K. G. (1986) Structure and properties of some vegetable fibres. Part 2. Pineapple fibre (*Anannus comosus*). *J Mater Sci*, 21, 51–56.

Palmario, M. S., Cueto, C. U., Imperial, Z. S., Tayco, S. A., Soriaga, R. P., Buenavenlure, R. V. and De Guzman, M. C. (1976) Pineapple fibres the retting process. *NSDB Technol J*, 1, 8–15.

Pavithran, C., Mukherjee, P. C., Brahmakumar, M. and Damodaran. A. D. (1987) Impact properties of natural fibre composites. *J Mater Sci Lett*, 6, 882–884.

Samal, R. K. and Ray, M. C. (1997) Effect of chemical modifications on FTIR spectra II. Physicochemical behaviour of pineapple leaf fibre (PALF). *J Appl Polym Sci*, 64, 2119–2130.

Satyanarayana, K. G., Sukumaran, K., Mukherjee, P. S. and Pillai, S. (1986) Materials science of some lignocellulosic fibers. *Metallography*, 19, 389–400.

Satyanarayana, K. G., Sukumaran, K., Mukherjee, P. S., Pavithran, C. and Pillai, S. G. K. (1990) Natural fibre-polymer composites. *Cement Concrete Comp*, 12, 117–136.

Sena Neto, A. R., Araujo, M. A. M., Souza, F. V. D., Mattoso, L. H. C. and Marconcini, J. M. (2013) Characterization and comparative evaluation of thermal, structural, chemical and morphological properties of six pineapple leaf fibers varieties for use in composites. *Ind Crop Prod*, 43, 529–537.

Sharma, U. (1981) Investigations on the fibers of pineapple [*Ananas comosus* (L). Merr.] leaves. *Carbohyd Res*, 97, 323–329.

Siregar, J. P., Sapuan, S. M., Rahman, M. Z. A. and Zaman, H. M. D. K. (2012) Effects of Alkali Treatments on the tensile properties of pineapple leaf fibre reinforced high impact polystyrene composites. *Pertanika J Sci Technol*, 20, 409–414.

Threepopnatkul, P., Thanaphat, K., Teerawattananon, W., Suriyaphaparkorn, K. and Kulsetthanchalee, C. (2013) Effect of flame retardants on performance of PALF/ABS composites. *Adv Mat Res*, 747, 351–354.

Venugopal, B., Pillai C. K. S. and Satyanarayana, K. G. (1990) Studies on flame retardance of needlefelt coir. *Res Ind*, 35, 108–110.

Yahya, S. A. and Yusof, Y. (2013) Comprehensive review on the utilization of PALF. *Adv Mat Res*, 701, 430–434.

Yu, C. (2001) Properties and processing of plant fiber. *Proceedings of the New Frontiers in Fibre Science 2001*, 23–25 May 2001. The Fiber Society, USA Raleigh, NC.

8

The use of banana and abaca fibres as reinforcements in composites

A. A. MAMUN and H. P. HEIM, University of Kassel, Germany, O. FARUK, University of Toronto, Canada and A. K. BLEDZKI, University of Kassel, Germany and West Pomeranian University of Technology, Poland

DOI: 10.1533/9781782421276.2.236

Abstract: The fibre extracted from banana trees is a by-product of banana plants, which can be found in all tropical countries. Unlike bananas, abaca is inedible and cultivated solely for fibre extraction purposes. Fibre properties depend on botanical type, growing condition and extraction methods. The abaca fibre is considered to be one of the strongest natural fibres and shows excellent mechanical and chemical properties. It is currently used for pulping, cordage and yarns and fabrics. Several suface modifications can minimize the drawbacks of these fibres with the aim of increasing the applications of these fibres in the automobile and packaging industries. This chapter will describe the processing and performance of composites and the latest economic and technical developments.

Key words: fibre extraction, chemical content and surface properties, enzyme modification, fibre length and load effect, mechanical properties.

8.1 Introduction

Bananas and abacas look similar and are both members of the Musaceae plant family. The fibre in question will be referred to as abaca fibre to avoid confusion. Abaca fibre is a cellulosic fibre obtained from the pseudo-stem of a plant. Sometimes the fibre used for industrial purposes is a cheap waste product of banana cultivation. In tropical countries, agricultural plants like abacas are readily available. The economical advantage of the abaca plant is that it does not require additional fertilizer, pesticides and water inputs for its cultivation (Goeltenboth and Muehlbauer, 2010).

The abaca plant was first found in the Philippine Islands. Abaca fibres are traditionally used in handicraft, ropes and clothing. It is the main material used for marine ropes due to its strength, lightness and water-resistance. It is also used in the manufacture of specialized papers including tea and coffee bags, sausage casing, electrolytic papers, currency notes, cigarette filters and medical and disposal papers. This is because of its long staple length, strength and high cellulose content (Anon, 2000). Abaca fibre-reinforced composites are currently gaining interest due to the innovative application of abaca fibre in under-floor protection for passenger cars made by DaimlerChrysler (Anon, 2005). The combination of polypropylene (PP) thermoplastics with embedded abaca fibres has been patented by DaimlerChrysler's researchers, and the manufacturing process (compression moulding process) was initiated by Rieter Automotive.

Abaca fibre has a high tensile strength, is resistant to rotting and has a specific flexural strength similar to that of glass fibre (Hintermann, 2005). Abaca is the first natural fibre to meet the stringent quality requirements for components used on the exterior of road vehicles, especially regarding resistance to stone strike, exposure to the elements and dampness. This chapter will describe the following in relation to current economic and technical developments: the cultivation process, the extraction process, the quality of the fibres, fibre properties, disadvantages of the fibre, the surface modification, processing and performance of composites.

8.2 Banana and abaca plants and their cultivation

Banana plants have been cultivated since 5000 BC and are one of the oldest plants in the world.There are about 700 varieties which can be cultivated around the world, mostly originatingfrom the Philippines archipelago. Only 20 varieties are used for common consumption. The most common genus is called *Musa* L. and contains seven species (Anon, 2013a), as shown in Table 8.1. Today, the National Abaca Research Center recommends very few species for further cultivation. Such a recommendation is made depending on the growth rate, cultivation cost and energy inputs (Moreno *et al.*, 2006).

Abacas and bananas are very alike but the stalk of the abaca plant is thinner and tapered from the base to the top (Sinon, 2008). The abaca fruits themselves and the corresponding bunches of them are much smaller than bananas and bunches of bananas. As early as 1880, the commercial usefulness

Table 8.1 Species of banana

Species	Name
Musa acuminate Colla	Edible banana
Musa balbisiana Colla	Banana
Musa nana sensu Parham, non Lour	
Musa xparadisiaca L.	French plantain
Musa textilis Née	Abaca
Musa troglodytarum L.	fe'i banana
Musa velutina H.	Hairy banana

Source: Anon, 2013a.

of these fibres was known and was referred to as 'Manila hemp' by the English (Batra, 1998). The Malays used abaca fibre for ropes, fishing nets, woven cloths, mats and bags.

The abaca plant is made up of a central soft core, which is wrapped in up to 30 sheaves that form a proto-stem. The stalk is usually about 2 to 6m tall and 9 to 30cm in diameter. The stem is up to 90% water, about 4% parenchyma tissues and about 5% fibre (Gonzal and Valida, 2003). A single plant can successively produce as many as about 25 stems. The first pseudo-stems are developed enough to be harvested when the plants are 18 months old. Depending on their rate of growth, four stems are then cut at intervals of 6 to 12 months (Franck, 2005). The leaves are classified into four groups in order to grade the fibres. The fibre strength, chemical composition and physical properties depend on the position of the leaf. The fibres from the leaf nearest to the centre are finer, whiter and softer (Catlin and Grayson, 1998).

Abaca grows in the hot humid tropics and is adapted to an average relative humidity of about 80%. The successful cultivation zone lies within the latitudinal range 15°N to 5°S. It is usually grown at elevations below 500m but can be grown in the tropics at altitudes between sea level and 1100m. The growing period is perennial and the plants grow for 1.5–3 years before being harvested for the first time (McGovern, 1990). Their economic life lasts 6 to18 years.

The required environmental conditions for commonly grown clones are listed below:

- Tangongon is usually grown on heavy clay soils and can easily be blown down. It is common in the Philippines and grows to about 4.5 to 5.5m tall. Its life span lasts 8 to 15 years.
- Bungulanon cannot be grown on stiff clay or dry sandy soils but requires muddy soil. It is common in Central America and its life span lasts 5 to 6 years.
- Maguindanao lives approximately 15 years or more. It cannot be grown in heavy clays. The root system is shallow and easily damaged by high winds (Anon, 2012). The annual fibre yield ranges from 0.30 to 1.70 tonnes/ha in the Philippines, while yield is between 1.5 and 2.5 tonnes/ha in Ecuador (Armecin *et al.*, 2005).

There are several advantages to planting abaca. It helps to minimize the need for biodiversity rehabilitation and may control erosion in rainforest areas, particularly when planted with coconut palms. It can minimize sedimentation problems in coastal areas which are important breeding places for seawater fish. Abaca improves the water holding capacity of the soil so that floods and landslides can also be prevented. Abaca waste materials can be used as organic fertilizer (Anon, 2012).

8.3 Fibre extraction

Fibre extraction takes place as soon as the stems are cut, while they are still moist and contain the gums which bind the fibres to each other. The first step, aimed at separating the fibre bundle from the leaf-sheaves, is called 'tuxying'. This is generally done manually. A sharp knife is inserted between the outer and middle layers of the leaf sheath. The outer layer is then gripped and pulled away from the rest of the leaf. This results in producing ribbons of fibres. The process of tuxying produces tuxies from the outer, middle and inner leaf-sheaths, which are separated for classification. The second step aims to remove gum, vascular tissue, parenchyma cells and any residual leaf matter remaining on the fibres. This is called stripping. Stripping is done manually, semi-manually or mechanically.

Stripping technologies are classified into fixed blade and moving blade. The fixed blade technique can be divided into hand and spindle stripping. Moving blade stripping is known as a decorticating technique. Detailed explanation of these techniques can be found in several studies, including Goeltenboth and Muehlbauer (2010), Sinon (2008) and Gonzal and Valida (2003). The different stripping techniques are summarized according to fibre quality, production rate and investment in Table 8.2.

In the hand stripping technique, the fibres are pulled by the farmer himself. The fibre production capacity on average is 15 kg. While fibre quality is often poor and includes impurities, the investment cost is very low. The technique is suitable for individual farmers. In spindle stripping, the fibres are pulled by a machine. The production capacity is about 100 kg per day and produces good-quality fibres. The fibres are fine and homogeneous and have low impurity content. The traditional moving stripper is too bulky, heavy and expensive for individual ownership. The fibres produced using decortication techniques are poor in quality and expensive. They are a mixture of primary and secondary fibres and not homogeneous.

Table 8.2 Fibre quality, production rate and investment

Technique	Capacity (kg/day)	Fibre quality	Investment cost (euro)
Hand stripping	10–20	Poor	4–6
Spindle stripping	100–200	Good	1700–2300
Multi-fibre decorticator	80–100	Poor	1200–1300
Auto-fed decorticator	600–800	Poor	6800–7000

Source: Sinon, 2008.

The fibre quality produced by the spindle stripping technique is thus better than that from other techniques. This technique is not attractive for individuals due to its high initial investment costs and even higher production costs. One alternative to help reduce the processing periods and produce pure high-quality fibres is to use enzyme technology in combination with stripping. This method has not yet been perfected.

The bundles or ribbons of fibre produced are up to 3 m long and have a moisture content of 55% to 60% after stripping or decorticating. The fibre bundles have to be dried immediately to prevent discolouration and damage. They can be hung outside to dry, weather permitting. Mechanical drying could be another option but is not feasible due to high investment costs. Once dried, they are graded and packed into bundles for shipping. The grading systems differ with the producing country. There are two existing grading systems, one used in the Philippines and one in Ecuador. The grading system for fibres produced in the Philippines focuses on the manner of extraction (see Table 8.3). The fibres produced in Ecuador are classified into grades 2 to 5 depending on their properties.

The world production of abaca by country is given in Table 8.4. The dominant producing countries are the Philippines and Ecuador, responsible for about 70% and 25% of total production respectively. In the past few

Table 8.3 Quality standards of abaca fibres

Grade	Cleaning	Colour	Leaf sheaths
Superior Current (AD)	Excellent	Ivory white to white	Inner
Mid Current (EF)	Excellent	Light ivory to very light brown	Middle
Streaky Two (S_2)	Excellent	Light ochre to very light brown	Next to outer
Streaky Three (S_3)	Excellent	Dark red to dark brown	Outer
Current (I)	Good	Light to very light brown	Inner and middle
Soft Seconds (G)	Good	Dingy white, light brown	Next to outer
Soft Brown (H)	Good	Dark brown	Outer
Seconds (JK)	Fair	Dull brown to dingy light brown	Inner middle
Medium Brown (M1)	Fair	Blackish	–
Coarse (L)	Coarse	Brownish	–
Residual (Y1, Y2, O, T, WS)	Loutish, damaged	Dark brown, black	All

Data were partially adapted with permission from F.G. Sinon, *Optimization of Stripping Technologies for the Production of High Quality Abaca Fibre*, Shaker Verlag GmbH, Aachen.

Table 8.4 Global production of abaca by country from 2000 to 2011

Country	2000	2005	2008	2011
Philippines	77,180	74,010	68,386	68,612
Ecuador	25,000	29,107	32,000	35,394
Costa Rica	1602	1100	1100	1609
Indonesia	618	498	503	515
Equatorial Guinea	398	363	276	230
Kenya	30	49	37	44

Source: Anon, 2013b.

years, global production has reached about 105,000 tonnes per year. In the early 2000s, it was 100,000 tonnes per year but it increased to 110,000 tonnes in 2005. It then decreased to its previous level in the year 2008 due to natural disaster in the Philippines. In the year 2011, the global production again increased to around 107,000 tonnes (Anon, 2013b).

8.4 Fibre structure and properties

Natural fibres, as well as abaca fibres, are composite materials, formed by the combination of three basic polymers:

- Cellulose
- Hemicelluloses
- Lignin.

A single cellulose molecule is formed by glucopyranosyl links between glucose units. It contains about 1500 glucose units, forming the primary structure of cellulose. This long-chain molecule lies side by side in bundles held together by hydrogen bonds, between the numerous neighbouring hydroxyl groups. The hydrogen bonds cause the cellulose bundles to be twisted to form rope-like structures which join together and form fibres. The cellulose bundles are embedded in lignin, which cements them to each other, producing the structure (Mathai, 2005). The cellulose and hemicellulose chains are integral parts of the cementing mechanism. Lignin plays an important role in the physical properties of the fibres. The hardness and stiffness of fibres depend on the lignin content and how it is embedded. The chemical composition of abaca fibres is shown in Table 8.5. The main constituent of abaca fibre is cellulose, making up about 65% of the total weight. It is also about 20% hemicelluloses, and the remainder consists of lignin, pectin, fat and wax (Batra, 1998; Franck, 2005; Lewin and Pearce, 1985).

The diameter of the abaca fibre bundles diameter is not uniform along the fibre, as shown in Fig. 8.1. It ranges from 60 to 250 μm. The fibre surface

Table 8.5 Chemical composition of abaca fibre

Component	Proportion (wt%)
Cellulose	63–68
Hemicellulose	19–20
Lignin	5–6
Pectin	<1
Fat and wax	<1
Water soluble	1–2

Sources: Batra, 1998; Franck, 2005; Lewin and Pearce, 1985.

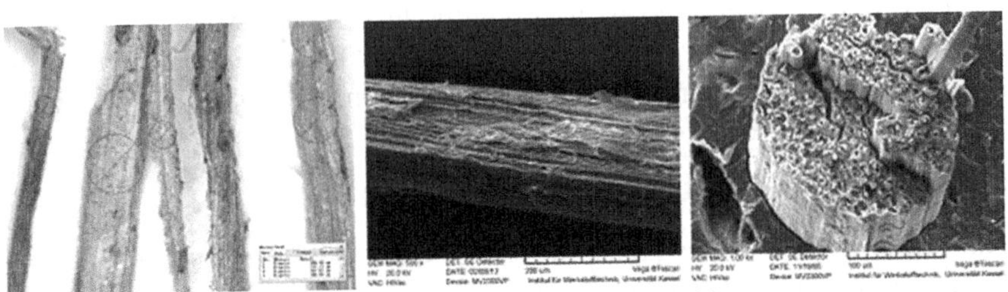

(a) (b) (c)

8.1 Abaca fibre: (a) longitudinal views, (b) surface, (c) cross-section.

is rough and damage is shown through the waxy and protruding parts (see Fig. 8.1). The cross-section of abaca fibre displays a compact structure with cylindrical holes orientated along the fibre axis. It is also evident that the bundles of polygonal shapes with round corners have a sufficiently large oval to circular lumen. In general, the abaca fibre bundles are surrounded by a layer of stegma cells filled with silica (Batra, 1998).

Energy dispersive X-ray spectroscopy analysis (EDX) was used to analyse the surface elements of abaca fibres. They consist mainly of cellulose (oxygen to carbon ratio, O/C = 0.83), hemicelluloses (polysaccharides with a range of monomers, where O/C ≈ 0.83), lignin (O/C = 0.35), pectin (O/C similar to hemicelluloses), and wax or lipids. The greater proportion of cellulose lies within the elementary fibre in the form of crystalline and amorphous cellulose. It can thus be predicted that the fibre surface will be enhanced in lignin and hemicellulose, which help to bind the fibres together, and in wax, which acts as a natural barrier to the environment for the plant stem. Figure 8.2 shows the spectra for abaca fibres. The relative atomic percent of the atoms was obtained from the peak area and corrected with an appropriate sensitivity factor. The spectra contained mainly carbon and oxygen as well

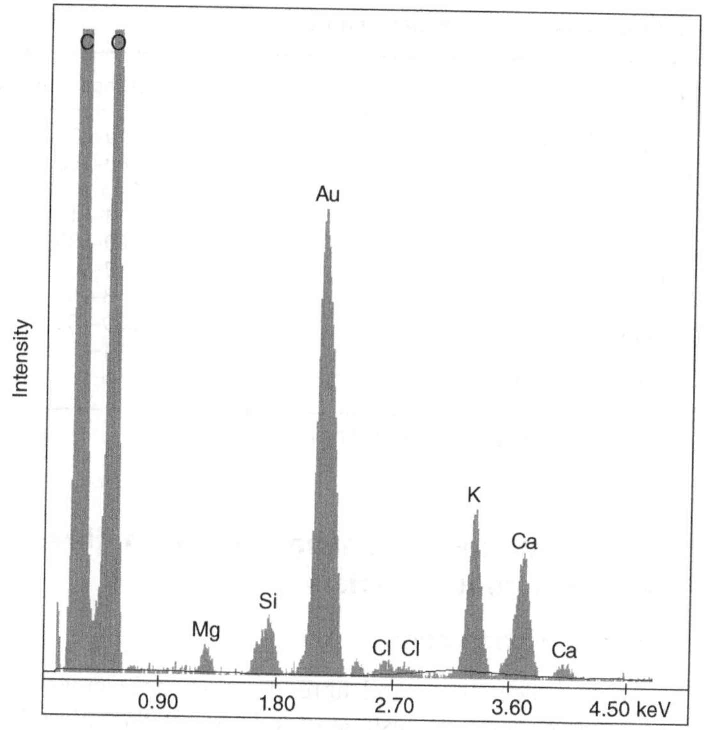

8.2 Energy dispersive X-ray spectroscopy of abaca fibre.

as small amounts of silicon, potassium, magnesium and calcium. The oxygen to carbon ratio (O/C) equalled 0.42, which is close to the O/C of lignin or fat and wax. Therefore, the abaca fibre surface could be covered by lignin or fatty material. The amount of fat or lipid on the fibre surface influences its hydrophobicity and surface tension. The greater the amount of lipid on the fibre surface, the more hydrophobic the fibres. The fibre surface layer is a relatively small portion of the overall fibre but plays an important role with regard to the wettability and surface tension.

There are many reports in the scientific literature on the physical, mechanical and thermal properties of abaca fibres. It is difficult to present all the properties in one chapter as the data differ for each citation depending on the different fibres, different conditions and testing methods used. Table 8.6 shows the properties of abaca fibres (Jarman, 1998; Biswas *et al.*, 2001). The density data represent the actual density, which is that of the whole fibre, not just the fibre cell wall. The diameter and linear density data of abaca fibres ranges very widely. This is because the fibres were separated using different methods and equipment. Abaca fibres thus have a wide range of mechanical properties depending on the exact fibre types.

Table 8.6 Properties of abaca/banana fibres

Property	Abaca/banana fibres
Length of commercial fibre (cm)	10–20
Diameter (mm)	60–250
Density (g/cm^3)	1.35
Linear density (Tex)	4–44
Tensile strength (MPa)	400–980
Elongation at break (%)	2–8
Elastic modulus (GPa)	8–20
Tenacity (MPa)	530–750
Micro-fibrillar angle (°)	11–29
Volume resistivity at 100 volts (Wcm × 10^5)	6.5–7

Sources: Jarman, 1998; Biswas *et al.*, 2001.

8.5 Disadvantages of banana and abaca fibres as reinforcement materials

8.5.1 Moisture absorption

One common problem of natural fibres is their rapid moisture uptake, which can lead to dimensional instability and rotting. Plant fibres like abaca fibres develop around a living cell which dies and creates a cavity or lumen. The fibre cell is thereforehollow and can transport moisture into a material during mixing. The moisture absorption property of abaca fibres depends on conditions such as temperature and relative humidity. The moisture absorption of S_2 grade abaca fibre was found to reach equilibrium after 90 days. The equilibrium moisture content (EMC) of this abaca fibre was 11% at 23°C with 65% RH and 17% at 50°C with 95% RH.

The transverse swelling in water of abaca/banana fibre has been reported by Satyanarayana and Wypych (2007) as 16 to 20%. Fibres change their dimensions with changing moisture content because the cell wall polymers contain hydroxyl and other oxygen-containing groups that attract moisture through hydrogen bonding. The hemicelluloses are mainly responsible for moisture absorption, but any accessible cellulose, noncrystalline cellulose and surface of cellulose also have an important effect. Water molecules absorbed by dry cellulose form a true cellulose hydrate. As the reaction is exothermic, it provides the driving force for further absorption. The primary layer of water bound directly to the $-CH_2OH$ group forms a relatively strong hydrogen bond and attracts other water molecules by weaker hydrogen bonding. This weaker bound layer may be thickened by several molecular layers and shows some order of moisture absorption (Pott, 2004; Young, 2007). This is a disadvantage for the industrial use of plant fibres.

However, it can be controlled by a number of surface modification treatments including heat treatments or chemical modification such as acetylation, although these are of course an extra cost.

8.5.2 Biological, heat and UV resistance

Fibres are degraded biologically because organisms recognize the carbohydrate polymers by the hemicelluloses in the cell wall. Biodegradation of the high molecular weight cellulose and hemicelluloses weakens the fibre cell wall. Fibre strength is lost as the fibre cellulosic polymer degrades due to oxidation, hydrolysis and dehydration reactions caused by the actions of fungi and/or bacteria (Huang and Edelman, 1995). Photochemical degradation takes place primarily in the lignin component, which is responsible for the characteristic colour changes. The lignocellulosic fibres burn with increasing temperature because the cell wall polymer undergoes pyrolysis reactions which give off volatile and flammable gas. The fat, starch and protein polymers are degraded by the heat long before the cellulose and hemicelluloses and their corresponding polymers are degraded by heat earlier than the lignin is (Mamun, 2011; Scott, 1995). Thus, the chemical composition and quantities clearly play a vital role in determining the fibre properties and composites properties when used as a reinforcement.

Figure 8.3 shows the thermal gravimetric (TG) and differential thermal gravimetric (DTG) analysis curves of abaca fibres that were measured in air and nitrogen at a heating rate of 10°C/min. The DTG curve for both conditions showed initial peaks between 40°C and 140°C, which correspond to the vaporization of water. After these peaks, both curves exhibit double decomposition steps. The decomposition peak temperatures in air were 334°C and 431°C. In nitrogen, the curve exhibits two decomposition steps and the decomposition peaks were at 431°C and at 614°C.

The incineration of natural fibre occurs in two steps. The decomposition peak at around 300°C is due to the thermal decomposition of hemicellulose and the glycosidic linkage of cellulose. The peak in the range 350–390°C is due to α-cellulose decomposition. Lignin degrades first but at a slower rate than the other constituents (Mamun, 2011). The peak corresponding to lignin degradation is slightly broader and appears in the range 190–500°C with a maximum value at about 350°C. Thus, it can be overlapped with the hemicellulose and cellulose peaks. The sharp decrease at about 600°C is caused by a sudden oxidation of the fibre contents in air. The thermal gravimetric analysis (TGA) programme was associated with two steps. The first step consisted of 25°C to 600°C in nitrogen, after which oxygen was supplied. The weight loss at temperatures up to 800°C was 98%. Therefore, the inorganic content of the fibre was close to 2%.

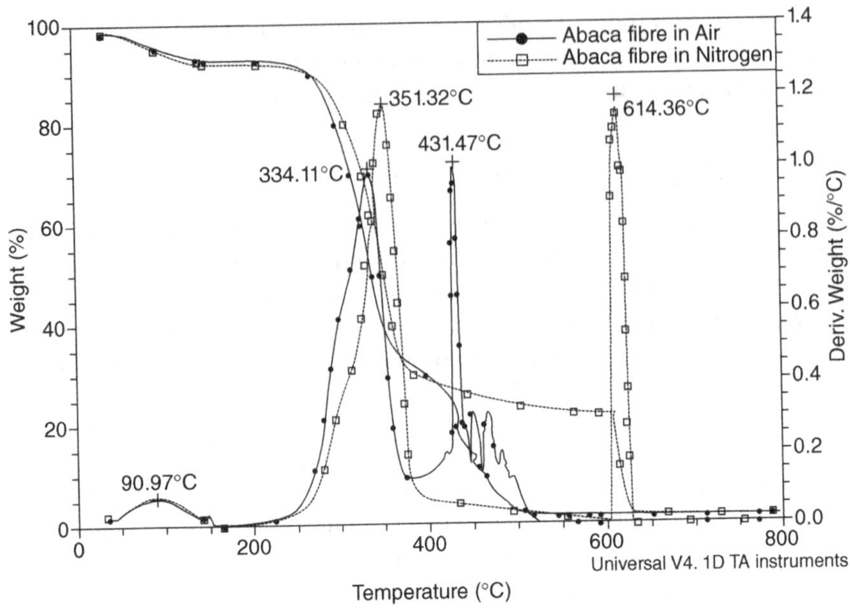

8.3 Thermal gravimetric and differential thermal analysis of abaca fibre.

8.6 Surface modification of fibres

The surface properties of natural fibre can be divided into two major groups: physical and chemical. Physical properties include the morphology, roughness, smoothness, specific surface area and permeability of natural fibres. Chemical properties include elemental, molecular, functional and group composition. Together these two groups determine the thermodynamics of the surface such as free energy, acid–base acceptor and donor numbers (Mamun, 2011). Natural fibre surface treatment generally involves altering the surface chemistry which changes the thermodynamic properties and creates beneficial microtopographical features without deliberately coating the surface (Tshabalala, 2005). In most cases, effective surface treatments not only remove native surface material and leave behind a more active functional group to promote wetting, but also roughen the surface to some degree. This increases the surface area, and can enhance mechanical interlocking. In terms of composite processing, the use of a surface treatment is desirable to promote wettability and increase the interfacial bonding between the matrix and the fibre (Rowell, 2001).

Unfortunately, the high moisture sensitivity of lignocellulosic fibre causes dimensional instability and limits the use of fibres as reinforcement in composite materials. Poor interfacial properties between the fibre and the polymer matrix often reduce their potential as reinforcing agents due to the hydrophilic nature of natural fibres. The three methods to alter the fibre surface, physical, chemical and biochemical, are explained below.

8.6.1 Physical methods

Physical methods include stretching, heat treatment, 'calendering' and the use of corona and plasma. These canalter surface polarity and other surface properties and improve the interface between the fibre and the matrix. This method is not commonly used for abaca fibre.

8.6.2 Chemical methods

Chemical modifications and chemical coupling agents are considered to optimize the interface of fibres. Chemical coupling agents are usually molecules with two key functions, firstly to react with hydroxyl groups of cellulose and secondly to react with functional groups of the matrix. The moisture absorbed by the fibres can be reduced by chemical modifications of the fibres such as acetylation, mercerization, methylation, cyanoethylation, benzoylation, permanganate treatment oracrylation.

Rahman *et al.* (2009) have reported that benzenediazonium salt can modify abaca fibre polypropylene composites, improving their mechanical properties. Teli and Valia (2013) explained that the property of abaca fibre oil absorption is increased by acetylation of abaca fibre. Pothan and Thomas (2003, 2004) investigated the chemical modification of abaca fibre-reinforced polyester composites in the compression-moulding process. The diffusion, sorption and permeability coefficient of various treated banana fibre-reinforced polyester composites are shown in Table 8.7. The equilibrium water uptake and diffusion coefficient were minimal for saline-treated fibre composites. The equilibrium water uptake of treated fibre composites increased with increasing temperature. Diffusion coefficient studies revealed that interfacial adhesion plays a vital role in the water-transport process.

Shibata *et al.* (2002, 2003) investigated treated and untreated abaca fibre-reinforced biodegradable polyesters in the injection moulding process. The mechanical properties can be improved regardless of the fibre treatment. The flexural strength of untreated and treated abaca fibre composites is shown in Fig. 8.4. The flexural strength of both abaca fibre composites increased with increasing fibre content in the range 0–10 wt%. Above this value, the flexural strength decreased for untreated fibre composites but only decreased slightly at 15 wt% fibre content and then increased again. The reason for this is not clear, although it may be related to the interfacial adhesion between matrix polymer and fibre, which is more important for the composite with higher fibre loading. The composite using 5h-BA abaca

Table 8.7 Diffusion,sorption and permeability coefficient of the various treated banana fibre-reinforced polyester composites

Sample	Temperature (°C)	Diffusion coefficient D (cm^2/s^{-1})	Sorption coefficient S (g/g)	Permeability coefficient P (cm^2/mm^{-1})
Untreated	30	0.123150	0.2261	0.0278
	50	0.6132	0.2431	0.1491
	70	0.052278	0.2388	0.01248
	90	0.19711	0.2982	0.0587
0.5% NaOH	30	0.1146	0.2117	0.0243
	50	0.1466	0.2321	0.0340
	70	0.055036	0.2250	0.01238
	90	0.2065	0.1811	0.0374
Saline A174	30	0.03554	0.2155	0.00766
	50	0.0562	0.2404	0.0135
	70	0.2611	0.2382	0.6219
	90	0.1188	0.1879	0.0223
Acetylated	30	0.09274	0.2558	0.02372
	50	0.3130	0.2115	0.0662
	70	0.10203	0.0172	0.00175
	90	0.13633	0.2392	0.0326

Sources: Pothan and Thomas, 2003, 2004.

had a higher flexural strength than the composite using untreated abaca. The abaca fibre treated with butyric anhydride/pyridine for 5hours is known as '5h-BA abaca'. The flexural properties of PHBV/5h-BA abaca composite were comparable to those of PHBV/GF composite, when compared with the same weight fraction of fibre.

8.6.3 Enzyme methods

Advances in biotechnology have had an overwhelming impact on society. Virtually all chemical reactions in biological systems are catalysed by macromolecules called enzymes. The reactions rarely occur at perceptible rates in the absence of enzymes while their rates increase as much as a million times in the presence of enzymes. The ability of enzymes to function outside a cell has greatly increased their use in a large variety of commercial products and reactions. The uses of enzyme technology in the processing of natural fibre and in the field of textiles, particularly for modification, are increasing substantially. A major advantage of this technology is that the application of enzymes is environmentally friendly and the reactions catalysed are very specific, causing focused results. Other potential benefits of enzyme technology include cost reduction, energy and water saving,

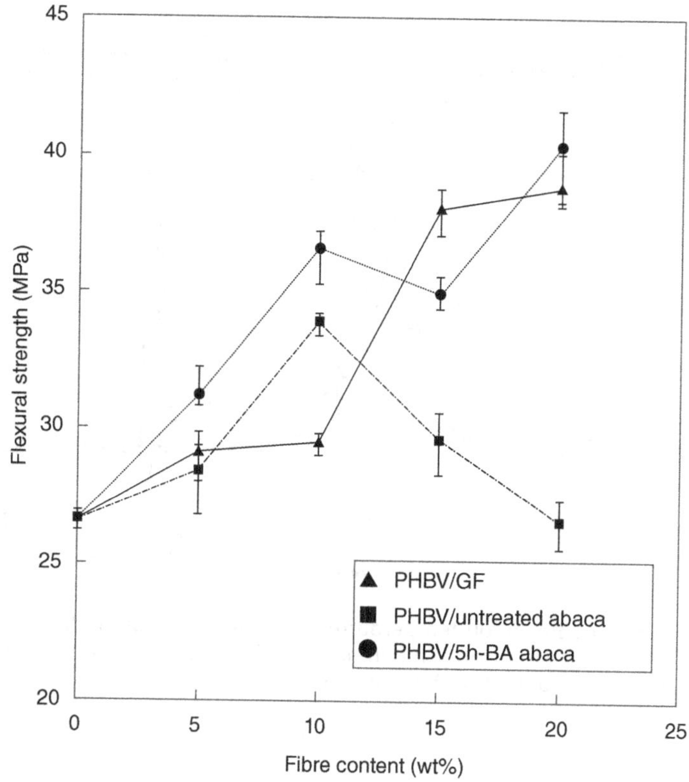

8.4 Flexural strength of modified abaca fibre and glass fibre PHBV composites.

improved product quality and potential process integration (Filer, 2003; Aehle, 2004).

The use of enzymes to modify the surface of chemical fibres and natural fibres such as hemp, flax, wool and cotton has been widely researched in industry. Most of the industrial applications aim to improve surface properties by removing adsorbed components, such as lignin, fats, waxes, proteins and non-crystalline parts. Not only are unwanted adsorbed materials removed but also the modification of the fibre surface can be accomplished using enzymes (Valchev *et al.*, 2009; Hansen, 2001). The effect of acetylation and enzyme modification of abaca fibres on their decomposition temperature is shown in Fig. 8.5. The main decomposition temperature of abaca fibres, explained above, is about 353°C. The decomposition temperatures increased by 24°C and 29°C due to acetylation and enzyme modification, respectively. The increase of the decomposition temperature of treated fibres is due to the removal of amorphous and smaller molecules (and thus an increase in molecular weight) from the fibre surface.

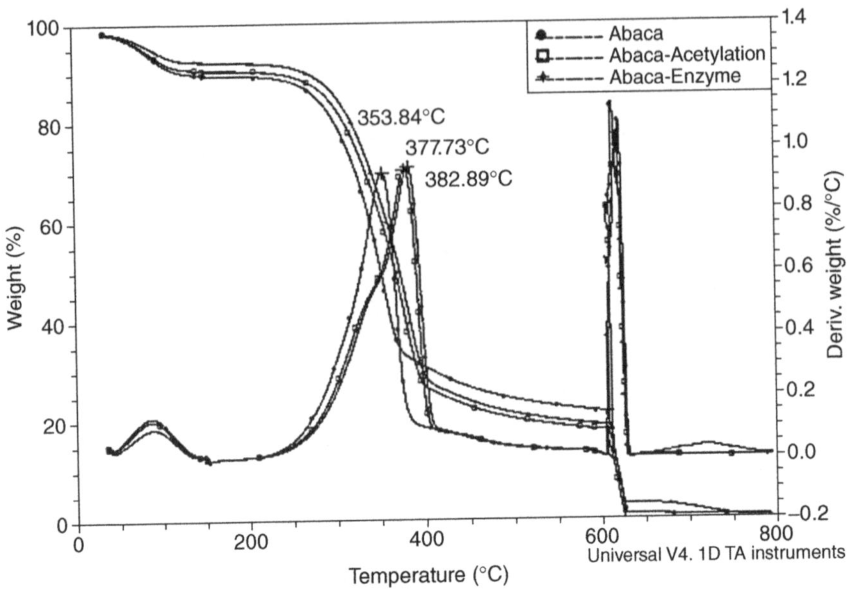

8.5 Decomposition temperature of modified and unmodified abaca fibre (reprinted from Shibata *et al.*, 2003, © 2002 Wiley-VCH Verlag GmbH & Co. KGaA, Weinheim, Germany).

8.7 Processing of banana/abaca fibre-reinforced composites

The nature, fibre types and processing parameters, processing method and tools geometry of the matrix materials have a strong influence on the processing and the final properties. Natural fibre as well as abaca fibre composites are of limited use in modern industrial applications. This is because the properties of abaca fibres vary considerably. The list below gives the factors affecting these properties:

- Fibre diameter
- Fibre type
- Thermal stability
- Structure (e.g. proportion of crystalline fibrils and non-crystalline regions, spiral angle of fibrils)
- Supramolecular structure (degree of crystallinity)
- Degree of polymerization
- Crystal structure: type of cellulose, defects, orientation of the chains of non-crystalline and crystalline cellulose
- Void structure: pore volume, specific interface, size of pores

- Whether the fibres are taken from the plant stem, leaf or seed and the corresponding growing conditions (Mamun, 2011; Bledzki and Gassan, 1999).

Secondly, this also due to the nature of the matrix material – whether it is polar or non-polar, melt rheology, etc. Finally, the properties of composite materials largely depend on processing parameters, methods and tool geometry. The use of the plastic technique for abaca fibres required a modification or optimization of the processing technique methods (Bledzki *et al.*, 2002). For example, new extruder screw design facilities reduced the thermal degradation of PLA and improvedthe compound's properties. The processing of biodegradable polymer (PLA) composites using existing compounds leads to thermal, oxidative and hydrolytic degradations during processing. This results in the cleavage of polymer chains, and hence in a decrease in molecular weight (Scherzer *et al.*, 2006).PLA (semi-crystalline) undergoes thermal degradation at temperatures above200°C due to the occurrence of hydrolysis, lactide reformation, oxidative main chain scission, and inter- or intra-molecular transesterification reactions.These degradation processes of PLA may also be influenced by the surface molecules of natural fibres. To overcome this problem the chain extension method could be used instead of an adaptation of the extruder screw geometry.The circled sections of the exiting screw are known as the mixing and kneading zones (Fig. 8.6). In the mixing or kneading zone, the screw elements are arranged with a specific angle or perpendicular to the screw direction. The shear moment generated in those zones during compounding is high and increases local temperature from 30°C to 50°C, which helps to degrade the PLA. These specific screw zones are modified without reducing the mixing ability. However, innovative technologies and process solutions must be carefully researched.

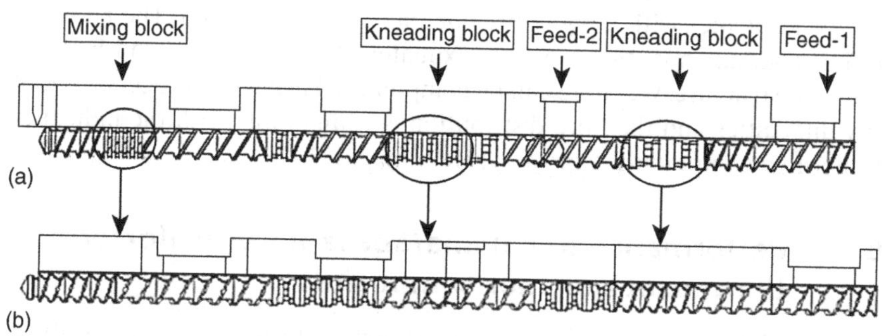

8.6 Extruder configurations: (a) exiting screw geometry, (b) modified screw geometry.

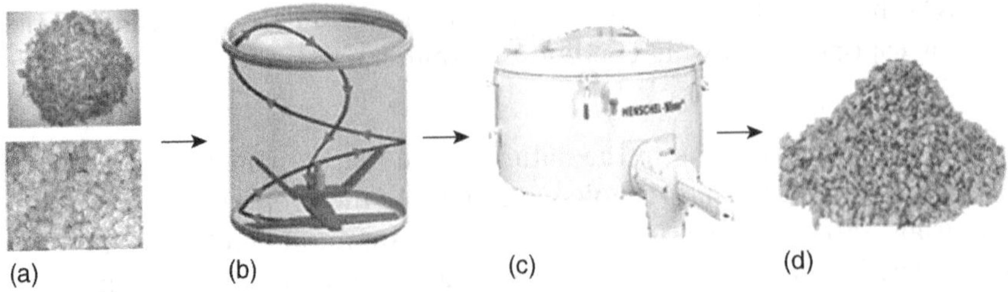

8.7 Processing steps of heating–cooling mixer: (a) fibre and matrix, (b) heating mixer, (c) cooling mixer, (d) agglomerate.

In the case of abaca fibre composites, the fibre types and forms and the nature and form of the matrix material are important in the selection of the processing technique. Short and long abaca fibres are available on the market. They can be compounded using the usual industrial techniques involving a heating–cooling mixer, extrusion and pultrusion, and compression and injection moulding. The heating–cooling mixer method is described below.

The heating–cooling mixer is the simplest way to produce granulates reinforced with short abaca fibres. This is a batch process, also known as a high-speed cascade mixer. The heating mixer is assembled with a special design impeller (Fig. 8.7), which is responsible for the friction and heating of the material. No external heating source is needed for this system. The abaca fibres and the matrix material are loaded into the heating mixer and the heat generated by the friction increases to the melting temperature of the matrix material. The mixing ability and the heating rate are controlled by the impeller speed. The hot agglomerate granules can be transferred directly from the heating mixer to the cooling mixer, where the agglomerate is cooled to room temperature. The size of the master batch mixers can vary from 10 litres to 1000 litres. The production rate ranges from 20 kg to 1000 kg per hour. In turn, the fibre content is adjustable within a wide range and concentrates with a high fibre content can be produced. The resulting pellets contain short fibres of approximately 1 mm in length and have an irregular structure. Wet fibre can be compounded using this process without any difficulties and the fibre distribution is comparable with the industrial twin screw compounding process (Mamun, 2011).

8.8 Performance of banana/abaca fibre-reinforced thermoset polymer composites

Natural fibre composites have been gaining importance in the furniture, packaging, automobile, building and construction industries. They have the potential to replace some existing methods in the construction industry. The

Table 8.8 Properties of kraft banana (B) and waste kraft *E. grandis* (EG) reinforced cements

Fibre content (%w/w)	Flexural modulus (GPa)	Flexural strength (MPa)	Fracture toughness (kJ m^{-2})	Water absorption (%w/w)	Density (g cm^{-3})	Permeable void volume (%v/v)
B-4	13.1 ± 1.5	15.5 ± 1.3	0.21 ± 0.03	16.5 ± 0.2	1.71 ± 0.02	28.2 ± 0.3
B-8	8.85 ± 0.81	19.5 ± 1.4	0.53 ± 0.08	18.4 ± 0.4	1.58 ± 0.02	29.0 ± 0.7
B-12	7.04 ± 1.22	20.1 ± 2.5	1.01 ± 0.15	21.4 ± 0.9	1.50 ± 0.04	32.1 ± 0.8
EG-4	15.3 ± 0.9	15.6 ± 0.8	0.29 ± 0.04	16.8 ± 0.8	1.78 ± 0.03	29.8 ± 0.8
EG-8	11.4 ± 0.9	21.4 ± 0.9	0.82 ± 0.11	20.7 ± 0.7	1.60 ± 0.02	33.3 ± 0.6
EG-12	8.04 ± 1.06	22.2 ± 1.3	1.50 ± 0.18	24.8 ± 0.8	1.47 ± 0.02	36.5 ± 0.6

Source: reprinted from Savastano *et al.*, 2000, © 2000, with permission from Elsevier.

abaca fibre is a particularly important natural fibre. The applications (short- or long-term) of abaca fibre composites have been considered according to physical, mechanical and chemical performance. Abaca fibre-reinforced cement composites have been researched by several scientists. Zhu *et al.* (1994) investigated banana fibre reinforced cement composites and reported that these materials could be suitable for use in building materials due to their moisture content of less than 25%.

Savastano *et al.* (2000) studied Brazilian fibre-reinforced cement-based composites. They prepared these using kraft pulps from banana waste (B) and from *Eucalyptus grandis* (EG) pulp mill residues. Cement composites with fibre mass fractions ranging from 4% to 12% were prepared in the laboratory using a slurry vacuum dewatering technique. The mechanical, water absorption, density and permeable void volume properties of these composites are shown in Table 8.8. The results show that the optimum performance of the various waste fibre-reinforced composites was obtained at a fibre content of around 12% by mass. The flexural strength values were about 20 MPa, fracture toughness values ranged from 1.0 to 1.5 kJ m^{-2} and density was approximately 1.5 g cm^{-3}. Experimental results showed that *E. grandis* is the best reinforcement for low-cost fibre-cement. These composites can only be used in low-cost housing construction.

Joseph *et al.* (2002) investigated and compared the mechanical properties of phenol formaldehyde (PF) composites reinforced with banana fibres and glass fibres. The incorporation of banana fibre induces an increase for the composites in their tensile strength of 400% and in their Young's modulus of 320%. Figure 8.8 shows the stress–strain behaviour of banana fibre and glass fibre PF composites under flexural loading. The glass fibre composites show high elasticity and lower extensibility. The high flexural strength of glass fibre is due to its own inherent properties. Banana fibre introduces

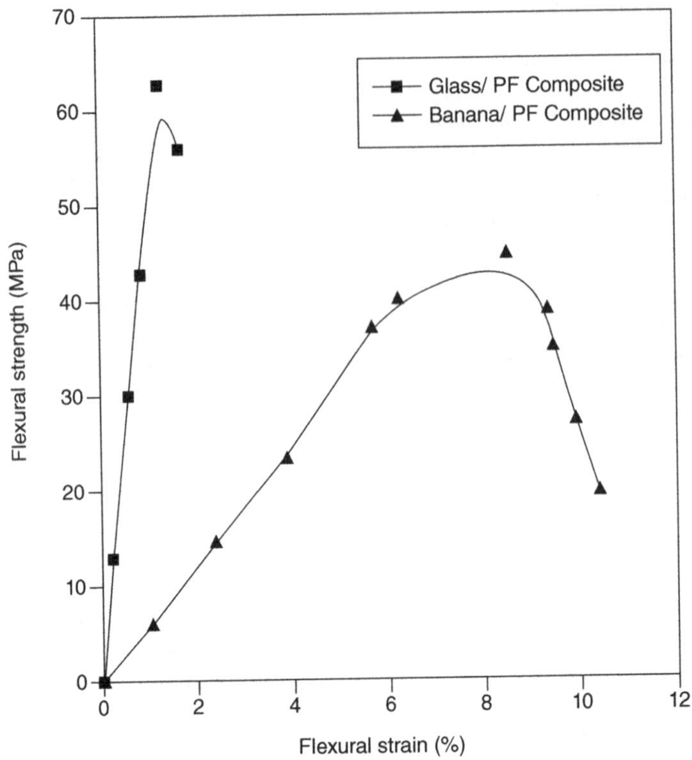

8.8 Flexural stress–strain curves of banana fibre and glass fibre reinforced composites (fibre loading 40%) (reprinted from Joseph *et al.*, 2002, © 2002, with permission from Elsevier).

plasticizing effects on the PF matrix. The banana and PF composites thus have higher toughness. Due to the high extensibility of the banana fibre, the composites can withstand the stress applied, which will prevent catastrophic failure.

Pothan *et al.* (2003) conducted dynamic mechanical analysis of banana fibre-reinforced polyester composites and found that the critical fibre loading value was 40 wt% fibre with 41 kJ per mol activation energy. At this value, the loss modulus peak broadens, reflecting the improved fibre/matrix adhesion. The glass transition temperature shifted positively after the addition of fibre. Cole–Cole plots show an imperfect semicircle which demonstrated the heterogeneity of the system as well as the good interfacial adhesion at a high fibre load. Chemical modification (such as the use of acetylation, NaOH and saline) can improve the storage modulus of banana fibre-reinforced polyester composites. Further, treatment with silane (A174) and NaOH caused the maximum increase of modulus values (Pothan *et al.*, 2006).

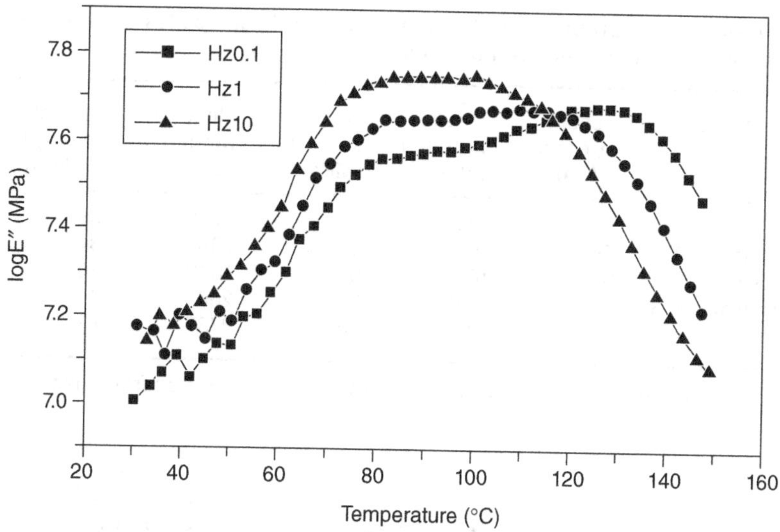

8.9 Effect of frequency on the storage modulus of saline (A174)-treated banana fibre reinforced polyester composites (40% fibre loading) (reprinted from Pothan*et al.*, 2006, © 2006, with permission from Elsevier).

Figure 8.9 shows the plot of the storage modulus G' as a function of temperature for saline (A174)-treated fibres at different frequencies (with fibre loading 40%). The viscoelastic properties of a material depend on the temperature and frequency of the applied stress. If a material is subjected to a constant stress its elastic modulus will decrease over a period of time. This is because the material undergoes molecular rearrangement in order to try to minimize the localized stresses. Storage modulus measurements performed at high frequency thus result in higher values, whereas measurements performed at low frequency result in lower values.

Idicula *et al.* (2005) investigated randomly orientated short banana and sisal hybrid fibre-reinforced polyester composites. They reported that 0.4 was the optimum volume fraction of banana and sisal fibre and the resulting composites showed their best properties. The mechanical properties of banana/sisal hybrid composites with different relative volume ratios of fibres are given in Table 8.9. A better fibre/matrix adhesion and stress transfer was found in composites with a relative volume fraction of banana and sisal of 3:1. These composites also showed the highest tensile strength and flexural modulus and lowest impact strength.

Idicula *et al.* (2006) have also studied the thermo-physical properties of polyester and banana/sisal fibre composites. They found that the inclusion of a banana/sisal fibre volume fraction ranging from 0.2 to 0.4 decreased the effective thermal conductivity of the composite from 0.181 to 0.153 W/

Table 8.9 Mechanical properties of banana/sisal hybrid composites possessing different relative volume ratio of the fibres

Volume ratio of fibres	Tensile strength (MPa)	Tensile modulus (MPa)	Flexural strength (MPa)	Flexural modulus (MPa)	Impact strength (kJ/mm²)
Banana	57	1352	53	2723	36
Banana : sisal = 3 : 1	59	1536	57	2981	34
Banana : sisal = 1 : 1	57	1601	62	2842	36
Banana : sisal = 1 : 3	55	1477	57	2661	41
Sisal	53	1079	56	2737	43

Source: reprinted from Idicula *et al.*, 2005, © 2005, with permission from Elsevier.

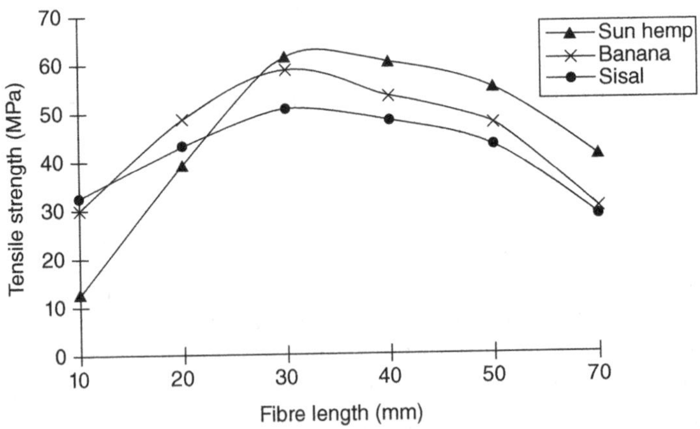

8.10 Effect of fibre length on tensile strength of various natural fibre-based composites (reprinted from Kiran *et al.*, 2007, © 2007, with permission from Elsevier).

mK. They concluded that the thermal conductivity of the polymeric matrix seems to be controlled by the banana/sisal volume fraction. The treated fibre showed a significant increase in thermal conductivity and density in both banana/sisal fibre composites.

Kiran *et al.* (2007) studied the tensile strength of banana fibre, sisal fibre and sun hemp reinforced polyester composites. Banana fibre (51 wt%) achieved the best tensile strength, about 59 MPa. The tensile strength of banana fibre, sisal fibre and sun hemp composites with respect to the fibre length are shown in Fig. 8.10. The tensile strength increases with the fibre length up to 30 mm. After this point, further increase in fibre length leads to a decrease in the tensile strength.

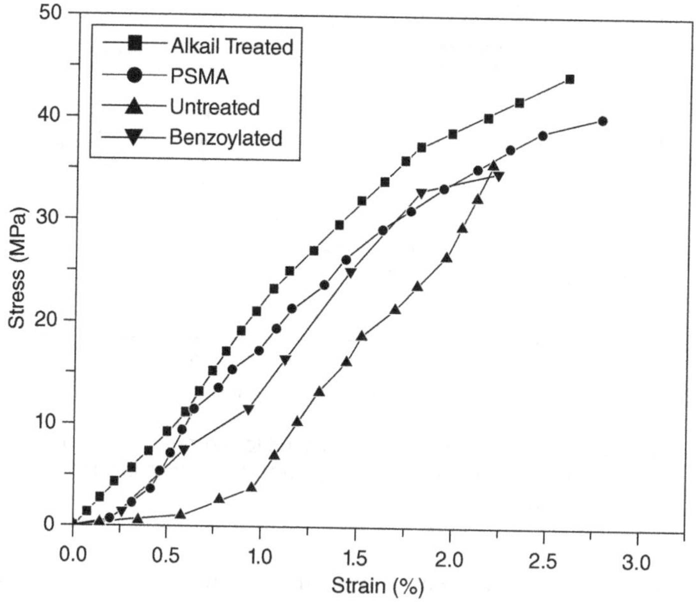

8.11 Tensile stress–strain curves of treated fibres (total fibre content 20% at 70/30 banana/glass concentrations) (reprinted from Haneefa *et al.*, 2008, © 2008, with permission from Elsevier).

Haneefa *et al.* (2008) researched the hybridization of banana fibre and glass fibre. They found that the increases in tensile strength and Young's modulus of the composites occur in correlation with an increasing volume fraction of glass fibre. This is because the glass fibre is more compatible with a polystyrene matrix. The tensile stress–strain curves of the hybrid composites are shown in Fig. 8.11. The tensile strength and modulus of hybrid composites with a fibre content of 20% at 70:30 banana to glass ratio were improved by chemical modifications (alkali treatment, benzoyl chloride and PSMA) of the banana fibre.

8.9 Performance of banana/abaca fibre-reinforced thermoplastic polymer composites

8.9.1 Effect of the fibre length

The tensile strength and modulus of abaca fibre–polypropylene (PP) composites depend on the fibre lengths and are illustrated in Fig. 8.12. The composites were produced using a heating–cooling mixer and then an injection–moulding process with 30 wt% fibre loading. The modulus and strengths increased with increasing fibre lengths from 5 mm to 40 mm. The

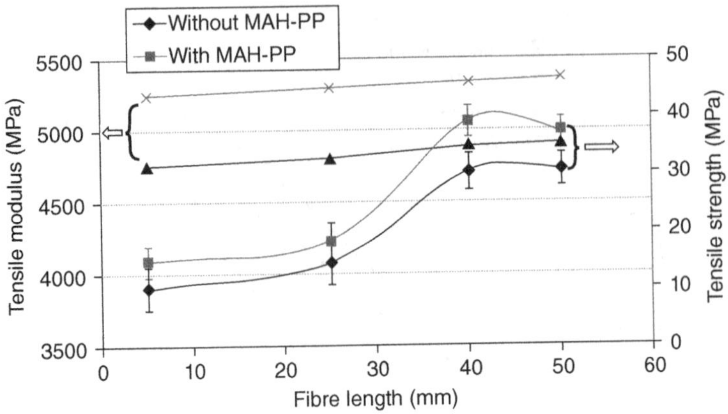

8.12 Influence of fibre length on tensile strength and modulus of abaca fibre–PP composites in injection-moulding process (fibre content: 30 wt%).

effect of the fibre ends was important to the fracture of short fibre composites. In order to achieve the maximum level of stress in the fibre, the fibre length must be at least equal to the critical fibre length required for the stress to reach the fracture stress of fibre. This is called the optimal fibre length. The stress level increases until the optimal fibre length and then decreases. This lowering of the stress value at higher fibre lengths is due to the entanglements formed at higher lengths. Modulus and strength increased until a fibre length of 40 mm, indicating that this is the optimum fibre length for this composite. The fibre length must be optimized for a particular matrix/fibre system so that the best properties can be achieved. However, another fibre length might be possible for a fibre content other than 30 wt%. It is also important to note that the length reduced randomly during composite processing. With the addition of MAH-PP, the tensile strength increased for all fibre lengths, ranging from 30% to 40%. The maximum increment of tensile strength was observed at fibre length of 40 mm (Bledzki *et al.*, 2008).

8.9.2 Effect of the fibre load

The tensile and the flexural strength depend on the different abaca fibre loads, as shown in Fig. 8.13. The tensile and flexural strengths increased until a fibre load of 40 wt% and then both decreased inversely with increasing fibre load. A fibre load around 40 wt% was shown to be the optimum fibre load for the abaca fibre-reinforced polypropylene composite. The SEM result (Fig. 8.14) confirmed that the abaca fibre was saturated with the polypropylene matrix at a fibre load of 40 wt%. The addition of MAH-PP

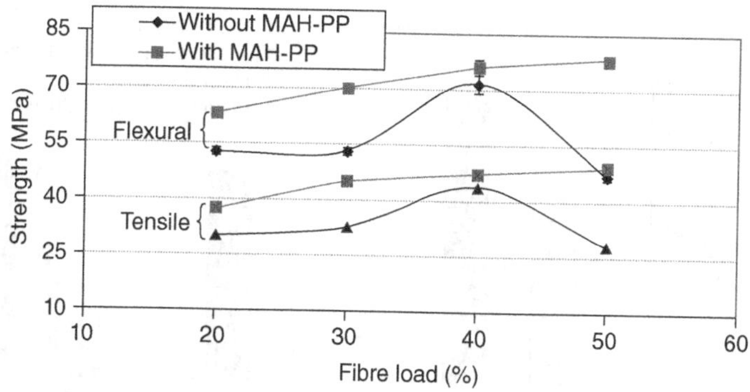

8.13 Influence of fibre load on the tensile and flexural strength of abaca fibre–PP composites.

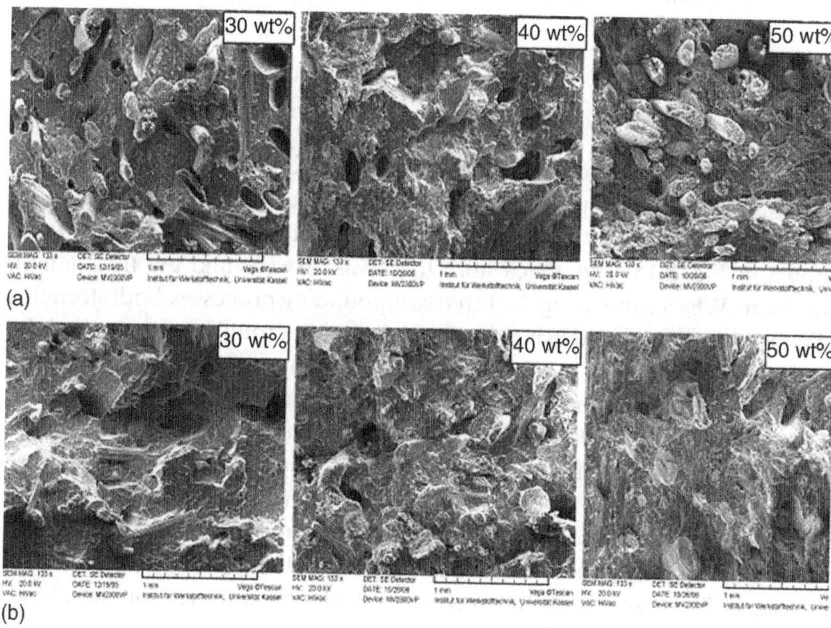

8.14 SEM micrographs of abaca fibre–PP composites with different fibre loads: (a) without MAH-PP, (b) with MAH-PP.

significantly reduced the occurrence of fibre debonding and pullouts. Moreover, the strength properties increased for all fibre loads. The addition of MAH-PP also meant that the strength properties were at their highest strength at a fibre load of 50 wt%. This may have occurred due to the strong equal portion of fibre and matrix interlinking via MAH-PP. Strength properties increased from 30% to 80% in all cases, compared with that of the fibre load at 40% (Bledzki *et al.*, 2008).

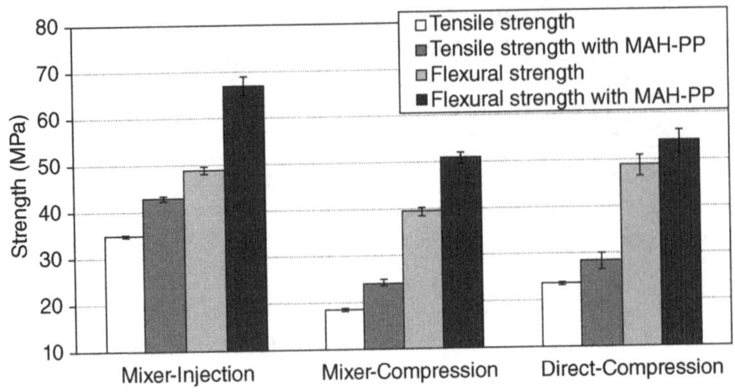

8.15 Influence of compounding processes on tensile and flexural strength of abaca fibre–PP composites (fibre length 5 mm, fibre content 30 wt%).

8.9.3 Effect of processing

The influence of compounding processes (including mixer-injection moulding, mixer-compression moulding and direct compression moulding) on the strength properties of the abaca fibre-reinforced PP composites is shown in Fig. 8.15. The abaca fibre load was 30 wt% and the fibre length was 5mm. When comparing the three compounding processes, both strengths of composites produced using the mixer-injection moulding process had the highest strength values of all the compounding processes. This may be due to the fibre distribution and alignment. It is notable that this property of the composites produced using the direct compression process was higher than that of the mixer-compression process. The shortening of the fibre length due to agglomeration could be the reason for the properties of the composites produced using the compression-moulding process. MAH-PP has a positive influence on both the tensile and flexural strengths, which increased 25% and 40% respectively, in the mixer-injection moulding process (Bledzki *et al.*, 2008).

Figure 8.16 depicts the notched Charpy impact strength of abaca fibre–PP composites in three different processes. The higher notched Charpy impact strength occurred in the direct compression-moulding process, and is 150% and 170% higher than that of the mixer-injection and mixer-compression-moulding processes, respectively. This is because the fibres were fractured into shorter lengths regardless of their initial length due to agglomeration in the mixing process. In the direct compression-moulding process, the fibres aligned with one another in composites according to their initial fibre length. This could be a reason for the higher Charpy impact strength. With the addition of MAH-PP, the notched Charpy impact strength increased in

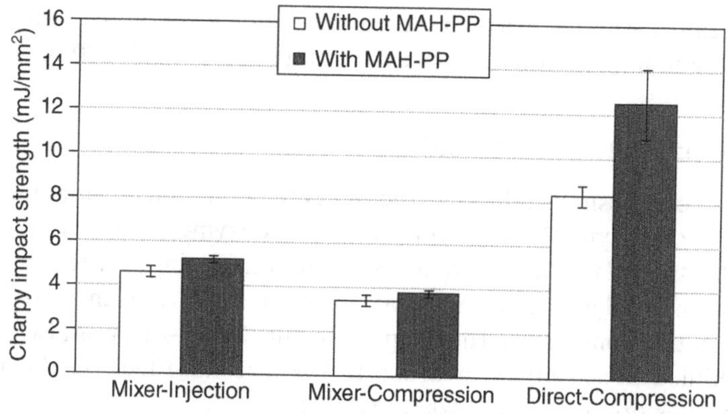

8.16 Notched Charpy impact strength of abaca fibre-reinforced PP composites in different processes (fibre length 5 mm, fibre content: 30 wt%).

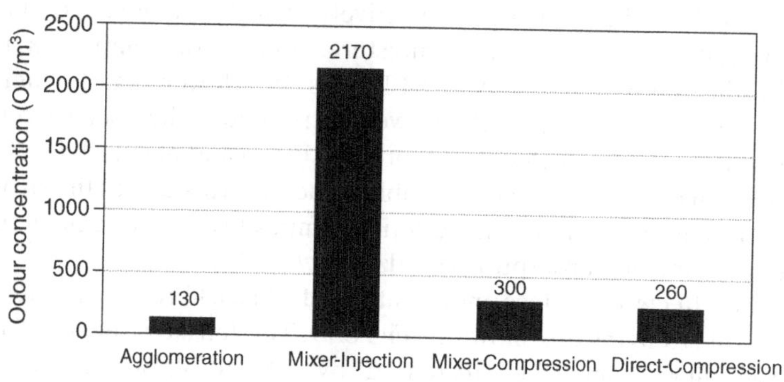

8.17 Odour concentrations of abaca fibre–PP composites after different processes (fibre length 5 mm, fibre load 30 wt%).

all three processes. However, in the direct compression-moulding process, this property increased 50% more than in composites without MAH-PP. The standard deviation in the direct compression-moulding process was relatively high, indicating the heterogeneity of the composites (Bledzki *et al.*, 2008).

The influence of the processing of abaca fibre–PP composites on odour emission is shown in Fig. 8.17. The odour emission induced by the mixer-injection moulding process was significantly higher than in the other processes. In the compression-moulding process there were relatively lower odour concentrations, which is favourable for the automotive sector. In the injection-moulding process, the fibre surface molecules decomposed instantly under high pressure and temperature. The increase of the temperature was due to a high shear force in the injection-moulding process, resulting in a higher odour concentration (Bledzki *et al.*, 2007).

8.9.4 Effect of modification

Prior to composite production, the fibres were modified using fungamix (a mixture of enzymes) and natural digested enzyme (NDS). The effects of modification of the fibre were assessed by considering moisture absorption and mechanical and environmental stress corrosion resistance properties of the resulting composites. The coupling agent (MAH-PP) was also used with unmodified abaca fibres in order to observe its effect on the resulting composites properties (Bledzki *et al.*, 2010).

The moisture absorption properties of modified and unmodified abaca fibre composites are shown in Fig. 8.18. Moisture absorption for all composites reached equilibrium after the 80-day conditioning period. The equilibrium moisture contents of fungamix and NDS modified abaca fibre composites were 45% and 20% lower, respectively, than for unmodified abaca fibre composites. The equilibrium moisture content of composites modified with the coupling agent (MAH-PP) was 35% lower than for unmodified composites. The enzymes' effects were attributed to the selective removal of surface extractives including fat, protein, lignin and hemicelluloses, without destroying the fibre surface and fibre structure. This causes the strong fibre/matrix adhesion found in the resulting composites and subsequent reduction of their moisture absorption (Bledzki *et al.*, 2010).

The effects of surface modification of abaca fibres on the tensile and flexural strengths are shown in Fig. 8.19. The tensile strength of modified abaca fibre composites increased by 5% to 45% due to different modifications. Modification of abaca fibre composites with natural digestive

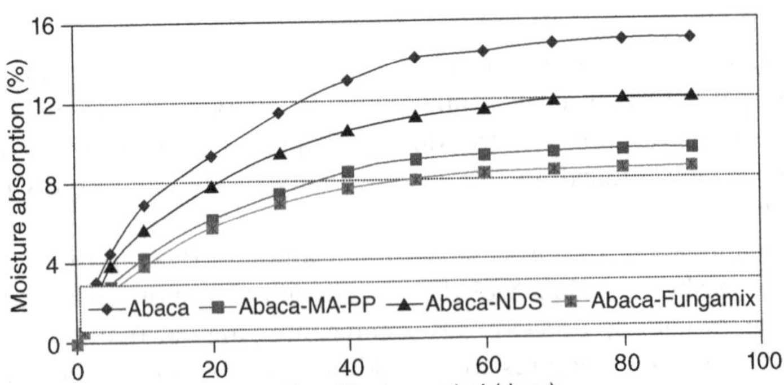

8.18 Moisture absorption of modified and unmodified abaca–PP composites (fibre length 5 mm, fibre load 30 wt%).

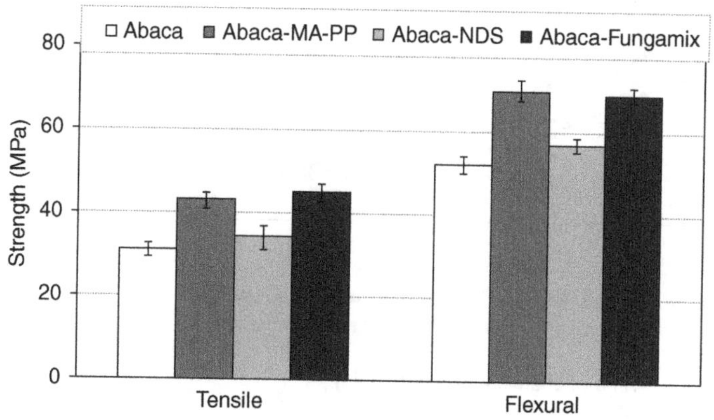

8.19 Tensile and flexural strength of modified and unmodified abaca–PP composites (fibre length 5 mm, fibre load 30 wt%).

system (NDS) caused little improvement. This may be because of the removal of the outer surface of fibre and simultaneous fibre damage during digestion. In contrast, the tensile strength of fungamix-modified composites increased by 45% when compared with unmodified fibre composites. The conventional coupling agent (MAH-PP) has a positive effect on the tensile strength. It improved tensile strength by 40% because an ester bond formed between the fibre and the matrix.

The flexural strength increased by 10% to 35% due to different modifications. Fungamix and MAH-PP modified abaca fibre composites both improved flexural properties by about 35% compared with unmodified fibre composites. The improvement of the tensile and flexural strengths of treated fibre composites is likely to be due to the removal of the outer surface. This increase in the effective surface area increases the interfacial adhesion and the number of physical and chemical changes induced by the fibre treatment. Fibrillation and diameter reduction of the abaca fibre also occur due to modification and may influence the strength properties of composites (Bledzki *et al.*, 2010).

Aggressive fluids lead to two types of failure mechanisms of cracking. The first and most obvious happens when these fluids irreversibly damage the polymer by inducing oxidation, chlorination and hydrolysis, and cause chemical modification in the form of chain scission and cross-linking (Mamun, 2011). The second type does not lead to a modification or degradation of the polymer. Instead it promotes cracking via a physical process which is generally understood to be due to the selective absorption of the fluid within a micro-yielded or stress-dilated zone. This embrittles the polymer in the presence of stress and reduces the yield strength. Local micro-yielding occurs before either ductile failure or brittle fracture.

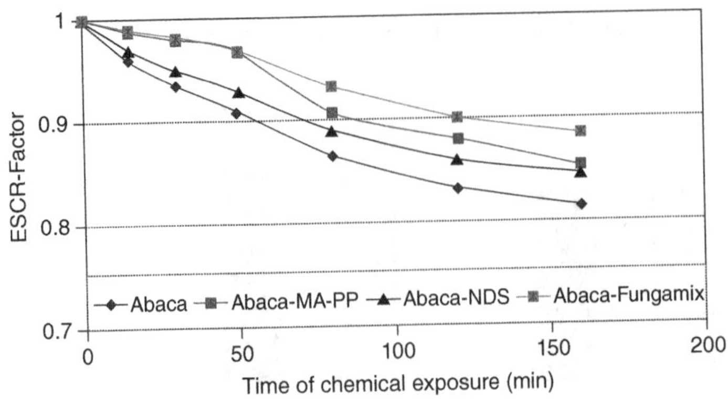

8.20 ESCR of modified and unmodified abaca–PP composites in HCl.

The failure mode is determined by the dynamics of initiation, growth and coalescence of individual zones. Initiation plays a very important role in both ductile and brittle fracture mechanisms (Wright, 1996). Environmental stress corrosion resistance (ESCR) depends on the system stress, strain, temperature, duration of exposure and concentration of an aggressive medium. When the stressed composite is exposed to a chemical environment, polymer resistance will vary with these parameters. Since micro-crazes, crazes or cracks weaken the chemically exposed polymer, it will usually take less time to rupture and have a lower stress value than the control medium.

Figure 8.20 shows the ESCR factor of modified and unmodified abaca fibre-reinforced polypropylene composites in an HCl medium. The ESCR factor decreased as the duration of the chemical exposure progressed. This behaviour can be expected of a viscoelastic material. There is a transition from a ductile to a brittle failure mode after 50 minutes of chemical exposure. The modified abaca fibre composites were more resistant to the transition than the unmodified abaca fibre composites. This could be due to the removal of surface impurities, leading to an increase in the interfacial adhesion. The fungamix-modified abaca fibre composites showed more resistance than NDS-modified abaca fibre composites. This could be because of a better interfacial adhesion than in NDS-modified abaca fibre composites. At the initial time of the chemical exposure period, MAH-PP modified composites showed more resistance than fungamix-modified composites because of the strong fibre/matrix adhesion caused by maleic anhydride. However, after a certain period of chemical exposure the resistivity of MAH-PP modified composites reduced swiftly (Bledzki *et al.*, 2010). It could be that this is caused by the fact that maleic anhydride reacts with water at room temperature to produce maleic acid. The high reactivity of maleic anhydride increases the local fluid concentration and thus reduces

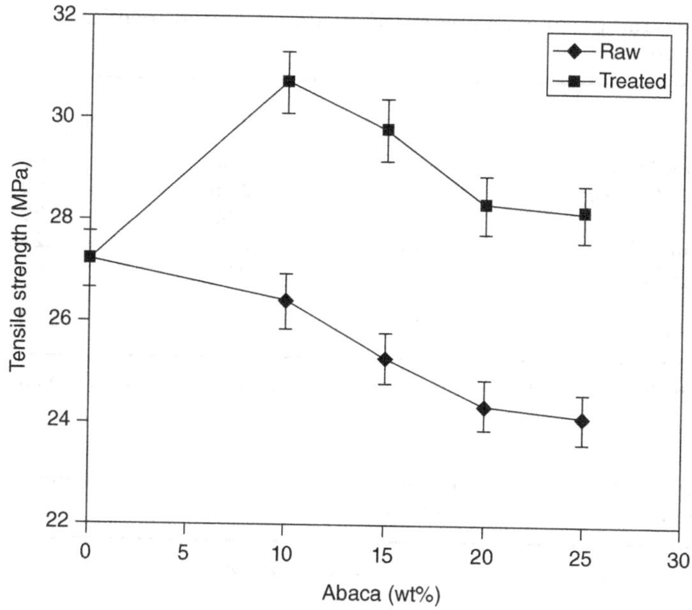

8.21 Variation of tensile strengths of abaca–PP composites with filler loading (reprinted from Rahman *et al.*, 2009, © 2009, with permission from Elsevier).

the local yield strength. The failure is preceded by absorption of the chemical medium within the micro-yielded zone. Further, the forming of micro-crazes increases the failure propagation (Mamun, 2011). Absorption of any fluid plasticizes the polymer and reduces its yield strength (Morris, 1994).

There are some relevant research studies into abaca fibre-reinforced thermoplastic composites which are discussed briefly below. In one, abaca was chemically treated with benzene diazonium salt and untreated and treated abaca fibres were used to prepare the composite. The tensile strengths of the composites of both untreated and treated abaca fibres decreased with increasing filler content. In fibre loading each type, the tensile strength of the treated abaca–PP composites was higher than that of composites containing untreated abaca fibres, as shown in Fig. 8.21. The Young's modulus, flexural strength, flexural modulus, impact strength and hardness values also increased in correlation with fibre loading. The values were higher for treated abaca–PP composites than for untreated fibre composites (Rahman *et al.*, 2009).

The chemical modification of banana fibre and its PP composites has been studied by Paul *et al.* (2010). The polarity parameters of chemically treated banana fibres are given in Table 8.10. The hydrogen bond accepting ability was lowest for treated fibre with 10% alkali. It was shown that the

Table 8.10 Kamlet–Taft polarity parameters of the untreated and chemically treated banana fibre

Sample	α	π^*	β
Untreated banana fibre	1.60	0.90	0.50
NaOH treated (10%) fibre	1.30	0.76	1.00
Benzoylatedfibre	1.50	0.89	0.50
Stearic acid treated fibre	1.40	0.64	1.20
Vinyl trimethoxysilane treated fibre	1.43	0.67	0.80
$KMnO_4$treated fibre	1.41	0.74	0.90

Source: reprinted from Paul *et al.*, 2010, © 2007, with permission from Elsevier.

Table 8.11 Tensile properties of composites reinforced with abaca strands and MAPP as coupling agent and of composites with fibreglass

Reinforce-ment (%)	Abaca strands–PP			Fibreglass–PP		
	Modulus (GPa)	Strength (MPa)	Elongation at break (%)	Modulus (GPa)	Strength (MPa)	Elongation at break (%)
20	3.2 ± 0.1	35.3 ± 0.4	3.7 ± 0.2	4.8 ± 0.1	50.9 ± 0.9	3.1 ± 0.1
30	5.0 ± 0.1	44.9 ± 0.7	3.1 ± 0.2	5.6 ± 0.2	58.5 ± 4.3	3.0 ± 0.2
40	6.0 ± 0.1	48.6 ± 1.0	2.6 ± 0.2	7.7 ± 0.2	67.1 ± 1.7	2.4 ± 0.1

Source: reprinted from Vilaseca *et al.*, 2010, © 2010, with permission from Elsevier.

fibre/matrix interactions were dependent on the polarity parameters of the modified banana fibre surface. Banana fibre with a lower polarity was more compatible with the non-polar PP matrix. This research also revealed that chemical treatments of the banana fibre improved the mechanical properties of composites. The banana fibre composites treated with 10% NaOH demonstrated the best mechanical properties (Paul *et al.*, 2010).

Vilaseca *et al.* (2010) investigated abaca strands and glass fibre-reinforced polypropylene composites. They reported that the tensile properties of composites reinforced with E-fibreglass were higher than those obtained for composites with abaca strands. This was observed despite the same reinforcement contents. However, similar tensile strength values can be achieved by increasing the abaca strands volume. The results are summarized in Table 8.11.

Ibrahim *et al.* (2010) extracted banana fibres using alkali pulping and steam explosion which were used to produce polyethylene composites. The addition of 20% maleated fibre results in better adhesion to the polymer matrix and thus leads to an improved tensile strength. However, the material seems to lose its flexibility, which leads to poor adhesion to the polymer matrix, if more fibres are added. It can be assumed that increased phase adhesion results in better properties regardless of the smaller particle size. In addition, the higher crystallinity of the steam-exploded banana fibre results in a higher modulus. Moreover, the higher lignin content facilitates fibre dispersion in the polymer matrix.

8.10 Performance of banana/abaca fibre-reinforced biodegradable polymer composites

Mechanical properties such as the tensile, flexural and impact strengths and the stiffness of synthetic cellulose and abaca fibre-reinforced PLA have been investigated. The test results of PLA composites werefound to be a significant improvement on those of the native PLA matrix. Synthetic cellulose fibres showed a much better potential as reinforcements for polylactides than abaca fibres. However, abaca fibres seem to produce a higher composite stiffness.The impact strengths of abaca and cellulose fibre-reinforced PLA and PP composites are shown in Fig. 8.22. The most significant improvement can be seen in the CharpyA-notch impact strength of PLA/cellulose 70/30, which increasedby a factor of 3.6 in comparison with the control PLA value. The impact strength of PLA/abaca fibre composites improved by a factor of 2.4. In the case of PP composites, the

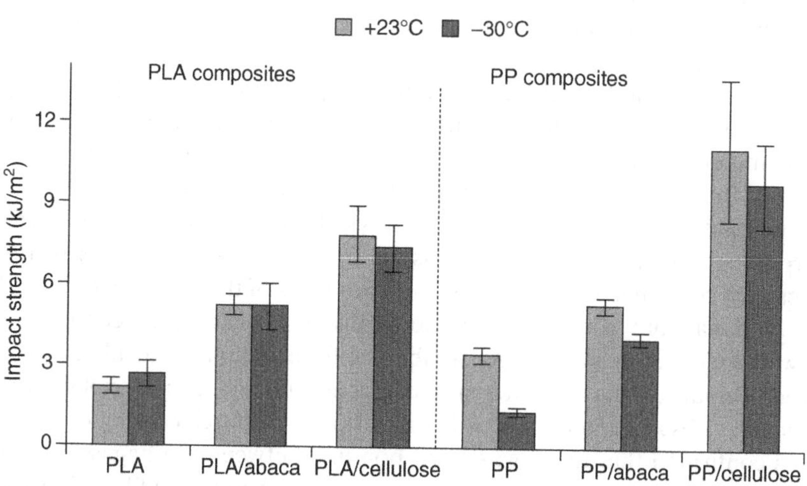

8.22 Notch Charpy impact strengths of PLA and PP composites.

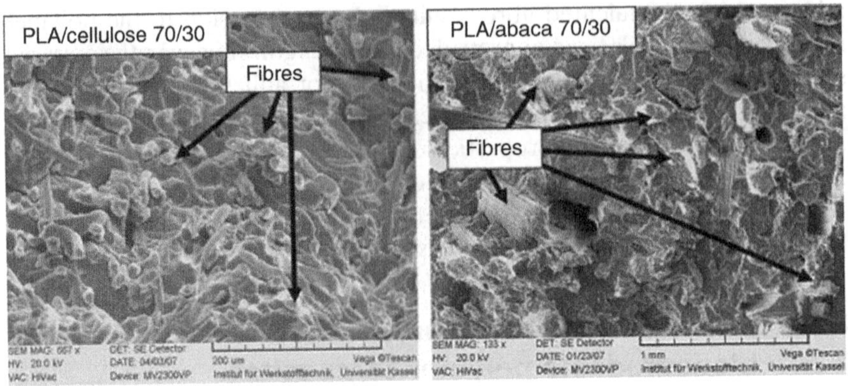

8.23 SEM photographs of PLA/cellulose (500×) and PLA/abaca (100×).

impact strength improved by factors of 1.5 and 3.2 due to the addition of the abaca and cellulose fibres, respectively.

Dynamic mechanical analysis of the composites showed that the incorporation of fibrous reinforcement causes a decrease in the polymer chain mobility.The storage modulus of PLA-based composites remained much higher than in unreinforced PLA. The glass transition temperature derived from the loss modulus was characterized by a slight shift to higher temperatures when compared with the matrix polymer. The morphology was evaluated using SEM and optical microscopy. Figure 8.23 shows the morphology of PLA/abaca and PLA/cellulose composites. Both SEM pictures showed several fibre pullouts. Abaca fibres seem to have a better physical bonding to the matrix than do synthetic cellulose fibres. This is probably due to the roughness of the fibre surface (Bledzki and Jaszkiewicz, 2009).

Starch/banana and starch/bagasse fibre composites can be prepared using compression moulding via two different processing methods. These both use commercial glycerol and crude glycerin as plasticizers. XRD studies showed that there were structural changes in both fibre composites, but without much variation in their crystallinity index (20–21%). The morphology of these composites was controlled by the properties of the fibres they contained rather than by the properties of the matrix. Compared with banana fibre composites, fractographs of bagasse fibre composites showed that they have a large number of fibre pullouts and fibres lying perpendicular to the fracture surface. These factors could explain the tendencies observed in their tensile properties. The starch coating on these fibres confirmed that there was good bonding between the bagasse fibres and the matrix, compared with the free surfaceof the banana fibres. Tensile testing of these composites indicated marginal improvements of strength properties compared with those of the matrix. The tensile properties displayed a strong correlation with the processing methods which automatically controlled the temperature. Furthermore, the applied pressure improved the properties for the same amount of fibre content, irrespective of the type of glycerol used. Crude glycerin produced more homogeneous composites, which could have better properties than those produced with commercial glycerol (Guimarães *et al.*, 2010).

8.11 Conclusions

The demand for abaca fibres in composites and their corresponding applications is increasing. However, there is still a great deal of work required to improve their performance to the desired level. The chemical content, crystallinity and surface properties of the fibres are very dependent on the growing conditions and the fibre extraction process. The thermal instability and sensitivity of abaca fibres to the environmental conditions make it difficult to design them for use as structural components. Bio-based and biodegradable matrix composites could be used for short-term applications. However, the development of a suitable compounding process for abaca fibres is an ongoing process. The proper optimization of each step required in processing long abaca fibre thermoplastic composites could produce not only a substitute for glass fibre-based composites but also an enhanced performance of that substitute.

8.12 References

Aehle, W. (2004), Industrial enzymes, in *Enzymes in Industry: Production and Application*, second edition, Wiley-VCH Verlag, Weinheim, Germany, 101–257

Anon. (2000), *Abaca Information*, Wigglesworth Fibres, London

Anon. (2005), DaimlerChrysler awarded for banana fibre use in Mercedes A class (http://www.netcomposites.com/news), Germany

Anon. (2012), *Future Fibres: Abaca*, Plant Datasheet, Food and Agriculture Organization of the United Nations, www.faostat.fao.org

Anon. (2013a), *Plants Database*, Natural Resource Conservation Service, USDA, http://www.usda.gov/plants

Anon. (2013b), *Production Data Sheet by Country*, FAOstat, Food and Agriculture Organization of the United Nations, www.faostat.fao.org

Armecin, R.B., Seco, M.H.P., Caintic, P.S. and Milleza, E.J.M. (2005), Effect of leguminous cover crops on the growth and yield of abaca, *Industrial Crops and Products*, 21, 317–323

Batra, S.K. (1998), Other long vegetable fibres, in Lewin, M. and Pearce, E.M., *Handbook of Fibre Chemistry*, Marcel Dekker, New York, 505–511

Biswas, S., Srikanth, G. and Nangia, S. (2001), Development of natural fibres in composites in India, *Proceedings of the Annual Convention and Trade Show of the Composite Fabricators Association*, Tampa, FL

Bledzki, A.K. and Gassan, J. (1999), Composites reinforced with cellulosic fibre, *Progress in Polymer Science*, 24, 221–274

Bledzki, A.K., Sperber, V.E. and Faruk, O. (2002), *Natural and Wood Fibre Reinforcement in Polymers*, Rapra Technology Ltd, Shrewsbury, UK, 28–30

Bledzki, A.K., Mamun, A.A. and Faruk, O. (2007), Abaca fibre reinforced PP composites and comparison with jute and flax fibre PP composites, *eEXPRESS Polymer Letters*, 1(11), 755–762

Bledzki, A.K., Faruk, O. and Mamun, A.A. (2008), Abaca fibre reinforced PP composites: Influence of fibre length and compounding processes on the mechanical properties, *Polimery*, 53(2), 35–40

Bledzki, A.K. and Jaszkiewicz, A. (2009), Mechanical properties of PLA composites with man-made cellulose and abaca fibre, *Composites: Part A*, 40, 404–412

Bledzki, A.K., Mamun, A.A., Jaszkiewicz, A. and Erdmann, K. (2010), Polypropylene composites with enzyme modified abaca fibre, *Composites Science and Technology*, 70, 854–860

Catlin, C. and Grayson, J. (1998), *Identification of Textile Fibres*, Archive Publications, London

Filer, K. (2003), Industrial production of enzymes for the feed industry, in Roussos, S., Soccol, C.R., Pandey, A. and Augur, C., *New Horizons in Biotechnology*, Kluwer Academic Publishers, Dordrecht, 1–15

Franck, R.R. (2005), *Abaca*, in Franck, R.R., *Bast and Other Plant Fibres*, Woodhead Publishing Limited, London, 315

Goeltenboth, F. and Muehlbauer, W. (2010), Abaca–cultivation, extraction and processing, in Muessig, J., *Industrial Applications of Natural Fibres: Structure, Properties and Technical Applications*, Wiley, Chichester, UK

Gonzal, L.R. and Valida, A.F. (2003), Musa textilis nee, in Brink, M. and Escobin, R.P., *Plant Resources of South-East Asia 17: Fibre Plant*, Backhuys, Leiden, 186–193

Guimarães, J.L., Wypych F., Saul, C.K., Ramos, L.P. and Satyanarayana, K.G. (2010), Studies of the processing and characterization of corn starch and its composites with banana and sugarcane fibers from Brazil, *Carbohydrate Polymer*, 80, 130–138

Haneefa, A., Bindu, P., Arvind, I. and Thomas, S. (2008), Studies of tensile and flexural properties of short banana/glass hybrid fiber reinforced polystyrene composites, *Journal of Composite Materials*, 42, 1471–1489

Hansen, C.A. (2001), Industrial uses of biotechnology, in *The Application of Biotechnology to Industrial Sustainability: Sustainable Development*, OECD Publishing, Paris, 17–24

Hintermann, M., (2005), Automotive exterior parts from natural fibres, *Inconference RIKO-2005*, Hannover, Germany

Huang, S.J. and Edelman, P.G. (1995), An overview of biodegradable polymers and biodegradation of polymers, in Scott, G. and Gilead, D., *Degradable Polymers: Principles and Applications*, Chapman & Hall, London, 18–24

Ibrahim, M.M., Dufresne, A., El-Zawawy, W.K. and Agblevor, F.A. (2010), Banana fibers and microfibrils as lignocellulosic reinforcements in polymer composites, *Carbohydrate Polymers*, 81, 811–819

Idicula, M., Malhotra, S.K., Joseph, K. and Thomas, S. (2005), Dynamic mechanical analysis of randomly oriented intimately mixed banana/sisal fiber reinforced with polyester composites, *Journal of Composite Science and Technology*, 65, 1077–1087

Idicula, M., Devi, U. and Thomas, S. (2006), Thermo physical properties of natural fiber reinforced polyester composite, *Journal of Composite Science and Technology*, 66, 2719–2725

Jarman, C. (1998), *Plant Fibre Processing: A Handbook*, ITDG Publishing, Rugby, UK

Joseph, S., Sreekala, M.S., Oommen, Z. and Thomas, S. (2002), A comparison of the mechanical properties of phenol formaldehyde composites reinforced with banana fibers and glass fibers, *Journal of Composite Science and Technology*, 62, 1857–1868

Kiran, C.U., Reddy, G.R., Dabade, B.M. and Rajesham, S. (2007), Tensile properties of sun hemp, banana and sisal fiber reinforced with polyester composites, *Journal of Reinforced Plastics and Composites*, 26, 1043–1050

Lewin, M. and Pearce, E.M. (1985), Fibre chemistry, in *Handbook of Fibre Science and Technology*, Volume 4, Marcel Dekker, New York

Mamun, A.A. (2011), *PP and PLA composites with grain by-products: Enzyme modification, characterization, processes optimization and properties*, PPH ZAPOL Dmochowski, Poland

Mathai, P.M. (2005), Coir, in Franck, R.R., *Bast and Other Plant Fibres*, Woodhead Publishing Limited, London, 275

McGovern, J.N. (1990), Vegetable fibers, in *Polymers–Fibers and Textiles*, University of Wisconsin Press, Madison, WI

Moreno, L.O., Parac, A.A. and Ocan, F.L. (2006), Fibre characteristics of promising abaca accessions in NARC Germplasm suited for specific industry end uses, NARC, Visayas State University, Baybay City, Philippines, 20

Morris, N. (1994), *Chemical Resistance*, Plastic Design Library, New York

Paul, S.A., Joseph, K., Gem Mathew, G.D., Pothen, L.A. and Thomas, S. (2010), Influence of polarity parameters on the mechanical properties of composites from polypropylene fiber and short banana fiber, *Composites: Part A*, 41, 1380–1387

Pothan, L.A. and Thomas, S. (2003), Polarity parameters and dynamic mechanical behaviour of chemically modified banana fibre reinforced polyester composites, *Composites Science and Technology*, 63, 1231–1240

Pothan, L.A., Oommen, Z. and Thomas, S. (2003), Dynamic mechanical analysis of banana fiber reinforced polyester composites, *Journal of Composite Science and Technology*, 63, 283–293

Pothan, L.A. and Thomas, S. (2004), Effect of hybridization and chemical modification on the water-absorption behavior of banana fiber-reinforced polyester composites, *Journal of Applied Polymer Science*, 91, 3856–3865

Pothan, L.A., Thomas, S. and Groeninckx, G. (2006), The role of fibre/matrix interactions on the dynamic mechanical properties of chemically modified banana fiber/polyester composites, *Composites: Part A*, 36, 1260–1269

Pott, G.T. (2004), Natural fibres with low moisture sensitivity, in Wallenberger, F.T. and Weston, N.E., *Natural Fibres, Plastics and Composites*, Kluwer Academic Publishers, New York, 105–120

Rahman, R., Huque, M., Islam, N. and Hasan, M. (2009), Mechanical properties of polypropylene composites reinforced with chemically treated abaca, *Composites: Part A*, 40, 511–517

Rowell, R.M. (2001), Modification of cellulose, in Hon, D.N.S. and Shiraishi, N., *Wood and Cellulose Chemistry*, Marcel Dekker, New York

Satyanarayana, K.G. and Wypych, F. (2007), Characterization of natural fibres, in Fakirov, S. and Bhattacharyya, D., *Handbook of Engineering Polymers*, HanserVerlag, Munich, 3–47

Savastano, J.H., Warden, P.G. and Coutts, R.S.P. (2000), Brazilian waste fiber as reinforcement for cement-based composites, *Journal of Cement and Concrete Composites*, 22, 379–384

Scherzer, D., Eipper, A., Weiss, C., Yamamoto, M., Skupin, G. and Witt, U. (2006), *Method for Compounding Polycondensates*, World Patent no. WO 2006087346

Scott, G. (1995), Photo-biodegradable polymers, in Scott, G. and Gilead, D., *Degradable Polymers: Principles and Applications*, Chapman & Hall, London, 169–183

Shibata, M., Takachiyo, K., Ozawa, K., Yosomiya, R. and Takeishi, H. (2002), Biodegradable polyester composites reinforced with short abaca fiber, *Journal of Applied Polymer Science*, 85, 129–138

Shibata, M., Teramoto, N., Ozawa, K., Yosomiya, R. and Takeishi, H. (2003), Biocomposites made from short abaca fiber and biodegradable polyesters, *Macromolecular Materials and Engineering*, 288, 35–43

Sinon, F.G. (2008) *Optimization of Stripping Technologies for the Production of High Quality Abaca Fibre*, Shaker Verlag, Aachen, 7–15

Teli, M.D. and Valia, S.P. (2013), Acetylation of banana fibre to improve oil absorbency, *Carbohydrate Polymers*, 92, 328–333

Tshabalala, M.A. (2005), Surface characterization, in Rowell, R.M., *Handbook of Wood Chemistry and Wood Composites*, CRC Press, Boca Raton, FL, 187–211

Valchev, I., Nenkova, S., Tsekova, P. and Lasheva, V. (2009), Use of enzymes in hydrolysis of maize stalks, *BioResources*, 4(1), 285–291

Vilaseca, F., Valadez-Gonzalez, A., Herrera-Franco, P.J., Pèlach, M.À., López, J.P. and Mutjé, P. (2010), Biocomposites from abaca strands and polypropylene. Part I: Evaluation of the tensile properties, *Bioresource Technology*, 101, 387–395

Wright, D.C. (1996), *Environmental Stress Cracking of Plastics*, Rapra Technology Limited, Shrewsbury, UK

Young, R.A. (2007), Wood and wood products, in Kent, J.A., *Handbook of industrial Chemistry and Biotechnology*, vol. 2, Springer, New York, 1234–1293

Zhu, W.H., Tobias, B.C., Coutts, R.S.P. and Langfors, G. (1994), Air-cured banana-fiber-reinforced cement composites, *Journal of Cement and Concrete Composites*, 16, 3–8

9
The use of palm leaf fibres as reinforcements in composites

D. KOCAK and S. I. MISTIK, Marmara University, Turkey

DOI: 10.1533/9781782421276.2.273

Abstract: Firstly, the importance of palm leaf fibres in different industries, and as a potential alternative to synthetic fibres, is outlined. The cultivation and uses of palm leaf fibres are then explored before a description of their various properties is outlined. Different methods of surface modification are explained before discussing how palm leaf fibres can be used as reinforcements in polymer nanocomposites.

Key words: palm leaf fibres, cellulose, natural fibres, mechanical properties, surface modification, composite, adhesion, nanocomposites.

9.1 Introduction

With the gradual worldwide depletion of petroleum resources in the last decade, fibres derived from plants have become important for the production of biocomposites, which have the advantages of renewability, biodegradability, sustainability and low cost, and are less hazardous to health, and thus have received considerable attention as a possible alternative to synthetic fibres based on petrochemical sources [1].

Palm trees are grown along coastal areas of tropical zones, and yield commercially important products such as fibres, which are mainly exploited for economic purposes. The advantages of palm leaf fibres include acceptable specific strength properties and low cost and the fact they are biodegradable [2].

Palm leaf fibres consist of 20 different species, depending on tree structure and the tropical zone in which they are grown. *Borassus flabellifer* (palmyra palm), *Corypha umbraculifera* (talipot palm), *Corypha taliera* (Roxb.) and *Raphia farinifera* (raffia palm) leaf fibre are the most commonly used types. Photos of palm leaf tree and fibre are shown in Fig. 9.1.

Palm leaves have been a popular writing material for over 2000 years in South and Southeast Asia. The use of palm leaves for recording literary and scientific texts has been reported from about the fifth century BC, with the oldest existing documents dating from about the second century AD. Palm leaf manuscripts are produced from two types of palms: palmyra and talipot [3].

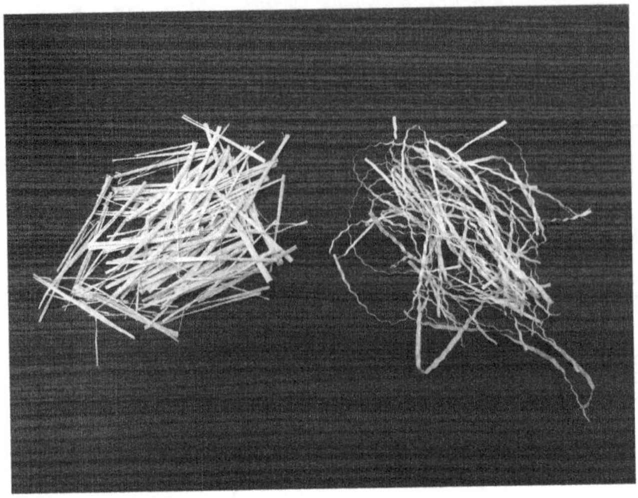

9.1 Palm leaf fibre.

The *Raphia* palm was identified in the nineteenth century. It is a genus of 20 species of palms native to tropical regions of Africa, Central and Southern America. The *Raphia* palm is characterised by its compound pinnate leaves, which are the largest leaves in the plant family.

The palm leaf has a maximum length of 65 feet. Research has shown that the *Raphia* palm can produce good volumes of fibre when grown in either hydromorphic or non-hydromorphic soils. However, the palm grows best in swampy soils, which experience high rainfall, high temperatures (22–33°C), sunshine of not less than five to seven hours daily, and high relative humidity. The morphology of *Raphia* palms is not identical, as also demonstrated in its vegetative growth. Like the African oil-palm (*Elaeis guineensis*), every single part of the *Raphia* palm is useful for economic and domestic purposes.

The palms grow up to 16 m tall and are remarkable for their compound pinnate leaves; dimensions up to 25.11 m long and 3 m wide have been recorded. The plants are either monocarpic, which flower once and then die after the seeds are mature, or hapaxanthic, where individual stems die after fruiting but the root system remains alive, sending up new stems [4].

9.2 Cultivation and uses of palm leaf fibres

Palm leaf fibres have many uses, especially in textiles and construction. In their local environments, they are used in the production of ropes, sticks, supporting beams and roof coverings. The membrane on the underside of each individual frond leaf is removed to create a long thin fibre that can be dyed and woven as a textile into products ranging from hats and shoes to decorative mats. The leaves and trunk are used for the construction of thatch houses and for making furniture. The hard fibre, piassava, is useful in making gunpowder, African clothes, carpets, baskets, and so on. On the other hand, the soft palm leaf fibre is used for making baskets, shoes, bags, and masquerade dresses, to name just a few uses. The raw materials utilised in paper manufacture are derived from the '*fibre*' pulp obtained from its trunk, bamboo and others.

A large palm tree grows leaves that can be stripped and dried to create palm leaf strands. Palm leaf, like jute or hemp twine, is a natural fibre that can be woven like straw, tied like silk ribbons, or packed like Styrofoam pills. The creamy-brown coloured lengths come from a specific palm tree that originally grew only on the island of Madagascar. *Raphia farinifera* actually has the largest leaves of palm trees. The fibrous leaves are cut off and torn apart in parallel lines to yield very long strips of palm leaf. The tree is cultivated specifically for harvest and export from East Africa.The palm leaf fibres are peeled from the fronds of the palm on both sides and then hung out to dry in the sun. They are then collected, classified, pressed and packaged into bales for shipment.

Date palm (*Phoenix dactylifera*) is a perennial with an average life of 150 years. There are approximately 62 million trees in the Middle East and North Africa and 100 million worldwide. Each year the trees are pruned to remove old, dead or broken leaves. This produces approximately 100,000 tons of date palm fronds (DPF) and 15,000 tons of date palm leaves in Saudi Arabia alone. Worldwide, an estimated 1,130,000 tons of date palm leaves are produced annually. The leaves are partially utilised in making crates, robes, baskets and mats but the bulk of the material is discarded as waste. More efficient utilisation of this natural resource would have a positive impact on our fragile environment and help preserve it [5].

Palm leaf also provides an important cultural drink, as the sap contains sugars. It is traditionally collected by cutting a box in the top of the palm from which is suspended a large gourd to collect the milky white liquid. Unlike oil palms, this process kills the tree. The sap from both the raffia and oil palms can be allowed to ferment over a few days. When first collected from the tree it is sweet and appears slightly carbonated. The sap is usually called wine. The raffia wine tends to be sweeter at any age when compared to oil palm wine. Both kinds of palm wine can also be distilled into strong liquors, such as Ogogoro. It is a tradition in many cultures for guests and spirits to be offered these drinks from the palm tree [6–8].

9.3 Properties of palm leaf fibres

Applications of the lignocellulosic materials vary according to their chemical composition and physical properties. Palm leaf fibre has the

Table 9.1 Comparative cost, density and moisture content of some raw materials

Raw materials	Cost/kg (Indian rupees)	Density (g/cm^3)		Moisture content (%)
		True	Apparent	
Date palm leaf	0.50	1.44	1.21	6.8
Jute stick	1.50	1.21	1.10	15.2
Bagasse	2.50	1.35	1.16	13.4
Hardwood	3.00	1.38	1.19	12.1
Softwood	3.50	1.30	1.18	13.8

highest α-cellulose content, and the matrix responds to chemical surface modifications rapidly because of its low lignin content.

Cellulose is the main structural component that provides strength and stability to the plant cell walls and the fibre. Fibres with higher cellulose content are preferable for textile, paper and other fibrous applications, whereas products with higher hemicellulose content are better suited for producing ethanol and other fermentation products because hemicellulose is relatively easily hydrolysable into fermentable sugars. Mechanically, hemicellulose contributes little to the stiffness and strength of fibres or individual cells and is more easily hydrolysed into sugars than cellulose. Therefore, fibres containing a higher proportion of hemicellulose are preferable for producing sugars, and eventually fuels such as ethanol. Hemicellulose has the highest water sorption.High moisture absorption of natural fibres leads to swelling and the presence of voids, which result in poor mechanical properties and reduce the dimensional stability of composites. Comparative cost, density and moisture content of some raw materials are given in Table 9.1 [1, 9–11]. The mechanical properties of palm leaf fibres are outlined in Table 9.2 and Fig. 9.2.

9.4 Surface modification of palm leaf fibres

Different chemical surface treatments have been experimented with to increase adhesion between natural fibres and matrices to produce a good composite. By performing surface treatment methods, the fibre surface becomes clean and thus more uniform due to the elimination of microvoids. The stress transfer capacity between the fibre and the matrix therefore improves. In addition to this, it reduces fibre diameter and thereby increases the aspect ratio (length/diameter). This treatment creates an effective fibre surface area for good adhesion with the matrix [12].

Table 9.2 Comparison of mechanical properties of various textile fibres

Fibres	Tensile strength (MPa)	Young's modulus (GPa)	Elongation at break (%)	Density (g/cm^3)	Specific strength (MPa)	Specific Young's modulus (GPa)
E-glass	2200–3600	65	2.5	2.54	850–1300	27
Carbon	3500	220–240	1.3–1.9	1.4	2750	165
Flax	365–1100	25.6	1.5–3.1	1.5	250–650	17.8
Sisal	490–635	8–21	2.1–2.7	1.5	335–430	7.2
Jute	370–760	24.5	1.3–2.2	1.3	310–625	15
Hemp	670	35–50	2.2–4.1	1.5–1.6	630	25
Cotton	290–490	5.0–12.0	6–7	1.5–1.6	191–310	5.1
Coir	160	4–5.5	15–35	1.2	146	0.2
Palm leaf fibre	140–650	12–35	2.2	1–4	180–850	40.1

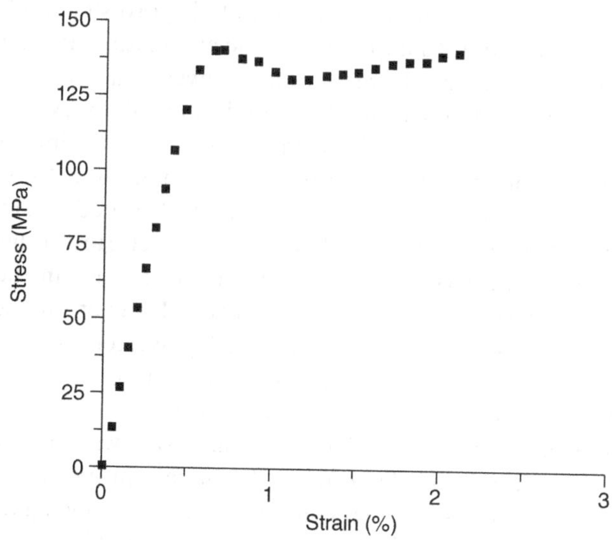

9.2 A typical stress–strain curve of palm leaf fibre after a room temperature tensile test at a strain rate of $3 \times 10^{-4} s^{-1}$.

Palm leaves were treated with different concentrations of NaOH, 0.5%, 1%, 2% and 5% (w/w). The results from this investigation indicated that female leaves have better tensile properties, which deteriorate with an increase of the alkali. Male leaves have lower tensile properties than female leaves and their mechanical properties are improved slightly by NaOH treatment. Female leaves have more pores than male leaves, which was proved and calculated by image analysis. Untreated female leaves have higher thermal stability (353°C) than male leaves (343°C). Both can be used as reinforcements in thermoplastic matrix materials whose processing temperatures are below 300°C. It was observed that alkali treatment increased the mechanical and thermal properties of palm leaf fibre-reinforced composites [13].

Deacetylation caused morphological changes in the fibre cells, and slowed movement of water through composites, with no detectable influence on bending stiffness or strength. The results suggest that high levels of natural acetylation are not a useful feature of the leaf fibres [14].

De Rosa *et al.* used different lengths of palm leaf fibre to reinforce the epoxy composite and observed that tensile strength and flexural modulus properties of the composites increased according to the fibre length [15]. Goulart *et al.* reported that palm leaf-reinforced polypropylene composite panels have better flexural strength and modulus properties than neat polypropylene polymer panels [16].

Short palm leaf fibre-reinforced unsaturated polyester and epoxy composites were produced. To improve interfacial adhesion, the surface of the fibre was treated with acetatewhen using a polyester and epoxy matrix. All the results lead to the conclusion that good interactions exist between the fibre and the epoxy matrix. These interactions can be improved by chemical treatment of the fibre. Improved thermomechanical properties, bending modulus, stress at break and maximum absorbed energy were reported. These strong interactions and/or reactions occurring at the epoxy/lignocellulosic fibre interface result in a significant increase in the glass–rubber transition temperature of the neat matrix. However, the adhesion level is very weak when using an unsaturated polyester as the matrix. Chemically treating the fibre with acetate does not improve the interfacial adhesion between the fibre and polyester resin [17].

Sugar palm leaf fibre-reinforced epoxy composite was produced by Bachtiar *et al.* in a study that determined the impact properties of sugar palm fibre-reinforced epoxy polymer composite after introducing the alkali treatment to the fibre. The alkali treatment modified the fibre surface and enhanced the interfacial bonding between fibre and matrix, thus contributing to the composite's ability to dissipate energywhen the İzod impact testing was complete. The ultimate impact strength of treated sugar palm fibre-reinforced epoxy composite took place at 0.5M NaOH solution with eight hours' soaking time, and higher impact strength was obtained from treated fibre epoxy composite than from untreated composite [18].

Thiruchitrambalam and Shanmugam produced palmyra leaf fibre-reinforced polyester resin composites. Before production, palmyra leaf fibres were mercerised, and treated with benzoyl chloride and permanganate to improve adhesion between the fibre and the matrix. The mercerised and benzoyl-treated fibre composites saw an improvement in tensile strength of around 60% while the tensile modulus increased by 37% and 60%, respectively. In the case of permanganate-treated fibre composites, the flexural strength increased by 70% and flexural modulus increased by 110% in comparison to the untreated composites. The impact strength for the mercerised and permanganate-treated fibre composite improved by 55% and 42% respectively. Chemical pre-treatment of fibres reduced water absorption of the composites. The benzoyl chloride-treated fibres absorbed less water in comparison to the untreated fibre composites [19].

By using date palm fibre (*Phoenix dactylifera*) and phenolic resin, composite construction panels were produced for insulation purposes. The orientation and size of the date palm fibre had no measurable effect on thermal conductivity. The compositesproduced were very stable to handle all the machining processes required as construction panels [20].

9.5 The use of palm leaf fibres as reinforcements in polymer nanocomposites

Generally, palm oil leaf fibres, pineapple leaf fibres and date palm leaf fibres are used in the production of nanocomposites. Patel *et al.* produced nano-biocomposite structures. In that study, the biocomposites of palm leaf were incorporated into polyvinyl alcohol (PVA)by using a solution cast technique. Structural and microhardness properties of pure PVA and palm leaf filled PVA biocomposites were determined by using FTIR and Vickers indentation technique respectively. The FTIR analysis reveals the presence of palm leaf moieties in PVA, which indicates the compatibility between palm leaf and PVA. The values of microhardness were increased in all compositions of palm leaf incorporated PVA films as compared to the pure PVA. This increment in microhardness is attributed to the excellent binding of palm leaf into PVA [21].

Phattaraporn *et al.* produced biodegradable rice starch films reinforced with different concentrations (10, 20, 30 and 40% glycidoxy propyl-triethoxysilane) of silane-treated palm-pressed fibre, and the properties of treated palm-pressed fibre were investigated. Higher tensile strength, water vapour permeability and thermal properties of biodegradable rice starch films were obtained when silane-treated palm-pressed fibre was applied. Increasing concentrations of silane and amount of silane-treated palm-pressed fibre resulted in increased tensile strength and water vapour permeability, but decreased elongation at break. Maximum improvements of rice starch films in the mechanical and thermal properties were obtained when 40% of silane-treated palm-pressed fibre was used. These results revealed that adhesion between the rice starch and the palm-pressed fibre increased with the silane treatment of palm-pressed fibres [22].

9.6 Conclusion

Adding palm leaf fibres into polymers increases the mechanical properties of the composites. In addition, surface-treated palm leaf fibre reinforcement improves the mechanical and thermal properties of the polymer composites due to the modified surface of the palm leaf fibre. Biocomposite structures can be produced by using palm leaf fibres and degradable polymers. Treated and untreated palm leaf fibre-reinforced polymer composites can be used, especially in the automotive and construction industries.

It is clear that surface treatment of palm leaf fibres increases adhesion between the fibre and the resin, especially for the polypropylene, epoxy and polyester resin polymer composites. Palm leaves are of significant economic and ecological importance and palm leaf fibres can be used as reinforcement for polymer composites. Palm leaf fibre-reinforced polymer composites are valued by the automotive and construction industries due to their beneficial mechanical and thermal properties.

9.7 References

1. Reddy N. and Yang Y. (2005) Biofibres from agricultural by-products for industrial applications, *Trends in Biotechnology* 23(2), 22–27.
2. Nawaz M.S., Chapatwala K.C. and Wolfram J.H. (1989) Degradation of acetonitrile by *Pseudomonas putida*, *Applied Environmental Microbiology* 55, 2267–2274.
3. Shi Z., Setlur S. and Govindaraju V. (2010) Digital enhancement of palm leaf manuscript images using normalization techniques, *International Journal on Electrical Engineering and Informatics* 2, 1.
4. Dransfield J. (2011) *Arecales (Palms)*, Wiley/Online Library.
5. Elenga R.G., Dirras G.F., Goma Maniongui J., Djemia P. and Biget M.P. (2009) On the microstructure and physical properties of untreated raffia textilis fiber, *Composites, Part A: Applied Science and Manufacturing* 40(4), 5.
6. South Africa plants: 'Raphia australis', http://www.plantzafrica.com/plantqrs/raphiaaust.htm
7. Fairchild Tropical Botanic Garden: 'Raphia', http://www.fairchildgarden.org/palmguide/genus.php?family=arecaceae&genus=Raphia
8. 'Raphia' articles, http://www.palmpedia.net/wiki/index.php/Palmpedia:Online_reference#raphia
9. Paster M. *et al.* (2003) *Industrial Bioproducts: Today and Tomorrow*, Report prepared for the US Department of Energy, Washington, DC.
10. Madsen B. (2004) *Properties of Plant Fibre Yarn Polymer Composites: An Experimental Study*, BYU-DTU Report R-082, Roskilde, Denmark, p. 75.
11. Maya J. and Sabu T. (2008) Biofibres and biocomposites, *Carbohydrate Polymers* 71, 343–364.
12. Abdal-hay A., Suardana N.P.G., Jung D.Y., Choi K.-S. and Lim J.K. (2012) Effect of diameters and alkali treatment on the tensile properties of date palm fiber reinforced epoxy composites, *International Journal of Precision Engineering and Manufacturing* 13(7), 1199–1206.

13. Al Maadeed M.A., Kahraman R., Noorunnisa Khanam P. and Al-Maadeed S. (2013) Characterization of untreated and treated male and female date palm leaves, *Materials and Design* 43, 526–531.
14. Newman R.H., Clauss E.C., Carpenter J.E.P. and Thumm A. (2007) Epoxy composites reinforced with deacetylated *Phormium tenax* leaf fibres, *Composites, Part A* 38, 2164–2170.
15. De Rosa I.M., Santulli C. and Sarasini F. (2010) Mechanical and thermal characterization of epoxy composites reinforced with random and quasi-unidirectional untreated *Phormium tenax* leaf fibers, *Materials and Design* 31, 2397–2405.
16. Goulart S.A.S., Oliveira T.A., Teixeira A., Miléo P.C. and Mulinari D.R. (2011) Mechanical behaviour of polypropylene reinforced palm fibers composites, *Procedia Engineering* 10, 2034–2039.
17. Kaddami H., Dufresne A., Khelifi B., Bendahou A., Taourirte M., Raihane M., Issartel N., Sautereau H., Gérard J.F. and Sami N. (2006) Short palm tree fibers – Thermoset matrices composites, *Composites, Part A* 37, 1413–1422.
18. Bachtiar D., Sapuan S.M. and Hamdan M.M. (2009) The influence of alkaline surface fibre treatment on the impact properties of sugar palm fibre-reinforced epoxy composites, *Polymer-Plastics Technology and Engineering* 48(3), 379–383.
19. Thiruchitrambalam M. and Shanmugam D. (2012) Influence of pre-treatments on the mechanical properties of palmyra palm leaf stalk fiber–polyester composites, *Journal of Reinforced Plastics and Composites* 31, 1400–1414.
20. Al-Sulaiman F.A. (2003) Date palm fibre reinforced composite as a new insulating material, *International Journal of Energy Research* 27, 1293–1297.
21. Patel A.K., Bajpai R., Keller J.M. and Saha A. (2011) Preparation and Characterization of palm leaf incorporated polyvinyl alcohol bio composites, *AIP Conference Proceedings*, 12 December 2011, Vol. 1393, Issue 1, 151.
22. Phattaraporn T., Waranyou S. and Thawien W. (2011) Effect of palm pressed fiber (PPF) surface treatment on the properties of rice starch films, *International Food Research Journal* 18, 287–302.

Part III

Seed fibers

10 The use of coir/coconut fibers as reinforcements in composites

D. VERMA, Indian Institute of Technology, B.H.U., Varanasi, India and P. C. GOPE, College of Technology, Pantnagar, India

DOI: 10.1533/9781782421276.3.285

Abstract: Composite reinforcement with natural fibers has recently gained attention due to low cost and easy availability. This chapter discusses the use of coir fiber in composites and the current status of research.

Key words: coconut, coir fiber, coconut cultivation, polymer, mechanical properties.

10.1 Introduction

Historically, the coconut was known as *Nux indica* (the Indian nut) and also the Nargil tree, the tree of life. Western literature has also mentioned the Malayalam name Tenga for the coconut palm which relates to Tamil 'Tennai', believed to be of Sri Lankan origin. Its geographical dispersion was aided by travelers and traders. Botanically, the coconut palm is a monocotyledon and belongs to the order Arecaceae, family Palmae and the species is known as *Cocos nucifera* Linn.

The Philippines, Indonesia, India and Sri Lanka account for 78% of coconut production. The most important and economically valuable product of coconut palm is its fruit popularly known as the 'nut'. It is made up of an outer exocarp, a thick fibrous fruit coat known as the husk and underneath lies the hard protective endocarp or shell. Lining the shell is a white albuminous endosperm or 'coconut meat' and the inner cavity is filled with a clear sweet liquid called 'coconut water'. In this chapter, utilization of coconut fiber/coir fiber in composites is discussed.

Scanning electron microscopy (SEM) is carried out to understand why the mechanical properties of composites prepared from treated and untreated coir are different and also to understand the effect of fiber treatment on the fiber–matrix interfacial adhesion.

10.2 The coconut plant and its cultivation

The coconut palm comprises a white 'meat' which has a percentage by weight of 28 surrounded by a protective shell and husk which have a

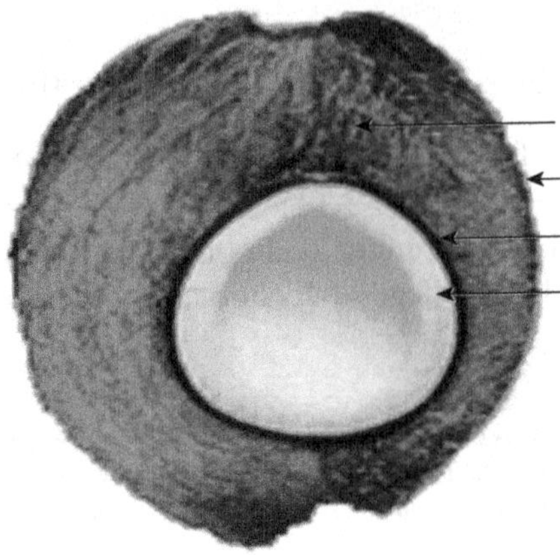

10.1 Coconut fruit.

Table 10.1 Chemical composition of coir fibers

Item	Percentage
Water soluble	5.25%
Pectin and related compounds	3.00%
Hemicellulose	0.25%
Lignin	45.84%
Cellulose	43.44%
Ash	2.22%

Source: Verma *et al.*, 2013.

percentage by weight of 12 and 35 respectively, as shown in Fig. 10.1. The husk from the coconut palm comprises 30% weight of fiber and 70% weight of pith material. Table 10.1 shows the composition of the coir fibers (Verma *et al.*, 2013).

The important properties of coconut fiber are:

- It is a renewable resource and CO_2-neutral material.
- The fiber is abundant, non-toxic, biodegradable, low density and low cost.
- The fiber has a high degree of water retention and is rich in micronutrients.

Coconut (*Cocos nucifera*) plays a significant role in the economy of India. Apart from the importance of copra and coconut oil, which is widely used in the manufacturing of soaps, hair oil, cosmetics and other industrial products, the husk is a source of fiber which supports a sizable coir industry. The tender nut also supplies coconut water.

Coconut is grown in more than 90 countries in an area of 14.231 million ha with a total production in terms of copra equivalent of 11.04 million MT. Indonesia (25.63%), the Philippines (23.91%), and India (19.20%) are the major coconut-producing countries of the world.

The coconut palm can tolerate a wide range of soil conditions, although a variety of factors such as drainage, soil depth, soil fertility and layout of the land have an influence on the growth of the palm. The major soil types that support coconut in India are laterite, alluvial, red sandy, loamy, coastal sandy and reclaimed soils with a pH ranging from 5.2 to 8.0.

Coconut palms require a continuous supply of water, which can be provided by rainfall of the order of 2000 mm per annum, or from groundwater (at a depth of 1–3 m), although they cannot tolerate waterlogging. The palms grow best at average temperatures of around 26–27°C, so they cannot grow above 750 m. Growth is stimulated by a sufficient supply of chlorine in the soil, and the coconut can withstand up to 1% salt in the soil. These conditions are generally found in tropical and subtropical coastal regions with little rainfall. Coconut palms can also grow on deep alluvial soil.

The quality of the seeds is very important to the yield from the palm. The seeds should originate from a healthy and productive stock plant. Two main groups are cultivated in the commercial sector: the tall plants of the Typica group, which generally need to be cross-fertilized, and dwarf types of the Nana group, where self-pollination is prevalent.

Stock plants that are suitable seed providers produce 100 nuts per year, 12–14 syncarpy of differing ages, and up to 180 g copra per nut. The fully ripened nuts which are intended to provide seeds are harvested after 11–12 months. The nuts germinate quicker at the lower end, or in the middle, of the syncarpy and should not be allowed to fall but should be cut down, and carefully lowered. Following the harvest, the produce should be stored for a short break in a covered and well-ventilated place.

Before sowing, the nuts are sorted and only those containing water are used. The shell is cut away on the germinating side of the nut to facilitate germination and then the nuts are soaked in water for 14 days before being sown in loose soil which can drain easily. The nuts are laid in the soil lengthways at a distance of 45 cm with the upper side still visible. Coconut fibers are used as mulching material between the rows. The nuts can also be sown in a glasshouse with 95% humidity. On smallholdings, the nuts are often set out in shaded areas, lightly dug in and then covered with organic material (Augstburger *et al.*, 2000).

Although coconut is grown in more than 90 countries, the Philippines, Indonesia, India and Sri Lanka contribute about 78% of the world's production. In 2007–08, India contributed 27.86% of global coconut production. Table 10.2 shows each country's area, production and productivity of coconut in 2006–07.

Table 10.2 Area, production and productivity of coconut in 2006–07

Country	Area (hectares)	Production (million tonnes)	Productivity (kg/ha)
Brazil	2,200	2,973,700	10,813
China	28,200	289,000	10,248
Dominican Republic	38,000	181,533	4,777
Ghana	55,000	315,000	5,727
India	1,940,000	15,840,000	8,165
Indonesia	2,660,000	16,289,000	6,123
Mozambique	70,000	265,000	3,785
Myanmar	41,000	350,000	8,536
Papua New Guinea	200,000	650,000	3,250
Philippines	3,253,927	14,344,920	4,408
Sri Lanka	447,000	1,950,000	4,362
Tanzania	310,000	370,000	1,193
Thailand	337,000	1,450,000	4,302
Vietnam	132,800	930,600	7,007

10.3 Preparation/extraction of coir fibers from coconut husk

The processes of fiber extraction are varied, and depend on the effectiveness of the wet processing such as bleaching and dyeing of the coir.

The traditional production of fibers from the husks is laborious and time-consuming. After separation of the nut, the husks are processed by various retting techniques generally in ponds of brackish water (for three to six months) or in backwaters or lagoons. This requires 10–12 months of anaerobic (bacterial) fermentation. By retting, the husks are softened and can be decorticated. The fiber is extracted by beating, which is usually done by hand. After hackling, washing and drying (in the shade) the fibers are loosened and cleaned. The remaining residual pith has recently found a new market as a peat moss substitute for horticultural production.

Retted fibers from green husks are the most suitable fibers for dyeing and bleaching. For the production of more coarse brown yarns, shorter periods of retting may be applied. These find an increasing outlet in geo-textile applications.

Mechanical processes using either defibering or decorticating equipment can process the husks after only five days of immersion in water tanks. Crushing the husk in a breaker opens the fibers. By using revolving 'drums' the coarse long fibers are separated from the short woody parts and the pith. The stronger fibers are washed, cleaned, dried, hackled and combed.

Husk defibering generally involves two processes:

1. *In the wet milling process*, the coconut husks are crushed between fluted rollers called husk crushers before they are soaked in the retting pond for a minimum of 72 hours in order to facilitate the penetration of water through the exocarp. After soaking, the fibers are extracted through specially constructed machines called drums. Mature coconut husks are usually processed through this method.
2. *The dry milling method* utilizes a special machine called the down decorticator. The husk segment is disintegrated by the use of metal beater bars revolving at high speed followed by the use of sifters to separate the non-fibrous matter from the fiber. This is considered the most efficient method of extracting coir and is well suited to areas where soaking facilities are limited or are not available.

Environmentally friendly methods for fiber production have the potential to produce a more constant quality of fibers. Novel developments by the Central Coir Research Institute using a biotechnological approach with specific microbial enzymes have reduced the retting time substantially to three to five days. By using specific (microbial) lignolytic enzymes (laccase/phenoloxidase), the fiber surface can be bleached or activated to react more easily with dyes.

The grades of coir are determined by three qualities: strength, cleanliness and color. The standard grades of coir are shown Table 10.3.

10.4 Surface modification of coconut fibers

Coir fibers can be modified by the following methods.

10.4.1 Alkaline treatment

Treatment of natural fiber by NaOH is widely used for composites. This treatment changes the orientation of highly packed crystalline cellulose and forms an amorphous region by swelling the fiber cell wall. This enables improved reception to penetration by chemicals. Alkali-sensitive hydrogen bonds existing among the fibers are agitated and new hydrogen bonds form between the cellulose molecular chains, increasing the surface roughness. The treatment removes the waxy substances on the fiber surface thereby improving the close contact of the fiber–matrix. Karthikeyan *et al.* (2013) reported that alkali-treated coir fiber–polyester composites, with a volume fraction ranging from 10% to 30%, show better properties than composites manufactured with untreated fibers, but the flexural strength (FS) of these composites was consistently lower than that of the bare matrix. A maximum value of 42.3 MPa is reported compared to a value of 48.5 MPa for polyester.

Table 10.3 Grades of coir fiber

Letter designation	Name of grade	Description
CH-1	Coir Good	Fiber (bristle) is of good cleaning, with little or no pulp content; color is light brown to almost dark brown; length is not less than 5 inches.
CH-2	Coir Fair	Fiber (bristle) is of fair cleaning; fibers are stuck together and considerable pulps are present; color ranges from dull brown to dark brown or black; length is not less than 5 inches.
CH-3	Coir Mixed	Mixture of bristle and mattress fibers, generally crumpled and tangled; of good and fair cleaning, must be free from coir dust and hard, undefibered portion of the husk; color ranges from light brown to dull brown.
CH-4	Coir Mattress	Consists mostlyof short crumpled fibers with an average length of not less than ½ inch; must be free from coir dust and hard, undefibered husk.
CH-W	Coir Waste	Consists of coir dust and fiber not fitted in any regular grades of coir, with length of less than 2½ inches.

Source: Fiber Industry Development Authority, Philippines.

10.4.2 Silane treatment

Coupling agents usually improve the degree of crosslinking at the interface. Silane coupling agents are effective in modifying the natural fiber–matrix interface. Silanols form in the presence of moisture and hydrolyzable alkoxy groups and react with the cellulose hydroxyl group of the fiber, improving fiber–matrix adhesion and stabilizing the composite properties (Bledzki *et al.*, 2008). Coupling agents such as toluene diisocyanate and triethoxyvinylsilane have been tested in fiber treatment processes in order to improve the interfacial properties. Silanes undergo hydrolysis, condensation and a bond formation stage. Silanols can form polysiloxane structures by reaction with the hydroxyl group of the fibers.

Abdullah and Ahmad (2012) used various concentrations of silane solution in fiber treatment, ranging between 0.25% and 1% w/w. For each concentration, silane was dilute with distilled water and stirred using a glass rod for 15 minutes to form an aqueous solution. Acetic acid was then added until the pH of the solution is 4, which optimizes the performance of the reinforcing material. The coir fiber was immersed in the solution for an hour, then washed with distilled water and dried at room temperature for two days.

10.4.3 Acetylation treatment

Plasticization of cellulose fibers can only be achieved by the esterification method, also known as acetylation treatment. This method involves the generation of acetic acid (CH_3COOH) which must be removed from the lignocellulosic material before the fiber is used. Acetic anhydride (CH_3–C(=O)–O–C(=O)–CH_3) which is used for chemical modification substitutes the polymer hydroxyl groups of the cell wall with acetyl groups, modifying the properties of the polymers so that they become hydrophobic (Sreekala *et al.*, 2000). Untreated coir fibers are immersed in 18% aqueous NaOH solution at 28°C for one hour. The fibers are then washed several times with cold water and finally with acidified water (0.1N HCl). The fibers are dried in an air oven andthen soaked in glacial acidic acid for one hour at the same temperature (Dixit and Verna, 2012).

10.4.4 Benzoylation treatment

Benzoyl chloride is used in benzoylation treatment to decrease the hydrophobicity of the fiber and to improve fiber–matrix adhesion leading to an increase in the strength of the composite. In benzoylation treatment, the alkali pretreatment is used to activate the hydroxyl groups of the fiber and then the fiber is soaked in benzoyl chloride solution for 15 minutes. Benzoyl chloride which is adhered to the fiber surface is removed by ethanol solution followed by washing with water and drying in an oven. Benzoyl chloride treatment on alkali pretreated sisal fiber exhibits higher thermal stability compared to untreated fiber composites (Joseph *et al.*, 1996).

In this process, the pretreated coir fibers are suspended in 10% NaOH solution and agitated with benzoyl chloride. The mixture is left for 15 min, filtered, washed thoroughly with water and then dried between filter papers. The isolated fibers are then soaked in ethanol for one hour to remove the benzoyl chloride and finally washed with water and dried in the oven at 800°C for 24 h (Chandra Rao *et al.*, 2012).

10.5 The properties of coir fiber-reinforced thermoset polymer composites

Two kinds of matrix can be used for the integration of the coconut fibers, thermosets and thermoplastics.

Romli *et al.* (2012) presented an analysis of the tensile strength (TS) of a coir-based thermoset matrix composite. Three parameters were used in this study, fiber volume fraction, curing time and compression load applied during fabrication. A total of 18 tensile test cases were performed which revealed that important factors in the tensile strength of the composites are coir fiber volume fraction and its interaction with the curing time. The contribution from compression loading during fabrication was found to be negligible.

As shown in Fig. 10.2, for 24-hour curing, the composite tensile strength initially increases with increase in coir fiber volume and compression loading, but decreases as coir fiber volume fraction and the compression load are increased further. On the other hand, some samples of the 48-hour curing time setting show an increase in maximum tensile strength with increase in coir fiber volume fraction and compression load, indicative of the effect of the fabrication process on the material properties.

Misra *et al.* (2007) investigated fire-retardant coir epoxy micro-composites. The coir fiber is treated with saturated bromine water and then mixed with stannous chloride solution. A 5% volume fraction of fire-retardant filler reduces the smoke density by 25% and the LOI value increases to 24%. The mechanical properties of the composites were not greatly affected by the incorporation of the fillers. The flexural strength and flexural modulus of the composites did, however, increase significantly.

Harish *et al.* (2009) investigated the mechanical properties of coir fiber composites. Scanning electron micrographs obtained from fractured

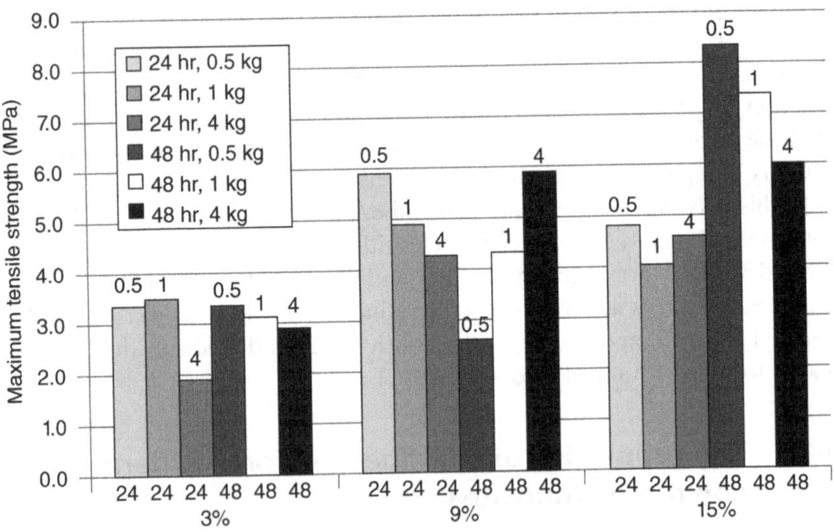

10.2 Experimental results on maximum tensile strength of the coir fiber composites (reprinted from Romli *et al.*, 2012, © 2012, with permission from Elsevier).

surfaces were used for a qualitative evaluation of the interfacial properties of the coir/epoxy and compared with glass fiber epoxy. These results indicate that coir can be used as a potential reinforcing material for fabrication of low-load-bearing thermoplastic composites. The tensile strength (TS), flexural strength (FS) and impact strength (IS) values of the coir and GFRP composites are tabulated in Table 10.4. All measures of the strength reported here show that the GFRP specimens are significantly stronger than the coir/epoxy specimens.

Chandra Rao *et al.* (2012) investigated the wear behavior of treated and untreated coir dust filled epoxy resin matrix composites. The effects of treated and untreated coir dust concentrations of 10%, 20% and 30%, varying loads of 10, 20 and 30 N, and varying velocities of 300, 400 and 500 meters per second on the abrasive wear rate of the composites were analyzed. The results showed that the treated fiber composites exhibit better wear resistance than the untreated composites. The abrasive wear rate decreases with increasing coir dust volume fraction.

Das (2012) bleached treated coir fiber and then cast them with epoxy resin using a handmade mold. XRD patterns confirmed that the degree of crystallinity decreases as a result of treatment of the fibers with H_2O_2. SEM images show the degree of roughness of the surface structure, confirming the increase in adhesion between the fibers and matrix after treatment. FTIR spectra confirm that the water content of the composites decreases due to intermolecular hydrogen bonding. The three-point bending test (Instron) shows that the strength of the fibers is increased by the bleaching process.

Mean tensile strength and its standard deviation employing a normal and a Weibull distribution are plotted in Fig. 10.3. The means of both distributions are nearly equal and the standard deviation is in a few cases slightly higher in the Weibull distribution. In Table 10.5 the Weibull parameters for the fiber species and for the different test lengths are given.

A shape factor (m) larger than 2 indicates that the failure rate increases at an increasing rate as the applied load increases and a shape factor of 3.4 yields a distribution that is similar to a normal distribution. Synthetic fibers usually have shape factors between 5 and 15, while for natural fibers, which display a larger variation in properties, the values vary typically between 1 and 6. Figure 10.3 shows that for all fiber species, the strength decreases as the test length increases because with larger test lengths there are more flaws in the fibers, making the chance of failure larger. The 5 mm test length is an exception and this may be related to a higher impact of process factors, for example due to the clamps.

Mithil Kumar *et al.* (2009) developed phenolic resin-based coir/glass hybrid composites using compression molding followed by hand lay-up. The mechanical properties of the composites were evaluated as a function of

Table 10.4 Textile, flexural and impact properties comparison

Composite type	Sample identification	Tensile strength (MPa)	Average tensile strength (MPa)	Flexural strength (MPa)	Average flexural strength (MPa)	Impact strength (kJ/m^2)	Average impact strength (kJ/m^2)
Coir	C1	18.110	17.86 ± 2.32	40.230	31.08 ± 6.01	10.551	11.49 ± 0.99
	C2	14.101		28.550		12.636	
	C3	20.481		24.261		12.332	
	C4	18.550		29.226		11.504	
	C5	18.073		33.131		10.448	
GFRP	GFRP1	83.453	85.35 ± 4.32	121.276	132.39 ± 11.85	53.695	52.66 ± 3.13
	GFRP2	86.548		132.578		54.334	
	GFRP3	85.887		120.800		55.607	
	GFRP4	91.329		138.661		47.520	
	GFRP5	79.547		148.643		52.151	

Reproduced with permission from Elsevier Ltd. (Harish *et al.*, 2009).

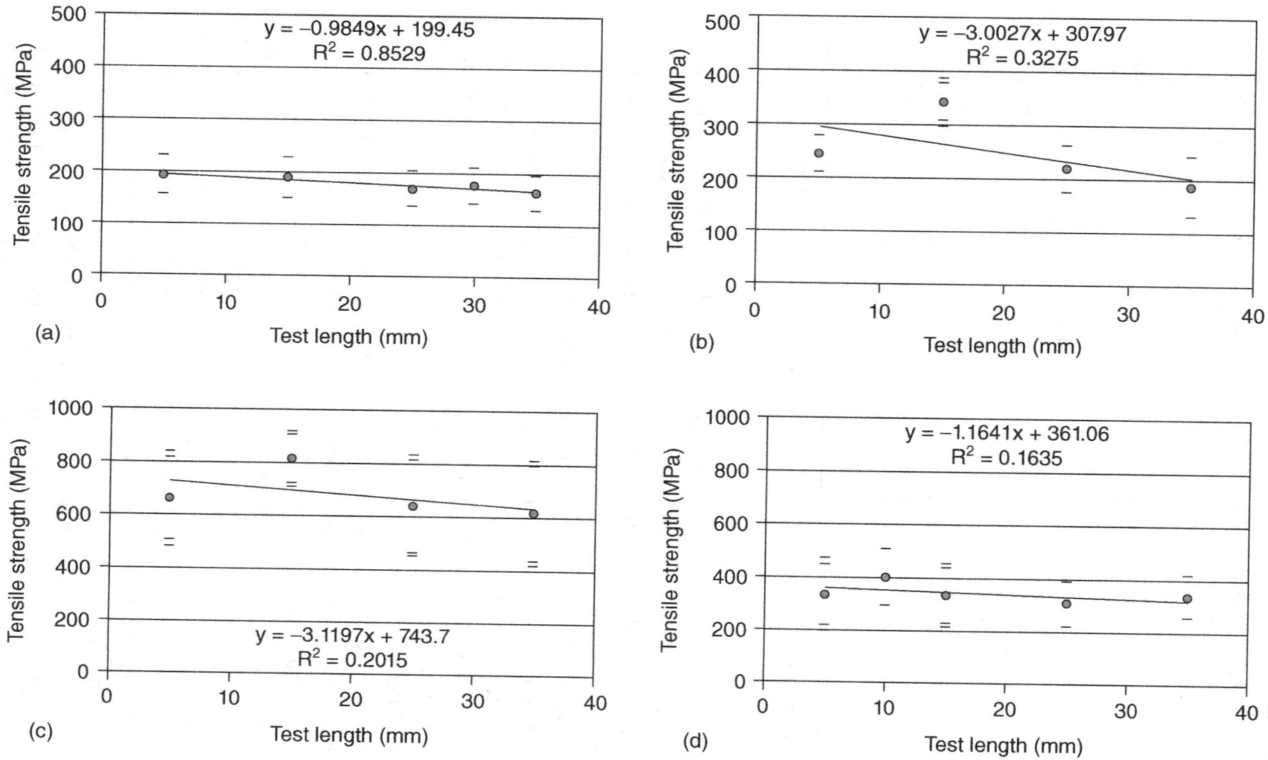

10.3 Tensile strength in function of the test length according to a normal (●, standard deviation: =) and Weibull (○, standard deviation: –) distribution for (a) white coir, (b) brown coir, (c) bamboo and (d) jute fibres; both distributions closely coincide. Reproduced with permission from Elsevier Ltd (Nele Defoirdt *et al.*, 2010).

Table 10.5 Weibull parameters calculated on the strength data of different test lengths

Test length	Parameter	White coir	Brown	Bamboo	Jute
5	*m* (l)	6.0	8.0	4.2	2.7
10		–	–	–	4.3
15		5.8	9.3	9.3	3.0
25		6.0	5.5	3.8	3.9
30		6.0	–	–	–
35		5.8	3.7	3.5	4.6
5	$\lambda = \sigma_0/N^{(1/m)}$ (MPa)	207	259	726	375
10		–	–	–	438
15		204	360	855	374
25		182	237	707	339
30		189	–	–	–
35		175	206	683	364

Reproduced with permission from Elsevier Ltd (Defoirdt *et al.*, 2010).

fiber content and fiber volume fraction. Results showed that by increasing fiber content the tensile properties of the composites increase. The effect of alkali treatment of fibers on the mechanical properties was also observed and found to improve the composites' performance significantly.

Wong *et al.* (2010) studied the impact strengths of E-glass, coir and oil palm as well as E-glass/coir and E-glass/oil palm hybrid polyester composites. The composites were reinforced with fiber volume fractions of 30%, 40% and 50% and fiber lengths of 3, 7 and 10 mm. Composite laminates reinforced with longitudinal and transverse coir fiber mats were also studied. The number of fiber mats varied from one to four layers for non-spaced fiber mats and from two to four layers for 1.5 mm spaced fiber mats. Coir–polyester composites with sand filler were also studied with 40%, 50% and 60% volume fractions and fiber lengths of 3, 7 and 10 mm. Test results showed that impact strength improves with fiber content and fiber length. In addition, it is also observed that longitudinal fiber mats always exhibit better impact toughness when compared to transverse fiber mats. Impact strength increases with the number of fiber layers but decreases with fiber spacing. For coir/polyester concrete, low fiber content and fiber length improve the impact strength.

Rout *et al.* (2001) studied the surface modifications of coir fibers using alkali treatment, bleaching and vinyl grafting with a polyester resin matrix. The mechanical properties of the composites such as tensile, flexural and impact strength increased as a result of surface modification. The bleached (65°C) coir–polyester composites showed the best flexural strength

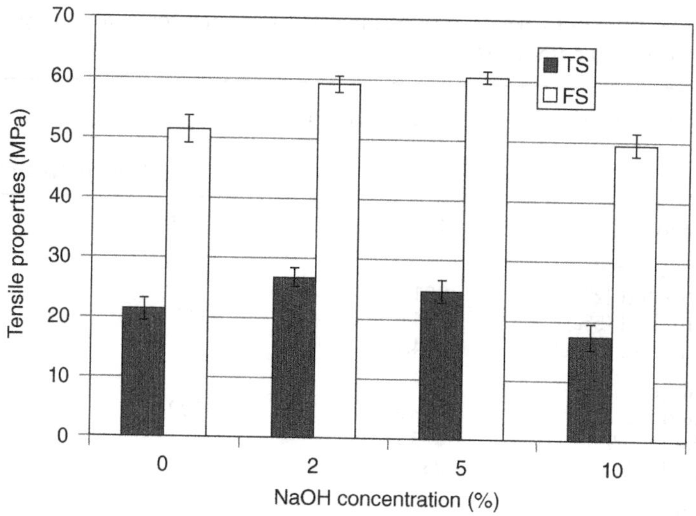

10.4 Effect of alkali treatment on TS and FS of coir–polyester composite (reprinted from Defoirdt *et al.*, 2010, © 2010, with permission from Elsevier).

(61.6 MPa) whereas 2% alkali-treated coir/polyester composites showed significant improvement in tensile strength (26.80 MPa).

Hybrid composites comprising of a glass fiber mat (7 wt%), a coir fiber mat (13 wt%) and a polyester resin matrix were also prepared. Hybrid composites containing surface-modified coir fibers show significant improvement in flexural strength and reduced water absorption. Scanning electron microscopy (SEM) investigations showed that the surface modifications improved fiber/matrix adhesion.

The effect of alkali treatment (2–10%) of coir fibers on the TS and FS of the composites (containing 17% fiber by weight) is shown in Fig. 10.4. NaOH (2%) treated fiber polyester composites showed improvements in the TS and FS of 26 and 15%, respectively when compared to the untreated fiber polyester composites. Increasing the NaOH concentration from 2 to 5% results in a reduction in the TS and an increase in the FS by 17% in comparison to the untreated fiber composite. At 10% NaOH concentration, the treated fiber composites exhibit a reduction in both the TS and the FS. The enhancement of mechanical properties using alkali treatment is attributed to the improved wetting characteristics of alkali-treated coir with the polyester matrix.

The impact strength of untreated and NaOH treated (2–10%) coir–polyester composites (17 wt% fiber loading) is shown in Table 10.6. With incorporation of 2% NaOH-treated fiber, the IS increases from 433.5 J/m

Table 10.6 Effect of surface modification on IS of coir–polyester and glass/coir–polyester hybrid composites

Surface modification	Impact strength	
	Coir–polyester	Glass/coir–polyester hybrid
Untreated	433.5	576
Alkali treatment:		
Alkali treated (2%)	521.9	–
Alkali treated (5%)	634.6	687.8
Alkali treated (10%)	494.7	–
Bleaching:		
Bleached (65°)	458.3	593.0
Bleached (75°)	425.2	–
Bleached (85°)	389.8	–
AN-grafting:		
AN-grafted (5%)	477	
AN-grafted (10%)	596.3	629.8
AN-grafted (15%)	189.7	–

Reproduced with permission from John Wiley & Sons (Rout *et al.*, 2001).

(untreated) to 521.9 J/m, and with a further increase in NaOH concentration to 5% the IS increases to 634.6 J/m. This large increase in IS is due to toughness of the coir fibers. At still higher NaOH concentration (10%), the IS decreases in comparison to 2% and 5% NaOH-treated fiber composites but is still higher than for untreated fiber composites.

The effect of bleaching of the coir fibers on the TS and FS of coir–polyester composites (17 wt% fiber) is shown in Fig. 10.5. The bleached coir–polyester composites show a slight reduction in TS in comparison to untreated fiber composites. This lowering of the fiber tensile strength is due to the reduction of the cementing material (lignin). However, the bleached fiber polyester composites showed a better FS than the untreated fiber composite. Coir fibers bleached at 65°C show an increase in FS of 20%. Fibers bleached at 75° and 85°C also show an improvement in FS in comparison to untreated fiber polyester composites, although the values are lower than the FS of fiber bleached at 65°C.

Figure 10.6 represents the effect of AN-grafting (varying from 5 to 15%) on the TS and FS of coir fiber composites (17 wt% fiber). In the case of 5% AN-grafted coir composites, both the TS and FS increased marginally, but with increased grafting (10%) both the TS and FS increased by 14% and 15%, respectively. A further increase to 15% reduces the TS and FS significantly.

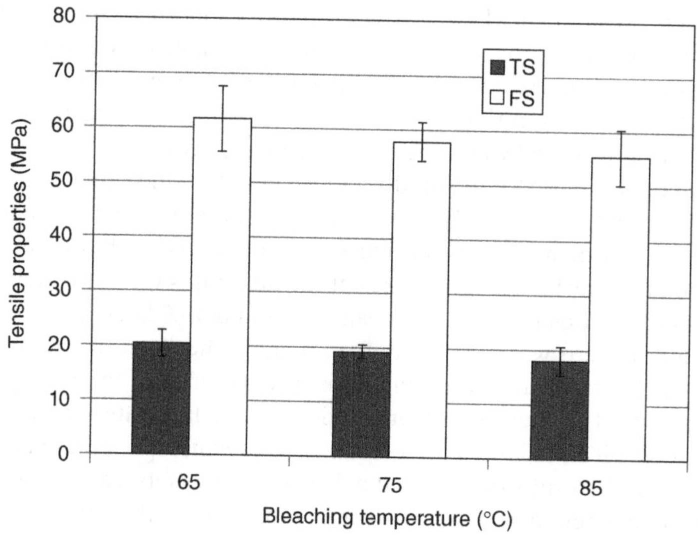

10.5 Effect of bleaching temperature (°C) on TS and FS of coir–polyester composite (Rout *et al.*, 2001, © 2001, John Wiley & Sons, Inc.).

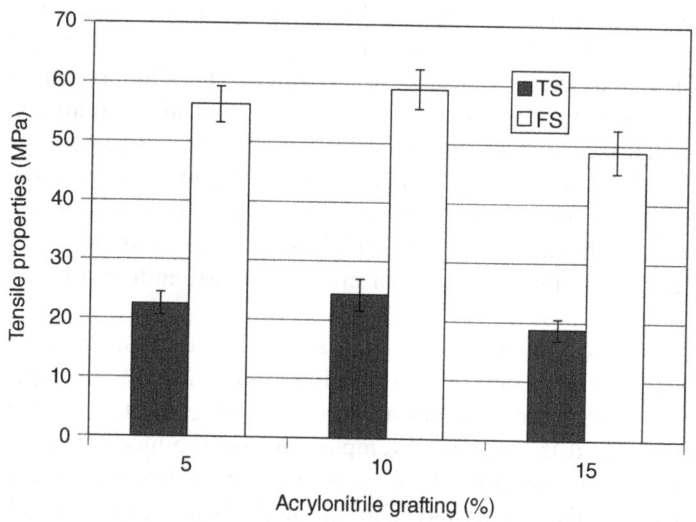

10.6 Effect of AN-grafting on TS and FS of coir–polyester composite (reproduced with permission from John Wiley & Sons, (Rout *et al.*, 2001, © 2001 John Wiley & Sons, Inc.).

The improved TS and FS of AN-grafted (5 and 10%) fiber composites may be due to improved wettability and adhesion between the grafted fibers and the matrix polymer. The decrease in TS and FS at the higher percentage of AN grafting (15%) is due to the fact that the fiber strength was reduced drastically and the fiber becomes porous.

Varma *et al.* (1985) fabricated hybrid laminates consisting of glass fiber mat, chopped bristle coir fibers and unsaturated polyester resin matrix. The effects of treatment of the coir fibers with alkali and dichloromethyl vinyl silane on the mechanical properties of the laminates were investigated. For hybrid composites having 66% resin (w/w) and 19.7% chopped coir fiber (1 cm long) there was a 35% improvement in the flexural strength and a 70% improvement in flexural modulus as compared to the pure polyester resin sheets. The mechanical properties of the laminates increased with increasing coir fiber length up to 1 cm, beyond which a decrease in strength was observed. Results showed that hybrid laminates fabricated from NaOH-treated coir fibers had better mechanical properties than those fabricated from untreated coir fibers. A decrease in thermal conductivity was observed with increasing weight fraction of coir fiber.

Rajini *et al.* (2012) used naturally woven coconut sheath as the reinforcement phase. The hybrid composites were developed with the combination of naturally woven coconut sheath/E-glass/nanoclay using an unsaturated polyester matrix, and the effect of nanoclay on the mechanical and free vibration behavior was investigated. It was observed that the addition of nanoclay improved the mechanical and damping properties of the composites. In addition, the effect of alkali and silane treatments on the fibers, and different stacking sequences, were also investigated. It was found that the alkali treatment of fibers improved the mechanical properties of the composites when cellulose fiber was present. The stacking sequence of the coconut sheath/coconut sheath/glass with alkali treatment of fibers provides the optimum increase in mechanical strength and free vibration characteristics.

Rajini *et al.* (2013) developed naturally woven coconut sheath/polyester composites by using compression molding and varying the weight percentage (1,2,3 and 5 wt%) of organically modified montmorillonite (MMT) nanoclay and investigated the effects of temperature on the mechanical properties of the resulting composite. Three classifications of fibers were considered: untreated, NaOH treated and silane treated. Dynamic mechanical analysis revealed a decrease in the storage modulus (E0) with an increase in temperature in the range 50°C to 110°C. Chemical treatment increases the loss modulus (E00) and damping peaks (tan δ) of coconut sheath. Silane-treated composites showed the maximum increase in E0 and E00 values. Incorporation of nanoclay increases the dynamic factors in all types of composites. A slight increase in the glass transition temperature value was noticed in association with the E00 peaks for silane- and NaOH-treated composites compared to untreated composites.

Rout *et al.* (2003) prepared composites using coir fiber (both continuous and short fiber) as the reinforcement phase with both thermosetting and thermoplastics as matrices. Untreated and chemically modified coir were used for the composite specimen. Chemical treatments such as defatted, PMMA grafted, PAN grafted and cyanoethylated were used to treat the coir fibers. Coir–polyester and glass/coir–polyester hybrid composites were prepared using a hand lay-up technique whereas the coir–BAK composites were prepared by the film-stacking method and the coir–PP composites were prepared by compression molding. The specimens made were subjected to tensile and impact tests and the fractured surfaces were observed under SEM. The SEM micrographs of the fractured surfaces of the composites showed fiber pull-outs under both failure modes.

Sapuan *et al.* (2005) fabricated coconut spathe and spathe-fiber-reinforced epoxy composites and evaluated the resulting tensile and flexural strengths. Samples were fabricated using the hand lay-up process (30:70 fiber and matrix ratio by weight) and the properties evaluated using the Instron material test system. Tensile and flexural strengths for the coconut spathe-fiber-reinforced composite laminates ranged from 7.9 to 11.6 MPa and from 25.6 to 67.2 MPa respectively, indicating that the tensile strengths of coconut spathe-fiber composites are inferior in comparison to other natural fibers such as cotton, coconut coir and banana fibers.

Girisha *et al.* (2012) prepared natural fiber (sisal and coconut coir) reinforced epoxy composites. In order to study the effects of water absorption on the mechanical properties, the composites were subjected to water immersion tests. The composites were fabricated using coconut coir (short fiber) and sisal (long fiber) in hybrid combination and fiber weight fractions of 20%, 30% and 40%. The specimens were immersed in a water bath at 250°C and 1000°C for varying time periods.

The tensile and flexural properties of water-immersed specimens were evaluated and compared with dry composite specimens. An increase in the fiber volume fraction increases the percentage of moisture uptake because of the high cellulose content of the fiber. The moisture uptake behavior of the specimens decreases the tensile and flexural properties of natural fiber-reinforced epoxy composite. At room temperature the water absorption pattern of these composites was found to follow Fickian behavior, whereas at high temperature it did not follow Fick's law.

Chandra Rao et al. (2012) studied the wear behavior of treated and untreated coir dust filled epoxy resin matrix composites. For wear analysis, treated and untreated coir dust concentrations of 10%, 20% and 30%, loads of 10, 20 and 30 N, and velocities of 300, 400 and 500 meters per second were employed. It was observed that the treated fiber composite shows better wear resistance than the untreated fiber composites and that increasing the coir dust concentration reduced the abrasive wear rate whereas the wear rate increased with load and velocity.

Salmah *et al.* (2013) studied the properties of untreated and treated coconut shell (CS) reinforced unsaturated polyester (USP) composites. The coconut shell was treated with 1% sodium hydroxide (NaOH). It was found that the addition of CS content increased the tensile strength, modulus of elasticity, flexural strength, flexural modulus and thermal stability, whereas elongation at break of the USP/CS composites decreased. As compared to the untreated composites the treated USP/CS composites exhibited a higher tensile strength, modulus of elasticity, flexural strength, flexural modulus and thermal stability. SEM results indicated better filler dispersion and adhesion between the CS and USP with alkali treatment.

Aireddy and Mishra (2011) conducted an experimental study to investigate the erosive and abrasive wear behavior of coir dust filled epoxy resin matrix composites. For wear test analysis, various coir dust concentrations, impingement angles(30°, 45°, 60°, 75°, and 90°), and impact velocities (34, 48, 60, 78, and 92 m/s) were used. Dry silica sand with a size distribution in the range 200–600 μm was used as the erodant. It was observed that the composite shows brittle-type failure and the maximum erosion rate was observed at a 90° impingement angle. Erosion wear rate was found to decrease with increasing coir dust concentration. Abrasive wear resistance was found to decrease with an increase in the normal load and also to increase with increasing fiber content.

10.6 The properties of coir fiber-reinforced thermoplastic polymer composites

Sindhu *et al.* (2007) analyzed the influence of different conditions on the mechanical properties of short coir fiber-reinforced polymer composites and glass fiber-reinforced polymer composites. They found that a fiber length of 20 mm yields the best performance characteristics. It was also observed that a filler loading of 30 phr was optimal.

Rimdusit *et al.* (2011) developed coconut fiber-filled composites based on PVC/ASA matrices. They compared composites fabricated from PVC/ASA/coconut fiber and PVC wood or the polyolefin wood and found that the impact strengths of the PVC/ASA/coconut fiber composites were higher than those of the PVC wood composites or polyolefin wood composites at the same fiber content. By increasing the fiber content, the storage modulus at room temperature and glass transition temperature (Ts g) of the composites are also increased. By slightly decreasing the fiber content, the tensile strength of the blend reached as high as 45 MPa.

Santos *et al.* (2009) prepared composites of PP and coir fibers in a mixer. Two coupling agents were added in order to improve the properties of the composites. The first coupling agent was a laboratory-made silane-functionalized-PP and the second was a commercial maleated-PP. The result showed that the coir fibers acted as reinforcing fillers and increased the static and dynamic moduli of the composites. Both coupling agents improved these properties further. The silanized-PP was found to exhibit superior properties and had a larger effect on the various moduli (Young, storage, and loss moduli). TGA analysis revealed an earlier degradation of the PP in the presence of the fibers and the couplers.

Haque *et al.* (2010) manufactured coir and abaca fiber-reinforced polypropylene composites using a single extruder and an injection molding machine. The raw coir and abaca were chemically treated with benzene diazonium salt. During composite manufacturing, both raw and treated coir and abaca fibers at different levels of fiber loading (10, 15, 20, 25 and 30 wt%) were utilized. They found that the chemically treated fiber-reinforced specimens displayed good mechanical properties when compared to the untreated composites and the coir fiber composites exhibited better mechanical properties than the abaca fiber-reinforced ones. Based on fiber loading, 30% fiber-reinforced composites had the optimum set of mechanical properties.

Sharma *et al.* (2006) developed glass or coir fiber-reinforced polypropylene (PP) composite specimens with 0, 15 and 30 wt% reinforcement and tested for weathering properties due to moisture and ultraviolet (UV) exposure. Short-duration soaking, long-duration soaking, accelerated (boiling water) soaking, and UV with moisture exposure tests were conducted on specimens prepared using a hot press designed for the purpose. The results showed that the PP/glass specimens were superior to the PP/coir specimens with respect to rate of water absorption and thickness swelling. Both deformation and delamination are observed during the soaking tests and discoloration and surface roughness were characteristic degradation features of the UV with moisture exposure test.

Ayrilmis *et al.* (2011) evaluated the physical, mechanical and flammability properties of coconut fiber-reinforced polypropylene (PP) composite panels. They selected four levels of coir fiber content (40, 50, 60 and 70% by weight) and the fibers were mixed with the PP powder and 3 wt% maleic anhydride grafted PP (MAPP) powder. By increasing coir content it was found that the water resistance and the internal bond strength of the composites were decreased. However, the flexural strength, tensile strength and hardness of the composites increased by increasing the coir fiber content up to 60 wt%. The flame retardancy of the composites improved with increasing coir fiber content.

Hatta and Akmar (2008) prepared polystyrene/polypropylene (PS/PP) reinforced coconut and jute fiber composites using the injection molding method. They designed composites by fixing the fiber content at 10 wt% with the ratios of 100/0, 75/25, 50/50, 25/75 and 0/100 by weight of the coconut fiber/jute fibers. They observed that the addition of 10 wt% of fiber (coconut and jute) increased the tensile properties of the composites. It was found that the tensile strength for composites reinforced with 10 wt% jute fibers displayed better performance characteristics than the composites reinforced with 10 wt% coconut fibers, although reinforcement was observed to reduce the impact strength of the materials.

Haque and Islam (2013) studied the influence of fiber surface treatment on the mechanical properties of coir fiber-reinforced polypropylene (PP) composites. In order to improve the compatibility between the fibers and the PP matrix the coir fiber was first treated chemically with sodium periodate and then with urea. It is found that the mechanical properties of treated coir–PP composites were better than those of the untreated ones, suggesting that improved fiber–matrix interfacial adhesion had occurred. Water absorption results showed that treated coir–PP composites absorb less water than the untreated ones, which indicates that the chemical treatment decreased the hydrophilic nature of the coir fibers. The scanning electron microscopy (SEM) results showed that the treated coir–PP composites possess better surface smoothness with less micro-voids and fiber agglomerates, together with better fiber.

Haydaruzzaman *et al.* (2010) prepared coir yarn-reinforced polypropylene (PP)-based unidirectional composites by compression molding. In this study it was observed that 20% coir yarn content showed improved mechanical properties. Jute yarns (20%/100%) were incorporated into the coir-based composites. It was found that 20% coir and 80% jute-reinforced PP matrix composites yeilded the best results. Gamma radiation from 400 to 1000 krad was used for all the materials (coir, jute and PP). The best mechanical properties were observed at 600 krad for PP-based composites (20% coir and 80% jute). Starch (210% in aqueous solution) was used as a stiffening agent for the yarns and improved the unidirectional positioning of the yarns in the composite.

Tran *et al.* (2009) developed short coir fiber-reinforced poly(butylene succinate) (PBS) biodegradable composites. They studied the effect of fiber content varying from 10 wt% to 50 wt% on the mechanical properties of short fiber coir/PBS composites. The effect of alkali treatment on the mechanical properties of the composites was also studied. They found that the mechanical properties of alkali-treated coir/PBS composites are better than those of the untreated coir fiber. The best mechanical strength of short coir/PBS composites was achieved at a fiber content of 20 wt%. The tensile and flexural modulus of the composites increased with increasing fiber content. Compared with untreated coir fiber, alkali-treated coir/PBS composites at 20 wt% fiber content showed increases in tensile strength (TS) by 29.6%, in tensile modulus (TM) by 32.6%, in flexural strength (FS) by 19.1% and in flexural modulus (FM) by 21.3%. The fractured surface of the tensile specimens exhibited an improvement in interfacial fiber–matrix bond for the alkali-treated coir/PBS composites.

Haque *et al.* (2009) manufactured palm and coir fiber-reinforced polypropylene biocomposites using a single extruder and injection molding machine. Benzene diazonium salt was used for treatment of the raw palm and coir to increase their compatibility with the polypropylene matrix. Five levels of fiber loading (15, 20, 25, 30 and 35 wt%) for both raw and treated palm and coir fiber were utilized during composite manufacture. The treated fiber-reinforced composites showed better mechanical properties when compared to the untreated fiber reinforced composites. It was also observed that the coir fiber composites had better mechanical properties than the palm fiber ones. Based on fiber loading, the optimum set of mechanical properties were achieved at 30 wt%. The tensile strength of the raw and treated palm and coir fiber-reinforced PP biocomposites at different fiber loadings are shown in Fig. 10.7.

For the untreated fiber-reinforced composites, when the fiber load increases the weak interfacial area between the fiber and matrix also increases, which results in a decrease in tensile strength. Treatment with benzene diazonium salt to increase the compatibility of the palm and coir fiber with the PP matrix resulted in an increase in the tensile strength of the 15 wt% treated fiber-loaded composites as compared to the PP matrix

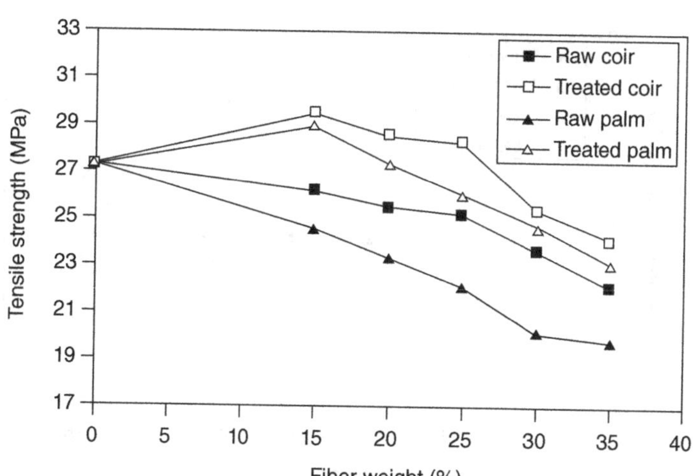

10.7 Variation of tensile strength at different fiber loading (reprinted from Haque *et al.*, 2009, © 2009, with permission from Elsevier).

itself. This is due to the fact that stress is transferred more effectively. The tensile strength of the treated composites decreased with fiber loading. Due to the lower cellulose content of the coir fiber compared to the palm fiber, the tensile strength of the untreated and treated coir fiber-reinforced PP composites was found to be higher than those of the raw and treated palm composites respectively. Chemical treatment and increasing fiber loading results in an improved Young's modulus, which was higher in the untreated and treated coir fiber-reinforced composites compared to the untreated and treated palm fiber-reinforced PP composites.

Increasing the fiber loading from 30 to 35 wt% results in fiber agglomeration and decreases the fiber/matrix adhesion, resulting in lower flexural strength values (as shown in Fig. 10.8). Again it is found that untreated and treated coir fiber-reinforced composites had higher flexural strength and modulus values compared to the untreated and treated palm fiber-reinforced PP composites.

Variation of the Charpy impact strength with fiber loading for both untreated and chemically treated fiber-reinforced composites is shown in Fig. 10.9. The impact strength of all the composites increases with increasing fiber loading. This is due to the fact that because of the strong interfacial adhesion between the fiber and matrix, the fiber was capable of absorbing energy.

Islam *et al.* (2010) investigated the physico-mechanical properties of coir-reinforced polypropylene (PP) composites. Coir was chemically treated with *o*-hydroxybenzenediazonium salt to attain improved mechanical

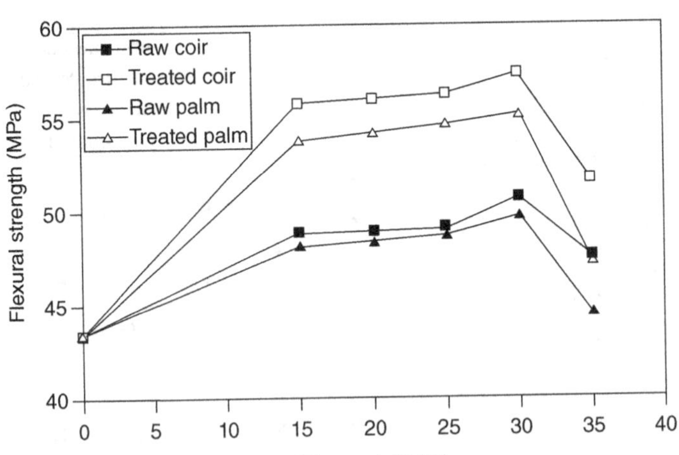

10.8 Variation of flexural strength at different fiber loading (reprinted from Haque *et al.*, 2009, © 2009, with permission from Elsevier).

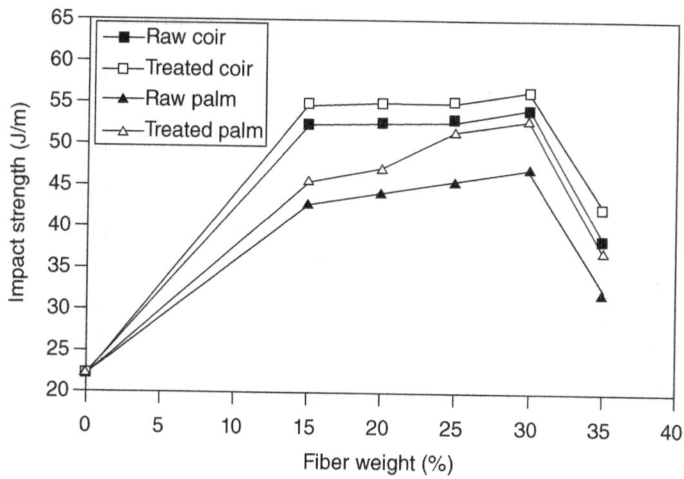

10.9 Variation of impact strength at different fiber loading (reprinted from Haque *et al.*, 2009, © 2009, with permission from Elsevier).

properties of the composites. Both untreated and treated coir samples were utilized for the fabrication of the composites. Results showed that the mechanical properties of the composites prepared from the chemically treated coir were better than those of the untreated ones. By increasing the filler content the tensile strengths of the composites of both untreated and chemically treated coir–PP composites were reduced.

Figure 10.10 shows the Young's modulus of the composites of raw and treated coir at different filler loadings. It is clear that the incorporation of the fiber significantly increases the modulus with increasing filler loading. Since coir also has a higher modulus than PP, incorporation of the filler into the matrix also improved the stiffness of the composites. Figure 10.11 shows the variation of tensile strength with filler loading for untreated and treated coir-reinforced PP composites.

The flexural strength and modulus of the composites of both raw and chemically treated coir-reinforced PP composites are shown in Figs 10.12 and 10.13, respectively. Addition of the fiber has increased the flexural strength of the composites as compared to pure PP. It was also observed that the values remain almost constant for 10% to 25% filler-loaded composites (Fig. 10.12). A significant improvement in the flexural strength is noticed for the composites of chemically treated coir fibers as compared to those of untreated ones. This is because of the better wetting of the treated coir with the PP matrix. It is also observed that the values of flexural modulus of both treated and untreated composites are found to increase with increasing filler content (Fig. 10.12).

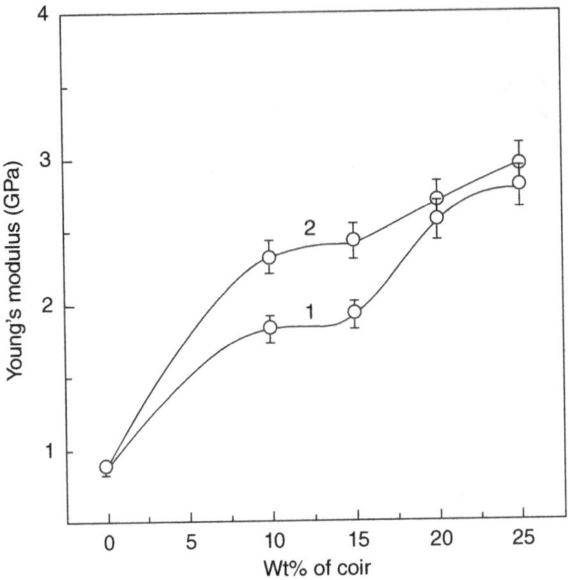

10.10 Variation of Young's modulus with filler loading for (1) untreated and (2) treated coir reinforced PP composites (reprinted from Islam *et al.*, 2010, © 2010, with permission from Elsevier).

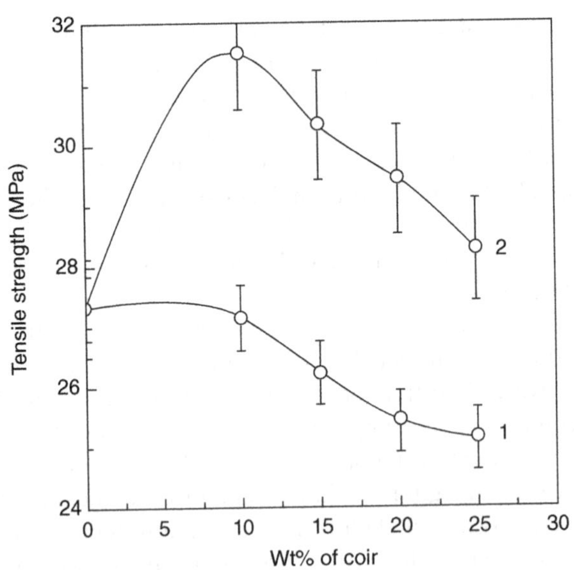

10.11 Variation of tensile strength with filler loading for (1) untreated and (2) treated coir-reinforced PP composites (reprinted from Islam *et al.*, 2010, © 2010, with permission from Elsevier).

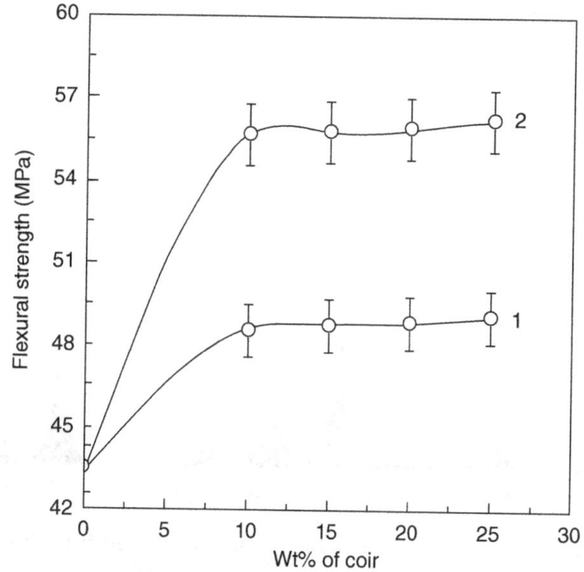

10.12 Plots of flexural strength at different filler loading for (1) untreated and (2) treated coir-reinforced PP composites (reprinted from Islam *et al.*, 2010, © 2010, with permission from Elsevier).

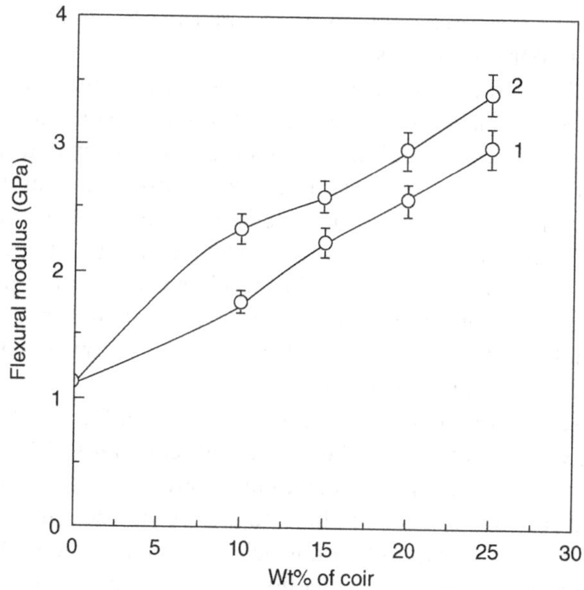

10.13 Plots of flexural modulus at different filler loading for (1) untreated and (2) treated coir-reinforced PP composites (reprinted from Islam *et al.*, 2010, © 2010, with permission from Elsevier).

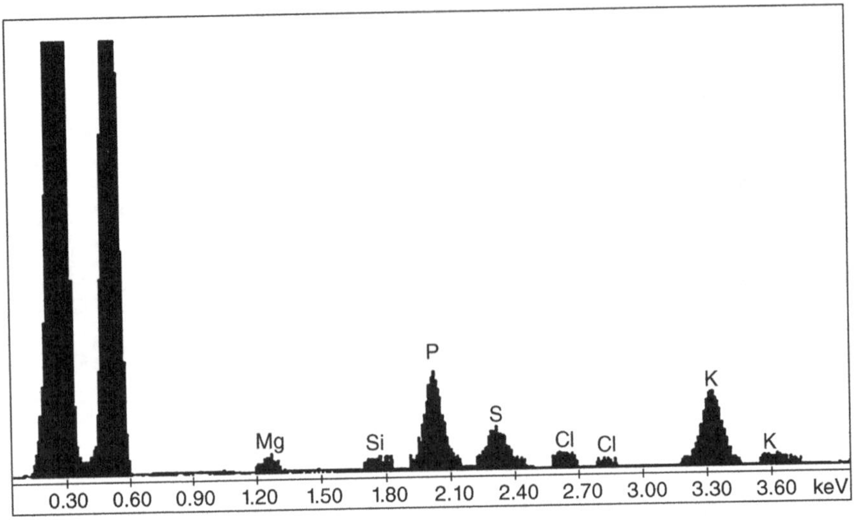

10.14 Elementary analysis of coconut shell (reprinted from Bledzki *et al.*, 2010, © 2010, with permission from Elsevier).

Bledzki *et al.* (2010) studied the potential of products such as barley husk and coconut shell as reinforcements for thermoplastics as an alternative or addition to wood fiber composites.

Figure 10.14 shows the spectra of fiber surface elements acquired for coconut shell fibers which contain mainly carbon, oxygen and a small amount of silicon, potassium, sulphur, phosphorus, magnesium and chlorine. The relative atomic percentages of the atoms were obtained from the peak area and corrected with an appropriate sensitivity factor. Coconut shell fibers show a higher proportion of carbon when compared to softwood fibers. The higher proportion of carbon in the fiber is due to the presence of the hydrocarbon-rich waxy coating on the cuticle of the fiber and the lignin present on the surface. Coconut shell fiber contains a lower proportion of silicon when compared to softwood.

The FT-IR spectrum of coconut husk fibers is shown in Fig. 10.15 and the assignment is given in Table 10.7. Peaks at 3401 cm^{-1} and 2911 cm^{-1} for coconut shell were observed. The peak at about 3400 cm^{-1} is due to hydrogen bonded O–H stretching. The hydrophilic tendency of coconut shell was reflected in the broad absorption band, which is related to the OH groups present in aliphatic or aromatic alcohol and present in their main components. The peak at around 2900 cm^{-1} is due to the C–H asymmetric and symmetric stretching from aliphatic saturated compounds. These two stretching peaks correspond to the aliphatic moieties in cellulose and

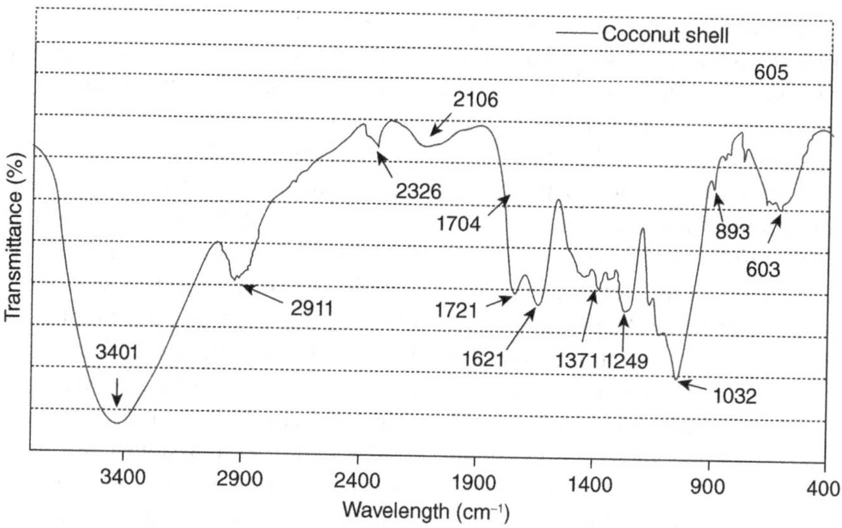

10.15 FT-IR of coconut husk (reprinted from Bledzki *et al.*, 2010, © 2010, with permission from Elsevier).

Table 10.7 FT-IR spectra of coconut husk

Peak location range (cm^{-1})	Assignment	Coconut shell
3460–3400	O–H stretching	3401
3000–2850	C–H asymmetric and symmetric stretching in methyl and methylene group	2911
2400–2300	P–H stretching and P–O–H stretching	2326
2200–2100	Si–H stretching	2106
1738–1700	C=O stretching in acetyl and uronic ester groups or in carboxylic group of ferulic and coumaric acids	1721
1650–1580	N–H bending in primary amine	1621
1375–1350	C–H rocking in alkanes or C–H stretching in methyl and phenolic alcohol	1371
1250–1200	Si-CH_2 stretching in alkane or C–C plus C–O plus C=O stretching	1249
1086–1030	C–O deformation in secondary alcohol and aliphatic ether	1032
900–875	C–H in plan deformation plus C–O deformation in primary alcohol	893
	C-1 group frequency or ring frequency	

Reproduced with permission from Elsevier Ltd (Bledzki *et al.*, 2010).

hemicellulose (Cherian *et al.*, 2008). The sharp peaks at around 2330 cm^{-1} are due to the P–H stretching and P–O–H stretching. The band at about 2112 cm^{-1} is reflected Si–H stretching. It could be summarized that the bands in this region were representing the multivalent inorganic hydride or oxide (Stuart, 1997). In the double bond region, a peak at 1721 cm^{-1} for coconut shell is assigned to the C=O stretching of the acetyl and uronic ester groups of hemicellulose or to the ester linkage of the carboxylic group of the ferulic and *p*-coumaric acids of lignin (Sun *et al.*, 2005). The sharp peaks at 1621 cm^{-1} for coconut shell are reflected for amide I. The amide I band represents 80% of the C=O stretching of the amide group, coupled to the in-plane N–H bending and C–N stretching modes. The exact frequency of this vibration depends on the nature of hydrogen bonding involving the C=O and N–H groups and the secondary structure of protein (Cherian *et al.*, 2008). The bands in the range 1375–1350 cm^{-1} were assigned from the C–H symmetric and asymmetric deformations in methyl and phenolic alcohol or C–H rocking in alkanes. The band region of 1250–1200 cm^{-1} represents Si–CH_2 stretching in alkane or C–C plus C–O plus C–O stretching and deformation bands in cellulose and lignin. Coconut shell showed absorbance peaks at 1050 cm^{-1} and 1032 cm^{-1}. The peak in the range 1086–1030 cm^{-1} is assigned to C–O deformation in secondary alcohol and aliphatic ether. A sharp peak at around 900 cm^{-1} which is indicative of the C-1 group frequency or ring frequency is characteristic of β-glycosidic linkages between the sugar units (Stuart, 1997). It has been also noticed that fibers content with different amounts have more or less similar chemical constituents but different peak area and sharpness.

10.7 Characterization of coconut/coir fiber-reinforced composites

Scanning electron microscopy can be used for the morphological characterization of coconut/coir fiber-reinforced composites. This is the most widely used method for the surface analysis of composite materials. In SEM, a highly focused scanning electron beam bombards the surface, causing large numbers of secondary electrons to be generated, the intensity of which is governed by surface topography. The method is suitable for all materials, but non-conductive materials must be given a thin conductive coating of sputtered gold which can alter or mask the true surface morphology.

The SEM image of untreated coir fiber is shown in Fig. 10.16(a) which indicates an uneven surface due to globular protrusions. The changes in surface topology of alkali-treated coir fiber are shown in Fig. 10.16(b). Alkali treatment removes the globular protrusions present on the surface

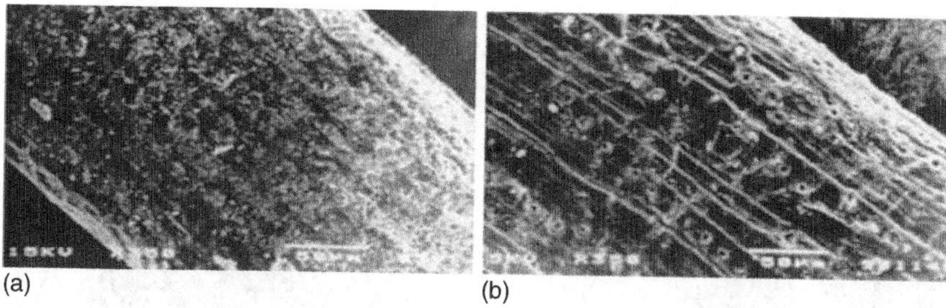

10.16 (a) Untreated coir fiber; (b) alkali-treated coir fiber (reprinted from Geethamma *et al.*, 2005, © 2005, with permission from Elsevier).

of the untreated fiber, leading to the formation of a large number of voids on the surface of the fiber. These voids promote better mechanical adhesion between the fiber and rubber matrix (Geethamma *et al.*, 2005).

In Fig. 10.17(a) a comparison is made between backscattered electron images of the short coconut fibers obtained from unaged specimens with those submitted to 25 cycles of wetting and drying. Microstructural characteristics of the interface of the composites showed that the matrix around the fibers of the unaged specimen (Fig. 10.17(a) and (b)) is relatively more porous and shows more cracks than those of the aged specimens (Fig. 10.17(c) and (d)). It can also be observed that the mortar zone delineated by the fiber surface and the crack around the fiber perimeter is increased in an unaged specimen when compared to the specimen submitted to cycles of wetting and drying. A weak matrix in the surrounding of the fiber is found to be good from the point of view of the ductility of the fiber composite. The migration of hydration products from the matrix to the fibers caused densening of the matrix around the fibers which ultimately results increase in crack strength and reduction in toughness.

To verify the migration of the cement products to the lumen, pores and spaces of the fibers, a dot mapping of chemical elements was carried out. The chemical elements were calcium (Ca), sulphur (S), silicon (Si), and sodium (Na). Fig. 10.18 presents the fiber–matrix interfacial zone where the dot mapping of the elements was carried out. Figure 10.19 shows the dot map of the chemical elements. The results show that there is a considerable transport of cement chemicals, mainly calcium products, to the fibers during the cycles of wetting and drying (Toledo Filho *et al.*, 2000).

The morphological features of tensile fractured surfaces of 20% coir-reinforced PP composites investigated by SEM are shown in Fig. 10.20. From Fig 10.20 it is observed that there are a number of agglomerations, resulting from strong fiber–fiber interaction and fiber pullout marks in the SEM images of the untreated coir–PP composite. A poor interfacial bonding

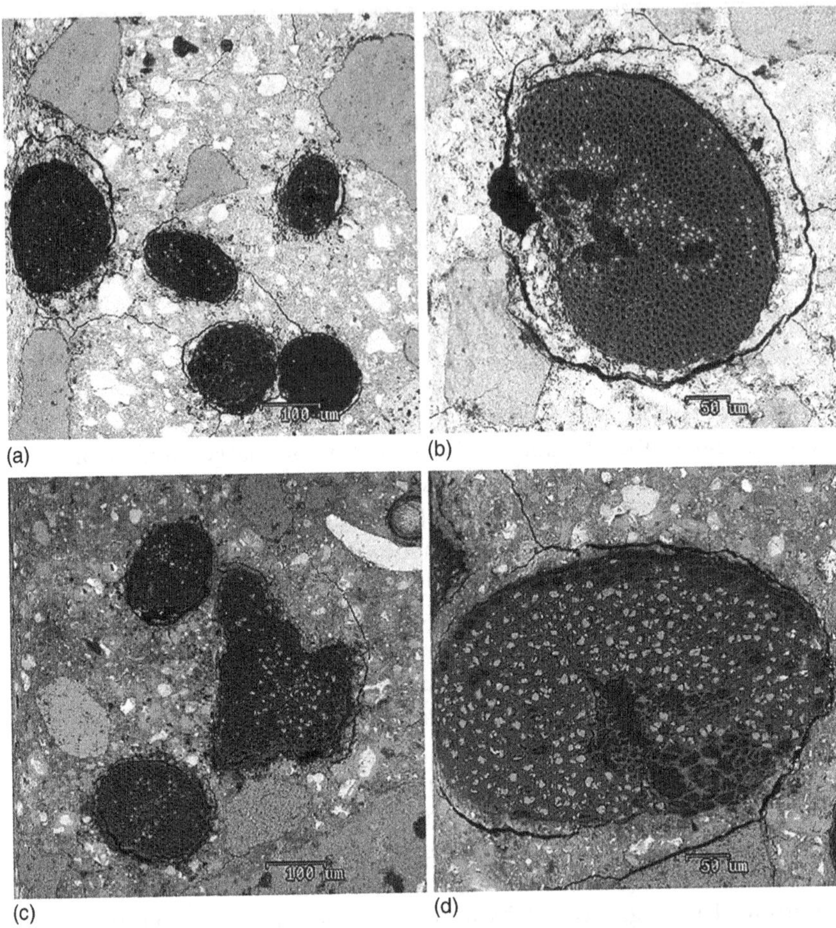

10.17 Backscattered images of cross-sections of coconut fibers: (a) images from a 28-day specimen; (b) higher magnification of the typical fiber–matrix interface before ageing; (c) images from a specimen submitted to 25 cycles of wetting and drying; (d) higher magnification of the typical fiber–matrix interface after 25 cycles of wetting and drying (reprinted from Toledo Filho *et al.*, 2000, © 2000, with permission from Elsevier).

exists between the matrix and the untreated filler (image (a)). On the other hand (image (b)), chemically treated coir–PP composites showed almost no signs of fiber agglomeration and micro-voids in the fractured surface of the composite. This result suggests that interfacial adhesion between the chemically treated coir and the PP matrix has become much more favorable upon treatment of the fiber with diazonium salt. Furthermore, a better filler–matrix interaction has occurred upon chemical treatment of coir.

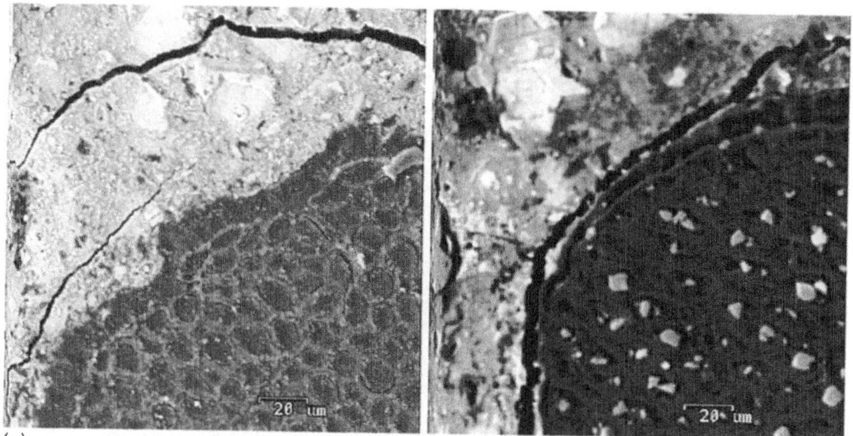

(a) (b)

10.18 Fiber–matrix interfacial zone where the dot mapping of elements was carried out: (a) coconut fiber–mortar interface zone for a specimen 28 days old; (b) coconut fiber–mortar interfacial zone for a specimen subjected to 25 cycles of wetting and drying (reprinted from Toledo Filho *et al.*, 2000, © 2000, with permission from Elsevier).

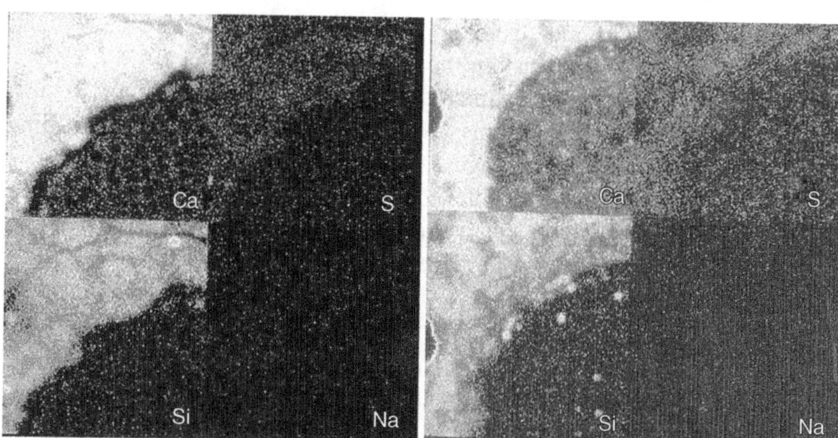

(a) (b)

10.19 Dot map of chemical elements in coconut fiber–mortar interface: (a) mapping of chemical elements for the specimen 28 days old (reference); (b) mapping of chemical elements for the specimen submitted to 25 cycles of wetting and drying (reprinted from Toledo Filho *et al.*, 2000, © 2000, with permission from Elsevier).

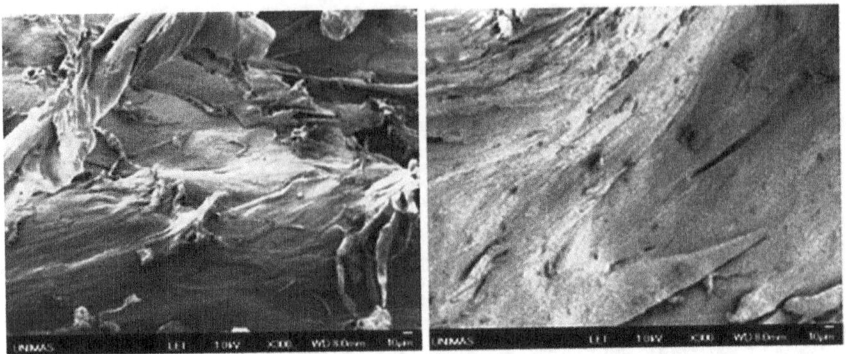

(a) (b)

10.20 SEM images of 20% coir-reinforced PP composites: (a) untreated coir–PP composite; (b) treated coir–PP composite (reprinted from Islam *et al.* 2010, © 2010, with permission from Elsevier).

10.8 Advantages of using coconut/coir fibers as reinforcement in composites

The advantages of coir/coconut fiber can be summarized as follows:

- The fibers are light and strong.
- The fibers can easily withstand heat.
- The fibers can withstand saline water.
- The use of coconut fibers seems to delay the plastic shrinkage controlling crack development at early ages.
- Coir is an abundant, renewable, cost-effective, and lignocellulosic fiber.
- The addition of coconut coir reduces the thermal conductivity of the composite specimens (Sen and Reddy, 2011).

10.9 Conclusions

The present chapter summarized the use of coir fiber as a filler in composite materials, especially those extracted from coconut husk. The mechanical properties of coir fibers in polymer composites were discussed. Surface modification of fibers (chemical modification) was also shown to play an important role in composite performance as it improves interfacial adhesion between the fiber and the matrix. A review of research in the field was given.

10.10 Acknowledgment

The authors are thankful and grateful to Elsevier Ltd for granting permission to reproduce figures and tables to include in this chapter from their journals.

10.11 References

Abdullah, N.M., Ahmad, I. (2012), 'Effect of chemical treatment on mechanical and water-sorption properties coconut fiber-unsaturated polyester from recycled PET', *ISRN Materials Science*, Article 134683, 1–8.

Aireddy, H., Mishra, S.C. (2011), 'Tribological behavior and mechanical properties of bio-waste reinforced polymer matrix composites', *Journal of Metallurgy and Materials Science*, 53, 139–152.

Augstburger, F., Berger, J., Censkowsky, U., Heid, P., Milz, J., Streit, C. (2000), *Organic Cultivation of Coco Palms*, 1st edition, Naturland e.V., Gräfelfing, Germany.

Ayrilmis N., Jarusombuti, S., Fueangvivat, V., Bauchongkol, P., White, R.H. (2011), 'Coir fiber reinforced polypropylene composite panel for automotive interior applications', *Fibers and Polymers*, 12, 919–926.

Bledzki, A.K., Mamun, A.A., Gabor, M. Lucka and Gutowski, V.S. (2008), 'The effects of acetylation on properties of flax fibres and its polypropylene composites', *Express Polymer Letters*, 2, 413–422.

Bledzki, A.K., Mamun, A.A., Volk, J. (2010), 'Barley husk and coconut shell reinforced polypropylene composites: The effect of fibre physical, chemical and surface properties', *Composites Science and Technology*, 70, 840–846.

Chandra Rao, C.H., Madhusudan S., Raghavendra G., Venkateswara Rao, E. (2012), 'Investigation in to wear behavior of coir fiber reinforced epoxy composites with the Taguchi method', *IJERA*, 2, 371–374.

Cherian, B.M., Pothan, L.A., Nguyen-Chung, T., Menning, G., Kottaisamy, M., Thomas, S. (2008), 'A novel method for the synthesis of cellulose nanofibril whiskers from banana fibres and characterization', *Journal of Agricultural and Food Chemistry*, 56, 5617–5627.

Das, S. (2012), 'Fabrication and characterization of raw and bleached treatment coir fiber reinforced polymer composite', MSc. Dissertation, National Institute of Technology, Rourkela, India.

Defoirdt, N., Biswas, S., De Vriese, L., Tran, L.Q.N., Van Acker, J., Ahsan, Q., Gorbatikh, L., Van Vuure, A., Verpoest I. (2010), 'Assessment of the tensile properties of coir, bamboo and jute fibre', *Composites: Part A*, 41, 588–595.

Dixit, S. and Verna, P. (2012). 'The effect of hybridization on mechanical behaviour of coir/sisal/jute fibres reinforced polyester composite material' *Research Journal of Chemical Sciences*, 2(6), 91–93.

Geethamma, V.G., Kalaprasad, G., Groeninckx, G., Thomas, S. (2005), 'Dynamic mechanical behavior of short coir fiber reinforced natural rubber composites', *Composites Part A: Applied Science and Manufacturing*, 36, 1499–1506.

Girisha, C., Sanjeevamurthy, Srinivas, G.R. (2012), 'Sisal/coconut coir natural fibers – epoxy composites: Water absorption and mechanical properties', *International Journal of Engineering and Innovative Technology (IJEIT)*, 2, 166–170.

Haque, M., Islam, N. (2013), 'A study on the mechanical properties of urea-treated coir reinforced polypropylene composites', *Journal of Thermoplastic Composite Materials*, 26, 139–155.

Haque, M.M, Hasan, M., Islam, M.S., Ali, M.E. (2009), 'Physico-mechanical properties of chemically treated palm and coir fiber-reinforced polypropylene composites', *Bioresource Technology*, 100, 4903–4906.

Haque, M., Rahman, R., Islam, N., Huque, M., Hasan, M. (2010), 'Mechanical properties of polypropylene composites reinforced with chemically treated coir and abaca fiber', *Journal of Reinforced Plastics and Composites*, 29, 2253–2261.

Harish, S., Michael, D.P., Bensely, A., Mohan Lal, D., Rajadurai, A. (2009), 'Mechanical property evaluation of natural fiber coir composite', *Materials Characterization*, 60, 44–49.

Hatta, N., Akmar, N. (2008), 'Mechanical properties of polystyrene/polypropylene reinforced coconut and jute fibers', *CUTSE International Conference*, 24–27 November 2008, Miri, Sarawak, Malaysia.

Haydaruzzaman, Khan, A.H., Hossain, M.A., Khan, M.A., Khan, R.A. (2010), 'mechanical properties of the coir fiber-reinforced polypropylene composites: Effect of the incorporation of jute fiber', *Journal of Composite Materials*, 44, 401–416.

Islam, N., Rahman, R., Haque, M., Huque, M. (2010), 'Physico-mechanical properties of chemically treated coir reinforced polypropylene composites', *Composites: Part A*, 41, 192–198.

Joseph, K., Thomas, S. and Pavithran, C. (1996), 'Effect of chemical treatment on the tensile ... sisal fiber-reinforced polyethylene composites', *Polymer*, 37, 5139.

Karthikeyan, A. and Balamurugan, K. (2012) 'Effect of alkali treatment and fiber length on impact behavior of coir fiber reinforced epoxy composites', *Journal of Scientific and Industrial Research*, 71, 627–631.

Misra, R.K., Kumar, S., Sandeep, K., Misra, A. (2007), 'Some experimental and theoretical investigations on fire retardant coir/epoxy micro-composites', *Journal of Thermoplastic Composite Materials*, 21, 1–32.

Mithil Kumar, N., Venkata Reddy, G., Venkata Naidu, S., Shobha Rani, T., Subha, M.C. (2009), 'Mechanical properties of coir/glass fiber phenolic resin based composites', *Journal of Reinforced Plastics and Composites*, 28, 2605–2613.

Rajini, N., Winowlin Jappes, J.T., Rajakarunakaran, S., Jeyaraj, P. (2012), 'Mechanical and free vibration properties of montmorillonite clay dispersed with naturally woven coconut sheath composite', *Journal of Reinforced Plastics and Composites*, 31, 1364–1376.

Rajini, N., Winowlin Jappes, J.T., Jeyaraj, P., Rajakarunakaran, S., Bennet, C. (2013), 'Effect of montmorillonite nanoclay on temperature dependence mechanical properties of naturally woven coconut sheath/polyester composite', *Journal of Reinforced Plastics and Composites*, 32, 811–822.

Rimdusit, S., Damrongsakkul, S., Wongmanit, P., Saramas, D., Tiptipakorn, S. (2011), 'Characterization of coconut fiber-filled polyvinyl chloride/acrylonitrile styrene acrylate blends', *Journal of Reinforced Plastics and Composites*, 30, 1691–1702.

Romli, F.I., Alias, A.N., Rafie, A.S.M., Majid, D.L.A.A. (2012), 'Factorial study on the tensile strength of a coir fiber-reinforced epoxy composite', *AASRI Procedia*, 3, 242–247.

Rout, J., Misra, M., Tripathy S.S., Nayak S.K., Mohanty A.K. (2001), 'The influence of fibre treatment on the performance of coir-polyester composites', *Composites Science and Technology*, 61, 1303–1310.

Rout, J., Misra, M., Mohanty, A.K., Nayak, S.K., Tripathy, S.S. (2003), 'SEM observations of the fractured surfaces of coir composites', *Journal of Reinforced Plastics and Composites*, 22, 1083–1100.

Salmah, H., Marliza, M., Teh, P.L. (2013), 'Treated coconut shell reinforced unsaturated polyester composites', *International Journal of Engineering & Technology*, 13, 94–103.

Santos, E.F., Mauler, R.S., Nachtigall, S.M.B. (2009), 'Effectiveness of maleated- and silanized-PP for coir fiber-filled composites', *Journal of Reinforced Plastics and Composites*, 28, 2119–2128.

Sapuan, S.M., Zan, M.N.M., Zainudin, E.S., Arora, P.R. (2005), 'Tensile and flexural strengths of coconut spathe-fibre reinforced epoxy composites', *Journal of Tropical Agriculture*, 43, 63–65.

Sen, T. and Reddy, H.N. Jagannatha (2011), 'Finite element simulation of retrofitting of RCC beam using coir fibre composite (natural fibre)' *International Journal of Innovation, Management and Technology*, 2(2).

Sharma, S.C., Krishna, M., Narasimhamurthy, H.N., Sanjeevamurthy (2006), 'Studies on the weathering behavior of glass coir polypropylene composites', *Journal of Reinforced Plastics and Composites*, 25, 925–932.

Sindhu, K., Joseph, K., Joseph, J.M., Mathew, T.V. (2007), 'Degradation studies of coir fiber/polyester and glass fiber/polyester composites under different conditions', *Journal of Reinforced Plastics and Composites*, 26, 1571–1585.

Sreekala, M.S., Kumaran, M.G., Joseph, S., Jacob, M. and Thomas, S. (2000), 'Oil palm fibre reinforced phenol formaldehyde composites: influence of fibre surface modifications on the mechanical performance', *Appl. Compos. Mater.*, 7, 295.

Stuart, B. (1997), *Biological Applications of Infrared Spectroscopy*, John Wiley & Sons, Chichester, UK.

Sun, X.F., Xu, F., Sun, R.C., Fowler, P., Baird, M.S. (2005), 'Characteristics of degraded cellulose obtained from stream exploded wheat straw', *Carbohydrate Research*, 340, 97–106.

Toledo Filho, R.D., Scrivener, K., England, G.L., Ghavami, K. (2000), 'Durability of alkali-sensitive sisal and coconut fibres in cement mortar composites', *Cement and Concrete Composites*, 22, 127–143.

Tran, N.H., Ogihara, S., Kobayashi, S. (2009), 'Mechanical properties of short coir/PBS biodegradable composites: Effect of alkali treatment and fiber content, *Proceedings of the 18th International Conference on Composite Materials*, 27–31 July 2009, Edinburgh, UK.

Varma, D.S., Varma, M., Varma, I.K. (1985), 'Coir fibres II: Evaluation as a reinforcement in unsaturated polyester resin composites', *Journal of Reinforced Plastics and Composites*, 4, 419.

Verma, D., Gope, P.C., Shandilya, A., Gupta A. (2013), 'Coir fibre reinforcement and application in polymer composites: a review', *Journal of Materials and Environmental Sciences*, 4, 263–276.

Wong, K.J., Nirmal, U., Lim, B.K. (2010), 'Impact behavior of short and continuous fiber-reinforced polyester composites', *Journal of Reinforced Plastics and Composites*, 29, 3463–3474.

11

The use of cotton fibers as reinforcements in composites

S. K. BAJPAI and G. MARY, Government Model Science College (Autonomous), India and
N. CHAND, Advanced Materials and Processes Research Institute (AMPRI) (CSIR), India

DOI: 10.1533/9781782421276.3.320

Abstract: This chapter discusses the properties of cotton fibers, and their use as a reinforcement phase in a number of polymer composites. Factors affecting the mechanical, and other, properties of the composites are detailed with reference to relevant studies, and the use of compatibilizers and their means of introduction into the composites are discussed. Particular reference is made to the biomedical uses of polymer/cotton fiber composites, requiring further the introduction of antimicrobial agents, and the chapter ends with a summary of the required properties of polymer/cotton fiber composites for medical application.

Key words: cotton fibers, acrylamide, N-vinyl-2-pyrrolidone, acrylic acid, grafting, chitosan, alginate, silver nanoparticles, FTIR, mechanical properties, SEM, antibacterial action.

11.1 Introduction

Cotton is the most widely used natural textile fiber in the world today. We use it every day, for example in the clothes we wear or the furniture in our homes. Cotton is comfortable to wear, easy to care for, and durable. It is also an important cash crop for many countries such as China, India and the United States. Cotton has been an important fiber since its cultivation in the United States by colonists in the sixteenth century. Cotton has since spread to more than 80 countries and makes up 56% of all fibers used for clothing and home furnishings in the United States. The US alone produces 3.8 million tons of cotton a year and consumes 1.7 million tons a year [1]. Cotton can be identified by visual inspection, burn tests, microscopy and solubility tests.

11.2 Physical properties of cotton fibers

11.2.1 Fiber length

Length is the most important attribute of cotton fiber. Fiber length is described [2] as 'the average length of the longer one-half of the fibers

Table 11.1 Classification of upland cotton based on length of cotton fiber

Upper half mean length (inches)	Quality
Below 0.99	Short
0.99–1.10	Medium
1.11–1.26	Long
Above 1.26	Extra long

Table 11.2 Length uniformity vs uniformity index

Length uniformity	Uniformity index (%)
Very high	>85
High	83–85
Intermediate	80–82
Low	77–79
Very low	<77

(upper half mean length)'. This measure is taken by scanning a 'beard' of parallel fibers through a sensing region. The beard is formed from the fibers taken from the sample, clasped in a holding clamp and combed to align the fibers. Typical lengths of upland cottons might range from 0.79 to 1.36 inches, as shown in Table 11.1.

Length uniformity or uniformity ratio is determined as 'a ratio between the mean length and the upper half mean length of the fibers and is expressed as a percentage' [2]. Typical comparisons are illustrated in Table 11.2. A low uniformity index indicates that there might be a high content of short fibers, which lowers the quality of a textile product.

11.2.2 Fiber strength

Fiber strength is measured in grams per denier. It is determined as the force necessary to break the beard of fibers clamped in two sets of jaws 1/8 inch apart [2]. Typical tensile levels are illustrated in Table 11.3. The breaking strength of cotton is about 3.0–4.9 g/denier, and the breaking elongation is about 8–10% [3].

11.2.3 Micronaire

Micronaire measurements reflect fiber fineness and maturity. A constant mass (2.34 grams) of cotton fibers is compressed into a space of known

Table 11.3 Fiber strength in grams per denier

Degree of strength	Fiber strength (g/denier)
Very strong	>31
Strong	29–30
Average	26–28
Intermediate	24–25
Weak	<23

Table 11.4 Micronaire measurement

Cotton range	Micronaire reading
Premium	3.7–4.2
Base range	4.3–4.9
Discount range	>5.0

volume, and air permeability measurements of this compressed sample are taken. These are indicative of micronaire values (Table 11.4).

11.2.4 Color

The color of cotton samples is determined by two parameters: the degree of reflectance (Rd) and the yellowness (+b). Degree of reflectance shows the brightness of the sample, and yellowness depicts the degree of cotton pigmentation. The appropriate area located in a Nickerson–Hunter cotton colorimeter diagram represents each color code. The color of the fibers is affected by climatic conditions, impact of insects and fungi, type of soil, storage conditions, etc. There are five recognized groups of color: white, gray, spotted, tinged, and yellow stained. As the color of cotton deteriorates the processability of the fibers decreases.

Work at the University of Tennessee has led to color measurement using both a spectrometer CIE-based average color measurement and a color uniformity measurement using image analysis to improve the accuracy and provide additional measurement for color grading [4]. The investigators have developed a two-color grading system using neural networks.

11.2.5 Trash

A trash measurement describes the amount of non-lint materials such as parts of the cotton plant in the fiber. Trash content is assessed from scanning the cotton sample surface with a video camera and calculating the percentage of the surface area occupied by trash particles. The values of trash content should be within the range from 0 to 1.6%. Trash content is highly correlated to the leaf grade of the sample.

11.3 Chemical and other properties of cotton fibers

Cotton is a natural cellulosic fiber and swells in a high-humidity environment, in water and in concentrated solutions of certain acids, salts and bases. The swelling effect is usually attributed to the sorption of highly hydrated ions. The moisture regain for cotton is about 7.1–8.5% and the moisture absorption is 7–8% [3].

Cotton is attacked by hot dilute or cold concentrated acid solutions. Acid hydrolysis of cellulose produces hydro-celluloses. The fibers show excellent resistance to alkalis. There are a few other solvents that will dissolve cotton completely, although included among them is a copper complex of cuprammonium hydroxide and cupriethylenediamine (Schweitzer's reagent [5]). Cotton degradation is usually attributed to oxidation, hydrolysis or both. Oxidation of cellulose can lead to two types of oxycellulose [6], depending on the environment in which the oxidation takes place.

Cotton fibers show double refraction when observed in polarized light. Even though various effects can be observed, second-order yellow and second-order blue are characteristic colors of cellulosic fibers [7]. A typical birefringence value, as shown in the table of physical properties, is 0.047.

Cotton fibers are very prone to attack by certain microorganisms, insects and fungi that cause functional, hygienic and aesthetic difficulties. Their tendency to promote bacterial growth is due to their moisture absorption tendency which provides a good environment in which microbes can develop.

11.4 Cultivation of and quality issues affecting cotton fibers

Prior to seed planting, the ground is plowed and divided into seed rows. Next the seeds, which have a size of about 0.5 cm, are inserted into the soil with a mechanical planter. Irrigation causes the seed to germinate.

Cotton is grown in large monocultures and is therefore susceptible to pests. After bloom the cotton balls become visible, each containing up to 25 seeds. After about eight weeks these balls tear open and the cotton emerges and harvesting can begin.

When harvesting the plants are sprayed with defoliant beforehand to minimize contamination with leaves and the freshly picked cotton is pressed into large modules. The cotton gin mechanically separates the fibers from the seed and turns it into ginned cotton, which is also called lint. The lint is pressed into large bales and transported to the textile mill. At the mill, the bales are picked apart and cleaned again before being fed into a carding-machine.

At this point, the fibers are laid side by side and made into an untwisted rope called a sliver. A spinning frame turns these slivers into yarn. In the next step, the yarn is dyed and looms are used to weave it into ready-to-use fabrics.

One method of analyzing individual fibers is to develop an optical sensing tool that can be used to characterize single cotton fiber quality (Thomasson *et al.*, [8]). Should cotton producers and processors employ the same method of characterizing cotton fiber quality, this would maximize the efficiency of the manufacture of cotton products.

Variations in the quality of cotton fibers exist between different areas of a cotton field, from plant to plant, and even within the same plant. Moreover, the quality of fibers on a single seed may vary in length, shape, thickness and maturity as described in Jost [9]. By studying individual cotton fibers from different parts of a seed, different plants and different locations on a field, it may be possible to determine the causes of the variation in cotton fiber quality (Cui *et al.*, [10]).

11.5 Processing of cotton fibers as reinforcements in composites

Cotton fibers are a good candidate for biomedical applications and in this chapter we describe some of the strategies that have been adopted to develop fiber/polymer composites and polymer-attached fibers for biomedical applications, with particular reference to antimicrobial efficacy. Synthetic and natural polymers can be attached to the cotton cellulosic backbone to yield polymer-grafted fibers. The following are examples of graft co-polymerization of vinyl monomers and natural polymers onto cotton cellulose:

- Grafting of monomer acrylamide (AAm) and *N*-vinyl-2-pyrrolidone (VP) on to cotton fibers
- Graft co-polymerization of acrylic acid onto cotton fibers
- Covalent attachment of biopolymer chitosan onto cotton fibers
- Preparation of calcium alginate attached cotton fibers.

The basic mechanism involved in graft co-polymerization onto a substrate is illustrated in Fig. 11.1. The initiator reacts with the appropriate functional groups of the substrate molecules to produce free radicals. Monomers then

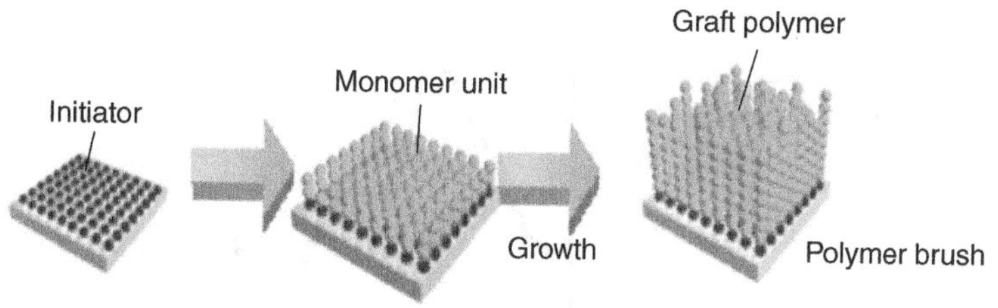

11.1 Grafting mechanism.

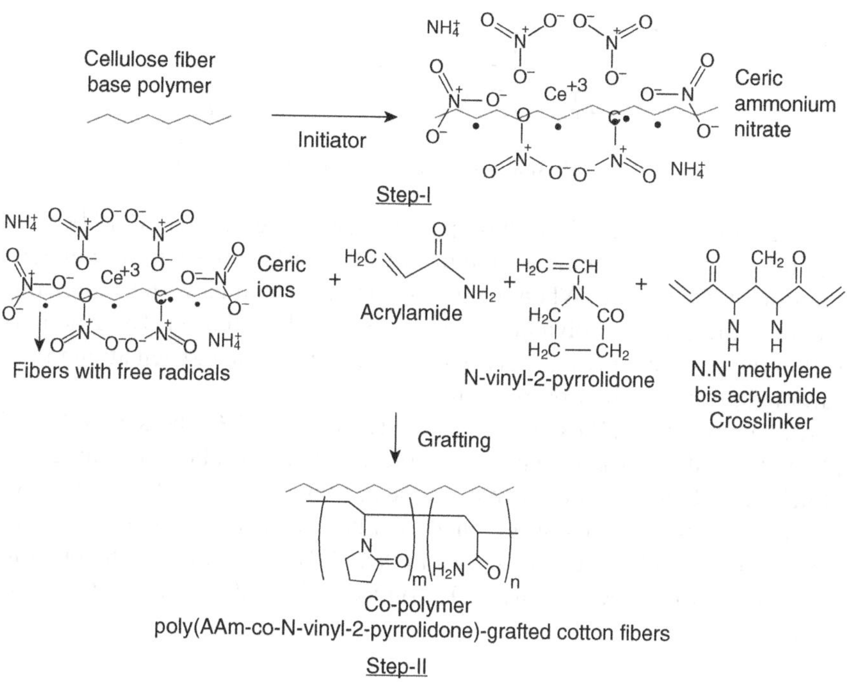

11.2 Grafting procedure.

attach to the free radical sites, forming macro-radicals. The build-up of monomer units in this manner is known as graft co-polymerization.

11.5.1 Grafting of AAm and VP onto cotton fibers

Cotton fibers, when treated with ceric ammonium nitrate, generate free radicals via proton abstraction. The free radicals are then treated with the monomers AAm and VP which attached to these active sites, thus forming graft chains of poly(AAm-*co*-VP) copolymer onto cotton cellulose fibers [11]. This is illustrated in Fig. 11.2.

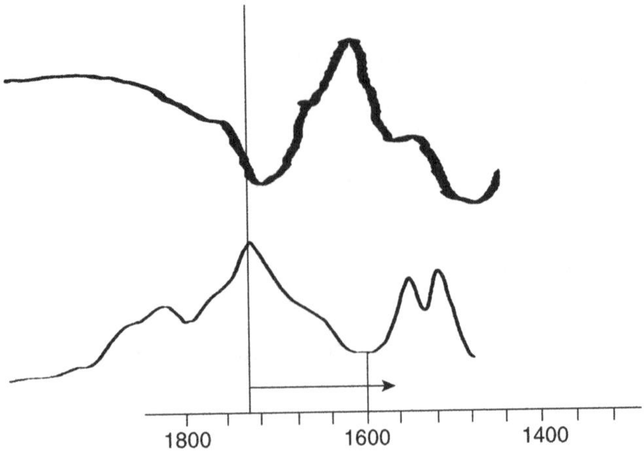

11.3 FTIR spectral analysis to confirm specific bindings [11].

Zinc ions are loaded into these polymer-grafted cotton (PGC) fibers by equilibrating an appropriate quantity of dry grafted fibers in an aqueous solution of Zn(II) ions. When the fibers come into contact with the aqueous medium, the grafted polymer network absorbs the solution. The zinc ions within the swollen polymer network may react with the oxygen atom in the carbonyl group of vinyl-pyrrolidone.

A comparison of FTIR spectra of Zn(II)-loaded PGC fibers with that of plain PGC fibers usually gives valuable information about the mode of attachment. It was observed that the peak appearing at 1705 cm^{-1} for CO group of *N*-vinyl-2-pyrrolidone in PGC fibers was shifted to 1633 cm^{-1} in Zn(II)-loaded PGC fibers (as shown in Fig. 11.3). This indicates the binding of Zn(II) with oxygen of the CO group of VP in Zn(II)-loaded grafted fibers.

11.5.2 Graft co-polymerization of acrylic acid onto cotton fibers and loading of gentamycin sulphate into composite fibers

An acidic monomer such as acrylic acid (AAc) can be grafted onto cellulose in order to load cationic antibiotic drugs such as gentamycin sulphate. These cationic drugs undergo ion-exchange with free H^+ ions present in the grafted polymer network which result from the ionization of –COOH groups of monomer AAc [12].

The percent grafting (PG) is calculated using the following expression:

$$PG = \frac{W_g - W_0}{W_0} \times 100$$

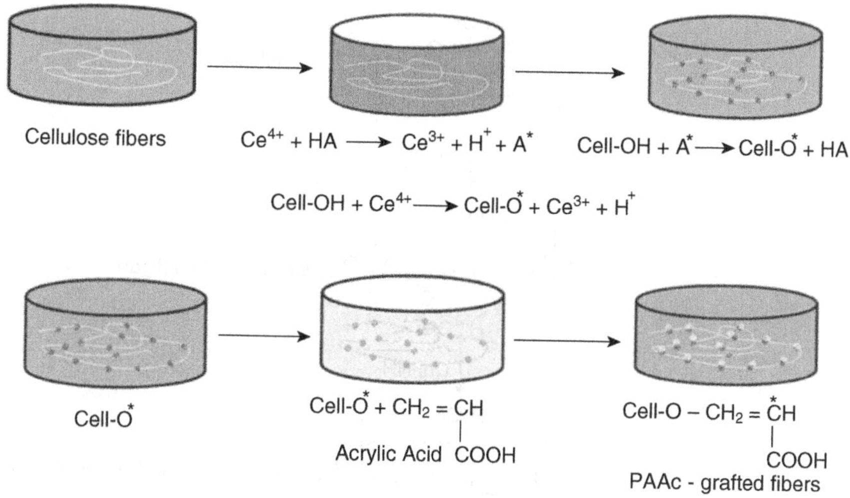

11.4 Polyacrylic acid grafted cotton fibers [12].

where W_0 = weight of cotton fibers before grafting procedure and W_g = weight of grafted cotton fibers or weight of cotton fibers after grafting procedure.

Figure 11.4 illustrates the grafting procedure of monomer acrylic acid onto cotton cellulose.

A brief protocol of procedure is as follows. Pre-weighed cotton fibers are placed in 10 ml of 20 mM CAN for 30 min, blotted with tissue paper to remove extra CAN, and then immersed in 25 ml of a solution containing pre-weighed quantities of monomer AAc and crosslinker MB. After the grafting reaction is over, each substrate is equilibrated in distilled water to remove unreacted salts. Finally, the grafted fibers are put in 50% methanol solution to remove any homopolymer formed [13] and placed at 40°C in a dust free chamber until the fibers are completely dry.

Poly(acrylic acid) grafted fibers can be readily loaded with Zn(II) ions by equilibration. Gentamycin sulphate (GS) is a water-soluble drug with a solubility of approximately 50 mg per ml H_2O. In aqueous medium it is present in the form of a cation:

$$G.SO_4 \rightleftharpoons G^{++} + SO_4^{--}$$

The poly (acrylic acid)-grafted fibers have carboxylic groups attached along the polyacrylic acid-grafted chains with free or counter H^+ ions:

$$\text{fibers–COOH} \rightleftharpoons \text{fibers–COO}^- + H^+$$

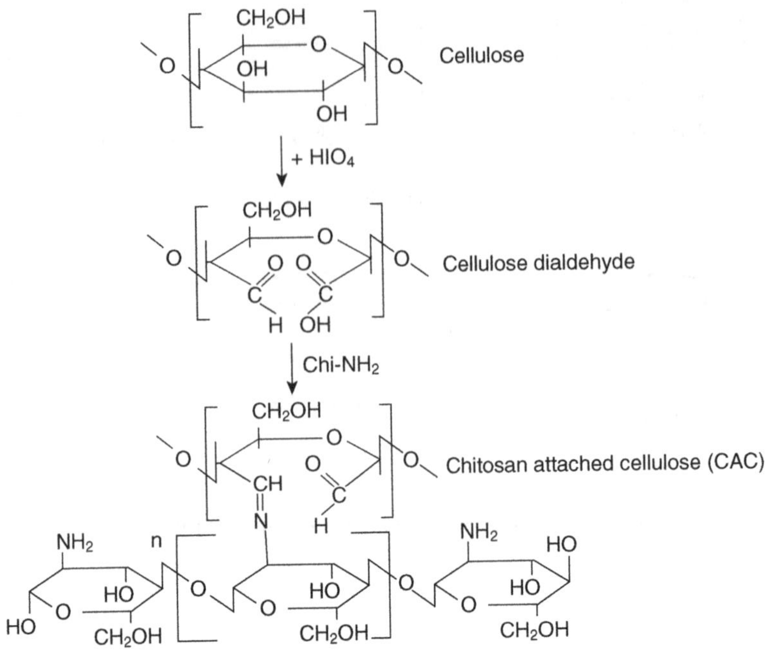

11.5 Chitosan-attached cotton cellulose fibers.

When grafted fibers are placed in an aqueous solution of the drug gentamycin sulphate, ion-exchange occurs between the counter H^+ ions attached along the fibers and the G^{++} ions present in the aqueous solution:

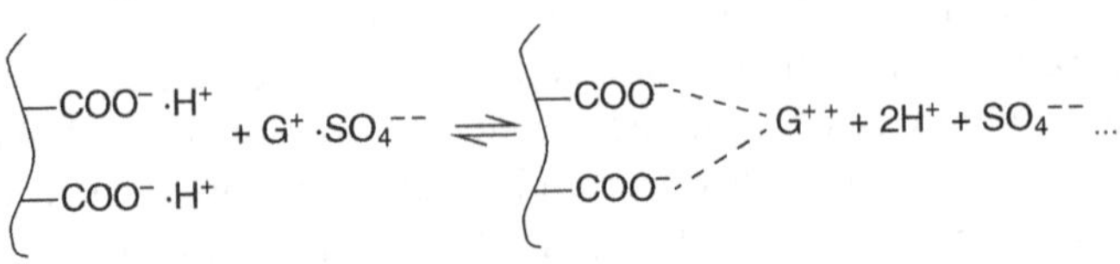

These fibers release G^{++} ions into a release medium and perform antibacterial action against antimicrobial infections.

11.5.3 Covalent attachment of biopolymer chitosan on to cotton fibers followed by loading of copper ions

The covalent attachment of a polymer onto the backbone of cotton cellulose fiber can also be achieved by oxidizing the cotton cellulose using an appropriate oxidant such as periodic acid to produce cellulose dialdehyde. These dialdehyde groups can be attached to a suitable polymer containing NH_2 groups via a coupling reaction. An example of such attachment is illustrated in Fig. 11.5.

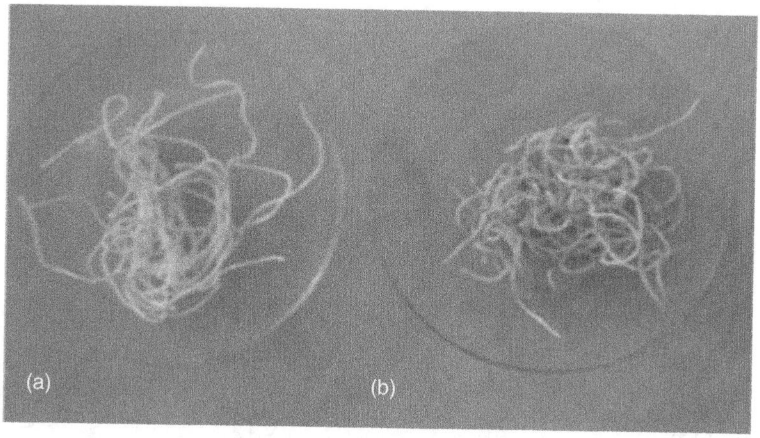

11.6 (a) Plain cotton fibers; (b) Cu(II) ion-bound chitosan-attached fibers [14].

The biopolymer chitosan exhibits antibacterial action against a number of microorganisms such as *E. coli*; however, in order to enhance its antibacterial action the chitosan-attached cotton fibers can be further modified by attachment of some strong antibacterial agent. For example, Cu(II) ions can be attached to the chitosan chains as Cu(II) has a strong tendency to bind with the NH_2 groups. Cu(II)-loaded chitosan-attached cotton cellulose fibers exhibit biocidal action against microorganisms. A photograph of plain and Cu(II)-loaded chitosan-attached cotton cellulose fibers is given in Fig. 11.6. These Cu(II) ions are released from the fibers when they are brought in contact with physiological fluid [14].

11.5.4 Preparation of calcium alginate-attached cotton fibers containing Cu nanoparticles

Bajpai and Thomas [15] attached chitosan macromolecular chains onto the cellulose cotton backbone via oxidation to cellulose dialdehyde by periodic acid, and the resulting cellulose dialdehyde was coupled with the chitosan chains with elimination of water. Placing the chitosan-attached cellulose fibers into an aqueous $AgNO_3$ solution allows Ag(I) ions to bind with the amino groups of the chitosan macromolecular chains. These Ag(I)-attached fibers are then reduced with sodium citrate to yield Ag nanoparticle-loaded chitosan-attached cotton cellulose fibers, as illustrated in Fig. 11.7.

These fabrics can be used for antibacterial applications wherever needed. The optical images of plain, chitosan-attached and Ag-loaded fabrics are shown in Fig. 11.8.

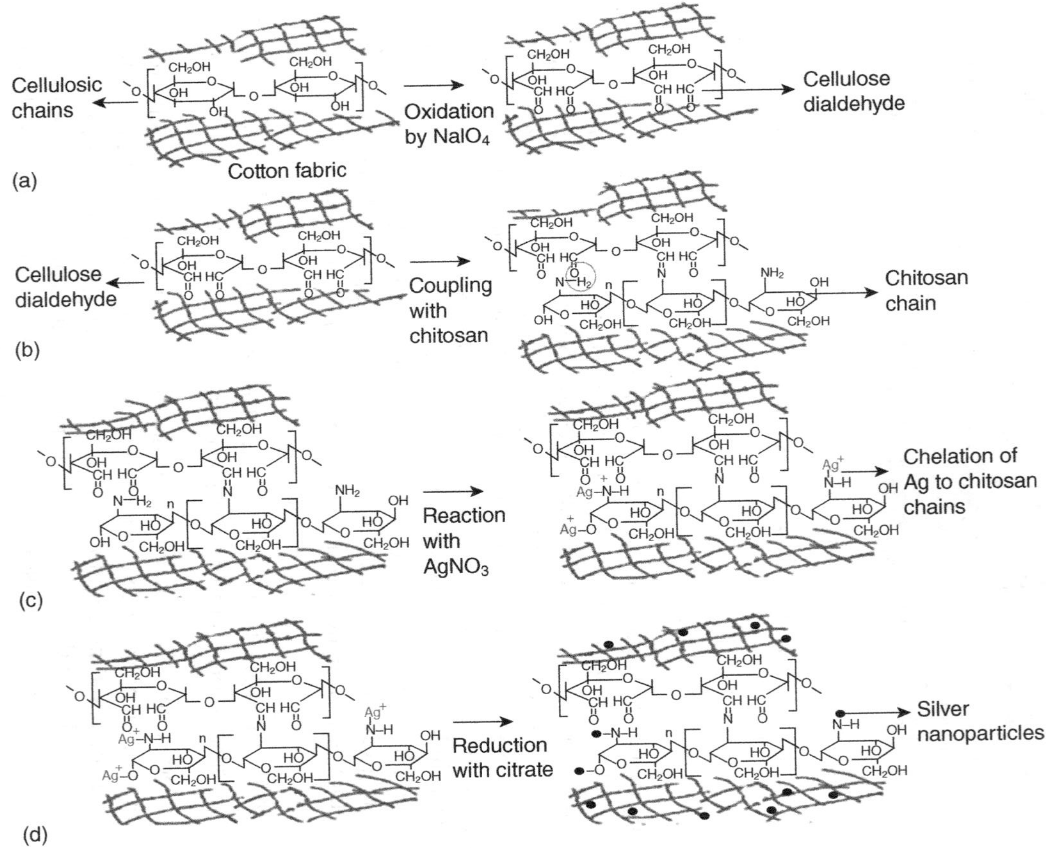

11.7 Matrix form of nanosilver-loaded fabric [15].

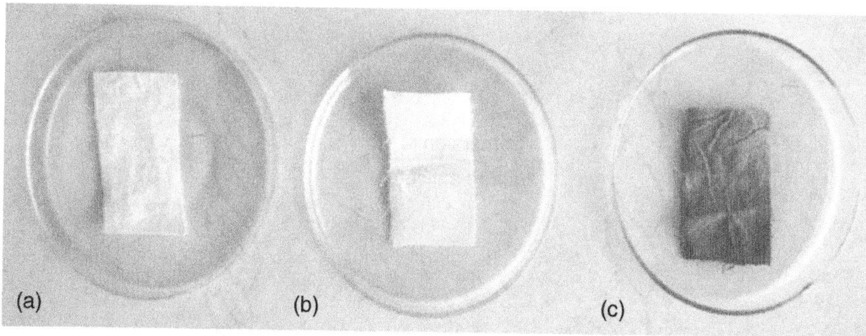

11.8 (a) Plain fabric; (b) chitosan-attached fabric; (c) nanosilver-loaded fabric.

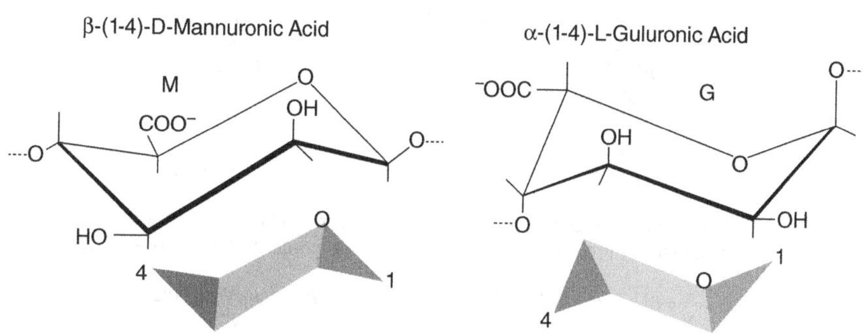

11.9 Structure of alginate.

The main drawback with the techniques described above is the loss of mechanical strength of the fibers due to cleavage of the C2–C3 bond, limiting their applications where high mechanical strength is required.

An alternative approach in which the mechanical strength of the fibers is not compromised involves absorption of a biopolymer sodium alginate into the ultra-fine pores of the cotton fibers, followed by ionotropic gelation of alginate using divalent Cu(II) ions to produce copper alginate/cellulose composite fibers. Here, the role of the Cu(II) is not only to crosslink the alginate chains but also to impart antibacterial properties to the resulting composite fibers [16] (see Fig. 11.9). This procedure is illustrated in Fig. 11.10 and the resulting fibers are shown in Fig. 11.11.

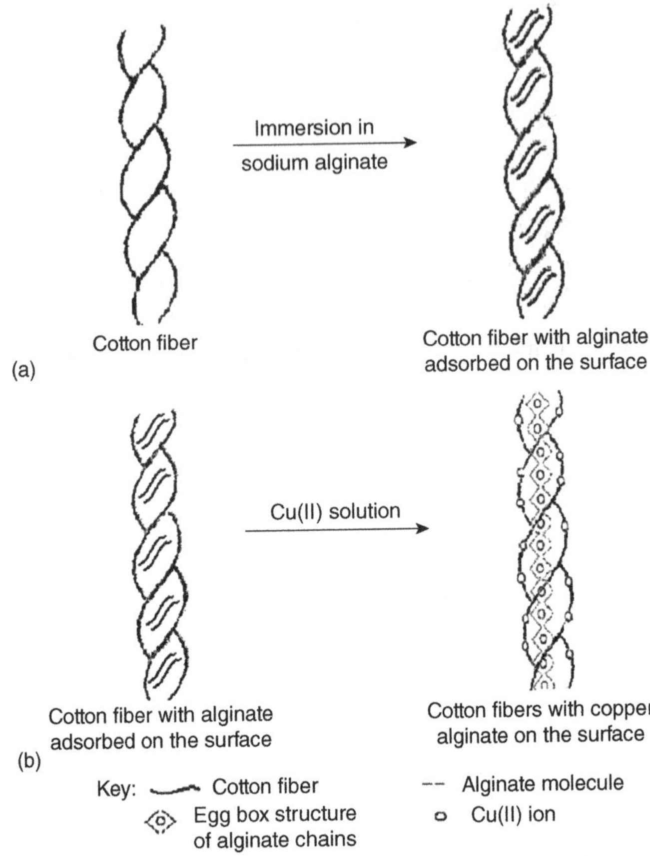

11.10 (a) Sorption of alginate chains into cotton cellulose fibers; (b) crosslinking of alginate chains by Cu(II) ions thus forming 'egg-box' cavities.

11.6 Assessing the antibacterial activity of biomedical composites reinforced with composite cotton fibers

Cotton fiber composites can exhibit antibacterial activity against bacteria, viruses, fungi, etc., on introduction of appropriate antimicrobial agents. These fibers/fabrics can be used as burn and wound dressing materials, and for other purposes in the biomedical fields. Figure 11.12, for example, compares the antibacterial action of Zn(II) ion loaded polymer-grafted fibers with polymer-grafted fibers as the control. It is clear that there is appreciable inhibition in the bacterial growth in the Petri plate supplemented with Zn(II) loaded fibers. Similar results are obtained with cellulose fibers loaded with alternative agents (see Figs 11.13–11.16).

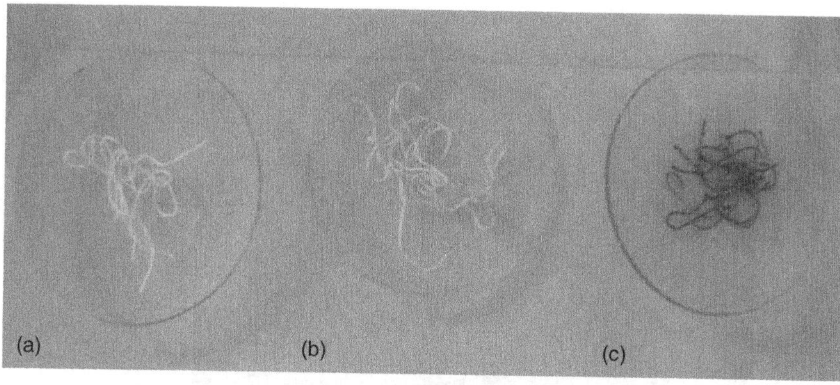

11.11 Photographs showing (a) plain cotton cellulose fibers; (b) copper alginate–cotton cellulose (CACC) fibers; (c) nanocopper-loaded alginate–cotton cellulose composite fibers.

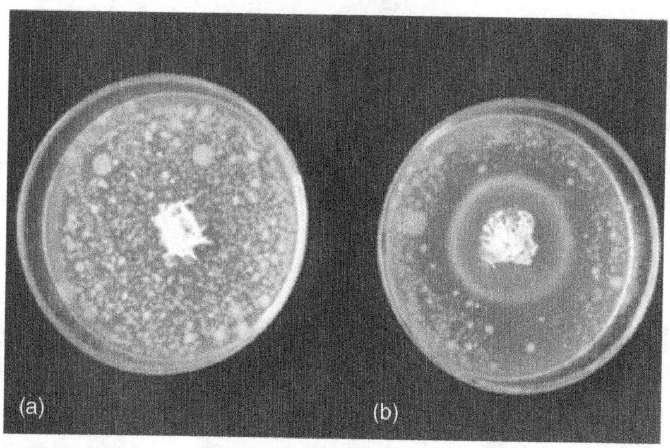

11.12 Photographs showing bacterial growth in the plates: (a) containing poly(acrylamide-*co*-vinyl pyrrolidone)-*g*-fibers, and (b) poly(acrylamide-*co*-vinyl pyrrolidone)-grafted-fibers containing Zn(II) ions.

11.7 Assessing the mechanical properties of biomedical and other composites reinforced with cotton fibers

Cotton fibers have also been incorporated into a polymer matrix as the reinforcement phase, yielding improved mechanical properties. In work by Way *et al.* [17], PLA composites with up to 25% w/w cotton linter (CL) loadings or up to 50% w/w maple wood fiber (WF) loadings were prepared by melt blending. The presence of the fibers in the PLA matrix was shown by TGA tests to improve the thermal stability of all of the PLA composites.

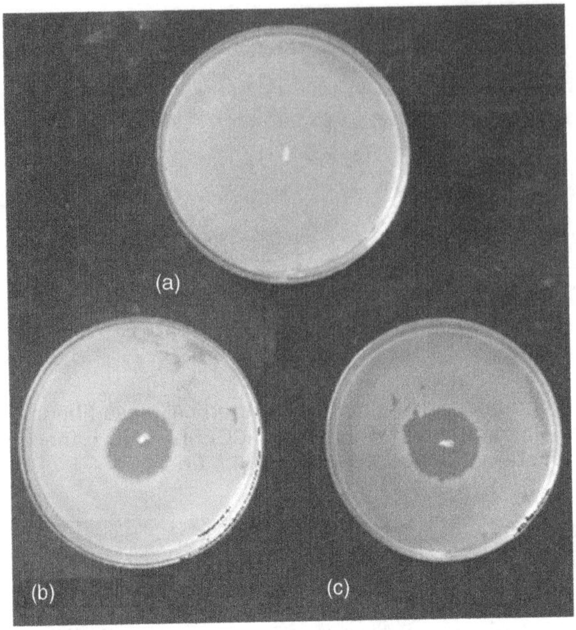

11.13 Antibacterial activity of drug-loaded grafted fibers: (a) plain fiber; (b) 2% drug-loaded fiber; (c) 4% drug-loaded fiber.

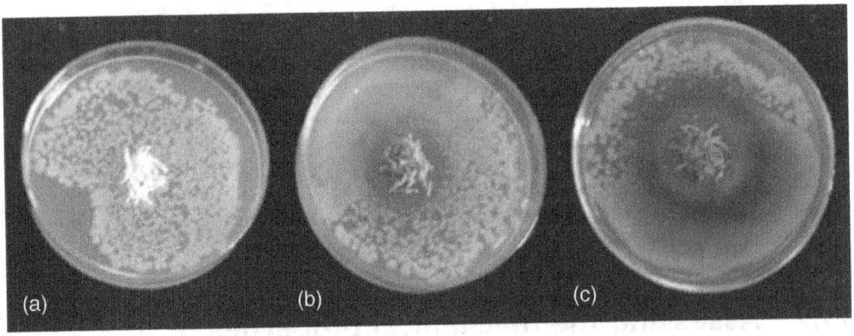

11.14 Biocidal action of (a) plain cellulose fibers; (b) chitosan-bounded copper attached cotton fibers (CBCAC)(2) and (c) CBCAC(4) fibers against *E. coli* as studied by the zone-inhibition method (the number in parentheses denotes the percent (w/v) concentration of aqueous solutions of Cu(II) ions used for loading).

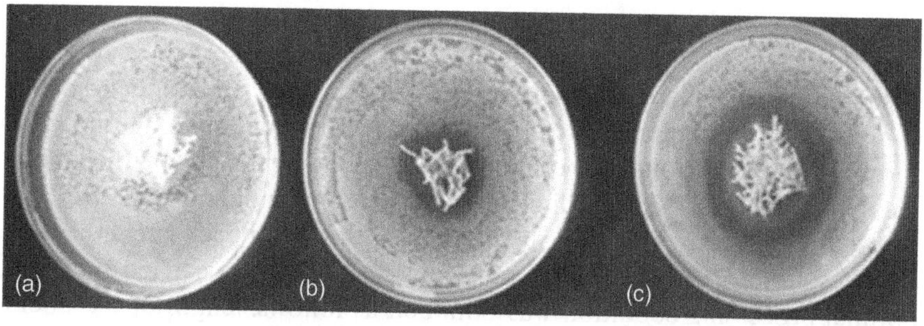

11.15 Biocidal action of (a) plain fibers; (b) CACC(2) and (c) CACC(4) fibers against *E. coli* as studied by the zone inhibition method (the number in parentheses denotes the percent (w/v) concentration of aqueous solutions of Cu(II) ions used for loading).

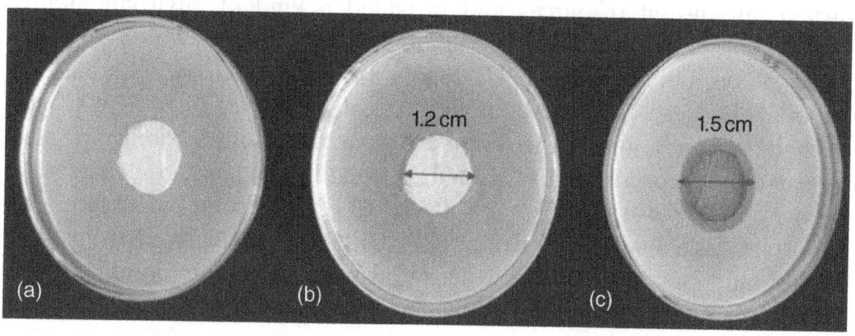

11.16 Photographs showing the inhibition zone of (a) plain cotton fabric disc; (b) chitosan-attached cotton fabric disc; (c) nanosilver-loaded 1% chitosan-attached fabric disc.

In contrast to many other natural fiber-reinforced thermoplastic composites, where the improvement in stiffness markedly reduces the toughness, the PLA CL and PLA WF composites offer excellent improvements in both stiffness (with increases of up to 51 and 98% in tensile modulus as well as 56 and 123% in flexural modulus respectively) and toughness (up to 72 and 56% increase in notched impact strength respectively). DMA results indicated that the mechanical loss factor (tan *d*) was decreased, suggesting better damping capabilities were achieved with the composites.

Because of the biodegradability of poly(lactic acid), it is the most frequently studied polymer for biomedical applications in last two decades, and attempts have been made, from time to time, to strengthen this polymer by incorporating suitable strengthening agents. The cotton stalks bast fiber is a new kind of natural cellulose fiber and it possesses properties similar to those of flax fibers. PLA is a class of biodegradable matrix material. As flax fibers can be reinforced and then compounded with PLA, cotton stalks bast fibers as reinforced materials and PLA as matrix can be used to make a kind of green and biodegradable composite in the experiment. With this basic idea, Wu *et al.* [18] have recently reported results of their research based on cotton stalk bast fibers reinforced polylactic acid biodegradable composites. The composites were prepared by mixing, molding and cold forming. An orthogonal experiment table including length of the fibers, mass fraction of PLA, molding temperature and molding pressure was devised. The tensile strength, flexural strength and impact strength were then tested and the analysis of range and variance was carried out. The optimal processing condition was obtained as follows: length of the fibers was 6 mm, mass fraction of PLA was 75%, and molding temperature was 175°C and molding pressure was 12 MPa. The result showed that the cotton stalks bast fibers reinforced PLA biodegradable composites had good mechanical properties. Through this study, waste cotton stalks basts were turned into useful resources and provided a kind of environmentally-friendly composite.

Nowadays, polyurethane (PU) foams are used extensively as cushion materials because of their superior physical properties (e.g., high tensile strength, abrasion and tear resistance, low temperature flexibility, oil and solvent resistance, printability) and high versatility [19]. With the increasing quantity of PU wastes, the recycling of PU has become an urgent and important task. Waste cotton fabric (WCF), which is derived from used clothes and surplus from textile industries, represents another type of environmental problem. Nowadays, most WCF is land filled or incinerated [20].

The PLA composites were also subjected to a composting environment in which increases in the onset of biodegradation were observed with increasing fiber content. This is due to the fact that the fibers are more readily mineralized than the PLA matrix. Conversely, after 90 days, the PLA with the lowest fiber content exhibited the highest degree of degradation due to mineralization of the matrix phase.

Wang *et al.* [21] undertook preparation of thermoplastic composites from reclaimed polyurethane (PU) foam and waste cotton fabric (WCF) with the aim of providing a cost-effective technology for recycling the two polymers. An activated WCF ultra-fine powder was successfully prepared through a mechano-chemical route and was used as a reinforcement to enhance the melt processability and mechanical properties of reclaimed PU foam. The relatively strong shearing and compression forces exerted by the pan mill destroyed the hydrogen bonds of the WCF cellulose and released reactive hydroxyl groups, which were beneficial to the interfacial cohesion between the WCF and the PU matrix and the mechanical properties of the PU/WCF composites were thus significantly improved. Similar results have been obtained using poly(ethylene tetraphthalate) PET/cotton composites (see Fig. 11.17).

11.17 SEM micrograph of the tensile fracture surface of polyester–cotton–kapok fabric composite [23].

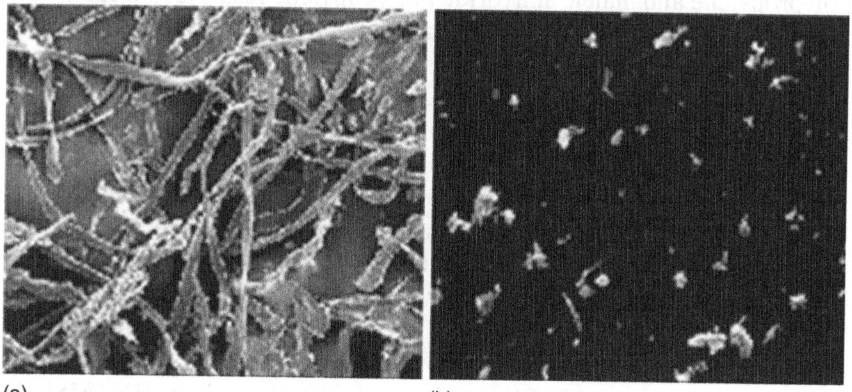

(a) (b)

11.18 SEM images of the WCF powders (a) before and (b) after 40 cycles of pan milling [22].

The heat shrinkage of the composites sharply decreased with increasing pretreated WCF content (Fig. 11.18). In this study, the mechano-chemically activated WCF was used as a low-cost but effective functional additive for the enhancement of the melt processability and mechanical properties of reclaimed PU foam (Zou *et al.*, [22]).

Plastic/fiber composites are widely used in many industries including the aircraft, automobile, leisure, electronic, and medical sectors. The potential advantages of natural fibers, aside from their environmental benefits, include the abundance of raw materials from renewable resources, and low cost. Furthermore, it is possible to obtain a higher loading of natural fibers in plastic/fiber composites than in conventional inorganic fillers because of the non-abrasive nature of the former.

In a work by Kim *et al.* [24] the mechanical properties of polypropylene (PP)/natural fiber composites were studied. For the natural fiber component of the composites, cotton fiber was compared with wood fiber. The effect of the melt index of PP on the mechanical properties of the composites was also investigated. In order to improve the poor interfacial interaction between the hydrophilic natural fibers and the hydrophobic matrix PP, maleic anhydride (MAH) grafted PP (PP-g-MAH) was used as a compatibilizer. The tensile strength of the PP/wood fiber composites was found to decrease with increasing wt% of the wood fibers, whereas that of the PP/cotton fiber composites showed higher tensile strength with the addition of 20 to 30wt% cotton fiber due to entanglement.

The use of PP-g-MAH was helpful in increasing the tensile and flexural strengths of the PP/cotton fiber and PP/wood fiber composites, due to the increased interaction between the fiber and PP matrixes. Similarly, Kapok/cotton fabric has also been used as reinforcement for conventional polypropylene and maleic anhydride polypropylene resins [25]. Treating the reinforcement with acetic anhydride and sodium hydroxide strengthens the fabric (fibers).

Many efforts have been made to develop biodegradable materials based on starch due to the environmental problems arising from petroleum-derived plastics. Due to its low cost and availability as a renewable resource, starch has been widely researched as an important raw material for packaging, agricultural and biomedical applications. In recent work [26] modification of thermoplastic rice starch (TPRS) was carried out using cotton fiber or low-density polyethylene (LDPE). It was found that the TPRS/cotton fiber and the TPRS/LDPE composites exhibited improved mechanical properties and lower water uptake. Use of maleic anhydride grafted-polyethylene (MAPE) and vinyl trimethoxysilane (VTMS) as compatibilizers in the TPRS/LDPE composite improved the tensile properties but slightly increased water absorption. The compatibilizers also reduced phase separation in the TPRS/LDPE composites (see Fig. 11.19).

In a recent study, Navin Chand *et al.* [27] found that nano cellulose fibers present in sisal fiber have been separated. The other components present in sisal fiber such as lignin, pectin and wax were removed in the process. These separated nano cellulose fibers were reinforced in some olefins namely low-density polyethylene (LDPE), linear low-density polyethylene (LLDPE) and polypropylene (PP) by the melt mixing technique.

Cellulose fibers obtained from various natural resources have also been utilized for reinforcement of polymers. For example, separated nanocellulose fibers derived from sisal were used to reinforce a number of olefins, namely low-density polyethylene (LDPE), linear low-density polyethylene (LLDPE) and polypropylene (PP) using the melt mixing technique. The mechanical properties of the composites exhibit improved strength up to

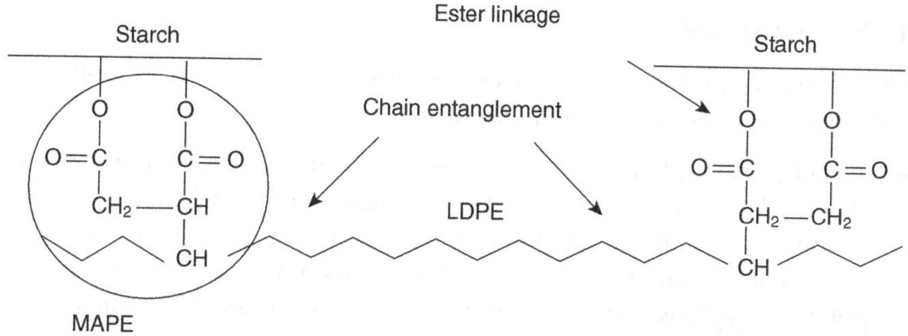

11.19 Schematic diagram of interaction in TPRS/LDPE/MAPE.

11.20 Optical micrographs of (a) neat PBS and (b) CSBF/PBS (20/80) composite.

certain loading of nanofibers in the LDPE and LLDPE cases, whilst the PP composites showed a reduction in strength with increased fiber content.

In recent decades, researches focused on biodegradable polymers have received more and more attention due to their promising applications and environmental considerations. One of the biodegradable polymers is poly(butylene succinate) (PBS) synthesized by succinic acid and butanediol, both of which can be derived from bio-based renewable resources. In a recent work by Tan *et al.* [28] biodegradable poly(butylene succinate) was reinforced with cotton stalk bast fibers (CSBF) which had been pretreated by the continuous steam explosion method. The non-isothermal crystallization kinetics, crystalline structure and spherulitic morphology of neat PBS and CSBF/PBS composites were investigated by differential scanning calorimetry (DSC), wide-angle X-ray diffraction (WAXD) and polarizing optical microscopy (POM). Moreover, the dynamic mechanical thermal properties of neat PBS and CSBF/PBS composites were studied by dynamic mechanical thermal analysis (DMA) (see Fig. 11.20).

11.8 Summary

The following properties are advantageous for application of polymer/cotton fiber composites in the medical sector:

- The fibers should possess reasonably good water absorption in order to absorb the exudates released from wounds. This can be achieved by introducing a hydrophilic polymer by means of grafting or some other technique. Synthetic polymers such as sodium acrylate or natural polymers such as starch or sodium alginate are possible candidates in this respect.
- The fibers should have good mechanical strength after attachment of the polymer and loading of the antibacterial agent. This is usually achieved by soaking the fibers in an aqueous solution of monomer/polymer.

11.9 References

1. R. Hegde, A. Dahiya, M.G. Kamath, *Cotton Fibers*, University of Tennessee, Knoxville, TN, April 2004.
2. *The Classification of Cotton*, USDA Agricultural Marketing Service, Cotton Division, Agricultural Handbook 566, September 1995.
3. J. Brandrup, E.H. Immergut, E.A. Grulke, *Polymer Handbook*, Wiley Interscience, New York, 1989.
4. K.E. Duckett, Color grading of cotton-measurement, *Beltwide Cotton Conference*, Orlando, FL, 5–8 January 1999.
5. P.B. Gove (ed.), *Webster's Third New International Dictionary*, G. & C. Merriam Company, Springfield, MA, 1963.
6. J.G. Cook, *Handbook of Textile Fibers, Part I. Natural Fibers*, Merrow Publishing, Watford, UK, 1968.
7. H.R. Mauersberger (ed.), *Matthews' Textile Fibers, Their Physical, Microscopic and Chemical Properties*, 6th edition, John Wiley & Sons, Inc., New York, 1954.
8. J.A. Thomasson, S. Manickavasagam, M.P. Mengüç, Cotton fiber quality characterization with light scattering and fourier transform infrared techniques, *Applied Spectroscopy*, 63, 321–330, 2009.
9. P. Jost, *Cotton Fiber Quality and the Issues in Georgia*, University of Georgia, 2005.
10. X. Cui, M.W. Suh, P. E. Sasser, Estimating single cotton fiber tensile properties from the load-elongation curves of slack bundles. *Textile Research Journal*, 73, 1066–1070, 2003.
11. S.K. Bajpai, N. Chand, G. Mary, A novel approach to prepare Zn(II)-loaded cotton fibers with antibacterial property, *Journal of Macromolecular Science, Part A*, 45(10), 1–9, 2008.
12. S.K. Bajpai, P. Das, Gentamycin-loaded poly(acrylic acid)-grafted cotton fibers, part-1: synthesis, characterization, and preliminary drug release study, *Journal of Applied Polymer Science*, 122(1), 366–374, 2011.

13. M.T. Taghizadeh, M.A. Darvish, Kinetics and mechanism of heterogenous graft polymerization of acrylonitrile onto polyvinyl alcohol initiated with ceric ammonium nitrate, *Iranian Polymer Journal*, 10(5), 283–292, 2001.
14. S.K. Bajpai, N. Chand, G. Mary, Copper(II) ions and copper nanoparticles-loaded chemically modified cotton cellulose fibers with fair antibacterial properties, *Journal of Applied Polymer Science*, 113(2), 757–766, 2009.
15. S.K. Bajpai, V. Thomas, In situ formation of silver nanoparticles within chitosan attached cotton fabric for antibacterial property, *Journal of Industrial Textiles*, 40(3), 229–245, 2011.
16. S.K. Bajpai, N. Chand, G. Mary, Copper alginate–cotton cellulose (CACC) fibers with excellent antibacterial properties, *Journal of Engineered Fibers and Fabrics*, 4(3), 24–35, 2009.
17. C. Way, D.Y. Wu, D. Cram, K. Dean, E. Palombo, Processing stability and biodegradation of polylactic acid (PLA) composites reinforced with cotton linters or maple hardwood fibres, *Journal of Polymers and the Environment*, 21(1), 54–70, 2012.
18. Z. Wu, C. Wei, Z. Tian, Preparation and mechanical properties of cotton stalk bast fibers reinforced polylactic acid biodegradable composites, *Advanced Materials Research*, 399–401, 1367–1371, 2011.
19. H. Koerner, J.J. Kelley, R.A. Vaia, Transient microstructure of low hard segment thermoplastic polyurethane under uniaxial deformation, *Macromolecules*, 41, 4709–4716, 2008.
20. R. Miranda, C. Sosa-Blanco, D. Bustos-Martinez, C. Vasile, Pyrolysis of textile wastes: I. Kinetics and yields, *Journal of Analytical and Applied Pyrolysis*, 80, 489–495, 2007.
21. M. Wang, X. Zhang, W. Zhang, D. Tian, C. Lu, Thermoplastic polyurethane composites prepared from mechanochemically activated waste cotton fabric and reclaimed polyurethane foam, *Journal of Applied Polymer Science*, 128(6), 3555–3563, 2013.
22. Y. Zou, N. Reddy, Y. Yang, Reusing polyester/cotton blend fabrics for composites, *Composites, Part B*, 42, 763–770, 2011.
23. L.Y. Mwaikambo, E.T.N. Bisanda, The performance of cotton–kapok fabric–polyester composites, *Polymer Testing*, 18, 181–198, 1999.
24. S.-J. Kim, J.-B. Moonb, G.-H. Kim, C.-S. Ha, Mechanical properties of polypropylene/natural fiber composites: Comparison of wood fiber and cotton fiber, *Polymer Testing*, 27, 801–806, 2008.
25. L.Y. Mwaikambo, E. Martuscelli, M. Avella, Kapok/cotton fabric–polypropylene composites, *Polymer Testing*, 19, 905–918, 2000.
26. J. Prachayawarakorn, P. Sangnitidej, P. Boonpasith, Properties of thermoplastic rice starch composites reinforced by cotton fiber or low-density polyethylene, *Carbohydrate Polymers*, 81, 425–433, 2010.
27. N. Chand, S.C. Prajapati, R.K. Singh, Development and characterization of sisal nano fiber reinforced polyolefin composites, *Journal of Scientific Research and Review*, 1(2), 26–32, 2012.
28. B. Tan, J.-P. Qu, L.-M. Liu, Y.-H. Feng, S.-X. Hu, X.-C. Yin, Non-isothermal crystallization kinetics and dynamic mechanical thermal properties of poly(butylenes succinate) composites reinforced with cotton stalk bast fibers, *Thermochimica Acta*, 525, 141–149, 2011.

12

The use of oil palm biomass (OPB) fibers as reinforcements in composites

M. D. H. BEG, M. F. MINA, R. M. YUNUS and A. K. M. MOSHIUL ALAM, Universiti Malaysia Pahang, Malaysia

DOI: 10.1533/9781782421276.3.342

Abstract: There are tremendous demands of lightweight materials for use in transportation and construction sectors. Natural fibers (NFs) in reinforced polymer composites are relatively lightweight. However, NFs have some limitations due to their moisture affinity, poor wettability and low thermal stability during processing with synthetic polymers. These drawbacks have been overcome by effective physical and chemical treatments of NFs. Among various NFs, oil palm biomasses (OPBs) are readily available in some tropical countries and have received intense attention in the composite industries. In this chapter, the performance of untreated and differently treated OPB fibers, reinforced thermoplastic and thermoset composites prepared through extrusion, compression and injection moldings have been discussed in detail.

Key words: natural fiber, oil palm biomass, polymer, physical and chemical treatment, reinforcement, composites, fiber-matrix interface, mechanical and thermal properties.

12.1 Introduction

An oil palm biomass (OPB) reinforced polymeric composite consists of a soft, ductile matrix that encapsulates fibers for reinforcement, transmitting stress to those fibers when under tension, and giving improved mechanical properties compared with the unreinforced matrix material. Interest in OPB-reinforced polymer composites has increased dramatically because of the advantages they offer over synthetic fibers, combining stiffness and strength, low densities, and non-toxicity with environmental friendliness, as well as being less abrasive (Mohanty et al., 2001; Luz *et al.*, 2008; Graupner *et al.*, 2009; Cheung *et al.*, 2009). OPB is, moreover, produced as waste by the palm oil industry. Many OPB-based polymeric composites have found advanced applications in the automotive and plastics industries, among others, because they are stronger than traditional, pure polymer materials (Jawaid *et al.*, 2011). They have also become popular as alternative materials for biocomposites and hybrid composites and for use in the pulp and paper industries (Abdul Khalil *et al.*, 2009, 2010a).

There are, however, also disadvantages that have limited the use of OPB-reinforced composites, in particular a tendency to absorb moisture, poor wettability, low thermal stability during processing, and poor adhesion with synthetic counterparts (Shinoj et al., 2011; Demir *et al.*, 2006; Son *et al.*, 2001). Their flexibility means they are not suitable for heavy-duty applications; and their environmental sensitivity and poor moisture resistance result in a degradation of strength and stiffness. Poor interfacial adhesion with polymer matrices during composite fabrication is caused by the hydrophilic nature of OPB.

There are treatments involving the surface modification of fibers that can effectively overcome most of these drawbacks. Chemical treatments can reduce the hydrophilicity of the fibers and improve the wettability of the polymer matrix (Khalid *et al.*, 2008). Various coupling agents have also been reported to improve the fiber–matrix interface and to enhance the mechanical properties of OPB composites (Xie *et al.*, 2010; Gao and Yan, 2004; Kim *et al.*, 2007), although in general the morphological, structural, mechanical and thermal properties are strongly influenced by the quality of the fibers themselves, their distribution within the polymer matrix, and the efficiency of stress transfer between these two components.

In this chapter, the properties described above will be discussed in detail for polymeric composites reinforced by untreated and variously treated oil palm empty fruit bunch fibers.

12.2 Oil palm biomass fibers

12.2.1 Production and availability

About 42 countries produce palm oil from *Elaeis guineensis* (the oil palm tree), which is cultivated across a total of about 27 million acres worldwide. The major oil palm cultivating countries are those of West Africa, as well as Malaysia, Indonesia, Thailand, Brazil and India (Joseph et al., 2006). Oil palm trees, shown in Fig. 12.1, have three main parts from which fibers are extracted: trunk, frond, and the empty fruit bunch. The trees have a life-span of about 25 years and grow to a height of ≈7–13 m with a girth of 45–65 cm, as measured 1.5 m above ground level (Abdul Khalil *et al.*, 2010a).

These trees generate a large amount of agricultural waste. The reported figures vary; for example, in Malaysia, even though oil palms have emerged there as a commercial crop, a 1 hectare (ha) plantation of trees produces about 55 tons of waste annually, while yielding 5.5 tons of oil (Hasamudin and Soom, 2002). It has also been reported that the total 6 million ha of plantation produce over 11.9 million tons of oil and 100 million tons of biomass (Abdul Khalil *et al.*, 2010b). In any case, Malaysia is the world's largest exporter of palm oil, accounting for approximately 60% of the

12.1 Photograph of an oil palm tree.

world's oil and fat production. The average amount of biomass produced by an oil palm tree is 231.5kg dry weight per year (Abdul Khalil et al., 2010c). Another estimate of 4.69 million ha of plantation and a production rate of dry oil-palm biomass of 20.34 tons per ha per year (Lim, 1998) indicated the production of 95.3 million tons of dry lignocellulosic biomass in 2009 by the Malaysian palm oil industry. Whatever the actual figure, it will, of course, increase if the total plantation area increases. Oil palm production has nearly doubled in the past 20 years (Abdul Khalil *et al.*, 2010c).

Oil palm biomass (OPB) residues are extracted from several parts of the tree, including trunk, frond, kernel shell and empty fruit bunch (EFB). Oil palm fronds account for 70% of the total residues produced, while the EFB accounts for 10% and the trunk accounts for only about 5% (Ratnasingam, 2011). Photographs of the trunk, EFB and frond and their fibers are shown in Figs 12.2 and 12.3. The EFB is the fibrous mass left behind after the fruits are separated from fresh fruit bunches (sterilized by steam treatment at 294kPa for 1h). The EFB can potentially yield 73% fibers (Wirjosentono *et al.*, 2004) and hence is preferable among the various fiber sources in the oil palm tree in terms of availability and cost (Rozman *et al.*, 2000). The palm oil industry disposes of approximately 1.1 tons of EFB fiber waste for every ton of oil produced (Karina *et al.*, 2008). Some of this highly cellulosic material is currently used as boiler fuel (Sreekala *et al.*, 1997), in the preparation of fertilizers, or as mulching material (Saheb and Jog, 1999), but a large part of it is left on the mill premises (Fig. 12.4), creating major environmental problems (Sreekala *et al.*, 1997; Law *et al.*, 2007).

12.2 Physical appearances of OPF: (a) felled trunk, (b) EFB and (c) frond.

12.3 Images of extracted fibers from trunk, EFB and frond.

12.4 EFB-fiber wastes in the mill premises.

12.2.2 Chemical composition

The chemical composition of natural fiber (NF) in general varies according to species, growing conditions, method of fiber preparation, and many other factors (Bledzki and Gassan, 1999). The chemical constituents of oil palm fiber specifically are known to depend on plant source (because of different soil conditions and climates), age, fiber extraction mechanism, and the presence of defects (Chew and Bhatia, 2008; Rowell *et al.*, 2000). The chemical composition of oil palm fiber has been the subject of several studies (Abdul Khalil *et al.*, 2008a; Killman and Lim, 1985; Law *et al.*, 2007; Abdul Khalil et al., 2009; Punsuvon *et al.*, 2005; Shinoj *et al.*, 2011; Chew and Bhatia, 2008; Mohamad *et al.*, 1985; Abdul Khalil and Rozman, 2004; Law and Jiang, 2001; Sreekala *et al.*, 2001). The major constituents in OPB fibers obtained from these studies are summarized in Table 12.1. Oil palm fiber is primarily composed of cellulose and hemicellulose reinforced in a lignin matrix (Raveendran *et al.*, 1995; Meier and Faix, 1999; Demirbaş 2000), as shown schematically in Fig. 12.5.

Oil palm EFB has high cellulose content, which along with its toughness makes it suitable for application in polymeric composites (Sreekala *et al.*, 2004; John *et al.*, 2008). The main chemical constituents of various EFB fibers are shown in Table 12.2. Ash, glucose and xylose are also present in oil palm fibers. Plant materials containing more than 34% cellulose are potentially useful for pulp and paper manufacture (Nieschlag *et al.*, 1960).

In oil palms, the highest percentage of lignin is found in the trunk fiber (24.51%), followed by the frond (20.48%) and EFB (17.84%). Mature tissues in the trunk accumulate higher amounts of metabolic product than the younger tissues in the frond (Ververis *et al.*, 2004). EFB fibers, with the lowest lignin content, are more easily bleached than trunk fibers. One study, by contrast, found large amounts of lignin in EFB fibers (Mohamad *et al.*, 1985), which may have been due to use of different methodologies and plant varieties. EFB fiber contains the highest percentage (3.4%) of ash, which

Table 12.1 Chemical compositions in different OPB fibers

Composition	EFB (wt%)	Frond (wt%)	Trunk (wt%)
Cellulose	43–65	40–50	29–37
Hemicellulose	17–33	34–38	12–17
Lignin	13–37	20–21	18–23
Ash	1–6	2–3	2–3

Sources: Abdul Khalil *et al.*, 2008a, 2008b; Killman and Lim, 1985; Law *et al.*, 2007; Punsuvon *et al.*, 2005; Shinoj *et al.*, 2011; Chew and Bhatia, 2008; Mohamad *et al.*, 1985; Abdul Khalil and Rozman, 2004; Law and Jiang, 2001; Sreekala *et al.*, 2001.

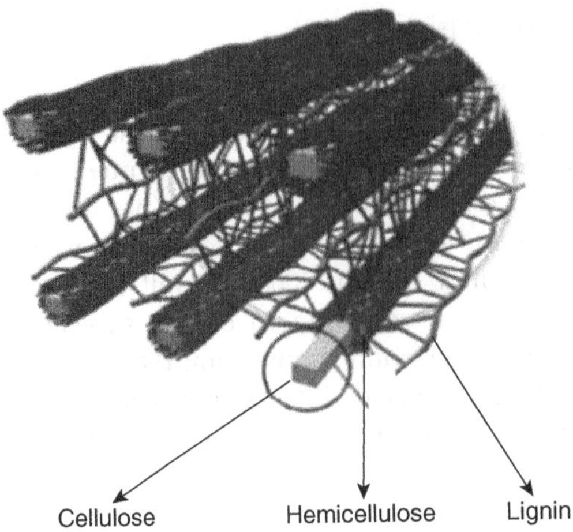

12.5 A scheme of cellulose, hemicellulose and lignin distribution in the natural fiber.

Table 12.2 Major chemical compositions in oil palm EFB fiber of different origins

Cellulose (wt%)	Hemicellulose (wt%)	Lignin (wt%)	Location	References
22	48	25	Malaysia	Hill and Abdul Khalil, 2000
30	50	18	Malaysia	Abdul Khalil *et al.*, 2008a
30	36	22	Indonesia	Minowa *et al.*, 1998
28	65	19	India	Sreekala *et al.*, 1997; Law *et al.*, 2007

consists of calcium and magnesium, but certain plants have high levels of silica.

Oil palm fibers contain 4.5% residual oil (Abu Bakar *et al.*, 2006), which is important because the ester components of oil residues may affect the coupling efficiency between fibers and the polymer matrix as well as the interaction between fibers and coupling agents (Rozman *et al.*, 2001a).

12.2.3 Morphological and physical characteristics

EFB fiber is a hard and tough multicellular fiber with a central portion called a 'lacuna', as shown in Figs 12.6(a) and (b). The porous surface morphology allows effective mechanical interlocking with the matrix resin during composite fabrication (Sreekala *et al.*, 1997). The cross-section of the fibers varies from rounded to polygonal in shape. A vascular bundle of EFB

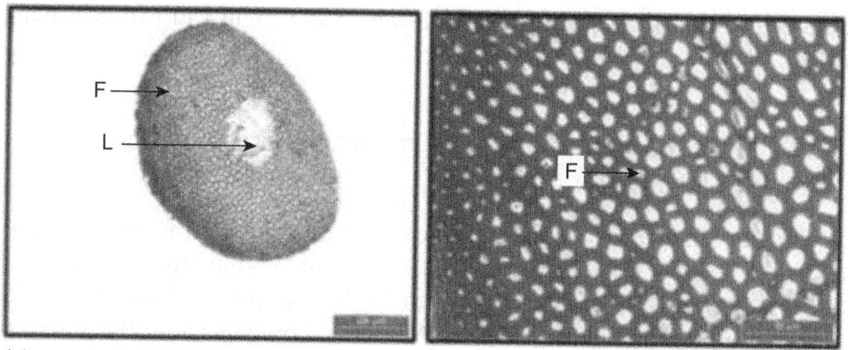

(a) (b)

12.6 Transverse sections of EFB fiber at low magnification (4×) after being stained with toluidine blue: (a) and (b) empty fruit bunch (EFB), F: fiber, L: lacuna.

Table 12.3 Physical characteristics of different OPB fibers

Fiber	Fiber length (mm)	Fiber diameter (μm)	Lumen width (μm)	Density (g/cm³)	Fibril angle (°)
EFB	0.89–142	8–300	8	0.7–1.6	46
Frond	0.59–1.59	11–19.7	8.2–11.7	0.6–1.2	40
Trunk	0.60–1.22	29.6–35.3	17.6	0.5–1.1	42

Sources: Mohamad *et al.*, 1985; Law and Jiang, 2001; Bismarck *et al.*, 2005; Amar *et al.*, 2005, Abdul Khalil *et al.*, 2008a, Hassan *et al.*, 2010; Ahmad *et al.*, 2010.

fibers is a simple bundle surrounded by a sheath consisting of a thickened cell layer, which functions to resist buckling under lateral compression forces (Dickison, 2000).

Physical characteristics of different OPB fibers are given in Table 12.3. Fibers with thicker cell walls resist collapse but contribute less to inter-fiber bonding (Reddy and Yang, 2005). Fiber length is a vital factor in determining bonding and stress distribution (Abdul Khalil *et al.*, 2008a). Oil palm EFB fiber length lies between hardwood and softwood fiber length (Hassan *et al.*, 2010). Microfibril angle, cell dimensions, and the chemical composition of fibers are the essential variables that determine their overall properties (John and Thomas, 2008); these and other physical aspects of oil palm fibers, such as diameter, aspect ratio (the ratio of fiber length to diameter, *l/d*), lumen width and density have a major effect on the physical and mechanical properties of composite materials (Hassan *et al.*, 2010; Shinoj *et al.*, 2011). For example, owing to their low specific gravity of 1.25–1.50 g/cm³, compared with the 2.6 g/cm³ of glass fibers, the lignocellulosic fibers of oil palms are able to give plastic materials a high strength-to-weight ratio (Abu Bakar *et al.*, 2005).

12.3 Surface modifications of empty fruit bunch (EFB) fibers

Several methods have been established for surface modification of natural fibers. For example, Norul Izani *et al.* (2013) demonstrated the treatment of empty fruit bunch (EFB) fibers with boiling water, sodium hydroxide (NaOH), and a combination of both NaOH and boiling water. By studying the morphological and mechanical properties of the treated EFB fibers, they established that alkali treatment improved most of the fiber properties, such as surface topography, mechanical properties and thermal stability.

Alam *et al.* (2012) have combined alkali and ultrasound techniques. They placed fibers separately in ultrasonic baths of water and of alkaline media and then used those fibers for the preparation of poly(lactic acid) (PLA) composites. It was found that the mechanical and interfacial properties of the composites were significantly increased by simultaneous alkali and ultrasound treatment. Sun *et al.* (1999) and Rosnah *et al.* (2010) have also described alkali treatment of oil palm fibers.

Arif *et al.* (2010) treated EFB fibers with alkali and saline to increase fiber loading in high-density polyethylene (HDPE) and improve the mechanical properties of HDPE and EFB composites. An enhancement of properties in EFB-reinforced polypropylene composites by the low-molecular-weight coupling agent MAPP has been reported by Ramli *et al.* (2011), and improvements in the mechanical properties of natural fiber-reinforced thermoplastic composites as a result of the inclusion of MAPP have been shown by several researchers (Godara *et al.*, 2009; Yang *et al.*, 2007; Kim *et al.*, 2007; Pracella *et al.*, 2006). Rozman *et al.* (2001b) used MAPP, PMPPIC and 3-(trimethoxysilyl)-propylmethacrylate (TMP) as coupling agents in a PP–EFB–glass fiber hybrid composite. They stated that MAPP and TMP improved the properties of the composite, but that PMPPIC was less effective.

12.4 Processing methods for EFB reinforced composites

Melt mixing, extrusion, injection molding and compression molding are all used in the preparation of EFB fiber-reinforced polymer composites, and a number of mixing methods for EFB fibers and polymer matrices have been demonstrated. Rozman *et al.* (2001b) compounded PP–EFB–glass fiber (GF) hybrid composites with a Haak Rheocord system consisting of a Haak

Rheodrive 5000 and Haak Rheomix 600 with roller blade. Mixing was at 175°C for 20 min at a rotor speed of 20 rpm. Abu Sharkh and Hamid (2004) compounded PP and EFB with a Brabender Measuring Head single-screw extruder. Alam *et al.* (2012) compounded PLA thermoplastic with short EFB fibers using a twin-screw extruder, with a temperature profile in the range 110–190°C and the screw rotating at 100–110 rpm. Extruded PLA composites were injection molded with 10–40 wt% EFB, and it was found that a 30 wt% composite showed the best mechanical properties. Abdul Khalil *et al.* (2010a) developed a new hybrid medium density fiberboard (MDF) of EFB and rubber wood using a steam-heated hotpress machine at 200°C and 3500 MPa. Jawaid *et al.* (2010) studied the mechanical performance of EFB–jute–epoxy hybrid composites prepared by a jute–EFB–jute sandwich technique and compression molding.

12.5 Effects of fiber treatments on the structures and properties of composites

12.5.1 Water absorption characteristics

This section describes the water absorption behavior of various thermoplastic composites (PP, PVC, etc.) reinforced by EFB fibers. A twin-screw extruder was used, with short EFB fibers (5–10 mm), variously untreated and treated. Certain additives were used, such as a coupling agent (MAPP) and flame retardant (magnesium hydroxide, $Mg(OH)_2$). Processing temperature and screw speed were dependent on matrix and fiber content. The compounded pellets were used to prepare test specimens with the injection-molding machine. With polyester used as the matrix phase, the formulation included a catalyst such as methyl ethyl ketone peroxide and exposure to a vacuum after thorough mixing. The process of resin impregnation was aided by the application of a hand roller. The resin-impregnated mats were transferred to a thick perspex sheet, which was placed in a cold press, squeezing out excess resin in the process. The press was left closed at room temperature overnight to allow a partial cure of the resin.

Figure 12.7 shows water absorption as a function of fiber loading. Water absorption increases as EFB fiber loading increases. Untreated fiber contains lignocellulosic materials (amorphous cellulose, hemicelluloses and lignin), which form hydrogen bonds with water through their easily accessible hydroxyl groups, thus giving the fibers their strongly hydrophilic character (Desch and Dinwoodie, 1989). EFB–PS composites are the most hydrophilic, followed by EFB-PVC and EFB-PP. This absorption of water into the structure of the cell wall causes swelling that leads to micro-cracks and thus severely degrades the mechanical properties of the composite (Peijs *et al.*, 1998). This degradation is not only limited to physical and

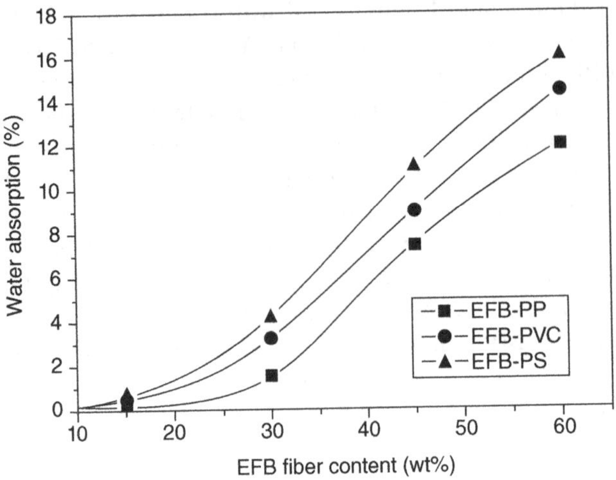

12.7 Water absorption of EFB-PP, EFB-PVC and EFF-PS composites at soaking time of 24h (data reported from Rozman *et al.*, 1999).

chemical modification of the polymer matrix but can also cause debonding of the fiber–matrix interface (Foulc *et al.*, 2005).

Figure 12.8 illustrates the absorption of water versus soaking-time for the EFB fiber–PP composite with various fiber loadings, in water at 25°C. Clearly, more fibers absorb more water. Water penetration involves the flow of water molecules along the fiber–matrix interface, followed by diffusion from the interface into the matrix and fibers (Andreopoulos and Tarantili, 1998). Specifically, water diffusion in polymeric composites is governed by three different mechanisms, as presented in a previous study (Lin *et al.*, 2002): diffusion of water molecules inside the micro-gaps between the polymer chains, capillary transport into the gaps and flaws at the fiber–matrix interface, and transport of micro-cracks in the matrix arising from the swelling of fibers (particularly in the case of natural fiber composites).

12.5.2 Chemical texturing

This section describes the chemical texturing of raw EFB (REFB) and treated EFB (TEFB) reinforced PLA thermoplastic composites. PLA was compounded with short EFB fibers by a twin-screw extruder, and test samples were prepared by an injection-molding machine. The EFB fibers were treated variously by an alkali medium and by ultrasound in both water and alkali media, and under different NaOH concentrations, exposure times and temperatures. The fibers thus treated are termed alkali-treated EFB (AEFB), ultrasound-treated EFB (UEFB), and simultaneous

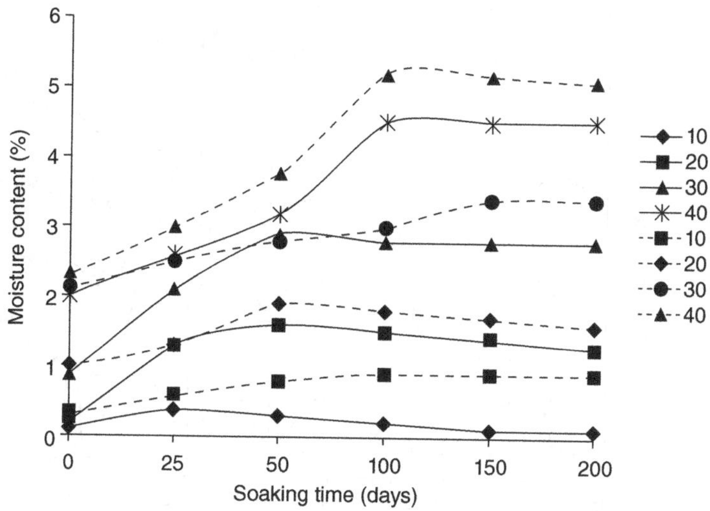

12.8 Plots of moisture content in EFB-PP composites with respect to soaking time at different wt% of fiber content with MAPP (dotted line) and without MAPP (solid line) at 25°C.

ultrasound- and alkali-treated EFB (UAEFB) fibers. AEFB, UEFB and UAEFB treated fibers under optimum treatment conditions are termed OAEFB, OUEFB and OUAEFB respectively. REFB, AEFB, UEFB and UAEFB fibers were loaded in PLA to fabricate composites termed REPC, AEPC, UEPC and UAEPC respectively; OAEFB, OUEFB and OUAEFB fiber-reinforced PLA composites are termed OAEPC, OUEPC and OUAEPC respectively.

Figure 12.9(a) presents the FTIR spectra of PLA and EFB fiber–PLA composites (PLA (a), REPC (b), OUEPC (c), OAEPC (d), and OUAEPC (e)). Considerable variations in the spectra are highlighted with the dashed box, circle and lines where the shape and the position of the absorption bands change, depending on the samples. The absorption peaks for PLA are located at 3500 cm^{-1} due to OH stretching, 2600 and 2100 cm^{-1} from the asymmetric and symmetric modes of C–H stretching respectively, 1700 cm^{-1} from C=O stretching, 1386 cm^{-1} from C–H deformation, and 1083 cm^{-1} due to the O–C asymmetric mode of the ester groups. The OH stretching in REPC is found to be different from that in PLA, indicating the formation of hydrogen bonds with the fibers and the PLA. A similar observation has been reported by other researchers for PLA–rice starch composites (Yew *et al.*, 2005). The spectra of OUEPC and OUAEPC are similar to the spectrum of REPC, except for the band shape around 1775 cm^{-1} (from C=O stretching), which is found to depend on fiber treatments.

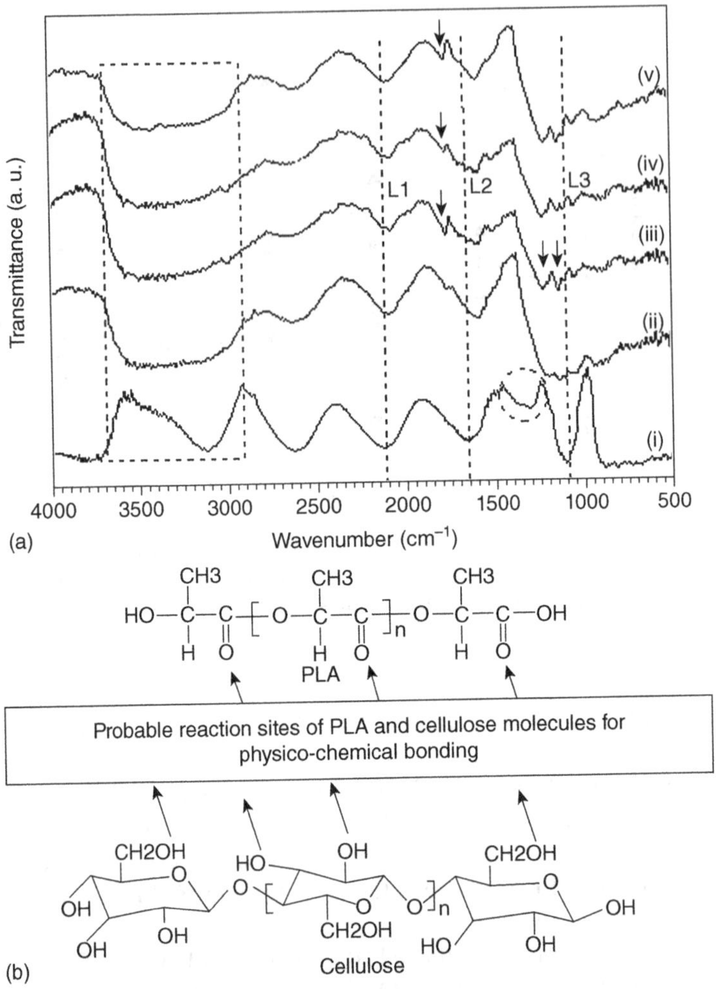

12.9 (a) FTIR spectra of (i) PLA, (ii) REPC, (iii) OUEPC, (iv) OAEPC and (v) OUAEPC. (b) Scheme for the probable reaction sites of PLA and EFB fibers.

Since cellulose is the main component of the EFB fiber, its OH groups can be bonded to the carbonyl (C=O) or carboxylic acid (COOH) groups of PLA. The possible sites of interaction between the cellulose of EFB fibers and PLA are depicted in Fig. 12.9(b). Since REFB fibers contain impurities (e.g. wax and pectin) on their surfaces, it is reasonable to assume that only a limited number of OH groups (from fibers) take part in bonding with the C=O groups of –COOH in the PLA. By contrast, the number of available OH groups for the alkali-treated fibers increase due to the removal of impurities from the surfaces. The increased number of OH groups on the

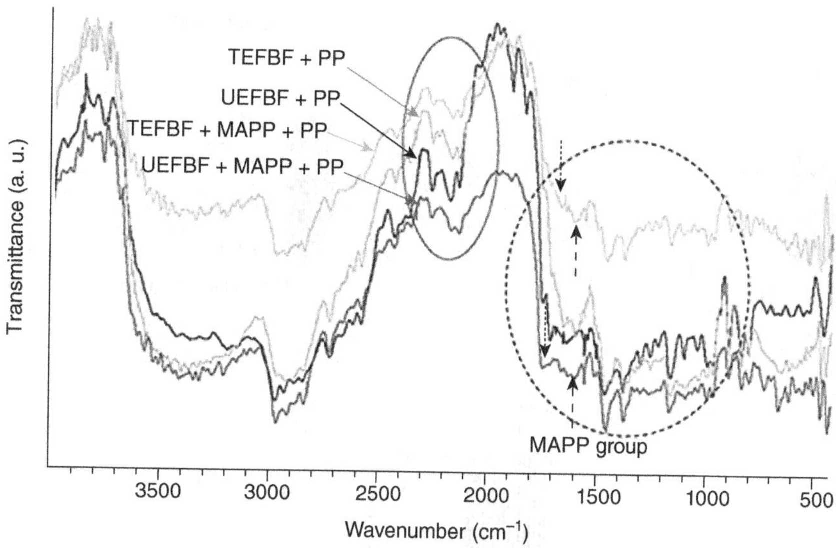

12.10 FTIR spectra of untreated and treated EFB-PP composites with and without MAPP.

fibers means that there is greater potential for hydrogen and covalent bonding with C=O and COOH groups in the PLA.

The FTIR spectra of the untreated EFB fiber (UEFBF) and treated EFB fiber (TEFBF) reinforced PP composites, both with and without MAPP, are shown in Fig. 12.10. Two enclosures (solid and dotted) are introduced to distinguish the important variations in the spectra for different samples. Arrows indicate the spectra for various samples at positions where noticeable variations can be observed. The absorption at around 1670–1600 cm^{-1} is principally associated with the absorbed water, since hemicelluloses usually have a strong affinity to water due to its disordered structures. The peaks at 1723 and 1515 cm^{-1} indicate the presence of the conjugated carboxyl group (C=O) of lignin. Bands around 1125–1000 cm^{-1} are typical of xylenes. The prominent bands at 1091 cm^{-1} and 1039 cm^{-1} are attributed to C–O and C–C stretching or C–OH bending in hemicelluloses, while those at 1510, 1468, 1379 and 1255 cm^{-1} correspond to C–O stretching and CH or OH bending in the hemicelluloses. The treated EFB fiber-reinforced composites give smaller, narrower peaks than the untreated ones.

A new, small peak appears around 1800–1750 cm^{-1} (downward arrows) due to the ester bonds between the fibers and MAPP. There is also increased absorption in the C–H region. Both the EFB fibers and the MAPP-loaded composites also show an increased absorption at 1630 cm^{-1} (upward arrows), which probably indicates the presence of C=C bonds in MAPP (Rozman

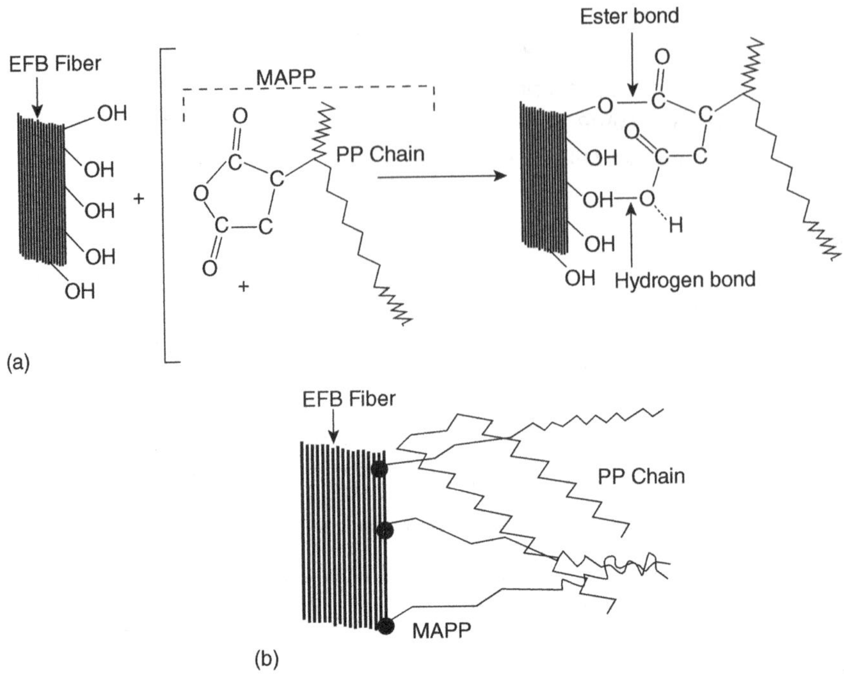

12.11 Illustrations of (a) bonding between MAPP and fiber and (b) molecular entanglement.

et al., 2003). A clear change in absorption bands in the range 2100–2300 cm^{-1} may indicate interaction between EFB fibers and MAPP. Thus, these FTIR analyses clearly demonstrate the formation of ester bonds and hydrogen bonds between fibers and the PP through the coupling agent (see Fig. 12.11).

12.5.3 Crystallization and crystallinity

Figure 12.12 represents the XRD profiles of REFB (a), OAEFB (b) and OUAEFB (c) fibers as well as PLA (d), REPC (e), OAEPC (f) and OUAEPC (g). The REFB fiber shows a diffraction peak (P1) at ~22.7°, which is characterized as the $(200)_f$ plane in the crystalline cellulose material of the raw EFB fibers (Nishiyama *et al.*, 2002). The $(200)_f$ peak appears in diffuse scattering (dashed line), defined here as ξ-scattering, with maximum intensity at $2\theta \approx 18°$. The ξ-scattering comes from the amorphous parts, such as hemicellulose, lignin, pectin, and other foreign impurities of EFB fibers. Its intensity decreases in the cases of OAEFB and OUAEFB fibers, which also show a comparatively less diffuse peak (P2) at ~16°, corresponding to the $(110)_f$ plane (Inagaki *et al.*, 2010). The appearance of 110 reflection and

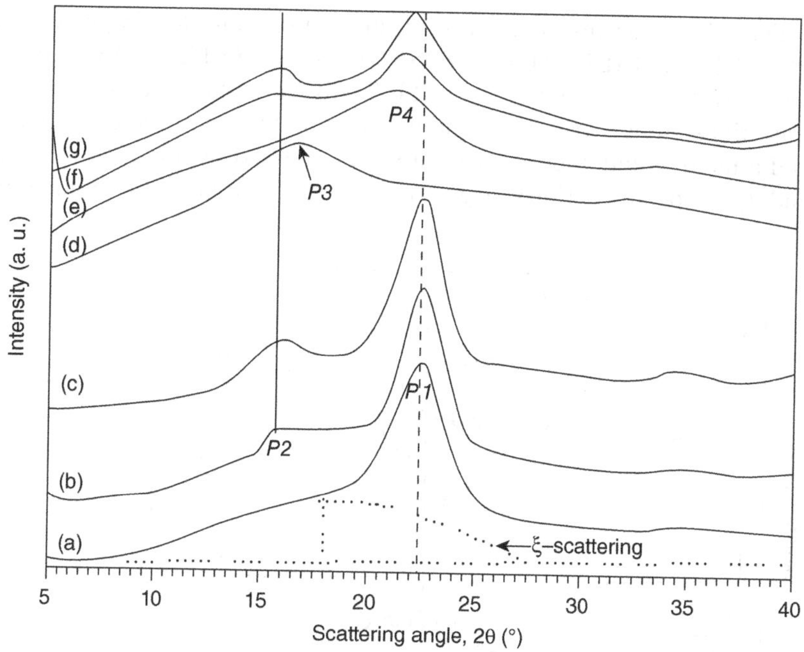

12.12 XRD profiles of (a) REFB, (b) OAEFB, (c) OUAEFB, (d) PLA, (e) REPC, (f) OAEPC and (g) OUAEPC.

the gradual decrease in the peak-width of $(200)_f$ indicates the removal of non-cellulosic components to leave well-ordered cellulose.

The diffraction patterns from PLA can be identified by a diffuse peak close to $2\theta \approx 16.6°$. The shape of this peak suggests that PLA chains are poorly ordered with a low degree of crystallinity due to the rapid cooling of the injection-molding process. The position of this peak for PLA corresponds to the $(200)_{p\alpha}$ plane of the α-phase (Wu and Wu, 2006). The diffraction patterns for REPC contain a very diffuse peak at $2\theta \approx 21°$ that suggests a very poor ordering of PLA molecules. When OAEFB and OUAEFB fibers are introduced in PLA, two distinct peaks appear. The $(110)_f$ peak intensity gradually increases and is shifted to the higher angle, though more for OUAEFB fibers than for OAEFB fibers. All of this indicates that the degree of PLA crystallization in the composites is increased more by OUAEFB fibers than by OAEFB and REFB fibers. The inter-planar spacing in the treated fibers also decreases. Hydrogen- and covalent-bonded interactions and mechanical interlocking between the fibers and PLA matrix are improved by surface modifications. These interactions can help the fiber surfaces to provide nucleation sites for better crystallization in the PLA.

DSC thermograms obtained from first and second heating courses for PLA, REPC, OAEPC and OUAEPC are shown in Figs 12.13(a) and (b) respectively. On heating, pure PLA displays three main successive transitions: a glass transition endotherm, a cold crystallization exotherm, and a melting endotherm. These are characterized by temperatures T_g, T_c and T_m, respectively (Suryanegara *et al.*, 2009). The injection-molded pure

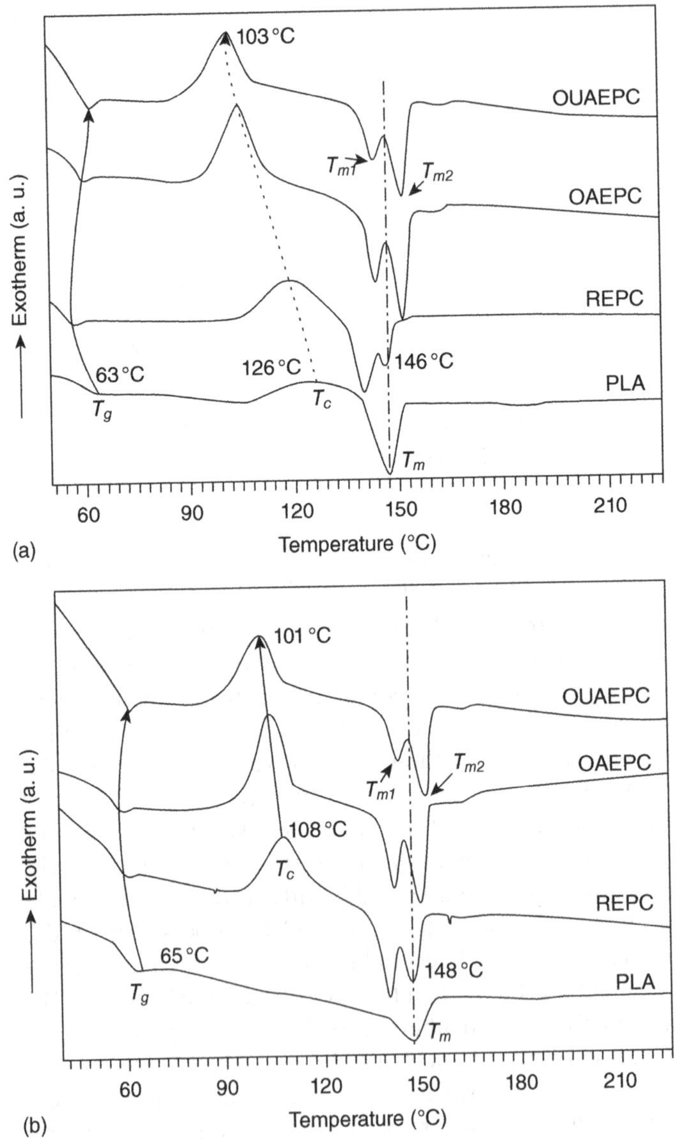

12.13 DSC thermograms at (a) first heating, and (b) second heating of the samples.

Table 12.4 The T_g, T_c, T_m, T_d and X_{dsc} values obtained from DSC and TGA thermograms

Sample	DSC				TGA	
	T_g (°C)	T_c (°C)	T_m (°C)	X_{dsc}(%)	T_d (°C) at 50% wt. loss	T_d (°C) at DTG peak
PLA	63.5	125.0	146.6	24.66	366.7	371
REPC	57.0	119.7	140.5 147.0	30.52	330.5	328
OAEPC	62	100.39	143.15 152.13	41	331	326
OUAEPC	59.7	105.0	143.9 151.5	43.11	350.7	364

PLA and composites all present three distinct peaks for T_g, T_c and T_m, although the composites show double melting endothermic peaks (T_{m1} and T_{m2}), where T_{m1} and T_{m2} are temperatures of the low- and high-temperature endotherms respectively. The observed T_c difference between OUAEPC and REPC is 16°C for first heating and 7°C for second heating. The T_g, T_c and T_m values and the corresponding enthalpies ΔH_g, ΔH_c and ΔH_m from first- and second-heating DSC runs as well as the $X_{dsc}(\%)$ values evaluated from the first-heating run are summarized in Table 12.4, which clearly distinguishes the effect of the two heating processes.

The T_g value is the lowest in REPC but slightly greater in OAEPC and greater still in OUAEPC. This progression in T_g depression is related to the chain mobility of PLA (Baratian *et al.*, 2001). A low T_g is commonly seen in a sample with a low degree of crystallinity, where the molecules can move easily. By contrast, in a sample with a more orderly crystalline state (and therefore higher density), the non-crystalline molecular chains are anchored to the immobile crystallites and constrained.

The variations in T_c values may be caused by factors in the solidification of the samples from the melt under the DSC and the injection-molding processes. Nucleating agents are commonly used to increase the crystallinity of a polymer by lowering the surface free energy barrier for nucleation and thus enabling crystallization at higher temperature upon cooling from the melt. Talc and montmorillonite (at ~5 wt%) effectively modify the crystallization rate of PLA (Baratian *et al.*, 2001), with the lowest crystallization induction period and maximum crystallization speeds observed around 100°C. In our study, if we compare the degrees of recrystallization in REPC, OAEPC and OUAEPC by injection molding (first-heating curve) and the controlled cooling DSC process (second-heating curve), the T_c values are observed to lie in the range 103–119°C and 101–108°C (Figs 12.13(a) and 13(b)) respectively. These differences in T_c suggest that the lowest crystallization induction temperature is similar to the reported value, almost irrespective of processing route. However, T_c of composites for both DSC heating processes is lowest for OUAEPC, greater for OAEPC, and greatest for REPC, which suggests that treated EFB fibers are favorable as active nucleating agents for PLA, enabling crystallization at lower temperatures on heating.

The reported melting points of PLA are 185°C for the α-structure and 175°C for the β-structure (Hoogsteen *et al.*, 1990). The observed low melting point (~148°C) of PLA is most likely due to its origin, because T_m depends on molecular weight, stereochemical structure and synthesis sources. Many different explanations have been given for the appearance of the double melting peaks in PLA. Di Lorenzo (2006) suggested that T_{m1} is due to the fusion of small and imperfect crystals of low thermal stability that are formed at T_c, and that T_{m2} is due to the melting of perfect crystals formed by structural reorganization of the imperfect crystals. A very similar explanation was proposed by Yasuniwa *et al.* (2004), who suggested that T_{m1} originates from small and imperfect crystals that form successively into more stable crystals through melting and recrystallization and that an exotherm between the two endothermic peaks is associated with recrystallization. Pan *et al.* (2007) have put forward more detailed proposals, describing T_{m1} as associated with both the phase transition and the melting of the original α-phase crystals, and T_{m2} as arising from the α-phase formed during the phase transition and melt-recrystallization (Yasuniwa *et al.*, 2004). Other explanations have involved the existence of dual crystal structures in the polymer or dual lamellae in the same crystalline structure (Di Lorenzo, 2006; Pan *et al.*, 2007).

12.5.4 Density and melt-flow index

The density of composites prepared with untreated EFB fiber (UEFBF) and treated EFB fiber (TEFBF), with and without MAPP and with magnesium hydroxide as the flame retardant (FR), is presented in Figs 12.14 (a) and (b) respectively. Density increases as the fiber content increases. This increase is more pronounced in TEFBF-reinforced composites than in those reinforced with UEFBF, which indicates better interfacial adhesion between EFB fibers and the PP matrix as compared with the untreated EFB fiber-loaded composites (Shinoj *et al.*, 2010). The more pronounced increase in density is due to the presence of MAPP, which encapsulates the surface of the EFB fibers and diffuses into the bulk of the PP matrix, reducing the formation of voids.

Figure 12.15 shows the effects of NaOH concentration, soaking time and treatment temperature on the density of EFB fiber-reinforced PLA

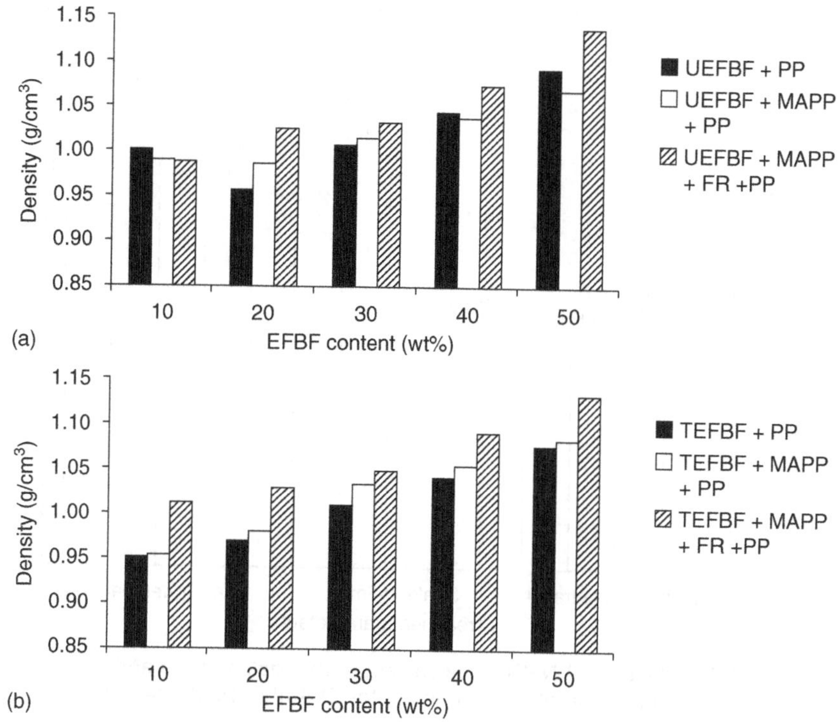

12.14 Density of untreated and treated EFB fiber-reinforced PP composites.

composites with fibers treated by ultrasound at a sonication power of 80%. The density increases with NaOH concentration up to 3 wt% and decreases thereafter. When fibers have been exposed to solution for between 30 and 120 min, the density increases with greater times up to 90 min and then decreases. The density of composites also increases with higher treatment temperatures (from 35°C to 80°C) and decreases at 100°C. These variations in density are due to changes in the compatibility between fibers and the polymer matrix (Hill and Abdul Khalil, 2000). Ultrasound treatment reduces the fiber treatment parameters and improves the melt-flow of the composites.

Table 12.5 shows a decrease in melt flow index (MFI) with increased fiber loading. This suggests that the addition of raw EFB fiber increases viscosity and ultimately reduces the capacity for flow during processing (Thattaiparthasarthy *et al.*, 2009). Figure 12.16 illustrates the MFI of virgin PLA, REPC, and variously treated EFB fiber (30 wt%) reinforced composites. The MFI reduces significantly with inclusion of raw EFB fibers in PLA, but increases with alkali treatment of fibers. This suggests that natural fiber greatly reduces the flow property of composites. By contrast, treated fibers can, under certain conditions, increase the MFI. This probably

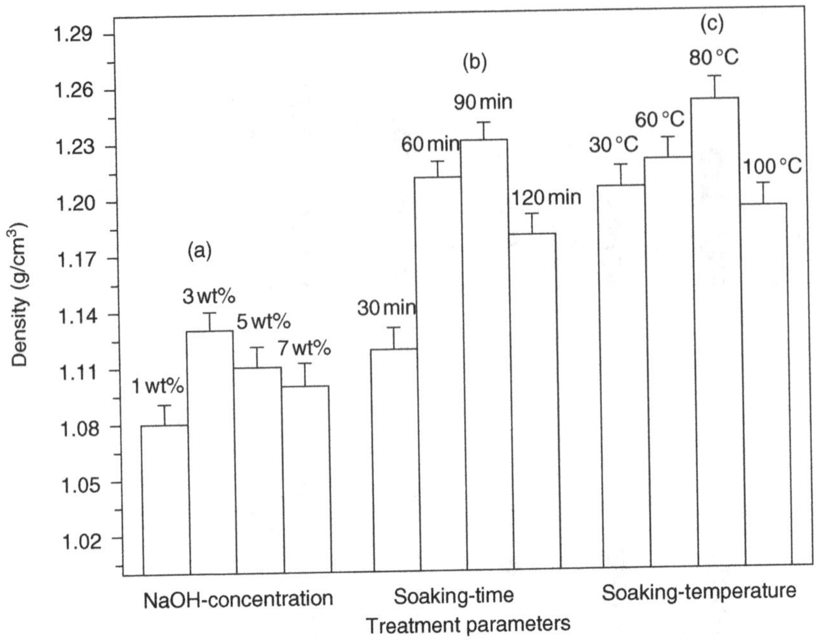

12.15 Effect of (a) NaOH concentration, (b) soaking time and (c) soaking temperature on density for EFB fiber-reinforced PLA composites.

Table 12.5 Melt flow at 230°C for treated EFBF composites with 3 wt% MAPP

Composition (wt%)	MFI (g/10 min)
0	45.00
10	35.80
20	27.60
30	19.58
40	14.22
50	8.46

occurs because of the increased compatibility between the fibers and PLA matrix after the removal of cementing components. The MFI values for UAEPC with fibers treated at room temperature in 3 wt% NaOH for 90 min and at 80°C in 3 wt% NaOH for 60 min are 3.31 and 3.55 g/10min respectively.

12.5.5 Mechanical properties

Some studies have indicated that the tensile strength (TS) of natural fibers and polyolefin composites decreases with increasing fiber content, while

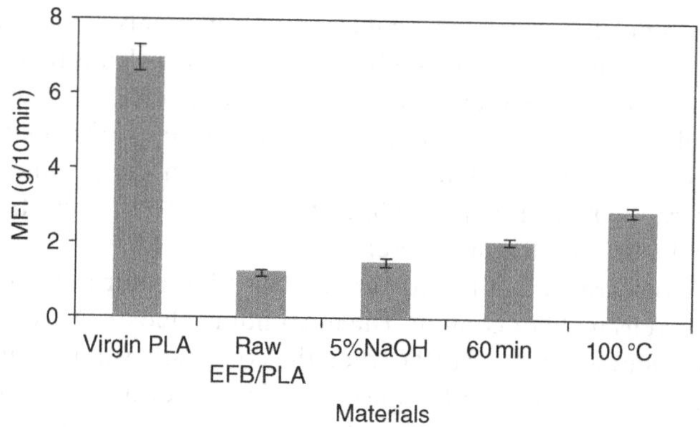

12.16 MFI of virgin PLA, raw EFB and Alkali-treated (5% NaOH, 60 min, 100°C) EFB reinforced PLA composites.

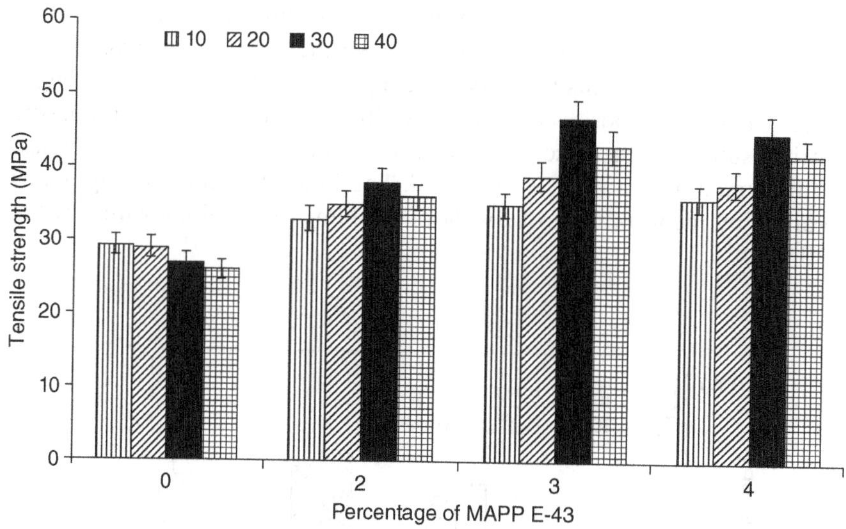

12.17 Tensile strength of EFB/PP composites in the presence of MAPP E-43 coupling agent (EFB fiber: 10–40 wt%).

other studies have shown an opposite trend (Beg and Pickering, 2006; Rozman *et al.*, 2003). Figure 12.17 displays the change in TS of EFB–PP composites with respect to coupling agent (MAPP, E-43). The TS of the EFB–PP composites is significantly affected by fiber incorporation, as compared with the virgin PP. The TS of virgin PP is 37 MPa, decreasing to 26 MPa with fiber loading up to 40 wt% without a coupling agent. This negative impact of fiber loading on the PP matrix is due to weak interfacial adhesion between the EFB fibers and the PP. The interfaces are the weakest part of the composite and serve as failure initiation points. Poor interfacial interaction or bonding between the fibers and the matrix leads to inefficient stress transfer when stress is applied to the tensile specimen (Gowda *et al.*, 1999). Irregularly shaped and shorter fibers like those of EFB may not be able to support the stresses that are transferred from the polymer matrix (Rahman *et al.*, 2009; Liu *et al.*, 2009; Botev *et al.*, 1999). Thus, there is a reduction in TS as fiber loading increases.

The TS of composites significantly increases with greater use of coupling agents and increasing fiber content. The maximum value of TS exhibited by the EFB–PP composite was 47 MPa, when the composite was prepared with 3 wt% MAPP and 30 wt% EFB fibers. Maleic anhydride reacts with the hydroxyl polar groups in the fibers and forms strong covalent ester bonds as well as hydrogen bonds (see Fig. 12.11).

Upon esterification, the exposed polyolefin chains are diffused into the PP matrix and entangled with PP chains during hot pressing. Chemical bonds are formed at the interface between the fiber and the PP matrix, thereby improving the compatibility between them and enhancing the TS.

Figure 12.18 represents the effect of coupling agents on flexural strength (FS) in EFB/PP composites. Generally, addition of fibers to the thermoplastic matrix results in an increase in FS. Overall, the FS shows significant increase as the percentage of fiber is increased. The highest flexural strength was obtained by incorporation of 3 wt% of coupling agent along with 30 wt% EFB fiber.

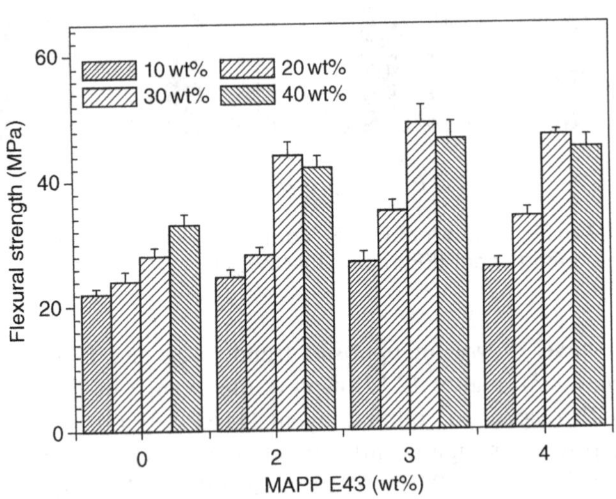

12.18 Flexural strength of EFB/PP composites in the presence of MAPP E-43 coupling agent (EFB, 10–40 wt% (Data reported in Ramli *et al.*, 2011).

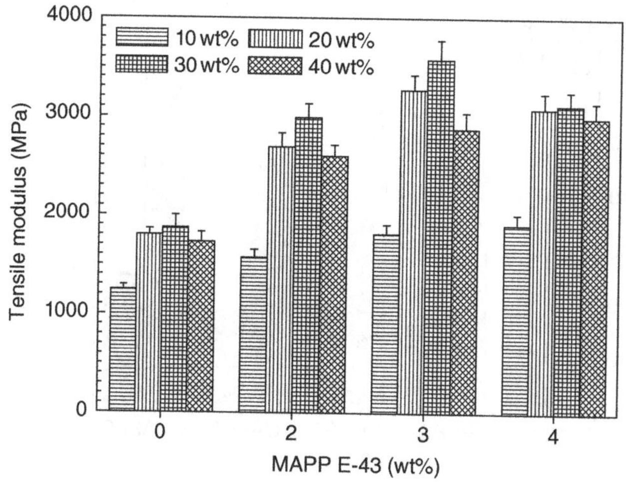

12.19 Tensile modulus of EFB/PP composites in the presence of MAPP E-43 coupling agent (EFB fiber: 10–40 wt%) (Data reported in Ramli *et al.*, 2011).

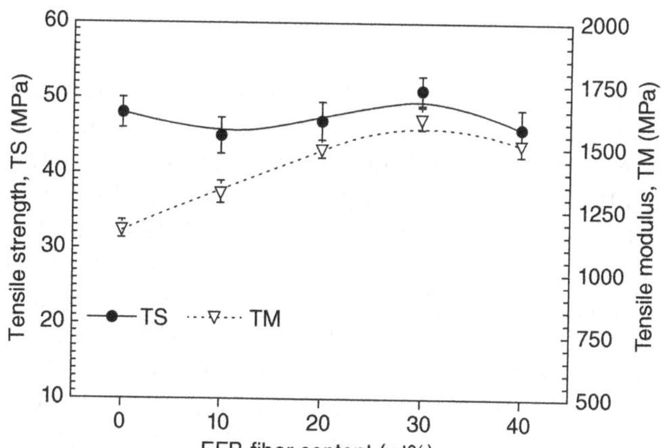

12.20 Tensile strength and modulus of EFB/PLA composites as a function of fiber content (data reported in Alam *et al.*, 2012).

Figure 12.19 shows the effect of coupling agents on the tensile modulus in EFB/PP composites. Generally there is a good effect on modulus regarding the fiber loading in composite materials, because the fiber is stiffer than the matrix. Modulus significantly increases with increase in the fiber content and the coupling agents.

The variation in tensile strength (TS) and tensile module (TM) with respect to raw EFB fiber content is plotted in Fig. 12.20, wherein the 30 wt%

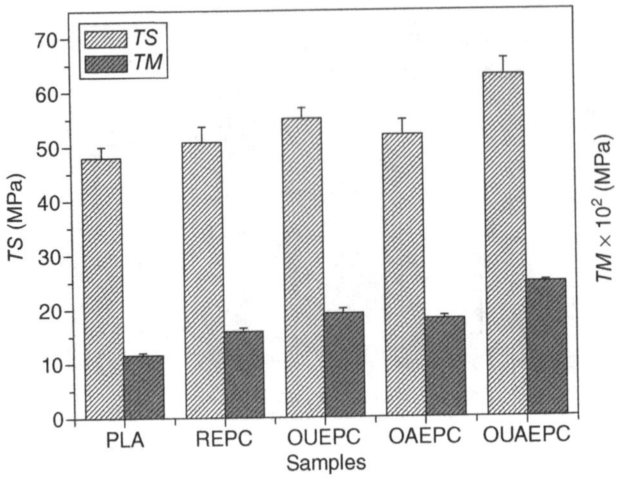

12.21 Tensile strength (TS) and tensile modulus (TM) plots for PLA, REPC, OUEPC, OAEPC and OUAEPC (data reported in Alam *et al.*, 2012).

EFB fiber-reinforced PLA composite exhibits the highest mechanical performance among all others. EFB fiber contents both higher and lower than 30 wt% in composites result in declining mechanical properties. This behavior of EFB/PLA composite is opposite to that of EFB/PP composite.

Figure 12.21 shows the TS of 48, 51, 55, 62 and 63 MPa as well as the TM of 1178, 1620, 1930, 1800 and 2470 MPa for PLA, REPC, OUEPC, OAEPC and OUAEPC, respectively. Clearly, OUAEPC shows the highest TS and TM values, which are immediately followed by OUEPC. The respective increases of TS and TM of OUAEPC from REPC are about 23 and 57%, those of OUEPC from REPC are 8 and 20%, and those of OAEPC from REPC are 2 and 12%, denoting a significant figure to enhance materials properties by fiber treatments. This better performance of the resulting composites can be attributed to the increased compatibility between EFB fibers and PLA because of the modification of the EFB fiber surface by both alkali and ultrasound treatments.

On the other hand, the increase of TS and TM in the composites merely for the inclusion of REFB fiber content may be assigned to the robust TS and TM values of the individual EFB fiber (Shinji, 2008). However, when EFB fiber content is greater than 30 wt%, the composites become inhomogeneous because of the fiber–fiber interaction whose effect may cause fiber agglomeration to develop in the composites, as reported elsewhere (Joseph *et al.*, 2006).

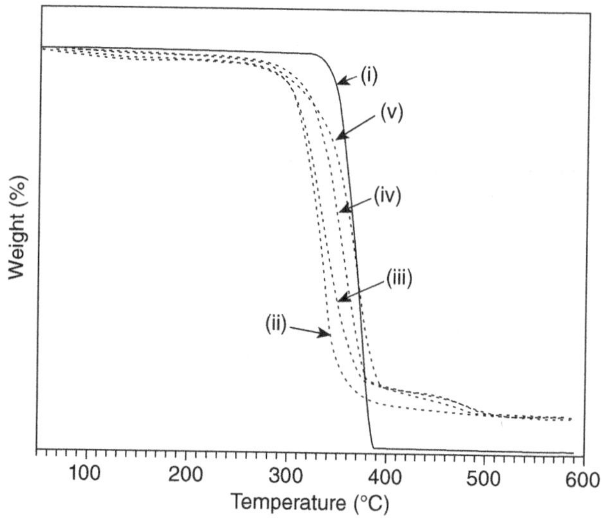

12.22 TGA thermograms of (i) PLA, (ii) REPC, (iii) OUEPC, (iv) OAEPC and (v) OUAEPC (data reported from Alam *et al.*, 2012).

12.5.6 Thermogravimetric properties

Figure 12.22 illustrates the weight loss of the samples investigated by TGA in a temperature range of 50–590°C. The TGA thermogram generally involves the release of absorbed water (if any) from a sample, the 'onset' of degradation of molecules in the sample, the steps of degradation and the presence of residual char. The thermal degradation of neat PLA takes place in one step in which the onset of weight loss occurs at about 330°C and finishes at 390°C. On the other hand, this degradation in composites commences at a relatively lower temperature than that found in PLA. The TGA traces of composites seem to fall at 100°C, indicating the emission of absorbed moisture, and the sharp fall, which occurs at 289°C for composites, looks quite distinct from that for PLA, showing a two-step process in the temperature ranges 275–375°C and 375–484°C. This may be connected to the decomposition behavior of the molecules of the EFB fibers in the variously treated composites.

The final decomposition temperature generally increases with the addition of fiber, which can be attributed to a barrier effect: the dispersion of fibers in the PP matrix hinders diffusion of volatile decomposition products (Abu Sharkh and Hamid, 2004). The degradation of natural fiber has been ascribed to the dissociation of C–C chains along with H-abstraction at the sites of dissociation (Joseph *et al.*, 2003).

Several authors have reported an increase in thermal stabilization of fiber-reinforced polymeric composites by means of coupling agents

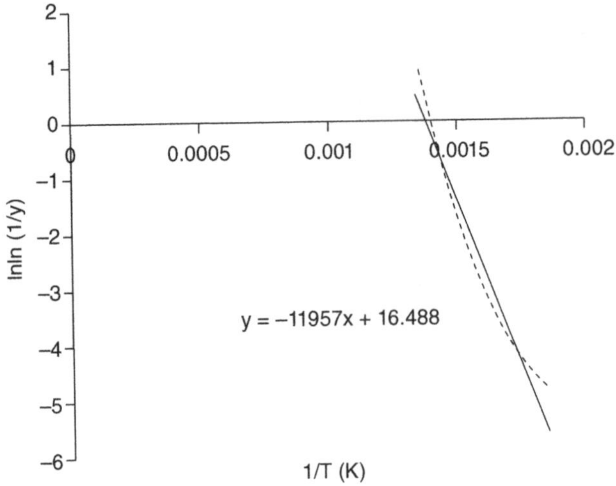

12.23 The lnln (1/*y*) versus 1/*T* plot for pristine PP with the best fit straight line.

(Mohanty *et al.*, 2006). This can be explained by additional intermolecular bonding, from a reaction between the EFB and the functional group of the coupling agents. The reaction can be understood through the kinetics of thermal degradation for EFB–PP composites at various stages, which have been determined from the TGA graphs using the following equation, given by Broido (1969):

$$\ln\left[\ln\frac{1}{y}\right] = -\frac{E_a}{RT} + \ln\left[\frac{RZ}{E_a\beta}T_{max}^2\right] \qquad [12.1]$$

where y is the fraction of nonvolatilized material not yet decomposed, T_{max} is the temperature at maximum reaction rate, β is the heating rate, Z is the frequency factor, and E_a is the activation energy. A plot of $\ln[\ln(1/y)]$ versus $1/T$ for various stages of the decomposition of the EFB–PP composite (Fig. 12.23) is found to be generally linear, suggesting good agreement with the Broido equation. The activation energy, E_a, determined from the slopes of these plots for various composites, is given in Table 12.6. It is found to increase with greater fiber content, but is comparatively higher for composites with MAPP. This higher activation energy is due to the bonding between the fiber and the coupling agent, as described above.

12.5.7 Electrical properties

Particular reports on the dielectric properties of EFB fiber composites based on PP, PLA, polyester and rubber are rare in the literature. Shinoj

Table 12.6 Thermal properties of EFB/PP composites

Formulation	Stage	Temperature range (°C)	T_{max} (°C)	Residue (%)	Activation energy (kJ/mol)
PP	1st	267–472	420	0.012	99.41
10% EFB + PP	1st	210–351	325	2.8	23.39
	2nd	377–490	464		64.62
40% EFB + PP	1st	214–361	331	5.78	65.37
	2nd	389–496	470		35.88
PP + 40% EFB + MAPP	1st	223–371	339	5.47	71.35
	2nd	392–499	472		40.06

Source: Ramli *et al*., 2011.

et al. (2010) produced a comprehensive study of the dielectric constant of EFB fiber–polyethylene composites as a function of loading with untreated and treated fibers. The dielectric constant was found to increase with fiber loading, as shown in Fig. 12.24, which was explained by the increase in effective dipole moment of the composites due to the polar groups in the filler material. A similar trend of variations in dielectric constant has been observed for coconut fiber–polypropylene composites (Lai *et al.*, 2005). Shinoj et al. (2010) showed that the dielectric constant of alkali-treated fiber composites changed slightly with fiber size. This fact was discussed by Shinoj *et al.* (2010), who explained that the effective dielectric constant decreases with increasing filler due to increased interface volume when filler with a smaller particle size is used at a given volume fraction. The smaller particle size creates more polarization in the interface surface. Chen *et al.* (2003) explained this as a result of increased moisture absorption due to the increased surface area provided by small filler particles.

Alkali-treated fiber composites exhibited a lower dielectric constant, particularly with a high degree of fiber loading. A similar trend was observed by Paul *et al.* (1997) in the case of sisal–LDPE composites. They explained this trend as due to alkali treatment reducing the water absorption capacity of the sisal fiber. In a study by Jacob *et al.* (2006) on the dielectric properties of OPF–sisal–natural rubber hybrid composites, it was also reported that chemical modification of fibers resulted in a lower dielectric constant, which was attributed to a reduction of the hydrophilicity of the treated fibers leading to a reduction of orientation polarization in those composites. Alkali treatment unlocked the hydrogen bonds, making them more reactive. In the untreated state, cellulosic OH groups form strong hydrogen bonds and so are relatively unreactive. Alkali treatment can also lead to fibrillation, i.e. the breaking down of large fibers into smaller ones. All these processes contribute to a larger surface area and thus to better mechanical interlocking

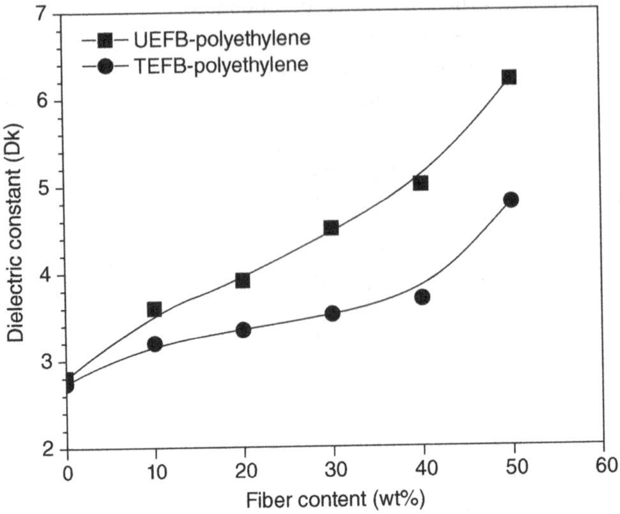

12.24 Dielectric constant versus fiber loading in EFB fiber–polyethylene composites with untreated and treated fibers (data replotted from Shinoj *et al.* 2010).

between the fiber and the matrix, which reduces water absorption. This in turn results in lowering overall polarity and hydrophilicity of the system, leading to a reduction of orientation polarization and consequently a reduction of the dielectric constant of the treated composites.

Ben Amor *et al.* (2010) measured the dynamic dielectric properties of short palm tree lignocellulosic fiber-reinforced polyester composites, with special reference to the effect of fiber orientation, in the temperature range 40–200°C and in the frequency range 1–100 kHz. Three relaxation processes were identified: orientation polarization attributed to the presence of polar water molecules in the palm fiber, the relaxation process associated with conductivity occurring as a result of carrier-charge diffusion (noted for high temperatures above the glass transition and low frequencies), and the interfacial or Maxwell–Wagner–Sillars relaxation that was attributed by these authors to the accumulation of charges at the palm fiber–polyester interface. They suggested that the orientation of the fiber can strongly influence dielectric properties and interfacial polarization processes in composites.

12.5.8 Degradation and stability

Degradation of a polymeric composite is a change in its properties (such as tensile strength, color or shape) due to one or more environmental

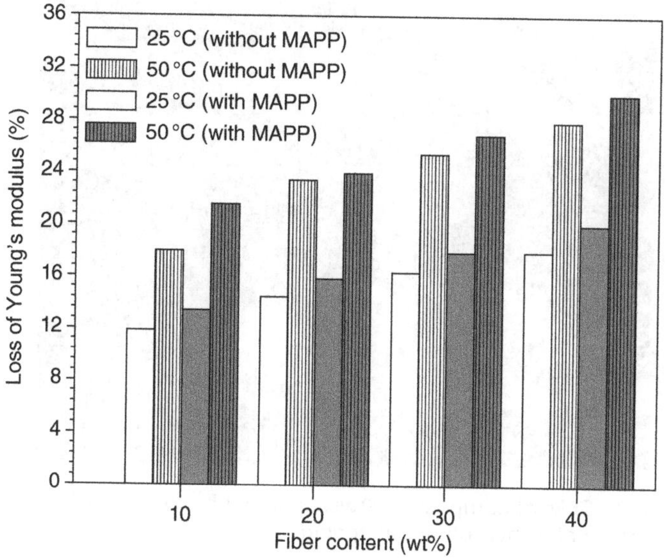

12.25 Loss of Young's modulus (%) of composites without and with MAPP before and after hygrothermal ageing at different temperatures.

factors (such as heat, light, or chemicals). It can be attributed to molecular degradation from processing, the natural environment, and aggressive chemical attack during the service lifetime of the composite.

Hygrothermal ageing lessens the reduction in Young's modulus for EFB fiber-based PP composites (Fig. 12.25). The reduction of Young's modulus is minimized for composites with high fiber content and high ageing temperature. The fall of Young's modulus without MAPP is higher than that with MAPP, and the maximum loss occurs at 40% fiber loading and 50°C. This is likely to be due to the degradation of fibers at the interface, as can be seen in SEM micrographs (Fig. 12.26).

Weathering similarly decreases the strength of composites. The flexural stress and flexural modulus of EFB–PP composites are 41.6 MPa and 3.85 GPa respectively, reducing to 27.4 MPa and 2.69 MPa after 12 months of soil exposure (Hill and Abdul Khalil, 2000). The loss of tensile properties in EFB–polyester composites due to degradation in soil has been quantified by Abdul Khalil and Ismail (2001). Tensile strength was reduced by 8%, 17% and 35% after soil exposures of 3, 6 and 12 months respectively. Impact strength was reduced by 6%, 18% and 43% after 3, 6 and 12 months respectively of soil exposure. Tensile strength, Young's modulus and elongation-at-break decreased from 35.1 to 34.6 MPa, from 3.29 GPa to 2.32 MPa, and from 3.75% to 2.48% respectively upon soil burial of EFB–polyester composites for 12 months (Hill and Abdul Khalil, 2000).

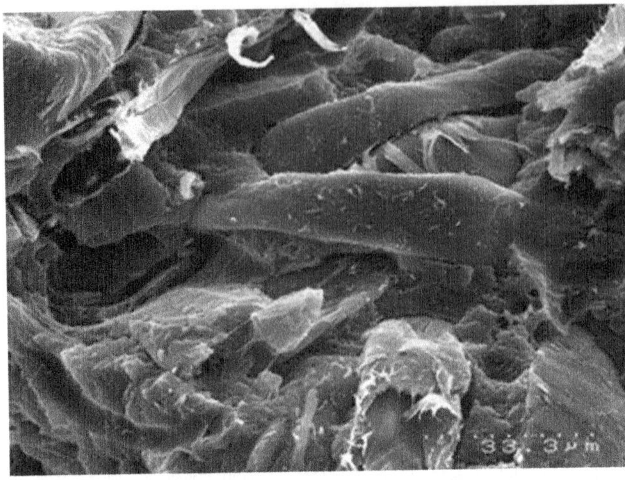

12.26 SEM of composite (30wt% fiber with 3wt% MAPP) fracture surface after hygrothermal ageing.

Chemical treatment of the fiber considerably reduced the mass-loss of OPF–polyester composites due to weathering (Hill and Abdul Khalil, 2000). The magnitude of mass-loss decreased in the following order: unmodified fiber > titanate treated > silane treated > acetylated. Chemical treatments could also conserve the mechanical properties of OPF–polyester composites upon ageing (Abdul Khalil *et al.*, 2000). Though further degradation was observed in the subsequent months, there was an increase in tensile strength and Young's modulus upon acetylation treatment of the fibers in the initial three-month period. Acetylation, silane and titanate treatment of fibers conserved the tensile properties of OPF–polyester composites even after soil exposure of many months (Abdul Khalil and Ismail, 2001). The loss in strength decreased in the same order as for mass-loss: unmodified fiber > titanate treated > silane treated > acetylated.

12.6 Applications of EFB fiber-based composites

Some 12.4 million tons of EFB fibers are discharged every year from palm oil refineries (Abdul Khalil *et al.*, 2010c). This EFB fiber has potential as a natural fiber resource, but the applications account only for a small percentage of the total biomass produced. Several studies have shown that EFB fibers could serve as an effective reinforcement in thermoplastics and thermosetting materials (Abdul Khalil *et al.*, 2008a; Hassan *et al.*, 2010; Shinoj *et al.*, 2011). In order to develop other applications, it will be necessary to extract EFB fibers from the waste using a retting process (Shuit *et al.*, 2009), mainly for soil conservation, erosion control, and ultimately the long-term benefit of nutrient recycling (Abu Hassan *et al.*, 1994); presently they

Table 12.7 Various applications of oil palm fibers

Oil palm biomass	Products	References
EFB fibers	Plywood	Abdul Khalil *et al.*, 2010c
	MDF	Abdul Khalil *et al.*, 2008b, 2010b
	Polymer biocomposite	Chai *et al.*, 2009
	Hybrid composite	Jawaid *et al.*, 2010
	Particle boards	Zaidon *et al.*, 2007
	Biofuel	Shuit *et al.*, 2009

12.27 Applications of natural fiber-reinforced plastics in building products (a) park bench, (b) railing, (c) cladding and siding, (d) decking, (e) fencing and (f) roofing.

are left rotting between the rows of palm trees. Research is underway into producing hybrid plywood, MDF, polymer composites, particle boards, paper, pulp, furniture and biofuels from these residues (see Table 12.7).

There is an increasing demand for natural and bio-based fibers from markets worldwide. In 1967, the demand for fillers from the plastics industry in the USA was 525,000 tons; it had grown to 1,925,000 tons by 1998 (Eckert, 1999) and the projected demand for 2010 was 4,250,000 tons, of which an estimated 350,000 tons (8%) were to be bio-based fibers. The major markets for natural fibers in plastic composites on a weight basis have been summarized elsewhere (Eckert, 2000). Based on that summary, the main application area of bio-based fiber-filled composites is in building products, where they are used in structural elements such as fencing, decking, roofing, railing, cladding and siding, park benches, etc., as shown in Fig. 12.27.

In the USA, the wood–plastic and natural fiber–plastic composites market has been dominated by rail and decking products, while in Europe more emphasis has been made on automotive applications, and most of the Asian countries use natural fibers as fuels. Automotive applications for these

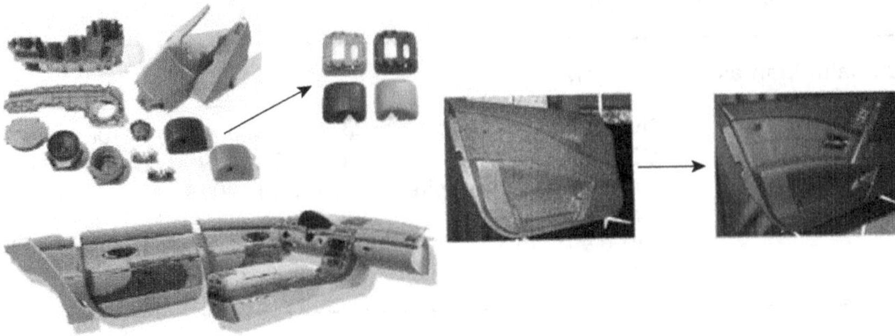

12.28 Interior automotive parts made of wood and natural fiber-reinforced plastics.

composites include interior panels, headliners, dashboards, car roofs, seat panels, parcel shelves and acoustic panels (Fig. 12.28).

12.7 Conclusions

EFB fiber is an agricultural by-product of the palm oil industry. It is left after oil extraction, is considered as waste, and creates environmental problems at mill premises. EFB accounts for 10% of the total biomass produced in Malaysia with a high cellulose content. It is a potential natural fiber resource, but its applications account for a small percentage of the total biomass produced. The properties of oil palm fibers are strongly influenced by chemical composition, fiber structure, microfibril angle, cell dimensions, and defects, and also by which part of the plant they come from and where that plant was grown. EFB fibers offer excellent specific properties and the potential to provide excellent reinforcement of a polymer matrix, and as such they are attractive alternative materials for use in bio composites in hybrid composites, and in the pulp and paper industries. They also hold promise for high-performance applications. Use of EFB fibers as reinforcement in various thermoplastics can give rise to significant changes in the structural, mechanical, thermal and electrical performance of the resulting composites. Most automobile manufacturers are now trying to replace synthetic fibers with natural fibers.

Extensive research is still required to develop EFB fiber-based composites. The microstructure of the interface between the oil palm fiber and matrix needs to be investigated and the interfacial properties studied with the single-fiber pull-out test, micro-bond test and single-fiber fragmentation test. Complementary techniques such as X-ray photoelectron spectroscopy and time-of-flight secondary ion mass spectrometry may reveal more information about the chemical composition of the surface, its wetting behavior, etc., which, when coupled with the mechanical assessment of the interface, will shed more light on the structural characteristics of the interface. The overall conclusion is that EFB fiber-reinforced polymeric composites will find application as cost-effective advanced composites, possessing (as required) superior mechanical properties, dimensional stability, stiffness, damping behavior and thermal stability. Future research on oil palm fiber-based hybrid composites will not be limited to its automotive applications but will also explore applications in aircraft components, the construction industry, rural housing and biomedicine.

12.8 References

Abdul Khalil, H.P.S. and Ismail, H. 2001. Effect of acetylation and coupling agent treatments upon biological degradation of plant fibre reinforced polyester composites. *Polymer Testing* 20: 65–75.

Abdul Khalil, H.P.S. and Rozman, H.D. 2004. *Gentian dan Komposit Lignoselulosik*, Universiti Sains Malaysia: Penang, Malaysia, 22.

Abdul Khalil, H.P.S., Rozman, H.D., Ahmad, M.N. and Ismail, H., 2000. Acetylated plant-fiber-reinforced polyester composites: a study of mechanical, hygrothermal, and aging characteristics. *Polymer Plastics Technology Engineering* 39: 757–781.

Abdul Khalil, H.P.S., Siti Alwani, M., Ridzuan, R., Kamarudin, H. and Khairul, A. 2008a. Chemical composition, morphological characteristics, and cell wall structure of malaysian oil palm fibers. *Polymer-Plastics Technology and Engineering* 47(3): 273–280.

Abdul Khalil, H.P.S., Nur Firdaus, M.Y., Anis, M. and Ridzuan, R.. 2008b. The effect of storage time and humidity on mechanical and physical properties of medium density fiberboard (MDF) from oil palm empty fruit bunch and rubberwood. *Polymer-Plastics Technology and Engineering* 47(10): 1046–1053.

Abdul Khalil, H.P.S., Kang, C.W., Khairul, A., Ridzuan, R. and Adawi, T.O. 2009. The effect of different laminations on mechanical and physical properties of hybrid composites. *Journal of Reinforced Plastics and Composites* 28(9): 1123–1137.

Abdul Khalil, H.P.S., Poh, B.T., Issam, A.M., Jawaid, M. and Ridzuan, R. 2010a. Recycled polypropylene–oil palm biomass: The effect on mechanical and physical properties. *Journal of Reinforced Plastics and Composites* 29(8): 1117–1130.

Abdul Khalil, H.P.S., Nur Firdaus, M.Y., Jawaid, M., Anis, M., Ridzuan, R. and Mohamed, A.R. 2010b. Development and material properties of new hybrid medium density fibreboard from empty fruit bunch and rubberwood. *Materials and Design* 31(9): 4229–4236.

Abdul Khalil, H.P.S., Nurul Fazita, M.R., Bhat, A.H., Jawaid, M. and Nik Fuad, N.A. 2010c. Development and material properties of new hybrid plywood from oil palm biomass. *Materials and Design* 31(1): 417–424.

Abu Bakar, A., Hariharan, and Abdul Khalil, H.P.S.. 2005. Lignocellulose-based hybrid bilayer laminate composite: Part I– Studies on tensile and impact behavior of oil palm fiber-glass fiber-reinforced epoxy resin. *Journal of Composite Materials* 39(8): 663–684.

Abu Bakar, A., Hassan, A. and Yusof, A.F.M. 2006. The effect of oil extraction of the oil palm empty fruit bunch on the processability, impact, and flexural properties of PVC-U composites. *International Journal of Polymer Materials* 55: 627–641.

Abu Hassan, O., Ishida, M., Mohamed Shukri, I. and Ahmad Tajuddin, Z. 1994. *Oil-Palm Fronds as a Roughage Feed Source for Ruminants in Malaysia*. Malaysia Agriculture Research and Development Institute (MARDI), *Kuala Lumpur*, Malaysia.

Abu Sharkh, B.F. and Hamid, H. 2004. Degradation study of date palm fibre/polypropylene composites in natural and artificial weathering mechanical and thermal analysis. *Polymer Degradation and Stability* 85: 967–973.

Ahmad, Z., Saman, H.M. and Tahir, P.M. 2010. Oil palm trunk fibre as a bio-waste resource for concrete reinforcement. *International Journal of Mechanical and Materials Engineering* 5(2): 199–207.

Alam, A.K.M.M., Beg, M.D.H., Reddy Prasad, D.M., Khan, M.R. and Mina, M.F. 2012. Structures and performances of simultaneous ultrasound and alkali treated oil palm empty fruit bunch fiber reinforced poly(lactic acid) composites. *Composites Part A: Applied Science and Manufacturing* 43: 1921–1929.

Amar, K.M., Manjusri, M. and Lawrence, T.D., (eds) 2005. *Natural Fibers, Biopolymers, and Biocomposites*. CRC Press, Taylor and Francis Group, Boca Raton, FL.

Andreopoulos, A.G., Tarantili, P.A. 1998. Water sorption characteristics of epoxy resin–UHMPE fibers composites, *Journal of Applied Polymer Science* 70: 747–755.

Arif, M.F., Yusoff, P.S.M.M. and Eng, K.K. 2010. Effect of reinforcement shape and fiber treatment on the mechanical properties of oil palm empty fruit bunch polyethylene composites, *AIP Conference Proceedings* 1217: 335.

Baratian S, Hall, E.S., Lin, J.S., Xu R. and Runt J. 2001. Crystallization and solid state structure of random polylactide copolymers: poly(L-lactide-*co*-D-lactide)s. *Macromolecules* 34: 4857–4864.

Beg, M.D.H. and Pickering, K.L. 2006. Fiber pre-treatment and its effects on wood fiber reinforced polypropylene composites. *Materials and Manufacturing Processes* 21: 303–307.

Ben Amor, I., Rekik, H., Kaddami, H., Raihane, M., Arous, M. and Kallel, A. 2010. Effect of palm tree fibre orientation on electrical properties of palm tree fibre-reinforced polyester composites. *Journal of Composite Materials* 44(13): 1553–1568.

Bismarck, A., Mishra, S. and Lampke, T. 2005. Plant fibres as reinforcement for green composites. In *Natural Fibres, Biopolymers and Biocomposites*, A.K. Mohanty, M. Misra and L.T. Drzal (eds). CRC Press, Taylor and Francis Group, Boca Raton, FL.

Bledzki, A.K. and Gassan, J. 1999. Composite reinforced with cellulose based fibres. *Progress in Polymer Science* 24(2): 221–274.

Botev, M., Betchev, A., Bikiaris, D. and Panayiotou, C. 1999. Mechanical properties and viscoelastic behavior of basalt fiber-reinforced polypropylene. *Journal of Applied Polymer Science* 74: 523–531.

Broido, A. 1969. A simple, sensitive graphical method of treating thermogravimetric analysis data. *Journal of Polymeric Science, Part A-2*, 7: 1761–1773.

Chai, L.L., Zakaria, S., Chia, C.H., Nabihah, S. and Rasid, R. 2009. Physico-mechanical properties of PF composite board from EFB fibres using liquefaction technique. *Iranian Polymer Journal* (English Edition) 18(11): 917–923.

Chen, Y., Chen Lin, H. and Lee, Y. 2003. The effects of filler content and size on the properties of PTFE/SiO_2 composites. *Journal of Polymer Research* 10: 247–258.

Cheung, H.Y., Ho, M.P., Lau, K.T., Cardona, F. and Hui, D. 2009. Natural fibre-reinforced composites for bioengineering and environmental engineering applications. *Composites Part B: Engineering* 40: 655–663.

Chew, T.L. and Bhatia, S. 2008. Catalytic processes towards the production of biofuels in a palm oil and oil palm biomass-based biorefinery. *Bioresource Technology* 99(17): 7911–7922.

Demir, H., Atikler, U., Balköse, D. and Tıhmınlıoğlu, F. 2006. The effect of fiber surface treatments on the tensile and water sorption properties of polypropylene–luffa fiber composites. *Composites Part A: Applied Science and Manufacturing* 37(3): 447–456.

Demirbaş, A. 2000. Mechanisms of liquefaction and pyrolysis reactions of biomass. *Energy Conversion and Management* 41(6): 633–646.

Desch, H.E. and Dinwoodie, J. M.1989. *Timber: Its Structure, Properties and Utilisation*, Chapter 9. Macmillan Press, London.

Dickison, W. 2000. *Integrative Plant Anatomy*. Harcourt Academic Press, New York, pp. 1–4.

Di Lorenzo M.L. 2006. Calorimetric analysis of the multiple melting behavior of poly (L-lactic acid). *Journal of Applied Polymer Science* 100: 3145–3151.

Eckert, C.H. 1999. Functional fillers for plastics: outlook to the year 2005. *Proceedings of the Fifth International Conference on Woodfiber-plastic Composites*. Proceedings No. 7263, Madison, WI, 26–28 May, 1999, pp. 10–22.

Eckert, C. 2000. Opportunities for natural fibers in plastic composites. *Proceedings of the Conference on Progress in Woodfibre-plastic Composites*, University of Toronto, Canada, 25–26 May 2000.

Foulc, M.P., Bergeret, A., Ferry, L., Ienny, P. and Crespy, A. 2005. Study of hygrothermal ageing of glass fibre reinforced PET composites. *Polymer Degradation and Stability* 89(3): 461–470.

Gao, C. and Yan, D. 2004. Hyperbranched polymers: from synthesis to applications. *Progress in Polymer Science* 29: 183–275.

Godara, A., Raabe, D., Bergmann, I., Putz, R. and Müller, U. 2009. Influence of additives on the global mechanical behavior and the microscopic strain localization in wood reinforced polypropylene composites during tensile deformation investigated using digital image correlation. *Composites Science and Technology* 69: 139–146.

Gowda, T.N., Naidu, A.C.B. and Chhaya, R. 1999. Some mechanical properties of untreated jute fabric reinforced polyester composites. *Composites Science and Technology* 30: 277–284.

Graupner, N., Hermann, A.S. and Mussig, J. 2009. Natural and man-made cellulose fibre-reinforced poly(lactic acid) (PLA) composites: An overview about mechanical characteristics and application areas. *Composites Part A: Applied Science and Manufacturing* 40: 810–821.

Hasamudin, W. and Soom, R.M. 2002. *Road making using oil palm fiber*. Malaysian Palm Oil Board Information Series, Kuala Lumpur, Malaysia, p. 171.

Hassan, A., Arshad A.S., Farid N.A. and Aznizam, A.B. 2010. A review on oil palm empty fruit bunch fiber-reinforced polymer composite materials. *Polymer Composites* 31 (12): 2079–2101.

Hill, C.A.S. and Abdul Khalil, H.P.S. 2000. Effect of fiber treatments on mechanical properties of coir or oil palm fiber reinforced polyester composites, *Journal of Applied Polymer Science* 78:1685–1697.

Hoogsteen, W., Postema, A.R., Pennings, A.J. and Brinke, G.T. 1990. Crystal structure, conformation, and morphology of solution-spun poly(L-lactide) fibers. *Macromolecules* 23: 634–642.

Inagaki, T., Siesler, H.W., Mitsui, K. and Tsuchikawa, S. 2010. Difference of the crystal structure of cellulose in wood after hydrothermal and aging degradation: A NIR spectroscopy and XRD study. *Biomacromolecules* 11: 2300–2305.

Jacob, M., Varughese, K.T. and Thomas, S. (2006). Dielectric characteristics of sisal oil palm hybrid biofibre reinforced natural rubber biocomposites. *Journal of Materials Science* 41: 5538–5547.

Jawaid, M., Abdul Khalil, H.P.S. and Abu Bakar, A. 2010. Mechanical performance of oil palm empty fruit bunches/jute fibres reinforced epoxy hybrid composites. *Materials Science and Engineering A* 527(29–30): 7944–7949.

Jawaid, M., Abdul Khalil, H.P.S. and Abu Bakar, A. 2011. Woven hybrid composites: Tensile and flexural properties of oil palm-woven jute fibres based epoxy composites. *Materials Science and Engineering A* 528(15): 5190–5195.

John, M.J. and Thomas, S.. 2008. Biofibres and biocomposites. *Carbohydrate Polymers* 71(3): 343–364.

John, M.J., Francis, B., Varughese, K.T. and Thomas, S. 2008. Effect of chemical modification on properties of hybrid fiber biocomposites. *Composites Part A: Applied Science and Manufacturing* 39(2): 352–363.

Joseph, P.V., Joseph, K., Thomas Pillai, C.K.S., Prasad, V.S., Groeninckx, G. and Sarkissova, M. 2003. The thermal and crystallization studies of short sisal fiber reinforced polypropylene composite. *Composites Part A: Applied Science and Engineering* 34(3): 253–266.

Joseph, S., Joseph, K. and Thomas, S. 2006. Green composites from natural rubber and oil palm fiber: Physical and mechanical properties. *International of Journal Polymer Materials* 55: 925–945.

Karina, M., Onggo, H., Abdullah, A.H.D. and Syampurwadi, A. 2008. Effect of oil palm empty fruit bunch fiber on the physical and mechanical properties of fiber glass reinforced polyester resin. *Journal of Biology Science* 8: 101–106.

Khalid, M., Ratnam, C.T., Chuah, T.G., Ali, S. and Choong, T.S.Y. 2008. Comparative study of polypropylene composites reinforced with oil palm empty fruit bunch fiber and oil palm derived cellulose. *Materials and Design* 29: 173–178.

Killman, W. and Lim, S.C. 1985. Anatomy and properties of oil palm stem. *Proceedings of the National Symposium on Oil Palm By-Products for Agro-Based Industries*, Malaysia.

Kim H.S., Lee, B.H., Choi, S.W., Kim, S. and Kim H.J. 2007. The effect of types of maleic anhydride-grafted polypropylene (MAPP) on the interfacial adhesion properties of bio-flour-filled polypropylene composites. *Composites Part A: Applied Science and Manufacturing* 38: 1473–1482.

Lai, C.Y., Sapuan, S.M., Ahmad, M. and Yahya, N. 2005. Mechanical and electrical properties of coconut coir fibre reinforced polypropylene composites. *Polymer-Plastics Technology and Engineering* 44: 619–632.

Law, K.N. and Jiang, X. 2001. Comparative papermaking properties of oil-palm empty fruit bunch. *TAPPI Journal* 84(1): 95–108.

Law, K.N., Daud, W.R.W. and Ghazali, A. 2007. Morphological and chemical nature of fiber strands of oil palm empty-fruit-bunch (OPEFB). *BioResources* 2(3): 351–362.

Lim, K.O. 1998. Oil palm plantations – A plausible renewable source of energy. *International Energy Journal* 20(2): 107–116.

Lin, Q., Zhou, X. and Dai, G. 2002. Effect of hydrothermal environment on moisture absorption and mechanical properties of wood flour-filled polypropylene composites. *Journal of Applied Polymer Science* 85(14): 2824–2832.

Liu, L.F., Yu, J.Y., Cheng, L. and Qua, W.W. 2009. Mechanical properties of poly(butylene succinate) (PBS) bio composites reinforced with surface modified jute fibre. *Composites Part A: Applied Science and Manufacturing* 40: 669–674.

Luz, S.M., Del Tio, J., Rocha, G.J.M., Goncalves, A.R. and Del'Arco, Jr. A.P. 2008. Cellulose and cellulignin from sugarcane bagasse reinforced polypropylene composites: Effect of acetylation on mechanical and thermal properties. *Composites Part A: Applied Science and Manufacturing* 39(9): 1362–1369.

Meier, D. and Faix, O. 1999. State of the art of applied fast pyrolysis of lignocellulosic materials–A review. *Bioresource Technology* 68(1): 71–77.

Minowa, T., Kondo, T. and Sudirjo, S.T. 1998. Thermochemical liquefaction of indonesian biomass residues. *Biomass and Bioenergy* 14(5–6): 517–524.

Mohamad, H., Zin Zawawi, Z. and Abdul Halim, H. 1985. Potentials of oil palm by-products as raw materials for agro-based industries. *Proceedings of the National Symposium on Oil Palm By-Products for Agro-Based Industries*, Malaysia.

Mohanty, A.K., Misra, M. and Drzal, L.T. 2001. Surface modifications of natural fibers and performance of the resulting biocomposites: an overview. *Composite Interfaces* 8: 313–343.

Mohanty, S., Nayak, S.K. and Verma, S.K. 2006. Dynamic mechanical and thermal properties of MAPE treated jute/HDPE composites. *Composites Science and Technology* 66(3–4): 538–547.

Nieschlag, H.J., Nelson, G.H., Wolff, J.A. and Perdue, R.E. 1960. A search for new fiber crops. *TAPPI Journal* 43(3): 193–201.

Nishiyama, Y., Langan, P. and Chanzy, H. 2002. Crystal structure and hydrogen-bonding system in cellulose Iβ from synchrotron X-ray and neutron fiber diffraction. *Journal of the American Chemical Society* 125: 14300–14306.

Norul Izani, M.A., Paridah, M.T., Anwar, U.M.K., Mohamed Nor, M.Y. and H'ng, P.S. 2013. Effects of fiber treatment on morphology, tensile and thermogravimetric analysis of oil palm empty fruit bunches fibers. *Composites Part B: Engineering* 45(1): 1251–1257.

Pan, P., Kai, W., Zhu, B., Dong, T. and Inoue, Y. 2007. Polymorphous crystallization and multiple melting behavior of poly(L-lactide): Molecular weight dependence. *Macromolecules* 40: 6898–6905.

Paul, A., Joseph, K. and Thomas, S. 1997. Effect of surface treatments on the electrical properties of low-density polyethylene composites reinforced with short sisal fibers. *Composites Science and Technology* 57: 67–79.

Peijs, T., Melick, V., Garkhail, S.K., Pott, G.T. and Baillie, C.A. 1998. Natural-fibre-mat reinforced thermoplastics based on upgraded flax fibres for improved moisture resistance. *ECCM-8 Conference*, Naples, Italy, pp. 119–126.

Pracella, M., Chionna, D., Anguillesi, I., Kulinski, Z. and Piorkowska, E. 2006. Functionalization, compatibilization and properties of polypropylene composites with hemp fibres. *Composites Science and Technology* 66(13): 2218–2230.

Punsuvon, P., Anpanurak, W., Vaithanomsat, P. and Tungkananuruk, N. 2005. Fractionation of chemical components of oil palm trunk by steam explosion. *31st Congress on Science and Technology of Thailand*, Suranaree University of Technology, Thailand.

Rahman, M.D.R., Huque, M.D.M., Islam, M.D.N. and Hasan, M. 2009. Mechanical properties of polypropylene composites reinforced with chemically treated abaca. *Composites Part A: Applied Science and Manufacturing* 40(4): 511–517.

Ramli, R., Yunus, R.M., Beg, M.D.H. and Alam, A.K.M.M. 2011, Effects of different coupling agents on the mechanical and interfacial properties of oil palm fiber reinforced polypropylene composites. *Journal of Reinforced Plastics and Composites* 30(4): 301–308.

Ratnasingam, J. 2011. Oil palm biomass utilization – Counting the successes in Malaysia. Available from http://www.woodmagmagazine.com/node/417

Raveendran, K., Ganesh, A. and Khilar, K.C. 1995. Influence of mineral matter on biomass pyrolysis characteristics. *Fuel* 74(12): 1812–1822.

Reddy, N. and Yang, Y. 2005. Biofibers from agricultural byproducts for industrial applications. *Trends in Biology* 23(1), 22–27.

Rosnah, M.Y., Ghazali, A., Wan Rosli, W.D. and Dermawan, Y.M. 2010. Influence of alkali peroxide treatment duration on the pulpability of oil palm empty fruit bunch. *World Applied Sciences Journal* 8(2): 185–192.

Rowell, R.M., Han, J.S. and Rowell, J.S. 2000. Characterization and factors effecting fiber properties, in *Natural Polymers and Agrofibres Composites*, E. Frollini, A.L. Leo and L.H.C. Mattoso (eds), Embrapa Instrumentacao Agropecuaria, San Carlos, Brazil, pp. 115–134.

Rozman, H.D., Lim, P.P., Abusamah, A., Kumar, R.N., Ismail, H. and Mohamed Ishak, Z.A. 1999. The physical properties of oil palm empty fruit bunch (EFB) composites made from various thermoplastics. *International Journal of Polymeric Materials* 44(1–2): 179–195.

Rozman, H.D., Lai, C.Y., Ismail, H. and Mahamed Ishak, Z.A. 2000. The effect of coupling agents on the mechanical and physical properties of oil palm empty fruit bunch–polypropylene composites. *Polymer International* 49: 1273–1278.

Rozman, H.D., Tay, G.S., Kumar, R.N., Abusamah, A., Ismail, H. and Mohamed Ishak, Z.A. 2001a. Polypropylene–oil palm empty fruit bunch–glass fiber hybrid composites: a preliminary study on the flexural and tensile properties. *European Polymer Journal* 37: 1283–1291.

Rozman, H.D., Tay, G.S., Abubakar, A. and Kumar, R.N. 2001b. Tensile properties of oil palm empty fruit bunch–polyurethane composites. *European Polymer Journal* 37(10): 1759–1765.

Rozman, H.D., Saad, M.J. and Ishak, Z.A. 2003. Flexural and impact properties of oil palm empty fruit bunch (EFB)-polypropylene composites – the effect of maleic anhydride chemical modification of EFB. *Polymer Testing* 22(3): 335–341.

Saheb, D. Nabi and Jog, J. P. 1999. Natural fiber polymer composites: a review. *Advances in Polymer Technology* 18(4), 351–363.

Shinji, O. 2008. Mechanical properties of kenaf fibers and kenaf/PLA composites. *Mechanics of Materials* 40: 446–452.

Shinoj, S., Visvanathan, R. and Panigrahi, S. 2010. Towards industrial utilization of oil palm fibre: Physical and dielectric characterization of linear low density polyethylene composites and comparison with other fibre sources. *Biosystem Engineering* 106: 378–388.

Shinoj, S., Visvanathan, R., Panigrahi, S. and Kochubabu, M. 2011. Oil palm fiber (OPF) and its composites: a review. *Industrial Crops and Products* 33(1): 7–22.

Shuit, S.H., Tan, K.T., Lee, K.T. and Kamaruddin, A.H. 2009. Oil palm biomass as a sustainable energy source: a Malaysian case study. *Energy* 34(9): 1225–1235.

Son, J., Kim, H.-J. and Lee, P.-W. 2001. Role of paper sludge particle size and extrusion temperature on performance of paper sludge–thermoplastic polymer composites. *Journal of Applied Polymer Science* 82(11): 2709–2718.

Sreekala, M.S., Kumaran, M.G. and Thomas, S. 1997. Oil palm fibers: Morphology, chemical composition, surface modifications and mechanical properties. *Journal of Applied Polymer Science* 66: 821–835.

Sreekala, M.S., George, J. Kumaran, M.G. and Thomas, S. 2001. Water-sorption kinetics in oil palm fibers. *Journal of Polymer Science, Part B: Polymer Physics* 39(11): 1215–1223.

Sreekala, M.S., Kumaran, M.G., Geethakumariamma, M.L. and Thomas, S. 2004. Environmental effects in oil palm fiber reinforced phenol formaldehyde composites: Studies on thermal, biological, moisture and high energy radiation effects. *Advanced Composite Materials: The Official Journal of the Japan Society of Composite Materials* 13(3–4):171–197.

Sun, R.C., Fang, J.M. and Tomkinson, J. 1999. Fractional isolation and structural characterization of lignins from oil palm trunk and empty fruit bunch fibers. *Journal of Wood Chemistry and Technology* 19(4): 335–356.

Suryanegara, L., Nakagaito, A.N. and Yano, H. 2009. The effect of crystallization of PLA on the thermal and mechanical properties of micro-fibrillated cellulose-reinforced PLA composites. *Composites Science Technology* 69: 1187–1192.

Thattaiparthasarthy, K.B., Pillay, S. and Vaidya, U.K. 2009. Rheological characterization of long fiber thermoplastics – Effect of temperature, fiber length and weight fraction. *Composites Part A: Applied Science and Manufacturing* 40: 1515–1523.

Ververis, C., Georghiou, K., Christodoulakis, N., Santas, P. and Santas, R. 2004. Fiber dimension, lignin and cellulose content of various plant materials and their suitability for paper production. *Industrial Crops Production* 19: 245–254.

Wirjosentono, B., Guritno, P. and Ismail, H. 2004. Oil palm empty fruit bunch filled polypropylene composites. *International Journal of Polymer Materials* 53: 295–306.

Wu, T.M. and Wu, C.Y. 2006. Biodegradable poly(lactic acid)/chitosan-modified montmorillonite nanocomposites: Preparation and characterization. *Polymer Degradation and Stability* 91: 2198–2204.

Xie, Y., Hill, C.A.S., Xiao, Z., Militza, H. and Maia, C. 2010. Silane coupling agents used for natural fiber/polymer composites: a review. *Composites Part A: Applied Science and Manufacturing* 41: 806–819.

Yang, H.S., Kim, H.J., Park, H.J., Lee, B.J. and Hwang, T.S. 2007. Effect of compatibilizing agents on rice-husk flour reinforced polypropylene composites. *Composite Structures* 77: 45–55.

Yasuniwa, M., Tsubakihara, S., Sugimoto, Y. and Nakafuku, C. 2004. Thermal analysis of the double-melting behavior of poly(L-lactic acid). *Journal of Polymer Science Part B: Polymer Physics* 42: 25.

Yew, G.H., Mohamed Yusof, A.M., Mohamed Ishak, Z.A. and Ishiaku, U.S. 2005. Water absorption and enzymatic degradation of poly(lactic acid)/rice starch composites. *Polymer Degradation and Stability* 90: 488.

Zaidon, A., Norhairul Nizam, A.M., Mohamed Nor, M.Y., Abood, F., Paridah, M.T., Nor Yuziah, M.Y. and Jalaluddin, H.. 2007. Properties of particleboard made from pretreated particles of rubberwood, EFB and rubberwood–EFB blend. *Journal of Applied Sciences* 7(8): 1145–1151.

Part IV

Grass, reed and cane fibers

13
The use of rice straw and husk fibers as reinforcements in composites

M. BASSYOUNI, King Abdulaziz University, Saudi Arabia and Higher Technological Institute, Egypt and S. WAHEED UL HASAN, King Abdulaziz University, Saudi Arabia

DOI: 10.1533/9781782421276.4.385

Abstract: Rice straw and rice husk fiber are underutilized agricultural residues with potential for use in reinforced polymer composites that would save existing wood and petroleum resources. With some brief description of rice plants and their cultivation, harvesting and milling, this chapter discusses methods of testing the properties of rice straw and rice husk fibers, fiber surface treatments and the use of coupling agents for the integration of fibers into a polymer matrix, the processing of thermoset and thermoplastic composites reinforced by rice straw and rice husk fibers, and the physical and mechanical properties of such composites. Differences in properties between rice straw composites and rice husk composites respectively are described. It is also noted that much research on coupling agents remains to be done.

Key words: coupling agent, fiber surface treatment, natural fiber, rice husk, rice straw, thermoset composite, thermoplastic composite.

13.1 Introduction

Rice straw and rice husk (RS/RH) belong to the family of non-wood biofibers. Rice straw is an agricultural production residue whereas rice husk, obtained off-field, is an agricultural processing residue. As RS/RH are agricultural residues with surplus availability, it is more economical to use them than wood fibers in the production of biocomposites. Rice residue is plentiful, which is boosting production of RS/RH composites, and yet it is still underutilized in many parts of the world. Rice residue is either disposed of as waste by being burned in the field or used as animal food and bedding material. Embedding rice straw and rice husk in polymer composites reduces environmental pollution and uses waste as a resource to generate revenue. A trend to used rice straw and rice husk fiber in place of wood will save existing wood and petroleum resources.

A composite is a combination of two or more materials held together by physical entanglement or a chemical matrix. Thermosets and thermoplastics are used as polymer matrices for RS/RH composites. Polyethylene and polypropylene are well-known thermoplastic matrices used with rice straw and rice husk. Major industrial uses of RS/RH are found in the paper and pulp industries, combustion for energy generation, housing, furniture and packaging. Cementitious materials, in which rice husk has its main use, serve as matrices in ceramic composites such as rice husk ash-filled cement and rice husk-filled gypsum wall and ceiling panels. The success of the rice straw composite industry is evidenced by an increasing use of rice straw particle board and medium and high density fiberboards; it is because of these applications that rice straw finds greater use than rice husk on an industrial scale.

Different trends are observed in natural fiber composite markets around the globe. In the United States and Asia, decking and rail products dominate the natural fiber composite market, whereas in Europe there is an emphasis on automotive parts such as dashboards and car roofs.

13.2 Cultivation and processing of rice straw and rice husk

The physical and chemical traits of rice plant varieties differ significantly. A study of various rice varieties confirms the unique sets of physical and chemical traits such as plant height, architecture and cell wall composition (Jahn *et al.*, 2011). Variation in the composition of cellulose, hemicellulose and lignin in rice straw and rice husk with different rice varieties results in different mechanical properties of the composites upon integration into the polymer matrix. Changes in the physical traits of straw are more prominent in different rice varieties than changes in chemical composition, which are only marginal. For a given rice variety, quality control of rice straw and husk is applied mainly at the fiber processing stage, as the harvesting methods are very basic in nature.

Rice is harvested mechanically or manually, the latter being more common in Asia. Although manual harvesting is labor intensive, it is very effective when the crop has fallen over. Mechanical harvesting is done by reaper or combine harvester. A combine harvester separates out husked rice from rice straw and piles the straw behind it as it moves forward. Balers collect the straw and gather it into round or rectangular bales.

The processing of rice straw begins with the opening of the bales. If the straw is not precut when harvested, it is cut to a length of around 400 mm when the bale twine is removed from the circular bales. There are three methods for opening a bale: guillotine, hopper feeder, and bale cutter. A hopper feeder accepts only rectangular bales with precut straw of around 400 mm length but delivers uniform straw with a steady output. A guillotine and a bale cutter accept both round and rectangular bales for opening but deliver uneven straw with a sporadic output.

In the guillotine, a hydraulically pressurized blade cuts the bales into segments without removing the bale twine. It has the highest capacity of all the three methods for bale opening but also requires more maintenance. The bale opener is a medium-capacity machine consisting of a bale-holding and turning mechanism, a self-sharpening and driven cutting tool, and a discharging conveyor. The bale-holding mechanism receives a bale from the feeding system and places it in a cutting position for the tool to cut strips of the required dimensions from the surface. The dimension of the strips determines the capacity of the bale cutter. The strips are carried out of the cutter on the discharge conveyor.

Rice straw bales may contain metallic objects and stones which might damage sensitive equipment in downstream parts of the process. Electromagnetic metal detectors are used to detect steel objects of various sizes, depending upon the sensitivity that has been set for the system. Magnetic separators can be used to separate ferrous objects from rice straw. Specially designed conveyors let the stones, loose seed, metallic objects and other dense impurities fall off under gravity, but around 25% of the stones are stuck inside the straw and carried along. Stones can also be detected electronically when a high level of purity is required, but the technology is expensive and rarely used.

The downstream machinery at a fiber processing plant requires a steady flow of rice straw, and hence a mass-flow metering mechanism. This consists of an inclined belt conveyor with a retaining rake above it which controls the thickness of the straw bed, thus ensuring a steady flow. In the next stage, the rice straw is ground to the desired particle size in a rotary knife cutter (discussed below). Rice husk is ground in the same way. After grinding, both rice straw and rice husk are washed with water to remove dust and impurities and dried in an oven at around 80°C.

13.3 Key fiber properties

13.3.1 Structure of rice straw and rice husk

Rice straw consists of jointed hollow culms, leaf blades and threshed-out panicles. The main rice culm, also called the rice stem, is made up of a series of round, hollow internodes separated by solid nodes. The leaves and buds originate from the node. The internode length and diameter change from the base to the top. The base internodes of the culm are short and strong where it grows into a thick, solid mass, and the lower internodes of the plant are thicker and greater in diameter than the upper internodes. The length of rice straw depends upon the rice species and cultivation method, the commercial varieties usually varying from 1 to 2 m.

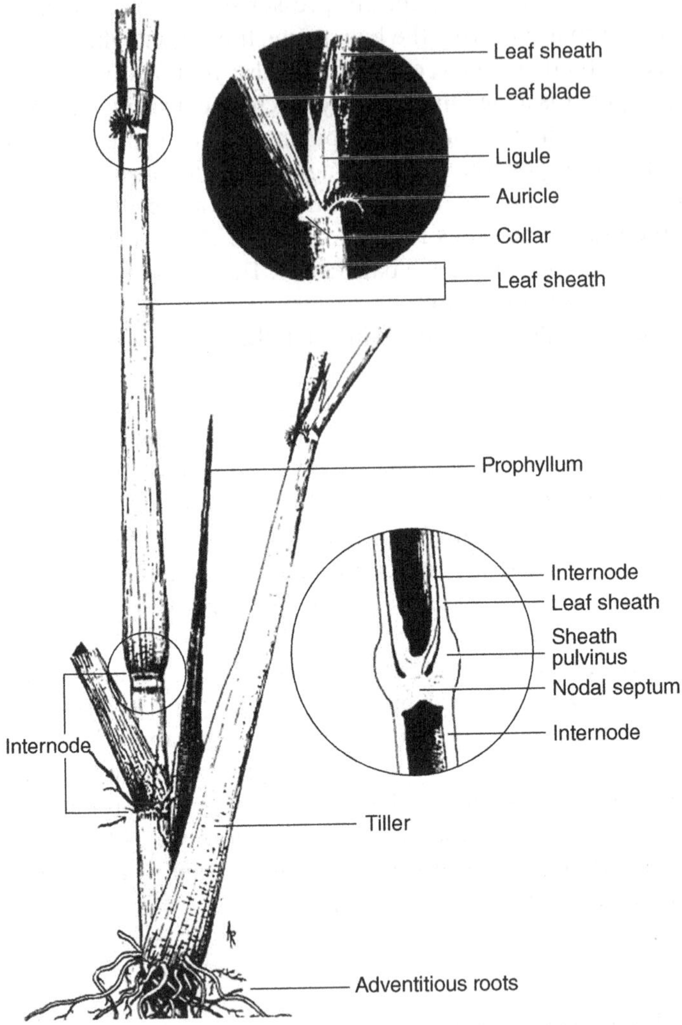

13.1 Rice plant structure (from Chang, T.-T. and Bardenas, E. A. 1965. *The Morphology and Varietal Characteristics of the Rice Plant.* Los Baños, Laguna, The Philippines: The International Rice Research Institute. Used with permission).

The branches of the rice plant originating from nodes of the main culm are called tillers. The primary tiller develops from the lower node of the stem, which may give rise to secondary and tertiary tillers, shown in Fig. 13.1. A tiller may or may not hold a panicle (the rice-bearing part of the plant). During the threshing process, rice grains along with husks are separated from panicles and the threshed-out panicles stay with the rice straw.

The main culm of the rice plant holds the greatest number of leaves. The leaves grow in two forms: the leaf blade and the leaf sheath. The leaf sheath originates from a node and envelops the internode above it and the succeeding portion of the culm, depending upon its length. The leaf blades are connected to the culm at a greater angle and are flat in shape. The shape and size of the leaf blades vary in different rice varieties. There exists a distinct leaf below the panicle which is different in shape, size and angle than the other leaves. This uppermost leaf is called the flag leaf.

During milling, rice grains are separated from rice husks. Scanning electron microscopy (SEM) micrographs of a well-dried rice husk are shown in Fig. 13.2. The epidermis contains well-structured corrugated cuticles, as shown in Fig. 13.2(a). When magnified, the outer surface reveals spiky trichomes, as shown in Fig. 13.2(b). These trichomes are mostly made of a form of silica called opaline silica, which is hydrated and amorphous in form. The silica content of the outer surface of the rice husk makes it abrasive and resistant to moisture. The high mineral content and type of lignin (different from that in wood) in rice husk reduces its biodegradability. The inner surface of husk is very smooth as shown in Fig. 13.2(c). SEM micrographs of traverse and longitudinal sections of a rice straw stem are shown in Fig. 13.3.

13.3.2 Physical, chemical and mechanical properties of rice straw and rice husk

The composition of washed, ground and screened (20–28 mm) rice straw and rice husk is given in Table 13.1.

The major constituent of rice straw and rice husk is cellulose, a long-chained polysaccharide consisting of 7000–15,000 glucose monomers. Glucose monomers go through a condensation reaction to form polysaccharide chains, staggering the bonding oxygen in the middle. Long stiff chains of polysaccharides associate together to form crystalline cellulose, shown in Fig. 13.4. Rice straw is a reinforced composite of cellulose fibers aligned to form microfibrils with a diameter of 3–4 nm. The microfibrils are bound together by hemicelluloses and lignin to form macrofibrils with a diameter of around 10–25 nm.

Table 13.1 Chemical composition (wt%) of rice straw and rice husk

Material	Cellulose	Hemicellulose	Lignin	Ash
Rice straw	38.3	28.0	14.9	18.8
Rice husk	45	19	19.5	15

Source: Reprinted from Zhang, Q. and Cai, W. 2008. Enzymatic hydrolysis of alkali-pretreated rice straw by *Trichoderma reesei* ZM4-F3. *Biomass and Bioenergy*, 32, 1130–1135. With permission from Elsevier.

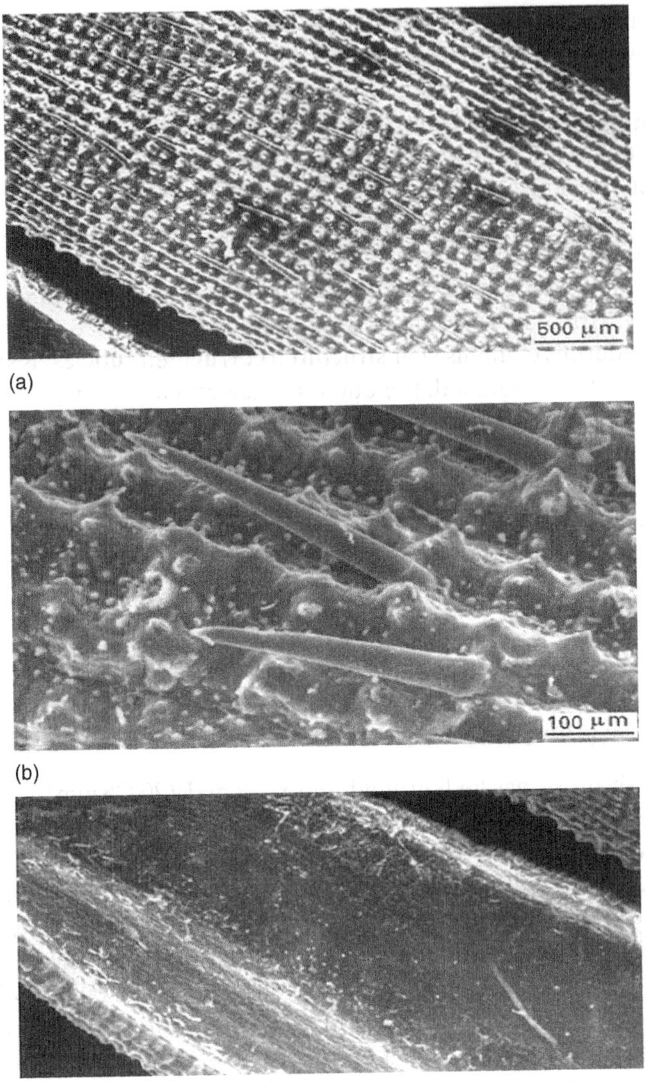

13.2 SEM micrographs of a rice husk: (a) and (b) outer surface and (c) inner surface (reprinted from Ryu, S.-E., Kim, T.-N. and Kang, T.-K. 1997. Pulverization of rice husks and the changes of husk densities. *Journal of Materials Science*, 32, 6639–6643. With kind permission from Springer Science and Business Media).

The microfibril structure is shown in Fig. 13.5. Lignin is an amorphous phenolic compound with a complex structure. Hemicellulose consists of 500–3000 monomers of amorphous polysaccharides with acidic groups, which is shorter in length compared to cellulose polymer. Table 13.2 presents

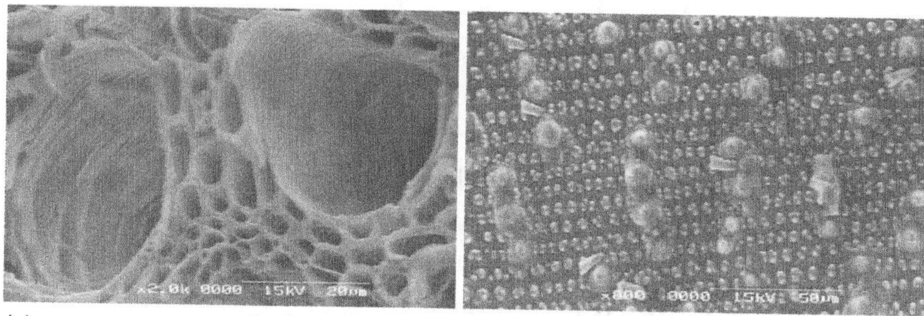

(a) (b)

13.3 SEM micrographs of a rice straw stem: (a) transverse section and (b) longitudinal section (reprinted from Zhang, Q. and Cai, W. 2008. Enzymatic hydrolysis of alkali-pretreated rice straw by *Trichoderma reesei* ZM4-F3. *Biomass and Bioenergy*, 32, 1130–1135. With permission from Elsevier).

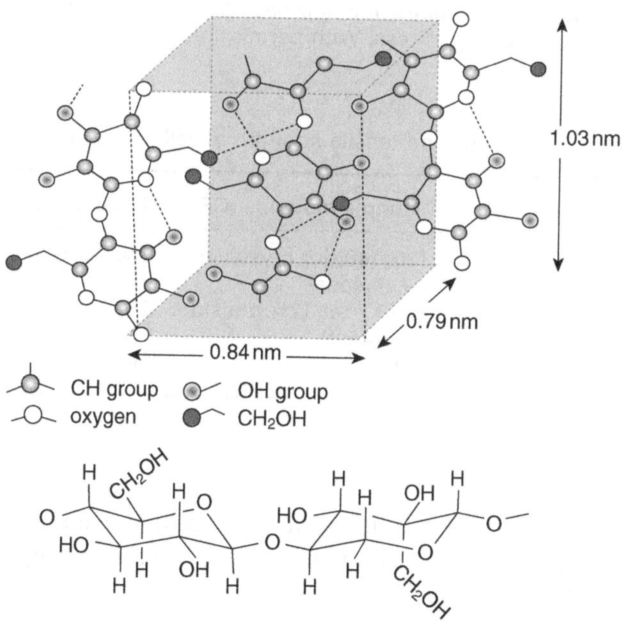

13.4 Structure of cellulose (reprinted from Gibson, L. J., Ashby, M. F. and Harley, B. A. 2010. *Cellular Materials in Nature and Medicine*. New York, Cambridge University Press. With permission).

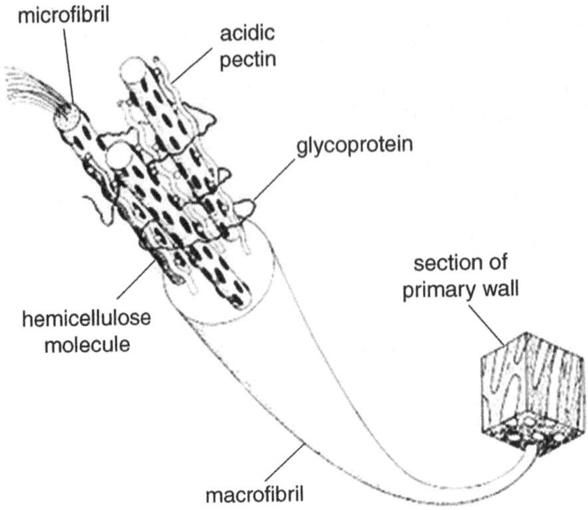

13.5 Structure of a macrofibril (reprinted from Niklas K. J. 1992. *Plant Biomechanics: an engineering approach to plant form and function*, University of Chicago Press. With permission).

Table 13.2 Young's modulus and tensile strength of cellulose and lignin

Material	Young's modulus (GPa)	Tensile strength (MPa)
Cellulose (along fiber length)	120–140 (for crystalline cellulose)	750–1080
Lignin	2.5–3.7 (at 12% moisture content)	25–75
Hemicellulose	5–8	N/A

Source: Klyosov, 2007.

the Young's modulus and tensile strength of cellulose and lignin, and Young's modulus data only for hemicellulose (tensile strength is not available). The Young's modulus data for cellulose correspond to strained and unstrained bleached ramie fibers and have been obtained through X-ray diffraction. The data for hemicellulose and lignin were obtained by performing unconfined compression tests on palletized, isolated powders of both species. The tensile strength of cellulose was obtained by performing tensile tests on high cellulose-concentration fibers.

Although the nature of the lignin present in rice husk differs significantly from that in wood, the chemical composition of rice husk is closer to that of wood and its ash consists mainly of silica. There is only a small amount

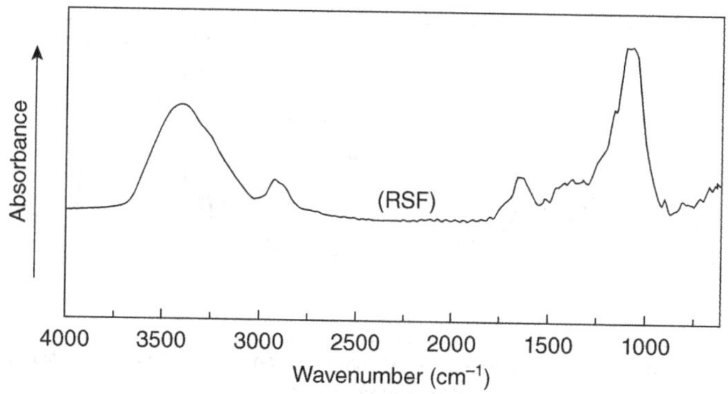

13.6 FTIR of rice straw fiber (adapted from Zhao, Y., Qiu, J., Feng, H. and Zhang, M. 2012. The interfacial modification of rice straw fiber reinforced poly(butylene succinate) composites: Effect of aminosilane with different alkoxy groups. *Journal of Applied Polymer Science*, 125, 3211–3220. With kind permission from John Wiley and Sons, Inc.).

of calcium present in rice husk; the amount of calcium attached to silica is around one hundredth by weight.

Fourier transform infrared spectroscopy (FTIR)

Figure 13.6 shows the Fourier transform infrared spectroscopy (FTIR) analysis of rice straw fiber (RSF). The broad absorption band from 3600 to 3200 cm^{-1} is a result of the O–H bond (stretching) and the OH…O hydrogen bond between the molecules. A steep absorption band around 1100 cm^{-1} derives from the vibrations of the glycosidic bond (C–O–C). This band is overlapped by C–OH (stretching vibrations) in cellulose, hemicellulose and lignin (C–O–C stretching at 1000–1160 cm^{-1}, C–O stretching of the primary hydroxyl group at 1000–1060 cm^{-1}, C–O stretching of the secondary hydroxyl group at 1070–1120 cm^{-1}, and C–O stretching of the tertiary hydroxyl group at 1050–1160 cm^{-1}).

Figure 13.7 shows the FTIR analysis of steam-treated rice husk. A medium-broad absorption band is found in the region of 3421 cm^{-1}, which is due to the stretching vibration of intermolecular hydrogen-bonded –OH groups in the cellulose fibers. An absorption peak around 1630 cm^{-1} is also due to –OH groups. Weak bands in the absorption region around 2919 cm^{-1} correspond to the vibration of carbon–hydrogen bonds superimposed on the –OH broad band around 3300 cm^{-1} (Ndazi *et al.*, 2007). Vibration of the carbonyl group within the carboxylic group in ester linkages or due to wax in natural fats is observed at 1735 cm^{-1}. The absorption peak around 1201–1156 cm^{-1} occurs in the vibration range of silica bonds.

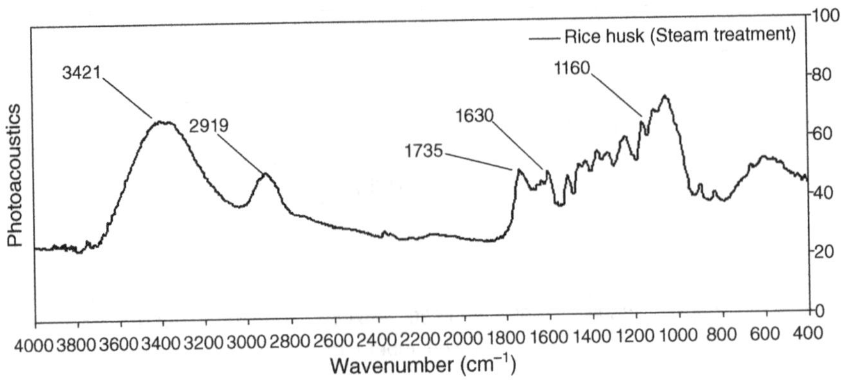

13.7 FTIR spectrum of steam-treated rice husk.

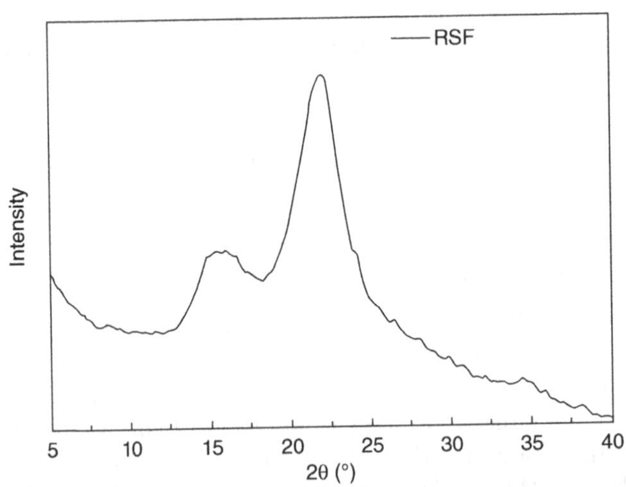

13.8 XRD profile of rice straw fibers (adapted from Zhao, Y., Qiu, J., Feng, H. and Zhang, M. 2012. The interfacial modification of rice straw fiber reinforced poly(butylene succinate) composites: Effect of aminosilane with different alkoxy groups. *Journal of Applied Polymer Science*, 125, 3211–3220. With kind permission from John Wiley and Sons, Inc.).

X-ray diffraction (XRD)

Chemical surface treatment of natural fibers might alter their crystallinity. The crystallinity of pure and silane-treated rice straw fibers was studied by Zhao *et al.* (2012) through X-ray diffraction (XRD) analysis, and it was found that the crystallinity index of silane-treated rice straw fibers does not differ significantly from that of untreated rice straw fibers. This confirms that the silanol does not percolate into the crystalline cellulose. Figure 13.8

shows an XRD profile of rice straw fiber. Distinct peaks are observed at 2θ values of 16°, 22° and 34.5°, representing the 110, 002 and 023 crystallographic planes respectively.

Thermal properties

The thermal properties of rice straw and rice husk are important as they limit the curing and extrusion temperatures for thermosets and thermoplastics. Figure 13.9 presents the thermogravimetric analysis (TGA) of rice straw and rice husk, along with the TGA of other natural fibers. The TGA of rice straw and rice husk are similar as both are lignocellulosic materials with almost the same chemical composition. In the rice straw TGA shown in Fig. 13.9(b), two mass-loss steps are observed. The first step represents the drying of the rice straw up to 150°C. The second step starts at 200°C and proceeds at a much higher rate, lasting until all the cellulose, hemicellulose and lignin have decomposed. A similar trend is observed in the rice husk TGA shown in Fig. 13.9(a). Rice husk is seen to be thermally more stable than rice straw, with thermal decomposition starting at around 230°C.

Physical dimensions of rice straw and rice husk

Rice husk is sold by industry as bulk, medium particles or fine particles, with around 10% moisture (Klyosov, 2007). The particle sizes used in composites vary from coarsely ground to fine flour. The different particle sizes are set by the rotating knife cutter that mills the rice straw and rice husk. Rice husk fed from the hopper feeder is cut against a set of stationary knives by high-speed rotary knives, as shown in Fig. 13.10. A screen at the base of the cutting chamber grinds and sieves the rice husk to the desired particle size. The same cutter can be used for rice straw during the processing of the composite, after the bales of rice straw have been cut to a predefined length using a bale cutter.

13.4 Composite processing: surface treatment

Strong adhesion between rice straw or rice husk and the matrix enhances stress transfer and load distribution all along the interface. Therefore, strategies have been devised for surface modification of the rice residue fibers to increase this interface adhesion and thus improve the mechanical properties of the composite. A coupling agent coats the fiber surface, bonding chemically with both the fiber and the matrix. This increases the strength of the composite. A treated surface binds better with the coupling agent. There are various methods available for surface modifications, such

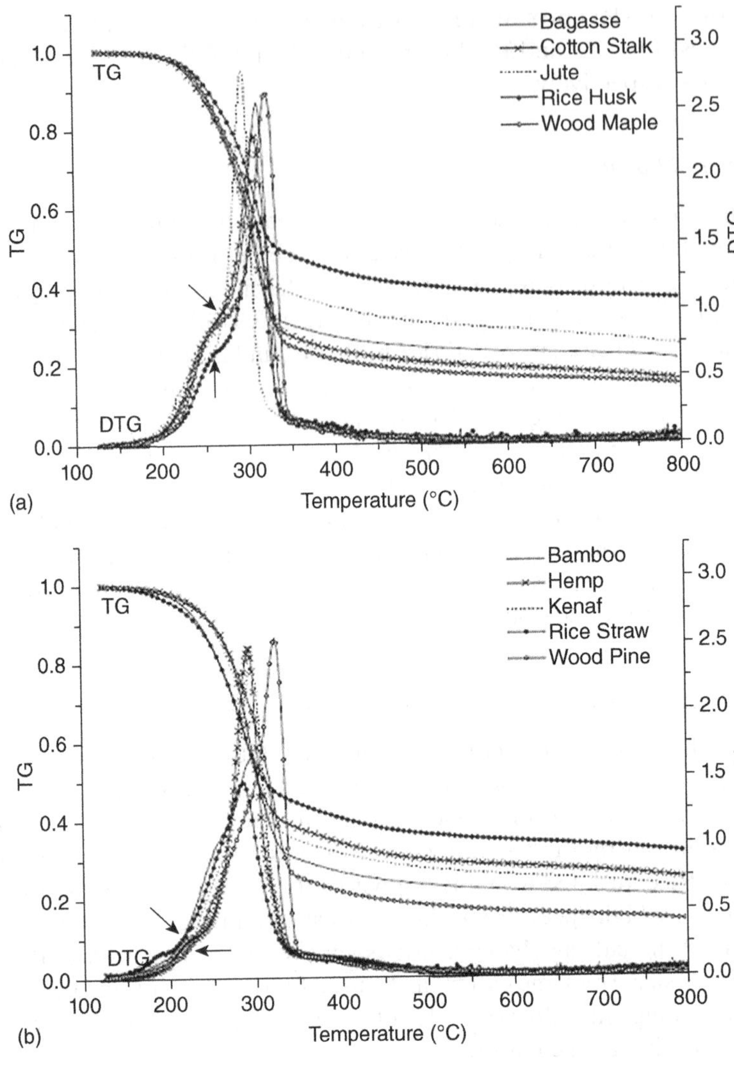

13.9 TGA of natural fibers including (a) rice husk and (b) rice straw (reprinted from Yao, F., Wu, Q., Lei, Y., Guo, W. and Xu, Y. 2008. Thermal decomposition kinetics of natural fibers: Activation energy with dynamic thermogravimetric analysis. *Polymer Degradation and Stability*, 93, 90–98. With permission from Elsevier).

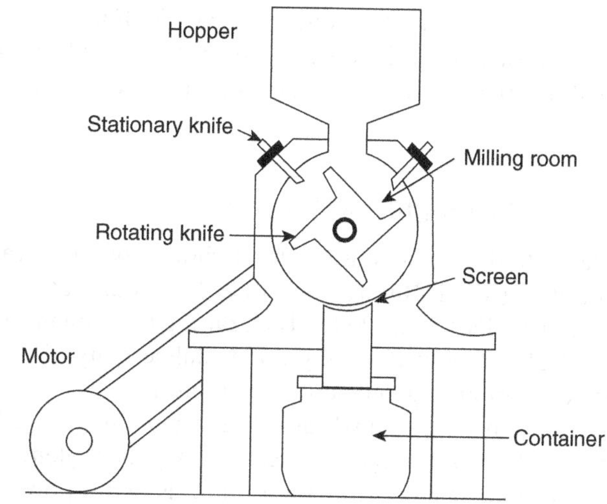

13.10 Rotating knife cutter (reprinted from Ryu, S.-E., Kim, T.-N. and Kang, T.-K. 1997. Pulverization of rice husks and the changes of husk densities. *Journal of Materials Science*, 32, 6639–6643. With kind permission from Springer Science and Business Media).

as physical and chemical treatments, mercerization, fiber impregnation, and silane treatments.

13.4.1 Mercerization

Mercerization is the most widely used surface treatment methodology for rice straw. Many studies report use of an alkali, usually NaOH, for surface modification of rice straw and husk (Liao *et al.*, 2011; Sherif *et al.*, 2009). Maleic anhydride is usually the coupling agent. The concentration of alkali, the duration of treatment, and the temperature during treatment determine the mechanical properties of the rice straw composite. Alkali treatment causes fibrillation of the straw by releasing lignin and hemicellulose. Increases in alkali strength and duration of treatment increase the tensile and flexural strength of the composite (Liao *et al.*, 2011). For production of rice husk composite panels, both NaOH and steam treatments are used. Steam explosion removes the fibrous matter from the rice husk, whereas dry grinding produces segmented particles only (Park *et al.*, 2004). Steam treatment is used mostly for production of straw particle boards and husk composite panels. Steam-treated phenol formaldehyde rice husk panels provide a better modulus of rupture and elasticity than alkali-treated rice husk, due to improved interfacial bonding (Ndazi *et al.*, 2007). For rice straw particleboards, steam-treated rice straw gives better results in comparison with straw treated chemically (e.g. with oxalic acid; Li *et al.*, 2011). Therefore, in both cases, of particleboards and of composite panels (Fuad *et al.*, 1995), rice straw and rice husk behave identically: steam-treated fiber leads to better mechanical properties of the composite than does chemically treated fiber.

13.4.2 Maleic anhydride-grafted polymers

Maleic anhydride-grafted polymers are a usual choice for rice straw and rice husk composites (Ming-Zhu *et al.*, 2011; Bassyouni *et al.*, 2012; Petchwattana *et al.*, 2012; Park *et al.*, 2004). The grafting reaction mechanism is shown in Fig. 13.11 for polyethylene-grafted maleic anhydride (PE-g-MAH) in the presence of an organic peroxide initiator.

Heating polyethylene in maleic anhydride solution generates polyethylene macroradicals by reaction of peroxide radicals with polyethylene and by reaction of polyethylene–maleic anhydride radicals with polyethylene chains (see Fig. 13.11). These macroradicals may set off the polymerization of maleic anhydride or join the developing chains of poly(maleic anhydride). The polypropylene-grafted maleic anhydride (PP-g-MAH) is formed by the same mechanism, only with the polymer chains being polypropylene instead. It acts as a bridge between the RS/RH and the polymer matrix where the polypropylene chains embed into the polymer matrix and maleic anhydride connects with hydroxyl groups in the fibers, as shown in Fig. 13.12.

13.4.3 Silane treatment

According to chemical bonding theory, silane molecules can act as a bridge between cellulose and resin through simultaneous chemical bonding with both of them. Siloxane bonds form with the surface of the cellulose, while organofunctional groups bond to the resin matrix, thus creating a link between the cellulose fibers and the polymer matrix (Martí-Ferrer *et al.*, 2006). Natural fibers such as rice straw and rice husk interact with silane in two steps. In the first step, silane monomers are converted into silanol reactive groups upon hydrolysis, liberating alcohol molecules. In the second step, silanol and natural fibers develop interlinking through self-condensation, adsorption and chemical grafting, as shown in Fig. 13.13. The choice of terminal functional groups of silane along with the processing conditions and fiber surface determine the strength of composite. Zhao *et al.* (2012) discussed the treatment of rice straw with four different silanes (see Table 13.3) for embedding into poly(butylene succinate) and found that 3-(2-aminoethylaminopropyl)-triethoxysilane-rice straw fiber–poly(butylene succinate) (AEAPTES-RSF-PBS) composite showed the best mechanical properties.

Figure 13.14 compares the FTIR spectra of untreated and silane-treated rice straw fiber samples. FTIR spectra of AEAPTMES-RSF and

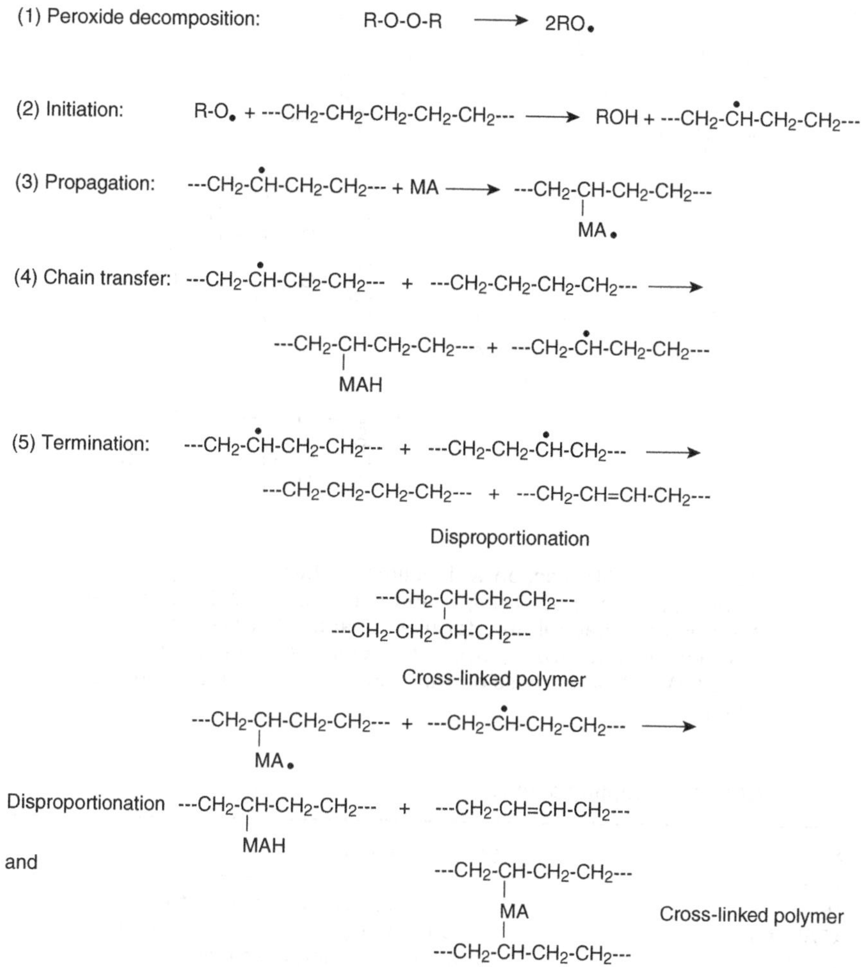

13.11 Grafting mechanism for polyethylene grafted maleic anhydride in the presence of a peroxide initiator (reprinted from Ghaemy, M. and Roohina, S. 2003. Grafting of maleic anhydride on polyethylene in a homogeneous medium in the presence of radical initiators. *Iranian Polymer Journal*, 12, 22–29. With permission).

APTMES-RSF show a clear peak split around 1100 cm^{-1}. The new peak, appearing at the lower wavenumber, arises from stretching vibrations of Si–O–Si bonds. The Si–O–Si peak becomes less prominent in APTES-RSF and almost indistinguishable in AEAPTES-RSF. This might be a result of increased chemical grafting of hydroxyl groups of rice straw fiber, leading to overlapping of C–O–C and Si–O–Si peaks by Si–O–C, which is in accordance with ξ-potential analysis.

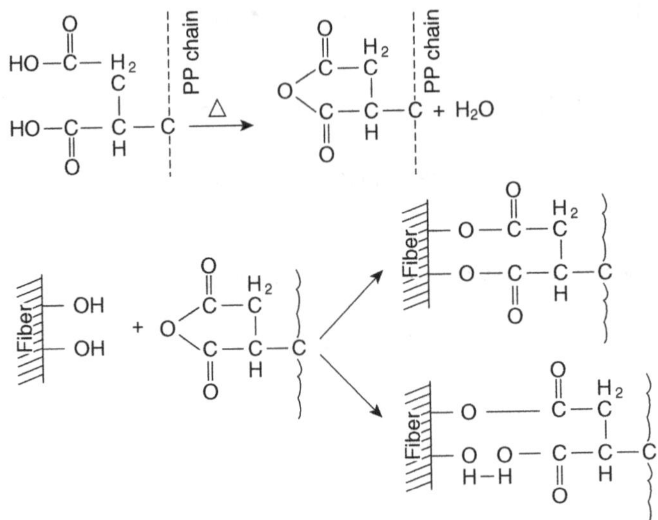

13.12 PP-g-MAH reaction with cellulose fiber hydroxyl group (reprinted from Li, X., Tabil, L. and Panigrahi, S. 2007. Chemical treatments of natural fiber for use in natural fiber-reinforced composites: a review. *Journal of Polymers and the Environment*, 15, 25–33. With kind permission from Springer Science and Business Media).

Table 13.3 Silane coupling agents

Silane	Chemical name
AEAPTES	3-(2-aminoethylaminopropyl)-triethoxysilane
AEAPTMES	3-(2-aminoethylaminopropyl)-trimethoxysilane
APTES	3-aminopropyltriethoxysilane
APTMES	3-aminopropyltrimethoxysilane

Source: Zhao *et al.*, 2012.

Silane is also used in rice husk composites. Park *et al.* (2004) studied the treatment of dry, ground and steam-exploded rice husk with γ-aminopropyltriethoxysilane and concluded that silane molecules in each case attached to rice husk by chemical reaction.

13.4.4 Acetylation

Acetylation is an esterification reaction which plasticizes natural fibers by-introducing acetyl functional groups (CH_3COO–). In this reaction, acetic acid (CH_3COOH) is produced as a by-product that has to be separated

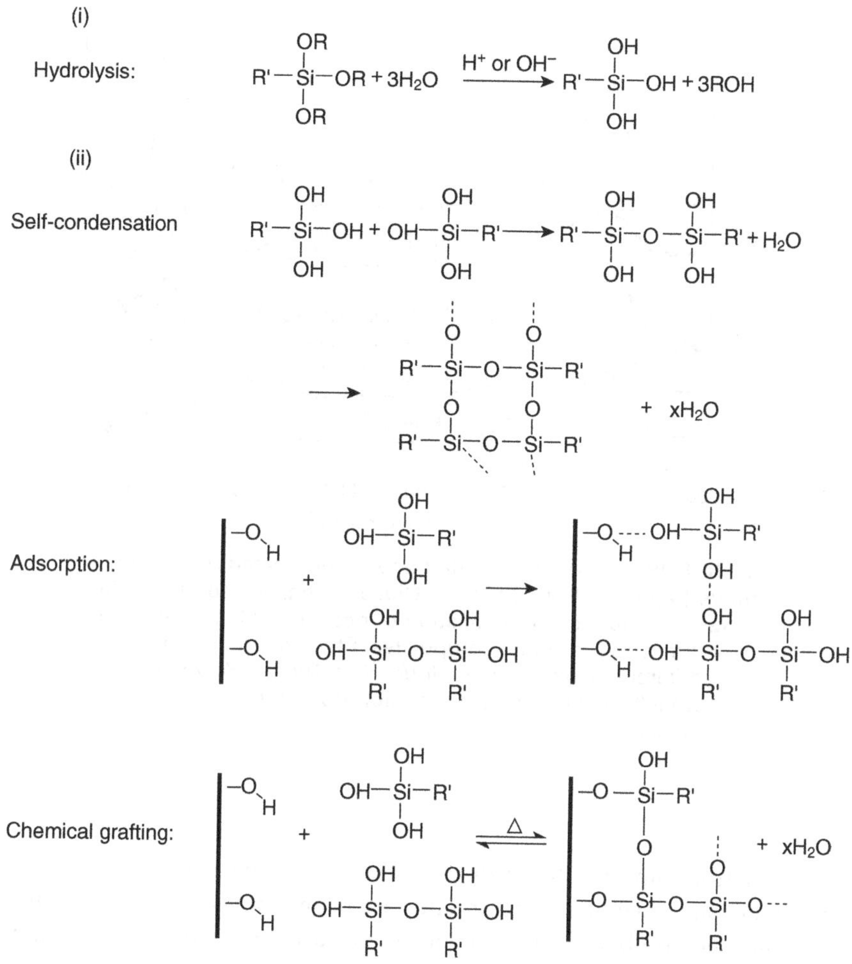

13.13 Grafting reaction mechanism of silane with cellulose hydroxyl group (reprinted from Zhao, Y., Qiu, J., Feng, H. and Zhang, M. 2012. The interfacial modification of rice straw fiber reinforced poly(butylene succinate) composites: Effect of aminosilane with different alkoxy groups. *Journal of Applied Polymer Science*, 125, 3211–3220. With kind permission from John Wiley and Sons, Inc.).

from the reaction mixture to prevent damage to the fibers. When RS/RH is treated with acetic anhydride (CH_3–C(=O)–O–C(=O)–CH_3), it replaces the hydroxyl groups present in the cell wall polymers with acetyl groups, making the fibers hydrophobic in nature (Hill *et al.*, 1998). The reaction of rice straw and rice husk with acetic anhydride is presented below:

$$\text{RS/RH–OH} + \text{CH}_3\text{–C(=O)–O–C(=O)–CH}_3 \rightarrow \text{RS/RH–OCOCH}_3 + \text{CH}_3\text{COOH}$$

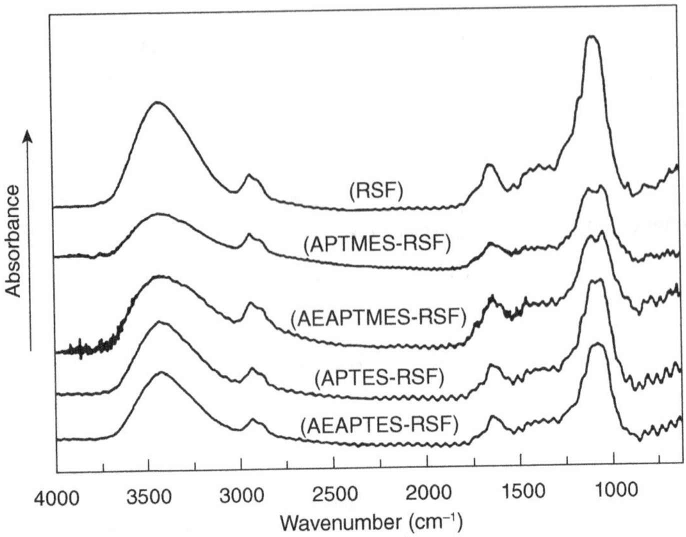

13.14 FTIR spectra of untreated and silane treated rice straw fibers (reprinted from Zhao, Y., Qiu, J., Feng, H. and Zhang, M. 2012. The interfacial modification of rice straw fiber reinforced poly(butylene succinate) composites: Effect of aminosilane with different alkoxy groups. *Journal of Applied Polymer Science*, 125, 3211–3220. With kind permission from John Wiley and Sons, Inc.).

Acetylation improves the water resistance and dimensional stability of the composite. Acetylene treatment has been studied mostly for flax, kenaf and hemp fibers and has been reported for rice straw by Barghini (2010).

13.4.5 Benzoylation

Benzoylation of RS/RH introduces benzoyl functional groups ($C_6H_5C{=}O$) at the fiber surface with the help of benzoyl chloride. It reduces the hydrophilic nature of the fibers and strengthens the linkage with the hydrophobic polymer matrix, increasing the water resistance of the composite. A schematic reaction of rice straw and rice husk fibers with benzoyl chloride is presented below:

$$\text{RS/RH–OH} + \text{NaOH} \rightarrow \text{RS/RH–O}^{-}\text{Na}^{+} + \text{H}_2\text{O}$$

$$\text{RS/RH–O}^{-}\text{Na}^{+} + \text{benzoyl–Cl} \rightarrow \text{RS/RH–benzoyl} + \text{NaCl}$$

13.4.6 Grafting of acrylic acid acrylonitrile

High-energy radiation generates free radicals in the cellulose of natural fibers, which initiates the acrylation reaction (Bledzki and Gassan, 1999). This has been a successful treatment for fiberglass and has been reported as a treatment of oil palm fibers (Sreekala *et al.*, 2000), but has not been evaluated for RS/RH composites. Acrylonitrile (CH_2=CHCN) is also used as a coupling agent for natural fibers, and reacts with surface hydroxyl groups of RS/RH as follows:

$$\text{RS/RH–OH} + \text{CH}_2\text{=CHCN} \rightarrow \text{RS/RH–OCH}_2\text{CH}_2\text{CN}$$

13.4.7 Permanganation (treatment with permanganates)

Permanganation introduces MnO_4 groups into cellulose fibers treated with alkali permanganate. MnO_3^- ion formation leads to the production of cellulose radicals, and highly reactive Mn^{3+} ions set off graft copolymerization. Permanganation is usually carried out by impregnating mercerized cellulose fibers in different concentrations of an acetone solution of potassium permanganate ($KMnO_4$) for 1–3 minutes (Paul *et al.*, 1997; Joseph *et al.*, 1996; Sreekala *et al.*, 2000).

13.4.8 Peroxide treatment

Organic peroxides (ROOR) decompose to form $RO^\bullet$ free radicals, which react with the hydrogen groups present in cellulose and polymers. A schematic reaction between cellulose fibers and polypropylene, initiated by an organic peroxide, is presented below for RS/RH:

$$\text{RO–OR} \rightarrow 2\,\text{RO}^\bullet$$

$$\text{RO}^\bullet + \text{PP–H} \rightarrow \text{ROH} + \text{PP}^\bullet$$

$$\text{RO}^\bullet + \text{RS/RH–H} \rightarrow \text{ROH} + \text{RS/RH}^\bullet$$

$$\text{PP}^\bullet + \text{RS/RH}^\bullet \rightarrow \text{PE–RS/RH}$$

For cellulose present in any natural fiber, there exist various other coupling agents and treatments, such as isocyanates, stearic acid, triazine derivatives and sodium chlorite, the choice of which depends upon the desired improvement in mechanical properties and resilience to water absorption in the relevant composites.

13.5 Critical issues for the integration of fibers into the matrix

13.5.1 Thermal stability of rice straw and rice husk

Degradation of RS/RH in the composite results in poor mechanical and organoleptic properties. RS/RH produce volatile organic compounds upon decomposition, and if this happens during processing, then the composite may form a porous and less dense structure with inferior mechanical properties. To avoid devolatilization of the fiber, the temperature during the processing of rice straw and rice husk composites therefore rarely exceeds 190°C.

13.5.2 Moisture content of the fiber

Rice straw and rice husk contain cellulose, which is hydrophilic in nature, whereas the polymer matrix is hydrophobic. Fibers with increased moisture content will therefore not embed effectively into the polymer matrix. Although the presence of silica in RS/RH prevents them from absorbing high amounts of moisture, the moisture content still ranges between 7 and 10%. This leads to inferior mechanical properties and dimensional inconsistency of the composite. Grafting RS/RH with acrylic or vinyl monomers can reduce the amount of moisture gain in the fiber. In the case of rice straw fiberboards, polymeric methylene diphenyl diisocyanate (PMDI) acts as a more efficient water resistant agent for rice straw fibers than various wax emulsions (Pan *et al.*, 2010).

13.5.3 Particle size

Fiber particle size plays an important role in determining the mechanical properties of the composite. RS/RH are used in various particle sizes in composites, from flour to coarse particles. Rice straw can be used as refined fiber, ground pellets and rice straw strands, as shown in Fig. 13.15. Refined rice straw fiber is obtained by steam explosion with a steam pressure of 1.2 bar and 80% moisture content, resulting in particles with a high aspect ratio (16.31) compared with ground rice straw (6.19) (Ming-Zhu *et al.*, 2011). Rice straw is normally used as ground particles in composites (Zhao *et al.*, 2012; Bassyouni *et al.*, 2012; Kalagar *et al.*, 2011) with average particle size in the range 100–1000 μm, and rarely exceeding 3000 μm. The use of rice husk particles in composites is reported in the range of 45 μm (Petchwattana and Covavisaruch, 2013) to 500 μm (Rozman *et al.*, 2003), rarely exceeding 2000 μm (Ndazi *et al.*, 2007).

At constant rice straw loading, increasing particle size increases the tensile strength of the composite, as reported by Zhao *et al.* (2012), because greater particle size more effectively transfers energy at the fiber–matrix

13.15 Rice straw of various particle sizes and aspect ratios: (a) rice straw, (b) refined rice straw fiber, (c) ground rice straw pellets and (d) rice straw strands (reprinted from Ming-Zhu, P., Chang-Tong, M., Xu-Bing, Z. and Yun-Lei, P. 2011. Effects of rice straw fiber morphology and content on the mechanical and thermal properties of rice straw fiber–high density polyethylene composites. *Journal of Applied Polymer Science*, 121, 2900–2907. With kind permission from John Wiley and Sons, Inc.).

interface. By contrast, increasing particle size at constant rice husk loading *decreases* the tensile strength of the composite (Petchwattana and Covavisaruch, 2013; Rozman *et al.*, 2003). This has been attributed to the exceptional amount of silica present in rice husk, which imparts toughness to the composite.

13.6 Processing of thermoset and thermoplastic composites incorporating rice straw/rice husk (RS/RH) fiber reinforcements

13.6.1 Thermosets

Fiberboards are probably the most widely used composites today, and use rice straw in addition to wood straw as the cellulosic component. These fiberboards are thermoset composites, usually making use of thermoset resins such as urea formaldehyde (UF), melamine urea formaldehyde (MUF) and PMDI. Other chemicals such as wetting and release agents, waxes and dyes are added to the mixture to make the boards resistant to water, fire and insects. Sometimes the term fiberboard is used interchangeably with particleboard. Particleboards are low-density fiberboards, which usually make use of urea formaldehyde. Medium- and high-density fiberboards (MDF, HDF) use polyurethane as the binding resin. HDF is more refined and denser in structure than MDF, and is used as a substitute for wood in many applications. It is ideal for pegboard and hardboard panels in household and point-of-sale fixtures. MDF is used mainly in furniture, cabinets and slatwalls. Particleboards are used mainly for non-display applications like shelves and inner wardrobe divisions.

The manufacturing of HDF is similar to that of MDF except that for HDF, ground wood and rice straw are subjected to steam explosion for fibrillation, whereas the straw is simply ground for MDF. Softwood logs are debarked and hacked into chips in a giant blender. The chips are graded and washed to remove impurities. Ground, graded and washed rice straw is mixed with chips in a defined proportion. To make HDF, the mixture is blasted with steam and mixed with resin, and then, once it has been dried again, it is formed into a giant, continuous fiber mat, which is compressed by a hot press at 260°C to one fortieth of its original thickness. In the case of MDF, the ground wood and rice straw mixture is dried, mixed with resin, and laid into a continuous fiber mat. The mat is conveyed to a hot press, which compresses it to a predetermined thickness between two heated steel belts. The pressed boards are cooled and transported into a sanding area where large belt sanders reduce the boards to the desired final thickness. After sanding, the boards are cut to size, bundled, and packed for delivery. Rice straw is used in cementitious composites and gypsum boards. Rice husk ash obtained below 600°C consists of amorphous silica and is used in pozzolanic cement.

13.6.2 Thermoplastics

RS/RH thermoplastic composites are processed similarly by extruding a proportionate dry blend of fibers, thermoplastics, coupling agents and processing aids to form pallets (Panthapulakkal, 2005; Rahman *et al.*, 2010) from a screw-type extruder (Bassyouni *et al.*, 2012; Petchwattana and Covavisaruch, 2013), as shown in Fig. 13.16.

The length-to-diameter ratio of the screw is around 36 (Ming-Zhu *et al.*, 2011) and it operates within a temperature range of 170–190°C (Petchwattana and Covavisaruch, 2013; Kalagar *et al.*, 2011). The extruded pallets are then

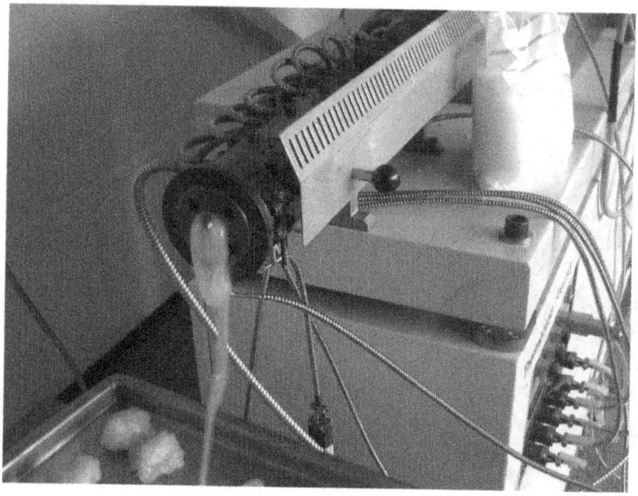

13.16 Brabender extruder.

ground to flakes with the help of a crusher. Rice straw–polypropylene flakes are shown in Fig. 13.17.

The flakes are injection-molded to a final form in an injection-molding machine. The injection-molding parameters for polypropylene–rice straw composite are shown in Table 13.4 (Bassyouni *et al.*, 2012).

13.7 Evaluating the performance of composites reinforced with RS/RH fibers

After the composite is processed, it is tested for its performance, i.e. its mechanical and physical properties. There is a continuous effort to improve these properties by manipulating fiber loading, size and shape of the fibers, coupling agents and other variables. Testing of mechanical properties of the composite involves impact, bending and tensile strength; testing of physical properties involves wettability, thermal properties (DSC, TGA), non-destructive testing (i.e. DMA), XRD, SEM and FTIR. RS/RH are mostly embedded into polyethylene and polypropylene matrices, with or without application of coupling agents; the mechanical properties of these composites are discussed below separately for rice straw and rice husk.

13.7.1 Density

The increase in the density of the composite with fiber loading follows the rule of mixtures. Bassyouni *et al.* studied the increase in density for rice

13.17 Rice straw–polypropylene flakes after pellets are crushed.

Table 13.4 Injection parameters for rice straw–polypropylene composite

Parameter	Value
Injection pressure (bar)	800
Holding pressure (bar)	80% of injection pressure
Mold temperature (°C)	40
Barrel zone temperature (°C)	30–170–175–180–190
Injection speed (cm^3/s)	22
Feeding volume (cm^3)	3
Cooling time (s)	10

straw–polypropylene composite with fiber loading. They found that the density increases linearly with rice straw loading up to 30% by weight, ranging from 0.9 to 1.01 g/cm^3, as shown in Fig. 13.18.

13.7.2 Water absorption

The water absorption of the composite is found to increase with fiber loading for rice straw reinforced polypropylene (RPP). The addition of a coupling agent decreases the water uptake by the rice straw fibers, as shown in Fig. 13.19 for 20% rice straw loaded polypropylene (20RPP) and 20% rice straw loaded maleic anhydride-grafted polypropylene (20RMAPP) (Bassyouni *et al.*, 2012). After 23 hours of immersion in water, RPP absorbs around 0.3% water whereas RMAPP absorbs 0.22% water, which makes evident the effect of the coupling agent on the wettability of the composite.

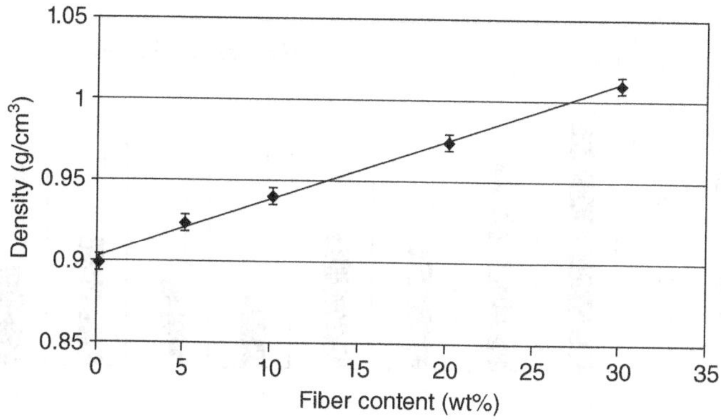

13.18 Increase in density of RPP with increase in fiber loading.

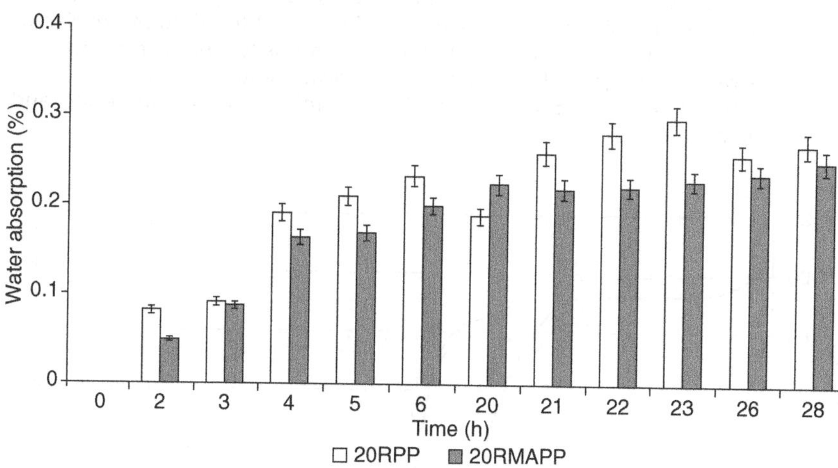

13.19 Water absorption characteristics of RPP and RMAPP loaded with 20% rice straw.

The addition of lignin as a coupling agent also reduces the wettability of the composite, because it is hydrophobic in nature. Kamel (2004) studied the effect of an increase in bagasse lignin on the percentage water gain (PWG) and swellability of mercerized rice straw–polyvinyl chloride (PVC) composites. It was found that the composite samples (1 to 8 in Fig. 13.20), made with 15% PVC and rice straw (treated with 15% NaOH at 80°C for 1 h), showed a decrease in PWG and percentage swellability with increase in lignin content from 0 to 10%.

13.7.3 Thermal stability

Figure 13.21 shows the effect of fiber loading (20 and 30% by mass of rice straw) on the thermal stability of RMAPP. Increased straw fiber loading

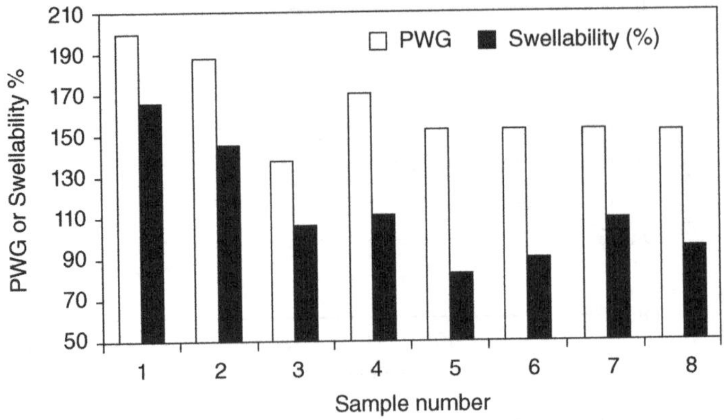

13.20 Decrease in PWG and percentage swellability of PVC–lignin–rice straw composite with increase in lignin content (reprinted from Kamel, S. 2004. Preparation and properties of composites made from rice straw and poly(vinyl chloride) (PVC). *Polymers for Advanced Technologies*, 15, 612–616. With kind permission from John Wiley and Sons, Inc.).

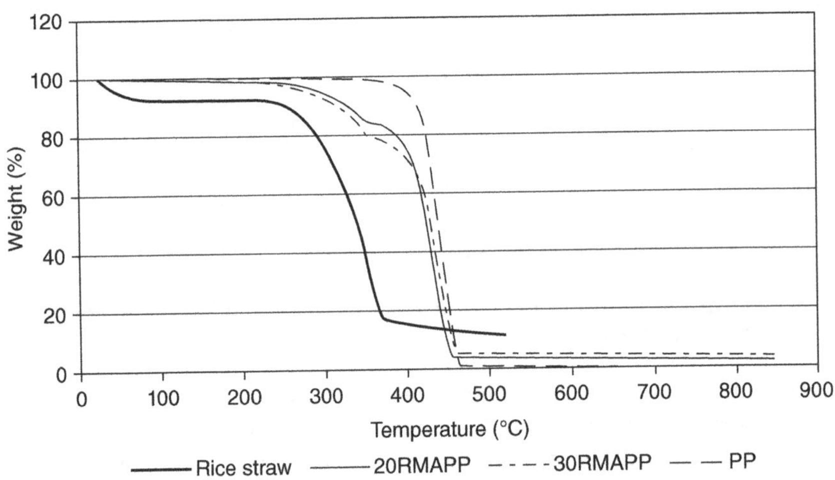

13.21 TGA of rice straw, PP, 20RMAPP, and 30RMAPP.

(30RMAPP) results in decreased thermal stability of the composite compared with 20RMAPP, although the rate of thermal decomposition is smaller.

Figure 13.22 shows that steam-treated rice husk underwent three stages of weight loss involving dehydration (60°C to 200°C) and hemicellulose degradation (150°C to 350°C), followed by cellulose degradation (275°C to

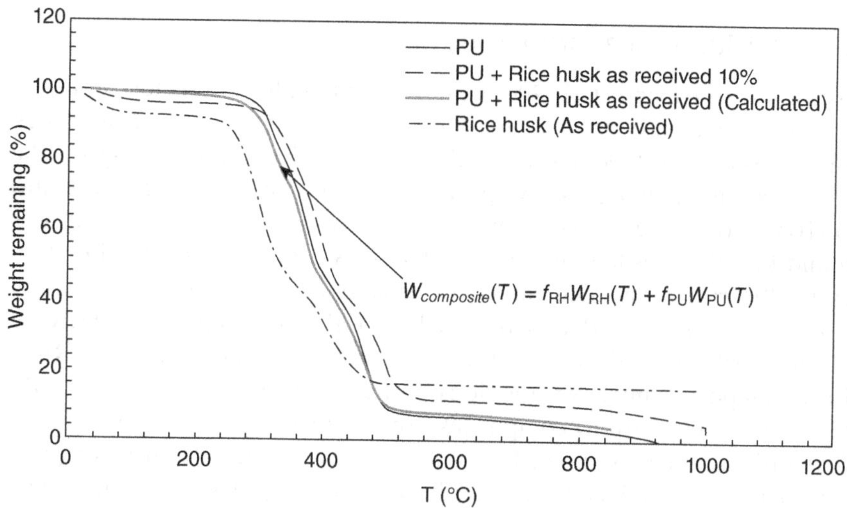

13.22 Calculated and experimental thermal degradation of PU–rice husk.

380°C) and then lignin decomposition (250°C to 550°C). Polyurethane showed greater thermal stability than untreated rice husk up to 459°C. The weight remaining from the rice husk above 550°C is due to its thermally stable silica content.

For nonreactive materials, the weight of the composite remaining at a certain temperature $W_{\text{composite}}(T)$ is normally calculated as follows:

$$W_{\text{composite}}(T) = f_{\text{RH}}W_{\text{RH}}(T) + f_{\text{PU}}W_{\text{PU}}(T) \qquad [13.1]$$

where:

f_{RH} = weight fraction of rice husk in the composite
$W_{\text{RH}}(T)$ = weight remaining of the rice husk at a certain temperature T
f_{PU} = weight fraction of polyurethane in the composite
$W_{\text{PU}}(T)$ = weight remaining of the polyurethane at a certain temperature T.

The theoretical thermal degradation of the PU–untreated rice husk composite (as received) was calculated based on equation 13.1. As shown in Fig. 13.22, the thermal degradation of the composite displayed greater stability than the rice husk and PU individually. The theoretical equation for nonreactive materials predicted a higher mass loss rate from 300 to 550°C. The experimental results indicated some form of interaction between the rice husk surfaces and the PU matrix.

13.7.4 Mechanical properties

According to Albano *et al.* (2001), the addition of fillers into thermoplastics can either increase or decrease the tensile strength of the composites. Buzarovska *et al.* (2008) studied the effect of rice straw loading on the tensile strength of rice straw–poly(hydroxybutyrate-*co*-hydroxyvalerate) (PHBV) copolymer, which contained 13 mol% hydroxyvalerate. It was found that the tensile strength of PHBV–RS composites containing 20% and 30% rice straw decreased by around 8% and 15% respectively, compared to the neat PHBV resin. Zhao *et al.* (2012) reported a decrease in the tensile strength of rice straw fiber–poly(butylene succinate) (RSF–PBS) composites made with various particle sizes of rice straw. RSF–PBS (30%/70%, 100–300 μm) composites prepared with increasing amounts of four different silanes showed an initial increase in tensile strength (with AEAPTES and APTES curves showing a sharp increase), and then decreased gradually when an excess amount of silane was used. The tensile strength of the composite can be improved by optimizing fiber dispersion in the matrix, which can be achieved through improved processing conditions or use of a coupling agent.

Similar results (of decreasing tensile strength with increasing fiber loading) have been reported by Bassyouni *et al.* (2012) in a study of the physicomechanical properties of RPP and RMAPP composites. It was found that an increase in rice straw loading decreased the tensile strength of both RPP and RMAPP and weakened the composite, which might be a result of fiber agglomeration causing poor stress distribution within the matrix. Figure 13.23 shows the tensile strength of RPP and RMAPP composites at various fiber loadings.

Albano *et al.* (2001) indicated that the tensile strength might increase with fiber loading, and this is supported by the studies of Petchwattana and Covavisaruch (2013). Figure 13.24 shows the increase in tensile strength of rice husk–PVC composites with fiber loading for a set of composite samples of various particle sizes. The tensile strength of 45 μm rice husk–PVC composite made with 60 parts per 100 resin (phr) of milled rice husk increased from 31.50 MPa (polymer matrix with additives only) to 44.70 MPa (composite with 60 phr milled rice husk).

This behavior might be attributed to silica-rich rice husk increasing the stiffness of the composites. However, a sharp decrease was observed in the tensile strength of the composites for all particle sizes when fiber loading was increased to 80 phr. This sharp decrease was a result of agglomeration of rice husk, as shown in micrographs of 250 μm rice husk–PVC composites for various fiber loadings in Fig. 13.25. Figure 13.25(a–c) shows that rice husk was homogeneously integrated into the PVC matrix up to a fiber loading of 60 phr, whereas at 80 phr loading, the rice husk made agglomerates

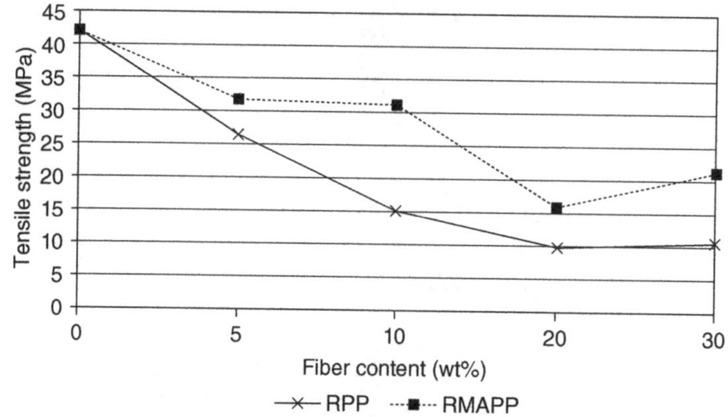

13.23 Tensile strength of RPP and RMAPP composites at various fiber loadings (reprinted from Petchwattana, N. and Covavisaruch, S. (2013). Effect of rice hull particle size and content on the mechanical properties and visual appearance of wood plastic composites prepared from poly(vinyl chloride. *Journal of Bionic Engineering,* 10, 110–117. With permission from Elsevier).

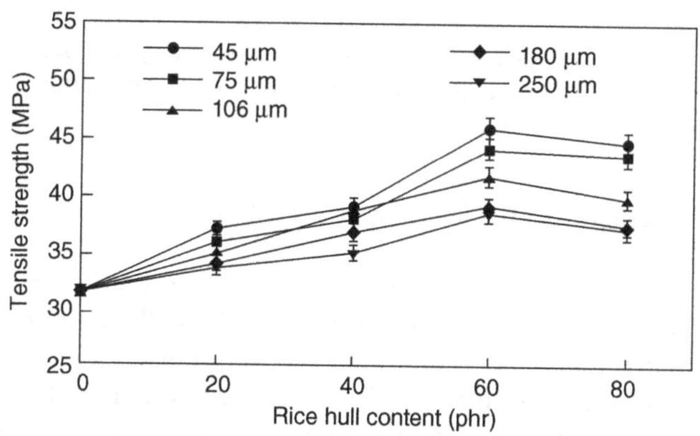

13.24 Tensile strength of rice husk–PVC composites with fiber loading (parts per hundred of resin) for various particle sizes (reprinted from Petchwattana, N. and Covavisaruch, S. (2013). Effect of rice hull particle size and content on the mechanical properties and visual appearance of wood plastic composites prepared from poly(vinyl chloride. *Journal of Bionic Engineering,* 10, 110–117. With permission from Elsevier).

of 500–700 µm, as shown in Fig. 13.25(d). It also shows that at constant fiber loading, an increase in particle size decreased the tensile stress in the composites, as larger particles leave a larger clearance at the fiber–matrix interface, reducing energy transfer at the interface. Another study by the same authors (Petchwattana *et al.*, 2012) gave similar results for the increase

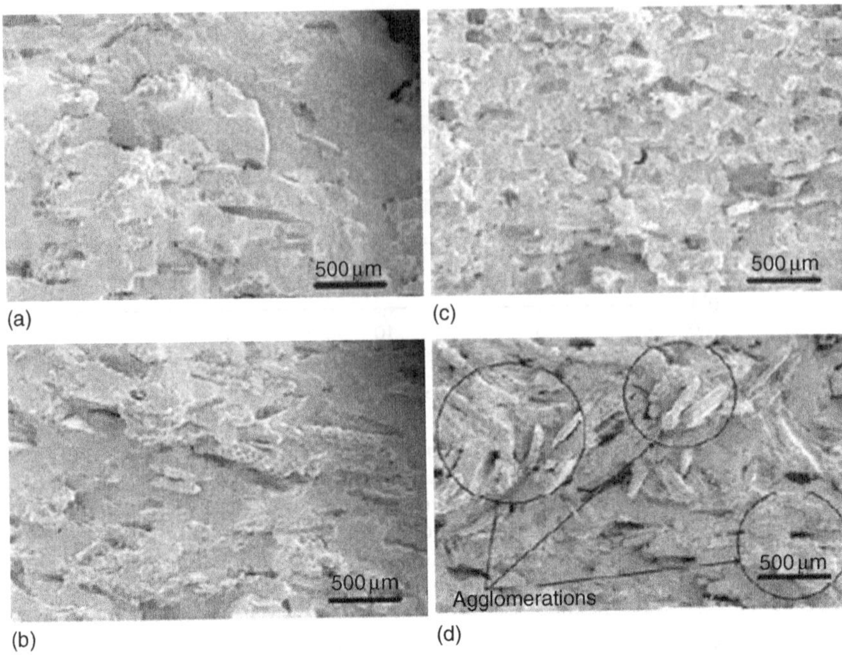

13.25 SEM micrographs of 250 μm rice husk–PVC composites with various fiber loadings: (a) 20 phr, (b) 40 phr, (c) 60 phr and (d) 80 phr.

in tensile strength with fiber loading of HDPE–rice husk composites for various particle sizes.

Kalagar *et al.* (2011) suggested that mercerization of rice straw for a sufficient time improves the tensile strength of RMAPP composites. They studied the effect of alkali concentration and duration of treatment of rice straw flour on the mechanical properties of a rice straw flour–polypropylene composite. Rice straw flour (40 mesh) was treated with two different concentrations of sodium hydroxide (5% and 10%) for two different durations (45 and 90 min), giving a total of four samples. Mercerized rice straw flour (30% by weight) was embedded in a polypropylene matrix (65%), using maleic anhydride (5%) as a coupling agent. It was found that the composite made from the rice straw flour, treated with a higher concentration of NaOH (10%) for a longer time (90 min), showed an increase of around 30% in tensile strength compared with untreated rice straw flour composite. The improvement of tensile strength with mercerization has also been reported by Kamel (2004) for rice straw–bagasse lignin–PVC composite. The effect of alkali concentration and treatment duration on the yield and composition of rice straw is summarized in Table 13.5 (Kamel, 2004).

Table 13.5 The effect of mercerization parameters on the yield and composition of rice straw

Sample no.	Treatment condition	Yield (%)	α-Cellulose (%)	Hemicellulose (%)	Lignin (%)	Ash (%)
1	Untreated	–	32.00	33.00	17.00	14.00
2	H_2O (130°C, 2 hr)	87.56	66.67	6.53	11.35	14.76
3	15% NaOH (130°C, 2 hr)	40.91	74.32	16.44	3.28	6.36
4	10% NaOH (100°C, 1 hr)	58.72	73.07	9.68	2.11	9.20
5	7.5% NaOH (100°C, 1 hr)	59.21	68.27	14.73	9.11	10.64
6	5% NaOH (100°C, 1 hr)	67.06	66.80	13.46	8.03	6.96
7	10% NaOH (80°C, 1 hr)	74.07	67.82	15.82	5.34	7.02
8	5% NaOH (80°C, 1 hr)	82.32	62.00	18.38	7.30	15.67

Source: Reprinted from Kamel, S. 2004. Preparation and properties of composites made from rice straw and poly(vinyl chloride) (PVC). *Polymers for Advanced Technologies*, 15, 612–616. With kind permission from John Wiley and Sons, Inc.

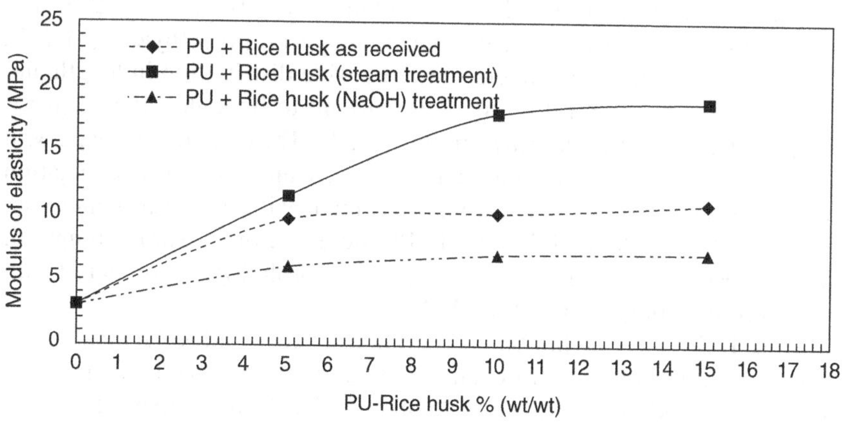

13.26 Young's modulus of rice husk–PU composites.

However, contrary results have been observed in the case of PU composites made with mercerized rice husk. Figure 13.26 compares the moduli of elasticity of rice husk–PU composites. Sodium hydroxide-treated rice straw composites with PU show a decrease in modulus of elasticity compared with untreated rice straw composite. Steam-treated rice husk shows a significant improvement in the modulus of elasticity.

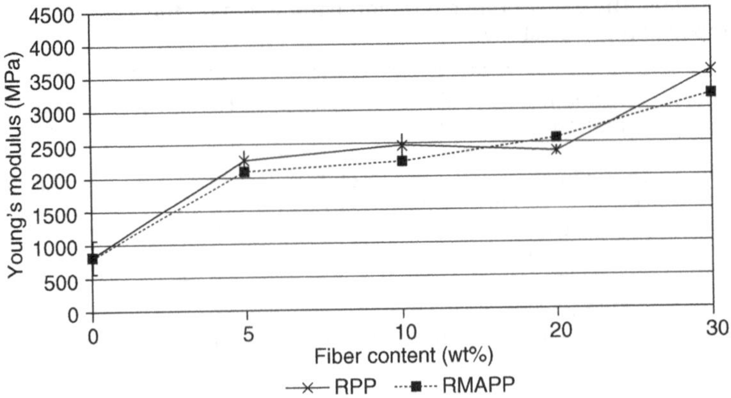

13.27 Young's modulus of RPP and RMAPP at different fiber loadings.

Ming-Zhu *et al.* (2011) studied the effect of rice straw fiber morphology and fiber loading on the mechanical properties of rice straw fiber–high density polyethylene (HDPE) composites. Three types of rice straw morphology (refined rice-straw fiber, rice straw pallet and rice straw strand; see Fig. 13.15) were prepared for embedding into HDPE, with 20% and 40% fiber loading in the presence of maleic anhydride polyethylene (MAPE; 5 wt%). Refined rice-straw fibers showed a wide distribution in length and width, and their aspect ratio (16.31) was found to be higher than that for rice pellets (6.19) and rice straw strands (14.48). The composites made with refined rice-straw fiber had more tensile strength and less tensile elongation at breaking compared with the composites made with rice straw pallets and rice straw strand. The addition of MAPE increased the flexural strength of the composites for all of the rice straw fiber morphologies, compared with the composites prepared without MAPE.

Figure 13.27, showing the Young's modulus of RPP and RMAPP, suggests that the addition of rice straw imparts greater stiffness to the composites up to a fiber loading of 30% (Bassyouni *et al.*, 2012). It is apparent that the addition of maleated polypropylene (MAPP) to polypropylene (PP) does not influence the stiffness of the composites, as the values of Young's modulus do not improve considerably for RMAPP. It is a common observation that there is a trade-off between tensile modulus and fracture strain, as depicted in Fig. 13.28 for RPP and RMAPP and cited by several researchers (Albano *et al.*, 2001; Taha *et al.*, 2008).

The RPP and RMAPP curves show a similar behavior in that the fracture strain rapidly falls from 500% to 6% as rice straw reinforcement increases from 0% to 5%. Any further addition of fiber lowers the failure strain of

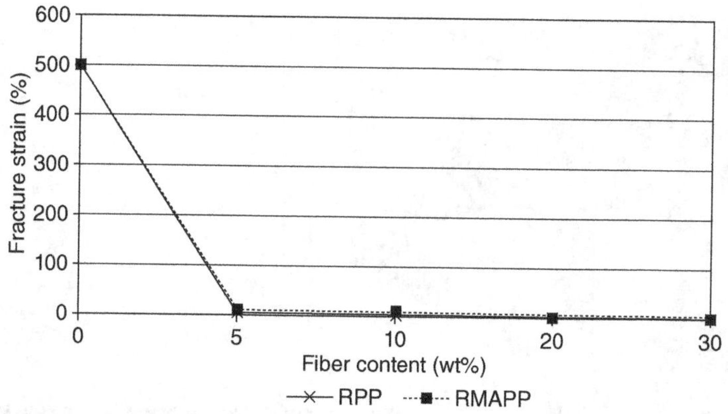

13.28 Percentage decrease in fracture strain with increased fiber loading.

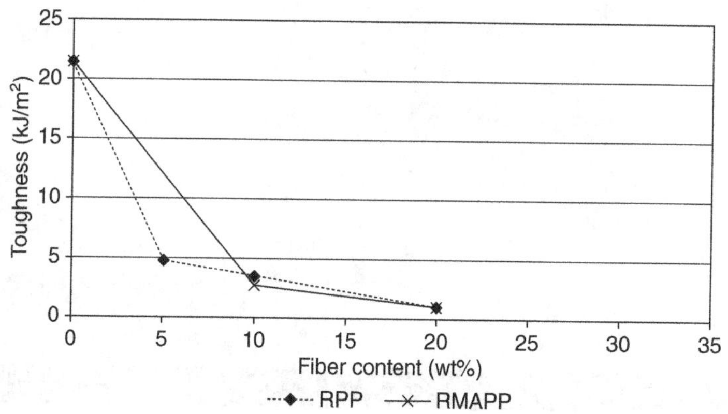

13.29 Decrease in impact strength of RPP and RMAPP with fiber loading.

the composites, but at a much lower rate. Such behavior is observed for other composites but with varying rates of decrease in failure strain depending upon the polymeric matrix. Therefore, any voids or impurities in the matrix will lead to composite stiffening.

Figure 13.29 shows the impact strength of both RPP and RMAPP decreasing with increased fiber loading in comparison with the pure polypropylene, and RMAPP having better impact strength than RPP (Bassyouni *et al.*, 2012).

The SEM micrographs in Fig. 13.30(a–f) show differences in tensile fracture surfaces of RPP and RMAPP. It can be observed in Fig. 13.30(a) that in the absence of the maleic anhydride coupling agent, there is a deficiency of polypropylene (PP) matrix around the rice straw, which results

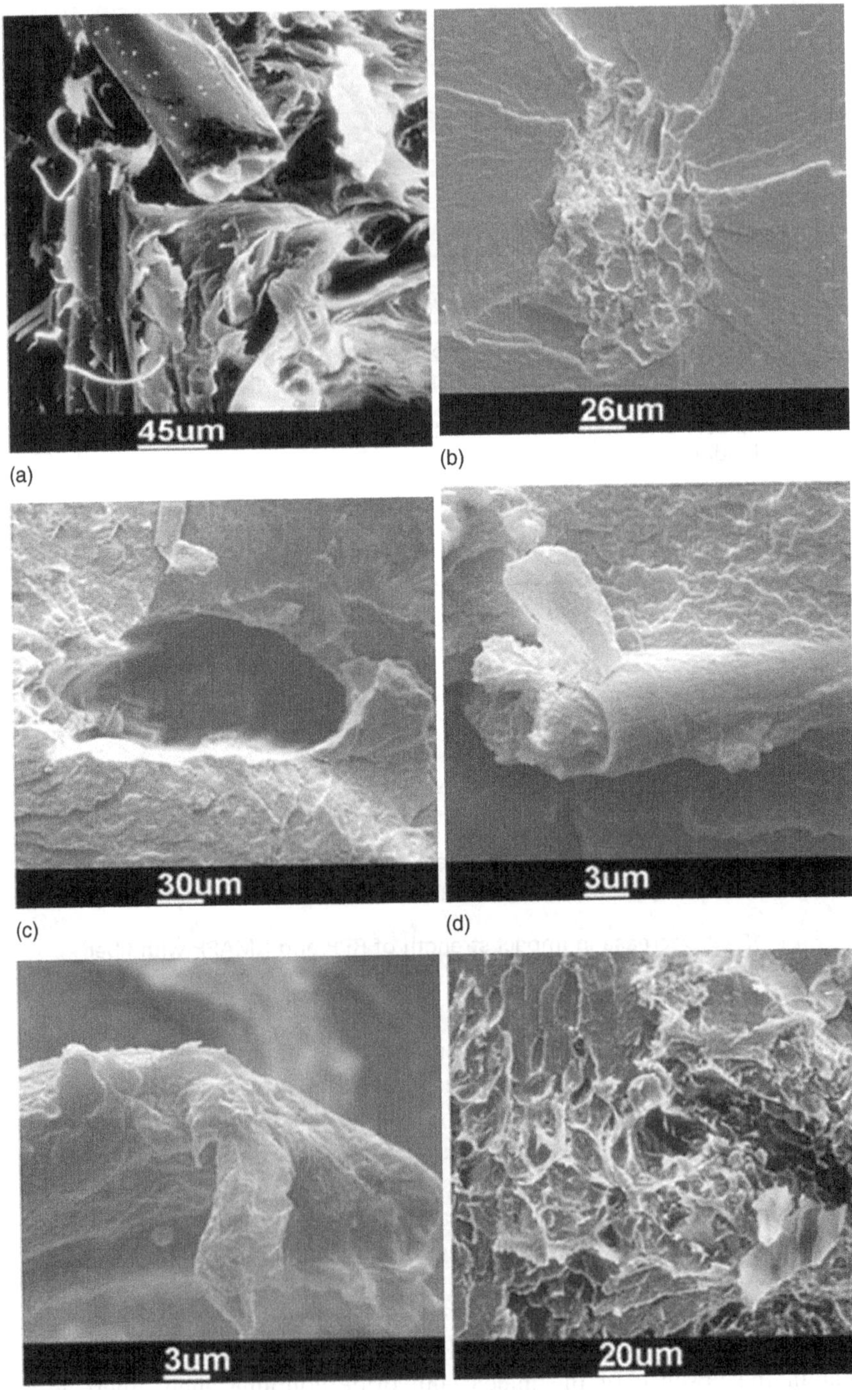

(a) (b) (c) (d) (e) (f)

13.30 SEM micrographs of RPP (a–c) and RMAPP (d–f).

in poor stress distribution within the matrix. The addition of maleic anhydride causes a film to form over the rice straw, which increases its affinity towards the hydrophobic PP, as shown in Fig. 13.30(e). The poor intermixing and debonding of rice straw from the polymer matrix, in the absence of maleic anhydride, is shown in Fig. 13.30(b–c). This is caused by fiber agglomeration, which reduces the effective surface available for adhesion. Figure 13.30(d) shows improved intermixing between rice straw and polypropylene in the presence of the coupling agent. The surface of rice straw is hydrophilic in nature, containing polar hydroxyl functional groups, while PP is hydrophobic in nature, due to polyolefins. In the presence of maleic anhydride-grafted PP, the anhydride groups connected to polypropylene link to –OH groups in the cellulose and lignin of the rice straw surface through the formation of a block copolymer, resulting in considerable reinforcement of the composite.

13.8 Conclusion

Rice straw and rice husk belong to the non-wood family of natural fibers and embedding them in polymer matrices improves the properties of the resulting composites and saves wood and petroleum resources. Rice straw is an agricultural production residue and a field product, whereas rice husk is obtained after milling operations and is thus an agricultural processing residue. The chemical composition of these fibers is sufficiently close to that of wood that they are a suitable alternative to wood fibers in composites. Rice husk contains an exceptional amount of silica, which makes it resilient to biodegradation and moisture penetration. The anomalous trend of increasing tensile strength with rice husk loading in composites is attributed to increased interfacial forces. FTIR studies prove the presence of hydrophilic hydroxyl groups on the surface of the fibers which are utilized by coupling agents to develop linkages with the hydrophobic polymer matrix. XRD studies show that chemical surface treatments do not alter the crystallinity of the cellulose fibers present in rice straw and rice husk. TGA studies show that rice husk starts decomposing around 180°C and thus has greater thermal stability than rice straw, which begins to degrade at 160°C.

For a given variety of rice plant, the quality of the rice straw and rice husk fibers depends on how well they are processed industrially, as cultivation and harvesting methods are very basic in nature and do not have much influence on the quality of the final fiber product. Mercerization (with NaOH) and steam explosion are widely used surface-treatment methodologies applied to rice straw and rice husk, with maleic anhydride used as a coupling agent. Coupling agents are reported to increase the water resistance and thermal stability of the composite. However, more coupling agents need to be evaluated. Rice straw and rice husk composites have been found to exhibit different mechanical properties. An increase in rice straw loading decreases the tensile strength in a manner similar to that found in other natural fiber composites, as high fiberloading results in fiber agglomeration. At a given fiber loading, increased rice straw particle size leads to increased tensile strength of the composite. Rice straw finds its applications in making particleboards and fiberboards, and car roofs and doors, whereas rice straw is mostly used in cementitious composites.

13.9 References

Albano, C., Ichazo, M., González, J., Delgado, M. and Poleo, R. 2001. Effects of filler treatments on the mechanical and morphological behavior of PP+wood flour and PP+sisal fiber. *Material Research Innovations*, 4, 284–293.

Barghini, A. 2010. *Ecocompatible, biodegradable polymers. Plastic items preparation & characterization*. Doctoral thesis, University of Pisa, Italy.

Bassyouni, M., Taha, I., Abdel-hamid, S. M.-S. and Steuernagel, L. 2012. Physico-mechanical properties of chemically treated polypropylene rice straw bio-composites. *Journal of Reinforced Plastics and Composites*, 31, 303–312.

Bledzki, A. K. and Gassan, J. 1999. Composites reinforced with cellulose based fibres. *Progress in Polymer Science*, 24, 221–274.

Bos, H. L. Van den Oever, M. J. A. and Peters, O. C. J. J. 2002. Tensile and compressive properties of flax fibers for natural fibre reinforced composites. *Journal of Materials Science*, 37, 1683–1692.

Buzarovska, A., Bogoeva, G. G., Grozdanov, A., Avella, M., Gentile, G. and Errico, M. 2008. Potential use of rice straw as filler in eco-composite materials. *Australian Journal of Crop Science*, 1(2), 37–42.

Chang, T.-T. and Bardenas, E. A. 1965. *The Morphology and Varietal Characteristics of the Rice Plant*. Los Baños, Laguna, The Philippines, The International Rice Research Institute.

Fuad, M. Y. A., Ismail, Z., Ishak, Z. A. M. and Omar, A. K. M. 1995. Application of rice husk ash as fillers in polypropylene: Effect of titanate, zirconate and silane coupling agents. *European Polymer Journal*, 31, 885–893.

Ghaemy, M. & Roohina, S. 2003. Grafting of maleic anhydride on polyethylene in a homogeneous medium in the presence of radical Initiators. *Iranian Polymer Journal*, 12, 22–29.

Gibson, L. J., Ashby, M. F. and Harley, B. A. 2010. *Cellular Materials in Nature and Medicine*, New York, Cambridge University Press.

Hill, C. A. S., Khalil, H. P. S. A. and Hale, M. D. 1998. A study of the potential of acetylation to improve the properties of plant fibres. *Industrial Crops and Products*, 8, 53–63.

Jahn, C. E., McKay, J. K., Mauleon, R., Stephens, J., McNally, K. L., Bush, D. R., Leung, H. and Leach, J. E. 2011. Genetic variation in biomass traits among 20 diverse rice varieties. *Plant Physiology*, 155, 157–168.

Joseph, K., Thomas, S. and Pavithran, C. 1996. Effect of chemical treatment on the tensile properties of short sisal fibre-reinforced polyethylene composites. *Polymer*, 37, 5139–5149.

Kalagar, M., Khademieslam, H., Bazyar, B. and Hejazi, S. 2011. Morphology and mechanical properties of alkali-treated rice straw flour–polypropylene composites. *BioResources*, 6, 4238–4246.

Kamel, S. 2004. Preparation and properties of composites made from rice straw and poly(vinyl chloride) (PVC). *Polymers for Advanced Technologies*, 15, 612–616.

Klyosov, A. A. 2007. *Wood-Plastic Composites*, Hoboken, NJ, John Wiley & Sons.

Li, X., Tabil, L. and Panigrahi, S. 2007. Chemical Treatments of natural fiber for use in natural fiber-reinforced composites: a review. *Journal of Polymers and the Environment*, 15, 25–33.

Li, X., Cai, Z., Winandy, J. E. and Basta, A. H. 2011. Effect of oxalic acid and steam pretreatment on the primary properties of UF-bonded rice straw particleboards. *Industrial Crops and Products*, 33, 665–669.

Liao, Z., Song, G., Shi, F., Yin, Z., Yang, Y., Niu, Z. and Tang, G. 2011. Preparation and characterization of PLA/rice straw fiber composite. *Applied Mechanics and Materials*, 71, 1154–1157.

Martí-Ferrer, F., Vilaplana, F., Ribes-Greus, A., Benedito-Borrás, A. and Sanz-Box, C. 2006. Flour rice husk as filler in block copolymer polypropylene: Effect of different coupling agents. *Journal of Applied Polymer Science*, 99, 1823–1831.

Ming-Zhu, P., Chang-Tong, M., Xu-Bing, Z. and Yun-Lei, P. 2011. Effects of rice straw fiber morphology and content on the mechanical and thermal properties of rice straw fiber–high density polyethylene composites. *Journal of Applied Polymer Science*, 121, 2900–2907.

Ndazi, B. S., Karlsson, S., Tesha, J. V. and Nyahumwa, C. W. 2007. Chemical and physical modifications of rice husks for use as composite panels. *Composites Part A: Applied Science and Manufacturing*, 38, 925–935.

Pan, M., Zhou, D., Ding, T. and Zhou, X. 2010. Water resistance and some mechanical properties of rice straw fiberboards affected by thermal modification. *BioResources*, 5, 758–769.

Panthapulakkal, S. 2005. Enhancement of processability of rice husk filled high-density polyethylene composite profiles. *Journal of Thermoplastic Composite Materials*, 18, 445–458.

Park, B.-D., Wi, S. G., Lee, K. H., Singh, A. P., Yoon, T.-H. and Kim, Y. S. 2004. X-ray photoelectron spectroscopy of rice husk surface modified with maleated polypropylene and silane. *Biomass and Bioenergy*, 27, 353–363.

Paul, A., Joseph, K. and Thomas, S. 1997. Effect of surface treatments on the electrical properties of low-density polyethylene composites reinforced with short sisal fibers. *Composites Science and Technology*, 57, 67–79.

Petchwattana, N. and Covavisaruch, S. 2013. Effects of rice hull particle size and content on the mechanical properties and visual appearance of wood plastic composites prepared from poly(vinyl chloride). *Journal of Bionic Engineering*, 10, 110–117.

Petchwattana, N., Covavisaruch, S. and Chanakul, S. 2012. Mechanical properties, thermal degradation and natural weathering of high density polyethylene/rice hull composites compatibilized with maleic anhydride grafted polyethylene. *Journal of Polymer Research*, 19, 1–9.

Rahman, W. A. W. A., Lee Tin, S., Rahmat, A. R., Isa, N. M., Salleh, M. S. N. and Mokhtar, M. 2010. Comparison of rice husk-filled polyethylene composite and natural wood under weathering effects. *Journal of Composite Materials*, 45, 1403–1410.

Rozman, H. D., Yeo, Y. S., Tay, G. S. and Abubakar, A. 2003. The mechanical and physical properties of polyurethane composites based on rice husk and polyethylene glycol. *Polymer Testing*, 22, 617–623.

Ryu, S.-E., Kim, T.-N. and Kang, T.-K. 1997. Pulverization of rice husks and the changes of husk densities. *Journal of Materials Science*, 32, 6639–6643.

Sherif, S. A., Sadek, M. A., Ashour, F. H. and Bassyouni, M. 2009. Effects of Surface treatment of ground rice husk on the polyurethane based on castor oil. *Polymers and Polymer Composites*, 17, 481–485.

Sreekala, M. S., Kumaran, M. G., Joseph, S., Jacob, M. and Thomas, S. 2000. Oil palm fibre reinforced phenol formaldehyde composites: influence of fibre surface modifications on the mechanical performance. *Applied Composite Materials*, 7, 295–329.

Taha, I., El-Sabbagh, A. and Ziegmann, G. 2008. Modelling of strength and stiffness behaviour of natural fibre reinforced polypropylene composites. *Polymers and Polymer Composites*, 16, 295–302.

Yao, F., Wu, Q., Lei, Y., Guo, W. and Xu, Y. 2008. Thermal decomposition kinetics of natural fibers: Activation energy with dynamic thermogravimetric analysis. *Polymer Degradation and Stability*, 93, 90–98.

Zhang, Q. and Cai, W. 2008. Enzymatic hydrolysis of alkali-pretreated rice straw by *Trichoderma reesei* ZM4-F3. *Biomass and Bioenergy*, 32, 1130–1135.

Zhao, Y., Qiu, J., Feng, H. and Zhang, M. 2012. The interfacial modification of rice straw fiber reinforced poly(butylene succinate) composites: Effect of aminosilane with different alkoxy groups. *Journal of Applied Polymer Science*, 125, 3211–3220.

14

The use of wheat straw fibres as reinforcements in composites

S. PANTHAPULAKKAL and M. SAIN, University of Toronto, Canada

DOI: 10.1533/9781782421276.4.423

Abstract: Wheat straw is an abundant and inexpensive lignocellulosic resource for manufacturing composites. Wheat straw-reinforced thermoplastic composites are generally prepared by melt blending followed by injection, extrusion or compression molding, and the properties of the composites vary depending on the type and form of the wheat straw used, and method of processing. Wheat straw can be used as an alternative to wood flour or other bast fibre composites with similar density, and could considerably reduce the cost of the product. Use of these inexpensive raw material in composites leads to socio-economic and environmental benefits through generating additional income to the farmers, generating cost-effective high-performing materials, and reducing the burning of the straw in the field.

Key words: wheat straw, wheat straw composites, surface modification, processing; wheat fibres; wheat nanofibres.

14.1 Introduction

Agricultural residues are readily available and inexpensive renewable lignocellulosic fibre resources that can be used as an alternative to woody lignocellulosic biomass. These residues have similar structure, composition and properties to those of other plant fibres and make them suitable for composite, textile and pulp and paper application. The demand for agro-residues as alternatives to woody cellulosic fibres has risen recently due to the increased awareness of global deforestation and environmental concerns of burning the residues, and also because of the low cost of the residues compared to other plant fibres. Further, use of agro-residues can alleviate the shortage of wood resources in countries where there are few forestry resources. The major lignocellulosic agricultural residues are wheat, rice, barley straw, corn stover, sorghum stalks, coconut husks, sugarcane bagasse, and pineapple and banana leaves. Not all of these agro-fibres are available across the world, but they grow in different regions depending on the climatic conditions required to grow food crops. Wheat straw is one of the agro-residues with the highest cellulose content of all agricultural fibres (see Agricultural residues: A promising alternative to virgin wood fibre;

www.wood consumption.org) and is one of the potential agro-fibre resources for the composites industry. In this chapter we try to explore the potential of wheat straw as a lignocellulosic resource for composites with its availability, structure and composition, and utilization in composites.

14.2 Worldwide availability and economics

Wheat straw (*Triticum aestivum* L.), the residue after wheat harvest, is produced in large quantities all over the world. The production of wheat crops in different regions of the world during 2011/12 is given in Table 14.1 (USDA database) and the total production of wheat in comparison with other crop residues is given in Table 14.2. According to USDA, world production of wheat in 2011/12 was 697.2 million tons, and at a conservative straw to grain ratio of 1:1 the production of wheat straw was probably of the order of 697 million tons also. Based on the data (Table 14.1), in North America, about 80 million tons of straw was available during the year 2011/12. As a matter of fact, about 30% of straw is required to practice good

Table 14.1 Worldwide production of wheat crop in the year 2011/12

Country/region	Total production (million tonnes)
North America:	
United States	54.4
Canada	25.3
Mexico	3.6
EU-27	137.3
Russia	56.2
Ukraine	22.3
Asia:	
China	117.4
India	86.9
Indonesia	–
Pakistan	25.0
Thailand	–
South America:	
Argentina	15.5
Brazil	5.8
Australia	29.9
South Africa	2.0
Others	96.7

Source: USDA database.

Table 14.2 Global production of wheat straw in comparison with other agro-residues in the year 2011/12

Crop	Cultivation area (million hectares)	Total production (million tonnes)
Wheat	221.29	697.2
Rice	159.22	465.81
Corn	169.64	883.25

Source: USDA database.

tillage soil conservation as determined by USDA (US Department of Agriculture). Despite leaving such a large fraction of these residues in the field to promote tillage and to supply nutrients, a large amount of straw is still available for other uses. For example, about 23 million tonnes of waste wheat straw is produced in Canada annually (Meissner and Shore, 2012). Only a small percentage is used in applications such as animal feed and bedding, construction of mud houses, and energy production, and the majority of the straw generated is left on the field to decompose. The general practice for disposing of the straw is to burn it in the fields and/or bury it in the fields in greater quantities than are required for soil conservation; however, both these practices have raised environmental concerns in many countries (Mohanty *et al.*, 2005). For example, in California about 56,000 tons of carbon monoxide is released to the atmosphere annually by burning these straws (see Agricultural residues: A promising alternative to virgin wood fibre, www.wood consumption.org). Legislation for mandatory reduction of burning is in place in many countries. These unused straws contain a large amount of fibre and have a great potential to replace wood for the production of a wide variety of industrial materials such as composites, pulp and paper, fuels and chemicals. Use of these fibres for industrial applications is a much more environmentally friendly practice and favorably contributes to the straw disposal problem as well as the overall CO_2 balance as a carbon sink.

Wheat straw is a low-cost agricultural residue and the price of the straw in Ontario was reported to be $0.10–0.15/lb based on an annual average production of wheat straw in 2008 (Ng, 2008). Mishra and Sain (2009) also reported that the straw is available from local farmers at a price of $0.02/lb. The price of this agro-residue is much lower than that of the other bast fibres such as jute, flax and hemp. For example, the price of jute fibre was about $1.45/kg in 2008 (www.World jute.com, 2008) and the price of wood flour and wood fibre varies in the range of $0.14–0.25/lb. Hence, the use of straw in composites provides substantial cost advantages over the currently used fibres. In assessing the economic potential of the utilization of wheat straw for composite manufacturing in comparison with wood flour composites, our group has performed an economic analysis of the production of wheat straw–thermoplastic composites in comparison with wood flour composites (Mishra and Sain, 2009). In this study, the total production cost of the compounded composite pellet preparation was calculated based on the capital cost involved and the operating cost at 80% operating time. The results indicated that the price for the production of the compounded Wiley milled wheat straw–PP composites ($0.63/lb) was less than that for wood flour–PP composite pellets with more or less similar mechanical properties ($0.66/lb). In short, use of these residues for value-added industrial applications provides a win–win scenario to the industry and to farmers as well as to the environment. Since wheat straw is a low-cost residue compared to other lignocellulosics, use of the straw reduces the price of the products, can support the farmers by providing an additional source of revenue for them, and provides environmental benefits without affecting the soil fertility (http://www.fao.org.DOCREP/004/Y1873E/y1873eob.htm).

14.3 Structure and composition of wheat straw

Physico-chemical and mechanical properties of any fibre depend on its structure and composition, and knowledge of these can shed light on the separation, processing and end-use of the straw. The details of structure and composition of wheat straw are described in the following sections.

14.3.1 Morphological structure

Macroscopically, stems of wheat straw are erect cylindrical structures, and comprised of internodes separated by nodes as shown in Fig. 14.1. Nodes are the hard points at which leaves are attached to the stem. The main botanical fraction of straw includes nodes, internodes and leaves. The stems of wheat straw usually have six internodes and, depending on the strain,

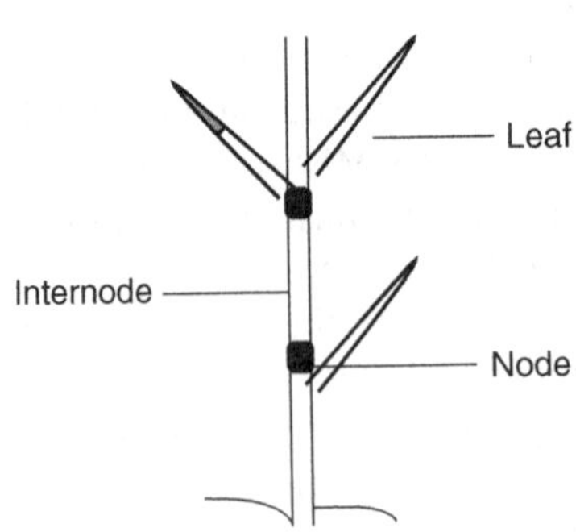

14.1 Sketch of a wheat stem.

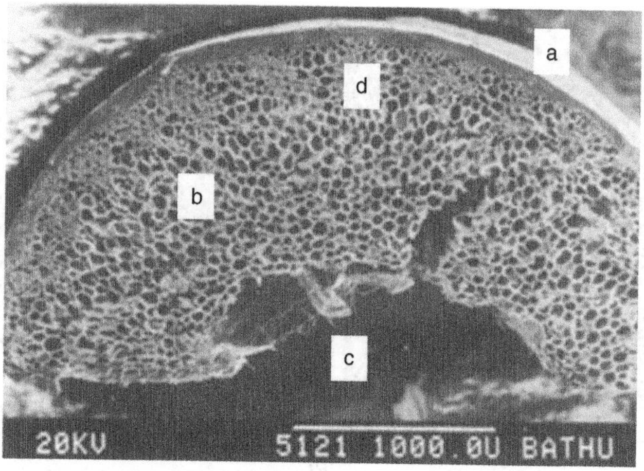

14.2 Polarized optical microscopic image of wheat stem cross-section showing different cell types: (a) epidermis, (b) parenchyma, (c) lumen and (d) vascular bundles (reprinted from White and Ansell, 1983, with kind permission from Springer Science and Business Media).

soil and climatic conditions, the stems grow from 0.5 to 1.5 m tall (Cooper *et al.*, 1999). The internodes are formed as concentric rings with void or lumen in the centre. The chemical composition of each fraction is different. For example, the epidermis is rich in cellulose and silica compared to other portions. The relative proportion of each fraction in the straw varies depending on the species, maturity at harvest, soil, climatic conditions, etc. Relative proportions of the fractions in a wheat straw were reported as follows: internodes 68.5%, leaves–sheaths 20.3%, leaves–blades 5.5%, nodes and fines 4.2%, and grains and debris 1.5% (Mckean and Jacobs, 1997).

Wheat straw stems are usually hollow and about 0.5 mm thick and consist of a heterogeneous population of cell types. This stem has a multilayer structure around the hollow lumen in the centre with an outer layer called the epidermis, and beneath the epidermis a loose layer of vascular and parenchyma bundles as shown in Fig. 14.2. The function of the epidermis is to give additional strength to the fibre and to prevent evaporation from the plant, whereas parenchyma bundles are responsible for the photosynthesis, storage of carbohydrates, and transport of water and solute throughout the plant (Hornsby *et al.*, 1997). Straw contains about 50% of bast fibres distributed in the epidermis and parenchyma cells.

In wheat straw, cellulose fibrils, which are responsible for the strength of the fibres, are present in the epidermis and parenchyma cells. However, it was found that the arrangement and morphology of cellulose microfibrils in these regions are different (Fig. 14.3). Knowledge of the structure of

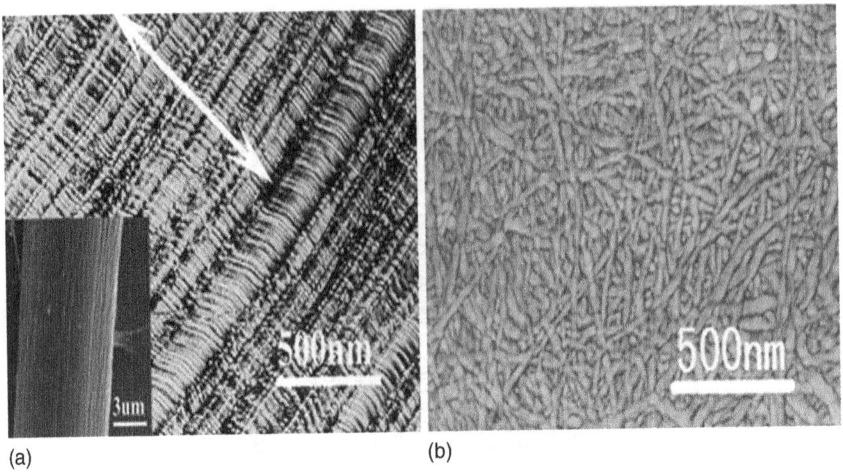

14.3 AFM photographs of cellulose microfibril arrangement in wheat straw: (a) epidermis and (b) parenchyma cells (insert in (a) is the SEM topography of the whole epidermal fibre) (reprinted from Yu *et al.*, 2008, © 2008, with permission from Elsevier).

cellulose microfibrils in the wheat straw can shed light on the deformation behavior of the cells and can provide information for the preparation of composites.

It was clear from the AFM photographs that cellulose fibrils in the epidermis layer are well arranged in the longitudinal direction (shown by arrows in Fig. 14.3(a)), whereas they do not have any preferred orientation in the parenchyma part (Liu *et al.*, 2005; Yu *et al.*, 2008). The fibrils in the epidermis layer were reported to be uniform and have a diameter of 20 nm. Thousands of microfibrils are packed in a well-ordered arrangement in the growth direction of wheat straw, leading to a crystal structure (Yu *et al.*, 2008). The diameter and length of the fibrils in the epidermis layer of the wheat stem are respectively 20 nm and 150–200 nm. On the other hand, in the parenchyma cells, the microfibrils are arranged in an interwoven network pattern, where the microfibrils interweave tightly and randomly without any particular orientation. The diameters of the microfibrils in this region are of the order of 20–30 nm and the length of the fibrils varies from 100 nm to several micrometres.

14.3.2 Ultra structure of wheat straw

Just like wood fibres, wheat straw fibre also consists of middle lamella (ML), primary wall (P), and three layers of secondary walls (S1-outer, S2-middle, and S3-inner) as shown in Fig. 14.4. The primary cell walls of wheat straw are composed of low amounts of pectic polysaccharides, non-cellulosic

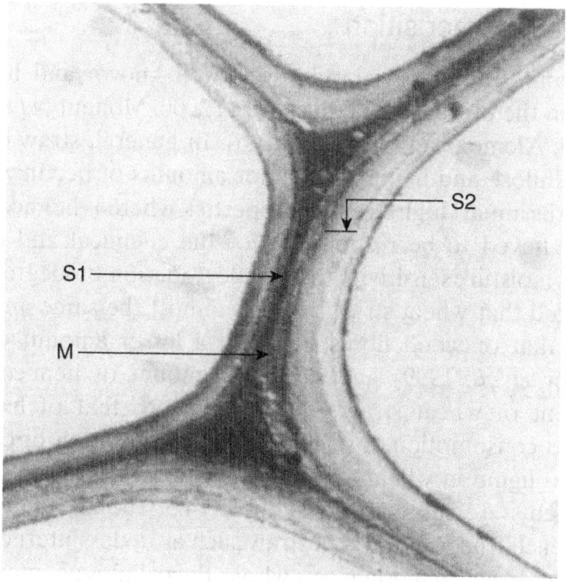

14.4 Transverse section of wheat straw, ×6000 (Zhai and Lee, 1989, reprinted by permission of the publisher, Taylor and Francis Ltd, http://www.tandf.co.uk/journals).

polysaccharides such as glucuronoarabinoxylans and β-1,3-β-1,4 glucans (Carpita and McCann, 2000), and hydroxycinnamic acids (Lam *et al.*, 1990, 1992), whereas the secondary cell wall is composed of cellulose, hemicelluloses and lignin.

The orientations of cellulose fibrils are different in various layers in wheat straw. The fibrils of the primary wall display a netlike structure, whereas in S1, the fibrils are oriented helically and almost perpendicular to the fibre axis. Fibrils with an angle of 20–30° can be seen in S2. It is reported that the thick S1 layer in wheat straw, with the fibrils oriented laterally in cross-helix, makes the defibrization of wheat straw difficult (Zhai and Lee, 1989).

The composition and composition distribution of wheat straw fibre cell walls vary widely with the cell origins and types of cells, and lead to different performances and properties. For example, the waxy and silica-rich epidermis layer of wheat straw is denser and is more resistant to water absorption than the lumen inside. The width of individual cells in wheat straw is smaller compared to wood cells and they can therefore form relatively finer fibres. Typical length and width of wheat straw fibre cells are 0.4–3.2 mm and 8–34 μm respectively (Reddy and Yang, 2005). However, their high *l*/*d* ratios favor their use in the manufacturing of paper and they can also be used for reinforcing polymers. Crystallinity of the fibres is reported in the range of 55–65% (Reddy and Yang, 2005).

14.3.3 Chemical composition

The chemical composition of wheat straw is well known and has been widely reported in the literature (Lequart *et al.*, 2000; Mohanty *et al.*, 2005; Foyle *et al.*, 2007; Alemdar and Sain, 2008a,b). In general, straw contains cellulose, hemicellulose and lignin with minor amounts of pectin, wax and ash. Cellulose fibrils impart high strength properties, whereas hemicelluloses and lignin are believed to be responsible for the chemical and physical properties such as moisture sensitivity, thermal degradation, biodegradability, etc. It was reported that wheat straw contains almost the same amount of holocellulose as that of wood fibres, but with a lesser amount of alpha cellulose (Cooper *et al.*, 1999) and a higher amount of hemicelluloses. The lignin content of wheat straw is comparable to that of hardwood fibres (www.wood consumption.org). The typical composition of cellulose, hemicellulose and lignin in wheat straw along with the structural features of these components are summarized in Table 14.3. Though the chemical compositions of each fraction of wheat straw such as nodes, internodes and leaves are different from each other, as a whole, the cellulose content varies from 33 to 42%, the hemicellulose content from 25 to 35%, and the lignin content from 14 to 23% (Lawther *et al.*, 1996; Sun *et al.*, 1998; Cooper *et al.*, 1999; Alemdar and Sain, 2008a). The extractive content is low, with about 1%, and straw has a relatively high ash content of 6–13%. The composition of wheat straw varies depending on the type of wheat straw, the climatic conditions, the geographical locations or places of cultivation, the age of the straw and the extraction conditions used to obtain the fibres. Table 14.4 demonstrates the variation in the chemical composition of straw reported by different authors.

Table 14.3 Structural characteristics of wheat straw

Component	Cellulose	Hemicellulose	Lignin
Weight fraction	33–52%	20–36%	10–23%
Structural features	About 40–60% crystallinity; cellulose chains with degree of polymerization of 500–1500 bonded into semi-crystalline microfibrils	About 70–90% xylan with substitutions of arabinan, uronic acid and acetyl groups. They are connected to lignin through ferulic and *p*-coumeric acid bridges	*P*-hydroxyphenyl-guacyl-syringyl phenolic monomers with a ratio respectively of 5%:49%:46%. Highly amorphous and branched polymer forming a protective shell around the cellulose–hemicellulose network

Table 14.4 Chemical composition of wheat straw from different sources

Source	% cellulose	% hemicellulose	% lignin	% ash	% silica	% extractives
Mckean and Jacobs, 1997	33.7	25.0	16–17	7.5–8.5	4.5–5.5	5.8
Cooper *et al.*, 1999	50–52	26–30	16–20	5–10	4–8	–
Mohanty *et al.*, 2005	39	36	10	–	6	–
Saha *et al.*, 2005	48.6	27.7	8.2	6.7	–	–
Perez *et al.*, 2008	37.6	24.7	17.4	4.8	–	–
Fu *et al.*, 2010	34.5	21.3	17.5	2.7	–	–

Other components in wheat straw are pectin, wax and ash. Pectin is a complex hetero-polysaccharide composed of methoxyl ester and sugar substituted galacturonic acid residues. Wheat straw pectins are low molecular weight and low viscosity polysaccharides and contain 5% of low-methoxy partially acetylated pectin with 25% galacturonic acid (Volyets and Dahman, 2011). These polymers are generally used in low-calorific, high-fibre beverages as they do not have gelling properties like hemicelluloses. Wheat straw also contains about 1% of wax by weight and is mainly present in the outer layer or epidermis of the straw (Deswarte *et al.*, 2006). Wax is primarily long-chain fatty acids and fatty alcohols, sterols and alkanes. Ash content represents the mineral content of wheat straw and varies from 6–13% depending on the soil and environmental conditions. Silica is the major component of wheat straw ash and accounts for 80%, the rest being metals such as sodium and potassium (Biricik *et al.*, 1999).

14.3.4 Applications

Wheat straw has traditionally been used for pulp and paper making and also for the production of regenerated cellulose fibres as an alternative to wood for cellulose-based materials (Staniforth, 1979; Focher *et al.*, 1998; Majumdar and Chanda, 2001). Since wheat straw is an excellent resource for cellulosic fibres, it can be used as reinforcements and/or fillers for non-structural and structural composites, and an immense amount of research in this direction has been observed during the last decades. This chapter also provides a short overview of the research on the use of wheat straw as a fibre resource for composites. Separation of cellulose microfibrils from wheat straw is also exploited in order to develop high-performing composites for high-end applications. They can also be used for other industrial applications such as production of fuels, chemicals, enzymes and food. Since wheat straw is rich in hemicelluloses, straw is another resource for polysaccharides and other fine chemicals derived from these polysaccharides. In short, wheat straw can be used for the manufacture of biopolymers such as polysaccharides, and fine chemicals. Some of the common products and chemicals from the cellulose and hemicelluloses from wheat straw include fibre for composites, particleboards, pulp and paper applications, adhesives and binders, surfactants and coagulants, furfural, ethanol and other alcohols (Reddy and Yang, 2005).

14.4 Wheat straw as a polymer composite reinforcement

The potential of any reinforcing lignocellulosic fibre depends on its chemical structure, chemical composition and mechanical properties. Though the cellulose content of wheat straw is less than that of wood fibres, our studies on the performance of wheat straw thermoplastic composites showed that straw can be used as an alternative to wood flour in composites (Panthapulakkal and Sain, 2006b). In general wheat straw in the cut or milled form is used in composites manufacturing. Milling can be done using a Wiley mill, a hammer mill or ball milling. Wheat straw fibres can also be prepared by mechanical and chemical refining. Details of the studies on wheat straw in different forms as reinforcements in various polymer matrices are given in Table 14.5. The properties of composites mainly depend on the amount of straw used, its chemical composition, its surface morphology and its mechanical properties.

The effect of chemical composition, surface morphology and mechanical properties of wheat straw fibres prepared using mechanical and chemical refining, and fungal treatment followed by a mechanical refining process on the reinforcing potential of these fibres in polypropylene, has been investigated (Panthapulakkal *et al.*, 2006; Panthapulakkal and Sain, 2006a). Chemical pulping was achieved through soda pulping using 15% NaOH solution. The pulping conditions used were 120°C at 1.5 kg/cm^2 pressure, with a solid to liquid ratio of 1:5 (w/v) for 450 g of fibres using a laboratory-scale M/K digester. Mechanical refining has been achieved using a laboratory-scale Sprout-Waldron refiner, by shearing the wet straw fibres between the grooved discs (passing through the refiner three times) with a gap width of 2–3 mm. The morphological differences between the fibres prepared using different processes are shown in Fig. 14.5. The mechanically defibrillated fibres were found to contain the cellular residues on the fibres, whereas alkaline pulped fibres did not have the surface residues and are smooth. Fungal retted fibres have lesser amounts of residues on the fibre

Table 14.5 Processing of thermoplastic composites of wheat straw in different forms and the properties studied by different researchers

Polymer	Wheat straw fibre size	Fibre content	Processing method	Properties studied	Reference
Polyethylene (MFI = 11 g/10 min) and polypropylene (MFI = 8 g/10 min)	50 mesh	35	Melt blending using Haake Buchler system (180°C for 10 min at 40 rpm) followed by granulation and injection molding (temperature: 185°C, pressure 3 MPa)	Thermal degradation properties and coefficient of thermal expansion	Zabihzadeh, 2010
Polypropylene (MFI = 12 g/10 min)	Average length 0.3 mm	20	Melt blending using Thermo Haake Rheomix (170°C for 7 min at 55 rpm) followed by granulation and compression molding (180°C for 4 min)	Tensile, dynamic mechanical and thermal properties, melt rheological properties and morphological analysis	Pan *et al.*, 2010
Polypropylene (MFI not reported)	Wheat fibre – type 1: length 20–40 mm; type 2: length 0.5–1.5 mm	40	Melt blending using a high-intensity K-mixer at 175°C followed by granulation and injection molding	Mechanical properties, melt flow, heat deflection temperature, coefficient of thermal expansion and water absorption	Mishra and Sain, 2009
Polypropylene (MFI = 12 g/10 min)	Wheat flour; wheat fibre 1.87 mm; fungal modified wheat flour	30, 40	Wheat flour: melt blending using high-intensity K-mixer at a preset temperature of 185°C. Wheat fibre: melt blending using a Brabender mixer at 180°C for 5 min at 60 rpm, followed by granulation and injection molding	Mechanical properties	Panthapulakkal and Sain, 2006a,b

Continued

Table 14.5 Continued

Polymer	Wheat straw fibre size	Fibre content	Processing method	Properties studied	Reference
Polypropylene (MFI = 12 g/10 min)	Mechanically refined wheat fibre (1.87 mm) and chemically pulped wheat fibre (2.70 mm)	30	Melt blending using Brabender mixing at 180°C, 60 rpm for 5 min	Mechanical properties	Panthapulakkal *et al.*, 2006
Linseed oil-based thermoset resin	Wiley milled wheat straw (0.5, 1 and 2 mm)	50–90	Compression molding at 140°C for 3 h, followed by 160°C for 3 h under a pressure of 400 psi, followed by post-curing at 180°C under atmospheric pressure for 2 h	Mechanical, thermal, dynamic mechanical and water absorption properties	Pfister and Larock, 2010
High-density polypropylene	Wheat flour of 0.25 mm	4–30	Dry mixing in a high-intensity mixer, and the blend melt blended using a single-screw extruder at 40 rpm at 170–200°C, followed by granulation, injection molded with foaming agents	Mechanical and thermal properties; morphological analysis	Mengeloglu and Karakus, 2012

High-density polypropylene (MFI = 16 g/10 min)	Wheat flour 50 mesh size	35	Melt blending using Haake Buchler system (180°C for 10 min at 40 rpm) followed by granulation and injection molding (temperature 185°C, pressure 3 MPa)	Tensile and impact strength; morphological analysis	Zabihzadeh and Dastoorian, 2010
Spunbonded PP (MFI = 38 g/10 min)	Wheat straw fibre: 1, 5 and 10 cm	40–80	Compression molding at a temperature of 185°C for a holding time of 80 s followed by a cooling to 35°C	Mechanical and acoustic properties	Zou *et al.*, 2010
Polypropylene powder (MFI = 12 g/10 min)	Wheat flour 35 mesh size and 2–5% clay	30–50	Melt blending using a Haake Minilab micro-compounder (co-rotating conical twin-screw extruder) at 190°C and 40 rpm followed by pelletization and injection molding (190°C at 100 psi)	Flexural properties and water absorption properties	Reddy *et al.*, 2010
PLA	Wheat straw fibres		Melt blending using microextruder under temperature conditions between 195 and 205°C, screw speed 100 rpm and cycle time 3 min, followed by granulation and injection molding	Bio-degradability	Pradhan *et al.*, 2010

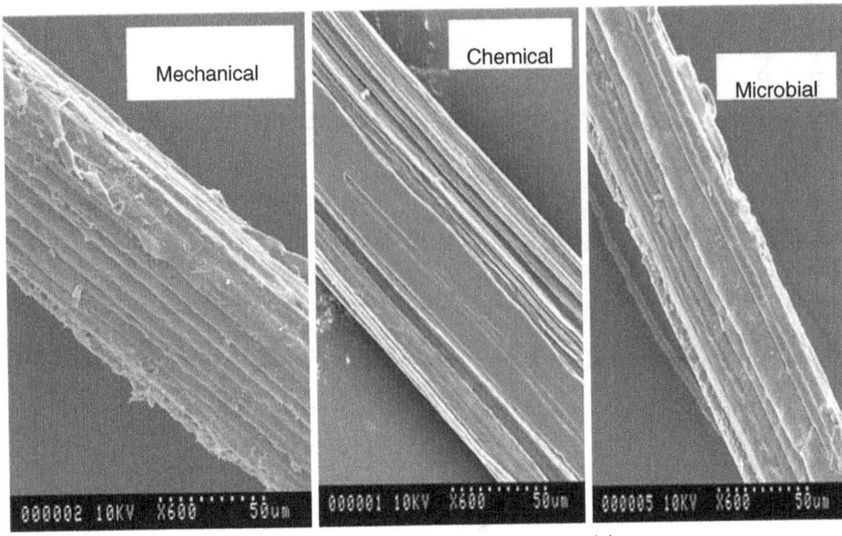

14.5 SEM of wheat straw fibres prepared by mechanical, chemical and microbial retting followed by mechanical refining: (a) mechanical, (b) chemical, (c) fungal retting followed by mechanical refining.

Table 14.6 Physico-mechanical properties of fibres prepared by different processes

Wheat straw fibres	Average diameter (mm)	Average length (mm)	Tensile strength (MPa)	E-modulus (GPa)
Mechanical processing	0.094 ± 0.03	1.87 ± 0.66	58.7 ± 51	3.7 ± 2.6
Chemical processing	0.076 ± 0.03	2.70 ± 0.74	146.3 ± 53	7.9 ± 3.7
Microbial retting	0.084 ± 0.02	1.49 ± 0.63	139.9 ± 53	4.80 ± 1.5

surface than mechanically refined fibres. The physico-chemical characteristics of these fibres are given in Table 14.6. Chemically refined fibres have lower diameter and have fewer points or defects in the fibres compared to fibres prepared by other processes, and are expected to have high strength properties. A large variation in the strength properties of these fibres is observed and is very characteristic of natural fibres.

The thermal properties of the fibres also showed an improved thermal stability for the chemically prepared fibres (Fig. 14.6 and Table 14.7) compared to microbial and mechanically refined fibres (onset of degradation 217°C versus 242°C), which is attributed to the higher removal of lignin and hemicelluloses during chemical pulping. Thermoplastic composites with these fibres were prepared with 30% fibre content in a polypropylene

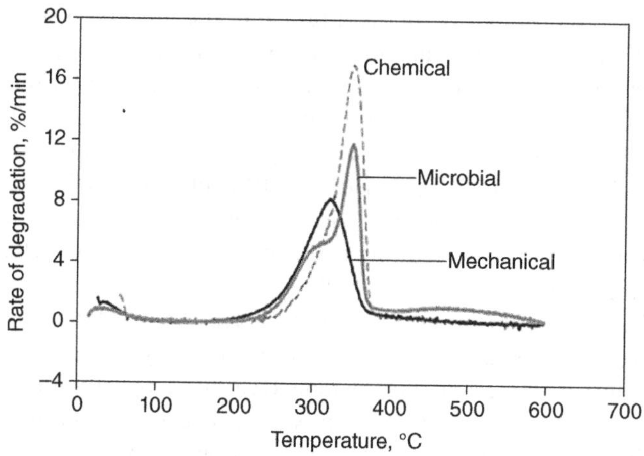

14.6 TGA curves of wheat straw fibres prepared by different processes.

Table 14.7 Thermal degradation characteristics of wheat straw fibres prepared using different methods

Wheat straw fibres	Moisture content (%)	Onset of degradation (°C)	Peak temperature (°C)	Rate of degradation (wt%/min)	Residue after 550°C (%)
Mechanical processing	5	217	324.7	8.2	25.8
Chemical processing	2	242	352.5	16.9	10.9
Microbial	4	231	351.5	11.6	9.3

matrix and it was found that tensile and flexural properties of composites increased compared to the neat polypropylene. However, despite having high strength, chemically prepared fibres exhibited lower performance compared to mechanically prepared fibres and this was attributed to the poor dispersion of the chemically treated fibres in the matrix. The pulp fibres were hard to disperse in the processing conditions used and the presence of agglomerates leads to poor mechanical properties of the composites (Panthapulakkal *et al.*, 2006).

Other than the fibre properties, interfacial interaction between the fibres and the polymer matrix is a critical factor in determining the physico-chemical properties of composites. The outer surface or epidermis of wheat straw is rich in silica and is composed of a layer of wax that degrades the adhesion between the straw and polymer resins. Physical, chemical or biological modification of straw fibres has generally been considered prior to the manufacture of composites to modify the surface properties (hydrophilic/hydrophobic) of the straw and thereby to improve the interfacial interaction between the fibre and the matrix. The available methods in the literature include ethanol/benzene extraction, steam explosion, chemical treatment (acid or alkali treatment), microbial and or enzyme modification and use of compatibilizers (Han *et al.*, 1999, 2009, 2010; Patil *et al.*, 1999; Schirp *et al.*, 2006; Panthapulakkal and Sain, 2006a; Jiang *et al.*, 2009; Zabihzadeh, 2010; Pan *et al.*, 2010; Pfister and Larock 2010). Extraction with an ethanol-benzene mixture was reported to improve the adhesion between the straw and the resins used in the wheat particleboard manufacturing process (Han *et al.*, 1999).

Pan *et al.* (2010) reported the surface modification of wheat straw fibres using alkalization (using 2% NaOH solution), acetylation (using acetic anhydride) and maleated polypropylene treatment (compounded wheat straw fibre and MAPP through a mixing process) for improving the interfacial interaction between the straw and polypropylene. The research emphasis was on the thermomechanical and rheological properties of the composites. Composites with alkali-modified wheat straw demonstrated high melt viscosity and this was attributed to the strong chemical interaction between the fibre and the polymer. The composites also exhibited a high storage modulus.

Surface modification of wheat straw by inoculating the straw with white rot fungi *Pleurotus ostreatus* for three months was reported to give comparable interfacial interaction between the straw and the HDPE matrix (Schirp *et al.*, 2006). It was also reported that the reduction in thermal stability of the straw did not influence the processability of the straw with the thermoplastic. Our results on the fungal modification of wheat straw using fungi isolated from the bark of elm trees indicated that the fungal treated fibres were stronger than the untreated fibres, due to the partial removal of extractives, hemicelluloses and lignin. Also, thermal degradation studies of these fibres suggested that these fibres were suitable for processing with thermoplastics (Panthapulakkal and Sain, 2006a). A study on the surface modification of wheat straw using lipases found that the enzymatic treatment could effectively remove the hydrophobic and lipophilic extractives and silica from the outer surface of the wheat straw without affecting the thermal stability and is expected to improve the interfacial interaction with the polymer resins (Jiang *et al.*, 2009).

Steam explosion is another technique for modifying straw fibre attributes such as morphology, wettability and silicon content. It was reported that, depending upon the treatment conditions, the morphology of the fibres changes, leading to varied surface wettability of the straw with the UF resin. During steam explosion, the straw is subjected to high temperature and pressure using saturated steam at temperatures typically in the range of 180–230°C for 2–10 minutes and the straw is quickly flashed to atmospheric pressure to get the explosion effect. During this process, water inside the substrate vaporizes and expands rapidly and leads to the destruction of the lignocellulosic network, and the fibres will get separated (Kokta and Ahmed, 1998). Morphological characteristics of the fibres after different pre-treatment conditions are different and are shown in Fig. 14.7. It is clear from the figure that explosion using higher temperature and time conditions

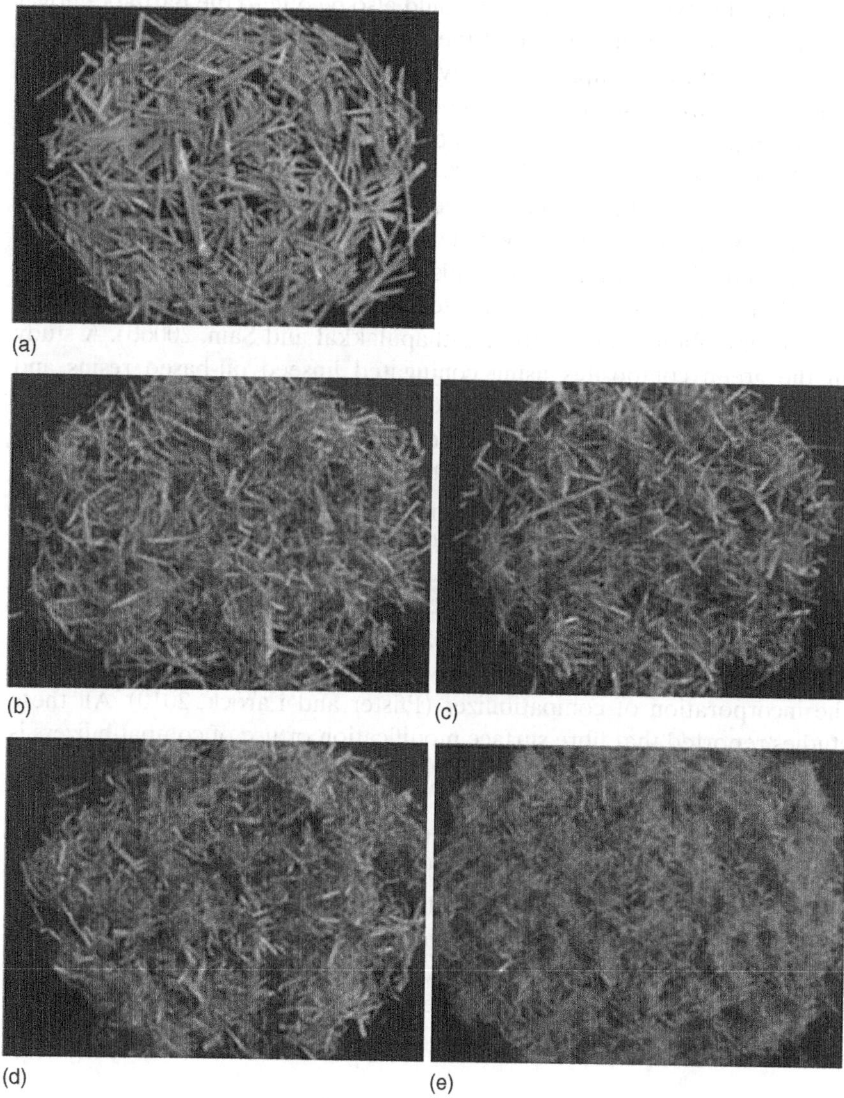

14.7 Forms of wheat straw after different steam explosion treatment conditions: (a) untreated straw, (b) treated at 190°C for 2 min, (c) treated at 190°C for 3 min, (d) treated at 200°C for 2 min, (e) treated at 200°C for (3 min) (reprinted from Han *et al.*, 2010, © 2010, with permission from Elsevier).

provides more fibres than the lower time–temperature combinations. The wettability of the straw with the resin depends on factors such as porosity, hygroscopicity and chemical composition of the straw, and the authors reported that the steam-exploded straw had improved wettability with the resin. This was suggested to be due to the porous structure of the straw as a result of the explosion effect and could also be due to the partial removal of extractives from the surface of the straw (Han *et al.*, 2009, 2010).

The addition of compatibilizers was found to increase the mechanical properties of wheat straw filled composites. We found that incorporation of compatibilizers (5% maleated PP and PE) in the wheat straw filled PP and HDPE composites led to a 24–46% improvement in tensile and flexural properties, which is believed to be due to the improved interaction between the wheat straw fibre and the polymer matrix (Panthapulakkal *et al.*, 2006). The amount of compatibilizer depends on the fiber loading in the composites, and different amount of compatibilizers (2–5%) were reported by different researchers (Zabihzadeh, 2010; Panthapulakkal and Sain, 2006b). A study on the green composites using conjugated linseed oil-based resins and wheat straw reported that incorporation of wheat straw in the resin increased the thermal and mechanical properties; however, higher amounts of wheat straw decreased the thermal degradation. About 80% of wheat straw in the resin increased the mechanical properties, whereas 90% decreased the performance. Incorporation of compatibilizer (maleic anhydride) improved the tensile strength of the composites but did not seem to affect the thermal properties of the composites. In this study about a 41–97% increase in water absorption was observed by incorporating 50–90% of wheat straw, and a small decrease was observed in the water absorption of the composites by the incorporation of compatibilizer (Pfister and Larock, 2010). All these studies reported that fibre surface modification or use of compatibilizers is required to get good compatibility between the fibres and the polymer matrices.

Patil *et al.* (1999) reported the effect of maleic anhydride grafting on wheat straw on the water absorption characteristics of wheat straw–Novolac resin in comparison with the maleated cane bagasse and teak wood and found that wheat straw-reinforced composites showed comparatively higher water absorption compared to the other fibres. The authors attributed this phenomenon to the loose criss-crossed chemical structure of cellulose and lignin in the wheat straw.

14.5 Processing of wheat straw fibre-reinforced polymer composites

Wheat straw filled or reinforced composites are materials in which the wheat straw in the particle or fibrous form is dispersed in polymer matrices.

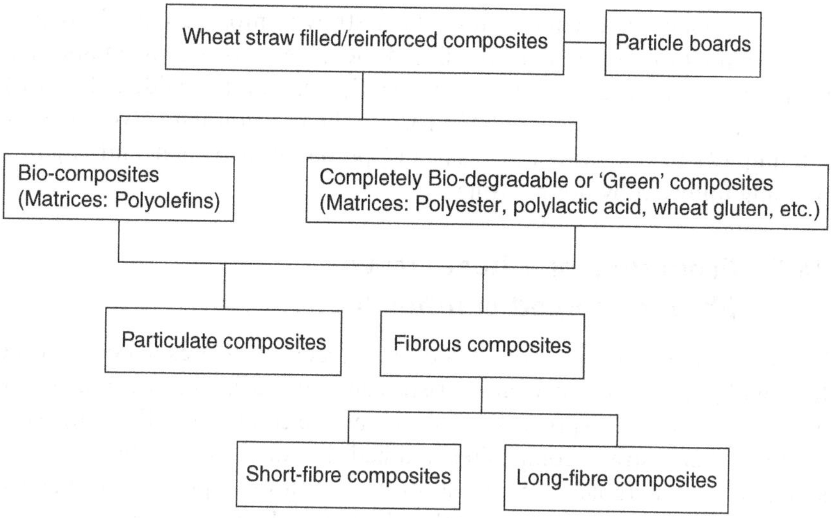

14.8 Classification of wheat straw composites.

A possible classification of wheat straw based composites is shown in Fig. 14.8 based on the fibre size and the type of polymer matrices used. Polymer matrices can be biodegradable or non-degradable. Bio-composites prepared using non-degradable polymer matrices are generally based on commodity-type plastics such as polypropylene and polyethylene (injection moldable or extrusion grade), whereas the biodegradable composites use degradable polymer matrices such as polyesters, polylactic acid, wheat gluten, etc. Depending on the fibre size wheat straw composites can be particulate-filled, short-fibre-reinforced or long-fibre-reinforced composites. The important factors that have to be considered during processing of the composites are processing temperature, moisture content of the straw and compatibility of the fibres and the matrix. Just as in any other lignocellulosic fibre processing, the moisture content has to be low in order to avoid defects in the polymer composites. Also, because of the thermal degradation of the straw above 200°C, usually thermoplastics with lower melt temperature, such as polyolefins, are preferred in the manufacture of the composites.

Commonly used manufacturing processes for wheat straw-reinforced composites are given in Table 14.5. Usually the procedure involves (1) size reduction of the straw, which includes either cutting, hammer-milling, Wiley milling and/or ball milling followed by screening to get uniform sized fibres; (2) melt blending of the thermoplastic polymer with the straw (untreated or pretreated) in the presence of compatibilizers; (3) granulation or pelletizing; and (4) injection molding or extrusion (using single screw or twin-screw extruder) or compression molding. Other than the above-mentioned composites, wheat straw-reinforced particle boards are manufactured using hot pressing of the fibre mats with formaldehyde resins such as urea-formaldehyde (UF), phenol-formaldehyde (PF), melamine-urea-formaldehyde (MUF), etc. The procedure involves (1) fibre refining, (2) mixing with the resin and preparation of mat and (3) pre-pressing followed by hot pressing.

14.6 Properties of wheat straw fibre-reinforced composites

The properties of wheat straw filled/reinforced composites depend on the fibre and polymer matrix characteristics and interfacial interaction, as well as the method used to process the fibres and the composites. The properties of the wheat straw composites studied by different researchers are summarized in Table 14.5. Most of the authors reported mechanical properties of the composites and the effect of polymers or compatibilizers on the mechanical properties, as these are important in directing the application of the composites. Other important properties in view of industrial applications of these composites include thermal, water absorption and rheological properties of the composites and were also studied by a few researchers (see Table 14.5). Some of the notable results are described below.

The mechanical properties of the composites mainly depend on the polymer matrix, fibre type, fibre form, dispersion of the fibres in the matrix and interfacial interaction between the fibre and the matrix. The effect of fibre size and processing conditions on the properties of wheat straw reinforced polypropylene composites were studied using two types of melt blending followed by injection molding (Panthapulakkal and Sain, 2006b). In one process, the melt blending was achieved using a high-intensity K-mixer (tip speed was 3200 rpm, preset temperature 185°C, and residence time about 2–3 minutes), whereas in the other process a low-shear Brabender mixer (185°C, 60 rpm for 5 minutes) was used. The mechanical properties of the composites indicated that wheat fibre-reinforced composites had comparatively higher strength than particulate composites, after melt blending using a low-shear mixer. However, there was no difference in the strength performance of the fibre and particulate composites after high-shear melt blending (Fig. 14.9). This similarity was attributed to the extensive fibre breakage that occurred during the high-shear mixing compared to the low-shear blending. Comparison of the strength properties of wheat straw with wood flour composites showed that wheat straw can be used as an alternative to wood fibres (Table 14.8).

Mechanical and water absorption properties of wheat straw polypropylene composites were reported by Kruger (2007). The variables used to study

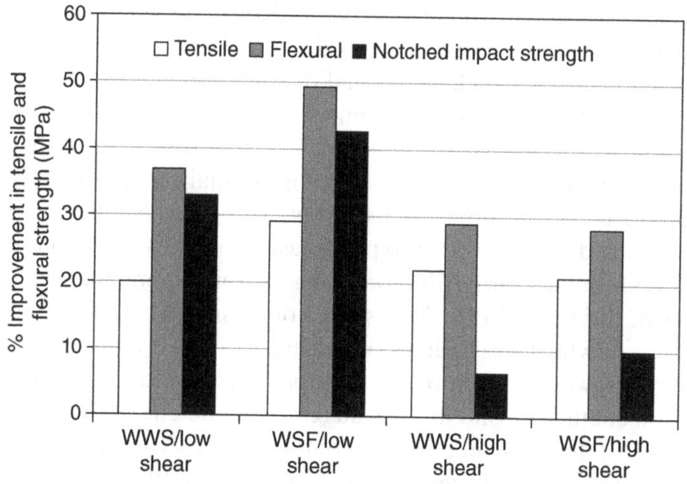

14.9 Mechanical properties of wheat straw–polypropylene composites prepared using different forms of wheat straw under different processing conditions (WWS: Wiley milled wheat straw, WSF: wheat straw in the fibre form).

Table 14.8 Comparison of mechanical properties of wheat straw–PP composites with wood flour–PP composites

Property	Wheat straw	Wood flour
Tensile strength, MPa	42.4 ± 2.2	42.7 ± 0.2
Tensile modulus, GPa	3.1 ± 0.2	2.9 ± 0.1
Flexural strength, MPa	79.9 ± 4.2	80.4 ± 1.1
Flexural modulus, GPa	4.0 ± 0.1	3.6 ± 0.2
Notched impact strength, J/m	28 ± 1	29 ± 2

the performance properties were filler content, processing at different conditions, and effect of additives. Particle size and processing parameters such as temperature and screw rotation were found to affect the final properties of the composites. Addition of compatibilizers increased the compatibility between wheat straw and PP and lead to increased strength and resistance to water absorption. Johnson *et al.* (1997) also reported the compounding of 30 wt% reinforced PP composites using a twin-screw extruder and processed using injection molding of the granulated pellets. The composites had a tensile and flexural strengths of approximately 40 MPa and 60 MPa respectively. The modulus value of the composites were twice the value of the base polymer. However, the impact strength of the composites considerably decreased.

Thermal properties such as thermal degradation and thermal expansion coefficients of the composites are important properties to be aware of for the practical application of wheat straw composites. A comparison of thermal degradation and the coefficient of thermal expansion of PP and PE wheat straw composites (65% wheat straw, 35% plastic) indicated that these properties are dependent on the type of plastics and the type of compatibilizer used (Zabihzadeh, 2010). Addition of compatibilizer enhanced the thermal stability of the PP and PE wheat straw composites. The thermal expansion coefficient of wheat straw filled composites was lower than that of the neat polymers; however, addition of compatibilizer increased the thermal expansion coefficient, and this was suggested to be due to the large thermal expansion of the polymer in the composite. PP-based composites were found to have lower thermal expansion than HDPE composites. This is because of the difference in the structure of the polymers as the thermal expansion of semicrystalline polymer and composites depends on degree of crystallinity, crystal size, processing parameters, volume fraction of fibres, thermal history, fibre orientation and distribution in the matrix.

Heat deflection temperature (HDT) and water absorption properties of composites are also important from the industrial application point of view and these properties of wheat straw–polypropylene composites are given in Table 14.9 in comparison with the virgin polymer. It is clear that HDT increased with the incorporation of the wheat straw and this is due to the increase in the stiffness of the polymer with the incorporation of the wheat straw. Water absorption is considerably higher for the wheat straw–PP composites and this is expected for the lignocellulosics due to their hydrophilic behavior (Mishra and Sain, 2009). More research in this direction is needed to manufacture composites with low water absorption. Hybrid composites using wheat straw in combination with clay have been developed to reduce the water absorption characteristics of composites intended for use in outdoor applications (Reddy *et al.*, 2010). However, the results

Table 14.9 Heat deflection temperature and water absorption of wheat straw–polypropylene composites

Material	Heat deflection temperature (°C)	Water absorption (%)
Polypropylene	121	0.10
40% wheat flour + compatibilizer	157	1.06

indicated that incorporation of the clay did not alter the water absorption resistance of the composites studied.

14.7 Potential applications of wheat straw fibre-reinforced composites

Traditionally, fibres from straw have been used in the pulp and paper industry as a fibre source, for manufacturing particleboards, as a building material, and for animal feed and bedding. Other applications of straw fibres include the housing and building sector, composite manufacturing, thermal insulation, and the energy sectors (Schirp *et al.*, 2006). Wheat straw-based composites can be used in semi-structural and structural applications in automotive interiors, where natural fibres are currently being used. Recently the Ford Motor Company announced the utilization of wheat straw-based composites in interior parts such as quarter-trim bin in their cars (Fig. 14.10). Besides the automotive industry, these composites can find use as an alternative to wood flour-reinforced composites in applications such as construction, housing, office panels and furniture. Utilization of wheat straw for industrial applications leads to substantial consumption of the straw, as well as reducing the cost of the product as the straw is available at low cost (0.04/kg, the cost of collection from the field) compared to other bast fibres such as jute ($1.45/kg) (Zou *et al.*, 2010).

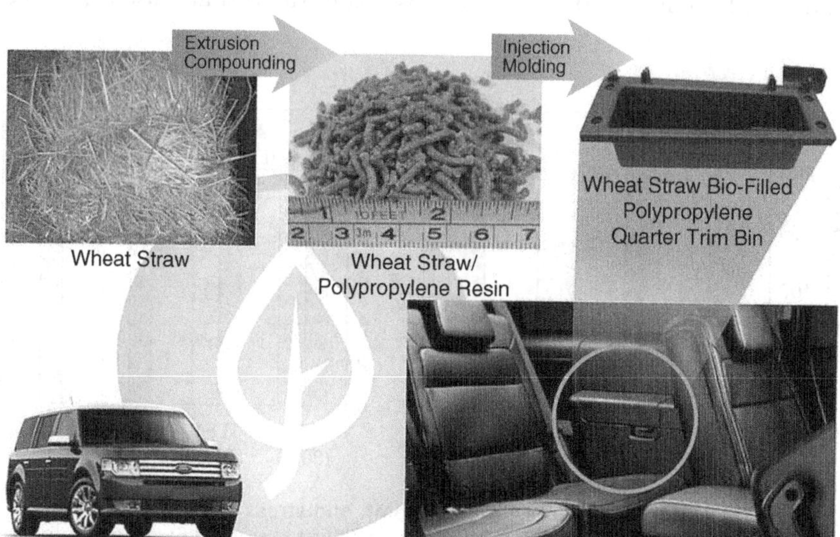

14.10 Wheat straw composite applications (adapted from Milewski, 2012, with permission from the Organizing Committee of the 12th International Conference on Biocomposites).

14.8 Future trends

Wheat straw is one of the non-wood lignocellulosic feedstocks for the bio-refinery industry, in which all of the major components of the straw can be utilized for industrial applications. As the straw contains more pentosans compared to woody counterparts, it can be an ideal candidate for hemicellulose extraction. The rest of the fibre which is rich in hemicelluloses and lignin can be used for making composites or for isolation of nanocellulose. For example, wheat straw fibres after fractionation and extraction of hemicellulose sugars using an acid hydrolysis, have been used to study the compatibility between these fibres with biodegradable and aromatic co-polyester (polybutyleneadipate-*co*-terephthalate, PBAT) for manufacturing a completely green composite (Le Digabel and Averous, 2006). The authors reported that depending upon the fractionation conditions, the amount of lignin content in the fibres, the size or granulometry distribution, and the surface characteristics of the fibres vary. Further, the authors could prepare composites with a large range of mechanical properties (Fig. 14.11) that can fulfill the requirements for various applications such as non-food packaging, or other short-lived applications in the agricultural and sports areas.

Recently, a complete green composite using wheat gluten and wheat straw fibres has been prepared by Montano-Leyva *et al.* (2013). Wheat gluten is a by-product of the wheat starch industry with good thermoplastic and film-forming properties and has been successfully used for the

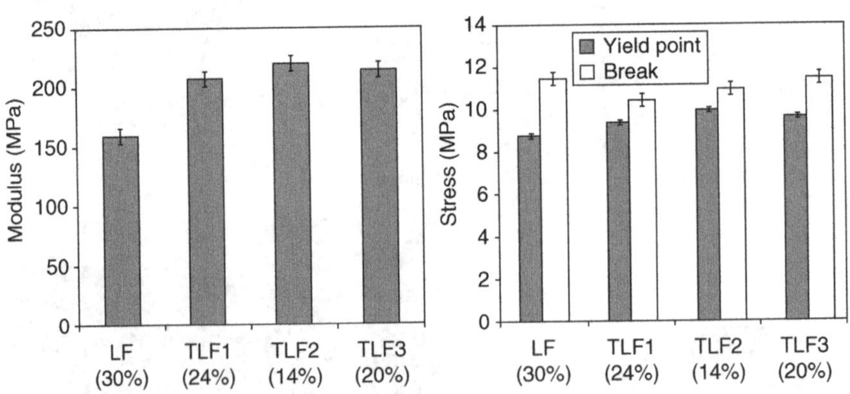

14.11 Modulus, stress at yield point, and stress at break of PBAT based bio-composites with different filler fractions (LF: lignocellulosic fibre, lignin content 30%; TLF1: treated lignocellulosic fibre, lignin content 24%; TLF2: lignocellulosic fibre, lignin content 4%; TLF3: lignocellulosic fibre, lignin content 20%) (reprinted from Le Digabel and Averous, 2006, © 2006, with permission from Elsevier).

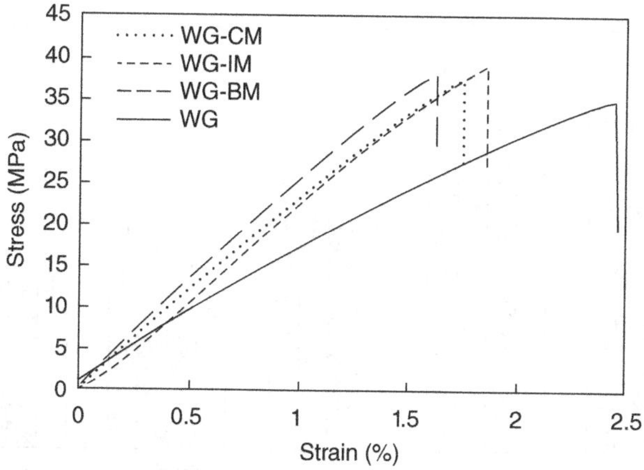

14.12 Typical stress–strain curves of the wheat gluten and wheat straw filled wheat gluten 'green' composites (11.1 vol% of wheat straw) (WG: wheat gluten; WG-CM: wheat gluten filled with cut-milled wheat straw; WG-IM: wheat gluten filled with impact-milled wheat straw; WG-BM: wheat gluten filled with ball-milled wheat straw) (reprinted from Montano-Leyva *et al.*, 2013, © 2013, with permission from Elsevier).

production of eco-friendly agro materials. Because of the lower cost of the straw, incorporation of wheat straw into wheat gluten could develop low-cost green industrial materials. The focus of the research was to get scientific knowledge about the effect of fibre characteristics on the mechanical properties of the composites and to investigate to what extent the wheat straw fibres can be incorporated to the wheat gluten to decrease the final cost of materials without changing the mechanical properties. It was reported by the authors that a fibre content of 11.1 vol% of cut-milled and impact-milled wheat straw or only 1.2 vol% of ball-milled wheat straw could produce equivalent or improved properties to wheat gluten-based materials. The stress–strain plots of the wheat gluten filled with wheat straw in different forms are shown in the Fig. 14.12. The results indicated that cut-milled or impact-milled wheat straw fibres can be incorporated to wheat gluten in large amounts to produce low-cost wheat gluten-based industrial materials. Other than thermoplastic composites, potential use of wheat straw in HDPE foams has been investigated by Mengeloglu and Karakus (2012) who found that 20% of wheat straw filled foams were the best in terms of density and with moderate mechanical properties.

In most of the reported work, wheat straw has been used in short fibre or ground or milled form to produce composites; however, recently, lightweight composites with long wheat straw have been prepared using

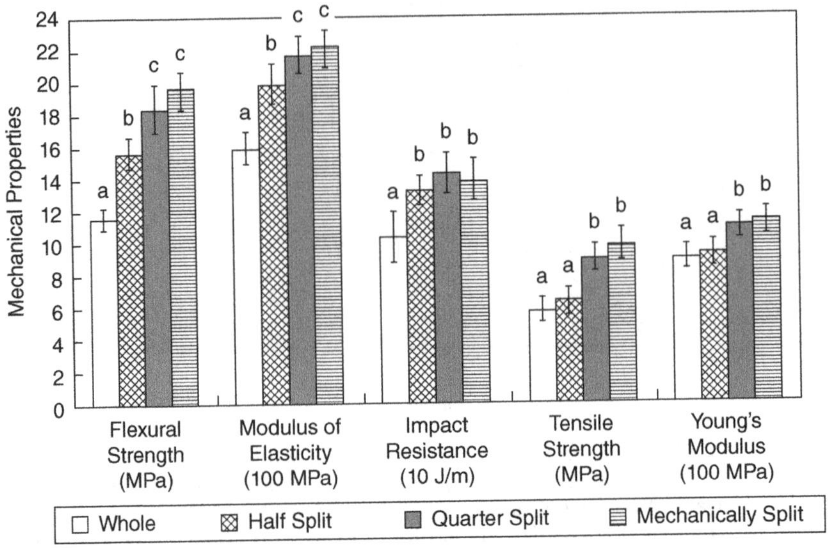

14.13 Mechanical properties of wheat straw–PP composites. The concentration and length of wheat straw were fixed at 60% and 5 cm, respectively. Letters a, b and c show the statistical significance of the properties: if any two data points of the same mechanical property showed totally different letters, the two data points were statistically different (reprinted from Zou *et al.*, 2010, © 2010, with permission from Elsevier).

mechanically split long wheat straw and nonwoven PP by compression molding (Zou *et al.*, 2010). Mechanical properties of 5 cm long wheat straw filled PP composites (60% wheat straw) are shown in Fig. 14.13. The results indicate that mechanical properties such as flexural strength, modulus of elasticity and impact strength increased significantly when the straw was split into halves and further into quarters. The authors attributed this to the increased surface area of the straw as well as the increased aspect ratio of the straw fibres. Further, mechanical properties of the mechanically split fibres reinforced composites were more or less similar to those of quarter-split wheat straw. This suggests that it would be desirable to split the straw using milling machines for large-scale manufacturing of straw-reinforced composites. In this study the authors also compared the properties of the straw–PP composites with those of jute–PP composites with similar density. The authors reported that the mechanically split wheat straw–PP composites had 114% higher flexural strength, 38% higher modulus of elasticity, 10% higher tensile strength, 140% higher Young's modulus, better sound absorption properties and 50% lower impact strength than the jute composites with similar density. This indicates that these composites can have an immediate opportunity in automotive interior applications where natural fibres are currently being used, with substantial cost advantages.

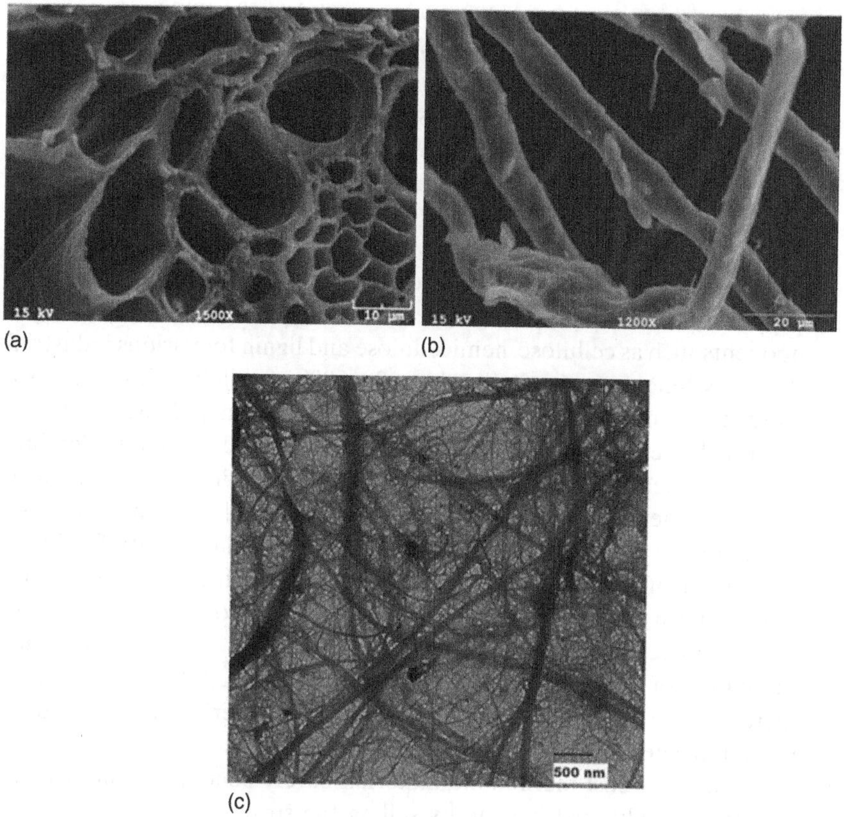

14.14 Changes in the diameter of the wheat straw during nanofibre isolation: (a) SEM images of the wheat straw cross-section, (b) wheat straw microfibres before mechanical fibrillation, and (c) TEM image (magnification ×15,000) of the wheat straw nanofibres (reprinted from Alemdar and Sain, 2008b, © 2014, with permission from Elsevier).

Another research area is the isolation of high-strength nanocellulose fibres from wheat straw for manufacturing nanocellulose composites having very high strength performance compared to macroscopic wheat straw composites. One of our group's research in this direction leads to high strength nanofibres from wheat straw using a chemico-physical process. Figure 14.14 shows the changes in the fibre dimension from the straw bundle to microfibre to nanofibre during the preparation of nanofibril isolation (Alemdar and Sain, 2008b). The fibre diameter distribution of nanofibrils is in the range of 20–120 nm with 60% of the fibrils having a diameter of 30–40 nm (Alemdar and Sain, 2008a). Wheat straw nanofibres have been used for reinforcing polymer matrices with low mechanical properties such as thermoplastic starch to enhance their performance properties. Alemdar and Sain (2008b) reported that 10 wt% of wheat straw nanocellulose could provide significant enhancement in the tensile strength and modulus (145%) of the thermoplastic starch. There are several potential applications of nanofibre composites including films, optical films, LED, surface coating applications, etc.

14.9 Conclusions

Wheat straw is an abundant and inexpensive raw material for the biorefinery sector. There is great research interest in the utilization of the major components such as cellulose, hemicellulose and lignin for various industrial applications. Since straw contains about 30–50% of cellulose, it is used as a reinforcement for thermoplastic as well as thermoset polymer matrices. Straw-based thermoplastics were generally prepared by melt blending followed by injection, extrusion or compression molding. The mechanical properties of the composites demonstrate their suitability as an alternative to wood flour or other bast fibre composites with similar density. Similarly, incorporation of wheat straw could considerably reduce the cost of the product and can be used as an immediate alternative to the expensive bast fibres. The processing temperature window and high water absorption are the limitation of the composites as in any other lignocellulosic composites. Research in this direction is needed to improve the performance of wheat straw composite materials. Converting wheat straw into novel high-performance, low-cost marketable materials surely benefits wheat farmers by generating additional income by selling the straw, benefits industry by generating cost-effective high-performing materials, and benefits the environment by reducing the burning of the straw as well as acting as a carbon sink.

14.10 References

Alemdar, A., Sain, M. (2008a) Isolation and characterization of nano fibres from agricultural residues – Wheat straw and soy hulls. *Biores. Technol.*, 99: 1664–1671.

Alemdar, A., Sain, M. (2008b) Biocomposites from wheat straw nanofibres: morphology, thermal and mechanical properties. *Comp. Sci. Technol.*, 68: 557–565.

Biricik, H., Akoz, F., Berktay, I., Tulgar, A.N. (1999) Study of pozzolanic properties of wheat straw ash. *Cement and Concrete Research*, 29: 637–643.

Carpita, N., McCann, M.C. (2000) The cell wall. In: B.B. Buchanan, G. Wilhelm and R.L. Jones (eds), *Biochemistry and Molecular Biology of Plants*. American Society of Plant Physiologists, Rockville, MD.

Cooper, P.A., Balatinecz, J.J., Flannery, S.J. (1999) Agricultural waste materials for composites: A Canadian reality. *Proceedings of Center for Management Technology Global Panel Based Conference*, Kuala Lumpur, 18–19 October.

Deswarte, F.E.I., Clark, J.H., Hardy, J.J.E., Rose, P.M. (2006) The fractionation of valuable wax products from wheat straw using CO_2. *Green Chem.*, 8: 39–42.

Focher, B. *et al.*, (1998) Regenerated and graft copolymer fibers from steam-exploded wheat straw: Characterization and properties. *J. Appl. Polym. Sci.*, 67: 961–974.

Foyle, T., Jennings, L., Mulcahy, P. (2007) Compositional analysis of lignocellulose materials: Evaluations of methods used for sugar analysis of waste paper and straw. *Biores. Technol.*, 98: 3026–3036.

Fu, D., Mazza, G., Tamaki, Y. (2010) Lignin extraction from straw by ionic liquids and enzymatic hydrolysis of the cellulosic residues. *J. Agric. and Food Chem.*, 58: 2915–2922.

Han, G., Umemura, K., Kawai, S., Kajita, H. (1999) Improvement mechanism of bondability in UF-bonded reed and wheat straw boards by silane coupling agent and extraction treatments. *J. Wood Sci.*, 45: 299–305.

Han, G., Cheng, W., Deng, J., Dai, C., Zhang, S., Wu, Q. (2009) Effect of pressurized steam treatment on selected properties of wheat straws. *Ind. Crops Prod.*, 30: 48–53.

Han, G., Deng, J., Zhang, S., Bicho P., Wu, Q. (2010) Effect of steam explosion treatment on characteristics of wheat straw. *Ind. Crops Prod.*, 31: 28–33.

Hornsby, P.R., Hinrichhsen, E., Trivedi, K. (1997) Preparation and properties of polypropylene composites reinforced with wheat and flax straw fibres. 1. Fibre characterization. *J. Mater. Sci.*, 32: 443–449.

Jiang, H., Zhang, Y., Wang, X. (2009) Effect of lipases on the surface properties of wheat straw. *Ind. Crops Prod.*, 30: 304–310.

Johnson, D.A., Jacobson, R., Maclean, W.D. (1997) Wheat straw as a reinforcing filler in plastic composites. *Fourth International Conference on Wood Fibre Composites*, Madison, WI, pp. 200–205.

Kokta, B.V., Ahmed, A. (1998) Chapter 6, Steam explosion pulping. In: R.A. Young and M. Akhtar (eds), *Environmentally Friendly Technologies for Pulp and Paper Industry*. John Wiley & Sons, New York.

Kruger, K. (2007) *Wheat Straw-Polypropylene Composites*, University of Waterloo, Ontario.

Lam, T.B.-T., Iiyama, K., Stone, B.A. (1990) Primary and secondary walls of grasses and other forage plants: Taxonomic and structural consideration. In: D.E. Akin, M. G. Ljundahl, R. J. Wilson, P. J. Hams (eds), *Microbial and Plant Opportunities to Improve Lignocellulose Utilization by Ruminants*. Elsevier, New York, pp. 43–69.

Lam, T.B-T., Iiyama, K., Stone, B.A. (1992) Cinnamic acid bridges between cell wall polymers in wheat and phalaris internodes. *Phytochemistry*, 31: 1179–1183.

Lawther, J.M., Sun R.C., Banks, W.B. (1996) Fractional characterization of alkali-labile lining and alkali insoluble lignin from wheat straw. *Ind. Crops Prod.*, 5: 291–300.

Le Digabel, F., Averous, A. (2006) Effect of lignin content on the properties of lignocelluloses-based biocomposites. *Carbohydr. Polym.*, 66: 537–545.

Lequart, C., Ruel, K., Lapierre, C., Pollet, B., Kurek, B. (2000) Abiotic and enzymatic degradation of wheat straw cell wall: A biochemical and ultrastructural investigation. *J. Biotechnol.*, 80: 249–259.

Liu, R., Yu, H., Huang, Y. (2005) Structure and morphology of cellulose in wheat straw. *Cellulose*, 12: 25–34.

Majumdar, P., Chanda, S. (2001) Chemical profile of some lignocellulosic crop residues. *Indian J. Agric. Biochem.*, 14(1&2): 29–33.

Mckean, W.T., Jacobs, R.S. (1997) *Wheat straw as a fiber source*. Clean Washington Center, Washington, DC.

Meissner, D., Shore, R. (2012) Step forward see waste wheat straw as the non-wood future of paper industry. *The Vancouver Sun*, 25 October 2012.

Mengeloglu, F., Karakus, K. (2012) Mechanical properties of injection molded foamed wheat straw filled HDPE biocomposites: The effects of filler loading and coupling agent contents. *Bioresources*, 7(3): 3293–3305.

Milewski, D. (2012) Greening and light weighting automotive: Ford's vision for a sustainable automotive industry. *12th International Conference on Biocomposites, Transition to Green Materials*, 6–8 May 2012, Ontario, Canada.

Mishra, S., Sain, M. (2009) Commercialization of wheat straw as reinforcing filler for commodity thermoplastics. *J. Natural Fibres*, 6: 83–97.

Mohanty, A.K., Misra, M., Drzal, L.T. (eds) (2005) *Natural Fibers, Biopolymers, and Biocomposites*. CRC Press, Boca Raton, FL.

Montano-Leyva, B., da Silva, G.H.D., Gastaldi, E., Torres-Chavez, P., Gontard, N., Angellier-Coussy, H. (2013) Biocomposites from wheat proteins and fibres: Structure/mechanical properties relationships. *Ind. Crops Prod.*, 43: 545–555.

Ng, Z.S. (2008) *Bulk Orientation of Agricultural Filler–Polypropylene Composites*, University of Waterloo, Ontario.

Pan M., Zhang, S.Y., Zhou, D. (2010) Preparation and properties of wheat straw fibre-polypropylene composites. Part-II. Investigation of surface treatments on the thermo-mechanical and rheological properties of the composites. *J. Comp. Mater.*, 44(9): 1061–1072.

Panthapulakkal, S., Sain, M. (2006a) Bioprocess preparation of wheat straw fibers and their characterization. *Ind. Crops Prod.*, 23: 1–8.

Panthapulakkal, S., Sain, M. (2006b) Injection molded wheat straw and corn stem filled polypropylene composites, *J. Polym. Environ.*, 14: 265–272.

Panthapulakkal, S., Zereshkian, A., Sain, M. (2006) Preparation and characterization of wheat straw fibres for reinforcing application in injection molded thermoplastic composites. *Biores. Technol.*, 97: 265–272.

Patil, Y.P., Gajre, B., Dusane, D., Chavan, S., Mishra, S. (1999) Effect of maleic anhydride treatment on steam and water absorption of wood polymer composites prepared from wheat straw, cane bagasse, and teak wood sawdust using Novolac as matrix. *J. Appl. Polym. Sci.*, 77(13): 2963–2967.

Perez, J.A., Ballesteros, I., Ballesteros, M., Saez, F., Negro, M.J., Manzanares, P. (2008) Optimizing liquid hot water pretreatment conditions to enhance sugar recovery from wheat straw for fuel-ethanol production. *Fuel*, 87: 3640–3647.

Pfister, D.P., Larock, R.C. (2010) Green composites from a conjugated linseed oil-based resin and wheat straw. *Composites, Part A*, 41: 1279–1288.

Pradhan, R., Misra, M., Erickson, L., Mohanty, A. (2010) Compostability and biodegradation study of PLA-wheat straw and PLA-soy straw based green composites in simulated composting bioreactor. *Biores. Technol.*, 101: 8489–8491.

Reddy, N., Yang, Y. (2005) Biofibres from agricultural byproducts for industrial applications. *Trends in Biotechnol.*, 23: 22–27.

Reddy, N.R., Sardshti, A.P., Simon, L.C. (2010) Preparation and characterization of polypropylene–wheat straw–clay composites. *Comp. Sci. Technol.*, 70: 1674–1680.

Saha, B.C., Iten, L.B., Cotta, M.A., Wu, Y.V. (2005) Dilute acid pretreatment, enzymatic saccharification and fermentation of wheat straw to ethanol. *Process Biochem.*, 40: 3693–3700.

Schirp, A., Loge, F., Aust, S., Swaner, P., Turner, G., Wolcott, M. (2006) Production and characterization of natural fiber-reinforced thermoplastic composites using wheat straw modified with the fungus *Pleurotus ostreatus*. *J. Appl. Polym. Sci.*, 102: 5191–5201.

Staniforth, A.R. (1979) Straw for pulping, paper and board. In: *Cereal Straw*, Clarendon Press, Oxford, pp. 116–123.

Sun, R.C., Fang, J.M., Rowlands, P., Bolton, J. (1998) Physico-chemical and thermal characterization of wheat straw hemicelluloses and cellulose. *J. Agric. Food Chem.*, 46: 2804–2809.

USDA database: http://usda01.library.cornell.edu/usda/current/worldag-production/worldag-production-06-12-2013.pdf

Volyets, B., Dahman, Y. (2011) Assessment of pretreatments and enzymatic hydrolysis of wheat straw as a sugar source for bioprocess industry. *Int. J. Energy Environ.*, 2(3): 427–446.

White, N.M., Ansell, M.P. (1983) Straw reinforced polyester composites. *J. Mater. Sci.*, 18: 1549–1556.

Yu, H., Liu, R., Shen, D., Wu, Z., Huang, Y. (2008) Arrangement of cellulose microfibrils in the wheat straw cell wall. *Carbohydr. Polym.*, 72: 122–127.

Zabihzadeh, S.M. (2010) Influence of plastic type and compatibilizer on thermal properties of wheat straw flour/thermoplastic composites. *J. Thermoplast. Comp. Mater.*, 23: 817–825.

Zabihzadeh, S.M., Dastoorian, F. (2010) Effect of MAPE on mechanical and morphological properties of wheat straw/HDPE injection molded composites. *J. Reinforced Plast. Comp.*, 29: 123–131.

Zhai, H.M., Lee, Z.Z. (1989) Ultra structure and topochemistry of delignification in alkaline pulping of wheat straw. *J. Wood Chem. Technol.*, 9(3): 387–406.

Zou, Y., Huda, S., Yang, Y. (2010) Light weight composites from long wheat straw and polypropylene web. *Biores. Technol.*, 101: 2026–2033.

15

The use of maize, oat, barley and rye fibres as reinforcements in composites

A. A. MAMUN and H. P. HEIM, University of Kassel, Germany and A. K. BLEDZKI, West Pomeranian University of Technology, Poland

DOI: 10.1533/9781782421276.4.454

Abstract: Lignocellulosic residues are waste materials that are abundant and comparatively cheap. The quality, fibre contents and chemical composition of lignocellulosic residues depend on the process of grain collection and on food processing. In this study, an investigation is conducted on the morphology, composition, surface chemistry and thermal degradation characteristics of maize, oat, barley and rye plant residues. The aim is to explore the viability of processing and producing composites. In addition, investigations are carried out to examine the potential of lignocellulosic residues as a possible reinforcement for polymeric materials.

Key words: fibre chemical content, surface and physical properties, enzyme modification, mechanical properties, chemical and dielectric properties.

15.1 Introduction

The challenge of the twenty-first century will be to utilize feedstocks such as lignocellulose and lignocellulosic waste materials as substitutes for existing fossil-based raw materials. The advantages of lignocellulosics as a feedstock material are that they are abundant, sustainable and carbon-balance neutral. There is also a strong push towards reduced greenhouse gas emissions. Fossil-based raw materials contribute over 25% of greenhouse gas emissions. It has been estimated that the utilization of plant/crop-based feedstock for the production of chemicals in the European Union (EU) could enable greenhouse gas emission reductions of over 6 million tonnes per annum in the next decade (Clark, 2008). Consequently, coordinated R&D strategies for the utilization of plant/crop-based products have been initiated across the globe. New functional waste material obtained from renewable resources has gained greater attention owing to the worldwide increasing demand for alternatives to fossil resources (Suramaythangkoor and Gheewala, 2011).

The use of cereal residues (husk, straw) or by-products as a filler or reinforcement in the production of plastic composites can alleviate the shortage of wood resources. Their use also has the potential to introduce a natural fibre industry to countries where few wood resources remain. The composite industries are looking into alternative, low-cost lignocellulosic sources, which can be used to decrease overall manufacturing costs and improve the properties of materials.

Cereal husk/straw raw material could be a potential alternative to wood for composites destined for use in automotive, packaging, household utilities and construction applications. In the last couple of years, cereal lignocellulosic raw materials (straw, husk, cornstalk, baggasses) have been used to make composites with polypropylene, polyethylene, polyester, polyvinyl acetate, polyurethane, poly(3-hydroxybutyrate-*co*-3-hydroxyvalerate), polylactic acid, polyethylene terephthalate and Novolac resin (Mamun, 2011).

The development of biodegradable packaging materials from renewable natural resources has received widespread government support in EU countries, and many national or international organizations have been established to facilitate development in this area. Apart from composite materials, particle-boards from cereal residues or by-products could be another potential application (Clark, 2008).

The main factors that restrict the use of cereal by-products (husk), straw and other agricultural residues in composites are problems associated with their collection, storage and transportation, as well as the economics related to the overall production of composites. However, research that focuses on the use of these materials could monitor a developing market, and lead to new market opportunities for these surplus, inexpensive, field crop residues (Donato *et al.*, 2011).

The final properties of composite materials depend on the incorporated fibre properties (morphology, surface chemistry, chemical composition and crystalline contents), as well as matrix properties (nature and functionality). Adhesion between the reinforcing fibres and the matrix in composite materials plays an important role. Wetting of the fibre is an integral step in the adhesion process. The interface depends on complex thermodynamics of the fibre and matrix. Fibre properties such as composition, surface roughness and surface polarity contribute greatly to fibre wettability and adhesion in composites.

There are, however, some contradictory results regarding the influence of the surface chemistry of reinforcements on the mechanical properties of composites. For instance, a natural fibre surface with a relatively high concentration of hydroxyl groups produces relatively low-performance composites when used as reinforcing filler in non-polar polymers. In contrast, removing lignin from wood fibres provides a higher number of free hydroxyl groups on the cellulose, which, in turn, results in better interfacial bonding with the polymer matrix. Moreover, lignin has been extracted from natural resources and used as a coupling agent for natural fibre composites. Stiffness and strength were enhanced to some extent, but the impact strength of resulting composites was reduced through the use of the lignin coupling agent (Graupner, 2008). Overall, it has thus been demonstrated that the surface chemistry of fibres has a direct influence on the mechanical properties of the associated composites. Therefore, it is important to determine the physical, chemical and thermal properties of a fibre before using it as reinforcement.

Wood fibre is the most widely used lignocellulosic natural fibre for the reinforcement of plastics. Considering economic and ecological aspects, wood fibre plastic has become established as a standard material, but unfortunately the availability of raw wood fibre is reducing. Scientists are therefore searching for new sources to provide a proper alternative to wood fibre.

As such an alternative, cereal residues or grain by-products are gaining attention in Asian, European and North American regions. They are generally waste products of food processing (milling) and have a good fibre value (cellulose content 25–60%). They are abundant, their by-products are eco-friendly, they are available on an annual basis, and they are cheap; however, they are also complicated in terms of cell geometry, morphology and chemical composition. In addition, their use can be associated with environmental issues such as fouling and attraction of pests.

Overall, proper utilization of these waste materials can provide cheap engineering materials, as well as offering a means for waste management. The quality, fibre content and chemical composition of cereal by-products depend on the grain collection process as well as the way in which the food is processed, as described below.

15.2 Types of reinforcing fibre

15.2.1 General information

Maize (*Zea mays* subsp. *mays* L.) is a large grain plant. It is known as 'corn' in some English-speaking countries. The leafy stalk produces ears containing the grain; its seeds are called kernels. Maize kernels are used as a starch in cooking. Maize is the most widely grown grain crop throughout the Americas, with 332 million metric tonnes grown annually in the United States. Maize spread to the rest of the world because of its ability to grow in diverse climates. Sugar-rich varieties called sweet corn are usually grown for human consumption, whereas field corn varieties are used for animal feed and as chemical feedstocks. Approximately 40% of the total world production is used for ethanol production. Physically, the maize plant is often 2.5 m high, although some natural strains can reach up to 12 m. The stem is commonly composed of 20 internodes, 18 cm long. A leaf grows from each node, and is generally 9 cm wide and 120 cm long (Ensminger, 1994).

Oat (*Avena sativa*) is a species of cereal grain. Oats are suitable for human consumption in the form of oatmeal and rolled oats. One of the most common uses for oat is as livestock feed, typically forming part of the daily diet of domestic horses and cattle. Oats are also used in some brands of dog food and chicken feed, and can be employed as industrial raw materials. Oats are annual plants; they are grown in temperate regions, particularly in areas with cool, wet summers (Zhou *et al.*, 1999). The different parts of the crop include the whole crop silage, straw and grain; grain derivatives are agro-residues and can be used for different industrial purposes (Welch, 1995).

Barley (*Hordeum vulgare*) is one of the common cereal grains, and is known as an important health food. It is also used as an animal feed crop. Recently, larger amounts have been used for malting beverages such as beer and whisky (Wrigley *et al.*, 2004). Barley husk is a lignocellulosic agro-waste, and comprises about 20% of the barley. The husk is used to a limited extent as a cattle food and fertilizer. Various parts of the barley plant (straw, grain and grain derivatives, hull) are considered to be agro-residues (Zohary and Hopf, 2000).

Rye (*Secale cereale*) is a member of the grass family. It is a diploid species of the tribe Gramineae, with two chromosomes. It is closely related to the wheat tribe (Triticeae). Rye grows more rapidly than wheat, oat, barley or various other annual grasses. Although it is usually regarded as a winter crop, several spring-sown varieties are available. It is grown extensively as a grain and forage crop. Rye grain is used for flour, rye bread, rye beer, whiskies, vodkas, and animal fodder. Non-food parts of rye form agro-waste, which comprises around 20–25 wt% of rye (Altpeter and Korzun, 2007).

15.2.2 Production of fibre

The production both worldwide and according to country of maize, oats, barley and rye is shown in Tables 15.1 to 15.4. Grains are an annual product and are available in abundant volume throughout the world. The production of grains worldwide and according to country in the year 2011 were estimated; the data shown in these tables were extracted mainly from the FAO database (FAO, 2013). The quantity of grain husk or other residues can be calculated by association with a given production of crops. The grain husk usually contains 15–20 wt% of grains. The quantity of other residues, including straw, is 1.5–6 times the grain amount, which is in turn dependent on the botanical type of grain and growing conditions.

Table 15.1 Global production of maize by country in 2011

Country	Production in 2011 (metric tons)
United States of America	313,918,000
China	192,904,232
Brazil	55,660,400
Argentina	23,799,800
Ukraine	22,837,900
India	21,570,000
Mexico	17,635,400
Indonesia	17,629,000
Romania	11,717,600
World	831,679,000

Table 15.2 Global production of oats by country in 2011

Country	Production in 2011 (metric tons)
Russian Federation	5,332,130
Canada	2,997,100
Poland	1,381,600
Finland	1,102,300
Spain	1,078,700
World	22,492,000

Table 15.3 Global production of barley by country in 2011

Country	Production in 2011 (metric tons)
Russian Federation	16,938,000
Ukraine	9,097,700
France	8,775,000
Germany	8,733,800
Turkey	7,600,000
World	122,607,000

Table 15.4 Global production of rye by country in 2011

Country	Production in 2011 (metric tons)
Russian Federation	2,970,750
Poland	2,600,680
Germany	2,521,000
Belarus	803,959
China	715,000
World	12,185,000

15.3 Fibre components and key properties

The chemical compositions of grain residues are presented in Table 15.5. The approximate cellulose content in corn (maize) stalk is 40 wt%, corn pith 25 wt%, oat straw 50 wt%, barley straw 47 wt%, barley husk 39 wt%, rye straw 50 wt% and rye husk 26 wt%, and for softwood it is 42 wt%. Thus, the cellulose contents of most grain residues are comparable with that of softwood fibre. Almost all grain residues contain a small percentage of silica, fat and ash. Grain husks contain 10–15 wt% starch. The analytical results are all approximate values and their standard deviation may be considered as equal to 5%.

15.3.1 Morphology

Figure 15.1 depicts the surface morphology of barley husk and rye husk. The surface of barley husk is relatively rough and covered with extractives, fat and loose, amorphous molecules. Defects, damage and internal cracks have been identified on the fibre surface of barley. The surface of rye husk, by contrast, is relatively smooth. A number of cracks, defects and some damage have also been observed on the fibre surface of rye husk (Mamun, 2011; Bledzki *et al.*, 2010a).

15.3.2 Surface properties

Figure 15.2 shows the inspection spectra of fibre surface elements acquired for barley husk and rye husk. Both husks exhibit spectra containing mainly carbon and oxygen and a small amount of silicon, potassium, sulphur and phosphorus. The elemental compositions and oxygen carbon atomic ratio of barley husk and rye husk are provided in Table 15.6 (Mamun, 2011; Bledzki *et al.*, 2010a). The relative atomic percentages of the atoms were obtained from the peak area and corrected using an appropriate sensitivity

Table 15.5 Chemical compositions (wt%) of maize, oat, barley and rye residues

	Cellulose	Hemicellulose	Lignin	Starch	Fat	Ash	Silica
Corn stalk fibre (Reddy and Yang, 2005a)	38–40	7–21	28	–	–	–	–
Corn pith (Yin *et al.*, 2007)	24.6	19.1	12.3	–	–	–	–
Oat straw	44–53	27–38	16–19	–	–	6–8	4–6.5
Barley straw	47–48	24–29	14–15	–	–	5–7	3–6
Barley husk (Bledzki *et al.*, 2010a)	39	12	22	11	4	2	–
Rye straw	50–54	27–30	16–19	–	–	2–5	1–4
Rye husk (Bledzki *et al.*,2010b)	26	16	13	17	7	11	<1
Softwood (Bledzki *et al.*, 2010b)	42	22	31	0.5	0.5	–	–

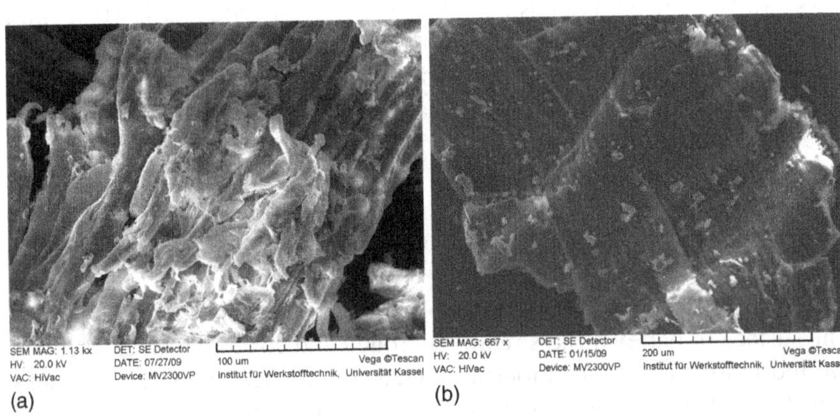

15.1 Scanning electron micrographs of grain by-products: (a) barley husk; (b) rye husk.

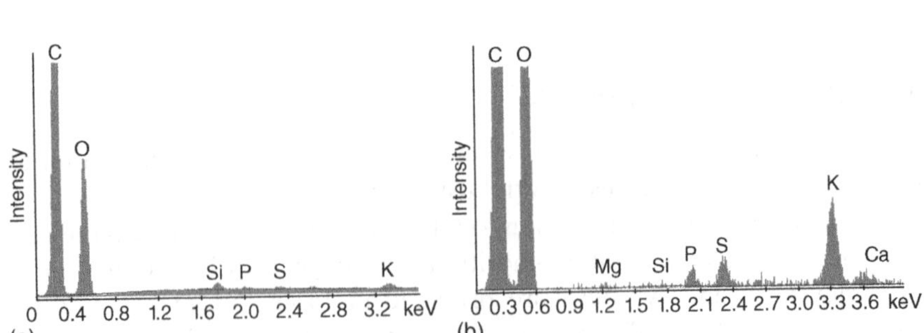

15.2 Elementary analysis of grain by-products: (a) barley husk; (b) rye husk.

Table 15.6 Elementary analysis (%) of barley husk, rye husk and softwood

Fibre	C	O	Si	K	S	P	O/C
Barley husk	75.9	23.5	0.4	0.2	–	–	0.31
Rye husk	75.6	18.9	0.2	3.3	1.3	0.7	0.25
Softwood	73.9	25.0	1.1	–	–	–	0.34
Cellulose (Dorris and Grey, 1978)							0.83
Lignin (Dorris and Grey, 1978)							0.35

factor. Barley husk and rye husk show a higher proportion of carbon atoms than softwood fibre. Moreover, barley husk shows a higher proportion of carbon atoms compared to rye husk. The higher proportion of carbon in the fibre can be attributed to the presence of a hydrocarbon-rich, waxy coating on the cuticle of the fibre and the lignin present on the surface. Barley husk and rye husk contain a small proportion of silicon in comparison with softwood. The silicon contained in the fibre may have an influence on the properties of fibre-reinforced composites. The oxygen/carbon ratios of barley husk and rye husk appear to be lower than that of softwood fibre, and the oxygen/carbon ratio of rye husk is lower than that of barley husk. This lower ratio indicates a higher proportion of aliphatic and aromatic carbons near the surface. Hence, a higher compatibility with non-polar polymers is anticipated, in addition to a large number of parameters related to the compatibility (Mamun, 2011).

15.3.3 Thermal properties

The results of the thermal analyses of barley husk, rye husk and softwood fibre are shown in Fig. 15.3 (Bledzki *et al.*, 2010a, 2010b). The derivative thermogravimetry (DTG) curve for all fibres shows initial peaks between 40°C and 130°C, which correspond with the vaporization of water. The curve for barley husk exhibits two decomposition steps. The decomposition peak temperatures are 299°C and 341°C. The curve for rye husk exhibits a single decomposition step and the decomposition peak temperature is 306°C. The DTG curve for softwood displays a single decomposition step and the decomposition peak temperature is 382°C. The temperature at which decomposition commences in all fibres is observed to be about 200°C. The decomposition peak at around 300°C is more likely attributable to thermal decomposition of the hemicellulose and the glycosidic linkage of the cellulose. The peak in the range 350–390°C is due to α-cellulose decomposition. Lignin degrades first and at a slower rate than the other constituents. The peak corresponding with lignin is slightly broader and appears in the temperature range 190–500°C, with a maximum value (peak) at about 350°C (Mamun, 2011); it therefore seems to overlap with the

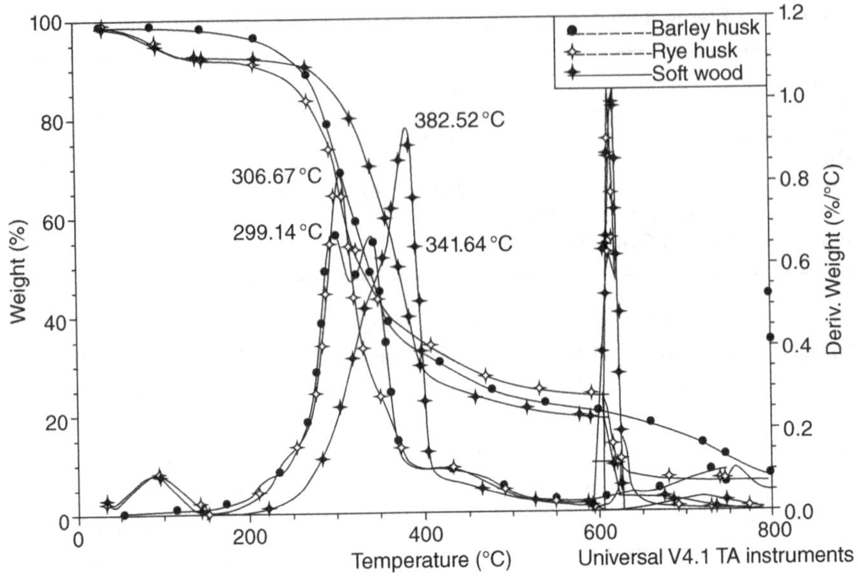

15.3 TGA and DTGA of barley husk, rye husk and softwood.

hemicellulose and cellulose peaks. The sudden drop at about 600°C is because of an abrupt oxidation of the fibre contents exposed to oxygen.

15.4 Surface modification of fibres

15.4.1 Chemical method

Chemical modifications of natural fibres result in the reduction of moisture absorption and an improved adhesion between the matrix and fibres. The chemical coupling agents are considered to optimize the interface of the fibres. Such modifications include acetylation, mercerization, methylation, cyanoethylation, benzoylation, permanganate treatment, and acrylation.

Alkaline treatment or mercerization is one of the most often used chemical treatments for natural fibres, as well as agro-residues. Alkali treatment leads to an increase in the amount of amorphous cellulose at the expense of crystalline cellulose, and can also sometimes result in changes to the cellulose configuration (structure). The removal of hydrogen bonding in the network structure is achieved when using this treatment. Alkali solution not only affects the cellulosic components inside the plant fibre, but also affects the non-cellulosic components such as hemicellulose, lignin and pectin. The concentration of alkali influences the degree of swelling and the degree of lattice transformation in the cellulose. The extreme delignification of natural fiber takes place when the alkali concentration is high, as a result of damage to the fibre surface (Li *et al.*, 2007).

Acetylation of natural fibres including agro-residues is a well-known esterification method causing plasticization of cellulosic fibres. The reaction of hydroxyl groups of the cell wall polymer with acetyl groups occurs, thereby modifying the properties of these polymers so they become hydrophobic; this can stabilize the cell wall against moisture, improving dimensional stability and resistance to environmental degradation. The reaction is known to proceed to full esterification of all three hydroxyls of anhydro-D-glucose when it is carried out in a homogeneous phase (i.e. when cellulose is dissolved), but in the case of fibres and wood the reaction is heterogeneous. The reacting hydroxyl groups are those of the minor constituents of the fibre, i.e. lignin, hemicelluloses and amorphous cellulose. This is because the hydroxyl groups in crystalline regions are completely inaccessible as a result of close packing and strong interlock bonding within the molecular structure (Bledzki *et al.*, 2008).

Huda and Yang (2008) extracted corn husks using a caustic agent and a mixture of enzymes. The caustic agent removes the non-cellulose materials. In addition, the enzyme treatment leads to an increase in the divisibility of fibres, finer fibre strands, and increased pore volume within the fibres. Alkaline/enzyme treatment of husks not only allows fibre extraction, but also changes the extracted fibres' surface properties, leading to increased interaction between corn husks and polypropylene Thus, surface modification can be achieved simultaneously with the extraction, without any additional processes. Table 15.7 shows that the cellulose content in corn husks increases as a result of the alkali treatment and enzyme treatments. The cellulose content in corn husk was 42.3%, in alkali-treated corn husk 64.5% and in alkali–enzyme-treated corn husk 83.5%. This can be attributed to the removal of non-cellulosic materials from the husk during treatment, mainly hemicellulose, lignin and ash. Lignin and ash are primarily removed during the alkaline process. The enzyme cellulase is used to remove short fibres from treated husk and pulpzyme is used to depolymerize hemicelluloses.

Table 15.7 Effect of chemical and enzyme modification on chemical composition and crystallinity

Constituent (%)	Corn husk	Alkali-treated corn husk	Alkali-enzyme-treated corn husk
Cellulose	42.3 ± 0.7	64.5 ± 0.5	83.5 ± 0.5
Hemicellulose	40.95	28.33	10.15
Lignin	12.6 ± 0.2	6.4 ± 0.4	5.9 ± 0.3
Ash	4.2 ± 0.3	0.7 ± 0.1	0.4 ± 0.1
Crystallinity	34.6 ± 0.9	39.4 ± 1.2	46.4 ± 1.4

Source: Adapted from Huda and Yang, 2008, © 2008 Wiley-VCH Verlag GmbH & Co. KGaA, Weinheim, Germany.

Furthermore, the process successively breaks the covalent bonds between lignin and carbohydrates in the corn husks. Crystallinity increases owing to the increase in cellulose content. In addition to the cellulose content, the microfibrillar angle and degree of polymerization of cellulose also have an effect on the fibre strength. Lower crystallinity and poor crystal orientation of corn husk fibre lead to a lower modulus and higher elongation of corn husk in comparison with jute fibres.

Corn husk has shown an average tenacity of 2.7 g/denier, an elongation of 15.3%, a modulus of 70 g/denier, and a work of rupture of 0.23 g/denier. This is comparable with values obtained for jute fibres: tenacity 3.4 g/denier, elongation 1.1%, modulus 195 g/denier, work of rupture 0.03 g/denier (Reddy and Yang, 2005b).

Bavan and Kumar (2012) treated maize fibre with alkali and acetylation, and reported an improvement in the morphological and thermal properties. Figure 15.4 shows surface and cross-sectional views of maize stalk and straw. Figure 15.4(a) and (b) reveals a cross-section of raw maize fibre; it has a thick layer of protective material, cellular deposits and also the presence of other constituents, such as the lumen, which can increase the absorbency of the fibres. An interesting feature of the fibre cells in corn stalk is the presence of a large lumen, as shown in Fig. 15.4(c); this is larger than the width of the cell wall in most cells, thereby reducing the density of the fibres. The surface of raw maize straw is shown in Fig. 15.4(d). Alkali- and acetylation-treated maize fibres are shown in Fig. 15.4(e) and (f), respectively. The surface morphology of the maize fibre becomes rough as a result of the chemical treatment.

The thermogravimetric analysis (TGA) curve for raw maize fibre is provided in Fig. 15.5(a). The initial and final decomposition temperatures are found to be 211°C and 434°C, respectively. The TGA curve of alkali-treated maize fibre is presented in Fig. 15.5(b). The initial and final decomposition occurred in the temperature range 227–444°C, which supports the increase in thermal stability.

Corn stalks have been modified using heat treatment, cooking with NaOH/anthraquinone for 30 and 60 minutes (Flandez *et al.*, 2012). This study reported on the thermal stability of corn stalks and modified corn stalks. The first weight loss occurred at 75°C and can be attributed to the water and volatile matter, related to the extractives present in corn stalk. Weight loss is also evident at temperatures ranging between 250°C and 260°C. The maximum weight loss in this step is found in corn stalk flour, and decreases progressively for chemically semi-modified corn stalks and chemically modified corn stalks. This weight change corresponds to the presence of lignin in the raw material, which is gradually eliminated by the different treatments. This observation is proven by the fact that the amount of lignin in chemically modified corn stalks fibres is almost zero.

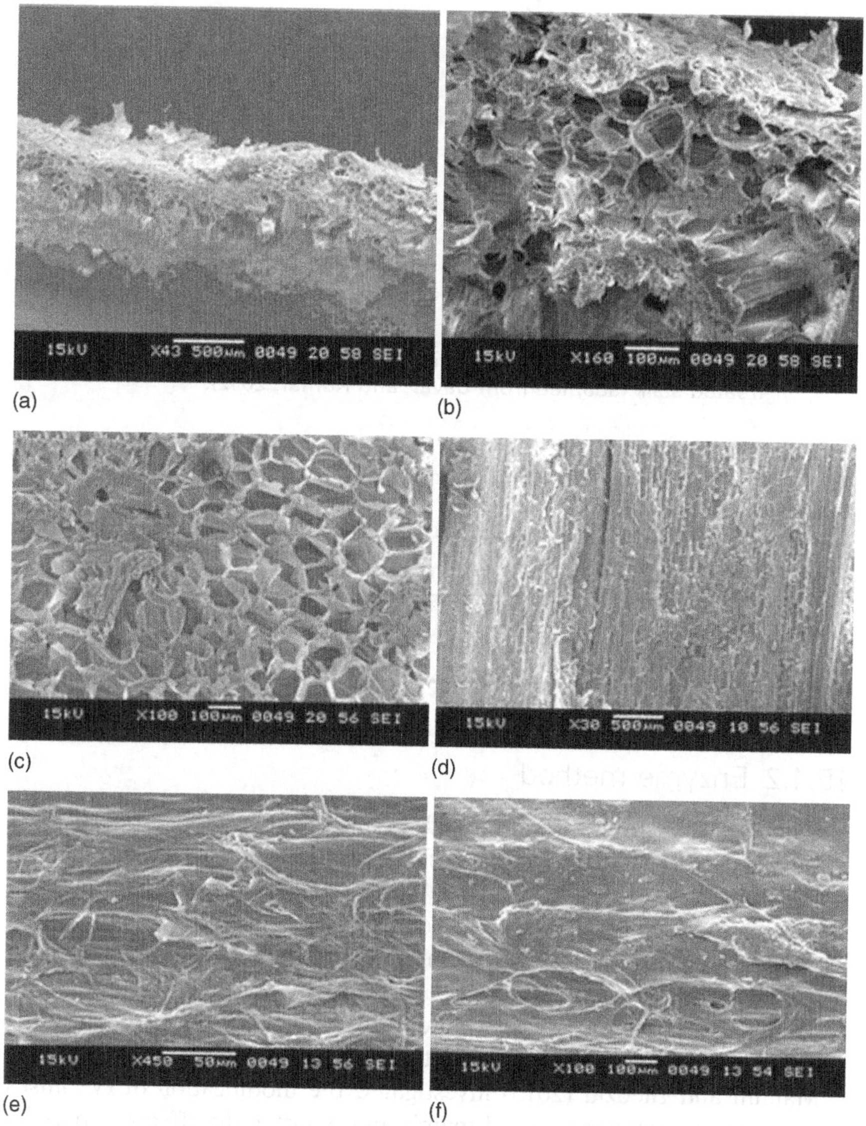

15.4 SEM micrographs of maize stalk fibre: (a) cross-section of raw maize fibre with cortex; (b) cross-section of raw maize fibre; (c) cross-section of maize stalk pith fibre; (d) surface of raw straw fibre; (e) alkaline-treated fibre; (f) acetylene-treated fibre (adapted from Bavan and Kumar, 2012).

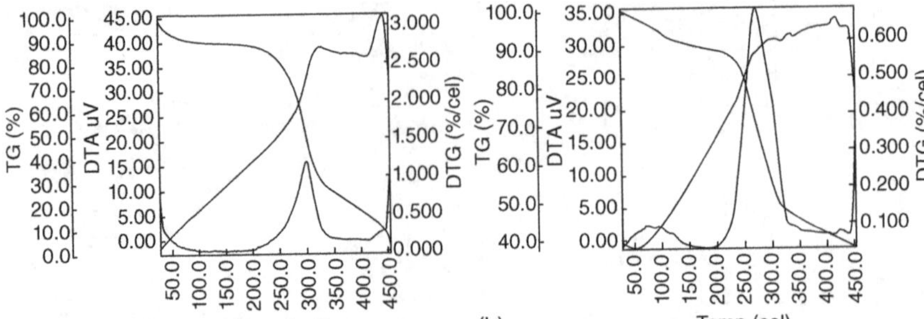

15.5 TGA thermographs of maize fibre: (a) raw stalk; and (b) alkaline-treated stalk (adapted from Bavan and Kumar, 2012).

Table 15.8 Parameter optimization of enzyme modification process for barley husks and rye husks (RH 95%, 23°C)

Fibre	Temperature (°C)	Dosing rate (wt%/L)	Treatment duration (h)
Barley husk	60	1.25	4
Rye husk	40	1.50	4

15.4.2 Enzyme method

Enzyme modifications are intended to improve surface properties by removing adsorbed components, such as lignin, fats, waxes, proteins and non-crystalline parts. This technology is environmentally friendly, and the reactions catalysed are very specific, so it is possible to control performance. The uses of enzyme technology for modification of natural fibres and in textile processing are increasing substantially. Such technology offers reductions in costs, energy and water use, as well as improved product quality and potential process integration.

Mamun and Bledzki (2013) investigated the modification of rye husks using an enzyme mixture called Lipamix and reported the effects on particle size, morphology and moisture absorption properties. Barley husks were also modified using Lipamix and the process parameters were optimized. Lipamix is a mixture of lipase, protease and amylase–xylanase enzymes. Prior to modification, the enzyme modification method was optimized in terms of the processing temperature, duration and dosing rate. The summarized results are shown in Table 15.8.

To observe the effects of modification, particle size distribution and surface morphology of both husk types were determined. The particle size distributions of modified and unmodified barley husk and rye husk can be

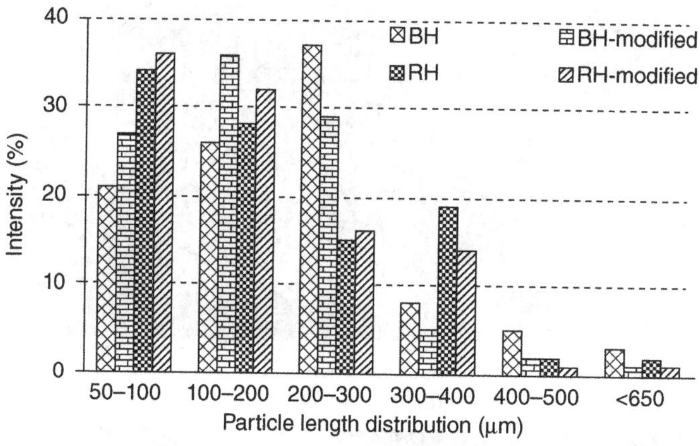

15.6 Particle size distributions of unmodified and modified barley husk and rye husk.

seen in Fig. 15.6 (Mamun and Bledzki, 2013). It can be observed that both husks were distributed in a wide range (50–650 μm), but almost 80% of the barley husk and 75% of the rye husk was distributed in the range 50–300 μm. After modification, the particle size distribution shifted to the lower range for both cases. This was caused by the removal of unwanted molecules on the surface (Mamun, 2011).

The surface morphologies of modified and unmodified barley husk and rye husk are shown in Fig. 15.7. The unmodified fibre surfaces (for both barley husks and rye husks) are full of cemented materials, defects and some damage. There are many knobs on the barley husk surface (Fig. 15.7(a)). The surface of rye husks is smoother and cemented by packing and waxy materials. This could be because fat and protein molecules have produced a thick coat on the fibre surface. The higher the amount of lipid found on the fibre surface, the more hydrophobic is the material. Also, the surface tension will be higher and the fibre surface smoother (Bledzki *et al.*, 2010b).

Loose and amorphous materials such as fat, protein, lignin and hemicelluloses are removed from the barley husk surface by means of modification; this can be seen in Fig. 15.7(b). The barley husk surface becomes porous, and defects and holes can also be observed. However, no fibre damage occurs as a result of the enzyme modification. In the case of rye husk, the removal of unwanted materials from the surface by enzyme modification resulted in an enhanced visibility of the microfibres. This effect can be seen in Fig. 15.7(d). In addition, some parts of the fibre surface became porous due to the removal of binding materials. Spring-like structures were visible on the surface of the fibre; these are assumed to be microfibrils of cellulose (Mamun, 2011).

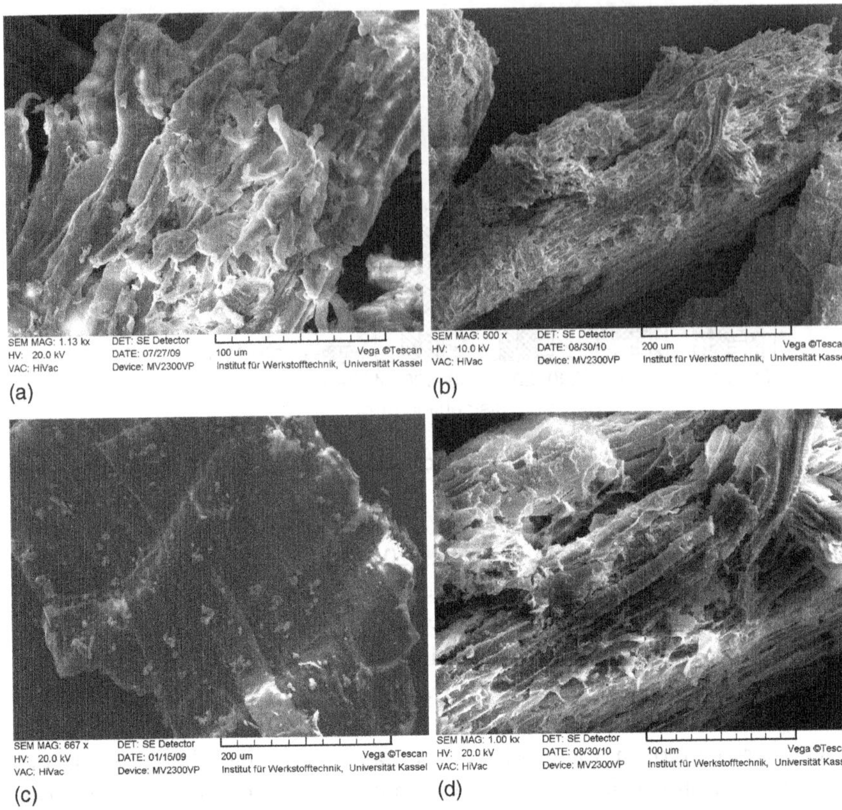

15.7 Surface morphology of grain by-products: (a) barley husk; (b) modified barley husk; (c) rye husk; (d) modified rye husk.

The moisture absorption properties of modified and unmodified barley husks and rye husks are illustrated in Fig. 15.8. The moisture uptakes of all types of husks were found to reach equilibrium after a conditioning period of 60 days. The equilibrium moisture contents of barley husks and rye husks were 18.6% and 21.8%, respectively. Barley husks and rye husks showed different moisture absorption properties; this is because their chemical compositions are different (Bledzki *et al.*, 2010a, 2010b). Moisture is mainly absorbed by hemicellulose, starch, non-crystalline cellulose, accessible cellulose and the surface of cellulose. After modification, the equilibrium moisture contents of barley husks and rye husks were found to reduce by 51 ± 3% and 33 ± 6%, respectively. The reduction of moisture can be attributed to the removal of hemicelluloses, amorphous cellulose and starch from the fibre surface. Additionally, hydroxyl groups in the cellulose and the rest of the hemicellulose were supposed to have been modified by sodium hydroxide during the enzyme deactivation process (Mamun, 2011).

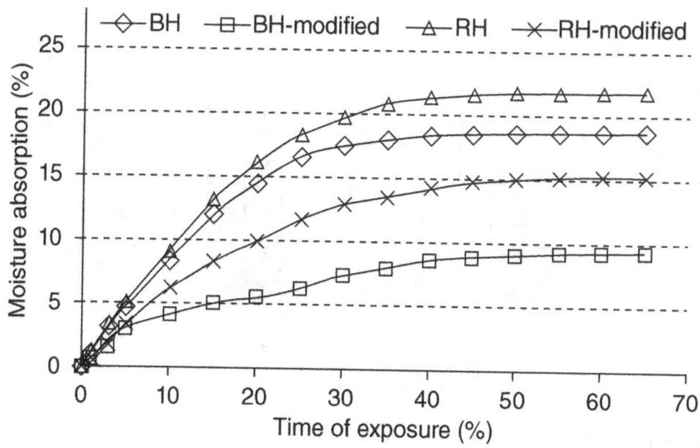

15.8 Moisture absorption of unmodified and modified barley husk (BH) and rye husk (RH).

15.5 Processing and performance: maize and oat flour composites

The choice of processing techniques depends on the forms and types of raw material being processed and the application. For grain residue composites, both the fibre type and form and the nature and form of the matrix material play an important role in the selection of the processing technique. Usually, grain residues are available in particle and powder form. Therefore, they can be compounded using conventional industrial techniques, such as a heating–cooling mixer, extrusion and pultrusion. They can also be moulded using compression and injection moulding techniques.

15.5.1 Maize flour composites

Derivatives from corn stalk waste have different chemical contents and are comparable to wood fibre. For example, the cellulose content of corn stalk and softwood fibre is 48% and 50%, respectively. It has been reported (Flandez *et al.*, 2012) that the tensile strength of PP reduced by about 20% with the addition of corn stalks (40 wt%). The addition of maleic anhydride-grafted polypropylene (MA-PP), however, to corn stalk–PP composites led to an improvement in tensile strength by 15%. The semi-chemical pulp from corn stalk fibre composites showed 90% improvement in tensile strength over PP alone. It was concluded that corn stalk wastes can be highly beneficial as raw materials for the preparation of plastic composites.

Flexural, impact resistance, tensile and sound absorption properties of composites formed from corn husk fibre (CHF) and PP were investigated by Huda and Yang (2008). The effects of the CHF length, CHF concentration and enzyme treatment of CHF on the mechanical properties were reported.

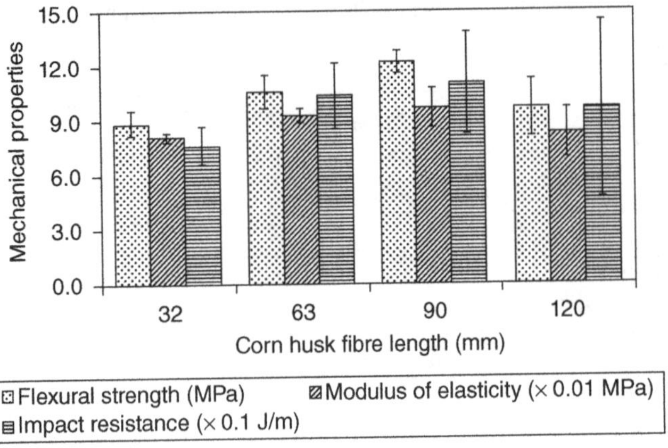

15.9 Effect of corn husk fibre (CHF) length on flexural and impact resistance properties of CHF/PP composites (fibre loading 40 wt% , holding temperature 185°C, holding time 80 s, thickness 3.2 mm) (adapted from Huda and Yang, 2008, © 2008 Wiley-VCH Verlag GmbH & Co. KGaA, Weinheim, Germany).

Table 15.9 Comparison of mechanical properties of CHF/PP and jute fibre/PP composites (fibre loading 40 wt%)

Material	Fibre extraction process	Flexural strength (MPa)	Flexural modulus (MPa)	Impact resistance (J/m)	Tensile strength (MPa)	Tensile modulus (MPa)
CHF-PP	Enzyme	12.0 ± 0.1	1073 ± 96	106.0 ± 12.5	10.0 ± 0.9	301.8 ± 22.7
CHF-PP	Caustic	10.6 ± 0.9	924 ± 40	103.3 ± 9.2	7.3 ± 0.5	223.5 ± 14.7
Jute-PP	Natural retting	8.4 ± 1.6	1280 ± 54	81.2 ± 8.0	10.6 ± 1.0	597.8 ± 30.1

Source: Adapted from Huda and Yang, 2008, Copyright © 2008 Wiley-VCH Verlag GmbH & Co. KGaA, Weinheim, Germany.

The effects of CHF length on flexural strength, modulus and impact resistance are shown in Fig. 15.9. Those properties increased until the fibre length of 90 mm was reached, and subsequently decreased. A comparison between the mechanical properties of CHF/PP and jute fibre/PP composites is shown in Table 15.9. CHF/PP composites (enzyme extraction, fibre loading 40 wt%) have a similar impact resistance, 30% higher flexural strength, 20% lower flexural modulus, the same level of tensile strength, 98% lower tensile modulus, and a slightly higher noise reduction coefficient than for jute/PP composites (natural retting, fibre loading 40 wt%).

A composite beam material was developed, composed of maize stalk and an unsaturated polyester resin polymer as matrix, with methyl ethyl ketone peroxide (MEKP) as a catalyst and cobalt octane as a promoter. The material was modelled to resemble a structural beam using suitable assumptions, and then analysed by means of the finite element method using ANSYS software to determine the deflection and stress properties. Morphological analysis and X-ray diffraction (XRD) analysis of this fibre were conducted. From the results, it was found that the finite element values are acceptable with proper assumptions, and the prepared natural fibre composite beam material could be used for structural engineering applications (Bavan and Kumar, 2013).

Green composites composed of long maize fibres and a poly ε-caprolactone (PCL) biodegradable polyester matrix were manufactured by means of thermo-mechanical processing (Dauda *et al.*, 2007). A variety of processing parameters, such as fibre area fraction, moulding temperature and forming pressure, were systematically controlled and their influence on the tensile properties was investigated. It was revealed that both the tensile strength and the elastic modulus of the composites increase steadily depending on an increase in fibre area fraction. This suggests a general conformity to the rule of mixtures (ROM), particularly up to 55% fibre area fraction. It was observed that the optimum processing conditions were a moulding temperature of 130°C and 10 MPa forming pressure.

Bavan and Kumar (2012) focused on the physical, chemical and morphological structure and the thermal properties of maize stalk fibre and analysed them using different characterization techniques. A simple hand layup method was employed to process the composite material. Chemical treatments of fibres were carried out to study the interaction of fibres with the matrix. The initial and final decomposition temperatures of maize fibre and alkali-treated maize fibre were 211°C and 434°C, and 227°C and 444°C, respectively. The spectrum corresponding to maize raw fibres displayed diffraction peaks for the amorphous and crystalline regions at 2θ angles of 22.58° and a high peak at nearly 29.46°. For alkali-treated fibres, the same peaks can be observed at 22.44° and 29.28°. Similarly, the peak of the crystalline region at 28.12° and that of the amorphous region at 20.36° were observed for chemically treated fibre with unsaturated polymeric resin (Bavan and Kumar, 2012). The position of the peak indicates an increase of interplanar distance in relation to the treated fibre. This occurs through the generation of disorder when the fibre is treated.

Thermo-mechanical maize-fibre-reinforced high-density polyethylene (HDPE) composites were processed, varying the fibre content. The incorporation of maize fibre induces a decrease in the thermal diffusivity and thermal conductivity. The thermal conductivity of the polymeric matrix seems to dominate over that of the incorporated maize fibres. Scanning electron microscopy (SEM) (Fig. 15.10) indicated that there were fewer voids on the surface and the maize fibres were well trapped by the HDPE

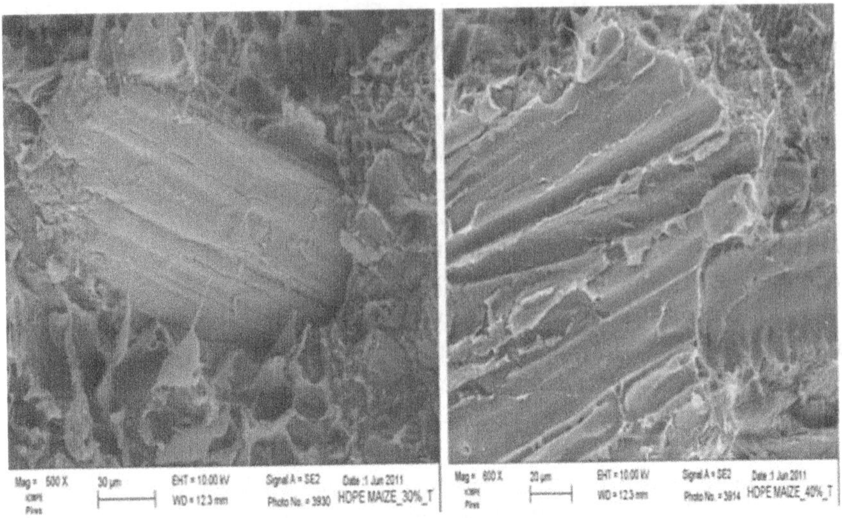

(a) (b)

15.10 SEM micrographs of HDPE–maize fibre composite: (a) 30 wt% fibre loading; (b) 40 wt% fibre loading (adapted from Trigui *et al.*, 2013).

matrix. Furthermore, the use of compatibilizer improved the affinity of the maize fibres to HDPE. The tensile strength, modulus and impact strength properties were enhanced by adding maize fibres up to a volume fraction of 35% (Trigui *et al.*, 2013).

The moisture absorption and mechanical properties of recycled HDPE composites reinforced by corn and wheat stalk flour were investigated by Thamae *et al.* (2009), who reported that the moisture absorption increased with increasing flour content for both types of composites. Composites made from the outer ring of corn stalk flour absorbed more moisture than composites made from whole corn stalk flour. The wheat stalk flour composites were found to absorb more moisture than corn stalk flour composites. The tensile strength of corn stalk flour composites decreased with increasing flour content. In contrast, the flexural strength of corn stalk flour composites improved according to flour content until the level of 30 wt% of flour was reached. Afterwards this property also decreased as the flour contents increased. Saline and sodium hydroxide treatments of corn stalk flour hardly affected the tensile and flexural properties.

15.5.2 Oat flour composites

Yellow poplar wood fibre-reinforced cellulose acetate butyrate composites were prepared by Onyeagoro (2012). Commercial oat fibres were mixed

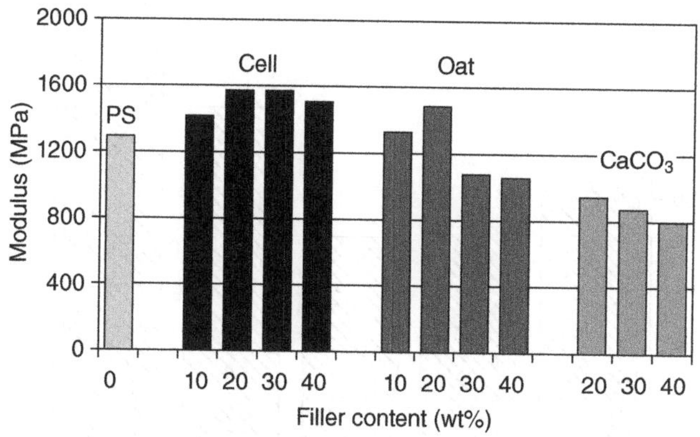

15.11 Effect of filler content on the elastic modulus for PS/cell, PS/oat and PS/$CaCO_3$ composites (adapted from Pacella *et al.*, 2010).

with the resulting composites. He recorded a slightly higher yield strain and an unchanged Young's modulus for composites with 40 wt% loading.

Composites consisting of polystyrene (PS) with cellulose (PS/cell) or oat (PS/oat) and $CaCO_3$ (PS/$CaCO_3$) were prepared using melt mixing (Pacella *et al.*, 2010). The main aim of the study was to evaluate the effects of the addition of PS copolymers containing reactive functional groups. The addition of polyethylene glycol (PEG) in PS/oat composites and styrene–ethylene–butylene–styrene-grafted maleic anhydride (SEBS-g-MA) in PS/$CaCO_3$ composites caused an improvement of the filler dispersion and interfacial interactions. The values of the tensile moduli for binary composites of PS with cell, oat and $CaCO_3$ are compared in Fig. 15.11. The filler contents ranged from 10 to 40 wt%. The moduli of PS/oat composites were better than those of PS/$CaCO_3$ composites. PS/cell and PS/oat samples showed rather brittle behaviour and similar values to those observed for PP/hemp.

Oat hull fibre-reinforced polypropylene/polylactic acid (9:1) composites were successfully manufactured by means of a melt processing method (Reddy *et al.*, 2013). TGA of the oat hull fibres clearly showed that the processing temperature of 190°C was acceptable. Oat hull fibre-reinforced composites displayed reductions in tensile and flexural properties when compared to the PP/PLA blend. The tensile and flexural properties of the composites significantly improved when compatibilizer was added. Heat deflection temperatures (HDT) of the PP, PP/PLA and PP/PLA/oat hull are shown in Fig. 15.12. The HDT of PP is 83°C and for PLA is 65°C. With the addition of compatibilizer a slight reduction in HDT was observed. The maleated ethylene–propylene copolymer affects the stiffness at higher temperature, hence a slight reduction in HDT was observed. Furthermore,

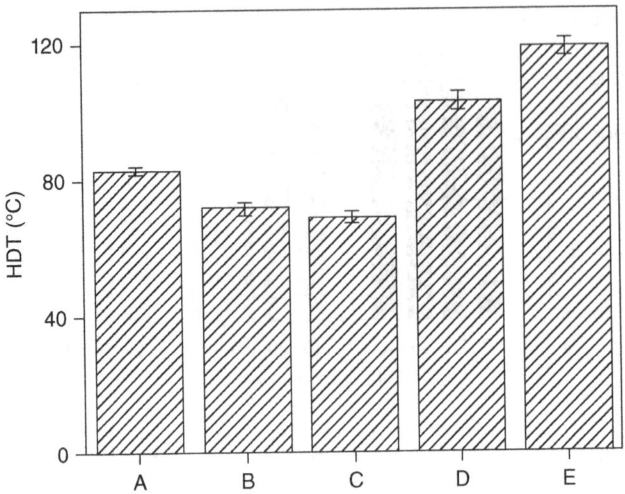

15.12 HDT of PP, PP/PLA, and PP/PLA/oat hull composites: (A) PP; (B) PP (90%) + PLA (10%); (C) PP (87%) + PLA (10%) + EP-g-MA (3%); (D) [PP (90%) + PLA (10%)] 70% + oat hull (30%); (E) [PP (87%) + EP-g-MA (3%) + PLA (10%)] 70% + oat hull (30%) (adapted from Reddy *et al.*, 2013).

HDT of the oat fibre-reinforced PP/PLA composites improved by 43% over that for the PP/PLA blend. Composites with 30 wt% oat hull loading showed improved storage moduli for both uncompatibilized and compatibilized systems (Reddy *et al.*, 2013).

The effects of the addition of compatibilizers (SEBS-g-MA, PS-*co*-MA, PEG) on the thermal, microstructural and morphological characteristics were examined for PS composites containing cellulose and oat (Pacella *et al.*, 2011). For all compatibilized systems, enhanced fibre dispersion and interfacial adhesion were observed. Moreover, the tensile properties were noticeably affected by the type, fibre content and concentration of the compatibilizer.

15.6 Processing and performance: barley and rye fibre composites

15.6.1 Mechanical properties

Bouhicha *et al.* (2005) produced soil composites with barley straw that had different fibre/soil ratios and fibre lengths. The results showed the positive effects of adding straw: decreased shrinkage, reduced curing time and enhanced compressive strength were seen if an optimal reinforcement ratio was used. Flexural and shear strengths were also increased and more ductile

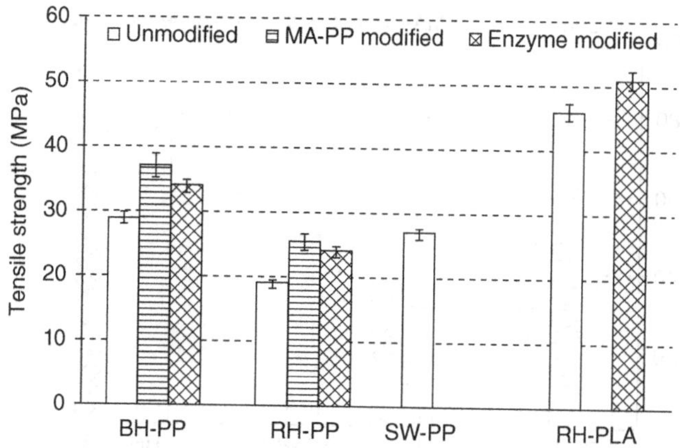

15.13 Tensile strength of modified and unmodified barley husk (BH) and rye husk (RH) reinforced PP or PLA composites: a comparison with softwood (SW) composites.

failure occurred for the reinforced fibres. The best water repellent treatment seems to be one that uses a cement render with an added polymer.

Barley husk (BH) and rye husk (RH) reinforced polypropylene or polylactic acid composites have been fabricated using a high-speed mixer followed by injection moulding with 40 wt% of fibre load. Mechanical, dielectric and chemical properties as well as odour emission of the resulting composites have been investigated and the results are described below.

The tensile strengths of barley husk and rye husk reinforced composites are displayed in Fig. 15.13. The tensile strength of barley husk PP composites was 8% better than that of softwood PP composites, but, when considering the standard deviation, the values are nearly the same. In contrast, rye husk PP composites showed a tensile strength 30% lower than that of softwood PP composites. With the addition of a coupling agent, the tensile strength improved by 30% and 35% for barley husk PP and rye husk PP composites, respectively. This improvement is caused by the formation of an ester linkage via MA-PP between the cellulosic filler and the polypropylene molecule. The tensile strength of softwood-reinforced PP composites also increased by 20% with the addition of a coupling agent (Bledzki *et al.*, 2010a, 2010b).

The increment in tensile strengths for different fibre composites are dissimilar, because the different fibres contain varying proportions of chemical components: cellulose (structural materials), starch, hemicelluloses, lignin and protein. Structural materials provide reinforcement of composites, with strong interlinking between the fibre and the matrix. Amorphous materials have less effect (positive/negative) on the composite properties. Fibre surface energy also plays an important role in the properties of the

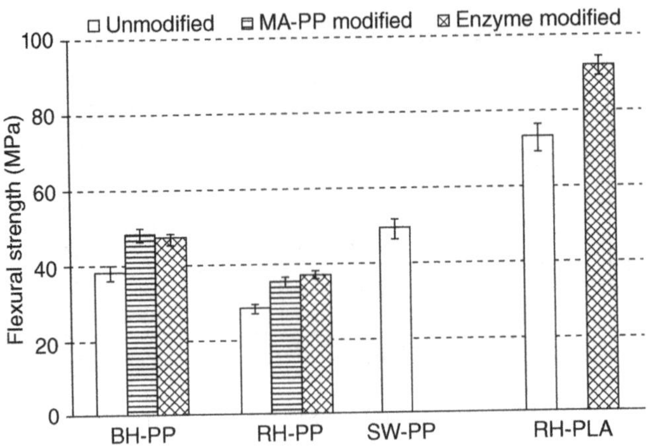

15.14 Flexural strength of modified and unmodified barley husk (BH) and rye husk (RH) reinforced PP or PLA composites: a comparison with softwood (SW) composites.

composite (Bledzki *et al.*, 2010a). As a result of enzyme modification, this property was found to improve by about 18% for barley husk PP composites and by 27% for rye husk PP composites. The tensile strength of rye husk PLA composites also improved by 11% as a result of enzyme modification. This is because of a reduction of the fibre particle size, an improved interface, and an increase in the average molecular weight of the fibre, which is achieved by the removal of amorphous and unwanted materials (Mamun and Bledzki, 2013).

The flexural strength of barley husk PP composites was 26% higher than that of rye husk composites (Fig. 15.14). The flexural strengths of barley husk PP and rye husk PP composites were 22% and 43% lower, respectively, than that of softwood PP composites. With the addition of MA-PP, the flexural strength of barley husk and rye husk composites increased by 27% and 25%, respectively. This property improved by 24% for barley husk PP composites, by 33% for rye husk PP composites and by 26% for rye husk PLA composites after enzyme modification (Mamun and Bledzki, 2013). The improvement in flexural strength of composites modified with enzyme is likely to be attributable to the removal of the outer surface. This increases the effective surface area and induces physico-chemical changes. As a consequence, an improvement in the interfacial adhesion and the shear stress ability also occurs. The effects of the modification include fibrillation and size reduction of the fibres, and this may have an influence on the strength properties of the composites (Mamun and Bledzki, 2014).

Notched Charpy impact strength tests were carried out according to EN ISO 179/1eA (instrumented) using a Zwick Charpy impact machine. Ten

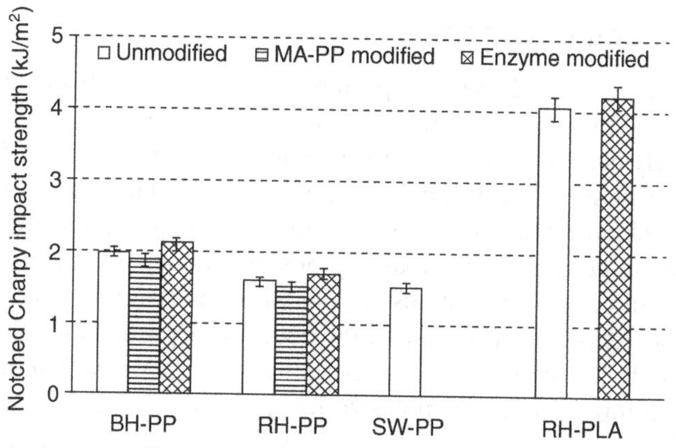

15.15 Notched Charpy impact strength of modified and unmodified barley husk (BH) and rye husk (RH) reinforced PP or PLA composites: a comparison with softwood (SW) composites.

specimens with 2 mm A type notches were used for this investigation. The notched Charpy impact strengths of barley husk (BH), rye husk (RH) and softwood (SW) composites are presented in Fig. 15.15. The Charpy impact test is a standardized high-strain-rate test which determines the amount of energy absorbed by a material during fracture. This absorbed energy is a measure of a given material's toughness and acts as a tool to study the brittle–ductile transition. The notched Charpy impact strength of barley husk composites was found to be 35% better than that of softwood composites. Rye husk composites showed a similar level of notched Charpy impact strength as softwood composites (Bledzki *et al.*, 2010a).

Particle size, shape and fibre surface properties all have an influence on this. The notched Charpy impact strength was reduced to some extent with the addition of a coupling agent. Damage initiation can be shifted to a higher force for composites with a strong fibre–matrix adhesion, whereas a smaller force is required to break down composites with a weak fibre–matrix adhesion. When composites do not contain a coupling agent, part of the impact energy is degraded in the fibre–matrix interface, for example by debonding and friction effects. In composites with no coupling agent, the damage saturation is less pronounced. This means that poor adhesion between the matrix and fibres leads to higher energy absorption, whereas relatively strong adhesion leads to lower energy absorption (Mamun and Bledzki, 2014). In the case of modified fibre composites, this property was found to improve slightly for PP and PLA composites as a result of enzyme modification. This is a consequence of an improvement in the fibre flexibility and, consequently, enhanced composite flexibility (Mamun and Bledzki, 2013).

15.6.2 Dielectric properties

Polymers were studied by comparing data gathered from dielectric, mechanical and nuclear magnetic resonance experiments on an homologous series of polymers, in order to investigate the occurrence of dielectric relaxation processes. The following types of molecular motion and their associated dielectric relaxation were identified: primary main chain motion in both crystalline and amorphous regions; secondary main chain motion in both crystalline and amorphous regions; side chain motion; and impurity motions. The relaxation processes that occur in a particular polymer are denoted as $\alpha, \beta, \gamma, \sigma$ in order of the increasing frequency at which they occur for a fixed temperature. The polarization process in polymers under an electric field takes place when units of the polymer chain tend to orient the dipole and strongly polarizable bonds in the direction of the electric field (Bal and Kothari, 2009).

The test specimens were conditioned in a climate conditioning cabinet at 40°C and 65% relative humidity and the properties were determined at predefined intervals. Rectangular specimens ($100 \times 100 \times 1.97\,mm^3$) were prepared using an injection-moulding process. The test samples were coated with silver paste on either side and copper wires were fixed on both sides of the samples as electrodes. The resistance was measured directly at room temperature using Haefely Capacitance and Tan Delta Bridge, Type 470, according to DIN 53483. The volume resistivity was measured using a TΩ Meter 5478, H+H High Voltage Technology, Tetley Instrument, according to DIN IEC.

Volume resistivity

The study of the specific volume resistivity of an insulating material is important, because the most desirable property of an insulator is its ability to resist the leakage of an electric current. Figure 15.16 shows the variation of volume resistivity with ageing. The volume resistivity was found to decrease with the addition of fibre in comparison with the control PP. The volume resistivity of neat PP remained unchanged as it aged (Bledzki *et al.*, 2009). This implies that the conductivity increased upon addition of lignocellulosic fibres. This could be due to the presence of polar groups, which facilitate the flow of current. It is well known for semi-crystalline polymers that most of the current flows through the crystalline regions, and the passage of currents in the amorphous regions is due to the presence of moisture (Jacob *et al.*, 2006). The specific volume resistivity of all composites reduced swiftly during the initial stage, and then slowly decreased as the ageing periods increased. The rapidly decreasing volume resistivity was associated with moisture uptake. In comparison to unmodified barley husk

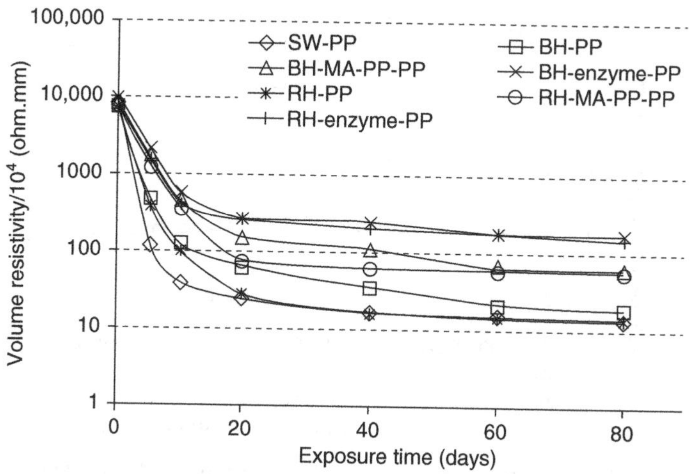

15.16 Volume resistivity of modified and unmodified barley husk (BH) and rye husk (RH) reinforced PP: a comparison with softwood (SW) composites (40°C, RH 65%).

composites, the volume resistivity of MA-PP modified composites was found to be 3.3-fold higher and for enzyme-modified composites ninefold higher at the end of the ageing period. In the case of rye husk composites, this property was fourfold higher for MA-PP modified composites and 11-fold higher for enzyme-modified composites compared to unmodified composites (Mamun *et al.*, 2010). These results are attributed to reduced moisture uptake due to modification. Unmodified barley husk and rye husk composites showed higher volume resistivity than softwood composites at all ageing times.

Figure 15.17 represents the volume resistivity of unmodified and modified rye husk PLA composites. The volume resistivity of the control PLA reduced very slowly in association with the exposure time. This result was caused by the slow moisture uptake of the PLA. The volume resistivity decreased due to addition of fibre in comparison with the control PLA. PLA composites exhibited similar tendencies to those observed for PP composites. In comparison to unmodified rye husk composites, the volume resistivity of enzyme-modified composites was better than that of unmodified composites (Mamun, 2011).

Dielectric constant

The dielectric constant of a material depends upon the polarizability of the molecules. The polarizability of non-polar molecules arises first from electronic polarization, in which the application of an electric field causes

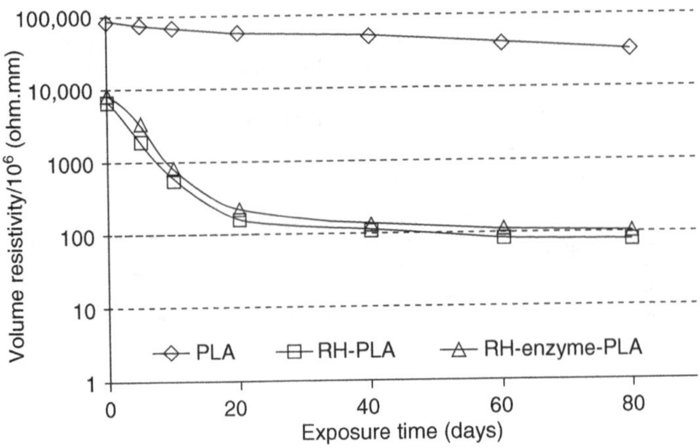

15.17 Volume resistivity of modified and unmodified rye husk (RH) reinforced PLA composites (40°C, RH 65%).

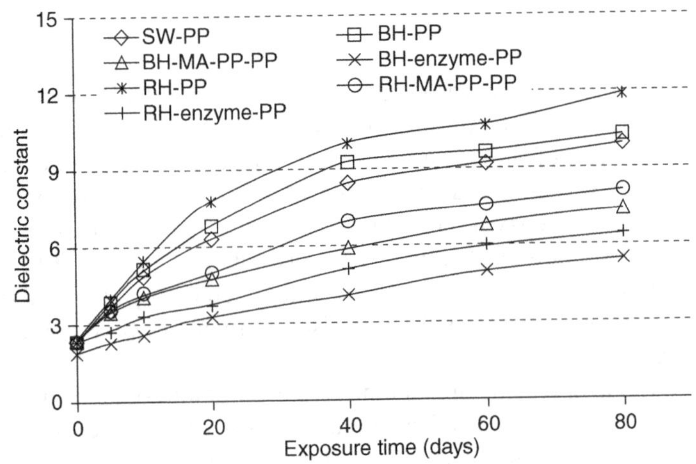

15.18 Dielectric constant of modified and unmodified barley husk (BH) and rye husk (RH) reinforced PP composites: a comparison with softwood (SW) composites (40°C, RH 65%).

a displacement of the electrons to the nucleus. Second, it can be caused by atomic polarization, which involves the application of an electric field, causing a displacement of the atomic nuclei relative to one another. In the case of polar molecules, a third factor is also involved, namely orientation polarization (in which the application of an electric field causes an orientation of dipoles).

Figure 15.18 shows the influence of modification of barley husk and rye husk on the dielectric constant values of polypropylene composites; this property was also compared with softwood fibre composites. In the initial stage, the dielectric constants of fibre-loaded composites were lower than those of the control polypropylene (Bledzki *et al.*, 2009). The addition of natural fibres to composite materials increases the air content of the composite because of the hollow space in the middle of each fibre, which is referred to as the lumen. The air trapped in the lumen results in a lower dielectric value (Jacob *et al.*, 2006). The dielectric constant of treated and untreated fibre composites was found to increase as the ageing time progressed. This was because of the absorption of moisture molecules, which are polar. They, in turn, increase the surface polarity of the composite materials. The MA-PP and enzyme modified barley husk and rye husk composites exhibited a lowering of the dielectric constant. This was due to the decrease of moisture absorption and the orientation polarization of composites containing treated fibres (Bal and Kothari, 2009). The enzyme treatment results in a reduction of the moisture absorption capacity of fibres owing to the reduction of interaction between the polar–OH groups of the fibres and water molecules. The decrease in the resulting hydrophilic nature of the fibres leads to a lowering of the orientation polarization and, subsequently, of the dielectric constant (Khan *et al.*, 1991).

The dielectric constant of the control PLA and rye husk composites is displayed in Fig. 15.19. The dielectric constant of the control PLA remained nearly constant or increased slightly as the ageing time progressed. This was because PLA absorbed a little bit of moisture. The dielectric constant of treated and untreated rye husk PLA composites increased as the ageing time progressed. In comparison to unmodified rye husk composites, the

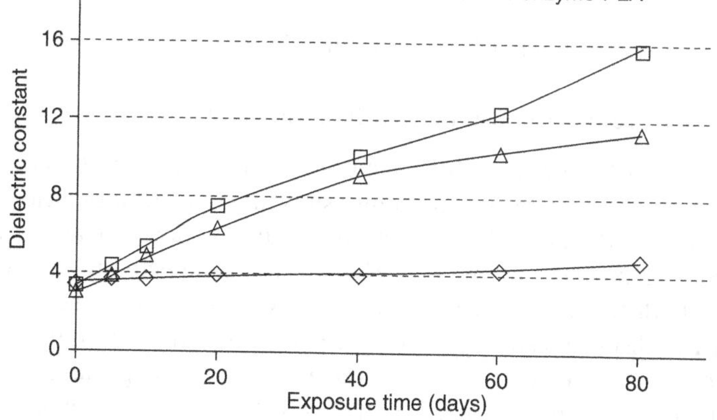

15.19 Dielectric constant of modified and unmodified rye husk reinforced PLA composites (40°C, RH 65%).

dielectric constant of enzyme-modified composites decreased by 30% at the end of the ageing period (Mamun, 2011).

15.6.3 Environmental stress cracking resistance

Environmental stress cracking resistance (ESCR) tests were carried out using a universal testing machine (Zwick 1440, Germany) according to the ISO standard 22088. Five flexural test specimens were used for this investigation. The specimens were bent using a three-point bending jig mounted in a chemical environment chamber. The investigations were carried out in a controlled medium (air) and a 0.25*N* hydrochloric acid medium. The load (F) was introduced to the sample until 1.25% elongation was reached, then fixed, after which 2 minutes (t_0) were allowed to stabilize the loading condition. Afterwards, the chemical was rapidly introduced into the chamber during loading. The variation of the load was recorded as follows: $t_1 = 15\,\text{min}$, $t_2 = 15\,\text{min}$, $t_3 = 20\,\text{min}$, $t_4 = 30\,\text{min}$, $t_5 = 40\,\text{min}$, $t_6 = 40\,\text{min}$. The test was carried out under laboratory conditions in an air-conditioned room with a controlled temperature of 23°C and relative humidity of 50%. The ESCR factor was calculated as the load in a corrosive medium divided by the load in a control medium at a certain time:

$$E_{\mathrm{T}}^{\mathrm{norm}}[-] = \frac{F_{\mathrm{T}}^{\mathrm{mM}}[\mathrm{N}]/F_{\mathrm{T}}^{\mathrm{oM}}[\mathrm{N}]}{F_{1}^{\mathrm{mM}}[\mathrm{N}]/F_{1}^{\mathrm{oM}}[\mathrm{N}]} \qquad [15.1]$$

where

$F_{\mathrm{T}}^{\mathrm{mM}}$ = force or load at a certain time with medium

$F_{\mathrm{T}}^{\mathrm{oM}}$ = force or load at a certain time without medium

F_{1}^{mM} = force or load at time t_1 with medium

F_{1}^{oM} = force or load at time t_1 without medium.

Figure 15.20 shows the ESCR factor of the modified and unmodified rye husk-reinforced polypropylene composites in a hydrochloric acid medium. It was observed that the ESCR factor was found to decrease as the duration of chemical exposure progressed. This was to be expected for a viscoelastic material. Both of the enzyme-modified rye husk-reinforced composites (PLA/PP) displayed stronger resistance properties than the unmodified rye husk composites. During the initial stage of chemical exposure, the MA-PP modified composites showed better resistance than enzyme-modified composites (Mamun, 2011). This can be attributed to the strong fibre–matrix adhesion associated with maleic anhydride. After a certain period

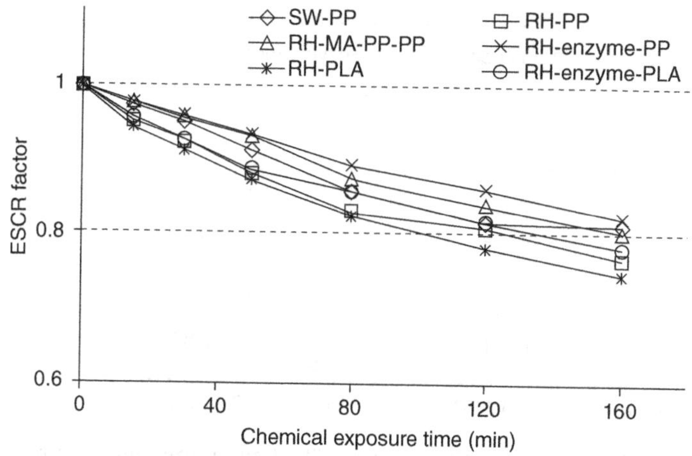

15.20 ESCR factors of modified and unmodified rye husk PP and PLA composites: a comparison with softwood (SW) composites (HCl, 0.25*N*).

of chemical exposure, the resistance of MA-PP modified composites reduced swiftly. This may be explained by the reactivity of maleic anhydride with water at room temperature. The high reactivity of maleic anhydride leads to an increase of the local fluid concentration and successively reduces the local yield strength. The failure is preceded by the absorption of the chemical medium within the micro-yielded zone, forming micro-crazes which lead to increase of failure propagation. Absorption of any fluid plasticizes the polymer and reduces its yield strength (Wright, 1996). PP composites showed better chemical resistance than PLA composites.

15.6.4 Odour emission

The odour measurements were performed with a T07 olfactometer produced by Odournet GmbH, Germany, and followed the standard method DIN EN 13725. The samples were stored at 80°C for 120 minutes in an oven prior to testing. In this method, the human nose is employed as a sensor. In olfactometry, the sample gas (odour sample) is diluted with neutral air at a defined ratio (1 g/litre). This dilution is represented in the panel lists as a smell sample. The panel lists display several levels of dilution. The odour concentration of the examined sample is the dilution factor at the detection threshold, and is expressed as multiples of one odour unit per cubic metre (OU/m^3) in standard conditions. The odour level describes the intensity of a sensation as a function of the logarithm of the amount of the stimulating quantity. The reference (butanol) quantity of the odorant concentration is 1 OU/m^3.

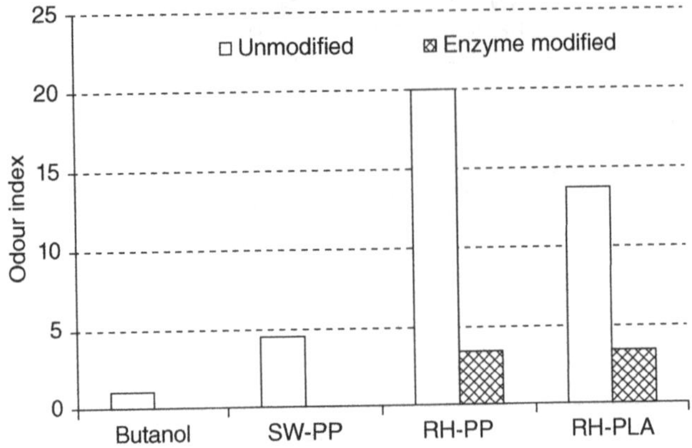

15.21 Odour emission from modified and unmodified rye husk PP and PLA composites: a comparison with softwood (SW) composites.

The odour emission intensity of rye husk composites is presented in Fig. 15.21 (Mamun, 2011). The odour emitted by unmodified rye husk PP and PLA composites showed 20 and 13.5 on the odour index, respectively. With thermal processing, the fibre surface molecules and thermally unstable components undergo thermal decomposition and emit odour. As a result of the modification, the odour emission was reduced by 83% for rye husk PP composites and by 75% for rye husk PLA composites. The reduction of odour emission could have been caused by the removal of extractives, fat, protein and other small molecules from the fibre surface. Both rye husk composites emit much higher levels of organic compounds than softwood composite. In contrast, enzyme-modified PP and PLA composites were found to emit fewer volatile organic compounds than softwood composites (Mamun, 2011).

15.7 Conclusion

The physical, chemical and thermal properties of lignocellulosic residues are more or less similar to those of standard natural fibres (mainly wood fibres). The composite processing conditions of these materials are not the same as for other natural fibres, but if the processing parameters are optimized, these materials could be used extensively in various applications. The enzyme modification of lignocellulosic residues or the addition of coupling agents results in improvement of the mechanical, chemical, dielectric and odour emission properties of composites. Thus, the proper utilization of these waste materials will provide cheap engineering materials and also assist in waste management. Moreover, they could be used as reinforcing agents in composite materials, in combination with wood fibre or as an alternative.

15.8 References

Altpeter, F. and Korzun, V. (2007), Rye, in Pau, E.C. and Davey, M.R. (editors), *Biotechnology in Agriculture and Forestry, vol. 59, Transgenic Crops IV*, Springer, Berlin, pp. 107–117

Bal, K. and Kothari, V.K. (2009), Measurement of dielectric properties of textile materials and their applications, *Indian Journal of Fibre and Textile Research*, 34, 191–199

Bavan, D.S. and Kumar, M.G.C. (2012), Morphological and thermal properties of maize fibre composites, *Fibers and Polymers*, 13(7), 887–893

Bavan, D.S. and Kumar, M.G.C. (2013), Finite element analysis of a natural fibre (maize)composite beam, *Journal of Engineering*, AID 450381, 1–7

Bledzki, A.K., Mamun, A.A., Lucka, M. and Gutowsk, V.S. (2008), The effects of acetylation on properties of flax fibre and its polypropylene composites, *eXPRESS Polymer Letters*, 2(6), 413–422

Bledzki, A.K., Lucka, M., Mamun, A.A. and Michalski, J. (2009), Biological and electrical resistance of acetylated flax fibre reinforced polypropylene composites, *BioResources*, 4(1), 111–126

Bledzki, A.K., Mamun, A.A. and Volk, J. (2010a), Barley husk and coconut shell reinforced polypropylene composites: The effect of fibre physical, chemical and surface properties, *Composites Science and Technology*, 70, 840–846

Bledzki, A.K., Mamun, A.A. and Volk, J. (2010b), Physical, chemical and surface properties of wheat husk, rye husk and soft wood and their polypropylene composites, *Composites: Part A*, 41, 480–488

Bouhicha, M., Aouissi, F. and Kenai, S. (2005), Performance of composite soil reinforced with barley straw, *Cement and Concrete Composites*, 27(5), 617–621

Clark, J.H. (2008), *Introduction to Chemicals from Biomass*, Wiley, Chichester, UK

Dauda, M., Yoshiba, M., Miura, K. and Takahashi, S. (2007), Processing and mechanical property evaluation of maize fiber reinforced green composites, *Advanced Composite Materials*, 16(4), 335–347

Donato, P.D., Fiorentino, G., Anzelmo, G., Tommonaro, G., Nicolaus, B. and Poli, A. (2011), Re-use of vegetable wastes as cheap substrates for extremophile biomass production, *Waste and Biomass Valorization*, 2, 103–111

Dorris, G.M. and Grey, D.G. (1978), The surface analysis of paper and wood fibres by ESCA II. Surface composition of mechanical pulp, *Cellulose Chemistry and Technology*, 12, 721–734

Ensminger, A.H. (1994), *Foods and Nutrition Encyclopedia*, 2nd edition, CRC Press, New York, p. 479

FAO (2013), *Production data sheet by country*, FAOstat, Food and Agriculture Organization of the United Nations

Flandez, J., González, I., Resplandis, J.B., Mansouri, N.E., Vilaseca, F. and Mutjé, P. (2012), Management of corn stalk waste as reinforcement for polypropylene injection moulding composites, *BioResources*, 7(2), 1836–1849

Graupner, N. (2008), Application of lignin as natural adhesion promoter in cotton fibre reinforced PLA composites, *Journal of Materials Science*, 43, 5222–5229

Huda, S. and Yang, Y. (2008), Chemically extracted corn husk fibres as reinforcement in light weight polypropylene composites, *Macromolecular Materials and Engineering*, 293, 235–243

Jacob, M., Varughese, K.T. and Thomas, S. (2006), Dielectric characteristics of sisal–oil palm hybrid biofibre reinforced natural rubber biocomposites, *Journal of Materials Science*, 41(17), 5538–5547

Khan, M.A., Ali, K.M.I. and Wang, W. (1991), Electrical properties and X-ray diffraction of wood and wood plastic composites, *Radiation Physics and Chemistry*, 38(3), 303–306

Li, X., Tabil, L.G. and Panigrahi, S. (2007), Chemical treatments of natural fiber for use in natural fiber-reinforced composites: A review, *Journal of Polymers and the Environment*, 15, 25–33

Mamun, A.A. (2011), *PP and PLA composites with grain by-products: Enzyme modification, characterization, processes optimization and properties*, PPH Zapol Dmochowski, Szczecin, Poland

Mamun, A.A. and Bledzki, A.K. (2013), Microfibre reinforced PLA and PP composites: Enzyme modification, mechanical and thermal properties, *Composites Science and Technology*, 78, 10–17

Mamun, A.A. and Bledzki, A.K. (2014), Enzyme modification of grain by-products and their biocomposites: Characterization, mechanical and thermal properties, *Macromolecular Materials and Engineering*, 299(2), 248–256

Mamun, A.A., Bledzki, A.K., Austermühl, A. and Claudi, A. (2010), Electrical and mechanical properties of modified grain by-product–PP composites, *8th Global WPC and Natural Fibre Composites Congress and Exhibition*, 22–23 June, 2010, Stuttgart, Germany

Onyeagoro, G.N. (2012), Effect of fibre acetylation on some mechanical properties of cellulose acetate butyrate/yellow poplar wood fibre composites, *International Journal of Academic Research*, 4(1), 55–63

Pacella, M., Haque, M. and Alvarez, V. (2010), Functionalization, compatibilization and properties of polyolefin composites with natural fibres, *Polymers*, 2, 554–574

Pacella, M., Pancrazi, C., Haque, M. and Alessio, A. (2011), Thermal and microstructural characterization of compatibilized polystyrene/natural fillers composites, *Journal of Thermal Analysis and Calorimetry*, 103, 95–101

Reddy, N. and Yang, Y. (2005a), Biofibers from agricultural byproducts for industrial applications, *Trends in Biotechnology*, 23(1), 22–27

Reddy, N. and Yang, Y. (2005b), Properties and potential applications of natural cellulose fibers from cornhusks, *Green Chemistry*, 7, 190–195

Reddy, J.P., Misra, M. and Mohanty, A. (2013), Injection moulded biocomposites from oat hull and polypropylene/polylactide blend: Fabrication and performance evaluation, *Advances in Mechanical Engineering*, AID 761840, 1–8

Suramaythangkoor, T. and Gheewala, S.H. (2011), Implementability of rice straw utilization and greenhouse gas emission reductions for heat and power in Thailand, *Waste and Biomass Valorization*, 1, 1–15

Thamae, T., Vaja, S., Shangguan, Y., Finoro, C., Stefano, N. and Baillie, C. (2009), Mechanical and moisture absorption of corn and wheat flour composites for developing countries, in Willems, F. and Moens, P., *Green Composites: Properties, Design and Life*, Nova Science Publishers, New York, pp. 1–16

Trigui, A., Karkri, M., Peña, L., Boudaya, C., Candau, Y., Bouffi, S. and Vilaseca, F. (2013), Thermal and mechanical properties of maize fibres–high density polyethylene biocomposites, *Journal of Composite Materials*, 47(11), 1387–1397

Welch, R.W. (1995), The chemical composition of oats, in *The Oat Crop: Production and Utilization*, Chapman & Hall, London

Wright, D.C. (1996), *Environmental Stress Cracking of Plastics*, Rapra Technology Limited, Shrewsbury, UK

Wrigley, C., Corke, H. and Walker, C.E. (2004), *Encyclopedia of Grain Science*, Elsevier Academic Press, St Louis, MO

Yin, Z.-D., Pan, Z.-L., Wang, C., Dong, Y.-P. and Ou, Y.-X. (2007), Composition, structure and mechanical properties of several natural cellular materials, *Chinese Science Bulletin*, 52(21), 2903–2908

Zhou, X., Jellen, E.N. and Murphy, J.P. (1999), Progenitor germplasm of domesticated hexaploid oat, *Crop Science*, 39, 1208–1214

Zohary, D. and Hopf, M. (2000), *Domestication of Plants in the Old World*, 3rd edition, Oxford University Press, Oxford, UK, p. 75

16

The use of bamboo fibres as reinforcements in composites

H. P. S. ABDUL KHALIL, M. S. ALWANI, M. N. ISLAM, S. S. SUHAILY, R. DUNGANI and Y. M. H'NG, Universiti Sains Malaysia, Malaysia and M. JAWAID, Universiti Putra Malaysia Malaysia

DOI: 10.1533/9781782421276.4.488

Abstract: Bamboos are of notable economic and cultural significance all over the world, especially in South Asia, Southeast Asia and East Asia, being used for building materials, as a source of food, as a decorative product and as a versatile raw product. Bamboo also has significant potential in composite making due to its high strength, environmentally friendly nature, rapid growing properties, low cost, availability and sustainability. This chapter summarizes production processes for bamboo fibres and their applications in composites. The production of nanocellulose from bamboo fibres is also briefly discussed. Overall, the chapter aims to show the versatility of bamboo fibre in various applications from construction to culinary.

Key words: bamboo fibres, nanocellulose, polymer composites, resin, surface treatment.

16.1 Introduction

As the worldwide demand for fibre grows, sustainable resource management and efficient industrial utilization must collaborate to develop a shared vision for both long-term management of bio-resources and sustainable economic development. Concerns about the environment and ecology have also sparked a new trend towards the use of environmentally friendly materials (Philippou *et al.*, 2001), thus interest in developing natural fibre-based composites is increasing rapidly. Various types of natural fibres have been studied for their application in composites, e.g. bamboo (*Bambusa* spp.) fibres, oil palm (*Elaeis* spp.) fibres, banana (*Musa* spp.) fibres, pineapple (*Ananas comosus)* leaf fibre, coconut (*Cocos nucifera*), coir (*Corchorus* spp.) fibres, jute, dhaincha (*Sesbania* spp.*)*, flax (*Linum usitatissimum*), kenaf (*Hibiscus cannabinus*), henequen (*Agave fourcroydes*) and hemp (*Cannabis sativa*) fibres (Islam *et al.*, 2010; Edyham and Hanafi, 2002; Pothan *et al.*, 1997; Devi *et al.*, 1997).

Bamboo, which belongs to more than 10 genera including about 1450 species ranging from small annuals to giant timber bamboo, is one of the oldest building materials used by mankind (Kamruzzaman *et al.*, 2008). It is the fastest growing woody plant in the world and is easily accessible globally. Around 64% of bamboo plantation originates from Southeast Asia, 33% is grown in South America and the rest comes from Africa and Oceania (Bonilla *et al.*, 2010). The bamboo culm, or stem, has been made into a diverse range of products from domestic to industrial applications. Examples of bamboo products are food containers, skewers, chopsticks, handicrafts, toys, furniture, flooring, pulp and paper, boats, charcoal, musical instruments, weapons, bicycles, dirigibles, windmills, scales, retaining walls, ropes and cables. With advances in science and technology and the restricted supply of timber, a wide range of technologies have been developed in recent years to process bamboo and make it more durable and usable as a building material.

Bamboo has also gained popularity as a 'green' fibre. It can be cultivated quickly and is a natural fibre (as opposed to popular synthetics like polyester) whose cultivation actually reduces greenhouse gases. Fibres can be made from the leaves, branches and trunks through chemical process, mechanical needling and scraping or through a steam explosion process. It is extremely resilient and durable as a fibre and has served as a foundation structure (Khanam *et al.*, 2011; Bhat *et al.*, 2011). In studies comparing it to cotton and polyester, it was found to have a high breaking tenacity, better moisture wicking properties and better moisture absorption. Cellulose from bamboo fibres is suitable for processing into viscose rayon, and viscose manufactured from bamboo is promoted as having environmental advantages over wood-pulp viscose. Bamboo is also used as an additive in biopolymers for construction and in many other applications.

To reduce any harmful destruction of the ecosystem and produce low-cost polymeric reinforced composites, researchers are working to manufacture composites using natural fibres which are entirely biodegradable, and consequently the use of bamboo fibres as a reinforcement in composite materials has increased tremendously in recent years. Its structural variation, mechanical properties, chemical modification and thermal properties have made it a versatile material (Amada *et al.*, 1996). The cost, availability, light weight, high specific strength and non-hazardous nature of bamboo fibres are its most attractive properties, encouraging researchers in composite technology to work on its development. Production of nanofibrillated cellulose (NFC) from bamboo is a recent development which has many different applications such as nanobiocomposites. There has been extensive research in many fields; engineering, biotechnological (genetic engineering) and cultivation, for example, all aiming to improve the use of bamboo fibres in the composites industry.

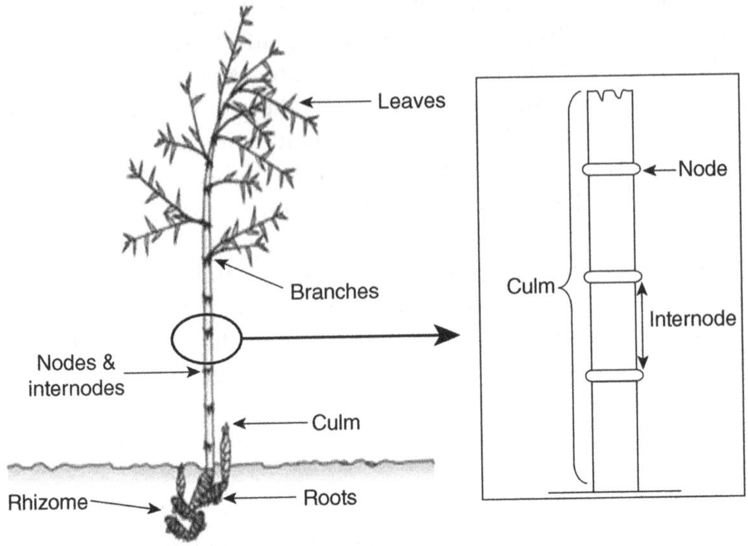

16.1 Structure of a bamboo (Sakaray *et al.*, 2012).

16.2 Structure of bamboo

The main parts of bamboo are the culm, node, internode, leaf and roots. The culm is a hollow stem likely to be cylindrical in shape, while the nodes can be found between the culms along the entire length of bamboo, acting as a disc inserted between each section of culms (Mwaikambo, 2006) (Fig. 16.1). These nodes play an important role in preventing buckling when the bamboo is bent (Amada *et al.*, 1996).

The structure of bamboo culm has been described over the years in great detail and most publications dealing with bamboo have focused on its anatomical structure (Parameswaran and Liese, 1980; Liese, 1998; Gritsch and Murphy, 2005; Jiang, 2007). Unlike wood, bamboo has no rings in its culm; it mainly consists of epidermis, cortex, ground tissue and lacuna. The vascular bundles which are made up of slerenchyma fibres, vessels and phloem are widely embedded in parenchymatous ground tissues (Londoño *et al.*, 2002) (Fig. 16.2). These bundles play the same role as reinforced fibres in composite materials. The fibres are characteristically thick-walled at maturity and the high tensile strength of bamboo tissue is attributed mainly to their multilayered cell wall structure (Gritsch *et al.*, 2004). The sieve tube and vessel which build up the vascular bundle are responsible for transporting nutrients and water. Jiang (2007) reported that the culm consists of about 52% parenchyma cells, 40% fibres and 8% conducting tissue. The vascular bundles are widely distributed from the periphery

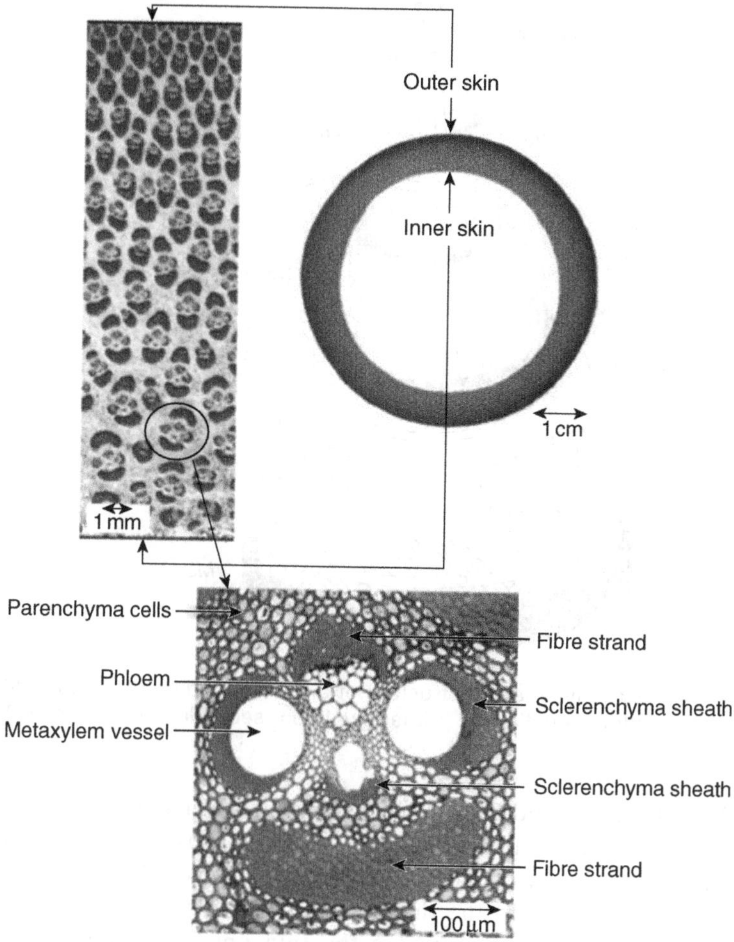

16.2 Crosssection of the bamboo culm (Chaowana, 2013).

towards the inner section of the culm; however, their concentration, size and shape vary. The fibre bundles are smaller and denser when closer to the periphery compared to the inner section of the culm (Grosser and Liese, 1971; Silva *et al.*, 2006). The fibres which are strongly lignified are mainly responsible for the mechanical properties of bamboo fibres (Alvin and Murphy, 1988). According to Liese (1998) the structure and anatomy of the bamboo culm has a large impact on its physical properties as well as its uses. For example, the mechanical properties of the culm are determined by its specific gravity, which depends mainly on the density and diameter of the fibres and the thickness of the fibre cell walls and makes it suitable for use as a building material.

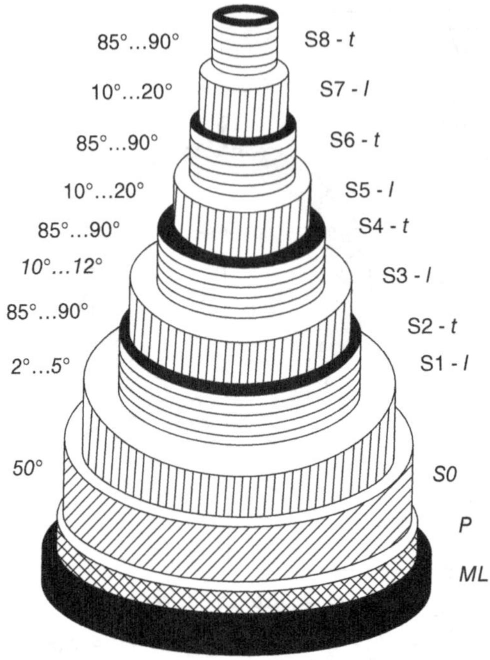

16.3 Polylamellate structure of a thick-walled bamboo fibre (ML = middle lamella, P = primary wall, S = secondary wall) (Liese, 1985).

Parameswaran and Liese (1980) reported that bamboo fibre consists of a multilayered or 'polylamellated' cell wall structure (Fig. 16.3). The lamellation consists of alternate broad and narrow layers with different fibrillar orientation. In the thick layers the fibrils are oriented at a slight angle to the fibre axis, whereas the thin ones generally show a more transverse orientation. This cell wall structure is very important because it determines the mechanical properties of the fibres (Osorio *et al.*, 2010). Fibre characteristics such as diameter, cell wall thickness and number of cell wall layers have been shown to vary according to their location in the culm and within vascular bundles, as well as to the maturity of the culm (Parameswaran and Liese, 1976; Murphy and Alvin, 1992).

16.3 Chemical properties of bamboo

Table 16.1 shows the compiled database for the chemical composition of bamboo. In general, the major chemical constituents of bamboo are cellulose, hemicelluloses and lignin, accounting for over 90% of the total mass, which is similar to wood (Jain *et al.*, 1992). According to Fengel and Wegener (1984) the α-cellulose content in bamboo is 40–50%, comparable with the

Table 16.1 Chemical composition of bamboo

Bamboo species	Holocellulose (%)	Cellulose (%)	Lignin (%)	Extractives (%)	Ash (%)	Reference
Phyllostachys makinoi	79.9	45.3	25.5	–	–	Fengel and Shao (1984)
Phyllostachys edulis (Riv.)	–	44.5	20.5	–	2.4	Amada *et al.* (1996)
Yunnanicus bamboo	70.3	52.7	25.5	5.7	2.3	Feng *et al.* (2003)
Whangee bamboo	70.2	52.4	23.2	7.2	1.8	Feng *et al.* (2003)
Dendrocalamus asper	74.0	–	28.5	5.5	1.5	Kamthai (2003)
Gigantochloa scortechinii	67.4	–	26.4	3.4	1.3	Kassim *et al.* (1992)
Phyllostachys pubescens	71.7	–	23.6	4.6	1.4	Li *et al.* (2007)
Schizostachyum zollingeri	71.6	–	21.4	2.5	–	Nor Aziha and Azmy (1991)
Bambusa vulgaris	67.1	50.2	23.9	–	1.3	Nahar and Hasan (2012)

reported α-cellulose contents of softwoods (40–52%) and hardwoods (38–56%), which makes bamboo a suitable raw material for the pulp and paper industry. According to Mwaikambo and Ansell (2001) cellulose is the most important component for composite fabrication as higher cellulose content leads to increased stiffness and therefore greater suitabilty for resin reinforcement. The high lignin content of bamboo contributes to its high heating value and its structural rigidity with the latter making it a valuable building material (Scurlock *et al.*, 2000). Lignin also facilitates reactivity, which allows a better response to chemical modification (Mwaikambo and Ansell, 2001). The high ash content for some bamboo species can adversely affect tool knife wear during machining operations and pulp processing (Mohmod, 1993). The ash content of bamboo is mostly silica along with metals such as calcium and potassium. According to Taj *et al.* (2007) its chemical makeup contributes to the overall properties of the fibre and the composition varies for different types of fibre. Usually the chemical content of bamboo changes with the age of the bamboo, the cellulose content in particular decreasing as the bamboo matures. Thus, it directly affects the chemical composition of bamboo fibre. Meanwhile, non-cellulosic components also contribute to fibre properties such as strength, flexibility, moisture and even density (Abdul Khalil *et al.*, 2012).

16.4 Mechanical properties of bamboo

The mechanical properties of plant fibres depend on their physical, chemical and morphological properties such as fibre orientation, cellulose content, crystal structure and diameter or cross-sectional area of the fibre (Munawar *et al.*, 2007). The mechanical properties of plant fibres are much lower compared to those of the most widely used equivalent reinforcing glass fibres. However, because of their low density, the specific properties (property-to-density ratio), strength and stiffness of plant fibres are comparable to the values for glass fibres (Taj *et al.*, 2007).

Bamboo is one of the most popular cellulose fibre resources that are abundantly available. Bamboo shows great potential as a sustainable structural material as well as for use in textiles due to its shorter maturity cycle and high cellulose content (Wan and Ko, 2009). Chaowana (2013) reported that the mechanical properties of bamboo fibres are extremely unstable due to its variable density and moisture content, it being a heterogeneous and anisotropic material. It is well established that the mechanical properties of bamboo culm improve with age (Liese, 1985; Li, 2004). Improved mechanical properties in mature bamboo are correlated with increased specific gravity, due to anatomical changes in the vascular bundles (Mohmod, 1993). The number of vascular bundles per unit area of the stem remains unchanged over its life cycle; however, the number of fibre cells with thickened secondary walls increases within the vascular bundles (Li *et al.*, 2007). The fibre cell wall consists of highly crystalline cellulose fibrils wound spirally in a matrix of amorphous hemicellulose, lignin and pectin which act mainly as bonding agents (Biagiotti *et al.*, 2004). The mechanical properties of some bamboo species selected from several studies are presented in Table 16.2.

16.5 Cultivation of bamboo, fibre extraction and surface modification

The major steps in producing natural fibres for use in plastics include:

- Harvesting the fibre-bearing plants
- Extracting the fibres
- Further processing the raw fibre to meet the required purity and performance for use in plastic composites.

The fibre extraction procedures depend on the type and portion of plant from which the fibres are derived (e.g. bast, leaves, wood) as well as the

Table 16.2 Mechanical properties of bamboo fibres

Species	Density (g/cm^3)	Diameter (μm)	Tensile strength (MPa)	Young's modulus (GPa)	Elongation at break (%)	Reference
Dendrocalamus membranaceus	1.38	366	639–813	33	2–7.5	Defoirdt *et al.* (2010)
Bambusa vulgaris	1.03–1.21	–	106–204	–	–	Monteiro *et al.* (2011)
–			575	27	–	Mwaikambo (2006)
Guadua angustifolia	1.44	156	811	46.4	1.9	Trujillo *et al.* (2010)
–	0.6–1.1	–	140–230	11–17	3.6–3.8	Kang and Kim (2011)
Bambusa vulgaris	–	–	882	33	3.9	Nahar and Hasan (2012)

required fibre performance and economics (Rao and Rao, 2007). Bamboos grow to their maximum height and diameter in a few months and are harvested at 4–6 years old (Wang, 2012). Different types of bamboo material can be produced during harvesting, including mat, sliver and veneer. Harvested bamboo should be split with a machete, hand splitting knives or a splitting machine, and bamboo strips are milled out from the bamboo wall. The removal of nodes is essential to ensure an even thickness of sliver. The knots (inner and outer) are removed manually using a sharp knife or width-sizing machine. This process is followed by either mechanical or chemical processing depending upon the subsequent use of the bamboo fibres (Abdul Khalil *et al.*, 2012).

All the previous reports regarding the extraction of fibres from bamboo have mainly concentrated on mechanical, chemical and combined chemi-mechanical procedures (Deshpande *et al.*, 2000; Das and Chakraborty, 2008). Chemical processing includes initial alkali hydrolysis to yield bamboo fibres. Most manufacturers use this process as it is less time consuming (Abdul Khalil *et al.*, 2012). However, bamboo fibre is also extracted mechanically. In the mechanical process, the woody parts of the bamboo plant are crushed and then natural enzymes are used to break the bamboo walls into a soft mass, enabling the natural fibres to be mechanically combed out and spun into yarn (Erdumlu and Ozipek, 2008). Conventional methods of compression moulding technique (CMT) and roller mill technique (RMT) were explored for mechanical separation (Deshpande *et al.*, 2000).

The average diameter of fibre extracted by RMT is finer than with CMT, but the latter technique produces fibres with a higher average strength (Deshpande *et al.*, 2000). Another method used to separate bamboo fibres is the steam explosion method. In steam explosion, the water contained in bamboo is heated under high temperature and pressure and the pressure is then rapidly released to the atmosphere, so that the water evaporates, shattering the parenchyma inside the bamboo (Ochi, 2012). This method is more environmentally friendly compared to the chemical process; however, it is rarely used because it is expensive. A combination of chemical and mechanical methods has also been used for the extraction of bamboo fibres. In recent research by Anyakora (2013) bamboo fibres were impregnated with 'white liquor' and the softened sample converted into fibre by mechanical action, followed by thorough washing, screening and drying. The extracted fibres were separated, re-washed and dried in a forced-air circulation type oven.

The ultimate quality of a bamboo composite mainly relies on a proper combination of reinforcement (bamboo fibres) and binder (matrix or resins). Surface treatment before fabricating the composite is essential to induce bridging between the fibre and polymer matrix (Chen *et al.*, 1998). Natural fibres are hydrophilic; hence several researchers have carried out studies on surface treatment of bamboo fibres. Generally, surface treatments include alkali or acid and silane treatment. A recent study by Ma and Joo (2011) is the most promising one to date, showing that the tensile strength of bamboo/polylactic acid (PLA) composite increased to 88.83 MPa with silane coupling after plasma (CAP) treatment, an increase of 71.2%, and the interfacial shear strength (IFSS) improved by 87.4%. Chen *et al.* (1998) claimed that acetylation treatment is the most effective way to increase the moisture resistance of bamboo fibre. Liu and Hu (2008) reported that the modification of bamboo fibres with sodium hydroxide concentration under 11% results in a slight increase in its crystallinity index. Biodegradable fibreboard was also prepared from bleached bamboo fibres modified with cationic guar gum by Han *et al.* (2013). An increase in tensile strength and strain were observed in the treated fibre as well as improved thermal stability compared to the pure fibre. Surface treatments for bamboo fibres are still limited and further detailed studies are needed. Table 16.3 shows the effect on bamboo fibre after surface treatment.

16.6 Properties of bamboo fibre-reinforced polymer composites

Bamboo fibre-reinforced polymer composites have been developed by various researchers using bamboo fibres in the form of strip, culm or woven mat or hybridizing bamboo with synthetic fibres such as epoxy, polyester,

Table 16.3 Effect of surface treatments on bamboo fibre

Treatment	Effect on bamboo fibre	References
Mercerization (alkali treatment)	De-waxing, remove the impurities such as matter that contained oil and grease to improve the wetting ability when spreading resin over bamboo fibre surface; reduce the weight of bamboo fibre; surface roughness of fibre increases; diameter of fibre decreases; percentage of crystallinity increased because of the removal of cementing materials which leads to better packaging of cellulose chain; lignin, fat, protein and non-water-soluble ingredients removed.	Deshpande *et al.* (2000), Mohanty *et al.* (2001), Das and Chakraborty (2008), Kang and Kim (2011)
Silane treatment	Silane coupling agent cuts down the amount of hydroxyl groups and improves the crosslinkage between fibre and matrix; increases the impact fatigue strength of composite; improves the thermal and mechanical properties of composite.	Kushwaha and Kumar (2010b), Bao *et al.* (2011), Kang and Kim (2011)
Silane coupling after plasma (CAP) treatment	Bamboo surface etching helps in increasing surface roughness and hence improves the binding of matrix.	Ma and Joo (2011)
Silane coupling during UV irradiation (CDU)	Alter the surface structure of the bamboo fibre and active binding site, which helps in improving the mechanical properties of composite.	Ma and Joo (2011)
Isocyanate silane treatment	Amino (NH_2) and cyano groups (CN) contained in amino and isocyanate silanes, which strengthen the crosslinking of cellulose and coupling agent in bamboo fibres.	Tung *et al.* (2004)
Acetylation	Moisture uptake ability decreases; alter the fiber surface structure by substituting the hydrogen atom on fibre surface with acetyl group to reduce the polarity and hence helps in fibre–matrix interaction; increase the micropores on the acetylated surface which helps to improve the bamboo/resin interaction.	Phuong *et al.* (2010), Chen *et al.* (2011)

Continued

Table 16.3 Continued

Treatment	Effect on bamboo fibre	References
Acrylation	Maleic anhydride treatment improves the tensile strength, modulus and impact strength of composite by proposing better bonding due to the crystallization which happens on the surface of bamboo fibre.	Chen *et al.* (1998)
Thermal treatment	Optimum temperature of thermal treatment ranging from 140–220°C increases the dimensional stability of bamboo in the existence of moisture meanwhile the mechanical properties are unaffected.	Colla *et al.* (2011)
Oxidization treatment	Potassium permanganate ($KMnO_4$) is used as an oxidant to form cellulose radical from MnO_3^- ions and increase the surface roughness of bamboo.	Chen *et al.* (2011)

unsaturated polyester, phenolic, vinyl ester, polypropylene, polylactic acid, natural rubber, Novalac and polyethylene resins. Table 16.4 sets out the work on bamboo fibre-based polymer composites.

16.6.1 Bamboo fibre-reinforced polymer composites

Gupta *et al.* (2011) studied potential uses of bamboo fibre in polymer composites (Fig. 16.4) and measured the effect of various parameters on the mechanical and erosion wear performance of bamboo fibre-reinforced epoxy composites. The results indicated that the erosion wear performance of the epoxy matrix improves quite significantly with the addition of bamboo fibres. Tensile, flexural and impact properties increased with fibre loading and displayed maximum values at 40 wt% fibre, 30 wt% fibre and 20 wt% fibre loading, respectively. The researchers recommended the potential use of bamboo fibre for low-cost housing components, pipes carrying coal dust, industrial fans and helicopter fan blades. In another study, alkali- and silane coupling agent-treated bamboo fibres were reinforced into the epoxy matrix to study the effect of surface modification on tensile and impact properties of bamboo–epoxy composites under controlled conditions (Lu *et al.*, 2013). Alkali-treated bamboo fibre-reinforced polymer composites displayed a 34% and a 31% increase in tensile strength and elongation at break respectively, while silane-treated bamboo fibre resulted in a 71% increase

Table 16.4 Reported work on bamboo fibre-reinforced polymer composites

Fibre	Matrix	References
Bamboo	Epoxy	Lu *et al.* (2013), Gupta *et al.* (2011), Rajulu *et al.* (2004)
	Polylactic acid (PLA)/ poly(butylene succinate) (PBS)	Lee and Wang (2006)
	Polyester	Wong *et al.* (2010), Prasad and Rao (2011)
	Natural rubber	Ismail *et al.* (2002)
	Polypropylene (PP)	Phuong *et al.* (2010), Lee *et al.* (2009), Chen *et al.* (1998), Okubo *et al.* (2004), Chattopadhyay *et al.* (2011a, b)
	PP, MA-grafted PP, polyvinylidene fluoride, and polyethylene terephthalate	Fuentes *et al.* (2013)
Bamboo husk	Epoxy	Shih (2007)
Bamboo culm	Epoxy	Verma and Chariar (2013)
Bamboo strip	Epoxy	Shin *et al.* (1989)
	Vinyl ester	Chen *et al.* (2009, 2011)
	Polyester	Das and Chakraborty (2009a)
	Novolac	Das *et al.* (2009), Das and Chakraborty (2007, 2009b)
Bamboo fibre/ strip mat	Epoxy	Nirmal *et al.* (2012), Kushwaha and Kumar (2009, 2010a, b)
	Polyester	Kushwaha and Kumar (2009, 2010a)
Woven bamboo mat	Polylactic acid (PLA)	Porras and Maranon (2012)
	Epoxy	Kumar *et al.* (2011), Kushwaha and Kumar (2011)
	Polyester	Kushwaha and Kumar (2011)

in tensile strength and a 53% increase in elongation at break as compared to untreated composites.

Short bamboo fibre-reinforced epoxy composites were fabricated with different fibre loadings and the void content, density and percent weight reduction of these composites were determined (Rajulu *et al.*, 2004). In an interesting work, Lee and Wang (2006) investigated the effect of lysine-based diisocyanate as a coupling agent on the tensile, water resistance and interfacial adhesion properties of bamboo fibre-reinforced poly(lactic acid) and poly(butylene succinate) biocomposites. The thermal degradation of both bamboo composites was lower than those of the pure polymer matrix,

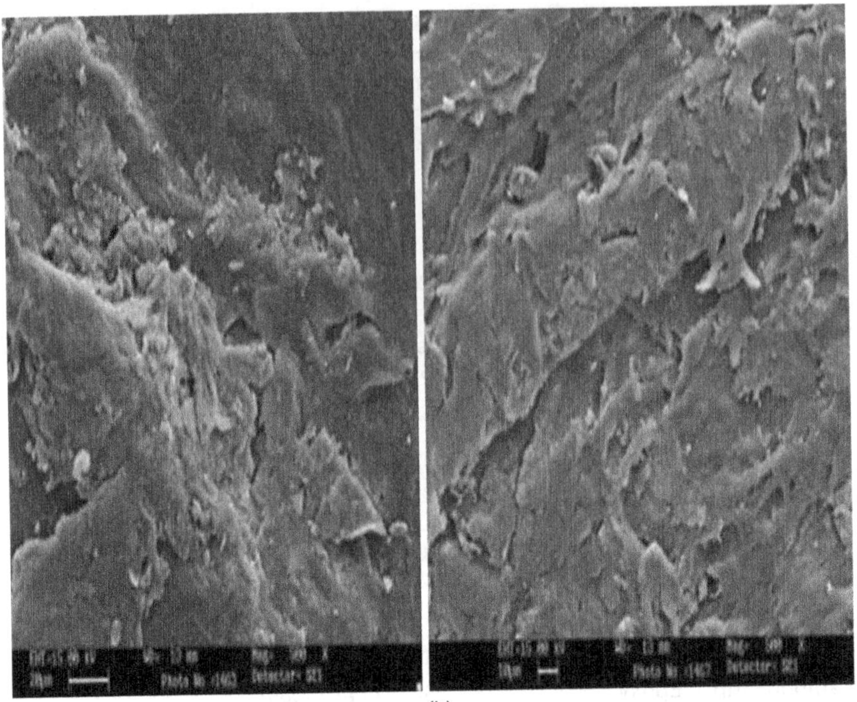

(a) (b)

16.4 SEM micrographs of bamboo fibre–polypropylene interface: (a) without maleic anhydride grafted polypropylene (MA-*g*-PP) and (b) with MA-*g*-PP for 50 vol% of bamboo fibre (Chattopadhyay *et al.*, 2011a, © 2010 Wiley Periodicals, Inc.).

but enzymatic biodegradability indicated that both composites could be quickly decomposed by enzyme, although the addition of lysine-based diisocyanate delayed the degradation.

In another study, the fracture and morphological behaviour of short bamboo fibre-reinforced polyester composites was investigated and the results indicated that the highest fracture toughness was achieved at 10 mm/50 vol% fibre-reinforced composite, with a 340% improvement compared to pure polyester (Wong *et al.*, 2010). Prasad and Rao (2011) conducted a study comparing bamboo to jowar and sisal fibres as reinforcement fibres in a polyester matrix and determined that the developed composite material can be used in different applications. Bamboo fibre-reinforced natural rubber composites were prepared and the mechanical properties of composites with and without bonding agent were studied (Ismail *et al.*, 2002). The results showed that interfacial bonding between bamboo fibre and natural rubber was enhanced by the addition of the bonding agent and their composites displayed enhancement in tensile modulus and hardness.

Bamboo fibre-reinforced polypropylene (PP) composites were prepared using compatibilizer (Chen *et al.*, 1998). The results indicated that 24% bamboo fibre loading-treated composites displayed better mechanical properties and that the newly developed composites were lighter, more water-resistant, cheaper, and with tensile properties more than three times higher than those of current commercial products. PP composites developed using steam-exploded bamboo fibres demonstrated better tensile properties compared to mechanically extracted bamboo fibre-based PP composites (Okubo *et al.*, 2004). Bamboo fibre-based PP composites were produced at 10, 15 and 50 vol% fractions of bamboo fibre and studied for their degree and rate of aerobic biodegradation by a specially designed experiment (Chattopadhyay *et al.*, 2011a). In another work, Chattopadhyay *et al.* (2011b) fabricated short bamboo fibre-reinforced PP composites by incorporating chemically modified bamboo fibres at various loading percentages. The results indicated that thermal stability and functionality of composites improved with modification of the bamboo fibres. Phuong *et al.* (2010) have reported on the fabrication of bamboo fibre-reinforced recycled polypropylene composites via direct melt blends using a twin-screw extruder. The effects on the mechanical, thermal, rheological and morphological properties of composites by alkaline and acetylation treatment of bamboo fibres were investigated.

In another study, the effects of coupling agent and filler loading on the mechanical and thermal properties of bamboo–PP biocomposites were studied and it was observed that the treatment of bamboo fibres with coupling agents affects their physico-mechanical, thermal and morphological properties (Lee *et al.*, 2009). Another work on bamboo–PP composites indicated that composites prepared with modified polypropylene resin showed drastic changes in their mechanical properties (Mohanty and Nayak, 2010; Sano *et al.*, 2002). Fuentes *et al.* (2013) used an integrated physical–chemical–mechanical approach to study the effect of adhesion on the mechanical strength of bamboo fibre-reinforced PP, MA-grafted PP, polyvinylidene fluoride and polyethylene terephthalate. Results indicated that physical adhesion improved interfacial and longitudinal strengthening of bamboo polyvinylidene fluoride composites as compared to the other thermoplastic matrices used in this study.

16.6.2 Bamboo husk fibre-reinforced epoxy composites

Morphological analysis of bamboo husk-reinforced composites reveals that modified fibre displays better fibre and matrix interfacial bonding compared to untreated fibres, and thermal resistance and mechanical properties are enhanced by the addition of coupling agent-treated fibres or untreated powders (Shih, 2007). Results also indicated that the storage moduli of

16.5 Bamboo laminated epoxy composite (reprinted from Verma and Chariar, 2013, © 2013 with permission from Elsevier).

epoxy were enhanced with the addition of 10% coupling agent-treated fibres and untreated powders.

16.6.3 Bamboo culm-based epoxy composites

Verma and Chariar (2013) fabricated four-layered laminated bamboo epoxy composites (LLBC) using thin laminas obtained from dry bamboo culm and epoxy resin. Experimental and theoretical values of stiffness and strength of LLCB were evaluated and compared with four-layered unidirectional laminated bamboo composites (HLLBCs). The results show that the stresses and strains obtained using a constitutive equation of laminate at macroscopic scale are lower than the experimental failure limit of LLBCs. An example of bamboo culm-based composite is shown in Fig. 16.5.

16.6.4 Bamboo strip-based polymer composites

Shin *et al.* (1989) and Corradi *et al.* (2009) used bamboo strip to fabricate unidirectional bamboo–epoxy laminates of varying laminae number and

16.6 Bamboo strip-based composite (Corradi *et al.*, 2009).

evaluated their tensile, compressive, flexural and interlaminar shear properties as shown in Fig. 16.6. The moisture sorption characteristics of bamboo strips and their influence on the interfacial shear strength of bamboo–vinyl ester composites were also investigated (Chen *et al.*, 2009). The results showed that relative humidity influenced the interfacial shear strength of composites and water immersion caused an accumulated loss of interfacial strength. A further study on bamboo strip-reinforced vinyl ester composites has been carried out to compare the effect of different chemical treatments (silane, alkali, oxidation and acetylation) on their ability to absorb moisture and the results showed that acetylation treatment was the most effective (Chen *et al.*, 2011). Das and Chakraborty (2009a) developed bamboo–polyester composites by hand lay-up technique using untreated strips and bamboo strips treated with alkali for 1 h at different concentrations, e.g. 0, 10, 15, 20 and 25%. The bamboo fibre strip and polyester composites with 20% caustic-treated strip reinforcements and 60% filler loading displayed the greatest improvement in mechanical properties.

In another study, bamboo strips were mercerized with varying concentrations of NaOH (10, 15, 20 and 25%) and made into bamboo strip based novolac composites (Das and Chakraborty, 2007). Das *et al.* (2009) investigated how alkali treatment of bamboo strips affected the thermal and weathering properties of the unidirectional bamboo strip Novolac resin composites. The results showed that the composites with treated bamboo strips showed better weathering and thermal stability compared to the untreated ones. In a further study, Das and Chakraborty (2009b) reported on the dynamic mechanical and thermal properties of untreated and treated

16.7 Three bidirectional roving bamboo fibre (Gupta *et al.*, 2011).

bamboo strip-reinforced Novolac composites. The results indicated that 20% alkali-treated fibre composites had the best dynamic mechanical properties, and differential scanning calorimetry analysis revealed that alkali treatment of the bamboo strip imparted better thermal stability to the composites as compared to the untreated one.

16.6.5 Bamboo fibre/strip mat-based polymer composites

The adhesive wear and frictional properties of bamboo fibre mat-reinforced epoxy composites were studied and results (see Fig. 16.7) showed that they had excellent wear resistance compared to neat epoxy (Nirmal *et al.*, 2012). Two sets of bamboo fibre mat-reinforced epoxy composites were fabricated in another study, one with silane-treated bamboo mats and the other with silane-treated mercerized bamboo mats (Kushwaha and Kumar, 2010a). It was observed that silane treatment improved the tensile and flexural strength of bamboo–epoxy composites but by contrast the addition of silane-treated mercerized bamboo led to a significant reduction in strength. Kushwaha and Kumar (2010b) also carried out a study to establish the effect of alkali and silane treatment on water absorption properties of bamboo matting-reinforced epoxy composites and discovered that both alkali and silane treatment resulted in a reduction of water absorption. The same researchers reported on the reinforcement of modified (alkali-treated) bamboo strip matting into epoxy and polyester matrix developed bamboo

16.8 Woven bamboo mat composite (Bäcklund, 2011).

fibre-reinforced plastic (BFRP) composites (Kushwaha and Kumar, 2009). Bamboo mats were treated with 1, 2, 5, 10, 15, 20 and 25% concentration of NaOH in distilled water for 30 min at 20°C (room temperature). This morphological study showed that 5% NaOH-treated bamboo strip mat exhibited better compatibility with the epoxy and polyester resins than the untreated bamboo, and treated bamboo-reinforced polyester composites displayed 69% higher tensile and 59% higher flexural strength. The same researchers also investigated the effect of acrylonitrile treatment of bamboo mats on bamboo-reinforced epoxy and polyester resin composites and observed that treatment of bamboo mats improved the tensile, flexural and water absorption properties of both composites (Kushwaha and Kumar, 2010a).

16.6.6 Woven bamboo mat-reinforced polymer composites

The dielectric behaviour of woven bamboo strip mats/epoxy composites (Fig. 16.8) was studied with dual fibre orientation (parallel and perpendicular) to the electric field (Kumar *et al.*, 2011). This study determined the effects of fibre alignment and alkali treatment on the dielectric properties of woven bamboo strip mats/epoxy composites and evaluated the structural performance of a standard laminating resin. The physical, thermal and mechanical properties of bamboo fabric-reinforced PLA composites were investigated by Porras and Maranon (2012). The results revealed that the tensile, flexural and impact properties of PLA increased when weft-direction bamboo fabric reinforcement was used, and bamboo fabric-reinforced PLA composites showed excellent ability to absorb energy, which could be exploited in structural engineering applications. In another recently published work, woven bamboo mats modified by maleic anhydride, permanganate, benzoyl chloride and benzyl chloride were used as reinforcements in epoxy and polyester matrices to estimate the physical, mechanical and morphological properties of bamboo-reinforced polyester/epoxy composites (Kushwaha and Kumar, 2011). The results obtained showed variations in the mechanical, physical and morphological properties of bamboo-reinforced polyester/epoxy composites.

16.7 Applications of bamboo composites

Design is a creative process that aims to establish the quality of a variety of objects, processes, services and systems throughout their life cycle. According to Pawlak (2008) design is a key factor of innovative humanization technology and an important factor in cultural and economic exchange. Research in material processes such as bamboo contributes to good design because it is concerned with factors such as safety, aesthetics, functionality, consumer acceptance and the potential impact on the environment. Indirectly it contributes to the development of products based on human values, whether in a physical or a cognitive sense (O'Grady and O'Grady, 2006). Most products are designed to take account of a diversity of factors such as safety and comfort without adversely affecting the ecology and without being more expensive (Rodgers and Milton, 2011).

This section describes the potential and explores the different types of biocomposites from various materials that are often associated with and used in design. The era of hybrid products formed over decades has continued to attract the world market until now when biocomposite material has proven its quality and importance to life cycle assessment (LCA) (Vogtländer *et al.*, 2010).

Compared to other materials, bamboo furniture is long-lasting and unique in general appearance but with similar strength to other wood-based furniture (Suhaily *et al.*, 2012). The growth rate of bamboo is an advantage as it is one of the fastest-growing plants and has a tensile strength comparable to that of mild steel. Normal bamboo will grow upright and can reach heights over 18.3 metres in only a few months. Bamboo is easily cultivated and does not affect the environment if cut frequently compared to trees grown for solid wood (Kowaluk *et al.*, 2011). Many researchers have already shown that modern furniture using biocomposite material such as bamboo fibre can create something unique, outstanding and biodegradable compared to solid wood materials (Brower *et al.*, 2009; Leao *et al.*, 2010). Wood consumption in line with population has increased every year. However, bamboo biocomposites are a new alternative material for wood-based industries and many researchers are developing panel products from bamboo fibre.

The integrated materials process determines the most suitable type of design material taking into account various factors such as its impact on nature and the community, technology, price, durability and sustainability. Bamboo is stronger when bent and glued under pressure, making bamboo plywood a very good deck and an ideal pilot material for furniture, interiors, transport and construction. Over the last few years, researchers have demonstrated that bamboo is an excellent material, capable of producing a variety of quality products as shown in Table 16.5 (Gabrielsen and Kristensen, 2004). Now, with the development of science and technology in the use of bamboo biocomposites, consumers need to be reassured about their quality and durability compared to other materials (Riley, 2003).

In little more than a decade, bamboo flooring has become a serious contender in the hardwood flooring market and some believe that bamboo plywood is next. Nowadays, bio composites in the production of various bamboo products for the construction industry and interior design have a high demand in the global market because both consumers and manufacturers realize their importance in supporting continuing efforts to encourage sustainability. Their success proves that the hybrid material bamboo can surpass other types of materials in various aspects, physical, mechanical and aesthetic. Lighter laminated bamboo can be used to create unusual effects and bamboo joinery can be bent or straightened by heating and clamping as demonstrated in the work of Colombian architects Simon Velez and Oscar Indalgo (Adams, 1998). Various types of bamboo-based products have been successfully commercialized, from ceilings, walls, floors, window frames and doors to stairs and home decor accessories.

For example, PlybooSport bamboo flooring has been designed using bamboo biocomposites for high-impact use in gym and basketball courts and is widely used in countries and regions such as the United States, Mexico, Europe and Canada (Smith and Fong, 2010). Now that bamboo flooring has grown beyond niche market status it is beginning to attract more scrutiny, especially in interior design. This is further enhanced by its excellence as an innovative material and has attracted recognition from various quarters with evidence of excellent designs made from bamboo material, for example the unique palm-shaped bamboo dome designed and built by Binh Duong and Vo Trong Nghia.

Bamboo furniture making is increasing gradually as the wood-based manufacturing industry is threatening the environment with its destruction of the forests. However, one major problem associated with bamboo is its huge wastage, especially through open burning, which creates its own

Table 16.5 Innovative design and applications from bamboo fibre biocomposites in various categories

Category	Name of product	Inventor/Designer	References
Interior design	Bamboo Flooring For Basketball Court 	Smith and Fong	Smith and Fong (2010)
	Bamboo Dome, Vietnam 	Vo Trong Nghia	Vo Trong Nghia (2010)
Building and construction	Kontum Indochine Café, Vietnam 	Vo Trong Nghia	Vo Trong Nghia (2010)
Furniture design	Infinity Bench 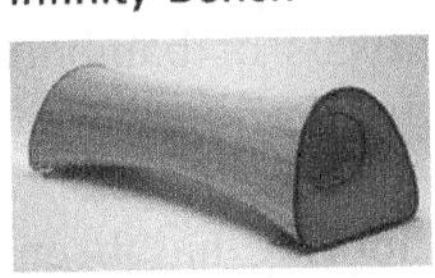	Andrew Williams and Tom Huang	Williams (2010), and Huang (2007)
	Hangzhou Bent Bamboo Stool 	Min Chen	Min Chen (2013)
Automotive components	Renault Mégane Bio-Concept car 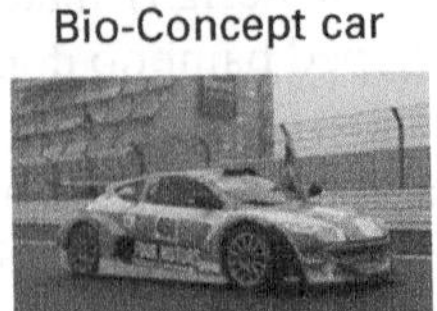	Cloth seats, floor mats, dashboard, door panels, etc.	Pro-Shift, (2010), Makinejad *et al.*, (2009)

burden on the environment. Bamboo has significant waste residues as both ends of bamboo poles cannot usually be used due to fungal attack and non-uniform diameter. It has been reported that in some cases about 15 cm must be cut and removed from each end. With advances in technological development, researchers have succeeded in producing composites from bamboo waste. This has enabled furniture designers to demonstrate that bamboo materials are not only resilient and pliable, but tremendously powerful in internal and external conditions.

For example, a piece of furniture called the Hangzhou Bent Bamboo Stool designed by Min Chen is made of 16 layers of thin bamboo veneer (Chen, 2013), the sections of bamboo being bent into an arc shape and glued together at the end, producing an effect like ripples on a water surface. It demonstrates the ability of each bamboo veneer layer to withstand a heavy load while offering flexibility to the user. Young designers Andrew Williams and Tom Huang have collaborated to create the Infinity Bench using bamboo strip construction (Williams, 2010; Huang, 2007). The Infinity has two hollow tubes, one large and one small, made with bamboo strips of the same size arranged closely together using adhesive on the outside and inside to ensure the strength of the design.

Building materials are selected based on three important criteria (van der Lugt *et al.*, 2005):

- Functionality
- Technical qualities
- Cost.

Sustainability has also been included as one of the criteria as the issue of environmental building materials has become important. Now, all organizations and institutions must comply with the construction and manufacturing criteria in each project. This is because the construction industry is also causing environmental damage, and recycling, sustainable production and use of renewable resources have to be considered in the development of new materials.

Unique architecture and distinctiveness apply where the whole structure of the building is made using bamboo, proving that bamboo materials have high strength comparable to metals. Moreover, the dynamic shape adds to the building's aesthetic quality. For example, the Kontum Café Indochine by Vo Trong Nghia Architects is a restaurant without walls, enabling a seamless view and maximizing the flow of air into the building in the summer, rendering air-conditioning unnecessary (Vo Trong Nghia, 2010). Built using a bamboo roof structure, it consists of 15 units of an inverted cone creating a space with a unique character.

The industrial revolution in transportation began in 1930 with the use of natural fibres in the construction of car interiors. Famous car inventor

Henry Ford led the way in this important era in the automotive manufacturing world because of his use of materials made from natural fibres. Studies have shown that the low cost of natural fibre materials makes them very suitable for use in automotive parts (Proemper, 2004; Makinejad *et al.*, 2009). The use of bamboo in the car industry continues to grow rapidly, especially in Europe and Southeast Asia. Now, bamboo-based innovation continues to be the choice of researchers globally because it promises excellent performance at minimal cost. In 2006, the European Union (EU) and Asian countries supported this by issuing guidelines for the global car manufacturing industry. The guidelines instruct all car manufacturers to produce automotive plastic reinforced with natural fibre. In addition, the European Union (EU) has set a target that recycled materials must make up 80% of manufactured products with that figure increasing to 95% in 2015.

Recent research has developed bamboo mat veneer composite (BMVC) hybrid in vehicles and trains. Natural fibre composites with thermoplastic and thermoset matrix have also been widely used in the manufacture of door panels, rear seats, headliners, package trays and dashboards by car manufacturers such as Audi, BMW, Peugeot, Volvo, etc. (Proemper, 2004). This is because natural fibres such as bamboo have been proven to reduce weight and cost and have the advantage of being recyclable. In aeronautics research, the kite and early aircraft were built using materials made from bamboo fibre because it is light, very strong and able to withstand wind resistance. The Philippines created a plane made entirely of bamboo during World War II (Proemper, 2004).

16.8 Sustainable and renewable products from bamboo composites

In the proliferation of the global bamboo manufacturing industry, three main factors should be given priority in relation to product life and sustainable development, as shown in Fig. 16.9:

- Ecological
- Economic
- Social–cultural.

There needs to be a balance between society's increasing demand for products, preservation of the forest, diversity of material resources and benefits, and the impact of culture. Previous research results show the use of bamboo composites can sustain economic and social-cultural factors, both locally and globally, and has a far less detrimental effect on the ecosystem. Consumer awareness can be achieved by evaluating the environmental impact of the entire life cycle of a product, from raw material through manufacture to final disposal.

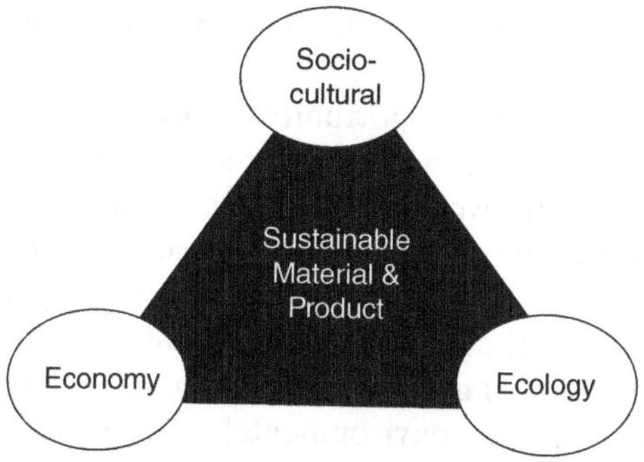

16.9 Three elements to support the sustainable product.

Like any other resource-based industry, the constant supply of raw material is important to ensure its continued existence. Over the years, the industry has observed a decline in the quantity and quality of bamboo. As a result, it has been forced to accept inferior quality bamboo which has led to problems in meeting consumers' specifications. The cultivation of bamboo through proper management is essential to ensure a continuous supply of high-quality raw material. The feasibility of planting bamboo under rubber and forest plantation has been studied. Either small-scale rubber holders or larger estates growing rubber could adopt the practice. However, growing bamboo in plantations still encounters management problems that need to be solved. Governments may have to play a part in land acquisition or long-term lease of land to make bamboo plantations more attractive. A form of the rural development scheme for rubber planting used in peninsular Malaysia may be a good model to follow.

The bamboo industry can improve the quality of the environment through reduced CO_2 emissions and almost zero net greenhouse gas emissions. Environmental issues are increasingly important and an issue of international debate. Bamboo materials can be an exciting challenge leading to new solutions through technology and research that are both economic and ecologically sustainable. Problems such as the lack of material resources due to forest fires, drought and an annually increasing world population are the main reasons why researchers have focused their attention on the development and design of new materials from bamboo fibre (Wagner *et al.*, 2010). The ability to see the potential of biocomposite bamboo products as part of the culture and heritage of the world could also lead to the development of innovative products (Zhang *et al.*, 2001). In addition to its product versatility, bamboo has other benefits such as being less susceptible to erosion, helping to reduce deforestation and being readily disposable without harming the environment compared to other materials.

Bamboo's natural strength is unmatched by modern materials such as steel and plastic.

The bamboo industry offers the best opportunity for sustainable economic development for biocomposite fibres, which are produced at minimal cost and will bring a new evolution in the world of supply chain and manufacturing (Zuo *et al.*, 2001). Bamboo cultivation requires minimal cost or care, bamboo being fully grown in 3–4 years compared to other types of plants that require 5–30 years to mature. The short period to maturity of bamboo ensures a cycle of optimal supply of good quality raw materials at all times.

Problems of waste disposal and environmental damage have raised concerns about the costs of commercial market production recently. Low-impact materials like bamboo are necessary to diversify the market by providing an alternative choice of material resource. This could increase potential new markets in developing sustainable global solutions (Ljungberg, 2005). To realize sustainable economic development, we need to take into account the costs associated not only with the goods such as the cost of the raw materials and production, but also other factors such as reputation, trends, environmental protection and natural resources. Bamboo also has a better variety of mechanical uses and anti-bacterial applications, making it an excellent resource for the development of a sustainable product (Kar and Jacobson, 2012).

Design is a practical activity and is part of our culture and current research. New product design can make an important contribution not only to products and services required by a creative community, but also for the development of materials (Lane and Flagg, 2010). Bamboo-producing countries of the world such as China and India have many sources of bamboo and various bamboo species (van der Lugt *et al.*, 2012). The world is faced with many serious problems such as global warming, acid rain, soil erosion, the financial crisis and extinction of flora and fauna habitat which are caused by profit-oriented manufacturing. Over the years, materials such as bamboo have proved to be innovative in global production, satisfying the current desire to develop sustainable products without forgetting traditional culture. In other words, materials and design are very much needed in maintaining quality of life because they reflect the values of our culture over the years, especially in the furniture manufacturing industry, construction of houses and buildings, appliances, jewellery and other industries (Steffen, 2007).

16.9 Future trends

A huge amount of research has been devoted to nanoscale-related technologies in recent years. This research revolution is significant, bringing about a new generation of composite processes and products with improved and unique properties. Technology is enabling the industry to manipulate matter at the atomic level so that its physical and chemical properties (e.g. stability, hardness, conductivity, reactivity, optical sensitivity, melting point, etc.) can be manipulated to improve the overall properties of conventional materials.

Cellulose nanofibres are found in the cell wall of bamboo fibres, acting as the framework of a biocomposite made of the matrix substances lignin and hemicellulose. These nanofibres have lateral dimensions of a few nanometres and comprise a bundle of cellulose molecular chains arranged parallel to the longitudinal direction of the nanofibres (Nakagaito *et al.*, 2011). Chang *et al.* (2012) produced bamboo nanofibres using hot-compressed water (HCW) treatment and disc milling. Bamboo nanofibres were also successfully prepared using chemical pretreatment combined with high-intensity ultrasonification (Chen *et al.*, 2011).

Over the years, the use of nanocellulose as reinforcing agents in paper and nanocomposites, membranes and films for filtration and packaging, stabilizing and texturing agents in cosmetics and food additives, wound dressings, artificial blood vessels, scaffolds for tissue engineering and drug delivery systems, among others, has been reported (Ioelovich and Figovsky, 2008; Taokaew *et al.*, 2013). However, applications of nanocellulose are still increasing due to its physical and mechanical properties. Cellulose nanocrystals, whiskers and nanofibres can be designed for a great variety of applications, ranging from medical to technological. Nanocomposites comprising nanoparticles and nanofibres are expected to be a major growth area in the plastics industry. According to Wagner *et al.* (2010) polymers reinforced with as little as 2–6 percent of nanoparticles exhibit dramatic improvements in properties such as the following:

- Thermomechanical properties
- Light weight
- Dimensional stability
- Barrier properties
- Flame retardancy
- Heat resistance
- Electrical conductivity.

Nanofibre composites are also used for making flexible circuits, solar panels and other electronics devices (Giri and Adhikari, 2013). This application potentially offers the benefits of reduced energy consumption together with more competitive pricing compared with conventional materials (Fan *et al.*, 2011).

Research developments have shown the possibility of increasing paper strength by the addition of nanocellulose particles or nanofibrillated cellulose (Ioelovich and Leykin, 2004). Henriksson *et al.* (2008) reported that nanopapers made from nanofibrillated cellulose have higher mechanical properties than conventional papers formed by the beating process. Recent research by Sehaqui *et al.* (2010) successfully developed a smooth and optical cellulose/inorganic hybrid nanopaper. They found that optical transparency and high tensile strength are demonstrated in 200 mm diameter nanopaper sheets, indicating well-dispersed nanofibrils.

Additionally recent advances in nanocellulose support different aspects of medical developments including the following (Kalia *et al.*, 2011):

- Skin transplants for burns and wounds
- Drug release systems
- Blood vessel growth
- Scaffolds for tissue engineering
- Stent covering
- Artificial organs
- Bone reconstruction.

Interestingly, a great variety of biomaterials have been developed recently (Fig. 16.10). Researchers from Chalmers University of Technology and the University of Gothenburg have shown that nanocellulose stimulates the formation of neural networks (Fig. 16.10(a)). This is the first step towards creating a three-dimensional model of the brain that could elevate brain research to a totally new level. Recently, MIT Chemical have discovered that arrays of billions of nanoscale sensors have unique properties that could help pharmaceutical companies produce drugs more safely and efficiently. Using these sensors, researchers were able to characterize variations in the binding strength of antibody drugs which hold promise for treating cancer and other diseases. A bionic ear (Fig. 16.10(d)) using nanoparticles has been developed by a scientist at Princeton University that can hear almost 1,000,000 times better than the normal human ear. The technology, which seamlessly interweaves biological components and sensitive electronic devices into a single bionic structure, could have a range of applications in regenerative medicine.

Currently, a team of researchers from the University of Pennsylvania has shown that nanoscale particles of the semiconductor cadmium selenide can be 'printed' or 'coated' on flexible plastics to form a flexible circuit (Fig. 16.10). This technology could pave the way for new kinds of devices and pervasive sensors which could have biomedical or security applications. Meanwhile, researchers at the Indian Institute of Technology in Madras have developed a $16 nanoparticle water filtration system (Fig. 16.10) which is the first to combine microbe-killing capacity with the ability to remove chemical contaminants such as lead and arsenic. The system can be customized to rid water of microbial contaminants, chemical contaminants or both, depending on the user's needs.

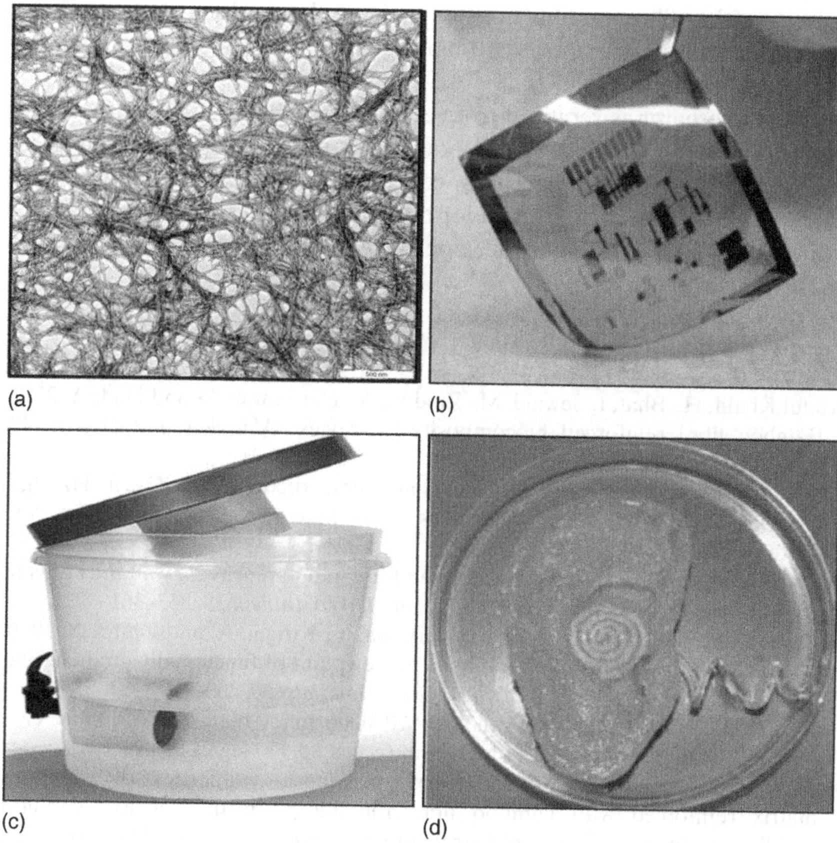

16.10 Examples of future applications of nanotechnology: (a) kenaf bast nanofibers for bionanocomposite applications; (b) flexible circuit; (c) water filtration system; and (d) bionic ear (Kim *et al.*, 2013; Gravotta, 2013; Anon, 2013).

16.10 Conclusions

The use of bamboo fibres in various applications has opened up new avenues for both academics and industrialists in designing sustainable building blocks. The fabrication of bamboo fibre-based composites using different matrices has developed cost-effective and eco-friendly biocomposites. These composites are likely to find more and more applications in the near future. To design such composites, a thorough investigation of the fundamental, mechanical and physical properties of bamboo fibres is necessary. Thus, this chapter has attempted to gather information on bamboo fibre processing and analyse its properties and applications for different composites. Current researches on bamboo fibre-based composites are in terms of either fibre modifications or their mechano-physical, thermal and other properties; however, the ultimate goal of fully utilizing the bamboo fibre is still some way off.

Further research is required to overcome the obstacles currently facing the advanced application of bamboo fibres in composites as well as in outdoor applications. Researchers worldwide are working to address and overcome these, and efforts to develop biocomposites and nanobiocomposites from bamboo fibres with improved performance for global applications are an ongoing process.

16.11 References

Abdul Khalil, H., Bhat, I., Jawaid, M., Zaidon, A., Hermawan, D. and Hadi, Y. 2012. Bamboo fibre reinforced biocomposites: a review. *Materials and Design*, 42, 353–368.

Adams, C. 1998. Bamboo architecture and construction with Oscar Hidalgo. *Designer/Builder Magazine*, Networks, 2405 Maclovia Lane, Santa Fe, NM 87505.

Alvin, K. and Murphy, R. 1988. Variation in fibre and parenchyma wall thickness in culms of the bamboo *Sinobambusa tootsik*. *IAWA Bulletin*, 9, 353–361.

Amada, S., Munekata, T., Nagase, Y., Ichikawa, Y., Kirigai, A. and Zhifei, Y. 1996. The mechanical structures of bamboos in viewpoint of functionally gradient and composite materials. *Journal of Composite Materials*, 30, 800–819.

Anon. 2013. A bionic ear that can hear 1000000 times better than human ears [online]. Available at (www.efytimes.com (accessed 15 August 2013).

Anyakora, A. N. 2013. Evaluation of mechanical properties of polyester matrix reinforced with bamboo fibre for the production of low strength building products. *International Journal of Engineering and Applied Sciences*, 2(2), 57–66.

Bäcklund, K. 2011. Bamboo and wheat straw as a green building composite material. Master's thesis, Chalmers University of Technology, Gothenburg, Sweden.

Bao, L., Chen, Y., Zhou, W., Wu, Y. and Huang, Y. 2011. Bamboo fibres @ poly(ethylene glycol) reinforced poly(butylene succinate) biocomposites. *Journal of Applied Polymer Science*, 122, 2456–2466.

Bhat, A., Abdul Khalil, H., Bhat, I. U. H. and Banthia, A. 2011. Development and characterization of novel modified red mud nanocomposites based on poly(hydroxy ether) of bisphenol A. *Journal of Applied Polymer Science*, 119, 515–522.

Biagiotti, J., Puglia, D. and Kenny, J. M. 2004. A review on natural fibre-based composites – Part I: Structure, processing and properties of vegetable fibres. *Journal of Natural Fibres*, 1, 37–68.

Bonilla, S. H., Guarnetti, R. L., Almeida, C. M. and Giannetti, B. F. 2010. Sustainability assessment of a giant bamboo plantation in Brazil: exploring the influence of labour, time and space. *Journal of Cleaner Production*, 18, 83–91.

Brower, C., Mallory, R. and Ohlman, Z. 2009. *Experimental Eco-Design*, RotoVision SA, Hove, UK.

Chang, F., Lee, S.-H., Toba, K., Nagatani, A. and Endo, T. 2012. Bamboo nanofiber preparation by HCW and grinding treatment and its application for nanocomposite. *Wood Science and Technology*, 46, 393–403.

Chaowana, P. 2013. Bamboo: an alternative raw material for wood and wood-based composites. *Journal of Materials Science Research*, 2, 90.

Chattopadhyay, S. K., Khandal, R., Uppaluri, R. and Ghoshal, A. K. 2011a. Bamboo fibre reinforced polypropylene composites and their mechanical, thermal, and morphological properties. *Journal of Applied Polymer Science*, 119, 1619–1626.

Chattopadhyay, S. K., Singh, S., Pramanik, N., Niyogi, U., Khandal, R., Uppaluri, R. and Ghoshal, A. K. 2011b. Biodegradability studies on natural fibre reinforced polypropylene composites. *Journal of Applied Polymer Science*, 121, 2226–2232.

Chen, H., Miao, M. and Ding, X. 2009. Influence of moisture absorption on the interfacial strength of bamboo/vinyl ester composites. *Composites Part A: Applied Science and Manufacturing*, 40, 2013–2019.

Chen, H. Y., Miao, M. H. and Ding, X. 2011. Chemical treatments of bamboo to modify its moisture absorption and adhesion to vinyl ester resin in humid environment. *Journal of Composite Materials*, 45, 1533–1542.

Chen, M. A. 2013. Hangzhou Bamboo Stool [online]. China. Available at http://chen-min.com (accessed 25 July 2013).

Chen, X., Guo, Q. and Mi, Y. 1998. Bamboo fibre-reinforced polypropylene composites: A study of the mechanical properties. *Journal of Applied Polymer Science*, 69, 1891–1899.

Colla, W. A., Beraldo, A. L. and Brito, J. O. 2011. Effects of thermal treatment on the physicochemical characteristics of giant bamboo. *Cerne*, 17, 361.

Corradi, S., Isidori, T., Corradi, M. and Soleri, F. 2009. Composite boat hulls with bamboo natural fibres. *International Journal of Materials and Product Technology*, 36, 73–89.

Das, M. & Chakraborty, D. 2007. Role of mercerization of the bamboo strips on the impact properties and morphology of unidirectional bamboo strips–Novolac composites. *Polymer Composites*, 28, 57–60.

Das, M. and Chakraborty, D. 2008. Evaluation of improvement of physical and mechanical properties of bamboo fibres due to alkali treatment. *Journal of Applied Polymer Science*, 107, 522–527.

Das, M. and Chakraborty, D. 2009a. The effect of alkalization and fibre loading on the mechanical properties of bamboo fibre composites, Part 1: Polyester resin matrix. *Journal of Applied Polymer Science*, 112, 489–495.

Das, M. and Chakraborty, D. 2009b. Processing of the uni-directional powdered phenolic resin–bamboo fibre composites and resulting dynamic mechanical properties. *Journal of Reinforced Plastics and Composites*, 28, 1339–1348.

Das, M., Prasad, V. and Chakraborty, D. 2009. Thermogravimetric and weathering study of Novolac resin composites reinforced with mercerized bamboo fibre. *Polymer Composites*, 30, 1408–1416.

Defoirdt, N., Biswas, S., Vriese, L. D., Tran, L. Q. N., Acker, J. V., Ashan, Q., Gorbatikh, L., Vuure, A. V. and Verpoest, I. 2010. Assessment of the tensile properties of coir, bamboo and jute fibre. *Composites Part A: Applied Science and Manufacturing*, 41, 588–595.

Deshpande, A. P., Bhaskar Rao, M. and Lakshmana Rao, C. 2000. Extraction of bamboo fibres and their use as reinforcement in polymeric composites. *Journal of Applied Polymer Science*, 76, 83–92.

Devi, L. U., Bhagawan, S. S. and Thomas, S. 1997. Mechanical properties of pineapple leaf fibre-reinforced polyester composites. *Journal of Applied Polymer Science*, 64, 1739–1748.

Edyham, M. and Hanafi, I. 2002. Effect of silane coupling agent on the mechanical properties of bamboo fibre reinforced natural rubber composites. *European Polymer Journal*, 38, 39–47.

Erdumlu, N. and Ozipek, B. 2008. Investigation of regenerated bamboo fibre and yarn characteristics. *Fibres and Textiles in Eastern Europe*, 16, 69.

Fan, M., Dai, D. and Yang, A. 2011. High strength natural fibre composite: Defibrillation and its mechanisms of nano cellulose hemp fibres. *International Journal of Polymeric Materials*, 60, 1026–1040.

Feng, W., Wang, Z. and Guo, W. 2003. A study on chemical compositions and fiber characteristics of two sympodial bamboos. *Chinese Forestry Science and Technology*, 2, 86–91.

Fengel, D. and Shao, X. 1984. A chemical and ultrastructural study of the bamboo species *Phyllostachys makinoi* Hay. *Wood Science and Technology*, 18, 103–112.

Fengel, D. and Wegener, G. 1984. *Wood: Chemistry, Ultrastructure, Reactions*. Walter de Gruyter, New York, 613, 1960–1982.

Fuentes, C., Tran, L. Q. N., van Hellemont, M., Janssens, V., Dupont-Gillain, C., van Vuure, A. and Verpoest, I. 2013. Effect of physical adhesion on mechanical behaviour of bamboo fibre reinforced thermoplastic composites. *Colloids and Surfaces A: Physicochemical and Engineering Aspects*, 418, 7–15.

Gabrielsen, G. and Kristensen, T. 2004. Some challenges to empirical studies of designing value creation. *Cat Fanciers' Association Annual Yearbook*, 22–27.

Giri, J. and Adhikari, R. 2013. A brief review on extraction of nanocellulose and its application. *Bibechana*, 9, 81–87.

Gravotta, L. 2013. Cheap nanotech filter clears hazardous microbes and chemicals from drinking water [online]. Available at www.scientificamerican.com (accessed 15 August 2013).

Gritsch, C. S. and Murphy, R. J. 2005. Ultrastructure of fibre and parenchyma cell walls during early stages of culm development in *Dendrocalamus asper*. *Annals of Botany*, 95, 619–629.

Gritsch, C. S., Kleist, G. and Murphy, R. J. 2004. Developmental changes in cell wall structure of phloem fibres of the bamboo *Dendrocalamus asper*. *Annals of Botany*, 94, 497–505.

Grosser, D. and Liese, W. 1971. On the anatomy of Asian bamboos, with special reference to their vascular bundles. *Wood Science and Technology*, 5, 290–312.

Gupta, A., Kumar, A., Patnaik, A. and Biswas, S. 2011. Effect of different parameters on mechanical and erosion wear behavior of bamboo fiber reinforced epoxy composites. *International Journal of Polymer Science*, 2011, 1–11.

Han, W., Zhao, C., Sun, Q., Yang, R. and Yu, D. 2013. Modification of bleached bamboo fibre using cationic guar Gum for fibreboard. *BioResources*, 8, 4421–4428.

Henriksson, M., Berglund, L. A., Isaksson, P., Lindström, T. and Nishino, T. 2008. Cellulose nanopaper structures of high toughness. *Biomacromolecules*, 9, 1579–1585.

Huang, T. 2007. TomHuangStudio.Com [online]. Available at http://www.tomhuangstudio.com (accessed 10 July 2013).

Ioelovich, M. and Figovsky, O. 2008. Nano-cellulose as promising biocarrier. *Advanced Materials Research*, 47, 50.

Ioelovich, M. and Leykin, A. 2004. Nano-cellulose and its application. *SITA*, 6, 17–24.

Islam, M. N., Rahman, M. R., Haque, M. M. and Huque, M. M. 2010. Physico-mechanical properties of chemically treated coir reinforced polypropylene composites. *Composites Part A: Applied Science and Manufacturing*, 41, 192–198.

Ismail, H., Edyham, M. and Wirjosentono, B. 2002. Bamboo fibre filled natural rubber composites: the effects of filler loading and bonding agent. *Polymer Testing*, 21, 139–144.

Jain, S., Kumar, R. and Jindal, U. 1992. Mechanical behaviour of bamboo and bamboo composite. *Journal of Materials Science*, 27, 4598–4604.

Jiang, Z. 2007. *Bamboo and Rattan in the World*. China Forestry Publishing House, Beijing.

Kalia, S., Dufresne, A., Cherian, B. M., Kaith, B., Avérous, L., Njuguna, J. and Nassiopoulos, E. 2011. Cellulose-based bio- and nanocomposites: a review. *International Journal of Polymer Science*, 2011, 1–35.

Kamruzzaman, M., Saha, S., Bose, A. and Islam, M. 2008. Effects of age and height on physical and mechanical properties of bamboo. *Journal of Tropical Forest Science*, 20, 211–217.

Kamthai, S. 2003. Alkaline sulfite pulping and ECF-bleaching of sweet bamboo (*Dendrocalamus asper* Backer). MS thesis, Kasetsart University, Thailand.

Kang, J. and Kim, S. 2011. Improvement in the mechanical properties of polylactide and bamboo fiber biocomposites by fiber surface modification. *Macromolecular Research*, 19, 789–796.

Kar, S. P. and Jacobson, M. G. 2012. NTFP income contribution to household economy and related socio-economic factors: Lessons from Bangladesh. *Forest Policy and Economics*, 14, 136–142.

Kassim, J., Ahmad, A., Jalil, A. A. and Mohmod, A. L. 1992. Variation in specific gravity of 1, 2 and 3 years old *Gigantochloa scortechinii* (Buloh Semantan). In: *Proceedings of the Seminar, National Bamboo Seminar, Forest Research Institute Malaysia (FRIM)*, Malaysia, 1992, 182–187.

Khanam, P. N., Khalil, H. A., Reddy, G. R. and Naidu, S. V. 2011. Tensile, flexural and chemical resistance properties of sisal fibre reinforced polymer composites: Effect of fibre surface treatment. *Journal of Polymers and the Environment*, 19, 115–119.

Kim, D., Diroll, B., Murray, C. and Kagan, C. 2013. Flexible and low-voltage integrated curcuits constructed from high-performance nanocrystal transistors. *Nature Communications*, 3, article 1216.

Kowaluk, G., Fuczek, D., , Beer, P., & Grzeskiewicz, A. M. 2011. Influence of the raw materials and production parameters on chosen standard properties for furniture panels of biocomposites from fibrous chips. *Bioresource Technology*, 6(3), 15.

Kumar, V., Kushwaha, P. K. and Kumar, R. 2011. Impedance-spectroscopy analysis of oriented and mercerized bamboo fibre-reinforced epoxy composite. *Journal of Materials Science*, 46, 3445–3451.

Kushwaha, P. K. and Kumar, R. 2009. Studies on water absorption of bamboo–polyester composites: Effect of silane treatment of mercerized bamboo. *Polymer-Plastics Technology and Engineering*, 49, 45–52.

Kushwaha, P. K. and Kumar, R. 2010a. Effect of silanes on mechanical properties of bamboo fibre–epoxy composites. *Journal of Reinforced Plastics and Composites*, 29, 718–724.

Kushwaha, P. K. and Kumar, R. 2010b. Studies on water absorption of bamboo–epoxy composites: Effect of silane treatment of mercerized bamboo. *Journal of Applied Polymer Science*, 115, 1846–1852.

Kushwaha, P. K. and Kumar, R. 2011. Influence of chemical treatments on the mechanical and water absorption properties of bamboo fibre composites. *Journal of Reinforced Plastics and Composites*, 30, 73–85.

Lane, J. P. and Flagg, J. L. 2010. Translating three states of knowledge – discovery, invention, and innovation. *Implement Science*, 5, 9.

Leao, A. L., , Souza, S. F., Cherian, B. M., Frollini, E., Thomas, S., Pothan, L. A. and Kottaisamy, A. M. 2010. Agro-based biocomposites for industrial applications. *Molecular Crystals and Liquid Crystals*, 522, 18–27.

Lee, S.-H. and Wang, S. 2006. Biodegradable polymers/bamboo fibre biocomposite with bio-based coupling agent. *Composites Part A: Applied Science and Manufacturing*, 37, 80–91.

Lee, S.-Y., Chun, S.-J., Doh, G.-H., Kang, I.-A., Lee, S. and Paik, K.-H. 2009. Influence of chemical modification and filler loading on fundamental properties of bamboo fibre reinforced polypropylene composites. *Journal of Composite Materials*, 43, 1639–1657.

Li, X. 2004. *Physical, Chemical, and Mechanical Properties of Bamboo and its Utilization Potential for Fibreboard Manufacturing*. Louisiana State University, Baton Rouge, LA.

Li, X., Shupe, T., Peter, G., Hse, C. and Eberhardt, T. 2007. Chemical changes with maturation of the bamboo species *Phyllostachys pubescens*. *Journal of Tropical Forest Science*, 19, 6.

Liese, W. 1985. Anatomy and properties of bamboo. In: *Proceedings of the International Bamboo Workshop*, 6–14 October 1985, Hangzhou, China.

Liese, W. 1998. *The Anatomy of Bamboo Culms. Brill*, Academic Publishers, Boston, MA.

Liu, Y. and Hu, H. 2008. X-ray diffraction study of bamboo fibres treated with NaOH. *Fibres and Polymers*, 9, 735–739.

Ljungberg, L. Y. 2005. Materials selection and design for development of sustainable products. *Materials and Design*, 28, 466–479.

Londoño, X., Camayo, G. C., Riaño, N. M. and López, Y. 2002. Characterization of the anatomy of *Guadua angustifolia* (Poaceae: Bambusoideae) culms. *Bamboo Science and Culture*, 16, 18–31.

Lu, T., Jiang, M., Jiang, Z., Hui, D., Wang, Z. and Zhou, Z. 2013. Effect of surface modification of bamboo cellulose fibres on mechanical properties of cellulose/epoxy composites. *Composites Part B: Engineering*, 51, 28–34.

Ma, H. and Joo, C. W. 2011. Influence of surface treatments on structural and mechanical properties of bamboo fibre-reinforced poly(lactic acid) biocomposites. *Journal of Composite Materials*, 45, 2455–2463.

Makinejad, M. D., Salit, M. S., Ahmad, D., Ali, A. and Abdan, K. 2009. A review of natural fibre composites in the automotive industry. *Research on Natural Fibre Reinforced Polymer Composites*, 16.

Mohanty, S. and Nayak, S. 2010. Short bamboo fibre-reinforced HDPE composites: Influence of fibre content and modification on strength of the composite. *Journal of Reinforced Plastics and Composites*, 29, 2199–2210.

Mohanty, A., Misra, M. and Drzal, L. 2001. Surface modifications of natural fibers and performance of the resulting biocomposites: an overview. *Composite Interfaces*, 8, 313–343.

Mohmod, A. L. 1993. Effects of age and height of three bamboo species on their machining properties. *Journal of Tropical Forest Science*, 5, 528–535.

Monteiro, S. N., Lopes, F. P. D., Barbosa, A. P., Bevitori, A. B., Da Silva, I. L. A. and Da Costa, L. L. 2011. Natural lignocellulosic fibres as engineering materials – an overview. *Metallurgical and Materials Transactions A*, 42, 2963–2974.

Munawar, S. S., Umemura, K. and Kawai, S. 2007. Characterization of the morphological, physical, and mechanical properties of seven nonwood plant fibre bundles. *Journal of Wood Science*, 53, 108–113.

Murphy, R. and Alvin, K. 1992. Variation in fibre wall structure in bamboo. *IAWA Bulletin (NS)*, 13, 403–410.

Mwaikambo, L. 2006. Review of the history, properties and application of plant fibres. *AJST*, 7, 121.

Mwaikambo, L. and Ansell, M. 2001. The determination of porosity and cellulose content of plant fibres by density methods. *Journal of Materials Science Letters*, 20, 2095–2096.

Nahar, S. and Hasan, M. 2012. Effect of chemical composition, anatomy and cell wall structure on tensile properties of bamboo fibre. *Engineering Journal*, 17, 61–68.

Nakagaito, A. N., Takagi, H. and Pandey, J. K. 2011. The processing and mechanical performance of cellulose nanofibre-based composites. *International Journal of Ocean System Engineering*, 1, 180–184.

Nirmal, U., Hashim, J. and Low, K. 2012. Adhesive wear and frictional performance of bamboo fibre reinforced epoxy composite. *Tribology International*, 47, 122–133.

Nor Aziha, M. and Azmy, H. 1991. Preliminary study on the four Malaysian commercial bamboo species. *India Bulletin*, 1, 6–10.

Ochi, S. 2012. Tensile properties of bamboo fibre reinforced biodegradable plastics. *International Journal of Composite Materials*, 2, 1–4.

O'Grady, J. V. and O'Grady, K. V. 2006. *A Designer's Research Manual*. Rockport Publishers, Quayside Publishing Group, 100 Cummings Center, Suite 406-L, Beverly, MA.

Okubo, K., Fujii, T. and Yamamoto, Y. 2004. Development of bamboo-based polymer composites and their mechanical properties. *Composites Part A: Applied Science and Manufacturing*, 35, 377–383.

Osorio, L., Trujillo, E., van Vuure, A.-W., Lens, F., Ivens, J. and Verpoest, I. 2010. The relationship between the bamboo fibre microstructure and mechanical properties. In: *Proceedings of the 14th European Conference on Composite Materials*, Budapest, Hungary.

Parameswaran, N. and Liese, W. 1976. On the fine structure of bamboo fibres. *Wood Science and Technology*, 10, 231–246.

Parameswaran, N. and Liese, W. 1980. Ultrastructure aspects of bamboo cells. *Cellulose Chemistry and Technology*, 14.

Pawlak, J. J. 2008. A sustainable economy. *Bioresource Technology*, 3, 2.

Philippou, J., Karastergiou, S. and Radoglou, K. 2001. Lignocellulosic materials from annual plants and agricultural residues as raw materials for composite building materials. In: *Proceedings of the International Conference: Forest Research: A Challenge for an Integrated European Approach* (ed. K. Radoglou), Thessaloniki, Greece, 817–822.

Phuong, N. T., Sollogoub, C. and Guinault, A. 2010. Relationship between fibre chemical treatment and properties of recycled PP/bamboo fibre composites. *Journal of Reinforced Plastics and Composites*, 29, 3244–3256.

Porras, A. and Maranon, A. 2012. Development and characterization of a laminate composite material from polylactic acid (PLA) and woven bamboo fabric. *Composites Part B: Engineering*, 43, 2782–2788.

Pothan, L. A., Thomas, S. and Neelakantan, N. 1997. Short banana fibre reinforced polyester composites: Mechanical, failure and aging characteristics. *Journal of Reinforced Plastics and Composites*, 16, 744–765.

Prasad, A. and Rao, K. 2011. Mechanical properties of natural fibre reinforced polyester composites: Jowar, sisal and bamboo. *Materials and Design*, 32, 4658–4663.

Proemper, E. 2004. New automotive international from natural fibre materials. In: *International AVK-TV Conference for Reinforced Plastics and Thermoset Molding Compounds*, Baden-Baden, Germany.

Pro-Shift, L. 2010. Renault Megane bio-concept car has first racing success with new PS3 [online]. Available at http://www.proshift.com (Accessed 1 August 2013).

Rajulu, A. V., Chary, K. N., Reddy, G. R. and Meng, Y. 2004. Void content, density and weight reduction studies on short bamboo fibre–epoxy composites. *Journal of Reinforced Plastics and Composites*, 23, 127–130.

Rao, K. and Rao, K. M. 2007. Extraction and tensile properties of natural fibres: Vakka, date and bamboo. *Composite Structures*, 77, 288–295.

Riley, N. 2003. *The Elements of Design*. Octopus, London.

Rodgers, P. and Milton, A. A. 2011. *Product Design: Portfolio*. Laurence King Publishing Ltd, London.

Sakaray, H., Togati, N. V. K. and Reddy, I. R. 2012. Investigation on properties of bamboo as reinforcing material in concrete. *International Journal of Engineering Research*, 2, 77–83.

Sano, O., Matsuoka, T., Sakaguchi, K. and Karukaya, K. 2002. Study on the interfacial shear strength of bamboo fibre reinforced plastics. *High Performance Structures and Materials*, 4, 147–156.

Scurlock, J., Dayton, D. and Hames, B. 2000. Bamboo: an overlooked biomass resource? *Biomass and Bioenergy*, 19, 229–244.

Sehaqui, H., Liu, A., Zhou, Q. and Berglund, L. A. 2010. Fast preparation procedure for large, flat cellulose and cellulose/inorganic nanopaper structures. *Biomacromolecules*, 11, 2195–2198.

Shih, Y.-F. 2007. Mechanical and thermal properties of waste water bamboo husk fibre reinforced epoxy composites. *Materials Science and Engineering A*, 445, 289–295.

Shin, F., Xian, X., Zheng, W. and Yipp, M. 1989. Analyses of the mechanical properties and microstructure of bamboo–epoxy composites. *Journal of Materials Science*, 24, 3483–3490.

Silva, E. C. N., Walters, M. C. and Paulino, G. H. 2006. Modeling bamboo as a functionally graded material: lessons for the analysis of affordable materials. *Journal of Materials Science*, 41, 6991–7004.

Smith & Fong. 2010. PlybooSport Bamboo Sport Flooring [online]. San Francisco: Smith & Fong Company. Available at http://www.plyboo.com/ (accessed 20 July 2013).

Steffen, D. 2007. *Product Language as a Reflection on Technical Innovation and Social–Cultural Change. Design Semantics of Innovation*. Department of Art and Design History, Bergische Universitat Wüppertal, Germany.

Suhaily, S. S., Jawaid, M., Abdul Khalil, H. P. S., Mohamed, A. R. and Ibrahim, F. 2012. A review of oil palm biocomposites for furniture design and applications: Potential and challenges. *BioResources.com*, 7, 4400–4423.

Suhaily, S. S., Abdul Khalil, H. P. S., Nadirah, W. O. W. and Jawaid, M. 2013. Bamboo based biocomposites: material, design and applications. In: Mastai, Y. (ed.), *Materials Science – Advanced Topics*. InTech Europe, Rijeka, Croatia.

Taj, S., Munawar, M. A. and Khan, S. 2007. Natural fibre-reinforced polymer composites. *Proceedings – Pakistan Academy of Sciences*, 44, 129.

Taokaew, S., Seetabhawang, S., Siripong, P. and Phisalaphong, M. 2013. Biosynthesis and characterization of nanocellulose–gelatin films. *Materials*, 6, 782–794.

Trujillo, E., Osorio, L., van Vuure, A.-W., Ivens, J. and Verpoest, I. 2010. Characterisation of polymer composite materials based on bamboo fibres. In: *14th European Conference on Composite Materials*, Budapest, Hungary: Budapest University of Technology and Economics.

Tung, N. H., Yamamoto, H., Matsuoka, T. and Fujii, T. 2004. Effect of surface treatment on interfacial strength between bamboo fibre and PP resin. *JSME International Journal Series A*, 47, 561–565.

Van der Lugt, P. V. D., van den Dobbelsteen, A. A. J. F. V. D., and Janssen, J. J. A. 2005. An environmental, economic and practical assessment of bamboo as a building material for supporting structures. *Construction and Building Materials*, 2005, 1–9.

Van der Lugt, P., Voglander, J. G., van der Vegte, J. H. and Brezet, J. C. 2012. Life cycle assessment and carbon sequestration; the environmental impact of industrial bamboo products. In: *Proceedings of the IXth World Bamboo Congress*, Antwerp, Belgium.

Verma, C. and Chariar, V. 2013. Stiffness and strength analysis of four layered laminate bamboo composite at macroscopic scale. *Composites Part B: Engineering*, 45(1), 369–376.

Vogtländer, J., Pablo, V. D. L. and Han, B. 2010. The sustainability of bamboo products for local and Western European applications. LCAs and land-use. *Journal of Cleaner Production*, 18, 1260–1269.

Vo Trong Nghia 2010. Vo Trong Nghia Architects [online], Vietnam. Available at http://www.votrongnghia.com (accessed 20 June 2013).

Wagner, S., Collet, C., Madsen, P., Nakashizuka, T., Nyland, R. D. and Sagheb-Talebi, K. 2010. Beech regeneration research: From ecological to silvicultural aspects. *Forest Ecology and Management*, 259, 2172–2182.

Wan, Y. and Ko, F. 2009. Hierarchical structure and mechanical properties of bamboo fibrils. In: ICCM-17, Edinburgh, Scotland, 27–31 July 2009.

Wang, L. 2012. Application of activated carbon derived from 'waste' bamboo culms for the adsorption of azo disperse dye: Kinetic, equilibrium and thermodynamic studies. *Journal of Environmental Management*, 102, 79–87.

Williams, A. 2010. Infinity Bamboo Bench [online], New York. Available at http://www.designaw.me (accessed 20 July 2013).

Wong, K., Zahi, S., Low, K. and Lim, C. 2010. Fracture characterisation of short bamboo fibre reinforced polyester composites. *Materials and Design*, 31, 4147–4154.

Zhang, Q., Jiang, S. and Tang, Y. 2001. *Industrial Utilization on Bamboo*. International Network for Bamboo and Rattan, Beijing.

Zuo, T., Wand, T. and Nie, Z. 2001. Ecomaterials research in China. *Materials and Design*, 22, 107-110.

17

The use of sugarcane bagasse fibres as reinforcements in composites

H. HAJIHA and M. SAIN, Centre for Biocomposites and Biomaterials Processing, Canada

DOI: 10.1533/9781782421276.4.525

Abstract: Currently, bagasse sugarcane, a waste product of the sugar industry, is mainly burned as fuel in sugar mill boilers. The low cost, low density and acceptable mechanical properties of bagasse fibre make it an ideal candidate to be considered for value-added applications such as reinforcement in plastic composites. Bagasse also has commercial potential as reinforcement in cement composites. The advantages of incorporating natural fibre as reinforcement in cement composites are related to their mechanical and thermal properties and reasonable cost. Different treatment techniques enhance the adhesion and compatibility between fibres and matrix, hence improving the mechanical properties of the composite. Several recent works on properties, processing and application of bagasse fibres have been reviewed in this chapter.

Key words: bagasse fibres, sugarcane, composite, surface treatment, mechanical properties.

17.1 Introduction

The use of natural fibre reinforcement for thermoplastics is beneficial due to its low cost, low density and utilization of locally renewable resources. Increasing regulations and emphasis on fuel efficiency and lighter-weight auto parts make the use of sugarcane bagasse and other natural fibres more important. The main disadvantage of natural fibre composites is their high moisture sorption and poor interface interaction with the matrix, but these can be improved with various surface treatment methods. Also, the use of coupling agents (which are molecules with two functional groups where one reacts with the hydroxyl group of natural fibre and the other reacts with the matrix) has enhanced the performance of composite.

Sugarcane is crushed in a series of mills with heavy rollers to extract the juice which is then processed for sugar production. The remaining crushed and broken cane stalks or bagasse fibres are the by-product (or waste) from sugar factories. Bagasse contains short fibres, water and small amounts of soluble solids. The bagasse/stalk ratio by mass is around 30% (Bilba and Arsène 2008). Sugarcane production in 2007 was over 1.4 billion tonnes worldwide and approximately 3 tonnes of wet bagasse are produced for every 10 tonnes of sugarcane in the milling process (Wirawan *et al.* 2011). Approximately 54 million dry tonnes of bagasse are produced annually worldwide (Mulinari *et al.* 2009a). While a tree takes years to grow, plants with short development cycles such as sisal, jute and sugarcane are very attractive for composite reinforcement and this is even truer of sugarcane bagasse, being a waste by-product of sugar plants.

Traditionally, bagasse has been used as a fuel for boilers in sugar factories and in smaller quantities for paper and board production (Huang *et al.* 2012). The low calorific power of bagasses makes burning them as a fuel an inefficient process (Verma *et al.* 2012) and it is anticipated that this application will reduce in future as boilers switch their fuel from SCB to gas (Vazquez *et al.* 1999). Currently, about 85% of bagasse is burnt as fuel and still there is an excess of bagasse, hence the need for new applications to utilize this by-product and reduce landfill waste. About 9% of bagasse is consumed in ethanol production but the efficiency of this process is also very low (Verma *et al.* 2012) Thus, it is necessary to boost the economic value of sugarcane bagasse (SCB).

17.2 Properties of sugarcane bagasse fibres

Bagasse is a vegetable fibre mainly composed of cellulose which has relatively high modulus. The dimension of the fibres affects its reinforcing properties. In general, bagasse has a length of 1.2 mm and a diameter of 15 μm, thus an aspect ratio of 80 (Luz and Gonçalves 2007). Figure 17.1 shows a sample of bagasse fibre.

17.1 Bagasse fibre (Acharya *et al.* 2009, © 2009 by Sage, reprinted by permission of Sage).

17.2.1 Chemical composition

SCB is composed of approximately 50% cellulose, 25% hemicellulose and 25% lignin (Huang *et al.* 2012, Xu *et al.* 2010). The high cellulose content makes SCB ideal as a composite reinforcement. Different researchers have recorded different chemical compositions for bagasse and some are listed in Table 17.1. Crystallinity in SCB is 47% and most of the cellulose in SCB is in a crystalline structure (Trindade *et al.* 2005).

Elemental analysis of bagasse shows that it is composed of 45.5 wt% C, 5.6 wt% H, 45.2 wt% O and 0.3 wt% N (Bilba *et al.* 2003). Moreover, the metal elemental analysis of bagasse shows that it contains 3.89 wt% Al^{3+}, 3.87 wt%; Ca^{+}, 1.32 wt% Mg^{+}, 0.97 wt% Na^{+} and 27.0 wt% Si^{4+} (Jústiz-Smith *et al.* 2008). It should be noted that the presence of these metals in the fibres increases their brittleness.

Table 17.2 summarizes the main bands observed in the FTIR spectrum of sugarcane bagasse cellulose with the corresponding functional groups and molecules.

17.2.2 Properties

Sugarcane bagasse has reasonable tensile strength and modulus, thus it has the potential to be used as reinforcement in composites. On average, sugarcane bagasse has a tensile strength in the range of 170–290 MPa and a modulus of elasticity in the range of 15–19 GPa (Wirawan *et al.* 2011). Values as high as 27.1 GPa for Young's modulus, with ultimate tensile strength of 222 MPa, and elongation at break of 1.1% have also been reported (Satyanarayana *et al.* 2007). Another researcher measured a tensile strength of 170–180 MPa and a modulus of elasticity of 17–19 GPa (El-Tayeb 2008).

17.3 Applications

Traditionally, bagasse has been used as a fuel for boilers in sugar factories and in smaller quantities for paper and board production. However, the use of sugarcane bagasse as a reinforcement and filler in polymeric composites is a more effective and attractive application, increasing the value of this agricultural waste. The high cellulose content of bagasse demonstrates its potential for industrial use. The matrix which binds the fibres together transfers loads to fibres and protects them against damage due to handling and the environment. Sugarcane bagasse's tensile strength (170–290 MPa) and modulus of elasticity (15–19 GPa) mean that it has the potential to be used as a reinforcement in polymer composites (Wirawan *et al.* 2011).

Table 17.1 Chemical composition of sugarcane bagasse fibre

Cellulose	Hemicellulose	Lignin	Protein	Fat and waxes	Ash	Saccharose	Silica	Glucose	References
50	25	25							Huang *et al.* 2012, Xu *et al.* 2010
40	24.4	15	1.8	0.6	5	14		1.4	Vazquez *et al.* 1999
40–43	28–30	9–11	8–9	2–2.5	5–6				Ramaraj 2007
46	24.5	19.5		3.5	2.4		2		Mulinari *et al.* 2009a
69.4	21.1	4.4		5.5	0.6				Habibi *et al.* 2008
41.8	28	21.8							Bilba *et al.* 2003
55.2	16.8	25.3			1.1				Trindade *et al.* 2005
56	6	29			7				Maldas and Kokta 1991
36.32	24.7	18.14							Vilay *et al.* 2008

Table 17.2 Infrared main transitions for sugarcane bagasse cellulose

Wavenumber (cm^{-1})	Vibration	Source
3300	O–H linked shearing	Polysaccharides
2885	C–H symmetrical stretching	Polysaccharides
1732	C=O unconjugated stretching	Xylans
1650–1630	OH (water)	Water
1335	C–O aromatic ring	Cellulose
1162	C–O–C asymmetrical stretching	Cellulose
670	C–OH out-of-plane bending	Cellulose

Source: Mulinari *et al.* 2009a, © 2009, with permission from Elsevier.

Bagasse fibres can also be used as reinforcement in cement composites. The advantages of incorporating natural fibre as reinforcement in cement composites are related to their mechanical and thermal properties and low cost. The advantages of cement/natural fibre composites are that they are environmentally friendly and reduce electricity consumption used in air-conditioning. The drawback for this application is that the lignin in natural fibre is attacked by alkaline cement resulting in the degradation of the reinforcing effect of fibres in composite performance. However, there are different ways to improve the ageing of natural fibre cement composites such as the use of pozzolans to reduce matrix alkalinity, carbonation of the matrix or treatment of the fibres (Onésippe *et al.* 2010, Bilba and Arsène 2008). The addition of fibres to cement results in a decrease in thermal conductivity from 0.62 W/m.K to 0.46 W/m.K for 1.5 wt% alkaline-treated bagasse cement composite. Treatment of the fibres with alkaline and heat did not affect the thermal conductivity for 3 wt% bagasse cement composite (Onésippe *et al.* 2010). Bilba *et al.* (2003) reported that the addition of bagasse fibre to cement led to delayed setting of the cement mixture which is probably due to the presence of some water soluble sugars. Another benefit for building materials is that the addition of 1 wt% heat-treated bagasse at 200°C resulted in reduction of the maximum hydration temperature of setting from 39.3°C to about 30°C (Bilba *et al.* 2003).

The use of sugar cane bagasse ash (SCBA) as pozzolans in the cement industry is possible due to the presence of reactive silica in ash. Ganesan *et al.* (2007) showed that addition of SCBA (5–30% in mass) can enhance the mechanical properties and durability of concrete. Fairbairn *et al.* (2010) carried out a case study simulating the use of SCBA on an industrial scale for south-east Brazil and showed that SCBA can reduce CO_2 emissions. The estimated emission reduction in the study was about 519 kilotonnes of CO_2 per year.

The removal of metals and colour from the solution is mainly achieved through adsorption. Rao *et al.* (2002) showed that raw bagasse can be an alternative to the common but expensive adsorbent of activated carbon. SCB has a porosity of 0.51% (compared to 0.78% for powdered activated carbon) and a surface area of 419.5 m^2/g (compared to 710 m^2/g for powdered activated carbon). The comparison at different pH values showed that powdered activated carbon had a greater adsorption capacity compared to bagasse, especially at lower PH values (2–6), but, bagasse is still a reasonable alternative to the costlier activated carbon. For example, maximum nickel removal by activated carbon was 96.2%, while for bagasse it was 69.7%. For bagasse, the optimum pH was found to be 6 for removal of Cr(VI) and 8 for removal of Ni(II). Moreover, reduction in particle size leads to increased removal of ions. Reducing the bagasse size from 4.75 mm to 75 µm led to an increase in the removal of Cr(VI) from 25% to 91% (Rao *et al.* 2002). Mall *et al.* (2006) studied some of the factors affecting the removal of colour from waste streams such as dosage, pH, initial dye concentration and contact time. A similar study was carried out by Gupta and Ali (2004) regarding the parameters affecting the removal of lead and chromium from wastewater using bagasse fly ash.

Some other uses of SCB are ethanol production and protein-enriched cattle feed, but these applications need further development as they are complex and uneconomic. In general, the application of SCB in composites is more beneficial (Huang *et al.* 2012).

17.4 Surface treatment techniques

The key to obtaining strong mechanical properties for biocomposites is good compatibility between the natural fibre and the matrix. Poor compatibility between hydrophilic fibres and the hydrophobic matrix leads to weak interfaces and poor mechanical properties. The main problem is that hydrophilic fibres tend to agglomerate and not disperse well in the matrix. Moreover, the high lignin content of bagasse (about 20% on average) reduces efficiency of wettability between fibre and matrix. Treating the fibre surface can reduce the lignin content and enhance its properties (Acharya *et al.* 2009). Different treatment techniques have been used to enhance adhesion and compatibility between fibres and matrix in order to enhance mechanical properties.

Some physical methods dealing with fibre treatment include mechanical activation treatment, processing techniques and varying the size of the bagasse fibres. Mechanical activation (MA) uses high friction, collision, shear and other mechanical actions to change the crystalline structure and properties of solids. It is usually carried out in high-energy milling and results in size reduction (increase in surface area) by distortion of the chemical bonds as the imposing force is greater than the chemical bonding energy. Compared to other methods, it is a simple and environmentally friendly method as it does not use any chemical solvent (Huang *et al.* 2012). SCB particles of 40–60 mesh size (0.25–0.38 mm) were milled using 6 mm diameter mill balls at 50°C for different milling times. Increasing the milling time resulted in an increase in flexural strength, tensile strength and Brinell hardness of 40 wt% SCB-PVC composite with the maximum values for milling time of 120 min (Huang *et al.* 2012). An increase of the amorphous phase and broken bonds results in activated radicals and functional groups, which make SCB more accessible to other reagents. Consequently, mechanical activation followed by surface modification can be more effective than surface modification alone.

Various chemical methods for fibre surface treatments are available for natural fibres, such as alkaline treatment, acetylation and silane treatment. In alkalization, the hydroxyl group of bagasse is ionized to alkoxide by the addition of aqueous sodium hydroxide. Alkalization eliminates some of the lignin, wax and oils covering the fibres and increases fibre defibrillation, exposing more contact area between fibre, matrix and sites for reaction. Moreover, surface roughness increases due to the disruption of hydrogen bonding, thus better mechanical intelocking is achieved between fibre and matrix, resulting in a higher mechanical resistance of the composites. In acetylation, the hydroxyl group of bagasse fibre is replaced with an acetyl group, making fibres hydrophobic and less polar. Some treatments involve the use of coupling agents, which are molecules having two functions: reacting with the OH group of the fibre and with the functional group of matrices. Common coupling agents are silane and isocyanates (Mohanty *et al.* 2005). The main variables in these treatments are the type and concentration of the chemical used and the time and temperature of the treatment.

17.5 Evaluation of fibre treatment techniques

Many researchers have studied the surface treatment of bagasse fibres in various resins. De Sousa *et al.* (2004) investigated the effect of the size of chopped bagasse fibre in polyester composites. It was noted that as the size of chopped fibres reduced, the flexural properties increased. The flexural strength of chopped bagasse of mesh size 20/polyester composite more than doubled in comparison to mesh size 10 fibre-reinforced composites. It was also observed that an increase in moulding pressure led to an increase in flexural properties. It is believed that higher moulding pressure results in closer contact between bagasse and resin and reduces entrapped voids, hence enhancing flexural properties. Overall, as the size of fibre reduces, surface area increases, hence the interfacial area available for fibre/matrix increases and flexural properties are improved.

El-Tayeb (2008) studied the effect of the length of sugarcane bagasse on the mechanical properties of SCB polyester composites. Chopped fibres of 1, 5 and 10 mm in length were added to the polyester matrix, resulting in a reduction of tensile strength in comparison to pure matrix. This is due to the weak interface between untreated natural fibres and matrix. The tensile strength was highest for 5 mm SCB composite (75 MPa) in comparison to 1 mm SCB composite (51 MPa) and 10 mm SCB composite (64 MPa).

Wirawan *et al.* (2011) studied the properties of bagasse polyvinyl chloride composites after treatment with sodium hydroxide, benzoic acid and a polymethylene polyphenyl isocyanate coupling agent. At a fibre content of 40 wt%, all treatments improved in tensile strength while the PMPPIC coupling agent resulted in the highest enhancement in mechanical properties. Surprisingly, properties of unwashed bagasse PVC composite, which contained sugar, surpassed those of the treated fibre composite. SEM micrographs showed that sugar fills the cavities in the unit cell of bagasse fibre, resulting in better load transformation from matrix to fibres. It is suspected that the sugar in unwashed SCB fibres may enhance the tensile strength and modulus of the composite and enable SCB to be used as a reinforcing agent without further chemical treatment (Wirawan *et al.* 2011). As shown in Table 17.3, the alkaline and benzoic acid treatment of SCB/PVC composite resulted in an initial increase in water absorption, which may be due to an increase in surface area, removal of hydrophobic lignin and conversion of cellulose I to more hydrophilic cellulose II. Unwashed SCB/PVC composites had a lower initial rate of water uptake compared to washed SCB/PVC composites. It is suspected that sugar covers the cell wall of SCB and reduces the volume of void space, resulting in slower water penetration and absorption. The use of a coupling agent also resulted in a

Table 17.3 Water absorption of sugarcane bagasse–PVC composites

Pre-treatment	Treatment	Initial rate of absorption, k (% $h^{-½}$)	Maximum weight gain, W_m (%)
Unwashed	Untreated	0.82	9.53
Washed (sugar-free)	Untreated	0.93	12.11
Washed (sugar-free)	Benzoic acid	0.99	11.92
Washed (sugar-free)	Alkali	1.25	13.45
Washed (sugar-free)	PMPPIC	0.76	11.76

Source: Wirawan *et al.* 2011, © 2011 by Sage, reprinted by permission of Sage.

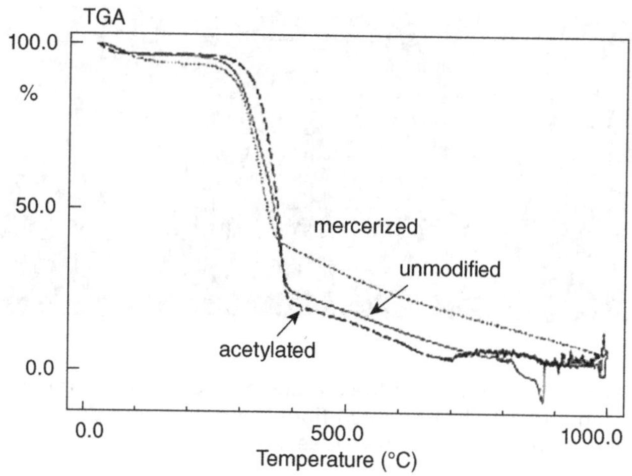

17.2 Thermogravimetric analysis data for the sugarcane fibres, unmodified, mercerized and acetylated (reprinted from Bertoti *et al.* 2009, © 2009, with permission from Elsevier).

more hydrophobic surface as the –N=C=O of isocyanate reacts with the –OH group of cellulose (Wirawan *et al.* 2011).

Bertoti *et al.* (2009) compared the effect of acetylation and mercerization on the properties of sugarcane fibres. Acetylated fibres were treated with acetic anhydride (with sulfuric acid as catalyst) in a reflux set up for 1 hour at 145°C, while mercerized fibres were treated with 10% NaOH for 1 hour at 0°C. Acetylated fibres showed higher thermal stability compared to mercerized fibres (Fig. 17.2). This is attributed to the small amount of water and acetic acid present in the fibres, as most of the acetic acid was removed by anhydride acetic vapour in this acetylation treatment. Also, acetylated fibres had less tensile strength due to fibrillation of fibres caused by loss of hemicellulose (Bertoti *et al.* 2009). Luz *et al.* (2008) also showed a slight reduction in tensile and flexural properties after acetylation of cellulose and cellulignin (fibre without hemicellulose) sugarcane bagasse reinforced polypropylene composites (10 and 20 wt%). They also found that an increase in fibre content led to a decrease in tensile strength.

Cao *et al.* (2006) investigated the effect of fibre content and alkali treatment on the performance of 20–75 wt% bagasse–polyester composites. SEM micrographs of treated and untreated fibres are shown in Fig. 17.3. They reported a tensile strength of 70.9 MPa for untreated bagasse and 83.4 MPa for alkali-treated bagasse. They also showed that the tensile and impact strength of a bagasse composite increased with an increase in fibre content up to a fibre content of 65 wt%. A 1% alkali treatment led to better results compared to 3% and 5% alkali treatments. Surface treatment with

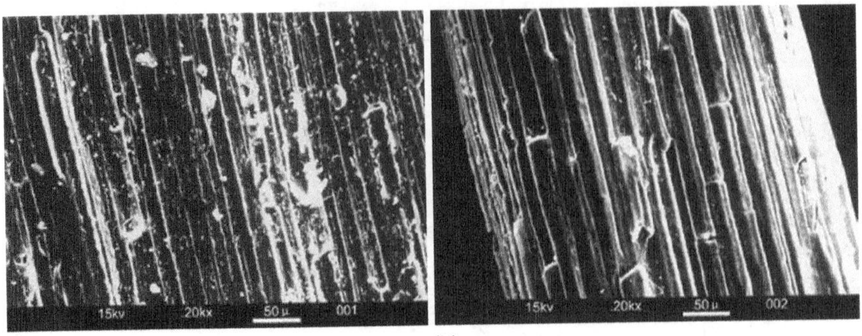

(a) (b)

17.3 SEM micrographs of the surface of bagasse fibre: (a) untreated fibre, and (b) 1% NaOH-treated fibre (reprinted from Cao *et al.* 2006, © 2006, with permission from Elsevier).

1% NaOH enhanced the tensile strength by 13%, flexural strength by 14% and impact strength by 30% (Cao *et al.* 2006). The mechanical properties of untreated and alkali-treated fibres are shown in Table 17.4.

Youssef *et al.* (2009) investigated the influences of bagasse content (alkali treated) and electron beam radiation of the matrix for LDPE and HDPE composites. SCB fibres with a diameter of 0.5–0.85 mm and a length of 2–10 mm were mixed with the matrix in a ratio of 20 to 80 wt%. Electron beam irradiation of the matrix was carried out in atmospheric air at ambient temperature at different doses of 10 to 70 kGy. The suggested mechanism of interaction between bagasse fibres and thermoplastic resins is shown in Fig. 17.4.

As fibre content increased from 20 to 80 wt% the modulus of rupture decreased (from 19 to 11 MPa for HDPE and from 10 to 8 MPa for LDPE) due to fibre agglomeration and poor interaction between hydrophobic matrix and hydrophilic fibre. An increase in fibre content also led to a reduction in impact strength (from 1.9 to 0.8 J/cm^2 for HDPE and from 1.4 to 0.8 J/cm^2 for LDPE). It was found that the deterioration in mechanical properties becomes more significant beyond 50 wt% bagasse fibre content, due to agglomeration of fibres. Also, the decrease in mechanical properties is more pronounced in the HDPE composites than in the LDPE composites. The optimum dose for electron beam irradiation was found to be 40 and 10 kGy for LDPE and HDPE respectively. Irradiation of the matrix led to enhanced modulus of rupture, reaching 130% and 160% for LDPE and HDPE composites, an increase of 130% and 137% for modulus of elasticity and an increase of 65% and 30% for impact toughness for LDPE and HDPE composites respectively. Moreover, SEM micrographs of LDPE composites showed enhanced adhesion between fibre and matrix for irradiated matrix (Youssef *et al.* 2009). The flexural strength (modulus of

Table 17.4 Mechanical properties of untreated and alkali-treated bagasse–polyester composite

Fibre (wt%)	Treatment of fibre	Tensile strength (MPa)		Flexural strength (MPa)		Flexural modulus (MPa)		Impact strength (KJ/m^2)	
		Mean value	S.D.	Mean value	S.D.	Mean value	S.D.	Mean value	S.D.
20	Untreated	16.52	0.78	31.19	2.18	1136.87	115.14	4.12	0.46
	Alkali treated	18.58	0.73	34.71	2.68	1321.72	56.69	6.07	0.79
35	Untreated	18.60	0.83	38.37	3.04	1451.81	157.64	6.82	0.37
	Alkali treated	21.35	0.69	43.96	3.04	1622.36	185.78	8.29	0.98
50	Untreated	21.09	0.89	40.16	2.36	1841.34	74.60	8.17	0.17
	Alkali treated	23.07	0.53	46.05	2.57	2031.37	155.15	9.52	0.31
65	Untreated	23.47	0.74	43.87	3.78	2292.02	174.52	8.82	0.57
	Alkali treated	26.77	0.75	50.86	3.79	2673.73	165.17	11.27	0.85

Source: Reprinted from Cao *et al.* 2006, © 2006, with permission from Elsevier.

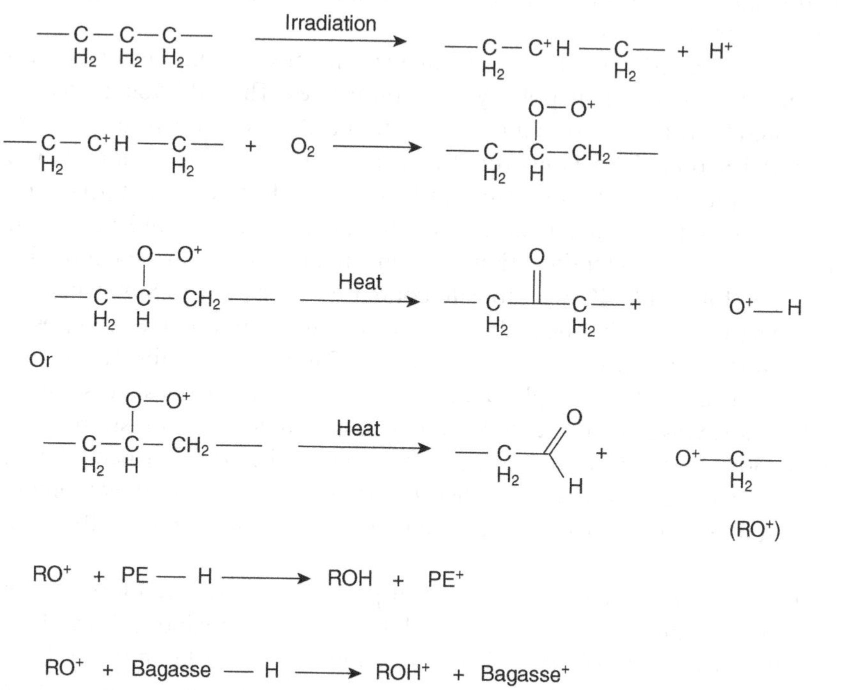

17.4 Suggested mechanism of interaction between bagasse fibres and thermoplastics (LDPE and HDPE) (Youssef *et al.* 2009, © 2009 by Sage, reprinted by permission of Sage).

rupture) was higher for HDPE compared to LDPE for all irradiation doses due to their higher degree of crystallinity.

Saini *et al.* (2010) evaluated the use of bagasse flour as a filler in the PVC matrix with a view to developing a value-added product from the waste of sugar mills and investigated the effect of particle size and alkali treatment. The addition of the SCB filler led to a reduction in the tensile strength, tensile modulus and impact strength of PVC composites. Particle size had a significant effect on the properties of the composite and a SCB filler with a particle size <50 μm had better properties compared to a filler with a particle size of 100–150 μm. The water uptake increased with larger amounts of filler and longer immersion time and was lower (3.63%) for smaller than for larger particles (7.98%). Alkali-treated filler led to an increase of 48% in tensile modulus, 10% in thermal stability and 14% in impact strength at a filler loading of 30 phr (per hundred rubber) (Saini *et al.* 2010).

Bilba and Arsène (2008) studied the silane treatment of bagasse fibre for reinforcement of cement composites for pyrolyzed and unpyrolyzed SCB/cement composites. Pyrolysis was done at 200°C for 2 hours and led to a reduction of hemicellulose and extractives content of bagasse fibres. This led to an increase in surface roughness, hence better adhesion and reduction in hydrophilicity. Moreover, the combination of pyrolysis and silane treatments led to lower porosity than silane treatment alone and much lower water absorption.

Rodrigues *et al.* (2011) carried out experiments to study esterification treatment of bagasse in polystyrene composites. The addition of 5 wt% untreated SCB fibres led to a reduction in tensile strength (from 19.7 MPa for pure PS to 11 MPa). However, it was observed that esterification of SCB led to an increase in tensile strength (from 11 to 12.5 MPa). Esterification was conducted using acetic anhydride, toluene, acetic acid and perchloric acid for 5 hours. This modification led to an enhancement of tensile modulus by 71.5% for 5 wt% SCB composite compared to the pure PS matrix.

Acharya *et al.* (2009) described the behaviour of SCB/epoxy composite where fibres were washed with acetone for 1.5 hours. Steam absorption was highest for unwashed samples and stabilized for acetone-washed samples at 40 hours. Washed and treated composite had higher flexural strength by about 50% compared to the unwashed composite. It is suspected that washing with acetone results in dissolution of hemicellulose, development of crystallinity and fibrillation, thus enhancing bonding between fibre and matrix.

Trindade *et al.* (2005) used thermoset phenolic matrices reinforced with 15% SCB (unmodified and oxidized by chlorine dioxide, followed by grafting furfuryl alcohol). After oxidization, the crystallinity of SCB increased from 47% to 52% due to the partial removal of lignin. Oxidization of SCB greatly reduced its tensile strength (from 222 MPa for unmodified

SCB to 126 MPa for oxidized SCB). However, further reaction with furfuryl alcohol enhanced the tensile strength (238 MPa) probably because of the introduction of a polymeric layer on the surface of the fibre which protects the fibre. Thermogravimetric analysis of SCB showed decomposition of polysaccharides near 300°C and of lignin near 480°C. As chemical modification mainly occurs at the surface, the TGA profiles were similar for the modified fibres as well. Modification of SCB led to a reduction in the Izod unnotched impact strength of phenolic composites reinforced with SCB. Oxidized SCB phenolic composite had impact strength of 15 ± 2 J/m, while after reacting with furfuryl alcohol it was 17 ± 2 J/m, in comparison to 28 ± 7 J/m for unmodified SCB phenolic composite. It was concluded that modification caused some fibre degradation that affected the mechanical properties of composites (Trindade *et al.* 2005).

Mulinari *et al.* (2009a) showed that modification of cellulose in SCB/HDPE composite with zirconium oxychloride improved adhesion between fibre and matrix and increased tensile strength and modulus from 1.54 MPa and 732 MPa to 18.2 MPa and 1233.1 MPa respectively. The modification of SCB cellulose by zirconium oxychloride resulted in a 15% reduction in composite elongation compared to untreated SCB/HDPE.

Maldas and Kokta (1991) examined the effect of adding two coupling agents, poly(methylene(polyphenyl isocyanate)) or PMPPIC and lignin, on particle boards made from bagasse and polymer matrix of PVC or PS. The total concentration of coupling agents was varied from 0 to 10 wt%, while the concentration of matrix (PVC or PS) was between 0 and 20 wt%. Bagasse was mixed with a roll mill at 175°C between 5 and 10 times. It was found that the lignin content of 3–5 wt% led to the best mechanical properties but overall its coupling action was less effective than the PMPPIC. To arrive at the best mechanical properties, the optimum condition for use of coupling agents was 10 wt% (5:5 of PMPPIC and lignin). Also, the PVC matrix led to better mechanical properties compared to PS and increased concentration of polymers in the particle board led to better mechanical properties, as shown in Fig. 17.5 (Maldas and Kokta 1991).

Habibi *et al.* (2008) investigated the use of a coupling agent for SCB/low-density polyethylene composites (0.8% maleic anhydride). A covalent bond can be formed between fibre and matrix in the presence of a coupling agent. Both the melting point and the degree of crystallinity of the matrix increased with the addition of the coupling agent (Table 17.5). The increase of fibre content (up to 30%) led to an increase in Young's modulus and stress at break of the maleated composite. A further increase of fibre content (up to 50%) led to a reduction in these values.

Vilay *et al.* (2008) compared two surface treatments of 1% alkali treatment for 3 hours and 1% acrylic acid treatment for 1 hour. Fibre treatment resulted in the enhancement of the tensile strength and modulus of fibres.

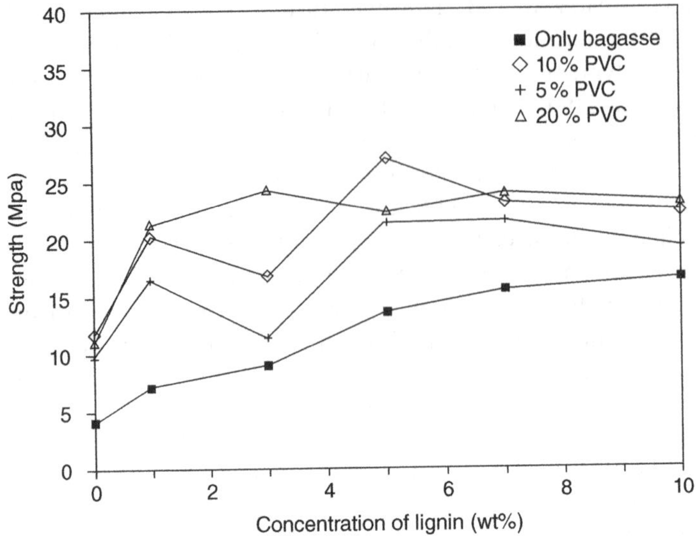

17.5 Effect of concentration of lignin and PVC on the tensile strength of bagasse PVC particle boards (reprinted from Maldas and Kokta 1991, © 1991, with permission from Elsevier).

Table 17.5 Melting characteristics of LDPE- and MLDPE-based composites reinforced with bagasse obtained from DSC curves

Sample	Fibre content (%)	LDPE			MLDPE		
		T_m (°C)	ΔH_m (Jg^{-1})	χ_c^a	T_m (°C)	ΔH_m (Jg^{-1})	χ_c^a
PELD	0	112.0	98.4	0.34	–	–	–
NPELD	0	–	–	–	122.4	121.1	0.42
Bagasse	10	112.2	104.2	0.40	122.5	129.8	0.50
	20	112.0	99.2	0.43	123.6	98.1	0.42
	30	112.3	88.9	0.44	124.8	83.5	0.41
	40	112.1	74.3	0.43	122.7	66.5	0.38
	50	112.1	65.6	0.45	124.6	64.6	0.44

Note: T_m: melting temperature, ΔH_m: heat of fusion, χ_c: degree of crystallinity.
Source: Habibi *et al.* 2008, © 2008, with permission from Elsevier.

The tensile strength increased from 96 MPa for untreated bagasse to 156 MPa for alkali-treated bagasse and 229 MPa for acrylic acid-treated bagasse, while the tensile modulus increased from 6.4 GPa to 7.1 GPa and 8.1 GPa respectively. Changing the fibre content from 0 to 20 vol% demonstrated that higher fibre content led to higher tensile and flexural properties in the polyester matrix. The tensile and flexural properties of

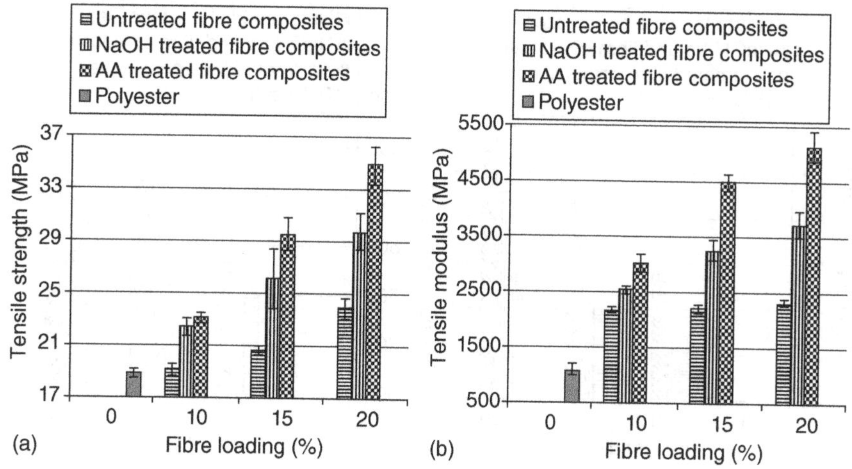

17.6 (a) Tensile strength and (b) tensile modulus of treated and untreated bagasse polyester composites (reprinted from Vilay *et al.* 2008, © 2008, with permission of Elsevier).

treated and untreated bagasse polyester composites are shown in Figs 17.6 and 17.7. Pure polyester was used as a control to study the effect of the addition of bagasse fibre to resin.

Mulinari *et al.* (2009b) studied the effect of adding sugarcane bagasse cellulose modified with zirconium oxychloride ($ZrOCl_2$ $8H_2O$) to HDPE matrix. The tensile strength of HDPE (16.7 MPa) reduced after the addition of 10 wt% SCB fibres (14.4 MPa). However, fibre surface modification enhanced the tensile strength (15.6 MPa). The addition of fibre to the matrix led to an increase in tensile modulus of the composite in comparison to pure HDPE (from 850 MPa to 880 MPa) while surface modification led to an enhancement of over 50% in tensile modulus of composite (1324 MPa).

17.6 Assessing composite performance

The performance of a fibre-reinforced composite depends on the fibre/matrix interface and the ability to transfer the load from matrix to fibre. The interface and stress transfer determine the mechanical properties and performance of the composite.

Tewari *et al.* (2012) investigated the effect of bagasse fibre content in the hybrid composite of bagasse–glass fibre epoxy composite. Hybrid composite was developed from 15–30 wt% bagasse fibre and 5% glass fibre of 2–3 mm with the addition of a hardener. SEM micrographs of the composite showed that bagasse fibres had an average length of 61 μm and a diameter of 13 μm. They also showed a uniform dispersion of both fibres. Increased fibre

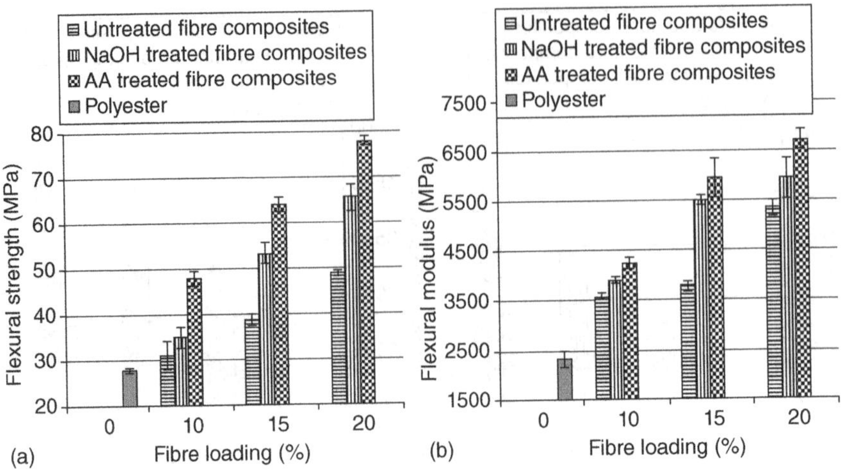

17.7 (a) Flexural strength and (b) flexural modulus of treated and untreated bagasse polyester composites (reprinted from Vilay *et al.* 2008, © 2008, with permission from Elsevier).

content of the reinforced components resulted in a decrease in the density of the composite. Water absorption capacity increased as the bagasse fibre content increased, due to the water-absorbing tendency of bagasse fibre and its voids. The addition of bagasse fibres led to a decrease in the ultimate tensile strength; however, the addition of glass fibre increased the ultimate tensile strength. Figure 17.8 shows a great difference in stress–strain behaviour of hybrid composites following the addition of different wt% of bagasse fibres with 5 wt% glass fibre in epoxy matrix (Tewari *et al.* 2012). The decrease in ultimate tensile strength (Fig. 17.8) with an increase in bagasse fibre content is due to voids present in the material. As bagasse content increased, the modulus of elasticity and ultimate compressive strength decreased. On the other hand, percentage elongation and impact strength increased with the increase of bagasse content in the hybrid composite, due to the more elastic nature of natural fibres in comparison to resin. The addition of bagasse fibre reduced the bending strength and hardness of the composite.

Vallejos *et al.* (2011) evaluated the potential use of fibrous material from ethanol–water fractionation of bagasse as a reinforcement for thermoplastic starch. The main drawbacks of starches are their brittleness, moisture absorption and low mechanical properties. The addition of natural fibre can improve their thermal and mechanical properties. Bagasse fibres were depitched and delignified in this study, with average fibre length of 667 μm, fibre diameter of 17.5 μm and length/diameter ratio of 38. The addition of 10 wt% bagasse fibres increased tensile strength of corn starches from 2.61

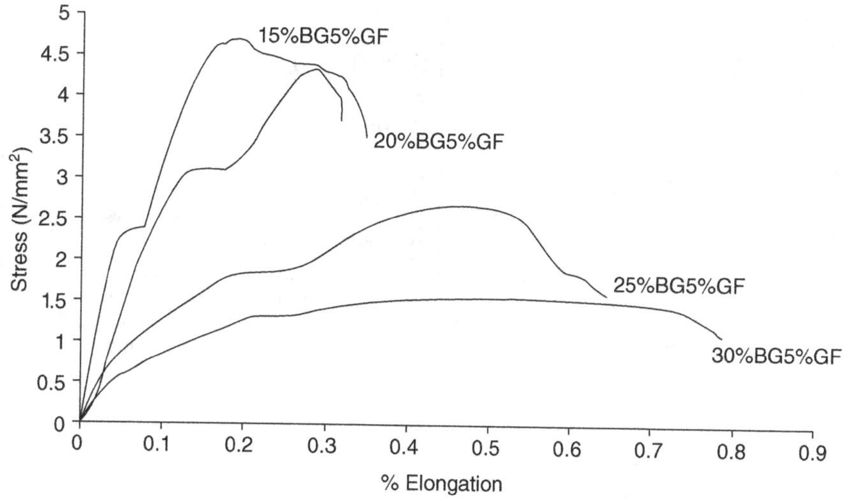

17.8 Stress–elongation (%) for different composite materials at 1 mm/min cross-head speed (Tewari *et al.* 2012).

to 3.75 MPa (47%) and from 2.21 to 3.24 MPa (44%) for cassava starches. Adding a higher amount of bagasse fibre did not result in an increase in tensile strength as fibres agglomerated, resulting in poor dispersion in the matrix. Reinforcement with 15 wt% bagasse resulted in a more than fourfold increase in the elastic modulus of corn and cassava starches. It was found that fibres tended to agglomerate at contents higher than 10 wt% in the composite, resulting in poor dispersion in the matrix. Hence, a lower content of bagasse fibre was more desirable as it led to better dispersion (Mulinari *et al.* 2009a).

Nadali *et al.* (2010) investigated the durability of bagasse/PP composite exposed to fungal decay, since deteriorating factors such as fungi can affect natural fibres. Composite specimens were exposed to rainbow fungus at 25°C and 75% relative humidity for 1–4 months. The bending strength, bending modulus and hardness reduced by about 30–50% due to fungal decay (Fig. 17.9). Fungal decay also resulted in an increase in water absorption of composites with the highest amount corresponding to the 4 months exposure.

Ramaraj (2007) examined the effect of SCB powder on the mechanical and thermal properties of the PP matrix by incorporating SCB powder (passing through a 1.5 mm size screen) in PP as a reinforcing filler. PP was compounded with 5, 10, 15 and 20% SCB filler in a co-rotating twin-screw extruder and injection which was moulded to create a specimen for testing. As the filler loading increased, tensile strength and tensile elongation reduced due to poor interaction between the hydrophilic filler and the

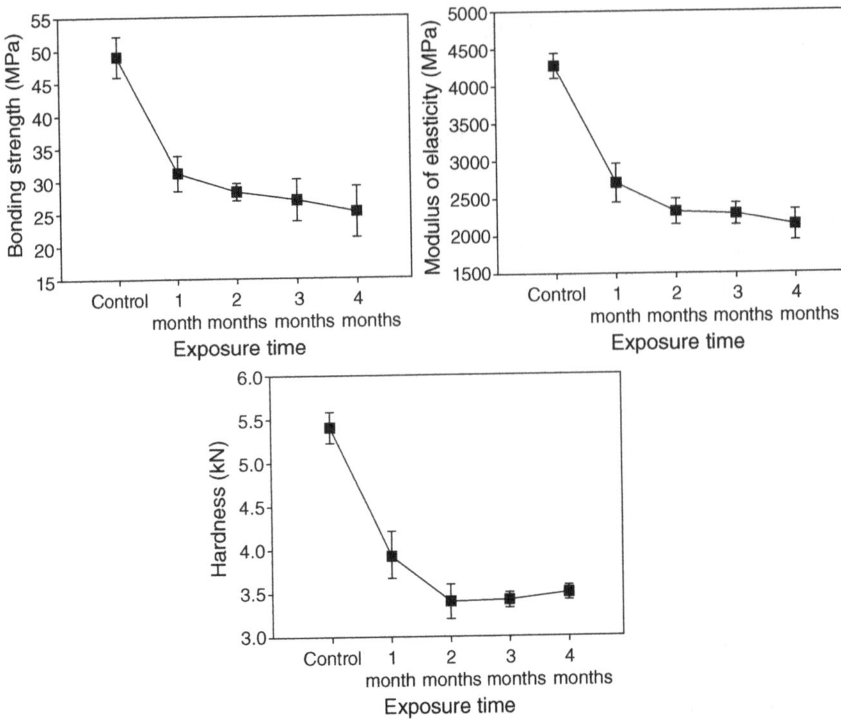

17.9 Effect of exposure time to fungus on the mechanical properties of the bagasse–PP composite (Nadali *et al.* 2010)

hydrophobic matrix (from 32.22 to 27.21 MPa and from 164.4% to 11.2% respectively with an increase in fibre loading from 5% to 20%). The impact strength increased with an increase in fibre loading, probably due to poor adhesion resulting in the fibre dissipating energy by mechanical friction. It was found that the Izod impact strength increased from 10.50 to 13.23 kg.cm/cm, the Charpy impact strength increased from 10.10 to 13.98 kg.cm/cm, the flexural strength increased from 23.66 to 26.84 MPa, and the HDT increased from 45.5°C to 66.5°C, with an increase in fibre loading from 5% to 20% in the PP matrix.

EVA (polyethylene-*co*-vinyl acetate) polymer can be a possible resin matrix as it has high flexibility (advantageous under impact conditions) despite low mechanical properties. Stael *et al.* (2001) evaluated and compared the impact properties of chopped SCB/EVA matrix composite to SCB/PP and PE matrix composites of varying size and volume fraction of bagasse used as filler. Three different chopped bagasse groups with lengths $L < 3$ mm, $3 < L < 5$ mm and $20 < L < 30$ mm were mixed with EVA composite at 20, 40, 60 and 70 wt%. The results showed that the impact behaviour of SCB/EVA is almost independent of the bagasse length, except in the case of 70 wt% SCB/EVA where impact strength increased for composites with longer fibres. The incorporation of SCB fibres resulted in a strong decrease in the impact strength from 20 kJ/m^2 to 3 kJ/m^2 for an increase in fibre loading from 20 to 70 wt%. Comparing impact strength of 60 wt% SCB/EVA composite (3.7 kJ/m^2) to SCB/PP (17.1 kJ/m^2) and SCB/PE (19.7 kJ/m^2) showed that EVA-based composites have lower impact properties, which may be due to their good interaction between the matrix and chopped bagasse which reduced the deformation capacity of the composite (Stael *et al.* 2001).

Acharya *et al.* (2008) investigated the weathering behaviour of SCB/epoxy composites under different environmental conditions such as subzero temperatures, steam and saline water. Changes in volume for SCB/epoxy composite exposed to steam stabilized after 24–48 hours, and as the bagasse content increased, changes in composite volume decreased. The reason for this is due to the fact that as fibre content increases, the exposed area decreases and fibres have less chance of fibre wetting and swelling, hence smaller changes in volume (Ramaraj 2007). A similar trend was observed for all weathering conditions, with steam having the highest volume change and subzero treatment having the lowest volume change. An increase in volume fraction of fibres led to a decrease in flexural strength (Acharya *et al.* 2008).

Shibata *et al.* (2005) studied the effect of fibre content and fibre length of flexural modulus of bagasse fibre biodegradable resin composite. Figure 17.10 shows an increase in flexural strength and modulus of the composite

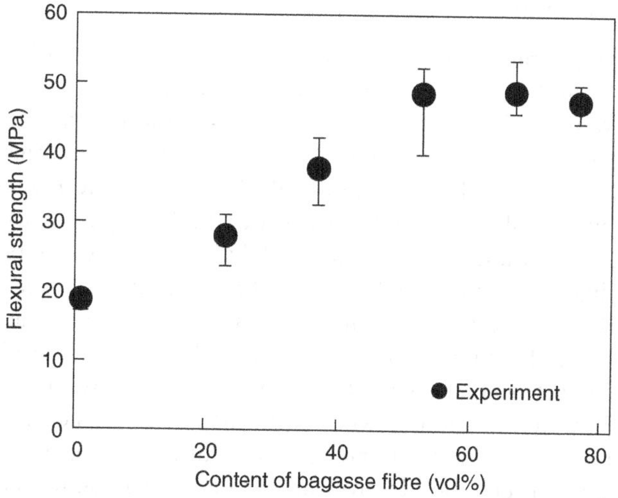

17.10 Variation of flexural strength with varying volume fraction of bagasse fibres (Shibata *et al.* 2005, © 2005 Society of Plastics Engineers).

of up to 65% volume fraction of fibres (flexural strength increased from 20 to 50 MPa as fibre volume fraction increased from 0 to 65%). The reduction of flexural modulus at 75% volume fraction of fibres can be attributed to the lack of resin to cover all fibre surfaces. Bagasse with various fibre lengths of 1.6, 3.2, 9.1 and 16.1 mm were used at 65% volume fraction in biodegradable composite. It was found that reducing the size of bagasse fibres below 3 mm (aspect ratio of 12) led to a decrease in the flexural modulus. The results from these studies were also in line with the predictions from Cox's model (Shibata *et al.* 2005).

Xu *et al.* (2010) replaced long-term creep experiments with shorter tests at higher temperatures for 50 wt% SCB/PVC (recycled and virgin) and SCB/HDPE (recycled and virgin) at 35–65°C for PVC and 35–85°C for HDPE. SCB/PVC showed higher stiffness than SCB/HDPE, probably due to the higher stiffness of the PVC matrix. Composites with recycled polymers had a slightly higher modulus than composites with virgin polymers. Also, SCB/PVC showed better creep resistance than SCB/HDPE at low temperature, but they were more temperature-dependent.

Mishra and Acharya (2010a, 2010b) carried out experiments to study the abrasive behaviour of bagasse epoxy composite by varying load, grit size of the abrasive papers, and orientation of the fibres in the composite. 20% fibre volume fraction composites were utilized in the composite in three directions:

- Parallel orientation
- Anti-parallel orientation (random)
- Normal orientation.

Specimens were abraded on silicon carbide abrasive papers with grit sizes of 150, 180, 320 and 400 at a sliding speed of 1 m/min, while four loads of 1, 3, 5 and 7 N were applied to them. The results showed that an increase of load and grit size led to an increase in the rate of abrasion. In terms of fibre direction and wear, parallel orientation of fibres led to the highest wear, followed by anti-parallel orientation (Mishra and Acharya 2010a). Mishra and Acharya also studied solid erosion of bagasse epoxy composites at different impingement angles of 30° to 90° at four different velocities of 48, 72, 82 and 109 m/s. The duration of the tests was 10 minutes and the erodent used was silica sand with a size of 150–250 μm. The results showed that an increase in impact velocity lead to an increase in erosion rate. Also, composites showed a brittle behaviour with maximum erosion rate at 90° impingement angle. It should be noted that addition of bagasse fibres to epoxy resin led to a reduction in erosion rate in comparison to pure epoxy matrix (Mishra and Acharya 2010b).

The processing technique and the condition of composite compounding greatly affect the performance of the composite as well. Luz and Gonçalves

(2007) compared compression and injection moulding processes on bagasse polypropylene composites to evaluate these common compounding methods. It was found that both techniques led to homogeneous distribution, but dispersion was random for injection moulding. In compression moulding, it was easier for shorter fibres to penetrate to the lower plate as the pressure was from up to down, hence a random dispersion was not achieved. The other problem with compression moulding in comparison to injection moulding was the appearance of blisters in the composite. Even using the better processing technique (injection moulding), inserting fibres (10 wt%) led to a 50% decrease in tensile strength compared to pure PP, which is explained by the agglomeration of fibres at various points. However, there was an increase in the flexural modulus of composites, as plastic became more rigid due to the presence of fibres (Luz and Gonçalves 2007).

Vazquez *et al.* (1999) reported that Kneader mixing of bagasse/PP composite decreased the average fibre length by more than 70%. On the other hand, fibre length was not affected in composite production with compression moulding of mat fibres between two PP sheets. However, fibre wetting was poor in the compression moulding technique, thus lower mechanical properties were obtained for the same fibre content compared to the Kneader-mixing technique. Tensile strength and the elongation at break of bagasse/PP composite decreased as the content of untreated bagasse fibre increased, showing that bagasse fibre acted only as a filler. However, isocyanate and mercerization treatment improved the tensile properties of the composite. Creep behaviour also improved for treated fibres, with the highest creep activation energy for mercerized fibres (Vazquez *et al.* 1999).

Aggarwal (1995) looked at the optimization of bagasse content (0–20% by mass), casting pressure (1–5 N/mm^2) and demoulding time (1–10 hour) for production of bagasse-cement composites. It was shown that increasing bagasse content led to an increase in water absorption while decreasing density and bond strength (Fig. 17.11). The optimum parameters were found to be a bagasse content of 12–16% by mass, a casting pressure of 2–3 N/mm^2 and a demoulding time of over 6 hours.

17.7 Future trends

Utilization of sugarcane bagasse as reinforcement and filler in polymeric composites is an effective application to increase the value of this agricultural waste. The high cellulose content of bagasse shows that it has the potential to be used in industry. As environmental regulations are encouraging the reuse or recycling of automotive materials in Europe (85%) and in Japan (95%) by 2015 (Satyanarayana *et al.* 2007) there will be an increase in the use of lignocellulosic fibres such as bagasse. Hybridization with some

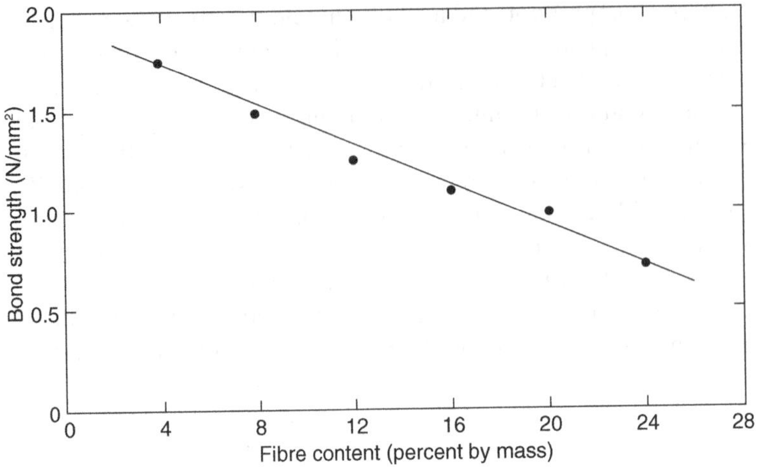

17.11 Bond strength vs. fibre content for bagasse–cement composite (reprinted from Aggarwal 1995, © 2009, with permission from Elsevier).

synthetic fibres makes bagasse a potential option for the automotive industry. Also, the use of carbonized bagasse as fillers in natural rubber will reduce environmental pollution (Osarenmwinda and Abode 2010).

Another trend which is a good potential for use of bagasse is bagasse fly ash metal and colour removal from wastewater. Bagasse fly ash is waste obtained from the bagasse-fired boiler stack of sugarcane mills. Currently, it is used as filler in building materials, but it also has the potential to be used to remove metals and dyes through adsorption on its surface and to replace commonly used activated carbon. The cost of the cheapest activated carbon available in India is about US$285 per ton, while waste bagasse fly ash costs about US$20 per ton, which makes bagasse fly ash an extremely attractive potential candidate for the removal of metals and colour from wastewater (Gupta and Ali 2004).

17.8 Conclusion

As reported by various researchers, it is beneficial to use bagasse as reinforcement and filler in plastic and cement matrices. The fibre surface treatment of bagasse reduces moisture absorption of fibres and enhances the wettability of the matrix, leading to better adhesion and interaction between fibre and matrix. The enhancement in the mechanical properties of bagasse composite as a result of a variety of surface treatment techniques has been reported by various authors.

17.9 References

Acharya, S., Mishra, P., Mehar, S.K. and Dikshit, V. 2008, 'Weathering behaviour of bagasse fibre reinforced polymer composite', *Journal of Reinforced Plastics and Composites*, vol. 27, no. 16–17, pp. 1839–1846.

Acharya, S., Mishra, P.P. and Mehar, S.K. 2009, 'The influence of fibre treatment on the performance of bagasse fibre-reinforced polymer composite', *Journal of Reinforced Plastics and Composites*, vol. 28, no. 24, pp. 3027–3036.

Aggarwal, L. 1995, 'Bagasse-reinforced cement composites', *Cement and Concrete Composites*, vol. 17, no. 2, pp. 107–112.

Bertoti, A.R., Luporini, S. and Esperidião, M.C.A. 2009, 'Effects of acetylation in vapour phase and mercerization on the properties of sugarcane fibres', *Carbohydrate Polymers*, vol. 77, no. 1, pp. 20–24.

Bilba, K. and Arsène, M. 2008, 'Silane treatment of bagasse fibre for reinforcement of cementitious composites', *Composites Part A: Applied Science and Manufacturing*, vol. 39, no. 9, pp. 1488–1495.

Bilba, K., Arsène, M. and Ouensanga, A. 2003, 'Sugar cane bagasse fibre reinforced cement composites. Part I. Influence of the botanical components of bagasse on the setting of bagasse/cement composite', *Cement and Concrete Composites*, vol. 25, no. 1, pp. 91–96.

Cao, Y., Shibata, S. and Fukumoto, I. 2006, 'Mechanical properties of biodegradable composites reinforced with bagasse fibre before and after alkali treatments', *Composites Part A: Applied Science and Manufacturing*, vol. 37, no. 3, pp. 423–429.

De Sousa, M., Monteiro, S. and d'Almeida, J. 2004, 'Evaluation of pre-treatment, size and molding pressure on flexural mechanical behavior of chopped bagasse–polyester composites', *Polymer Testing*, vol. 23, no. 3, pp. 253–258.

El-Tayeb, N. 2008, 'Abrasive wear performance of untreated SCF reinforced polymer composite', *Journal of Materials Processing Technology*, vol. 206, no. 1, pp. 305–314.

Fairbairn, E.M.R., Americano, B.B., Cordeiro, G.C., Paula, T.P., Toledo Filho, R.D. and Silvoso, M.M. 2010, 'Cement replacement by sugar cane bagasse ash: CO_2 emissions reduction and potential for carbon credits', *Journal of Environmental Management*, vol. 91, no. 9, pp. 1864–1871.

Ganesan, K., Rajagopal, K. and Thangavel, K. 2007. 'Evaluation of bagasse ash as supplementary cementitious material', *Cement Concr. Compos*, vol. 29, pp. 515–524.

Gupta, V.K. and Ali, I. 2004, 'Removal of lead and chromium from wastewater using bagasse fly ash – a sugar industry waste', *Journal of Colloid and Interface Science*, vol. 271, no. 2, pp. 321–328.

Habibi, Y., El-Zawawy, W.K., Ibrahim, M.M. and Dufresne, A. 2008, 'Processing and characterization of reinforced polyethylene composites made with lignocellulosic fibres from Egyptian agro-industrial residues', *Composites Science and Technology*, vol. 68, no. 7, pp. 1877–1885.

Huang, Z., Wang, N., Zhang, Y., Hu, H. and Luo, Y. 2012, 'Effect of mechanical activation pretreatment on the properties of sugarcane bagasse/poly(vinyl chloride) composites', *Composites Part A: Applied Science and Manufacturing*, vol. 43, no. 1, pp. 114–120.

Jústiz-Smith, N.G., Virgo, G. and Buchanan, V.E. 2008, 'Potential of Jamaican banana, coconut coir and bagasse fibres as composite materials', *Materials Characterization*, vol. 59, no. 9, pp. 1273–1278.

Luz, S. and Gonçalves, A. 2007, 'Mechanical behavior and microstructural analysis of sugarcane bagasse fibres reinforced polypropylene composites', *Composites Part A: Applied Science and Manufacturing*, vol. 38, no. 6, pp. 1455–1461.

Luz, S., Del Tio, J., Rocha, G. and Gonçalves, A. 2008, 'Cellulose and cellulignin from sugarcane bagasse reinforced polypropylene composites: Effect of acetylation on mechanical and thermal properties', *Composites Part A: Applied Science and Manufacturing*, vol. 39, no. 9, pp. 1362–1369.

Maldas, D. and Kokta, B. 1991, 'Studies on the preparation and properties of particle boards made from bagasse and PVC: II. Influence of the addition of coupling agents', *Bioresource Technology*, vol. 35, no. 3, pp. 251–261.

Mall, I.D., Srivastava, V.C. and Agarwal, N.K. 2006, 'Removal of Orange-G and Methyl Violet dyes by adsorption onto bagasse fly ash – kinetic study and equilibrium isotherm analyses', *Dyes and Pigments*, vol. 69, no. 3, pp. 210–223.

Mishra, P. and Acharya, S. 2010a, 'Anisotropy abrasive wear behaviour of bagasse fibre reinforced polymer composite', *International Journal of Engineering, Science and Technology*, vol. 2, no. 11.

Mishra, P. and Acharya, S. 2010b, 'Solid particle erosion of Bagasse fibre reinforced epoxy composite', *International Journal of Physical Sciences*, vol. 5, pp. 109–115.

Mohanty, A.K., Misra, M. and Drzal, L.T. 2005, *Natural Fibres, Biopolymers, and Biocomposites*, CRC Press, Boca Raton, FL.

Mulinari, D.R., Voorwald, H.J.C., Cioffi, M.O.H., Pinto da Silva, M.L.C. and Luz, S.M. 2009a, 'Preparation and properties of HDPE/sugarcane bagasse cellulose composites obtained for thermokinetic mixer', *Carbohydrate Polymers*, vol. 75, no. 2, pp. 317–321.

Mulinari, D.R., Voorwald, H.J.C., Cioffi, M.O.H., Pinto da Silva, M.L.C., da Cruz, T.G. and Saron, C. 2009b, 'Sugarcane bagasse cellulose/HDPE composites obtained by extrusion', *Composites Science and Technology*, vol. 69, no. 2, pp. 214–219.

Nadali, E., Karimi, A., Tajvidi, M. and Naghdi, R. 2010, 'Natural durability of a bagasse fibre/polypropylene composite exposed to rainbow fungus (*Coriolus versicolor*)', *Journal of Reinforced Plastics and Composites*, vol. 29, no. 7, pp. 1028–1037.

Onésippe, C., Passe-Coutrin, N., Toro, F., Delvasto, S., Bilba, K. and Arsène, M. 2010, 'Sugar cane bagasse fibres reinforced cement composites: Thermal considerations', *Composites Part A: Applied Science and Manufacturing*, vol. 41, no. 4, pp. 549–556.

Osarenmwinda, J.O. and Abode, S.I. 2010, 'Potential of carbonized bagasse filler in rubber products', *Journal of Emerging Trends in Engineering and Applied Sciences*, vol. 1, no. 2, pp. 157–160.

Ramaraj, B. 2007, 'Mechanical and thermal properties of polypropylene/sugarcane bagasse composites', *Journal of Applied Polymer Science*, vol. 103, no. 6, pp. 3827–3832.

Rao, M., Parwate, A. and Bhole, A. 2002, 'Removal of Cr^{6+} and Ni^{2+} from aqueous solution using bagasse and fly ash', *Waste Management*, vol. 22, no. 7, pp. 821–830.

Rodrigues, E., Maia, T. and Mulinari, D. 2011, 'Tensile strength of polyester resin reinforced sugarcane bagasse fibres modified by estherification', *Procedia Engineering*, vol. 10, pp. 2348–2352.

Saini, G., Narula, A., Choudhary, V. and Bhardwaj, R. 2010, 'Effect of particle size and alkali treatment of sugarcane bagasse on thermal, mechanical, and morphological properties of PVC–bagasse composites', *Journal of Reinforced Plastics and Composites*, vol. 29, no. 5, pp. 731–740.

Satyanarayana, K., Guimarães, J. and Wypych, F. 2007, 'Studies on lignocellulosic fibres of Brazil. Part I: Source, production, morphology, properties and applications', *Composites Part A: Applied Science and Manufacturing*, vol. 38, no. 7, pp. 1694–1709.

Shibata, S., Cao, Y. and Fukumoto, I. 2005, 'Effect of bagasse fibre on the flexural properties of biodegradable composites', *Polymer Composites*, vol. 26, no. 5, pp. 689–694.

Stael, G., Tavares, M. and d'Almeida, J. 2001, 'Impact behavior of sugarcane bagasse waste–EVA composites', *Polymer Testing*, vol. 20, no. 8, pp. 869–872.

Tewari, M., Singh, V., Gope, P. and Chaudhary, A.K. 2012, 'Evaluation of mechanical properties of bagasse–glass fibre reinforced composite', *Journal of Materials and Environmental Science*, vol. 3, no. 1, pp. 171–184.

Trindade, W., Hoareau, W., Megiatto, J., Razera, I., Castellan, A. and Frollini, E. 2005, 'Thermoset phenolic matrices reinforced with unmodified and surface-grafted furfuryl alcohol sugar cane bagasse and curaua fibres: properties of fibres and composites', *Biomacromolecules*, vol. 6, no. 5, pp. 2485–2496.

Vallejos, M.A., Curvelo, A.S., Teixeira, E.M., Mendes, F.M., Carvalho, A.J., Felissia, F.E. and Area, M.C. 2011. 'Composite materials of thermoplastic starch and fibers from the ethanol-water fractionation of bagasse', *Industiral Crops and Products*, vol. 33, pp. 739–746.

Vazquez, A., Dominguez, V. and Kenny, J. 1999, 'Bagasse fibre–polypropylene based composites', *Journal of Thermoplastic Composite Materials*, vol. 12, no. 6, pp. 477–497.

Verma, D., Gope, P.C., Maheshwari, M.K. and Sharma, R. 2012, 'Bagasse fibre composites – a review', *Journal of Materials and Environmental Science*, vol. 3, pp. 1079–1092.

Vilay, V., Mariatti, M., Mat Taib, R. and Todo, M. 2008, 'Effect of fibre surface treatment and fibre loading on the properties of bagasse fibre–reinforced unsaturated polyester composites', *Composites Science and Technology*, vol. 68, no. 3, pp. 631–638.

Wirawan, R., Sapuan, S.M., Yunus, R. and Abdan, K. 2011, 'Properties of sugarcane bagasse/poly(vinyl chloride) composites after various treatments', *Journal of Composite Materials*, vol. 45, no. 16, pp. 1667–1674.

Xu, Y., Wu, Q., Lei, Y. and Yao, F. 2010, 'Creep behaviour of bagasse fibre reinforced polymer composites', *Bioresource technology*, vol. 101, no. 9, pp. 3280–3286.

Youssef, H., Ismail, M., Ali, M. and Zahran, A. 2009, 'Studies on sugarcane bagasse fibre – Thermoplastics composites', *Journal of Elastomers and Plastics*, vol. 41, no. 3, pp. 245–262.

Part V
Wood, cellulosic and other fibers

18
Isolation and application of cellulosic fibres in composites

R. A. SHANKS, RMIT University, Australia

DOI: 10.1533/9781782421276.5.553

Abstract: Cellulose fibres have high modulus and outstanding strength and are the most abundant organic material. This chapter outlines cellulose resources from plants to products. Cellulose sources are summarised and separation of fibres from plant materials, purification of fibres by extraction, bleaching and hydrolysis are reviewed. Techniques for combining fibres by weaving or felting, and short fibre dispersion are described. Dissolving and reconstitution is preferred for shaping, though more processable derivatives such as esters and ethers are found in many applications. Most cellulose materials are composites, including plasticisers, surface modifiers, coupling agents and lubricants. The chapter concludes with information sources.

Key words: biocomposite, renewable material, cellulose purification, cellulose chemistry, sources, fibre processing.

18.1 Introduction

Cellulose is the skeletal material of plants. Plant structures are natural composites of cellulose fibres in conjunction with lignin, pectin and hemicelluloses. Cellulose is widely used as timber and in derivatised form in plywood, particle-board and medium-density fibreboard (MDF), packaging materials and as a filler in many thermoplastics and thermosets. In these products cellulose is not separated from the other plant constituents. Recycling of cellulose occurs into generally lower-grade or lower-performance products. Modification of cellulose is carried out by acetylation, methylation, ethylation, nitration, bonding with coupling agents such as silanes, stearic acid–calcium stearate, and lubricants/processing aides. These and other cellulose derivatives can be purified by separation from other plant materials.

Cellulose products assist the environment through carbon dioxide binding and storage during product lifetime. Wood has always been the principal building material for dwellings, boats and many tools and utensils. Cellulose has been used over thousands of years for textiles, e.g. cotton, flax and hemp. A problem with natural fibres is that they displace food crops from fertile land in a highly populated world. However, fibres can be obtained from agricultural waste where food and industrial fibres can coexist. Fibres obtained from fast-growing crops can be used for textiles and as composites in structural materials instead of wood.

Cellulose fibres have amazing chemistry and stereochemistry where puckered ring geometry is arranged into approximately parallel sheets. The closely packed cellulose sheets constitute hard, strong and tough crystalline structures that optimise intermolecular interactions. Equatorial monomer to monomer hydrogen bonds occur through substituent hydroxyl groups. Orientation of the hydroxyl groups provides a combination of intramolecular chain stiffening, intermolecular close packing and crystallisation enhanced by the hydrogen bonds. Though cellulose is highly polar it is only soluble in special solvents and it cannot be melted, thus limiting separation and processing. The cellulose to cellulose interactions are so strong that most solvents cannot compete. This is the challenge in purifying and shaping cellulose by methods typically used for most other polymers.

To meet increasingly demanding environmental regulatory requirements, the automotive, civil engineering and other industries are exploring the opportunity of replacing current synthetic materials used in composite matrices or reinforcement with natural materials. Cellulosic fibres such as flax, ramie and hemp can be used to replace glass fibres as reinforcements in non-structural components in automobiles, for example.

The advantages of using natural materials include their origin from renewable sources, their biodegradability, their relatively low cost of production, low density and adequate tensile properties. Disadvantages include variability in fibre geometry and physical properties that determine mechanical properties, poor interfacial adhesion and incompatibility with hydrophobic matrix resins (Kozlowski *et al.*, 2005; Netravali, 2005). Cellulosic fibres are susceptible to mechanical degradation during processing and vulnerable to moisture and fungal attack that reduce their performance (Schloesser, 2004).

18.2 Types of cellulosic fibre reinforcement and their properties

Cellulose is the principal structural component of plants. A plant cell wall is a natural composite of cellulose combined with lignin, pectin and hemicelluloses, providing strength and rigidity. Cellulosic fibres include cotton, bast fibres (such as flax, hemp, jute and ramie) and wood. Wood, for example, contains 30–40% cellulose, depending on the species.

Cellulose molecules in plants form into fibrils or bundles of molecular chains that combine in groups to form fibres. Different parts of the fibre may have cellulose molecules lying parallel to each other whilst other parts are characterised by a more random molecular arrangement. The parts of

Table 18.1 Mechanical properties of cellulose fibres compared to e-glass, aramid and carbon fibres

Fibre	Density (g/cm^3)	Elongation at break (%)	Tensile strength (MPa)	Young's modulus (GPa)	Specific modulus (GPa cm^3/g)
Flax	1.40–1.50	2.7–3.2	343–1035	27–80	19–53
Jute	1.30–1.50	1.4–3.1	187–773	3–55	2–37
Ramie	1.50	3.6–3.8	400–938	44–128	29–85
Hemp	1.40–1.50	1.4–4.7	580–1110	3–90	2–60
Abaca	1.50	10–12	980	72	48
E-glass	2.50–2.55	2.5	2000–3500	73	29
Aramid	1.40–1.45	3.3–3.7	3000–3150	63–67	45–48
Carbon	1.40–1.75	1.4–1.8	4000	230–500	164–171

Source: after Biagiotti *et al.* (2004).

the fibre where molecules lie side by side are characterised as crystalline. If the molecules, as well as lying side by side, are parallel to the longitudinal axis, there is a high degree of molecular orientation. High orientation and crystallinity usually imply strength, rigidity, low elongation and low pliability. The strength of cellulose fibres is influenced by the degree of polymerisation (X_n) as well as by the molecular arrangement or conformation. The higher the X_n, the stronger the fibre. A typical X_n for a cellulose plant fibre is about 10,000 repeat units. Comparing the resulting properties of cellulosic with synthetic fibres, Biagiotti *et al.* (2004) found that bast fibres have lower tensile strength but comparable specific modulus compared with E-glass (see Table 18.1). The low density of cellulosic fibres gives them a high strength to weight ratio, termed specific strength.

Because of the hydroxyl (–OH) groups in cellulose, cellulose fibres have a high attraction for water. Cellulose swells in liquids such as alkali and salt solutions, causing plasticisation of cellulose. Swelling has been correlated with liquid molar volume, basicity and polarisability, as well as cellulose crystal size, surface area and whether chains are parallel or anti-parallel (Fidale *et al.*, 2008). Microorganism growth is encouraged by the equilibrium water content of cellulose, which is typically 10 wt% under typical ambient humidity. Cellulose can therefore be vulnerable to fungal attack.

Cellulose is not stable at temperatures over 200°C. Above this temperature cellulose loses water and becomes brittle and the fibre structure degrades. Water is lost progressively from free water, then bound water, then finally dehydration reactions of the anhydroglucopyranose repeat units. This limits the type of thermosetting polymer that can be used as the matrix phase, as well as the cure temperature if a thermoset is used as the matrix material. Cellulose fibres also tend to burn readily.

Cellulose monomers are linked by acetal groups that are inherently resistant to alkali, though hydrolysed by acid. Cellulose is usually decomposed by acid solutions, especially strong mineral acids, but has excellent resistance to alkaline solutions. Alkali, however, promotes oxidation of cellulose when accompanied by oxidising agents such as chlorine (Shanks, 2014).

These general properties of natural fibres are significantly affected by cultivar, the regions where they are produced, methods of cultivation and the resulting fibre geometry (Schloesser, 2004). This makes the process of selecting appropriate fibres for a particular application more complex. The properties of wood as a cellulose fibre, for example, are influenced by a wide range of factors, ranging from species to whether trees are from plantation forests, regrowth, native or old-growth forests.

Depending on the source, natural fibres can be supplied as long fibre bundles (bast and cotton fibres) or as short staple fibres (such as wood). After suitable processing, these fibres can be converted into various preform shapes (Kim, 2012):

- Fibre mats (non-axial 2D arrangement)
- Unidirectional pre-impregnation sheets (monoaxial 2D arrangement)
- Woven structures (biaxial, triaxial and multiaxial)
- Knit structures (multiaxial 2D arrangement).

The most common types made using natural fibres are fibre mats and woven fabrics. Many polymer matrix composites using natural fibre reinforcements are made using short fibre nonwoven preforms. Fibre mat preforms are constructed in parallel-laid, cross-laid or random-laid (air-laid) webs. The resulting webs are then consolidated by needle punching. The mechanical properties of fibre mats made using natural fibre reinforcements are mostly determined by the fibre volume fraction, followed by fibre length and fibre orientation distribution (Aziz and Ansell, 2005). The maximum fibre volume fraction achievable with randomly orientated short fibre assemblies (nonwoven mats) is around 40% (Shah *et al.*, 2012). Parallel fibre assemblies such as staple fibre rovings and yarns can achieve fibre volume fractions of up to 60%.

18.3 Cultivation and fibre separation processes

Cellulose fibres can be derived from a wide range of sources, e.g. wood or bast fibres such as flax, hemp, ramie, jute, kenaf, coir or sisal. Agricultural cellulosic waste such as palm bunch fibre and sugarcane bagasse is also used as a source of cellulose fibre. Quality problems are reduced with monoculture crops, since the harvested plants are more uniform and can be more easily and cost-effectively processed by purpose-built machinery.

Tree plantations, for example, may be of species such as pine, native trees, and crop forests such as rubber trees that are no longer productive. The advantage of plantations is that they are a monoculture with only one species planted, usually all at the same time. This means that the wood is more consistent in quality and the harvesting can be mechanised for maximum efficiency. However, pests and diseases may be difficult to arrest without diversity of species to decrease susceptibility. Forest plantations also have a long lag between planting and harvesting.

Kelp and other materials that are used for food, provide cellulose waste, or direct materials for extraction of carbohydrate gels and cellulose. Marine plant materials are often rapidly growing and abundant, although an underutilised resource. They offer potential for separation and extraction of many complex carbohydrates. An advantage is that raw materials for composites and other uses can be harvested without occupying productive agricultural land that could be used for food production.

Plankton, bacteria, and vascular plant tissues are carbohydrate sources. Glucose yield enhancement factors confirmed predominance of α-cellulose-poor marine polysaccharides and increased levels of α-cellulose-rich vascular plants (Cowie and Hedges, 1984). Soluble and insoluble dietary fibre content of seven marine algae (sea vegetables) were determined by thermogravimetry. These seaweeds were rich in dietary fibre. Marine algae appear to be suitable sources of fibres with significant chemical, physic-chemical and rheological diversities (Lahaye, 2006).

Bacterial cellulose contains many very fine, nano-fibres. Bacteria can be grown in nutrient tanks rapidly producing fibres for specialised purposes. The fibres are desirable for their uniform and very fine structure. They have less potential as industrial fibres for commodity polymer composites. However, their characteristics make them suitable for high performance applications such as in biomedical products.

Efficiency of native and surface-modified bacterial cellulose nanocrystals as reinforcing agents was studied. Bacterial cellulose enhanced thermal, mechanical and viscoelastic properties of cellulose acetate butyrate (CAB) matrix. Significant improvement of the properties was due to adhesion between the cellulose and CAB (Grunert and Winter, 2002).

Deformation micromechanics of bacterial cellulose (BC) and microfibrillated cellulose (MFC) networks were investigated. Raman spectroscopy was used to evaluate orientation of the fibres and stress transfer from matrix to fibre. The effective moduli of single fibrils of BC and MFC in the networks were estimated to be 79–88 and 29–36 GPa, respectively. The model used to fit the data required use of a negative Poisson ratio (Tanpichai *et al.*, 2012).

After harvesting the fibres must be removed from the extraneous plant materials and further purified or modified prior to use in composites or any other product. Value-added microcrystalline cellulose (MCC), cellulosic nanofibres (CNF) and cellulosic nanocrystals (CNC) can be prepared by further processing using selective hydrolysis of amorphous and less crystalline regions of the structure, leaving the higher-performance crystalline materials.

There are various techniques for separating cellulose fibres from plants:

- Mechanical
- Biological
- Chemical.

A simple and effective means of separating cellulose fibres from plant materials is by mechanical disruption. Mechanical techniques include milling, decortication and steam explosion. Milling is a grinding process while decortication involves shredding by passing between embossed rollers. Steam explosion is less destructive because it relies on force from steam rather than mechanical force. Steam is equilibrated within the fibres under pressure and is absorbed into the plant material. Sudden removal of the pressure causes the water inside plant material to rapidly vaporise, disrupting internal structures and releasing the fibres (Jacquet *et al.*, 2011). Where mechanical disruption is used to release the fibres from the plant fibre matrix, damage may result, such as shortening of long fibres or thermal and chemical degradation due to lack of selectivity in removing lignin, pectin and hemicelluloses from the crystalline cellulose fibre.

Biological processes include retting where the plant material is left in stagnant water and natural microorganisms degrade the plant materials until fibres are released. The fibrous materials are removed from the water, washed and physically separated from the partially degraded waste materials. A more selective method is to immerse the plant materials in specific enzymes such as pectinase. The medium should be buffered to optimum pH and brought to a temperature of about 38°C. The enzymes perform their reactions under these controlled conditions and selectively remove components that are not desired in the resulting cellulose fibres (Ouajai and Shanks, 2006; Duchemin and Staiger, 2009).

Chemical methods include extraction with organic solvents such as acetone and ethanol. Extraction using sodium hydroxide solution is widely used to remove pectins, lignin and some hemicelluloses. Cellulose swells in concentrated alkali, such as 10 wt% sodium hydroxide. These extraction processes can be chemically enhanced by oxidative treatments. As an example sodium chlorite with sodium hydroxide provides an oxidative–alkaline environment to more aggressively remove non-cellulosic components. An alternative method uses nitric acid and acetic acid to provide an oxidative acidic environment for cellulose purification from wood sources. After treatment the acids are extracted from the cellulose with ethanol, water and acetone (Brendel *et al.*, 2000; Patrício *et al.*, 2013).

Acid hydrolysis of cellulose fibres depends upon adsorption of the acid into the materials. Thus amorphous structures such as pectins, hemicelluloses and some cellulose are progressively hydrolysed into soluble low molar mass sugars. As hydrolysis progresses the remaining cellulose crystals will be more perfect, though the amount of cellulose remaining will be less and the cellulose fibres or crystals will become smaller (Morán *et al.*, 2008). The remaining cellulose fibres can be separated from the soluble hydrolysis products by filtration, dialysis or centrifugation.

Cellulose can be chemically bound to synthetic polymers by grafting to or from the cellulose chain. The grafting may be used to increase the compatibility of cellulose with a matrix polymer or to modify the properties of the cellulose, particularly water sorption. Polypropylene–cellulose composites have been compatibilised with maleic anhydride grafted polypropylene. Bonding is then formed between the maleic anhydride residues on polypropylene (Ouajai *et al.*, 2004).

18.4 Fibre processing

There are two main methods of processing cellulose into fibres:

- Dispersion in a polymer
- Dissolving the cellulose and reconstituting it as a fibre or other shape.

Fibre surfaces need purification primarily through solvent extraction using acetone or ethanol to remove waxes. This may be followed by oxidative treatment for surface etching. Physical surface treatments include additives such as stearic acid, dialkyltitanates that facilitate wetting and lubrication during processing. Reactive surface treatments, often referred to as coupling agents, consist of siloxanes with amino propyl trimethoxy siloxane being widely used. Graft polymerisation using acrylates, styrene and acrylonitrile has also been used to modify cellulose surfaces. Nonionic surfactants can be used such as poly(ethylene oxide) (PEO) and related substances such as poly(ethylene glycol-*co*-propylene glycol) (PEG). Other plasticisers include triethyl citrate and glyceryl triacetate (Park *et al.*, 2006).

Plasticisers can improve composite performance and service life. An example of processing fibres is natural fibre–biopolymer composites prepared from flax and poly(hydroxybutyrate) (PHB) or poly(lactic acid) (PLA). The flax was vacuum dried, followed by absorption of a plasticiser such as tributyl citrate, triethyl citrate, glyceryl triacetate or other glycerides, replacing the lost water to prevent embrittlement (Wong *et al.*, 2002, 2003). Many materials are exposed to non-critical continuous stress where they may gradually deform over time and fluctuations in temperature, i.e. creep.

Comparative interpretation of creep behaviour between composites with different additives has been used to identify factors that determine creep. Thiodiphenol-formed hydrogen bonds at the interface of polymer and fibre decreased creep in flax–poly(hydroxybutyrate) composites. Creep increased when tributyl citrate plasticiser was added to these same composites. Thiodiphenol reduced the viscous flow component of creep while increasing the elastic and viscoelastic moduli. Alternatively, addition of a hyperbranched polyester increased permanent strain and reduced recoverable strain. The hyperbranched polyester did not introduce chain entanglements due to its compact molecular shape (Wong and Shanks, 2008).

Biocides need to be added to any cellulose materials to prevent the growth of microorganisms that degrade cellulose. Cationic antimicrobial polymers function similarly to membrane-active antimicrobial magainin peptides. Biological activity can be obtained from either biocidal side-groups or polymers that release biocidal substances (Siedenbiedel and Tiller, 2012). Fibre-forming polymers, that are *N*-halamine precursors demonstrate biocidal activity. *N*-halamines are compounds with halogen bonded to nitrogen, such as those formed from imides, aminos and amines by chlorination. Poly(*m*-phenylene isophthalamide) (Nomex) can form *N*-halamine by reaction with chlorine by soaking the fibres in sodium hypochlorite solution (Lee *et al.*, 2007).

Organic composite materials such as cellulose are a potential fire hazard. Cellulose can be combined with various phosphates and polyphosphates, such as melamine phosphate. Other fire-retardant additives can be dispersed into the composites along with the polymer matrix. An example is ammonium polyphosphate (APP) which can be mixed in aqueous dispersion (Fox *et al.*, 2012). Nanoparticles such as montmorillonite clay and microcrystalline cellulose enhance thermal stability and flame retardance (Shanks and Genovese, 2009). A metal hydroxide fire retardant, magnesium hydroxide or aluminium hydroxide, can significantly reduce the heat release rate through absorption of heat during conversion to its metal oxide. Another type of fire retardant is zinc borates which undergo dehydration and vitrification or glass formation with increasing temperature (Genovese and Shanks, 2007).

Flame retardant urea formaldehyde board made from cellulose fibres was investigated with flame-retardants including boric acid and borax that were mixed with the cellulose fibres (Nagieb *et al.*, 2011).

There are several ways of integrating cellulosic fibres into a matrix. Short fibres are typically co-mingled and dispersed into a liquid resin that is then cured to form a flat sheet. Fibres dispersed in thermosetting resins can be applied by a spray gun. The resin should include a thixotropic agent such as fumed silica or microcrystalline cellulose to restrict flow or draping between spraying and gelation. The fibres can be partially dissolved in an ionic liquid for example, such as 1-butyl-3-methylimidazolium chloride (Shibata *et al.*, 2013) or N-methylmorpholine-N-oxide (Ouajai and Shanks, 2009) and the dissolved cellulose is precipitated to form the matrix phase of the composite.

Short cellulose fibres can be dispersed into thermoplastics using an internal batch mixer or various types of extruder, such as a twin-screw extruder with customised screw elements. Many thermoplastics will require a compatibiliser additive or pre-treatment for cellulose fibres. Thermoplastics suited to cellulose fibres are those with processing temperatures less than 200°C to avoid cellulose degradation. Polymers include polypropylene, polyethylene, poly(lactic acid) and other bio-polyesters. Maleated forms of the same polymers have been frequently used as compatibilisers. The polymers have moderate mechanical properties, making cellulose fibres more than adequate for reinforcement. They are often used for lower-cost, less technically demanding applications.

Extruded and compression-moulded sections of thermoplastic–cellulose fibre composites can be shaped using heat and pressure or vacuum to conform with end-product requirements. This reshaping of composites using bast fibres is easier than with wood panels or sections where steam, heat and pressure are required. Thermoset cellulose fibre–resin composites cannot be thermally reshaped. A typical composite sheet will consist of polymer-dispersed fibres with continuous polymer phase surfaces. An alternative composite is where the fibres are exposed on one surface. This is known as a carpet and the continuous phase surface is called the carpet backing. The carpet style composite is formed from woven or felted mats impregnated or interspersed with thermoplastics. These carpets can be thermoformed in the same way as a full composite. Carpet thermoforming is used mainly for automotive fitting.

Pre-dispersed and formulated thermoplastic–cellulose fibre compounds are converted into integral product parts that can be three-dimensional using injection moulding. Short fibre composites are suitable for this process, though fibre length reduction, and thus reduced aspect ratio, is a problem due to shear during the moulding process. In general, short cellulose fibres still retain a suitable aspect ratio for reinforcement. Cellulose has high modulus and strength together with well-defined functionality for compatibilisation or reaction with a coupling agent.

18.5 Assessing performance

Material selection for fibre reinforcements is typically based on testing mechanical properties such as tensile strength, modulus, elongation and impact strength. In practice, composites are rarely subjected to a single steady deformation in a stable environment. Since the published values of the mechanical properties of composites are generated from laboratory tests conducted under standard conditions, a thorough understanding of mechanical properties, test conditions and the effect of a range of environmental factors on mechanical properties over time is critical to predict the performance and service life of a component (Kim, 2012).

Tensile tests are prescribed in ASTM D3039, EN 61 and ISO 3268, setting out key parameters such as specimen preparation, dimensions, conditioning and environmental conditions such as temperature and humidity. The most common test of mechanical performance is a stress–strain test. This can determine tensile strength (ultimate strength), tensile modulus (Young's or elastic modulus) and elongation (ultimate strain). Key test conditions include specimen preparation and storage prior to the test, prevailing temperature and strain rate. This method can be used to test the tensile properties of the composite, particularly short fibre composites where the matrix polymer is under load and transfers the stress to the fibres. If the fibres are long compared with the specimen length, then it is predominantly the strength of the fibres that is being tested. In composites with oriented fibres, the tension can be applied perpendicular to the fibres so fibre–matrix adhesion will be tested. Single cantilever, dual cantilever or three-point bend modes test both the polymer matrix and the fibres together.

Flexural strength is the ability of a material to withstand bending forces perpendicular to its longitudinal axis. The resulting stresses are a combination of compressive and tensile stresses. If a composite component is a beam subject to bending, a flexural test is more appropriate. Test methods are reported in ASTM D790 and ISO 178. The two main methods are method 1 (three-point bending) and method 2 (four-point bending). Repetitive application of flexing is used to measure fatigue failure.

Dynamic mechanical analysis (DMA) and thermomechanical analysis (TMA) are a group of techniques where a controlled temperature environment is used, with generally small test specimens and deformations. Small deformations mean that a material can be tested with minimal damage to its structure or morphology. These tests are able to measure stress transfer across the interface to assess polymer–filler interaction or adhesion. DMA is able to distinguish between elastic and viscoelastic responses. Elastic responses are immediate and reversible, where strain is in-phase with the stress. Viscoelastic responses are time-dependent and reversible, where strain lags behind stress and the energy loss is dissipated or damped in the material. Natural fibre composites tend to be energy damping, hence their ability to be thermal and sound insulators.

An important requirement for all materials is impact resistance, especially in such applications as automotive composites. Test methods are set out in ASTM D256 and EN ISO 180: 2000. Standard impact tests include Izod and Charpy tests (ASTM D256) where a swinging pendulum of adjustable mass causes impact fracture and the energy loss to the pendulum momentum is taken as a measure of the impact resistance. Another test geometry is impact by a falling or propelled dart whose momentum is retarded by the impact and fractures resulting. The ability to resist impact can be measured by other techniques such as the fracture mechanics approach using the critical stress intensity factor (K_{Ic}) as a measure of resistance to crack propagation through a material. Another indirect measure is to use the damping factor from DMA as a measure of the ability of the material to absorb applied stress.

Impact resistance depends upon rapid transfer of stress from matrix to fibres. Long fibres, or those with high aspect ratio, are more effective in resisting impacts. Entangled or felted and woven fibres will be more effective than independently dispersed fibres. This is partly due to felted fibres crossing the mat and providing reinforcement to the thickness dimension, a process often called z-knitting. Impact strength or toughness usually increases in inverse proportion to tensile strength, though two-phase materials may combine both properties as the stress is transferred to a much more resistant fibre network, limiting the need for both toughness and strength in the polymer. Flax–PLA composites were found to be toughened by blending PLA with hyperbranched polyesters. The main contributor to high fracture toughness was enhanced wetting of the fibres by the matrix when the hyperbranched polyesters were present. This increased matrix ductility and caused ductile tearing along the fibre–matrix interface during crack propagation (Wong *et al.*, 2004).

Interfacial shear strength (IFSS) is evaluated by single-fibre pull-out tests. Scanning electron microscopy images of fracture surfaces can be used to confirm test results. It has been found that additives can significantly increase the IFSS of composites of PLA or poly(hydroxybutyrate) and flax fibre (Wong *et al.*, 2007). Long-term properties such as abrasion resistance can be tested by continuous rubbing with an abrading substance.

Cellulose-containing materials are moisture absorbing. The emergence of natural fibre and biopolymer materials has led to accessory humidity generators for DMA and TMA instruments becoming available. Humidity scans are parallel to temperature or time scans. The rate of change in humidity must be slow compared with the rate of change in temperature due to slow equilibration with the test specimen. Other tests can be conducted with the specimen stored in water and/or immersed in water throughout the test.

Tests of flammability include the cone calorimeter (ISO 5660/ASTM E1354), the flame spread index (ASTM E162) and the smoke chamber test (ASTM E662). The most common laboratory tests are the limiting oxygen index (LOI) test (ASTM D2863), the vertical burn test (ASTM D568 and D3801) and the horizontal burn test (ASTM D635). The LOI test is the most accurate laboratory test, determining the minimum oxygen content required to sustain burning. A high LOI index is associated with lower flammability.

18.6 Applications

After the civil engineering sector, the automotive industry was one of the first to use long cellulose fibres in composites. The lower density of cellulosic fibres in comparison with glass and other mineral fibres helps to reduce vehicle weight and therefore fuel consumption. Cellulosic fibres have other advantages such as good thermal and acoustic properties (Miao and Finn, 2008). Puglia *et al.* (2004) have reviewed the use of natural fibre composites in automotive components. Bast fibres have been used to produce non-load bearing components such as dashboards, door panels, floor panels, insulation material and trunk/boot components.

Extrusion of thermoplastics reinforced with cellulose fibres is used to produce beams, planks and sheets of many types for building applications. Extruded products may include surface embossing to mimic wood for visual appeal or to provide slip resistance to surfaces intended to be walked upon. Internal structures may be designed to replicate wood grains and textures that will facilitate saw cutting and fastening with nails or screws. The internal structures can be aligned along the extrusion direction and created by partial foaming or extrusion of strands that are pressed together in the die. These internal textures, surface embossing and suitable pigmentation combine to provide products difficult to distinguish from wood. Advantages of a polypropylene–cellulose fibre composite include water resistance, no warping upon external exposure, colour stability and weather resistance.

Wider commercial uses need to be developed for current material developments into cellulose and its derivatives. Successful commercialisation depends upon the new materials being accepted into industry and regulatory standards. The use of nanotechnology in the form of the cellulose and other filler components added to composites, such as hybrid composites of cellulose and nanoparticles in a polymer, will provide new materials with significantly different properties.

Most polymers are produced from oil-derived chemicals. With increased demand for synthetic polymers and their likely decreasing availability and increasing cost, a challenge is to find replacement materials prepared from renewable resources such as cellulose. Few polymers are derived directly from cellulose or related renewable resources. Significant potential exists for the production of polymers from cellulose-derived materials. Many different materials are available by purification and/or derivatisation of cellulose. Sources of cellulose with rapid growth or as waste materials make it a promising material for the future. Currently polymers derived either directly from cellulose processing and regeneration or by derivatisation form the basis of novel polymers and co-polymers (Rose and Palkovits, 2011).

The world population is growing rapidly and placing additional demands for food. Arable land is limited and cropland per person will decrease, resulting in a food crisis. Demand for textiles and other fibre uses is increasing. Cotton production cannot expand significantly because of limited availability of suitable land. Approximately a third of textile fibres consist of cellulose because of its desirable properties such as absorbency and moisture management. Growth in demand for fibres will need to be through synthetic fibres. Properties of some cellulose fibres (cotton, viscose, Tencel, Lyocell and Modal), especially their physiological performance, cannot be replicated by synthetics (polyester, polyamide, polypropylene, and polyacrylonitrile). Cellulosic fibres are likely to provide 33–37% of fibres. Regenerated cellulose fibres are extremely sustainable because they are ecological fibres (Hämmerle, 2011).

18.7 Conclusions

Currently the use of cellulosic fibres in composites is restricted to non-load-bearing applications. Future research needs to focus on such areas as reducing variability and improving predictability of properties of raw fibres through better understanding and control of variables in cultivation, fibre surface treatment to control such problems as humidity/moisture sorption, improving interfacial adhesion for better load transfer (e.g. by continued developments in chemical modification or the use of coupling agents) and better control of fibre orientation for optimum configuration and performance.

18.8 Sources of further information and advice

18.8.1 Books

Research and commercial interest in application of biofibres as reinforcement in composites has increased. This can be construed as a return to traditional materials, since throughout history biofibres have been the basis for textiles and composites. Many books are available, too many to be listed here without emphasising some books unnecessarily.

18.8.2 Journals

There are many journals with emphasis on polymer composites and most polymer journal publish papers on composites. A group of journals highlight biomaterials, bioresources and specific natural materials such as cellulose and starch. A selection of journals can be viewed in the references for this chapter and some of the following journals feature biofibre composites, though the list is not conclusive:

Biological Macromolecules
Biomacromolecules
Biomaterials
Biopolymers
Bioresources
Carbohydrate Polymers
Cellulose
Composite Interfaces
Composites Part A Applied Science and Manufacturing
Composites Part B Engineering
Composites Science and Technology
eXpress Polymer Letters
Food Hydrocolloids
International Journal of Carbohydrate Chemistry
International Journal of Polymer Science
Journal of Applied Polymer Science
Journal of Biomaterial Applications
Journal of Biomaterials Science – Polymer Edition
Journal of Composite Materials
Journal of Reinforced Plastics and Composites
Macromolecular Bioscience
Macromolecular Materials and Engineering
Polymer Composites
Polymers and Polymer Composites
Starch – Starke

18.8.3 Industry standards

Cellulose materials in many forms are produced and marketed in products and forms that meet required standards for measurement of composition, properties and performance. Confidence in these data requires standardisation by international and national standards organisations.

The International Organization for Standardization (ISO) (http://www.iso.org/iso/home.html) has many standards for testing and performance of composites, though not specifically biocomposites. American Standard Test Methods (ASTM) (http://www.astm.org) has many standards for mechanical properties and performance of polymer composites.

The British Standards Institution (BSI) and NCS International (NCSI) (http://www.bsigroup.com/BSIandNCSI), Standards Australia (http://www.standards.org.au), the German: Deutsches Institut für Normung (DIN) (http://www.din.de) and the Canadian Standards Association (CSA) (www.csa.ca) are among and many other national standards organisations.

18.8.4 Patents

Biocomposites with potential commercial application may be patented. Patents describe unique methods for materials modification, specific composite compositions and novel preparation or manufacturing processes. Patents may be searched online at the US Patents Office (www.uspto.gov/patents/), the European Patents Office (http://www.epo.org) and many other national patents offices. Patents provide detailed information on compositions and manufacturing that are not revealed in the scientific literature. They can assist researchers in finding openings in technology and in finding research and development pathways that are blocked by prior claims.

18.8.5 Agricultural, fibre and textile agencies

Forestry and agriculture in most countries are government regulated. Governments and international organisations produce many studies on the management and uses of these resources both as a means of controlling and regulating, and to promote the use of these resources in commerce and trade. Industry associations exist to solve common problems, generically market and standardise product types and present the industry views and policies to the public. There are many such organisations; for instance, forestry organisations in the USA can be found at http://www.forestryusa.com (source 22 July 2013). These organisations are mostly concerned with managing forestry resources; however, an increasing interest is in adding value to resources through conversion of cellulose to microcrystalline cellulose, cellulose nanofibres, cellulose nanocrystals and development of higher-performance polymer composites using these forestry products.

The fibre and textiles industry has controlling bodies to control product specification, recommend suitability for application, assist with imports and exports, and lobby governments to the value and needs of the industry when policies are being formed. Reports are available in various countries or regions that promote use of bioresources. While some applications are in decline, the industry seeks new applications and particularly those with higher value. Some of the materials from forestry and crops are harvested specifically for use in materials, while others are waste materials from agriculture. New applications for waste materials reduce disposal costs and create value from materials that are otherwise a liability.

18.9 References

Aziz, S. H. and Ansell, M. P. (2005). Optimising the properties of green composites, in Baillie, C. (ed.), *Green Composites: Polymer Composites and the Environment*, Woodhead Publishing Limited, Cambridge, UK.

Biagiotti, J., Puglia, D. and Kenny, J. M. (2004). A review of natural fibre-based composites. Part 1: Structure, processing and properties of vegetable fibres, *Journal of Natural Fibres*, **1**(2): 37–68.

Brendel, O., Iannetta, P. P. M. and Stewart, D. (2000). A rapid and simple method to isolate pure alpha-cellulose, *Phytochemical Analysis*, **11**: 7–10.

Cowie, G. L. and Hedges, J. I. (1984). Carbohydrate sources in a coastal marine environment, *Geochimica et Cosmochimica Acta*, **48**: 2075–2087.

Duchemin, B. and Staiger, M. P. (2009). Treatment of Harakeke fiber for biocomposites, *Journal of Applied Polymer Science*, **112**: 2710–2715.

Fidale, L. C., Ruiz, N., Heinze, T. and Seoud, O. A. E. (2008). Cellulose swelling by aprotic and protic solvents: What are the similarities and differences? *Macromolecular Chemistry and Physics*, **209**: 1240–1254.

Fox, D. M., Temburni, S., Novy, M., Flynn, L., Mauro Zammarano, M., Kim, Y. S., Gilman, J. W. and Davis, R. D. (2012). Thermal and burning properties of poly(lactic acid) composites using cellulose-based intumescing flame retardants, in *Fire and Polymers VI: New Advances in Flame Retardant Chemistry and Science, ACS Symposium Series*, **1118**: Chapter 16: 223–234.

Genovese, A. and Shanks, R. A. (2007). Structural and thermal interpretation of the synergy and interactions between the fire retardants magnesium hydroxide and zinc borate, *Polymer Degradation and Stability*, **92**: 2–13.

Grunert, M. and Winter, W. T. (2002). Cellulose nanocrystal reinforced cellulose acetate butyrate nanocomposites, *Polymeric Materials Science and Engineering*, **86**: 367–368.

Hämmerle, F. M. (2011). The cellulose gap (the future of cellulose fibres), *Lenzinger Berichte*, **89**: 12–21.

Jacquet, N., Quivy, N., Vanderghem, C., Janas, S., Blecker, C., Wathelet, B., Devaux, J. and Paquot, M. (2011). Influence of steam explosion on the thermal stability of cellulose fibres, *Polymer Degradation and Stability*, **96**: 1582–1588.

Kim, Y. K. (2012). Natural fibres for automotive applications, in Kozlowski, R. M. (ed.), *Handbook of Natural Fibres. Volume 2: Processing and Applications*, Woodhead Publishing Limited, Cambridge, UK.

Kozlowski, R., Baraniecki, P. and Barriga-Bedoya, J. (2005). Bast fibres (flax, hemp, jute, ramie, kenaf, abaca), in Blackburn, R. S. (ed.), *Biodegradable and Sustainable Fibres*, Woodhead Publishing Limited, Cambridge, UK.

Lahaye, M. (2006). Marine algae as sources of fibres: Determination of soluble and insoluble dietary fibre contents in some 'sea vegetables', *Journal of the Science of Food and Agriculture*, **54**: 587–594.

Lee, J., Broughton, R. M., Worley, S. D. and Huang, T. S. (2007). Antimicrobial polymeric materials; cellulose and m-aramid composite fibers, *Journal of Engineered Fibers and Fabrics*, **4**: 25–32.

Miao, M. and Finn, N. (2008). Conversion of natural fibres into structural composites, *Journal of Textile Engineering*, **54**(6): 165–177.

Morán, J., Alvarez, V., Cyras, V. and Vázquez, A. (2008). Extraction of cellulose and preparation of nanocellulose from sisal fibers, *Cellulose*, **15**: 149–159.

Nagieb, Z. A., Nassar, M. A. and El-Meligy, M. G. (2011). Effect of addition of boric acid and borax on fire-retardant and mechanical properties of urea formaldehyde saw dust composites, *International Journal of Carbohydrate Chemistry*, **2011**: article ID 146763.

Netravali, A. N. (2005). Biodegradable natural fibre composites, in Blackburn, R. S. (ed.), *Biodegradable and Sustainable Fibres*, Woodhead Publishing Limited, Cambridge, UK.

Ouajai, S. and Shanks, R. A. (2006). Solvent and enzyme induced recrystallization of mechanically degraded hemp cellulose, *Cellulose*, **13**: 31–44.

Ouajai, S. and Shanks, R. A. (2009). Preparation, structure and mechanical properties of all-hemp cellulose biocomposites, *Composites Science and Technology*, **69**: 2119–2126.

Ouajai, S., Hodzic, A. and Shanks, R. A. (2004). Morphological and grafting modification of natural cellulose fibers, *Journal of Applied Polymer Science*, **94**: 2456–2465.

Park, H. M., Mohanty, A. K., Drzal, L. T., Lee, E., Mielewski, D. F. and Misra, M. (2006). Effect of sequential mixing and compounding conditions on cellulose acetate/layered silicate nanocomposites, *Journal of Polymers and the Environment*, **14**: 27–35.

Patrício, P. S. de O., Pereira, F. V., dos Santos, M. C., de Souza, P. P., Roa, J. P. B. and Orefice, R. L. (2013). Increasing the elongation at break of polyhydroxybutyrate biopolymer: Effect of cellulose nanowhiskers on mechanical and thermal properties, *Journal of Applied Polymer Science*, **127**: 3613–3621.

Puglia, D., Biagiotti, J. and Kenny, J. M. (2004). A review of natural fibre-based composites. Part 2: Application of natural reinforcements in composite materials for the automotive industry, *Journal of Natural Fibres*, **1**(3): 23–63.

Rose, M. and Palkovits, R. (2011). Cellulose-based sustainable polymers: State of the art and future trends, *Macromolecular Rapid Communications*, **32**: 1299–1311.

Schloesser, T. (2004). Natural fibre reinforced automotive parts, in Wallenberger, F. T. and Weston, N. E. (eds), *Natural Fibres, Plastics and Composites*, Kluwer Academic Publishers, Norwood, MA.

Shah, D. U., Schubel, P. J., Licence, P. and Clifford, M.J. (2012). Determining the minimum, critical and maximum fibre content for twisted yarn reinforced plant fibre composites, *Composites Science and Technology*, **72**: 1909–1917.

Shanks, R. A. (2014). Processing cellulose for cellulose fiber and matrix composites, in Thakur, V. K. (ed.), *Green Composites from Natural Resources*, CRC Press, Boca Raton, USA, CRC Press.

Shanks, R. A. and Genovese, A. (2009). Fire-retardant properties of polymer nanocomposites, in Thomas, S., Zaikov, G. E. and Valsaraj, S. V. (eds), *Recent Advances in Polymer Nano-composites*, Koninklijke Brill NV, Leiden, The Netherlands: 439–454.

Shibata, M., Yamazoe, K., Kuribayashi, M. and Okuyama, Y. (2013). All-wood biocomposites by partial dissolution of wood flour in 1-butyl-3-methylimidazolium chloride, *Journal of Applied Polymer Science*, **127**: 4802–4808.

Siedenbiedel, F. and Tiller, J. C. (2012). Antimicrobial polymers in solution and on surfaces: Overview and functional principles, *Polymers*, **4**: 46–71.

Tanpichai, S., Quero, F., Nogi, M., Yano, H., Young, R. J., Lindstrom, T., Sampson, W. W. and Eichhorn, S. J. (2012). Effective Young's modulus of bacterial and microfibrillated cellulose fibrils in fibrous networks, *Biomacromolecules*, **13**: 1340–1349.

Wong, S. and Shanks, R. A. (2008). Creep behaviour of biopolymers and modified flax fibre composites, *Composite Interfaces*, **15**: 131–145.

Wong, S., Shanks, R. A. and Hodzic, A. (2002). Properties of poly(3-hydroxybutyric acid) composites with flax fibres modified by plasticiser absorption, *Macromolecular Materials and Engineering*, **287**: 647–655.

Wong, S., Shanks, R. A. and Hodzic, A. (2003). Poly(L-lactic acid) composites with flax fibers modified by plasticizer absorption, *Polymer Engineering and Science*, **43**: 1566–1575.

Wong, S., Shanks, R. A. and Hodzic, A. (2004). Mechanical behavior and fracture toughness of poly(L-lactic acid)–natural fiber composites modified with hyperbranched polymers, *Macromolecular Materials and Engineering*, **289**: 447–456.

Wong, S., Shanks, R. A. and Hodzic, A. (2007). Effect of additives on the interfacial strength of poly(L-lactic acid) and poly(3-hydroxy butyric acid)–flax fibre composites, *Composites Science and Technology*, **67**: 2478–2484.

19

The use of biobased nanofibres in composites

S. BANDYOPADHYAY-GHOSH and S. B. GHOSH, Birla Institute of Technology and Science, Pilani, India and University of Toronto, Canada and M. SAIN, University of Toronto, Canada

DOI: 10.1533.9781782421276.5.571

Abstract: Biobased nanoscale fibres are unique with respect to their high aspect ratio, remarkable strength, renewability and biodegradability, which make them excellent candidates as reinforcements in producing nanocomposites. The present chapter reports the various techniques of isolation and surface modification of biobased nanoreinforcements, in particular cellulose nanoreinforcements. Strategies related to integration and efficient dispersion of these engineered nanoscale biofibres into common matrices have been highlighted. The chapter also reviews the properties, applications and fundamental challenges related to biobased nanocomposites and the current research efforts that are being directed towards solving these problems and in widening the potential application areas of these materials.

Key words: biobased nanomaterials, biobased nanoreinforcements, nanocellulose, cellulose nanofibre, cellulose nanoparticle, bio-nanocomposite, surface modification, nanotoxicity.

19.1 Introduction

The words nanomaterials and nanotechnology come from the Greek prefix 'nano' meaning 'dwarf'. In modern scientific parlance, a nanometre is one billionth of a metre, about the diameter of 10 atoms placed side by side. It was on 29 December 1959 that physicist Richard Feynman introduced the ideas and concepts behind nanoscience and nanotechnology at an American Physical Society meeting with a talk entitled 'There's plenty of room at the bottom' [1]. Over a decade later, in his explorations of ultra-precision machining, Professor Norio Taniguchi coined the term 'nanotechnology' [2]. In October 2010, the International Organization for Standardization (ISO) published the *ISO TS 8004* series of standards [3], in which a 'nanomaterial' was defined as 'material with any external dimension in the size range of approximately 1 nm to 100 nm (nanoscale) or having internal structure or surface structure in the nanoscale'. However, in that standard [3], it was also acknowledged that nanomaterials may not abruptly end at dimensions of

100 nm and that the actual size range of nanomaterials may fall outside the precise boundaries.

Today's scientists and engineers are finding a wide variety of ways to deliberately make materials at the nanoscale to take advantage of their enhanced properties such as higher strength, lighter weight, etc. Materials engineered at the nanoscale often exhibit novel or improved chemical, thermal, mechanical, electronic and biological properties when compared with the same materials in bulk form. In addition, entirely new classes of materials have been created at the nanoscale, offering unique properties not otherwise achievable. Combining these unique properties with their remarkable recognition capabilities has resulted in systems with significantly improved performance and novel applications. An important use of nanomaterials is in composites, materials that combine one or more separate components and which are designed to exhibit overall the best properties of each component. This multi-functionality applies not only to mechanical properties, but extends to optical, electrical and magnetic ones. Like conventional composites, nanocomposites too consist of a bulk material, known as the 'matrix', and nanoscale reinforcement material. Recent progress in the field of nanoreinforcements has shown the potential to advance composite properties to a high level of performance. Polymer nanocomposites, in particular, represent a new alternative to conventional polymer matrix composites (PMCs). Because of their nanometre sizes, nanoreinforcement-based composites exhibit markedly improved properties when compared to the pure polymers or their traditional composites, and often over that expected from continuum mechanics predictions [4–6]. This occurs because nano-size reinforcements have much higher surface area compared to conventional reinforcements, and the number of defects might be reduced at the nano level [7].

19.2 Biobased nanoreinforcements

Biobased reinforcement materials have attracted significant attention during the past decade, mainly due to ecological and climatic factors along with their superior strength, low density, non-abrasiveness, low cost, and biodegradable properties [8]. However, biodegradable reinforcement materials that are produced from naturally occurring substances lack certain characteristics when compared to synthetic reinforcements. The use of biobased nanoscale reinforcements allows the avoidance of such drawbacks. The production of biobased nanoreinforcements and their application in composite materials have gained increasing attention due to their high strength and stiffness combined with low weight, biodegradability and renewability. The interfacial area of the reinforcing particles is very large due to the small size of the particles, and the reinforcing particle will thus to a great extent interact with the polymer, further enhancing the effectiveness of the reinforcement [9–13]. Depending on the chemical nature of the natural building blocks, biobased nanoreinforcements can be categorized into the following groups.

19.2.1 Cellulose nanoreinforcements

Cellulose is the major component of wood and most natural fibres such as cotton, flax, hemp, jute, ramie and sisal. This natural polymer represents about a third of the plant tissues and can regenerate through photosynthesis. Apart from plant-derived cellulose, cellulose is also produced by a family of sea animals called tunicates (sea squirts), by several species of algae, and by some species of bacteria, amoebae and fungi [14]. Nanoscale cellulosic elements have attracted a great deal of interest as reinforcement [9–13, 15–22]. The reinforcing potential of cellulose in the nanoscale dimension combines the benefits of biobased materials and nanotechnology in a synergistic manner. By extracting cellulose at the nanoscale, the majority of the defects associated with the hierarchical structure can be removed, and a new cellulose-based 'building block' becomes available for the next generation of cellulose-based composites. Cellulose nanoreinforcements (CNs) have high aspect ratio, low density (around $1.6\,g/cm^3$), very low coefficient of thermal expansion (CTE) and a reactive surface of –OH side groups that facilitates grafting chemical species to achieve different surface properties (surface functionalization) [23–25].

19.2.2 Polysaccharide-based nanoreinforcements

Nanosized starch crystals obtained by hydrolysis of native starch granules have been used as reinforcing fillers in glycerol plasticized starch (GPS) matrix [26, 27]. The nanocrystals consist of platelet-like particles with a thickness of 6–8 nm, a length of 40–60 nm, a width of 15–30 nm and a density of $1.55\,g/cm^3$. Another method of preparation of starch nanoparticles has been reported by Ma *et al.* [28] using ethanol and citric acid. In another study, pea starch nanocrystals (StNs) were incorporated into a soy protein isolate (SPI) matrix to produce a class of fully biodegradable nanocomposites [29]. The StN with low loading level (2 wt%) showed a predominant reinforcing function, resulting in an enhancement in strength and Young's modulus [29]. The water uptake behaviour of the nanocomposites was found to be similar to that of neat SPI material and was attributed mainly to the strong interfacial interaction between the StN filler and the SPI matrix.

Chitin (also known as 'animal cellulose') is the most abundant source of polysaccharides found in nature [30–32]. Chitins are the main components in crab and shrimp shells, in the outer skins or cuticles of other arthropods, and in the molluscan shells of squid. They are chemically identical to cellulose except that the secondary hydroxyl group on the carbon atom of the cellulose molecule is replaced by an acetoamide group. Iftekhar *et al.* [33] reported extraction of chitin nanofibres having a uniform width of 20–30 nm from crab shells using a simple mechanical technique under acidic conditions. The nanofibres were small enough to retain the transparency of neat acrylic resin. Interestingly, chitin acrylic resin nanocomposite exhibited much higher transparency than cellulose acrylic resin films owing to the close affinity between less hydrophilic chitin and hydrophobic resin. Furthermore, the authors also reported that the incorporation of chitin nanofibres contributed to the significant improvement of the thermal expansion and mechanical properties of the neat acrylic resin [33].

19.2.3 Protein-based nanoreinforcements

Scientists at the National Center for Agricultural Utilization Research in Peoria, IL, USA , have developed processes to produce soy protein-based nanoparticles [34] with potential uses in the reinforcement of rubber/plastic materials and as adsorbing/releasing carrier for chemicals. These soy protein nanoparticles were produced with a microfluidizer and characterized in terms of particle size, size distribution, morphology, rheological properties, and aggregate structure. Wheat proteins were also used in forming protein-based nanoparticles. These nanoparticles strongly adhered onto the surface of hydrophobic materials.

19.2.4 Lignin-based nanoreinforcements

Various synthetic carbon precursors such as graphite powders, petroleum pitch, carbon-rich synthetic polymers and other kinds of liquid/gaseous hydrocarbons have been extensively used for the fabrication of carbon nanoparticles [35, 36]. However, there is a need for alternative carbon sources for the synthesis of carbonaceous materials due to increasing oil prices, depleting petroleum resources, their negative environmental impacts, and increasing demand for bioresource-based carbon nanomaterials. Hence, renewable carbon resources have been explored for the fabrication of carbon nanostructures [37, 38].

Among the various renewable precursors, lignin receives great attention [39, 40]. Lignin, which is obtained as a co-product of pulp and second-generation cellulosic ethanol industries, is released as industrial waste discharge, or burned as fuel, or for the production of some low-value chemicals. Lignin is a major component of plant fibre raw materials in nature reserves and is second only to cellulose. It is a three-dimensional network structure having a natural aromatic polymer. Since the carbon content of lignin is high (generally between 55% and 66%), it can be used as a carbon precursor Material. As a result, lignin has shown the potential to open up a new use as raw material of carbon nanofibre. However, since the industry-average relative molecular mass of lignin is low and it is composed of relatively rigid molecules, its fibre-forming performance is poor. In one attempt, Lallave *et al.* [41] reported use of ethanol lignin at a mass ratio of 1:1 and of coaxial electrospinning technology to generate lignin nanofibre, of diameters ranging between 400 nm and ~2 μm. This fibre was subjected to stabilizing treatment at 200°C for 24 h, and then carbonized at 900°C to obtain nanoscale carbon fibre. In another invention, lignin-based carbon nanofibres were prepared by following five steps such as lignin pretreatment, preparation of spinning solution, electrostatic spinning, pre-oxidation and carbonization [42]. Pretreatment was carried out on alkali lignin and acetic acid lignin (or formic acid lignin) to remove carbohydrate and inorganic salt impurities. As the obtained lignin had good spinnability, the nanofibre could be spun and further processed into the carbon nanofibre without involving synthetic macromolecules. The electrostatic spinning was carried out by maintaining a distance of 35 cm between the grounded collection plate, with a spinneret diameter of 0.6 mm, an electrostatic generator voltage of 50 kV, and a solution feed rate of 200 μL/h. Lignin nanofibres were then pre-oxidized in air by heating them from room temperature to 300°C at the heating rate of 20°C/min and incubating at this temperature for 2 h. Pre-oxidized lignin nanofibres were heated in a nitrogen atmosphere at 15°C/min heating rate for 2 h up to a temperature of 900°C to obtain lignin-based nanocarbon fibre. The diameter of the obtained carbon nanofibre was within 50 to 300 nm. Lignin nanoparticles (50–500 nm) have also been synthesized from lignins obtained from wheat straw and grass by hydroxymethylation or by epoxidation reaction.

Among the different types of biobased nanoreinforcement, nanoscale cellulosic elements from plant-based lignocellulosic materials is one of the most studied topics in the literature today and unarguably the most used biobased nanomaterials in composite application. In the following sections, we will therefore focus primarily on the use of plant-based cellulose nanoreinforcements and the terms biobased nanoreinforcements, cellulose nanoreinforcements, nanocellulose and cellulose nanofibre will be used interchangeably throughout the chapter.

19.3 Ultrastructure of cellulose nanoreinforcements

19.3.1 Chemical structure

Cellulose is a linear fibrous, semicrystalline, polydispersed linear homopolysaccharide which consists of glucose (d-glucopyranose) linked by

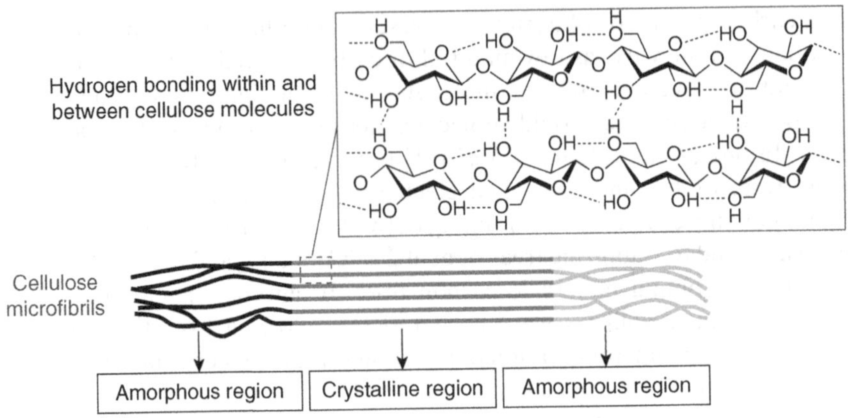

19.1 Scheme of interaction between cellulose molecular chains within the crystalline region of cellulose microfibrils (adapted from [45]).

glycosidic β(1-4) linkages (Fig. 19.1). In this connection, the hydroxyl at C-1 (anomeric carbon) of one molecule of glucose is in the β conformation and reacts with the hydroxyl at C-4 of the second molecule. As a result of the thermodynamically preferred 4C1 conformational orientation of its glucan units, cellulose is an extended, linear-chain polymer with a large number of hydroxyl groups (three per anhydroglucose unit) [43]. The non-reducing end of a cellulose chain terminates with a C4–OH group, while the other reducing end terminates with a C1–OH group. Hydroxyls present in the structure of cellulose are distributed equatorially in a quasi-planar arrangement that allows the formation of a linear chain, in which adjacent chains are aligned to form crystalline structures stabilized by hydrogen bonds inter- and intramolecularly [44].

19.3.2 Crystalline structure

The crystalline structure of cellulose I (native cellulose) is present in two different forms, namely one-chain triclinic structure I_α and two-chain monoclinic structure I_β whose ratio depends on the origin of the cellulose. In addition to the thermodynamically less stable cellulose I, cellulose may occur in other different polymorphs (cellulose II, IIII, IIIII, IV_I and IV_{II}) with the possibility of conversion from one form to another, of which cellulose II is the most stable structure owing to the anti-parallel orientation of cellulose chains. Alkali (NaOH) treatment or dissolution of cellulose and subsequent precipitation converts cellulose I into cellulose II and forms the basis of structures of technical and commercial interest. The amorphous regions act as structural defects and are responsible for the transverse cleavage of the microfibrils into short monocrystals, or whiskers [45]. The pore structure also has considerable importance as it influences the accessibility of reactants and enzymatic biodegradation, and by controlling the variation of pore structure it is possible to tailor the properties for specific applications.

19.3.3 Hierarchical structure

Cellulose biosynthesis is a multistep process that is highly specific to the organism producing the cellulose. Variations in this natural process dictate the morphology, aspect ratio, crystallinity and crystal structure of the resulting microfibril. As shown in Fig. 19.2, the first stage of cellulose biosynthesis from natural plant resources involves the self-assembly of the cellulose chains producing a 'minisheet' of ordered cellulose chains. The second stage is the self-assembly of these cellulose 'minisheets'. The final stage involves the assembly of cellulose fibrils into either micro- or macrofibrils, depending on the organism (plant, alga, tunicate, bacterium). Each microfibril can be considered as a string of crystallites linked along the microfibril by amorphous domains and having a modulus close to that

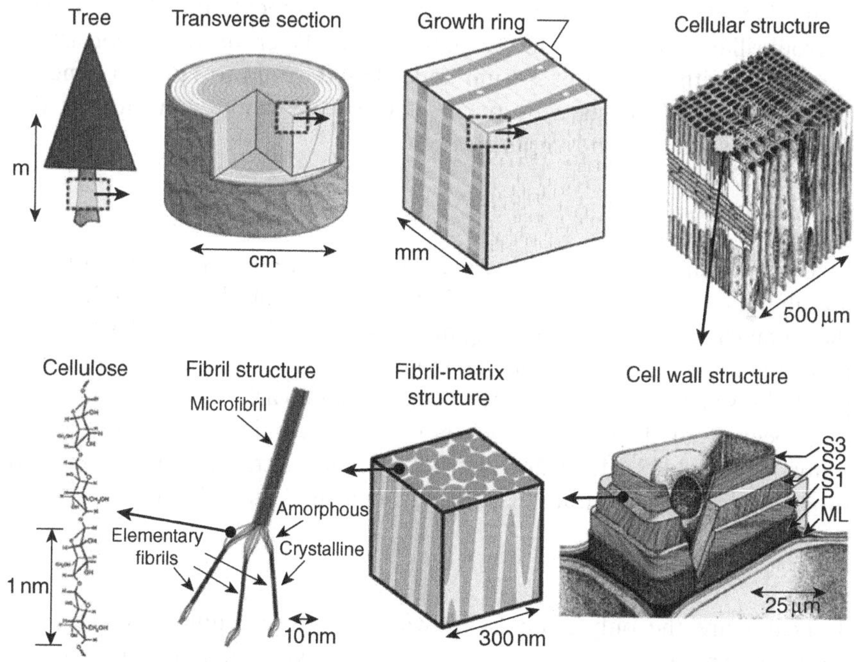

19.2 Schematic of the hierarchical structure from tree to the CNs. ML = middle lamellae, P = primary cell wall, S1, S2, S3 = cell wall layers (adapted from [47]).

of the perfect crystal of native cellulose (~150 GPa) and a strength that should be in the order of 10 GPa [46]. The 'reverse engineering' of cellulose hierarchical morphology with a top-down approach, along with a deeper understanding of cellulose structure, forms the basis of producing nanoscale cellulose reinforcements from various sources.

19.4 Source materials for cellulose nanoreinforcements

19.4.1 Wood

Wood as a cellulose source material has many advantages. Most significantly, it is abundant. The annual production of cellulose in nature is about 1 trillion tons. In addition, an existing infrastructure is in place for harvesting, processing and handling, i.e., the pulp and paper, packaging and pharmaceutical industries. Cellulose chains are biosynthesized by enzymes and are deposited in a continuous fashion primarily within the cell walls. The cell wall comprises primary, secondary and tertiary layers. The secondary layer is further divided into S1, S2 and S3 layers. The S2 layer is the predominant one and contains bundles of microfibrils running parallel to each other and oriented at an acute angle with respect to the fibre axis. The microfibrillar bundles in the secondary (S2) layer of wood consist of nanofibrils with a width of 3–5 nm and a length of up to 1 μm. In general, extraction of CNs starts with 'purified' wood, i.e. wood with most of the lignin, hemicellulose and impurities removed.

19.4.2 Plants

Like wood, plants are an attractive cellulose source primarily because they are abundant. Cellulose is present in the cell wall of plant fibres. Besides, the technical knowhow of isolating micron-sized cellulose particles through retting/pulping and product processing is already established. The length of plant fibres varies from species to species. A wide variety of plant materials have been studied for the extraction of CNs, including cotton [32, 48–52], ramie [53], sisal [32, 51, 54–56], flax [13, 51], wheat straw [50, 57–59], potato tubers [46], sugar beet pulp [32, 48], soybean stock [60–62], etc.

19.4.3 Animals

Tunicates are the only animals known to produce cellulose microfibrils. Tunicates are a family of sea animals that have a mantle consisting of cellulose microfibrils embedded in a protein matrix. Most researchers have used a class of tunicates that are commonly known as 'sea squirts' (Ascidiacea).

19.4.4 Bacteria

Bacterial cellulose (BC) is a three-dimensional reticular network of fine fibrils (2–4 nm in diameter) coalesced into a ribbon (~100 nm in width, 1–9 μm in length). The most studied species of bacteria that produces cellulose is generally called *Gluconacetobacter xylinus* (reclassified from *Acetobacter xylinum*) [63]. Under special culturing conditions the bacteria secrete cellulose microfibrils, producing a thick gel composed of cellulose microfibrils and ~97% water, called pellicle, on the surface of the liquid medium [64]. Because of the extensive hydrogen bonding within the reticulated network, it can hold up to 700 times its dry weight in water. The crystallinity index is above 70% and a separate fibril can have a theoretical Young's modulus of 173 GPa and tensile strength in the order of 2 GPa [65]. These unique properties emphasize the potential of BC as a nanoreinforcement material for use in composites [65–68]. The advantage of cellulose microfibrils derived from bacteria is that it is possible to adjust culturing conditions to alter the microfibril formation and crystallization [69].

19.4.5 Algae

Cellulose is found, although in small quantities, in all of the brown algae (Phaeophyta), most of the red algae (Rhodophyta) and most of the golden algae (Chrysophyta, or chrysophytes) [70]. In green algae, cellulose, xylan or mannan may serve as structural cell wall polysaccharides [70].

19.5 Classification of cellulose nanoreinforcements

Depending on the degree of elementarization and the physical nature, cellulose nanoreinforcements are obtained in various forms, which typically differ from each other based on cellulose source materials and the particle extraction method. Each reinforcement/particle type is distinct, having a characteristic size, aspect ratio, morphology, crystallinity, crystal structure and properties. Unfortunately, the diversity of terminology used to describe cellulose nanoreinforcements leads to misunderstanding and ambiguities, as different terminologies have been and are still used to describe a given set of nanocellulose particles. The term cellulose nanoreinforcements (CN) is used in this chapter to broadly refer to several of the particle types that have at least one dimension in the nanoscale. For comparison purposes, micron-sized scaled particles (MCC) are also defined. Figure 19.3 shows characteristic images of each particle type.

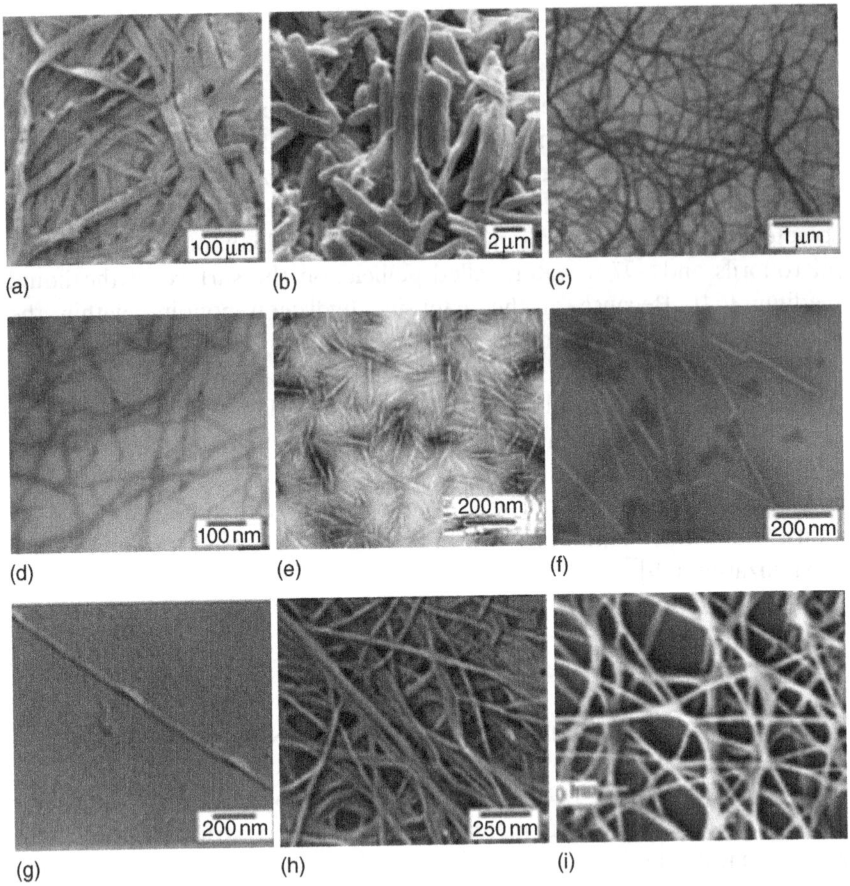

19.3 Several cellulose particle types: (a) SEM image of wood fibre, (b) SEM image of MCC, (c) TEM image of MFC, (d) TEM image of TEMPO-NFC, (e) TEM image of CNC, (f) TEM image of t-CNC, (g) TEM image of AC, (h) SEM image of BC, (i) SEM image of electrospun cellulose (adapted from [79, 89–91]).

19.5.1 Microcrystalline cellulose (MCC)

Microcrystalline cellulose (MCC), also known as hydrocellulose, is available commercially as white odourless, crystalline powder under the brand names Avicel™ (FMC Corporation, Philadelphia, USA), Emcocel™ (Edward Mendell Co. Inc., Carmel, USA), etc. MCC is generated by chemical treatment (mild acid hydrolysis) of different plant fibres, which is often accompanied by a mechanical milling operation [71]. The resulting particles are porous, 50–300 µm in diameter, with a high cellulose content and a higher crystallinity (80–85%) and are composed of aggregate bundles of multi-sized cellulose microfibrils that are strongly hydrogen bonded to each other. They have an aspect ratio of about 1–3. Usually the MCC aggregates are broken up into smaller micron-sized rod-like particles prior to use in composites.

19.5.2 Microfibrillated cellulose (MFC)

The term microfibrillated cellulose (MFC) was introduced by Turbak [72, 73]. Strong mechanical treatment that imposes high shear forces to cellulose fibres allows the extraction of microfibrils and microfibril aggregates with high aspect ratio (lateral dimensions in the 10–100 nm range and length generally in the micrometre scale), which form highly entangled networks. MFC has an aspect ratio of 50–100. Constituting elements typically have a wide size distribution down to nanoscale, and usually a low degree of crystallinity. Production of MFC is generally carried out by passing a liquid suspension of cellulose through a small-diameter orifice in which the suspension is subjected to a high pressure drop and a high-velocity shearing action followed by a high-velocity decelerating impact. MFC is now a commercial product available from various companies such as Daicel (Japan), Rettenmaier (Germany), Innventia AB (Sweden), UPM Kymmene and VTT (Finland) and Borregaard Chemcell (Norway) [6, 13].

19.5.3 Nanofibrillated cellulose (NFC)

The differentiation of NFC from MFC is based on the fibrillation process that produces finer particle diameters. NFC particles are finer cellulose fibrils produced when specific techniques to facilitate fibrillation are incorporated in the mechanical refining of wood fibre and plant fibre [74, 75]. NFCs are also prepared by enzymatic or chemical treatments of MCC or fibres with homogenization. NFCs are reminiscent of elementary fibrils in the wood and plant cellulose biosynthesis process and are considered to consist of cellulose chains arranged in a I_β crystal structure. They have a square cross-section, have a high aspect ratio (4–20 nm wide, 500–2000 nm in length), are ~100% cellulose and contain both amorphous and crystalline regions. Innventia AB (Sweden) inaugurated a pilot plant in 2011 with production capacity of ~100 kg/day of NFC.

19.5.4 Cellulose nanocrystal (CNC)

Cellulose nanocrystals (CNC) are nanoreinforcements which have been grown under controlled conditions that lead to the formation of high-purity single crystals [76]. These are stiff rod-like entities, usually obtained from natural fibres such as sisal [54], ramie [77], cotton stalks [78], wheat straw [59, 79], bacterial cellulose [80], sugar beet [81] and potato pulp [46, 82]. Different terminologies are often used with CNC, such as cellulose nanowhiskers (CNW) and nanocrystalline cellulose (NCC), to describe the form of stable dispersion of cellulose nanoparticles in colloidal suspension. The main characteristics of the whiskers are their high aspect ratio (3–5 nm wide, 50–500 nm in length) and the facts that they are ~100% cellulose, highly crystalline (54–88%) and contain a high fraction of I_β crystal structure (68–94%).

19.5.5 Tunicate cellulose nanocrystals (t-CNC)

Particles produced from the acid hydrolysis of tunicates are called t-CNCs. They have been distinguished from other CNCs because of differences in particle morphology, crystal structure and mechanical properties. The ribbon-like shaped t-CNCs have a width of ~20 nm and a length of 100–4000 nm, are ~100% cellulose, are highly crystalline (85–100%) and contain a high fraction (76–90%) of I_β crystal structure.

19.5.6 Algae cellulose (AC)

Algae cellulose (AC) particles are the microfibrils extracted from the cell wall of various algae by acid hydrolysis and mechanical refining. The resulting microfibrils are microns in length and have a large aspect ratio (greater than 40) with a morphology depending on the algae. A typical example is *Valonia* microfibrils with a square cross-section (~20 nm by ~20 nm) terminating surfaces, with primarily I_α crystal structure.

19.5.7 Bacterial cellulose (BC)

Gram-negative bacteria *Gluconacetobacter xylinus* have long been studied for their ability to produce cellulose, also called 'bacterial cellulose' (BC), as a biosynthetic product of strain. The biosynthesis of BC occurs at a cellulose-synthesizing complex in the bacterial cell. The cellulose chain exits the cell as a so-called *elementary fibril* through pores at the bacterium surface. A recent model of the BC structure in the never-dried state was given by Fink and co-workers [83]. The resulting microfibrils are microns in length and have a large aspect ratio (greater than 50) with a morphology depending on the specific bacteria and culturing conditions. Typically, BC microfibrils have a rectangular cross-section (6–10 nm by 30–50 nm) and have primarily I_α crystal structure.

19.5.8 Electrospun nanocellulose

'Electrospun' cellulose nanofibres are obtained in continuous form as nonwoven mats either directly from cellulose or from cellulose derivatives. Depending on the operating conditions and solution system, cellulose fibres of diameters ranging from 90 nm to 10 μm have been achieved [84–87]. In electrospinning a high voltage is applied between the orifice and a 'target' which can be just a conductive plate (which produces a nonwoven mat of fibres), or the fibre can be wound onto a spool. The surface characteristics of electrospun cellulose fibres are strongly dependent on the solvent used to dissolve the cellulose [88].

Recently, Sneck [92] proposed another type of classification of cellulosic nanoreinforcements based on size, dimensions, branching, charge and mass. This classification is shown in Fig. 19.4.

19.6 Synthesis/isolation of cellulose nanoreinforcements

The isolation of nanocellulose is generally carried out in two steps. The first stage consists of pretreatment of the raw material to obtain 'purified' individual cellulosic fibres that can be further processed. Depending on the source of the raw material, different pretreatments are applied. According to the desired size and morphology of the final nanocellulose, the second stage (generally called 'fibrillation') concerns the transformation of the individual cellulosic fibres into microfibrils, MFCs or NCCs. Processes typically used for this transformation can be used separately or combined.

19.6.1 Pretreatment

In the case of wood, plants and agricultural residues, the most used pretreatment processes are as follows.

Chemo-mechanical processes

As developed in the pulp and paper industry, this pretreatment is used to obtain bleached or chemically dissolved pulps by extracting lignin, hemicelluloses and other secondary products via partial hydrolysis of lignocellulosic material. The different typical steps are (1) chemical treatment of wood chips/plants in batch reactors (e.g. as per classical kraft or sulfite processes), (2) mechanical treatment of the cooked chips/plants to separate the fibres, and (3) bleaching of the resulting pulp to remove the residual lignin and other remaining by-products.

Steam explosion

Steam explosion was introduced and patented as a biomass pretreatment process in 1926 by Mason [93]. It is a thermomechanical process to break down the structural components of cellulose. In general, steam explosion is

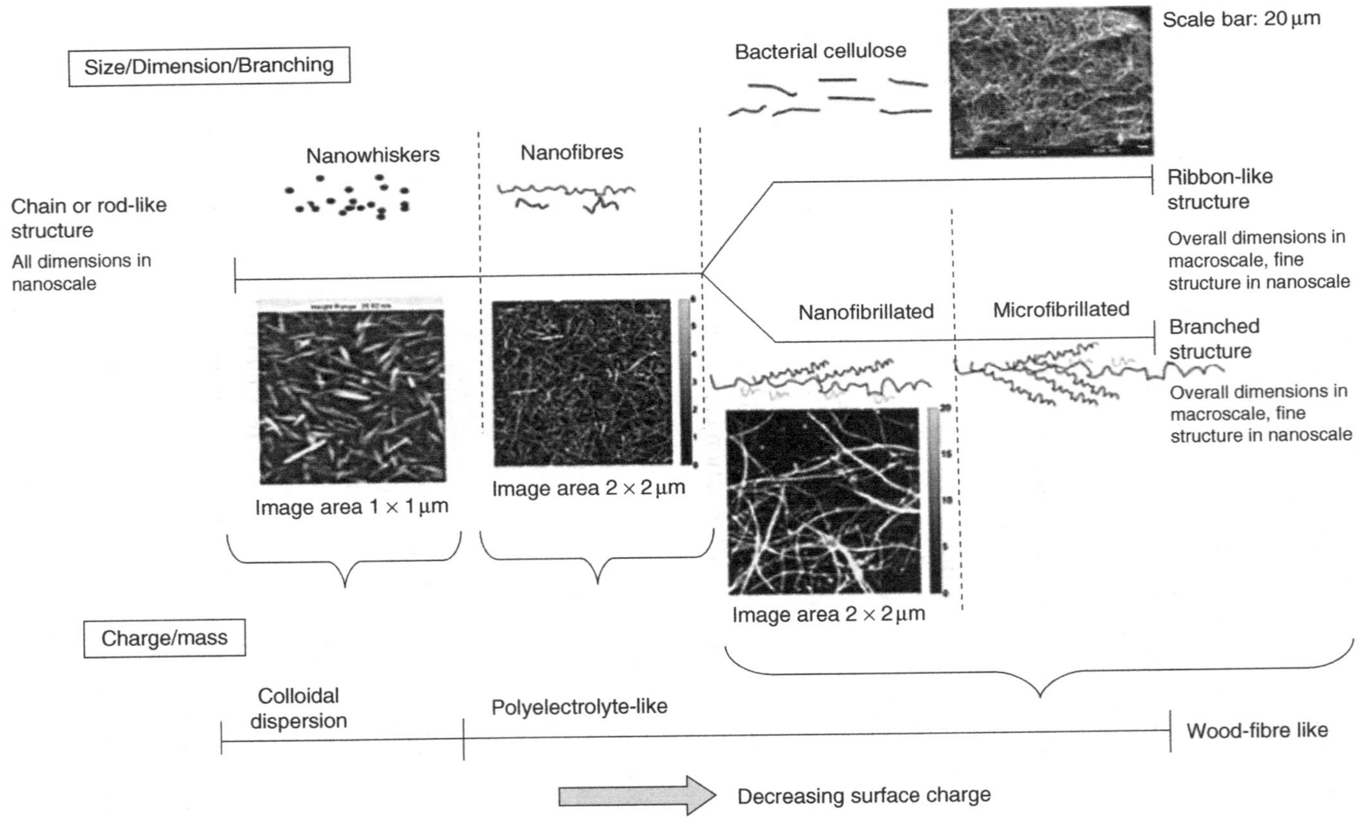

19.4 Nanocellulose grade classification – suggested property interdependencies (adapted from [92]).

a process in which biomass is treated with hot steam (180 to 240°C) under pressure (1 to 3.5 MPa) followed by an explosive decompression of the biomass to atmospheric pressure that results in a rupture of the biomass fibres rigid structure, changing the starting material into a fibrous dispersed solid. The sudden release of pressure generates shear force which hydrolyzes the glycosidic bond and hydrogen bonds between the glucose chains [94]. Steam explosion processes can be run in continuous or batch mode. A batch reactor is usually used for laboratory scale pretreatment while continuous systems are commonly used for large-scale, industrial processes.

A wide variety of plants have been treated using these two pretreatments (chemo-mechanical and steam explosion) including cotton [74, 95], bamboo [96], ramie [53], sisal [54–56] wheat straw [57, 59, 79], etc., to generate nanocellulose.

Other pretreatments

For tunicates, the pretreatment involves the isolation of the mantle from the animal (*Styela clava*) and the isolation of individual cellulose fibrils with the removal of the protein matrix. The cellulose mantles are separated from the rest of the organs by soaking in an alkaline solution (KOH), followed by washing, bleaching (NaClO) and further washing till neutrality [97]. The detailed operating conditions of the pretreatments applied on different raw materials (e.g. tunicates [97–100], algae [101–103], bacterial cellulose [69, 104], etc.) are well described in the literature.

19.6.2 Fibrillation

Chemical methods

The chemical process comprises the treatment of pretreated cellulosic mass with the required amount of alkali for delignification, organosolvation (with acetic acid, aqueous methanol or ethanol), and acid treatment for the hydrolysis. The chemical process also involves oxidation with oxidizing agent.

Acid hydrolysis

Stable aqueous suspensions of CNs can be prepared by acid hydrolysis. Acid hydrolysis of the biomass involves the use of mineral acids like H_2SO_4, HCl and H_3PO_4 [105]. Acid hydrolysis leads to the isolation of micro- and nanofibres with a high degree of crystallinity by removing the amorphous regions of the raw cellulose material. The time and temperature of the hydrolysis reaction as well as the acid concentration play important roles concerning the morphology and the dimensions of the obtained fibres. Many researchers have successfully used this method, alone or in combination with other methods, managing to obtain cellulose structures with nanoscale dimensions starting from different cellulose sources [55, 78, 106, 107]. Bondeson *et al.* treated microcrystalline cellulose (MCC) with sulfuric acid in concentration of 63.5% (w/w) in order to isolate CNC with an aspect ratio between 20 and 40 [108]. Lee *et al.* reported the obtaining of CNC by acid hydrolysis of MCC by testing two different concentrations of hydrobromic acid (1.5 M and 2.5 M) [109]. Rosa *et al.* obtained cellulose whiskers with diameters as low as 5 nm and aspect ratio of up to 60 by sulfuric acid hydrolysis from coconut husk fibres [110]. In all cases, the raw material had been previously submitted to a pretreatment process in order to facilitate the isolation of cellulose nanowhiskers [111].

Alkaline hydrolysis

Alkaline hydrolysis allows the partial separation of the cellulose fibres from the cell wall [55, 62, 112]. These treatments are usually made using diluted solutions of NaOH (1–10%) at low or high temperatures and of concentrated NaOH solutions at low temperatures. NH_4OH and anhydrous NH_3 (gas or liquid) are also used to activate the organic materials. Zuluaga *et al.* [113] reported that by applying four different alkaline treatments (peroxide alkaline, peroxide alkaline–hydrochloric acid, 5 wt% potassium hydroxide and 18 wt% potassium hydroxide) cellulose fibres with average diameters between 3 nm and 5 nm were obtained.

Oxidation

The oxidation of cellulose using 2,2,6,6-tetramethyl-1-piperidinyloxyl radical (TEMPO) as a catalyst allows the formation of carboxyl groups in the C_6 position from hydroxyl groups present on the surface of the fibres. Since the dissociated carboxyl groups are negatively charged, there are repulsive forces between microfibrils, as a result of which, nanocellulose with a diameter of 3–4 nm may arise from the simple mechanical treatment of the oxidized material. The reaction can be carried out under alkaline or acidic conditions with the addition of various reagents as oxidants [114, 115].

Mechanical methods

Under the influence of mechanical forces a local 'cracking' phenomenon occurs, forming some critical tension centre in the fibrous material. Mechanical processes can be carried out in different ways: compression mechanical technique, cryocrushing, homogenization, friction grinding, etc.

Compression mechanical technique

In this technique, delignified fibres of cellulosic materials are placed in a bed of stripes placed between the two plates (or rollers) and are subjected to a constant load of about 10 tons for 10–15 seconds [116].

Refining and high-pressure homogenization

This treatment consists of refining followed by a high-pressure homogenizing [7, 117] in which a diluted cellulosic suspension is forced through a gap between a rotor and the stator disk of a refiner. The disks surfaces are grooved and fitted with bars to subject the fibres to repeated cyclic frictional stresses. During homogenization, refined fibres are passed through a valve at high pressure (20,000 psi) and exposed to a pressure drop to atmospheric condition when the valve is released, resulting in high shear force on the fibre surface [116, 118, 119]. The non-homogenized product enters into the valve area at high pressure and low velocity. The homogenizing effect is caused by the product entering the valve inlet at high pressure. As it passes through the minute gap, the velocity quickly increases while the pressure rapidly decreases to atmospheric pressure. The homogenized product impinges on the impact ring and exits at a sufficient pressure for moving to the next processing stage. The ensuing defibrillated fibres are cooled at room temperature.

Microfluidization

The microfluidizer (developed by Microfluidics Inc., Newton, MA, USA) equipment also allows the defibrillation of cellulosic pulps. The fibre suspension is placed in an inlet reservoir, and then with a pump intensifier generating high pressure, the slurry is accelerated and led into the interaction chamber. When the pressurized product enters into the interaction chamber and passes through geometrically fixed micro-channels, very high velocities are achieved. Thin Z-shaped chambers with different sizes are used for the defibrillation process. The lower the chamber size, the higher is the degree of defibrillation. At the end of the process, a heat exchanger cools down the product stream to ambient temperature.

Cryocrushing

Rigid polymers like cellulose can be ground down to particles with diameters of 1–3 μm. In the 'cryocrushing' technique, water-swollen cellulose material is immersed in liquid nitrogen followed by the crushing of the material using a mortar and pestle. The high shear and impact forces acting on the fibres turn them to powder comprising microfibrils. This method has already been used in the case of kraft fibres after refining [13, 58, 120].

Ultra-fine friction grinding

Recently, several research groups have successfully used 'supermasscolloider' (Masuko Sangyo Ltd, Japan) series of friction grinding machines to produce CN [24, 25, 33, 121, 122]. The supermasscolloider ultrafine grinders feature two ceramic nonporous grinders which are adjustable between the upper and lower grinder. The grinder device first proposed and developed by Masuko© (Tokyo, Japan), involves the breakdown of the cell wall structure, thanks to the shearing forces generated by two grinding stones with countersense rotation [123]. The pulp is passed between a static grinding stone and a rotating grinding stone revolving at about 1500 rpm. Jin *et al.* [124] used never-dried bleached birch kraft pulp to access native cellulose nanofibres. The pulp suspension was at first diluted to 3% consistency, and cellulose nanofibres were disintegrated using an ultrafine friction grinder (Masuko Supermasscolloider, model MKZA 10-15J). The grinding leads to disintegration of the macroscopic cellulose fibres into nanofibres.

Physico-mechanical methods

Ultrasonication

Because of the large requirements of energy necessary in mechanical diminution, ultrasonication has been used alone or in combination with other methods (e.g. acid hydrolysis) in order to obtain cellulose nanofibres [50, 125]. Filson and Dawson-Andoh applied ultrasonication treatment combined with acid hydrolysis to obtain nano-fibres with an average diameter between 21 and 23 nm [50]. It was observed that the dispersion of nanocellulose depends on several parameters such as amplitude and intensity at the tip of the probe, temperature and concentration of the disperser, pressure, volume and shape of the used vessel.

Microwave

Using microwave is another way to obtain cellulose fibres disintegrated down to nanoscale. The main disadvantage of the method is that the obtained material is highly degraded and the nanofibre strength characteristics are low.

Electrospinning

Electrospinning is one of the simplest and most effective methods for producing micro- and nanofibres with an extremely high surface area to volume ratio on a subcellular scale. Until now, many synthetic and natural biopolymers have been electrospun into ultrathin fibres; however, electrospinning of some natural biopolymers such as cellulose is still a challenge [126]. Cellulose nanofibres have been produced by dissolving cellulose in solvents, with a salt selected from the group consisting of potassium thiocyanate, potassium iodide and mixtures thereof, the salt being present at their saturation points [127]. Several cellulose solvents and ionic liquids that have low volatility and do not evaporate completely between the electrospinning spinneret and the collector, have been adopted for the electrospinning process, including NMMO (*N*-methylmorpholine-*N*-oxide)/H_2O [128]. Other solvents, including LiCl/DMAc (dimethylacetamide) [129] and ethylene diamine/salt [87] are multicomponent systems containing both a nonvolatile salt and a volatile component. Petersson and Oksman [8] reported the swelling of the cellulose fibres into a solvent system (*N,N*-dimethylacetamide and lithium chloride) in order to facilitate isolation of the cellulose nanofibres. Trifluoroacetic acid (TFA) is one of the nonaqueous solvents for cellulose, and can dissolve both native and regenerated celluloses at room temperature. TFA is currently considered one of the most promising candidate solvents for electrospinning cellulose [130–132].

Combined methods

Researchers from the University of Toronto reported the combined chemical treatment, mechanical refining, homogenization, and crushing of the water-soaked material in the presence of liquid nitrogen in order to obtain cellulose nanofibres [13]. Through a combination of chemical and mechanical treatments, Jonoobi *et al.* [133] obtained nanofibres from unbleached and bleached kenaf pulp. A method based on a combination of ball milling, acid hydrolysis and ultrasound was developed by Qua *et al.* [51] in order to obtain cellulose nanofibres starting from flax fibres. Roohani *et al.* [32] reported the preparation of CNC with an average diameter of approximately 15 nm, by combination of acid hydrolysis and ultrasound using cotton linters as raw material. Disintegration of cellulose wood pulps by enzyme or acid hydrolysis treatments in combination with mechanical shearing was carried out by Henriksson *et al.* [134].

Biological methods

Degradation of the cellulose substrate occurs in the presence of microorganisms (fungi, bacteria) or, directly, with cellulose enzyme preparations. Enzymatic treatment was performed by Henriksson *et al.* who reported that such kinds of treatment facilitated the obtaining of microfibrillated cellulose nanofibres [134]. In another study, endoglucanase enzyme was used to hydrolyze recycled pulp to produce cellulose nanocrystals with widths of 30 nm to 80 nm and lengths of 100 nm to 1.8 μm [135]. In some cases, multi-enzyme preparations are usually applied which contain enzymes degrading both cellulose and other accompanying polymers, such as pectin, hemicellulose and lignin. The advantage of enzymatic hydrolysis, in contrast to acid hydrolysis, is the fact that with enzymatic hydrolysis the fibre surface is not esterified, e.g. by sulfate groups. CN thus obtained is a biocompatible material and can be used to produce biomedical and pharmaceutical products [52]. Besides, enzymatic processes are widely considered to be 'clean', i.e. environmentally friendly, unlike conventional methods of acid hydrolysis. Enzymatic hydrolysis compared to acid hydrolysis provides longer nanofibres characterized by a greater number of connections between nanofibrils.

19.7 Surface modification of cellulose nanoreinforcements

Three distinct surface modifications of cellulose nanofibres and nanoparticles are generally used: (1) via processing during CN extraction, (2) via adsorption of molecules to the surface of the particles, and (3) via chemical modifications including covalent bonding of molecules onto the surface.

19.7.1 Modification during extraction

Sulfuric acid hydrolysis during extraction of cellulose nanoreinforcements from source material promotes the formation of sulfate esters with high surface acidity. Hydrochloric acid hydrolysis results in hydroxylated surfaces [136]. Sulfuric acid treatment is the most commonly used method as the resulting high sulfates tend to stabilize CN dispersion. TEMPO-mediated oxidation, already described, is also used for modification during nanofibre extraction, yielding a carboxylic acid-rich surface.

19.7.2 Modification by adsorption

The overall objective is to reduce surface energy of CNs in order to improve their dispersibility/compatibility with nonpolar media. Noncovalent surface modifications are typically made via adsorption of surfactants such as stearic acid [137], cetyltetramethylammonium bromide [138], xyloglucan, xyloglucan block copolymers, and mono- and diesters of phosphoric acid with alkylphenol tails [139,140]. Most commonly, a layer-by-layer deposition technique is used in this method [141]; thus, the obtained surfactant-coated nanocellulose disperses well in nonpolar solvents [139].

19.7.3 Chemical modification

Nanocelluloses possess an abundance of hydroxyl groups on their surface, providing the possibility of extensive chemical modifications. Most of these modifications aim to improve (1) the dispersibility of nanocellulose by reducing the hydrogen bonds which cause the reaggregation and (2) the compatibility of nanocellulose in different solvents or matrices that are suitable in the production of nanocomposites. Figure 19.5 describes the common chemical modification strategies. The main chemical modifications, discussed in more detail in the following sections, are esterification, silylation, cationization, fluorescent labeling, carboxylation by TEMPO-mediated oxidation, and polymer grafting [142–145]. Although those operations decrease the surface energy and polar character of the nanocellulose, thereby improving the adhesion with a nonpolar polymeric matrix, a detrimental effect is also reported concerning the low mechanical performance of the composite.

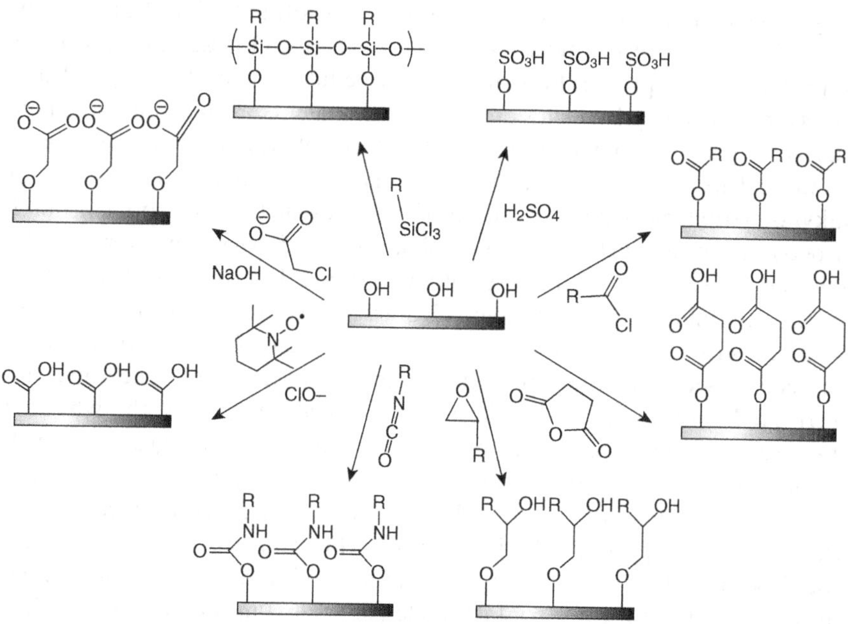

19.5 Common modification chemistries of CN surfaces. Clockwise from top right: sulfuric acid treatment provides sulfate esters, carboxylic acid halides create ester linkages, acid anhydrides create ester linkages, epoxides create ether linkages, isocyanates create urethane linkages, TEMPO-mediated hypochlorite oxidation creates carboxylic acids, halogenated acetic acids create carboxymethyl surfaces, and chlorosilanes create an oligomeric silylated layer (adapted from [146]).

Esterification

The hydrophobization of the cellulose surface is commonly achieved through the esterification process, which basically uses carboxylic acid, acid anhydrides or acetyl chlorides as reacting agents. The esterification reaction introduces an ester functional group (O–C=O) onto the surface of cellulose by condensation of the previous reagents with a cellulosic alcohol group, while the acetylation reaction introduces an acetyl functional group CH–C(=O)– onto the surface of cellulose. Several methods have been employed to achieve the surface acetylation [147–149] and the esterification processes [150, 151] of CNs. Avoiding swelling conditions, the reactions only occur on the cellulose chains located at the surface. The nature and extraction treatment appear to be important factors, controlling the extent of esterification and the final morphology of nanocellulose.

Silylation

Silylation consists of the introduction of substituted silyl R_3Si– groups onto the surface of cellulose nanoreinforcements. Gousse and coworkers reported on the preparation of silylated cellulose microfibrils [152] and whiskers, especially by using etherification reagents as halogenated silane derivatives (e.g., isopropyl dimethyl chloro silane). Cellulose whiskers from acid hydrolysis of tunicate were silylated partially with a series of alkyl dimethylchlorosilanes, with alkyl moieties such as isopropyl, *n*-butyl, *n*-octyl and n-dodecyl. When mild silylation conditions were applied, the microfibrils retained their morphology and could be dispersed in a non-flocculating manner into organic solvents. For example, partially silylated whiskers with degree of substitution (DS) between 0.6 and 1 could readily be redispersed in medium-polarity organic solvents, such as acetone and THF. At DS of less than 0.6, the morphological integrity of the whiskers was preserved; however, it was disrupted when the DS was greater than 1. Andresen *et al.* [153] hydrophobized NFC via partial surface silylation using the same silylation agent and reported that when silylation conditions were too harsh, partial solubilization of NFC and loss of nanostructure could occur. Films prepared from modified cellulose by solution casting showed a very high water contact angle (117–146°) in this case. In another study, the surface trimethyl silylation of NCC from bacterial cellulose and their reinforcement characteristics in nanocomposites were also investigated. In one patent, researchers surface-treated cellulose nanofibres with an alkoxysilane in order to improve the chemical bonding between the reinforcements and the resin to which they were added. In another invention [154], the preparation of a silylated NFC material containing bisacylphosphine oxide (BAPO) initiator groups was conducted by refluxing the dry nanofibrils in the solution of the BAPO-silane reagent dissolved in toluene for about 20 hours. The functional nanofibrillated product was isolated by consisted in washing out the unreacted BAPO with solvent, followed by drying.

Cationization

A one-step method has been developed to introduce positive charges onto the surface of NCC through the grafting of epoxypropyltrimethyl ammonium chloride [145]. This process, carried out under mild operating conditions, reverses the surface charge and leads to stable aqueous suspensions of NCC with thixotropic gelling properties while preserving the original morphology and maintaining the integrity of the crystal.

Fluorescence

Fluorescence techniques have been used significantly to study the cellular uptake and biodistribution of nanoparticulate delivery systems, by tracking the fluorophores. An interesting method to label NCC with fluorescein-5-isothiocyanate for fluorescence bioassay and bioimaging applications has been described [155]. The process was carried out in three steps. First, epoxy functional groups were fixed at the surface of NCC via reaction with epichlorohydrin. Then the epoxy ring was opened with ammonium hydroxide to introduce primary amino groups. Finally, the primary amino group was reacted with isothiocyanate group to form a thiourea.

Polymer grafting

The methods for polymer grafting onto NCC surfaces are based on two approaches, named 'grafting-onto' and 'grafting-from' [156–160]. 'Grafting-onto' consists in directly grafting an existing polymer on the NCC surface with the use of a coupling agent. The main advantage of this technique is that the polymer can be fully characterized before grafting and then it is possible to control the properties of the resulting material. The main drawback is due to steric hindrance and high viscosity of the medium. 'Grafting-from' consists in mixing NCCs with a monomer and an initiator. The polymer grows directly on the surface of the nanoparticles. The main advantage of this approach is that the reaction is fast and easy because there is no steric hindrance and the viscosity of the reaction medium remains low. The main drawback is that the grafted polymer is not initially fully characterized. In these approaches, nanocelluloses are never dried before grafting but solvent is exchanged from water to organic solvent. The 'grafting-onto' approach was used to graft onto NCCs with various polymers such as polycaprolactone via an isocyanate-mediated coupling reaction [53], pre-synthesized waterborne polyurethane [161], thermo-responsive polymers via a peptidic coupling reaction [162], maleated polypropylene [163] and DNA oligomers [164]. The 'grafting-from' approach was used to grow polymeric chains from the NCC surface via atom transfer radical polymerization (ATRP). This technique allows for very precise control over the grafting process that produces well-defined monodisperse particles [165]. Ring-opening polymerization was also used in the 'grafting-from' approach to graft polycaprolactone polymers onto the NCC surface [156], where ($Sn(Oct)_2$) was used as a grafting and polymerizing agent. To enhance the grafting efficiency, similar grafting processes via microwave irradiation were also studied [166, 167].

19.8 Characterization of cellulose nanoreinforcements

Appropriate measurement and characterization standards are essential for efficient introduction of CN into the marketplace. Some of these basic characterization methods are expected to be done on a daily basis for characterization, testing and quality control purposes [168]. Different types of nanocellulose need different characterization methods, for example, The characterization of nano- and microfibrillated celluloses is more difficult than the characterization of NCC, owing to their higher aspect ratio and branching. Light optical microscopy can be useful as a quick method for visualizing macrostructure and the amount of unfibrillated cellulose fibres present. Scanning Electron Microscopy (SEM) has been found suitable in obtaining a detailed fibril structure. Besides, thickness of fibrils can be measured, and their length and branching can be estimated. Other microscopic techniques such as Atomic Force Microscopy (AFM) and Transmission Electron Microscopy (TEM) have also been used to successfully characterize nanocellulose. X-ray techniques such as wide-angle X-ray scattering (WAXS), small-angle X-ray scattering (SAXS) and X-ray diffraction (XRD) have provided useful information about nanocellulose. From conventional bright-field transmission electron microscopy, it was possible to identify individual whiskers, which enabled determination of their sizes and shape. Although atomic force microscopy generally overestimates the width of the whiskers due to the tip-broadening effect, it can be a powerful technique to identify and even determine the nature of agglomeration and shape of cellulose whiskers.

Based on shape, concentration, colour and difference between refractive indices of the particle and the sample fluid, the transmission technique has

Table 19.1 Important properties of nanocellulose and suggested methods for their characterization

Property	Method(s)
Particle morphology: • Size (length, width, size, aspect ratio) • Distribution • Degree of branching	• Ultraviolet–visible spectroscopy (UV/Vis) • Atomic Force Microscopy (AFM), Scanning Electron Microscopy (SEM), Transmission Electron Microscopy (TEM) • Light microscopy (with/without fractionation by centrifugation) • Viscosity (for aspect ratio)
Surface chemistry	• Functional group – Fourier Transform Infrared Spectroscopy (FT-IR), Nuclear Magnetic Resonance Spectroscopy (NMR), Raman Spectroscopy • Inverse Gas Chromatography (IGC) • Charge determination: – Zeta potential – Conductimetric titration
Amount of nanomaterial	• Fractionation by centrifugation • Field flow fractionation
Rheology: • Viscosity, gel point	• Low shear viscosity (modified Brookfield) • Drainage time
Dissolved (colloidal) Substances (amount and quantity)	• Capillary electrophoresis (CE) • Size exclusion chromatography (SEC) • High-performance liquid chromatography (HPLC) • AFM/SEM/TEM
Crystallinity: I-Alpha, I-Beta, II	• Wide-angle X-ray scattering (WAXS) • Small-angle X-ray scattering (SAXS) • Raman spectroscopy • FT-IR • Solid-state NMR • Thermogravimetric analysis (TGA)
Specific surface area	• SAXS

Source: adapted from Ref. 168.

been used to characterize nanocellulose samples. For example, individual nanofibrils are transparent and have slight colour shade, while microfibrils and branched fibrils are white and opaque in nature. Highest transmittance, as a result, indicates finest structure in general. A UV-Vis spectrophotometer is generally used for this purpose. The performance of this method depends on size fractions of the sample. Viscosity measurement is another technique to characterize cellulose nanoreinforcements. Viscosities change owing to change of particle size, size distribution, shapes and charges. Table 19.1 shows the different characterization techniques related to CN.

19.9 Matrices

Both hydrophilic and hydrophobic polymers are used as the matrices for biobased nanocomposites. Matrix parameters such as solubility, dispersibility and degradation strongly influence the processing conditions. The various types of matrices that are used with nanoscale cellulose reinforcements are described below.

19.9.1 Hydrophilic matrices

Due to their strong hydrophilic character and high aspect ratio, CNs tend to flocculate through hydrogen bonding. As a result, the main challenge is related to their homogeneous dispersion within a polymeric matrix. Water could be considered as the preferred processing medium because of the high stability of aqueous CN dispersions. To ensure a good level of dispersion, this implies the use of either hydrophilic polymers or aqueous dispersed polymers, i.e., latexes as a matrix. In this context a majority of patents [169–175] have described CN-based nanocomposites produced in aqueous media.

Non-biodegradable matrices

A variety of non-biodegradable matrices containing hydrophilic groups, such as phenol-formaldehyde resin, epoxy resin, acrylic resin, melamine-formaldehyde resin, ethylene-vinyl alcohol and latexes, have been used to prepare nanocomposites with cellulose nanoreinforcements. Commonly used thermoset matrices, for example epoxies, phenolics, acrylic and phenol-formaldehyde resin, have been used through a fibre impregnation process [122, 176, 177]. Nakagaito and Yano [7, 178] impregnated microfibrillated kraft pulp with a phenol-formaldehyde resin and then compressed the resulting material under high pressure to produce high-strength cellulose nanocomposites. Henriksson and Berglund [179] produced nanocomposite films of MFC and melamine formaldehyde (MF) resin as a potential material for use in loudspeaker membranes. The MF-MFC nanocomposites showed average Young's modulus as high as 16.6 GPa and average tensile strength as high as 142 MPa. Other researchers have also reported use of different resins including different types of acrylic and epoxy resins [7, 24, 180, 181]. Not only were all the composites significantly stiffer and stronger than the unmodified resins, but other advantages were also reported such as high composite transparency, low thermal expansion coefficient (CTE), etc.

Poly(styrene-*co*-butyl acrylate) latex was used by Malainine *et al.* as a matrix with aqueous suspension of cellulose microfibrils to prepare nanocomposites by solution casting [182]. These researchers observed a significant reinforcing effect of cellulose microfibrils on the matrix. The tensile modulus increased from 0.6 MPa to 34.5 MPa when adding 10 wt% filler and the thermal stability of the composite was significantly enhanced. More recently, Dalmas *et al.* [183] dispersed cellulose nanofibrils obtained from sugar beet pulp in the same poly(styrene-*co*-butyl acrylate) latex. The mixture was cast and then hot pressed, and it was reported that formation of a rigid nanofibril network linked by strong hydrogen bonds took place, leading to high mechanical reinforcement and thermomechanical stability [183].

Fernandez *et al.* [184] prepared solvent-cast films from ethylene vinyl alcohol copolymers (EVOH29 and EVOH44) with 2 wt% nanofibre (MFC) content using isopropanol:water (70:30) as a solvent. Polarized light optical microscopy and AFM confirmed that the MFC microfibrils were well dispersed across the sample thickness.

Considering that polyurethanes (PU) are relatively polar polymers and can interact with the polar groups of cellulose, good interfacial adhesion can potentially be achieved, which is essential for enhanced mechanical properties. PU-MFC composite materials were prepared recently using a film stacking method in which the PU films and nonwoven cellulose fibril mats were stacked and compression moulded [185]. The thermal stability and mechanical properties of the pure PU were significantly improved by MFC reinforcement.

Biodegradable matrices

Among the biodegradable polymers of interest is poly(vinyl alcohol) (PVOH) which is water-soluble, has excellent chemical resistance and is biocompatible. Wan *et al.* [186] and Millon and Wan [187] tested BC as a potential reinforcing material in PVOH for medical device applications. Zimmermann *et al.* [188] dispersed MFC into PVOH and generated fibril-reinforced PVOH nanocomposites (fibril content 20 wt%) with up to three times higher E-modulus and up to five times higher tensile strength when compared to the reference polymer. Similar results have been reported by other researchers, where cellulose nanoreinforcements from different sources have been used with PVOH, with a range of nanocellulose contents (0–90 wt%), and it was also reported that tensile strength increased linearly as a function of filler content [13, 62, 180, 189]. However, Lu *et al.* [190] observed no further improvement of mechanical properties when MFC was applied above 10 wt%.

Starch obtained from a variety of crops such as corn, wheat, rice and potatoes is a another hydrophilic biodegradable polymer, which has immense potential for use as a matrix in composites, owing to its abundant availability and low cost. However, considering that native starch is very sensitive to moisture and tends to be brittle, it is often converted to thermoplastic starch (TPS). TPS is obtained after disruption and plasticization of native starch by applying thermomechanical energy in a continuous extrusion process.

Application of nanofillers, such as cellulose whiskers or microcrystalline cellulose [191–195], has been reported. MFC and BC have also been reported as promising candidates for starch reinforcement [196–198]. Dufresne and Vignon [82] used potato starch as matrix to prepare nanocomposites through addition of MFC, using glycerol as plasticizer. The tensile modulus was found to be ~7 GPa at 50 wt% cellulose content compared to ~2 GPa for unreinforced samples (0% MFC). Addition of MFC to the plasticized starch matrix also resulted in a decrease of both water uptake at equilibrium and the water diffusion coefficient. In a different approach, starch was added to the culture medium of cellulose-producing bacteria (*Acetobacter* sp.) to introduce the granules into the forming network of cellulose [196]. The application of such a bottom-up technique allowed the preservation of the natural ordered structure of cellulose nanofibres. The BC–starch mats were hot pressed to obtain nanocomposite sheets. Several other researchers have used thermoplastic starch (TPS) from different sources (potato, tapioca) to prepare nanocomposites from cellulose nanofibres by either the solution casting or the melt compounding method [58, 197, 199], The tensile strength and modulus were significantly enhanced in the nanocomposite films, owing to the strong interactions between the starch matrix and the high-aspect-ratio nanofibres.

Other major biodegradable hydrophilic matrices that have been used with cellulose nanoreinforcements are amylopectin [200–202], poly(ethylene oxide) (PEO) [203, 204] and chitosan [205–209].

19.9.2 Hydrophobic matrices

In general, the homogeneous incorporation in most common hydrophobic thermoplastic polymers is challenging. A non-homogeneous dispersion of the nanofillers in the polymer matrix is often obtained with such matrices, thus decreasing the final mechanical properties of the nanocomposite material [11]. However, many thermoplastic polymer nanocomposites were developed with CN dispersed in adequate organic medium using surfactants [10, 139] or organic solvents [210–214], with thermoplastic matrices such as poly(lactic acid) under specific compounding conditions [215–217].

Non-biodegradable matrices

Among the non-biodegradable hydrophobic matrices, poly(ethylene) (PE) and poly(propylene) (PP) are the most common ones used with cellulosic nanoreinforcements. Commercial MFC, for example, has been used with PP to fabricate composites by melt mixing or compression moulding techniques

[61,218]. Although some improvements in tensile properties were observed, the enhancements were not significant and SEM results indicated lack of sufficient adhesion at the fibre/matrix interface. Ljungberg *et al.* [219], on the other hand, prepared nanocomposite films of cellulose whisker-reinforced isotactic polypropylene by solvent casting from toluene, dispersed with surfactant, and reported that the linear mechanical properties above the glass–rubber transition were found to be drastically enhanced. These effects were attributed to a mechanical coupling between the polypropylene crystallites and filler–filler interactions. In another study, polypropylene (PP)–microcrystalline cellulose (MCC) composites were prepared containing poly(propylene-graft-maleic anhydride) (PP-g-MA) and MCC treated with silicone oil, stearic acid or alkyltitanate coupling agent to promote matrix–filler dispersion and compatability [137]. The researchers reported that tensile modulus, storage modulus, loss modulus and glass transition temperature increased with MCC concentration due to effective interaction between PP and MCC.

Biodegradable matrices

Among the biodegradable hydrophobic matrices, poly(lactic) acids or polylactides (PLA) have gained a lot of attention as matrices for use with biobased nanoreinforcements. As with other commodity polymers, PLA actually refers to a large family of compounds that includes copolymers with other monomers. CN-reinforced PLA composites are attractive because both the reinforcement and the matrix are obtained from renewable resources. PLA can be produced from renewable biobased resources such as starch and sugars. It is one of the most promising biodegradable polymers owing to its mechanical property profile, thermoplastic processibility and biological properties, such as biocompatibility and biodegradability. Microcrystalline cellulose (MCC) has been used as a whisker reinforcement with PLA as a matrix [220]. However, within the PLA matrix, MCC exists as aggregates of crystalline cellulose entities, and basically poor or no adhesion was obtained between MCC particles and PLA.

So far, most of the studies reported in the literature refer to either solution casting or vacuum filtering followed by hot pressing to produce MFC reinforced biocomposites. Only a few reported studies used industrially relevant methods. In one such study, a twin-screw Brabender mixer has been used to disperse cellulose microfibrils (MFC) in water solution in a PLA matrix and the resulting compound was then hot pressed at 190°C [199]. Microscopic images revealed uniform dispersion of MFC in the PLA matrix, indicating the potential of this preparation method. However, Mathew *et al.* [221] reported a nonuniform dispersion of cellulose fillers in PLA matrix when nanocomposites of PLA with 5 wt% cellulose nanowhiskers and MFC were prepared by twin-screw extrusion. They used water as the pumping medium during extrusion. It was revealed that fast evaporation of the water led to reaggregation of the fillers in the polymer matrix and resulted in poor mechanical properties of the nanocomposites. In another study, a papermaking-like process was adopted to prepare PLA nanocomposites. Other researchers have also reported the use of PLA with nanocellulose in preparing nanocomposites via a solvent exchange method [215], either by acetone or by acetone followed by dichloromethane. It was reported that significantly more uniform filler distribution could be achieved compared to that possible by direct introduction of MFC to the molten PLA matrix.

Another biodegradable polymer that has gained some attention in preparing biobased nanocomposites is poly(ε-caprolactone) or PCL. It is an oil-derived biodegradable and semicrystalline polyester. Its low melting point and low viscosity make it easily processable using conventional melt-blending technologies. However, due to the hydrophobic nature of PCL, good dispersion of CNs in the matrix requires some modification of the cellulose. In one study, PCL-grafted MFC films were hot-pressed together with a PCL film to produce a laminate [222]. In another study, *N*-octadecyl isocyanate was used as a grafting agent for modifying the surface of two types of cellulose nanofillers (MFC and CNW), and it was observed that grafted MFC-reinforced PCL composites possessed higher modulus and lower elongation at break at a given loading level compared to cellulose whisker-reinforced nanocomposites. The differences were mainly attributed to the capability of MFC to form an entangled network.

More recently, there have been several attempts to develop new biobased thermoset resins such as those based on polyfurfuryl alcohol, pine oil and soy oil. Pranger and Tannenbaum [223] established the feasibility of using furfuryl alcohol as a matrix for CNW reinforcements. They found sulfonic acid residues at the CNW surface, which were left over from the hydrolysis treatment and acted as catalysts for the curing reaction of the furfuryl alcohol polymer [184].

19.10 Incorporation of biobased nanoreinforcements into matrices

19.10.1 Strategies

The efficient incorporation and homogeneous dispersion of CN in the polymer matrix is considered as the main challenging step to manufacture nanocomposites with desired properties. The nano entities can be either in the form of powder or dispersed in solution. In both cases CNs may attach together and form agglomerates due to their high adhesive forces, which makes homogeneous dispersion into the matrix very difficult. There are different approaches to incorporating CN in a polymer matrix to obtain well-dispersed and separated nano entities. In most cases, more than one method is used. In such cases, the respective advantages of each method combine synergistically.

In-situ *polymerization*

In this approach nanoparticles are produced directly (*in-situ*) in the matrix during the polymerization process. This method is mainly used for thermoset matrices. In this method, nano entities are first dispersed in the thermoset resin precursor (monomer or oligomer). At this stage, the resin has a low viscosity, so an even dispersion can be achieved easily. Once the nano entities are well dispersed in the resin precursor, the polymerization reaction is initiated by heating, adding hardener, or subjecting to UV light, depending on the chosen curing mode. Often an organic solvent of the monomer is added to the system to dilute the mixture and reduce viscosity. Generally solvents with lower boiling point are used because it is easy to remove them from the system and they can be recovered and reused. Zeng [181] dispersed cellulose nanowhiskers in epoxy resin via organic solvent EGME (ethylene glycol monomethyl ether) and the mixture was rotoevaporated to remove EGME from the nanocomposite. The sol–gel technique can also be used to produce the nanoparticles directly in the matrix during the polymerization step [224].

Mechanical mixing

In this approach, dry CN is disposed in a liquid polymer resin using mechanical mixing in which high shear forces are applied during dispersion to break up agglomerates and give a homogeneous dispersion of CN. For thermoset nanocomposites, there are two main types of mechanical mixing methods, namely mechanical stirring and high shear mixing. Mechanical stirring is mainly used for distributive mixing purposes, since its dispersive capacity is poor and it is done by a simple magnetic or electric agitator. High shear mixing is a dispersive and distributive mixing method in which high shear forces are applied during the dispersion process to break up the agglomerates and to separate and distribute the individual nano entities homogeneously in the resin. The high local shear forces can be realized using several techniques such as rotating disks (dissolver), a rotor–stator system, the grinding effect of moving ceramic balls (ball mill), or three-roll mill. The choice of dispersive technique is determined by the properties of the matrix and the nanomaterial, such as viscosity and polarity [225].

Surface treatment

Another method is to apply a special surface treatment before the mechanical dispersion process. This technique aims to increase the compatibility between the matrix and CN. Considering the importance of surface modification of cellulose nanoreinforcements, it is elaborated separately in the following section.

Ultrasonic mixing

In ultrasonic mixing, ultrasound energy is used. Sonication is considered to be a very effective mixing process for dispersing nanomaterials in water, organic solvents and resins. In the sonication process, a probe generates ultrasound with a frequency above 19 kHz that propagates into the liquid media by alternating high- and low-pressure cycles that cause ultrasonic cavitations. Dispersion by sonication uses liquid jet streams resulting from ultrasonic cavitations to break the secondary adhesive forces (hydrogen bonds, van der Waals forces) between nanoparticles without breaking the covalent bonds within the particles. This separates the agglomerated particles as individual particles and improves dispersion and the stability of suspension. However, the use of ultrasound results in a rapid heating of the suspension and thus the duration of the treatment should be limited to several minutes in an ice bath.

Solvent exchange

In the solvent exchange method, water is substituted by a low-boiling-point organic solvent, which is more compatible with the hydrophobic matrix. The aqueous suspension is progressively solvent-exchanged in liquids of decreasing polarity by several successive centrifugation and redispersion operations after each solvent exchange step to avoid aggregation [213]. The modification exhibits a large effect on the individualization of the cellulose nanofibres due to reduced hydrogen bonding and fewer polar interactions between the individual particles.

19.10.2 Characterization

Characterization of nanoreinforcement dispersion is important, as it is closely related to the realization of efficient reinforcement and the resultant properties of the reinforced products. Besides the choice of a suitable processing technique, managing the optimized viscosity during processing can also be controlled through characterization. For example, the particle coalescence process during solvent evaporation, especially when the matrix is in the latex form, can lead to inhomogeneous nanocomposite film. Processing time and temperature also have important roles to play in the dispersion of nanoreinforcements. Naked eye examination of the film's surface can often indicate inhomogeneity of the sample and entrapped air bubbles. Various other characterization techniques such as scanning electron microscopy (SEM), wide-angle X-ray scattering (WAXS) and polarized light microscopy have been used to demonstrate the effect of processing on dispersion of nanocellulose from the morphology of cellulose nanocomposites [226]. The birefringence obtained from polarized light microscopy has also been useful in investigating inhomogeneity of cellulose nanocomposite films through the existence of different coloured domains [226].

Fractography technique using SEM can reveal valuable structural details such as homogeneity of the composite, presence of voids, dispersion level of the nanoscale reinforcements within the continuous matrix, presence of aggregates and sedimentation. The fracture surface of a thermoplastic starch (TPS) and a nanocomposite film filled with 10 wt% of cellulose nanofibres is shown in Fig. 19.6. From the image, it is clear that the nanofibres are well dispersed and covered by the matrix. No fibre pull-out or debonding was observed because of the good adhesion between the nanofibres and the polymer matrix [13, 58]. Small-angle neutron scattering (SANS) has also been used to inspect dispersion of nanocellulose and homogeneity of the resulting composite [48].

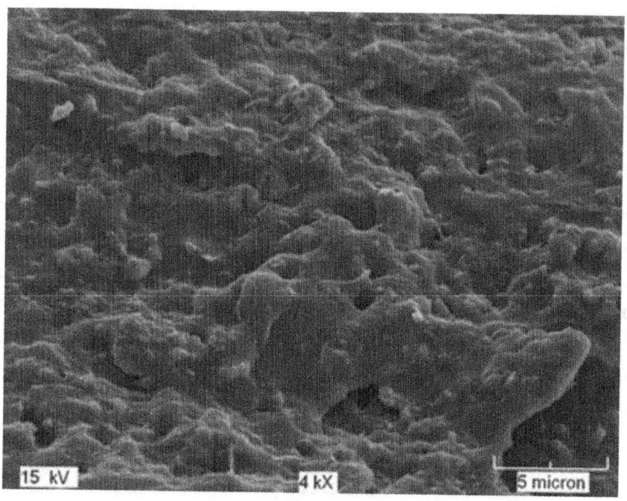

19.6 Scanning electron micrograph of nanocomposite filled with 10 wt% of cellulose nanofibres (from [58]).

19.11 Nanocomposites

A composite is defined as a material that combines two or more distinct constituents or phases, where one or more of the discontinuous phases (reinforcements) are dispersed in another continuous phase (matrix) in order to obtain tailor-made characteristics and properties. Nanoscale reinforcements have the advantages of increased interfacial interactions between the reinforcement and the matrix phase due to the exceptionally high interfacial area and improvement in composite properties for a relatively small amount of reinforcements. Due to the nanoscale dimensions of the reinforcements, the fracture-initiating defects are also expected to be smaller in size, thereby delaying the failure of the composites, also known as nanocomposites. These nanocomposites, in particular, are two-phase materials in which one of the phases has at least one dimension in the nanometre range. The advantages of nanocomposite materials when compared with conventional composites are their superior thermal, mechanical and barrier properties at low reinforcement levels (e.g., B5 wt%), as well as their better recyclability, transparency and low weight [118, 227–229]. Due to their abundance, biodegradability and relatively low price, there is a significant history of using cellulose fibres from plants as reinforcement in composite materials [230–236]. However, the application of nanoscale cellulose fibre for this purpose is a relatively new research area and a number of researchers have therefore explored the concept of fully bioderived nanocomposites or partially biodegradable nanocomposites as a route to development of bioplastics or bioresins with better properties.

19.11.1 Reinforcement mechanisms

The unusually high reinforcing effect observed for cellulose nanofibre-reinforced composites motivated some researchers to investigate the underlying mechanism(s). Indeed, classical prediction models for short-fibre composites cannot be applied, as the observed experimental values were much higher than those obtained from these classical models [226]. Since the 1990s, various analytical modeling approaches have investigated the effect of CN fillers within polymer matrices on effective nanocomposite properties. Typically, mean field and percolation approaches have been used to understand 'unusual' reinforcing effects at low concentrations of CN [237].

Mean-field theory

Most prior modeling efforts for CN composite properties have used the so-called mean-field theory. This theory is a special case of elementary bounds but imposes some restrictions on geometry and therefore gives tighter bounds. It is a building-block approach, which was described by Halpin and Kardos [238]. The main idea of the mean field theory is to focus on one particle and assume that the most important contribution to the interactions of such particle with its neighbouring particles is determined by the mean field due to the neighbouring particles. In this approach, the modulus and the geometry of the fibres are accounted for, but no interaction between the fibres is assumed. Indeed, this model is based on the concept that the material is made of short fibres, homogeneously dispersed in a continuous matrix. In this approach, the composite is assimilated to a four-ply laminate. The mechanical properties of each ply are derived from the micromechanical Halpin–Tsai equations [239]. However, the predicted results could be improved by replacing the Halpin–Tsai equation with a concentric cylinder model, which gives refined upper bounds and can evaluate the role of CN aspect ratio and CN/matrix interface. Experimental results that exceed any of these upper bounds should be treated with skepticism.

Percolation approach

Percolation is a random probabilistic process which exhibits a phase transition. Studying the statistics of the clusters helps to identify the critical value of density when formation of infinite or long-range connectivity in random systems first occurs. This is called the percolation threshold [240]. In other words, the concept of percolation deals with the development of a connected network in a multiphase system [221, 241, 242]. At low concentrations, no network is formed. As the concentration of the percolating phase increases, it eventually reaches the percolation threshold defined as the first development of a connected network. Its significance in composites is that some effective properties will increase dramatically and rapidly for concentrations above the percolation threshold. Its significance in nanocomposites, including a fibrous phase, is that the percolation threshold decreases as the fibre aspect ratio increases [243]. In particular, with cellulose nanocomposites, the strong reinforcing effect of the whiskers is generally attributed to the formation of a percolating network structure above the percolation threshold resulting from hydrogen bonding between nanoparticles [244].

In a nanocomposite, the interaction between nanofillers and polymer media is at the molecular level and it is known that the properties of the matrix and the nanoreinforcements and the amount and distribution of the nanoreinforcement have a strong impact on the interactions between the nanoreinforcements and the matrix. Of particular interest, therefore, is the percolation threshold, i.e. the minimum loading of nanofillers required to form a cluster that spans the whole system [245]. Not surprisingly, this would provide opportunities to manipulate the percolation threshold to produce a nanocomposite with optimized properties and as low as possible a loading of nanoreinforcements, so that the percolation threshold and the subsequent rapid increase in properties can be shifted to very low concentrations by using high-aspect-ratio nanoreinforcements.

For nanocomposites, the bounding methods give insight into the role of percolation in composite properties. The observation of experimental results exceeding Halpin–Kardos [238] mean-field calculations is a common basis for claiming a percolation effect. A critical evaluation of that model, however, shows that the percolation concept predicts an approximate lower bound result, and thus exceeding it is the expected result. Further, a good agreement between experimental and predicted data was reported when using the series-parallel model of Takayanagi *et al.* [246] modified to include a percolation approach.

Although, over the last several decades, a tremendous amount of work has gone into the modeling of the percolation phenomenon, forming a comprehensive picture of the influence of the material parameters on the percolation threshold of nanofibre dispersions in composites remains an ongoing process. Compared to analytical methods, computer simulations such as Monte Carlo simulations have been used more successfully to model the percolation phenomenon in composite materials.

19.11.2 Processing

Four main processing techniques have been utilized to produce CN composites: casting-evaporation [11, 247, 248], electrospinning [214, 249–251], melt compounding [247, 248, 252] and solid-phase compounding [61, 62].

Casting-evaporation process

Casting-evaporation is one of the most common processes used to produce nanocomposite films. In general, CN is dispersed within a given medium (0.05–5 wt% solids), typically water, but various organic media have also been used and then polymer solutions are mixed with the CN dispersion. To achieve good reinforcement, the dispersibility of CN in both the polymeric matrix and processing solvents is critical. Composite films can then be produced from this mixture via casting on a suitable surface followed by evaporation. For example, nanocomposite film based on ethylene oxide/epichlorohydrin and CN was produced by dispersion-casting of CN fillers in THF/water mixtures [253].

Electrospinning process

This technique, already described (see Section 19.6.2) is a fast and simple process to produce nanofibres and nanocomposites and has been widely studied[214, 249–251]. In general, CN is dispersed within a given medium (typically 0.05–5 wt% solids) and then polymer solutions are mixed with the CN dispersion. Using this technique, the solvent evaporates as the fibre moves between the source and the collector and the polymer coagulates, forming a composite fibre. The CN reinforced fibres can be further dried and/or go through additional treatments (e.g. heating to crosslink the matrix polymer).

Melt-compounding process

A melt-compounding technique is used in the production of hydrophobic polymeric nanocomposites. In general, melt-compounding processes involve the incorporation of CN into thermoplastic polymers by using thermal–mechanical mixing, extrusion of the melt mixture, and optional compression-moulding into specific test specimen geometries and configurations. In one invention, a concentrated suspension of cellulose nanowhiskers and a plasticizer liquid were pumped at the same time into an extruder, giving a partially molten matrix, to produce nanocomposites [215]. The extrusion process operating parameters must be carefully controlled to minimize CN and matrix degradation resulting from shear stresses or temperature. In one study, the preparation of cellulose whiskers-reinforced poly(lactic acid) (PLA) nanocomposites by melt extrusion was carried out by pumping the suspension of nanocrystals into the polymer melt during the extrusion process [227]. However, this process of adding the aqueous-phase fibre suspension into the polymer matrix directly has limitations and is not commercially feasible. But even if the fibres are dried before adding them into the polymer mass, as in conventional melt-mixing, the hydrophilic nature of cellulose causes irreversible agglomeration during drying and aggregation in nonpolar matrices, resulting in poor mechanical properties of the resulting composites. Therefore only a few studies [227, 254] have been reported concerning the processing of CN-reinforced composites by extrusion methods.

Solid-phase compounding process

A solid-phase compounding technique was used to mix the nanofibres isolated from a soybean source with PE or PP [61, 62]. 2.5% and 5% by weight of the coated nanofibres were added to the molten polymer in a laboratory compounder at 170°C. After the samples were well mixed, samples were compressionmoulded with hot press into sheet form at 180°C and under pressure of 50 MPa. Similarly, freeze-dried CNs have been mixed with poly(hydroxy butyrate) (PHB) in a high-intensity kinetic mixer at 150°C followed by compression moulding at 180°C [61, 62]. In other techniques, freeze-dried CNWs and supercritical dried CNW aerogels were used without any solvent for nanocomposite preparation.

Layer-by-layer (LBL) process

A different method for producing nanocomposites was presented in a patent by using the layer-by-layer (LBL) technique [172]. Glass microscope slides were used as a substrate for LBL assembly. The glass slides were sequentially immersed into an aqueous poly(diallyldimethylammonium chloride) (PDDA) solution, and then into a cellulose nanocrystal suspension. This cycle was repeated up to 10 times to produce a multilayer nanocomposite.

19.11.3 Properties

Mechanical properties

The mechanical properties of cellulose nanocomposites are of great interest as more and more potential application areas are now being considered for these materials. Both static and dynamic mechanical properties provide useful information about the mechanical performance of cellulose nanocomposites. The mechanical properties of cellulose nanocomposites, like those of any other composite materials, depend on several factors, including the specific behaviour of each phase, the composition, volume fraction, morphology (spatial arrangement of the phases), orientation of reinforcements, moisture content, degree of crystallinity of the reinforcement, matrix and the interfacial properties. Although cellulose crystals have high axial properties (Young's modulus ~150 GPa, tensile strength ~7000 MPa), this has yet to be fully exploited in CN neat films and composites, in which the properties are typically well below the theoretical upper bounds. Processing methods also strongly influence the mechanical properties of the resulting nanocomposite. For example, it has been found that solvent-casting of nanoscale reinforcements, especially with an aqueous latex dispersion matrix, produces superior results compared to melt-extruded composites, owing to the better dispersion of cellulose nanoscale reinforcements and the possibility of forming hydrogen bonds among reinforcements and matrix material [21].

Static mechanical properties

Incorporation of nanoreinforcements usually increases stiffness, strength and toughness of the polymer matrix. Nanocomposites prepared by adding

5 wt% of swollen MCC to cellulose acetate butyrate (CAB) and poly(lactic acid) (PLA) showed improvements in mechanical performance for the materials [255]. The PLA nanocomposite showed 12% improvement in tensile strength, while with CAB composites a 13% increase in tensile strength and a 135% increase in elongation at break were observed [200]. The property increment was higher with CAB composite due to better dispersion of MCC within the CAB matrix compared to PLA. The tensile modulus, however, was not improved in any of the nanocomposites. The toughness of the CAB nanocomposite increased approximately 300% compared to pure CAB, while the PLA nanocomposites maintained the high toughness of pure PLA. When 10% NFC was used as reinforcements in poly(vinyl alcohol) (PVA), the tensile strength of the resultant film approximately doubled as compared with a non-reinforced pure PVA film. Strong interfacial bonding between hydroxyl groups (–OH) of nanoreinforcements and hydrophilic PVA polymer resulted in an increase in tensile strength of the composite film. A four- to fivefold increase in Young's modulus was observed in this nanofibre-reinforced composite film compared to non-reinforced polymer.

Degree of fibrillation also has a strong influence on mechanical properties of the nanocomposite. With increasing degree of fibrillation (30 passes through the refiner), 5 wt% wood pulp-based MFC-reinforced nanocomposites showed remarkable improvement in mechanical properties, such as higher tensile strength (300 MPa), compared to the result following only 16 passes (200 MPa) [256]. With higher degree of fibrillation, the area of possible contact points per fibre increased which, in turn, led to the possibility of forming more hydrogen bonds and stronger intermolecular forces. Besides, microfibrillation eliminates crack-initiating defects or weaker parts of the original fibres. A blend containing 10 wt% cellulose nanofibres obtained from various sources, such as flax bast fibres, hemp fibres, kraft pulp or rutabaga, and 90 wt% poly(vinyl alcohol) (PVOH) was used for making nanofibre-reinforced composite material by a solution-casting procedure [13]. Both tensile strength and Young's modulus were improved compared to neat PVOH film, with a pronounced four- to fivefold increase in Young's modulus observed. Similarly, Wang and Sain [61, 62] reported that soybean stock-based nanofibre-reinforced PVOH films (up to 10% nanofibre content) demonstrated a doubling of tensile strength when compared to films without filler.

As is evident in naturally occurring composites, the orientation and alignment of the reinforcing phase (CNs) in the composite matrix is a key feature responsible for the improved properties of the final product. Cellulose nanowhiskers have been reportedly oriented by various methods, such as magnetic field [257–261], mechanical drawing [262], shearing forces [263], and electric fields [264]. Kvien and Oksman [260] applied a magnetic

Table 19.2 Average mechanical properties of various nanocellulose reinforced composites. Values in parentheses correspond to the reported values for different matrices.

Material	Nanocellulose content (wt%)	Modulus of elasticity (GPa)	Tensile strength (MPa)	Elongation to break (%)
PLA/MCC	5	4.1 (3.6)	31.9 (28.5)	1.8 (2.4)
PLA/MFC	5	2.9 (2.7)	71.1 (65.5)	2.8 (4.2)
TPS/NFC	5	0.2 (0.1)	5.9 (4.5)	–
CAB/MCC	5	0.2 (0.3)	39.5 (30.3)	40.0 (17.0)
PVA /NFC	5	6.2 (2.3)	102.6 (64.8)	1.7 (1.3)
iPP/CNW	6	–	27.0 (19.0)	12.0 (8.0)
PU/MFC	8.5	107 (25)	11 (5.2)	–

PLA = poly(lactic) acid; TPS = thermoplastic starch; CAB = cellulose acetate butyrate, PVA = poly (vinyl) alcohol), iPP = isotactic polypropylene, PU = polyurethane, MCC = microcrystalline cellulose, MFC = microfibrilated cellulose, NFC = nanofibrilated cellulose, CNW = cellulolose nano whisker.
Source: adapted from Refs 13, 220 and 265.

field (7 tesla) during solution casting of a CNC/PVA suspension (2 wt% CNC), and were able to lock in the preferential alignment of the CNCs. It was anticipated that with the high axial properties of crystalline cellulose, producing composites with highly aligned CNs would increase properties in the aligned direction [262]. Mechanical properties of some of the cellulosic nanocomposites are shown in Table 19.2.

Dynamic mechanical properties

Dynamic mechanical analysis (DMA) is carried out to understand the viscoelastic properties of materials in a wide range of temperatures. In general, both storage modulus (E') and loss modulus (E'') of the nanocomposites are found to increase with increasing nanofibre content and there is a continuing decrease in both moduli as a function of temperature due to glass transition. The temperature dependence of E' and E'' may not change significantly with introduction of nanofillers [220]. DMA results on wheat straw nanofibre-reinforced thermoplastic starch (TPS) polymer showed that the storage modulus was increased from 112 MPa for the pure TPS to 308 MPa for the TPS composite with 10 wt% nanofibre [258]. Tan δ ($\tan \delta = E''/ E'$), also known as damping, is related to glass transition (characterizing the micro-Brownian motions of main polymeric chain segments) and its peak has been found to shift to higher temperatures for nanocomposites compared to the tan δ peak for pure polymer. For example, the tan δ peak of pure CAB was found to occur at approximately 125°C, while for 0.1 wt% loading of native cellulose nanocrystals, the temperature of the peak value increased to 131°C. In another study, with CAB nanocomposites [266] reinforced with 5 wt% cellulose whiskers from MCC, it was reported that the tan δ peak of the matrix was affected by the presence of the cellulosic filler. A slight decrease of the peak temperature from 158 to 152°C was reported. Similar effects were also observed for other nanocomposites such as MCC-reinforced PLA composites [220].

Iwatake *et al.* [215] used a solvent-exchange technique to premix MFC with PLA using acetone and the mixture was kneaded after the removal of the solvent to attain uniform dispersion. As a consequence of uniform distribution the storage modulus of the composites was kept constant above the glass transition temperature of the matrix polymer. More recently, Suryanegara *et al.* [217] applied the same method but then exchanged acetone for dichloromethane and showed that the resulting MFC–PLA nanocomposites had improved storage modulus when compared to neat PLA. The presence of surfactant (20 wt%) in the nanocomposite was also evaluated [267]. Higher shift and increased magnitude of the tan δ peak was observed and was attributed to the presence of the surfactant. The surfactant was able to hinder the segmental movements of PLA chains.

Thermal properties

Glass transition

The glass transition temperature (T_g) is an important parameter, indicative of thermal behaviour of the material. Values of T_g obtained from differential scanning calorimetry (DSC) and dynamic mechanical analysis (DMA) influence mechanical behaviour, matrix chain dynamics and swelling behaviour of the cellulose nanocomposite. Amounts of nanoreinforcement content usually do not affect the glass transition temperature regardless of the nature of the polymer matrix. This has been found to be true with tunicin [268–271], bacterial [17] and wheat straw [259] cellulose nanofibre-based composites. However, with plasticized starch, the level of loading of cellulose nanofibres has been found to affect the T_g. For example, with sorbitol (hexane-1,2,3,4,5,6-hexol) plasticized starch with up to about 15% loading of cellulose nanofibres, the T_g increased, while the plasticization effect decreased for higher loading [271].

Heat capacity

Change in heat capacity of the matrix upon incorporation of fillers can be realized from DSC experiments. With bacterial cellulose-derived nanocrystal-reinforced CAB matrix, a vertical shift of the cooling DSC curves was observed which signified a decrease in specific heat capacity of the matrix [20]. With native crystals, the decrease was abrupt at 2.5 wt% loading and small changes were observed after subsequent loading, while with surface-modified nanocrystals the specific heat capacity decreased gradually. However, the experimental heat capacities of the nanocomposites were found to be smaller than the predicted values calculated from the weighted averages of the components. The discrepancies have been ascribed to the change in heat capacity of the matrix in the interphase.

Melting

Melting temperature (T_m) is another important thermal parameter, especially with semicrystalline polymer matrix. T_m values were found be nearly independent of the filler content in plasticized tunicin whiskers-reinforced starch [192,268,271], in poly(oxyethylene) (POE) [19] composites and with bacterial cellulose whiskers-reinforced cellulose acetate butyrate (CAB) composites. However, with native nanocrystal reinforced CAB, the melting temperature remains constant, while with increasing amounts of trimethylsilylated nanocrystal, the melting point of CAB matrix increases, owing to the stronger filler–matrix interaction in the case of chemically modified whiskers [20].

Thermal degradation

Thermogravimetric analysis (TGA) is carried out to investigate the thermal performance of nanocomposites. Figure 19.7 shows TGA thermograms of thermoplastic starch (TPS) and nanocomposite filled with 5 wt% nanofibre [58]. It is known that starch starts to degrade at around 275°C. The degradation temperature for the nanofibres was around 296°C. TGA thermograms show that the degradation temperatures of the polymer matrix and the nanocomposites are close to each other and lower than those of each component.

Dynamic mechanical thermal analysis (DMTA) of PLA/MCC composites was performed to investigate whether the addition of microcrystalline cellulose (MCC) would improve the thermal properties, such as maximum use temperature, for PLA [220]. It was noted that addition of MCC in PLA increased the softening temperature from 57 to 60°C. The results also indicated that small a improvement in thermal stability was obtained by the addition of nanocellulosic reinforcements to the PLA matrix. The nanocomposite also exhibited an improvement in heat distortion/deflection temperature (HDT) compared to pure PLA. Thermogravimetric analysis of cellulose nanocrystal-reinforced polysulfone composites revealed that at 2 wt% loading of the nanoreinforcements, the broad degradation step associated with cellulose nanocrystal shifted to higher temperature [18].

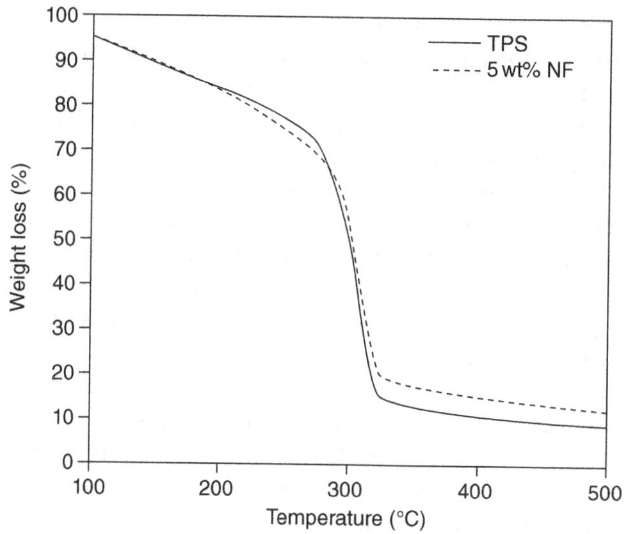

19.7 TGA thermograms of the thermoplastic starch (TPS) and the nanocomposite filled with 5 wt% of cellulose nanofibres (adapted from [57]).

This indicated that the reinforcements and the matrix were associated, thereby altering the thermal stability of the cellulose nanocomposite.

Thermal expansion

Bacterial cellulose (BC) nanocomposites are characterized by unusually reduced thermal expansion properties [256]. The coefficient of thermal expansion coefficient (CTE) of BC/epoxy composite was measured to be 6×10^{-6}/K, which is very low compared to the CTE of epoxy matrix (120×10^{-6}/K). With BC/phenol-formaldehyde, the CTE is found to be even lower, of only 3 ppm/K, comparable to that of silicon crystal.

Crystallinity

An X-ray diffraction (XRD) technique is employed to understand change in crystalline structure of the matrix after the addition of reinforcements. Incorporation of tunicin whiskers was found to increase the crystallinity of sorbitol-plasticized starch [271]. A similar effect could be observed for poly(oxyethylene) (POE) based composites [19], cellulosic whisker-reinforced isotactic polypropylene (iPP) [219] and medium-chain-length poly(hydroxyalkanoate) (mcl-PHA) [272]. While aggregated or surfactant-modified whiskers displayed two crystalline forms (α and β) in the nanocomposites owing to the nucleating effect of the fillers, neat iPP matrix and maleated polypropylene-grafted whisker-reinforced iPP only crystallized in the α-form, indicating that the appearance of the β-phase is favoured if the whisker surface was more hydrophilic. It is anticipated that when semicrystalline polymer such as mcl-PHA is used as the matrix, the formation of the cellulose network is hindered owing to the presence of a transcrystalline region around the whiskers. Cellulose whiskers probably act as nucleating agents for PHA, producing a transcrystalline region around the cellulose whisker. Transcrystallization is the preferential nucleation of polymer melts at crystalline surfaces. All these observations seem to indicate that the nucleating effect is mainly governed by the chemical nature of the filler surfaces. Preferential crystallization of the amorphous polymeric matrix chains during cooling has also been observed with tunicin whiskers-filled glycerol (propane-1,2,3-triol) plasticized starch due to accumulation of plasticizers around the filler/amylopectin interface, thus promoting crystallization of amylopectin chains [269].

Optical properties

Nanoreinforcements are unique in that they do not significantly affect the clarity of the polymer matrix. The nanocomposites appear transparent because the dimensions of the nanoparticles are smaller than the wavelength of visible light, thereby preventing the scattering of light [20, 255, 256]. Optically transparent composites have been produced using bacterial cellulose reinforcements [7, 176, 256]. The fibre contents in these nanocomposites were rather high (70 wt%) with mechanical strength about five times that of engineered plastics. Both epoxy resin (EP) and phenol-formaldehyde (PF) resin were used as the matrix. In the range of wavelengths from 500 to 800 nm, the BC epoxy nanocomposites transmitted more than 80% of the light (surface reflection included), a reduction of less than 10% compared to neat epoxy resin. This is also significant considering the fact that the refractive indices of reinforcing elements and matrix do not exactly match. For MFC reinforcements, at a wavelength of 600 nm, grinder fibrillated cellulose/acrylic nanocomposites with 70 wt% fibre content transmitted 80% of the light, a transmission reduction of just 20%. Nogi *et al.* [24] obtained transparent composites by reinforcing various acrylic resins with BC at loadings up to 70 wt%. Yano *et al.* [25] also reported high light transmittance with bacterial cellulose-reinforced epoxy and acrylic resin composites (70% fibre content), with only less than 10% loss in transparency due to the nanofibre network compared to neat epoxy resin. The nanofibre network led to significantly higher mechanical strength and significantly low thermal expansion coefficient, while retaining the flexibility [26]. CAB-cellulose nanocomposite films prepared by the solvent-casting method also appeared mostly clear [20]. Slight opaqueness was observed with increasing filler content, which, however, disappeared upon annealing. The opaqueness was attributed to the matrix crystallinity.

Barrier properties

The barrier properties, most importantly the oxygen transmission rate (OTR) and water vapour transmission rate (WVTR), are critical for application of any material in the packaging industry. Figure 19.8 shows the barrier properties (OTR versus WVTR) of several materials including nanocellulose. As seen in the figure, nanocelluloses demonstrate low OTR values, while their WVTR values are high. Pure nanocellulose films have even been reported to have an oxygen permeability of 0.07 $cm^3\,\mu m\,m^{-2}\,d^{-1}\,kPa^{-1}$ at 50% relative humidity [273]. The low OTR values of nanocellulose films are due to the low porosity of the well-packed and ordered material (cellulose). On the other hand, the inherent affinity of cellulosic materials to water vapour leads to a high WVTR, via moisture diffusion through cellulose. Other researchers have also found similar observations where different forms of nanocellulose (MFC, NCC) films recorded high OTR values [274, 275] .

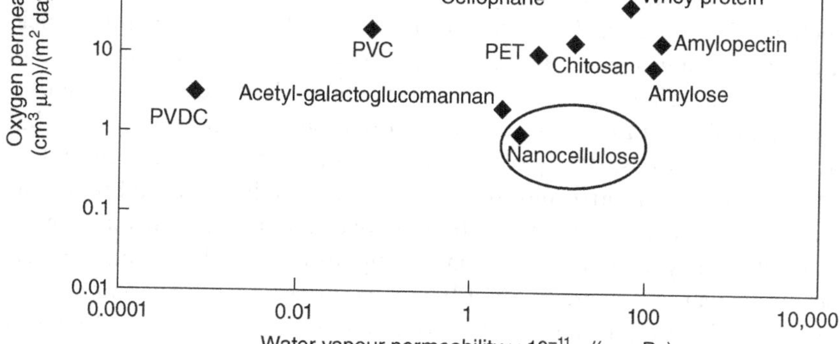

19.8 Oxygen and water vapour permeability of various materials (adapted from [273]).

Biodegradability

Biodegradation is the degradation caused by biological activity, especially by enzymatic action, leading to a significant change in the chemical structure of the exposed material and resulting in the production of carbon dioxide, water, mineral salts (mineralization) and new microbial cellular constituents (biomass) [276]. Currently most of the polymer nanocomposites are prepared with synthetic polymers and nanoscale fillers such as nanosilicates and, as a consequence, their biodegradability is much more limited. Cellulose nanocomposites are advantageous in that respect as cellulose is an abundant natural biodegradable polymer. However, very few researchers have investigated the biodegradability of cellulose nanocomposite. In one investigation on PLA-based nanocellulose-reinforced composites, biodegradation studies showed that PLA started degrading by 3–4 weeks while the composites started degrading rapidly during 4–8 weeks.

19.11.4 Applications

Despite the major advantages of cellulose nancomposites, their use is still restricted to a niche of specialized applications. The main cause can be found in the limited availability of the nanoreinforcement, but also in their tendency to aggregate, which hinders the homogeneous dispersion in the polymer matrix. The major potential applications of cellulose nanocomposites are described below.

Automotive

Cellulose micro- and nanobased materials have received attention in the automobile industry because of their abundance, renewability, low density, environmentally benign nature (non-toxicity) and outstanding mechanical properties [277–280]. Considering that markets for vehicle safety, performance and fuel efficiency provide a demand for new and advanced materials in automotive applications, lower-density cellulose-based nanofillers in composites can address lighter-weight materials that can meet or surpass the performance requirements of the automotive industry. The automotive industry's adoption of cellulose-based materials has therefore been gradually accelerating over the last several years and is mainly driven by regulatory, economic, company policy-related and product-specific concerns. In one study, PE nanocomposites were prepared with various CN loadings using a PVA carrier system by melt blending, followed by injection moulding [281]. The authors claimed that the commercial application for this method would be in automotive applications such as side panels, dashboards, etc. More recently, cellulosic nanofibres have been incorporated and dispersed in soy-based rigid PU foam to explore their potential as a supernucleating agent for development of automotive parts [282]. It was anticipated that in cellulosic nanofiller-reinforced PU foam, development of interconnected micro-bubbles would lead to better heat dissipation in cyclical stress, hence giving superior compressive strength properties. The authors reported enhancement in mechanical properties and that the developed soy PU foam had significant odour reduction compared to commercial foam. All these properties, they reported, can prove to be useful for automotive part applications.

Biomedical

Unique characteristics like nano sizes, superior mechanical properties, and compatibility make nanocelluloses indispensable biomaterials in the health area. The nanocellulose composite scaffolds have been found to be biocompatible with less rejection with cellular contact and blood contact cells interaction. They are considered to be promissory biomaterials, suitable for cell adhesion/attachment, and can be used for wound dressing or as tissue-engineering scaffolds. Researchers showed that nanocellulose stimulated the formation of neural networks [283]. In their investigations, it was found that when the nerve cells were attached to the scaffold they began to develop and generate contacts with one another, so-called synapses [283]. A neural network of hundreds of cells was produced. The researchers could use electrical impulses and chemical signal substances to generate nerve impulses, that spread through the network in much the same way as they do in the brain (Fig. 19.9). From the composition point of view, it is

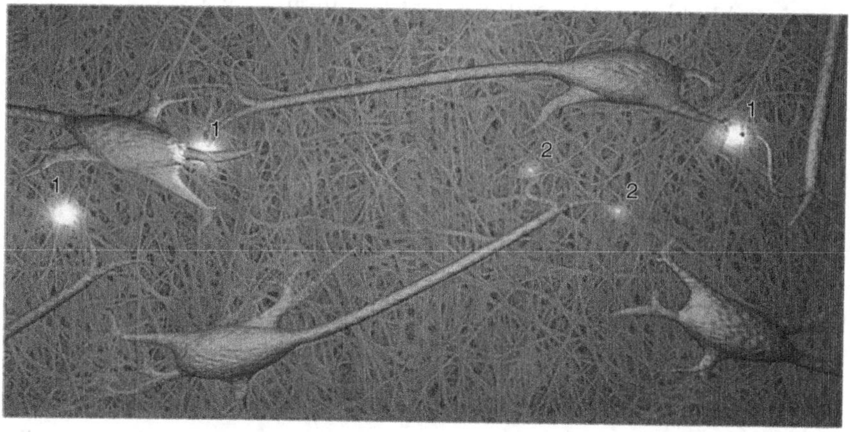

19.9 Nerve cells growing on a three-dimensional nanocellulose scaffold. 1 = functioning synapses and 2 = synapses that have been destroyed (from [283]).

important to note that cardiovascular tissues are composite materials with elastin (an elastic protein structure in connective tissue) and collagen (high-strength fibrous structural proteins) as the main load-bearing components. It was demonstrated that the isotropic PVA–BC nanocomposites were able to closely match the mechanical properties of cardiovascular tissues, such as aorta and heart valve leaflets in selective directions, while anisotropic (poly(vinyl alcohol)) PVA–BC nanocomposite allowed a broader range of control of mechanical properties along with aortic tissue replacement [186, 187, 284].

Nanocomposites containing hydroxyapatite ($Ca_5(PO_4)_3(OH)$, a mineral form of calcium apatite), with structural features close to those of biological apatite, are attractive for applications as artificial bones. A novel class of hydroxyapatite–bacterial cellulose nanocomposite was prepared by Wan *et al.* [285]. Modifying bacterial cellulose with chitosan (a linear polysaccharide produced by deacetylation of chitin) during its biosynthesis and incorporation of antibacterial agents (glucosamine and *N*-acetylglucosamine) into the cellulose chain resulted in a nanocomposite material which was characterized by a number of valuable features: good mechanical properties in the wet state, high moisture-keeping properties, bacteriostatic activity and bactericidal activity [209]. These features make such composite materials an excellent dressing material for treating burns, bedsores, skin ulcers, hard-to-heal wounds and wounds requiring frequent changes of dressing.

Nanocomposite films have also been developed by incorporating cellulose nanocrystals into a polysulfone matrix for potential use as a microchannel device for separation technologies such as kidney dialysis [18]. By adding nanocellulose fillers in a small amount (2 wt%) in the currently used polysulfone polymer, the nanocomposite holds the potential of being used as a portable bioseparation device which will allow dialysis operation at home, resulting in significant improvements in treatment and patient lifestyle. Immersion of nanocellulose into PVA has yielded hydrogels having a wide range of mechanical properties of interest for cardiovascular implants, In the ophthalmology area, Huia *et al.* [286] explored the potential of nanocellulose as a scaffold in engineering cornea tissue.

Other biomedical applications of cellulose nanocomposites include nasal and dental reconstruction, artificial vascular implants, medical devices like biocompatible drug delivery systems, blood bags, temporary skin substitutes, etc. [68, 287–290].

Electrical

There has been a surge of interest in the field of ion-conducting solid polymer electrolytes because of their potential application in rechargeable batteries, fuel cells, light-emitting electrochemical cells, electrochromics, and many other electrochemical devices [291, 292]. Cellulose whiskers have been used as mechanical reinforcing agents of low-thickness polymer electrolytes for application in lithium batteries application [293, 294]. High-performance solid lithium-conducting nanocomposite polymer electrolytes have been prepared from lithium salts such as lithium trifluoromethyl sulfonyl imide (LiTFSI) and polymers such as high-molecular-weight poly(oxyethylene) (POE) and ethylene oxide–epichlorohydrin copolymers (EO–EPI) with addition of high-aspect-ratio cellulose nanocrystalline whiskers. The main effect of the whiskers is reinforcement and thermal stabilization of the storage modulus of composites above the melting point of the polymer–lithium salt complex, while retaining a high level of ionic conductivity with respect to unfilled polymer electrolytes. Using these techniques, the electrolyte thickness could be reduced by a factor of 100 without compromising the conductivity or safety.

Electronic

High composite transparency can be important for some applications (e.g., in the optoelectronics industry). The low thermal expansion coefficient (CTE) of nanocellulosics combined with high strength, high modulus and transparency make them a potential reinforcing material in roll-to-roll technologies (e.g., for fabricating flexible displays, solar cells, electronic paper and panel sensors) [295]. Organic light-emitting diode (OLED) materials were prepared with wood–cellulose nanocomposites [24, 25, 33, 121, 176]. Okahisa *et al.* [121] succeeded in depositing an OLED on flexible, low-CTE and optically transparent wood–cellulose nanocomposites (Fig. 19.10). Sain and his group also developed transparent and flexible nanocomposite composed of bacterial cellulose and polyurethane (PU) based resin, which was used as a substrate for OLED. The nanocomposite had appealing features of flexibility, optical transparency up to 80%, and dimensional stability in terms of CTE as low as 18 ppm/K.

Acoustic

Another practical use of CN-reinforced composites is related to acoustic diaphragms. CN has been found to bear two essential properties: high sonic velocity and low dynamic loss. In fact, the sonic velocity of pure nanocellulose film was found to be almost equivalent to those of aluminum and titanium along with the warm, delicate sound that a paper diaphragm provides [296]. Jonas and Farah [296] stated that Sony had already been using CN-reinforced composites in headphone diaphragms. The nanocomposite diaphragms were developed by dehydration and compressed to a thickness of 20 μm in a diaphragm die.

19.10 Luminescence of an organic light-emitting diode deposited onto a flexible, low-CTE and optically transparent wood–cellulose nanocomposite (from [121]).

Magnetic

Magnetic nanocomposites based on bacterial cellulose substrates containing large quantities of magnetite particles (Fe_2O_3) have been prepared [297]. In BC membranes, needle-like lepidocrocite (γ-FeOOH) was formed along the cellulose fibrils, using the crystalline surface as a nucleation site. Spherical magnetite particles subsequently formed around the needles. The treated BC composite membranes were super-paramagnetic at room temperature.

Biopacking

Cellulose is a renewable and nontoxic biopolymer with biocompatibility with other substances. As a result, cellulose nanocomposite materials have a huge potential for a wide range of applications in the food industry, including innovative active food packaging with biofunctional properties such as antimicrobial packaging [298]. The acceptable structural integrity and barrier properties, along with functional and film-forming properties of cellulose nanocomposite, could be the primary driving force in the development of biopackaging applications.

Smart materials

One of the promising applications of cellulose nanocomposites currently being explored is solidified liquid crystals for optical applications such as security papers [299, 300]. Integration of cellulose nanocrystals in biobased foams can produce smart materials which have the advantages of low density and biodegradability and effective control of orientation of nanocrystals to integrate functional properties such as directional functionality. Because nanocellulose composite foams are ultra-light, they can also be used for biomimetic sensor/actuator devices, and micro-electromechanical systems (MEMS).

19.12 Challenges

19.12.1 Raw materials

Plant-based resources are attractive in producing CNs, but the high levels of variability inherent in these materials, sourced directly from nature, pose concerns over product consistency/quality. Additional problems may arise from uncertainty in quality, from prices of raw materials, and from stocking raw materials for extended periods of time, due to the possibility of degradation, biological attack from fungi, mildew, etc. The extraction of CNs from a given source material also has the inadvertent effect of altering the percent crystallinity and particle morphology, and possibly introducing new defects within the CN. Another factor that has been proved to be a barrier for many potential applications of nanocellulose is the tendency of cellulosic raw material to absorb moisture [11, 146].

19.12.2 Dispersion

As discussed previously, owing to the polar and hydrophilic nature of biobased nanoreinforcements, their dispersion in nonpolar media in order to use them with hydrophobic matrices is a significant challenge. As a result, the majority of reported nanocomposites have been prepared by mixing water-compatible matrices [32, 49, 227, 268]. Another challenge, as with any other nanomaterial, is the uniformity of the dispersion within the polymeric matrices, owing to the nanoscale sizes and aggregation. This is even compounded by the abundance of inter- and intramolecular hydrogen bonds present in nanocellulose. Possible solutions to overcome this challenge have been described earlier in this chapter.

19.12.3 Processing

A reliable recipe for producting CNs of uniform size, aspect ratio and surface chemistry is necessary in order to provide more control in CN suspensions, and in the design and processing of nanocellulose-based composites. Developing extraction processes with tighter controls on CN particle size distributions (length, width, aspect ratio) will also provide more control in CN suspensions, CN-surface functionalization, CN-polymer blends, and the design and processing of CN composites. Thermal stability of cellulose materials is among the other major issues that limit CN applications. The onset of thermal degradation of CN, which typically occurs at around 200°C, provides an upper limit to applications and processing, especially for thermoplastics because their processing temperature often exceeds 200°C.

19.12.4 Nomenclature

Unified and universally accepted terminology and nomenclature are essential to successful communication among all groups working with similar materials. Harmonized terminology and nomenclature are also the essential steps before developing international standards in other areas such as measurement and characterization, environmental health and safety (EHS) and materials specifications. The cellulose particle nomenclature has not been standardized and because of this, there is an inconsistent use of terms in the literature to describe a given set of cellulose particles [9, 63, 118, 146, 168, 301, 302]. While there is a particular urgency to develop terminology and nomenclature for plant-based nanocellulose, international standards for cellulose nanomaterials should also include nanocellulose from bacteria, algae, tunicates and other sources. For example, cellulose whiskers have also been called cellulose monocrystals, cellulose microcrystals, cellulose crystallites, cellulose microcrystallites, microcrystalline cellulose, and more recently cellulose nanowhiskers, cellulose nanocrystals, nanocrystalline cellulose, nanorods, rod-like cellulose microcrystals, rod-like cellulose crystals, nanowires, etc. TAPPI has recently been engaged in developing standardization of cellulose nanomaterial definitions (TAPPI WI 3021: Standard Terms and their Definition for Cellulose Nanomaterials), but their results have not yet been published. In this standard, definitions are structured using terminology as defined by existing ISO standards for nanotechnologies and nanostructured materials.

19.12.5 Scaling-up

Despite the major advantages of nanocellulose, its use so far has been in niche applications. One of the bottlenecks of nanocellulose research is the lack of sufficient quantities, as nanocelluloses have a low yield. Production of nanocellulose has up to now been on the laboratory scale, in kilogram batches. However, there is a worldwide research effort to increase the production to an industry scale. A number of manufacturing facilities have been or are being built that will increase production to upwards of multiple tons per day. Some pilot plant facilities are in the design stage for nanocellulose processing, for example FP Innovation (Canada), Domtar (USA), BioVision Technology Inc. (USA) and USDA-Forest Service Forest Production (USA). It is estimated that nanocellulose production has already increased by approximately 1000% in the last two years and will likely increase by a further 500% at least by 2017 [303].

19.12.6 Cost

Currently, the critical barrier impeding utilization of the full potential of biobased nanoreinforcements is the inability to economically manufacture these materials in sufficient quantities with adequate control and precision. The cost of a kilogram of nanocellulose, like other nanomaterials, is currently incompatible with such cases. Simply put, the potential range of applications of nanomaterials is enormously compromised by the difficulties of manufacturing at competitive cost and in sufficient quantity. As Zafeiropoulos and his co-workers reported, a need to surface-treat cellulose nanomaterials for certain applications in composite fabrication can have negative economic impacts [304]. Successfully overcoming such challenges is critical to realizing the potential of nanomaterials and enabling the social, economic, environmental and technological promises of nanocellulose. For example, by using suitable pretreatments, the number of 'passes' through mechanical refiners can be decreased. Similarly, it has been found that enzymatically pretreated NFC production is an energy-intensive process, though with high yield and low water consumption; while chemically pretreated NFC consumes less energy, but more water is required for the process and it is behind in yield. A comprehensive study, therefore, on energy use, water consumption and yield along with the possibility of use of new catalysts and/or recovery processes, will hold the potential of breakthroughs in lowering CN production costs.

19.12.7 Characterizations and measurements

While the cellulose crystal structures are generally well known, the characterization of CNs is lacking as the small size scale pushes the limits of sensitivity of current methodologies. Improved structural understanding of CNs will facilitate the development of models for the design of improved products. There is a need for the development of standardized measurement methods and reporting for CN mechanical properties. More work is also needed to develop the characterization tools and then to characterize CN–matrix interfaces and the role of the hierarchical structure on the resulting macroscopic properties.

19.12.8 Environmental, health and safety risks

The unique properties of engineered nanoparticles are not only used to achieve novel products and properties, but also have led to the possibility of manipulating biological structure and function in a fundamentally new way, along with the possibility of introducing new types of hazards at the nano–bio interface. Some problems have already been identified with conventional synthetic nanoreinforcements. Exposure to carbon nanofibres and nanotubes has been associated with cell death, DNA damage, epithelial, lung carcinoma, cell/tissue inflammation and fibrosis in human and other animal cells [305–338]. Cellulosic resources (wood pulp, lignocellulosic plant fibres, cellulose powder, etc.), on the other hand, are generally considered safe and can be used, for example, as food additives. However, as with other nanomaterials, the biological effects of nanocellulose cannot be predicted solely from the chemical nature of cellulose. The size, shape, aggregation properties, degree of branching and specific surface properties, among other still poorly understood factors, may affect the interactions of nanocellulose with cells and living organisms. Because of their unique physicochemical properties, biobased nanoreinforcements, nanocellulose in particular, are quite different from and potentially more complex than conventional nanomaterials, chemicals or medicinal agents. Consequently, these biobased nanomaterials produce a new set of challenges for academia, industry and regulatory agencies to develop appropriate safety assessment and risk management strategies.

So far only a handful of studies on the safety of nanocelluloses have been carried out [339–343]. In one study, the cytotoxic and genotoxic properties of nanocelluloses were addressed using well-characterized *in-vitro* model systems commonly applied in toxicity testing protocols [343]. The biological effects were assessed using *in-vitro* bioassays. Cytotoxicity was studied in human or animal cell lines and sub-ethical effects were tested on cultured human cells. The ability to damage DNA was assessed using a bacteriological assay. The samples were characterized using a combination of analytical techniques including scanning electron microscopy (SEM), viscosity, transmittance measurements, and nanoparticle-tracking analysis. The nanocellulose samples did not indicate any eco-toxicological concern.

19.13 Future trends

Growing awareness of sustainability, eco-efficiency and green chemistry has driven a search for renewable raw materials which are more environmentally friendly and can be produced from sustainable resources and processes. This, along with ground-breaking research on cellulose and advancement of nanotechnology, has triggered enormous interest in

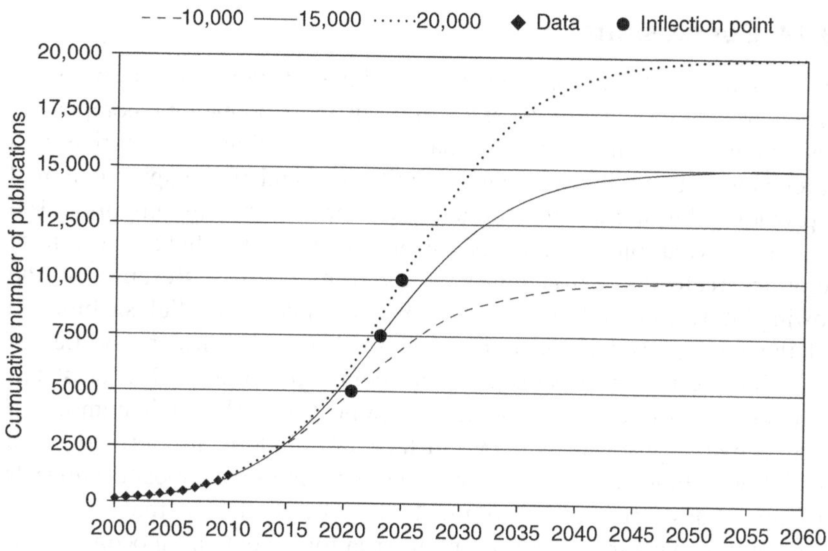

19.11 Extrapolation growth curve from nanocellulose scientific publications (adapted from [344]).

cellulose nanocomposites. The study of nanocellulose is a fascinating field of research, largely because of the potential for it to be used as a material with nanoscale features that is sustainably produced from natural resources. With current trends in miniaturization and 'green products', the use of cellulose nanomaterials is predicted to increase. In a recent study [344], scientific and technological developments on nanocellulose were investigated by using bibliometric indicators obtained from scientific publications and patent documents from 2001 to 2010, as shown in Fig. 19.11. The extrapolation curves suggest there is a long period until the publications and patent documents initiate the maturation stage. This finding means that there are several nanocellulose-related discoveries and technologies to be developed and it is the moment to invest in nanocellulose research.

The future development and growth of biobased nanoreinforcements, however, is based on the understanding of how natural materials work. This type of research will finally lead to the development and mass production of these novel materials. The strategies should include improving the understanding of natural nanostructures, understanding synthesis, adaptation and healing strategies used by nature to build, maintain and repair these usually very complex materials, a comprehensive study on the effect of these nanomaterials on the environment and health and finally the use of this knowledge base to create next generation biobased nanoreinforcements.

19.14 Conclusions

The present chapter reports the use of biobased nanoreinforcements, in particular cellulose nanoreinforcements, in developing nanocomposites. Development of nanocomposites based on nanocellulosic materials is a rather new but rapidly evolving research area and the chapter is aimed at providing knowledge to stimulate further research in this area. The present contribution summarizes various works in the field of cellulose nanoreinforcements, extracted from different sources. Because of the growing interest in the bioconversion of renewable lignocellulosic biomass and the unsurpassed quintessential physical and chemical properties of nanocellulosic materials, substantial academic and industrial interest has been directed toward their potential applications. The high number of reactive hydroxyl groups on the surface of cellulose also provides the possibility for fabricating a wide range of functionalized composite materials for future advanced applications. It has been shown that cellulose nanomaterials have an exciting potential as reinforcements in nanocomposites. Nanoscience and nanotechnologies are leading to the production of biobased nanocomposite materials and devices such as scaffolds for cell and tissue engineering, and sensors that can be used for monitoring aspects of human health. Nevertheless, some drawbacks limit its use, such as aggregation and low compatibility with hydrophobic polymeric matrices, for instance. Moreover, it was found that different terminologies are used to classify cellulose nanoreinforcements, leading to some misunderstanding and ambiguities around the subject rather than unifying and clarifying comprehension.

Despite the constantly growing scientific production in the field of nanocomposites, the practical transition from laboratory scale to industrial scale and the wider use of biobased nanoreinforcements requires the development of suitable technologies with interdisciplinary contributions from scientists and engineers. For example, use of cellulose nanocomposites in vehicle body components is a promising approach to weight reduction, but it is important not to compromise safety, cost or other desired attributes. In the longer term it is hoped that the synergistic advancement of nano and bio-based technologies will enable more efficient approaches to manufacturing biobased nanoreinforcements, with reduced resource use and environmental pollution. This in turn will produce a host of cost-effective multifunctional materials in a socially responsible manner with reduced resource use, environmental and health concerns.

19.15 References

1. R. P. Feynman, '*There's plenty of room at the bottom*', in *Annual Meeting of the American Physical Society*, 1959: California Institute of Technology, USA.

2. N. Taniguchi, '*On the basic concept of nano-technology*', in *International Congress on Production Engineering*, 1974, Japan Society of Precision Engineering, Tokyo, Japan.
3. International Organization for Standardization, '*Nanotechnologies – terminology and definitions for nano-objects – nanoparticle, nanofibre and nanoplate*', ISO/TS 27687: 2008.
4. Y. Kojima, A. Usuki, M. Kawasumi, A. Okada, T. Kurauchi, and O. Kamigaito, '*Synthesis of nylon 6–clay hybrid by montmorillonite intercalated with ε-caprolactam*', Journal of Polymer Science Part A: Polymer Chemistry, 1993. **31**: pp. 983–986.
5. Y. Fukushima, S. Inagaki, O. Kamigaito, M. Kawasumi, T. Kurauchi, A. Okada, S. Sugiyama, and A. Usuki, '*Composite material and process for manufacturing same*', US Patent: US 4,739,007 A: 1988.
6. M. Kawasumi, '*The discovery of polymer-clay hybrids*', Journal of Polymer Science Part A: Polymer Chemistry, 2005. **42**: pp. 819–824.
7. A. N. Nakagaito, and H. Yano, '*Novel high-strength biocomposites based on microfibrillated cellulose having nano-order-unit web-like network structure*', Applied Physics A: Materials Science and Processing, 2005. **80**: pp. 155–159.
8. L. Petersson, and K. Oksman, '*Biopolymer based nanocomposites: comparing layered silicates and microcrystalline cellulose as nanoreinforcement*', Composites Science and Technology, 2006. **66**: pp. 2187–2196.
9. G. Siqueira, J. Bras, and A. Dufresne, '*Cellulosic bionanocomposites: A review of preparation, properties and applications*', Polymers, 2010. **2**: pp. 728–765.
10. I. Kvien, B. S. Tanem, and K. Oksman, '*Characterization of cellulose whiskers and their nanocomposites by atomic force and electron microscopy*', Biomacromolecules, 2005. **6**: pp. 3160–3165.
11. M. A. Hubbe, O. J. Rojas, L. A. Lucia, and M. Sain, '*Cellulosic nanocomposites: A review*', BioResources, 2008. **3**: pp. 929–980.
12. D. Bondsen, I. Kvien, and K. Oksman, '*Strategies for preparation of cellulose whiskers from microcrystalline cellulose as reinforcemement in nanocomposites*', in *Cellulose Nanocomposites – Processing, Characterization and Properties*, K. Oksman and, M. Sain, Editors, 2006. ACS Publication, Washington DC, USA.
13. A. Bhatnagar, and M. Sain, '*Processing of cellulose nanofiber-reinforced composites*', Journal of Reinforced Plastics and Composites, 2005. **24**: pp. 1259–1268.
14. H. Charreau, M. L. Foresti, and A. Vazquez, '*Nanocellulose patents trends: A comprehensive review on patents on cellulose nanocrystals, microfibrillated and bacterial cellulose*', Recent Patents on Nanotechnology, 2013. **7**: pp. 56–80.
15. D. Bondeson, K. Oksman, and P. Syre, '*Nanocomposites based on cellulose whiskers and cellulose plastics*', US Patent: US20080108772 A1: 2008.
16. G. Siqueira, J. Bras, and A. Dufresne, '*Cellulose whiskers versus microfibrils: Influence of the nature of the nanoparticle and its surface functionalization on the thermal and mechanical properties of nanocomposites*', Biomacromolecules, 2008. **10**: pp. 425–432.
17. M. Grunert, and W. T. Winter, '*Nanocomposites of cellulose acetate butyrate reinforced with cellulose nanocrystals*', Journal of Polymers and the Environment, 2002. **10**: pp. 27–30.

18. S. Noorani, J. Simonsen, and S. Atre, '*Polysulfone-cellulose nanocomposites*', in *Cellulose Nanocomposites – Processing, Characterization and Properties*, K. Oksman, and M. Sain, Editors. 2006, ACS Publication, Washington DC.
19. M. A. S. A. Samir, F. Alloin, J. Y. Sanchez, and A. Dufresne, '*Cellulose nanocrystals reinforced poly(oxyethylene)*', Polymer, 2004. **45**: pp. 4149–4157.
20. M. Roman, and W. T. Winter, '*Cellulose nanocrystals for thermoplastic reinforcement: Effect of filler surface chemistry on composite properties*', in *Cellulose Nanocomposites – Processing, Characterization and Properties*, K. Oksman, and M. Sain, Editors. 2006, ACS Publication, Washington DC.
21. V. Favier, G. R. Canova, J. Y. Cavaille, H. Chanzy, A. Dufresne, and C. Gauthier, '*Nanocomposite materials from latex and cellulose whiskers*', Polymers for Advanced Technologies, 1995. **6**: pp. 351–355.
22. X. Li, E. Ding, and G. Li, '*A method of preparing spherical nanocrystals cellulose with mixed crystalline forms of cellulose I and II*', Chinese Journal of Polymer Science, 2001. **19**: pp. 291–296.
23. B. M. Novak, '*Hybrid nanocomposite materials – between inorganic glasses and organic polymers*', Advanced Materials, 1993. **5**: pp. 422–433.
24. M. Nogi, K. Handa, A. N. Nakagaito, and H. Yano, '*Optically transparent bionanofiber composites with low sensitivity to refractive index of the polymer matrix*', Applied Physics Letters, 2005. **87**: pp. 1–3.
25. H. Yano, J. Sugiyama, A. N. Nakagaito, M. Nogi, T. Matsuura, M. Hikita, and K. Handa, '*Optically transparent composites reinforced with networks of bacterial nanofibers*', Advanced Materials, 2005. **17**: pp. 153–155.
26. H. Angellier, S. M. Boisseau, P. Dole, and A. Dufresne, '*Thermoplastic starch–waxy maize starch nanocrystals nanocomposites*', Biomacromolecules, 2006. **7**: pp. 531–539.
27. U. Funke, W. Burgthaller, and M. G. Lindhauer, '*Processing and characterization of biodegradable products based on starch*', Polymer Degradation and Stability, 1998. **59**: pp. 293–296.
28. X. Ma, R. Jian, P. R. Chang, and J. Yu, '*Fabrication and characterization of citric acid-modified starch nanoparticles/plasticized-starch composites*', Biomacromolecules, 2008. **9**: pp. 3314–3320.
29. H. Zheng, F. Ai, P. R. Chang, J. Huang, and A. Dufresne, '*Structure and properties of starch nanocrystal-reinforced soy protein plastics*', Polymer Composites, 2009. **30**: pp. 474–480.
30. K. G. Nair, and A. Dufresne, '*Crab shell chitin whisker reinforced natural rubber nanocomposites. 1. Processing and swelling behavior*', Biomacromolecules, 2003. **4**: pp. 657–665.
31. K. G. Nair, A. Dufresne, A. Gandini, and M. N. Belgacem, '*Crab shell chitin whiskers reinforced natural rubber nanocomposites. 3. Effect of chemical modification of chitin whiskers*', Biomacromolecules, 2003. **4**: pp. 1835–1842.
32. M. Roohani, Y. Habibi, N. M. Belgacem, G. Ebrahim, A. N. Karimi, and A. Dufresne, '*Cellulose whiskers reinforced polyvinyl alcohol copolymers nanocomposites*', European Polymer Journal, 2008. **44**: pp. 2489–2498.
33. S. M. Iftekhar, S. Ifuku, M. Nogi, T. Oku, and H. Yano, '*Fabrication of optically transparent chitin nanocomposites*', Applied Physics A, 2011. **102**: pp. 325–331.
34. L. Jong, '*Characterization of soy protein nanoparticles prepared by high shear microfluidization*', Journal of Dispersion Science and Technology, 2013. **34**: pp. 469–475.

35. Y. Wang, S. Serrano, and J. Santiago-Aviles, '*Conductivity measurement of electrospun pan-based carbon nanofiber*', Journal of Materials Science Letters, 2002. **21**: pp. 1055–1057.
36. S. L. Hu, K. Y. Niu, J. Sun, J. Yang, N. Q. Zhao, and X. W. Du, '*One-step synthesis of fluorescent carbon nanoparticles by laser irradiation*', Journal of Materials Chemistry, 2009. **19**: pp. 484–488.
37. M. Sharon, '*Carbon nanomaterials and their synthesis from plant-derived precursors*', Synthesis and Reactivity in Inorganic, Metal-Organic and Nano-Metal Chemistry, 2006. **36**: pp. 265–279.
38. K. Sudo, and K. Shimizu, '*A new carbon fiber from lignin*', Journal of Applied Polymer Science, 1992. **44**: pp. 127–134.
39. Suhas, P. Carrott, and M. R. Carrott, '*Lignin – from natural adsorbent to activated carbon: A review*', Bioresource Technology, 2007. **98**: pp. 2301–2312.
40. J. Kadla, S. Kubo, R. Venditti, R. Gilbert, A. Compere, and W. Griffith, '*Lignin-based carbon fibers for composite fiber applications*', Carbon, 2002. **40**: pp. 2913–2920.
41. M. Lallave, J. Bedia, R. Ruiz-Rosas, J. Rodríguez-Mirasol, T. Cordero, J. C. Otero, M. Marquez, A. Barrero, and I. G. Loscertales, '*Filled and hollow carbon nanofibers by coaxial electrospinning of alcell lignin without binder polymers*', Advanced Materials, 2007. **19**: pp. 4292–4296.
42. S. Fangeng, F. Yu, and Z. Xuesong, '*Lignin carbon nanofiber and preparation method thereof*', Chinese Patent: CN101768799 B: 2010.
43. D. Klemm, B. Heublein, H. P. Fink, and A. Bohn, '*Cellulose: Fascinating biopolymer and sustainable raw material*', Angewandte Chemie International Edition, 2005. **44**: pp. 3358–3393.
44. P. Zugenmaier, '*Conformation and packing of various crystalline cellulose fibers*', Progress in Polymer Science, 2001. **26**: pp. 1341–1417.
45. C. Zhou, and Q. Wu, '*Recent development in applications of cellulose nanocrystals for advanced polymer-based nanocomposites by novel fabrication strategies*', in *Nanocrystals – Synthesis, Characterization and Applications*, S. Neralla, Editor. 2012, InTech Publisher.
46. A. Dufresne, D. Dupeyre, and M. R. Vignon, '*Cellulose microfibrils from potato tuber cells: Processing and characterization of starch-cellulose microfibril composites*', Journal of Applied Polymer Science, 2000. **76**: pp. 2080–2092.
47. M. Postek, A. Vladar, J. Dagata, N. Farkas, B. Ming, R. Wagner, A. Raman, R. J. Moon, R. Sabo, T. H. Wegner, and J. Beecher, '*Development of the metrology and imaging of cellulose nanocrystals*', Measurement Science and Technology, 2011. **22**: p. 024005.
48. L. Chazeau, J. Y. Cavaillé, G. Canova, R. Dendievel, and B. Boutherin, '*Viscoelastic properties of plasticized PVC reinforced with cellulose whiskers*', Journal of Applied Polymer Science, 1999. **71**: pp. 1797–1808.
49. Y. Lu, L. Weng, and X. Cao, '*Biocomposites of plasticized starch reinforced with cellulose crystallites from cottonseed linter*', Macromolecular Bioscience, 2005. **5**: pp. 1101–1107.
50. P. B. Filson, and B. E. Dawson-Andoh, '*Sono-chemical preparation of cellulose nanocrystals from lignocellulose derived materials*', Bioresource Technology, 2009. **100**: pp. 2259–2264.
51. E. H. Qua, P. R. Hornsby, H. S. S. Sharma, G. Lyons, and R. D. McCall, '*Preparation and characterization of poly(vinyl alcohol) nanocomposites made*

from cellulose nanofibers', Journal of Applied Polymer Science, 2009. **113**: pp. 2238–2247.

52. P. Satyamurthy, P. Jain, R. H. Balasubramanya, and N. Vigneshwaran, '*Preparation and characterization of cellulose nanowhiskers from cotton fibres by controlled microbial hydrolysis*', Carbohydrate Polymers, 2011. **83**: pp. 122–129.
53. Y. Habibi, and A. Dufresne, '*Highly filled bionanocomposites from functionalised polysaccharide nanocrystals*', Biomacromolecules, 2008. **9**: pp. 1974–1980.
54. N. L. G. de Rodriguez, W. Thielemans, and A. Dufresne, '*Sisal cellulose whiskers reinforced polyvinyl acetate nanocomposites*', Cellulose, 2006. **13**: pp. 261–270.
55. J. I. Morán, V. A. Alvarez, V. P. Cyras, and A. Vázquez, '*Extraction of cellulose and preparation of nanocellulose from sisal fibers*', Cellulose, 2008. **15**: pp. 149–159.
56. G. Siqueira, J. Bras, Y. Habibi, and A. Dufresne, '*Sisal and Luffa cylindrica cellulose whiskers reinforcing phase in nanocomposites*', in *Recent Advances in Research on Biodegradable Polymers and Sustainable Composites*, A. Jimenez, and G. E. Zaikov, Editors. 2008, Nova Publishers pp. 197–205.
57. A. Alemdar, and M. Sain, '*Isolation and characterization of nanofibers from agricultural residues – wheat straw and soy hulls*', Bioresource Technology, 2008. **99**: pp. 1664–1671.
58. A. Alemdar, and M. Sain, '*Biocomposites from wheat straw nanofibers: morphology, thermal and mechanical properties*', Composites Science and Technology, 2008. **68**: pp. 557–565.
59. W. Helbert, J. Y. Cavaillé, and A. Dufresne, '*Thermoplastic nanocomposites filled with wheat straw cellulose whiskers. Part I: Processing and mechanical behavior*', Polymer Composites, 1996. **17**: pp. 604–611.
60. N. Y. Uesu, E. A. G. Pineda, and A. A. W. Hechenleitner, '*Microcrystalline cellulose from soybean husk: Effects of solvent treatments on its properties as acetylsalicylic acid carrier*', International Journal of Pharmaceutics, 2000. **206**: pp. 85–96.
61. B. Wang, and M. Sain, '*Dispersion of soybean stock-based nanofiber in a plastic matrix*', Polymer International, 2007. **56**: pp. 538–546.
62. B. Wang, and M. Sain, '*Isolation of nanofibers from soybean source and their reinforcing capability on synthetic polymers*', Composites Science and Technology, 2007. **67**: pp. 2521–2527.
63. M. Iguchi, S. Yamanaka, and A. Budhiono, '*Bacterial cellulose – a masterpiece of nature's arts*', Journal of Materials Science, 2000. **35**: pp. 261–270.
64. P. Gatenholm, and D. Klemm, '*Bacterial nanocellulose as a renewable material for biomedical applications*', MRS Bulletin, 2010. **35**: pp. 208–213.
65. S. Yamanaka, K. Watanabe, N. Kitamura, M. Iguchi, S. Mitsuhashi, Y. Nishi, and M. Uryu, '*The structure and mechanical properties of sheets prepared from bacterial cellulose*', Journal of Materials Science, 1989. **24**: pp. 3141–3145.
66. D. C. Johnson, and A. R. Winslow, '*Bacterial cellulose has potential application as new paper coating*', Pulp and Paper, 1990: pp. 105–107.
67. E. Cannon, and S. M. Anderson, '*Biogenesis of bacterial cellulose*', Critical Reviews in Microbiology, 1991. **17**: pp. 435–447.
68. J. D. Fontana, A. M. Souza, C. K. Fontana, I. L. Torriani, J. C. Moreschi, B. J. Gallotti, S. J. Souza, G. P. Narcisco, J. A. Bichara, and L. F. X. Farah, '*Acetobacter*

cellulose pellicle as a temporary skin substitute', Applied Biochemistry and Biotechnology, 1990. **24–25**: pp. 253–264.

69. H. Yamamoto, and F. Horii, '*In situ crystallization of bacterial cellulose I. Influences of polymeric additives, stirring and temperature on the formation celluloses I_α and I_β as revealed by cross polarization/magic angle spinning (CP/MAS) ^{13}C NMR spectroscopy*', Cellulose, 1994. **1**: pp. 57–66.
70. B. M. Cherian, A. L. Leão, S. F. de Souza, G. M. de Olyveira, L. M. M. Costa, C. V. S. Brandão, and S. S. Narine, '*Bacterial nanocellulose for medical implants*', in *Advances in Natural Polymers, Composites and Nanocomposites*, S. Thomas, P. M. Visakh, and A. P. Mathew, Editors. 2013. Springer-Verlag.
71. M. Janardhnan, and M. Sain, '*Isolation of cellulose microfibrils – an enzymatic approach*', Bioresource Technology, 2004. **1**: pp. 176–188.
72. A. F. Turbak , '*Recent developments in cellulose solvent systems*', TAPPI Journal, 1984. **67**: pp. 94–96.
73. A. F. Turbak, F. W. Snyder, and K. R. Sandberg, '*Microfibrillated cellulose, a new cellulose product: Properties, uses and commercial potential*', in *Proceedings of the 9th Cellulose Conference. Applied Polymer Symposia*, 1983, Wiley: New York.
74. T. Saito, Y. Nishiyama, J. L. Putaux, M. Vignon, and A. Isogai, '*Homogeneous suspensions of individualized microfibrils from TEMPO-catalyzed oxidation of native cellulose*', Biomacromolecules, 2006. **7**: pp. 1687–1691.
75. R. K. Johnson, A. Zink-Sharp, S. H. Renneckar, and W. G. Glasser, '*A new bio-based nanocomposite: Fibrillated TEMPO oxidized celluloses in hydroxypropylcellulose matrix*', Cellulose, 2009. **16**: pp. 227–238.
76. J. V. Milewski, '*Whiskers*', in *Concise Encyclopedia of Composite Materials*, A. Kelly, Editor. 1994, Pergamon: New York.
77. R. L. Whistler, and J. N. BeMiller, '*Carbohydrate Chemistry for Food Scientists*', 1997. Eagan Press, St Paul, MN.
78. M. El-Sakhawy, and M. L. Hassan, '*Physical and mechanical properties of microcrystalline cellulose prepared from agricultural residues*', Carbohydrate Polymers, 2007. **67**: pp. 1–10.
79. W. Helbert, J. Y. Cavaillé, and A. Dufresne, '*Thermoplastic nanocomposites filled with wheat straw cellulose whiskers. Part II: Effect of processing and modeling*', Polymer Composites, 1997. **18**: pp. 198–210.
80. M. Stromme, A. Mihranyan, and R. Ek, '*What to do with all these algae?*', Materials Letters, 2002. **57**: pp. 569–572.
81. M. Samir, F. Alloin, M. Paillet, and A. Dufresne, '*Tangling effect in fibrillated cellulose reinforced nanocomposites*', Macromolecules, 2004. **37**: pp. 4313–4316.
82. A. Dufresne, and M. R. Vignon, '*Improvement of starch film performances using cellulose microfibrils*', Macromolecules, 1998. **31**: pp. 2693–2696.
83. H. P. Fink, H. J. Purz, A. Bohn, and J. Kunze, '*Investigation of the supramolecular structure of never dried bacterial cellulose*', in *2nd Japanese/German Seminar on Future Development of Polysaccharides – Fundamentals and Applications*, 1996, Sapporo, Japan.
84. H. Q. Liu, and Y. L. Hsieh, '*Ultrafine fibrous cellulose membranes from electrospinning of cellulose acetate*', Journal of Polymer Science Part B Polymer Physics, 2002. **40**: pp. 2119–2129.
85. M. S. Khil, H. Y. Kim, Y. S. Kang, H. J. Bang, D. R. Lee, and J. K. Doo, '*Preparation of electrospun oxidized cellulose mats and their in vitro degradation behavior*', Macromolecular Research, 2005. **13**: pp. 62–67.

86. C. W. Kim, D. S. Kim, S. Y. Kang, M. Marquez, and Y. L. Joo, '*Structural studies of electrospun cellulose nanofibres*', Polymer, 2006. **47**: pp. 5097–5107.
87. A. Frenot, M. W. Henriksson, and P. Walkenström, '*Electrospinning of cellulose-based nanofibers*', Journal of Applied Polymer Science, 2007. **103**: pp. 1473–1482.
88. D. J. Gardner, G. S. Oporto, R. Mills, and M. A. S. A. Samir, '*Adhesion and surface issues in cellulose and nanocellulose*', Journal of Adhesion Science and Technology, 2008. **22**: pp. 545–567.
89. J S. J. Hanley, J. F. Revol, L. Godbout, and D. G. Gray, '*Atomic force microscopy and transmission electron microscopy of cellulose microfibrils from Micrasterias denticulata; Evidence for a chiral helical microfibril twist*', Cellulose, 1997. **4**: pp. 209–220.
90. T. Saito, S. Kimura, Y. Nishiyama, and A. Isogai, '*Cellulose nanofibers prepared by TEMPO-mediated system*', Biomacromolecules, 2007. **8**: pp. 2485–2491.
91. S. Ifuku, M. Nogi, K. Abe, K. Handa, F. Nakatsubo, and H. Yano, '*Surface modification of bacterial cellulose nanofibers for property enhancement of optically transparent composites: Dependence on acetyl-group DS*', Biomacromolecules, 2007. **8**: pp. 1973–1978.
92. A. Sneck. '*New approach to classification of cellulose fibrils and suitable methods for their characterization*', in *TAPPI International Conference on Nanotechnology for Renewable Materials*, 2011, TAPPI Press, Washington, DC.
93. W. H. Mason, '*Process and apparatus for disintegration of wood and the like*', US Patent: US 1578609: 1926.
94. B. M. Cherian, A. L. Leăo, S. F. de Souza, S. Thomas, L. A. Pothan, and M. Kottaisamy, '*Isolation of nanocellulose from pineapple leaf fibres by steam explosion*', Carbohydrate Polymers, 2010. **81**: pp. 720–725.
95. S. Elazzouzi-Hafraoui, Y. Nishiyama, J. L. Putaux, L. Heux, F. Dubreuil, and C. Rochas, '*The shape and size distribution of crystalline nanoparticles prepared by acid hydrolysis of native cellulose*', Biomacromolecules, 2008. **9**: pp. 57–65.
96. B. S. L. Brito, F. V. Pereira, J. L Putaux, and B. Jean, '*Preparation, morphology and structure of cellulose nanocrystals from bamboo fibers*', Cellulose, 2012. **19**: pp. 1527–1536.
97. D. Zhang, Qing Zhang, Xin Gao, and Guangzhe Piao, '*A nanocellulose polypyrrole composite based on tunicate cellulose*', International Journal of Polymer Science, 2013, pp. 1–6.
98. S. Kimura, and T. Itoh, '*New cellulose synthesizing complexes (terminal complexes) involved in animal cellulose biosynthesis in tunicate Metandrocarpa uedai*', Protoplasma, 1996. **194**: pp. 151–163.
99. W. Helbert, Y. Nishiyama, T. Okano, and J. Sugiyama, '*Molecular imaging of Halocynthia papillosa cellulose*', Journal of Structural Biology, 1998. **124**: pp. 42–50.
100. S. Kimura and T. Itoh, '*Cellulose synthesizing terminal complexes in the ascidians*', Cellulose, 2004. **11**: pp. 377–383.
101. J. F. Revol, '*On the cross-sectional shape of cellulose crystallites in Valonia ventricosa*', Carbohydrate Polymers, 1982. **2**: pp. 123–134.
102. J. Sugiyama, H. Harada, Y. Fujiyoshi, and N. Uyeda, '*Lattice images from ultrathin sections of cellulose microfibrils in the cell wall of Valonia macrophysa kütz*', Planta, 1985. **166**: pp. 161–168.

103. S. J. Hanley, J. Giasson, J. F. Revol, and D. G. Gray, '*Atomic force microscopy of cellulose microfibrils: Comparison with transmission electron microscopy*', Polymer, 1992. **33**: pp. 4639–4642.
104. C. Tokoh, K. Takabe, M. Fujita, and H. Saiki, '*Cellulose synthesized by Acetobacter xylinum in the presence of acetyl glucomannan*', Cellulose, 1998. **5**: pp. 249–261.
105. X. Qian, Y. Y. Lee, P. Pettersson, and R. W. Torget, '*Heterogeneous aspects of acid hydrolysis of α-cellulose*', in *Biotechnology for Fuels and Chemicals*, B. Davison, J. Lee, M. Finkelstein, and J. McMillan, Editors. 2003, Humana Press. pp. 505–514.
106. Y. Chen, C. Liu, P. R. Chang, X. Cao, and D. P. Anderson, '*Bionanocomposites based on pea starch and cellulose nanowhiskers hydrolyzed from pea hull fiber: Effect of hydrolysis time*', Carbohydrate Polymers, 2009. **76**: pp. 607–615.
107. J. Zhang, T. J. Elder, Y. Pu, and A. J. Ragauskas, '*Facile synthesis of spherical cellulose nanoparticles*', Carbohydrate Polymers, 2007. **69**: pp. 607–611.
108. D. Bondeson, A. Mathew, and K. Oksman, '*Optimization of the isolation of nanocrystals from microcrystalline cellulose by acid hydrolysis*', Cellulose, 2006. **13**: pp. 171–180.
109. S.-Y. Lee, D. J. Mohan, I.-A. Kang, G.-H. Doh, S. Lee, and S. O. Han, '*Nanocellulose reinforced PVA composite films: Effects of acid treatment and filler loading*', Fibers and Polymers, 2009. **10**: pp. 77–82.
110. M. F. Rosa, E. S. Medeiros, J. A. Malmonge, K. S. Gregorski, D. F. Wood, L. H. C. Mattoso, G. Glenn, W. J. Orts, and S. H. Imam, '*Cellulose nanowhiskers from coconut husk fibers: Effect of preparation conditions on their thermal and morphological behavior*', Carbohydrate Polymers, 2010. **81**: pp. 83–92.
111. N. Wang, E. Ding, and R. Cheng, '*Thermal degradation behaviors of spherical cellulose nanocrystals with sulfate groups*', Polymer, 2007. **48**: pp. 3486–3493.
112. E. S. Abdel-Halim, H. E. Emam, and M. H. El-Rafie, '*Preparation and characterization of water soluble poly(acrylic acid)–hydroxypropyl cellulose composite*', Carbohydrate Polymers, 2008. **74**: pp. 783–786.
113. R. Zuluaga, J. L. Putaux, J. Cruz, J. Vélez, I. Mondragon, and P. Gañán, '*Cellulose microfibrils from banana rachis: Effect of alkaline treatments on structural and morphological features*', Carbohydrate Polymers, 2009. **76**: pp. 51–59.
114. A. Isogai, T. Saito, and H. Fukuzumi, '*TEMPO-oxidized cellulose nanofibers*', Nanoscale, 2011. **3**: pp. 71–85.
115. S. Fujisawa, Y. Okita, H. Fukuzumi, T. Saito, and A. Isogai, '*Preparation and characterization of TEMPO-oxidized cellulose nanofibril films with free carboxyl groups*', Carbohydrate Polymers, 2011. **84**: pp. 579–583.
116. S. Wolfgang, and A. R. Sanadi, '*Preparation and characterization of cellulose nanofibers from two commercial hardwood and softwood pulps*', Industrial and Engineering Chemistry Research, 2009. **48**: pp. 11211–11219.
117. P. Stenstad, M. Andresen, B. S. Tanem, and P. Stenius, '*Chemical surface modifications of microfibrillated cellulose*', Cellulose, 2008. **15**: pp. 35–45.
118. I. Siró and D. Plackett, '*Microfibrillated cellulose and new nanocomposite materials: A review*', Cellulose, 2010. **17**: pp. 459–494.
119. S.-Y. Lee, S.-J. Chun, I.-A. Kang, and J.-Y. Park, '*Preparation of cellulose nanofibrils by high-pressure homogenizer and cellulose-based composite films*', Journal of Industrial and Engineering Chemistry, 2009. **15**: pp. 50–55.

120. A. Chakraborty, M. Sain, and M. Kortschot, '*Cellulose microfibrils: A novel method of preparation using high shear refining and cryocrushing*', Holzforschung, 2005. **59**: pp. 102–107.
121. Y. Okahisa, A. Yoshida, S. Miyaguchi, and H. Yano, '*Optically transparent wood–cellulose nanocomposite as a base substrate for flexible organic light-emitting diode displays*', Composites Science and Technology, 2009. **69**: pp. 1958–1961.
122. A. N. Nakagaito, and H. Yano, '*The effect of fiber content on the mechanical and thermal expansion properties of biocomposites based on microfibrillated cellulose*', Cellulose, 2008. **15**: pp. 555–559.
123. K. Missoum, M. N. Belgacem, and J. Bras, '*Nanofibrillated cellulose surface modification: A review*', Materials, 2013. **6**: pp. 1745–1766.
124. H. Jin, M. Kettunen, A. Laiho, H. Pynnoönen, J. Paltakari, A. Marmur, O. Ikkala, and R. H. A. Ras, '*Superhydrophobic and superoleophobic nanocellulose aerogel membranes as bioinspired cargo carriers on water and oil*', Langmuir, 2011. **27**: pp. 1930–1934.
125. S. Wang, and Q. Cheng, '*A novel process to isolate fibrils from cellulose fibers by high-intensity ultrasonication, part 1: Process optimization*', Journal of Applied Polymer Science, 2009. **113**: pp. 1270–1275.
126. S. Xu, J. Zhang, A. He, J. Li, H. Zhang, and C. C. Han, '*Electrospinning of native cellulose from nonvolatile solvent system*', Polymer, 2008. **49**: pp. 2911–2917.
127. M. Frey, and Y. L Joo, '*Cellulose solution in novel solvent and electrospining thereof*', US Patent: US2005247236A1: 2005.
128. P. Kulpinski, '*Cellulose nanofibers prepared by the N-methylmorpholine-N-oxide method*', Journal of Applied Polymer Science, 2005. **98**: pp. 1855–1859.
129. C.-W. Kim, M. W. Frey, M. Marquez, and Y. L. Joo, '*Preparation of submicron-scale, electrospun cellulose fibers via direct dissolution*', Journal of Polymer Science Part B: Polymer Physics, 2005. **43**: pp. 1673–1683.
130. K. Ohkawa, S. Hayashi, A. Nishida, H. Yamamoto, and J. Ducreux, '*Preparation of pure cellulose nanofiber via electrospinning*', Textile Research Journal, 2009. **79**: pp. 1396–1401.
131. M. Hasegawa, A. Isogai, F. Onabe, and M. Usuda, '*Dissolving states of cellulose and chitosan in trifluoroacetic acid*', Journal of Applied Polymer Science, 1992. **45**: pp. 1857–1863.
132. M. W. Frey, '*Electrospinning cellulose and cellulose derivatives*', Polymer Reviews, 2008. **48**: pp. 378–391.
133. M. Jonoobi, J. Harun, A. Shakeri, M. Misra, and K. Oksman, '*Chemical composition, crystallinity, and thermal degradation of bleached and unbleached kenaf bast (Hibiscus cannabinus) pulp and nanofibers*', BioResources, 2009. **4**: pp. 626–639.
134. M. Henriksson, G. Henriksson, L. A. Berglund, and T. Lindström, '*An environmentally friendly method for enzyme-assisted preparation of microfibrillated cellulose (MFC) nanofibers*', European Polymer Journal, 2007. **43**: pp. 3434–3441.
135. P. B. Filson, B. E. Dawson-Andoh, and D. Schwegler-Berry, '*Enzymatic-mediated production of cellulose nanocrystals from recycled pulp*', Green Chemistry, 2009. **11**: pp. 1808–1814.

136. S. Beck-Candanedo, M. Roman, and D. G. Gray, '*Effect of reaction conditions on the properties and behavior of wood cellulose nanocrystals suspensions*', Biomacromolecules, 2005. **6**: pp. 1048–1054.
137. S. Spoljaric, A. Genovese, and R. A. Shanks, '*Polypropylene–microcrystalline cellulose composites with enhanced compatibility and properties*', Composites Part A: Applied Science and Manufacturing, 2009. **40**: pp. 791–799.
138. S. Padalkar, J. R. Capadona, S. J. Rowan, C. Weder, Y. H. Won, L. A. Stanciu, and R. J. Moon, '*Natural biopolymers: Novel templates for the synthesis of nanostructures*', Langmuir, 2010. **26**: pp. 8497–8502.
139. L. Heux, G. Chauve, and C. Bonini, '*Nonflocculating and chiral-nematic self-ordering of cellulose microcrystals suspensions in nonpolar solvents*', Langmuir, 2000. **16**: pp. 8210–8212.
140. L. Heux, and C. Bonini, '*Microfibrillated and/or microcrystalline dispersion, in particular of cellulose, in an organic solvent*', International Patent: WO 2000/077088: 2000.
141. E. D. Cranston, and D. G. Gray, '*Morphological and optical characterization of polyelectrolyte multilayers incorporating nanocrystalline cellulose*', Biomacromolecules, 2006. **7**: pp. 2522–2530.
142. C. Goussé, H. Chanzy, G. Excoffier, L. Soubeyrand, and E. Fleury, '*Stable suspensions of partially silylated cellulose whiskers dispersed in organic solvents*', Polymer, 2002. **43**: pp. 2645–2651.
143. Y. Habibi, H. Chanzy, and M. Vignon, '*TEMPO-mediated surface oxidation of cellulose whiskers*', Cellulose. **13**: pp. 679–687.
144. B. Braun, and J. R. Dorgan, '*Single-step method for the isolation and surface functionalisation of cellulosic nanowhiskers*', Biomacromolecules, 2009. **10**(2): pp. 334–341.
145. M. Hasani, E. D. Cranston, G. Westman, and D. G. Gray, '*Cationic surface functionalisation of cellulose nano-crystals*', Soft Matter, 2008. **4**: pp. 2238–2244.
146. R. J. Moon, A. Martini, J.Nairn, J. Simonsen, and J. Youngblood, '*Cellulose nanomaterials review: Structure, properties and nanocomposites*', Chemical Society Reviews, 2011. **40**: pp. 3941–3994.
147. J. Sassi, and H. Chanzy, '*Ultrastructural aspects of the acetylation of cellulose*', Cellulose, 1995. **2**: pp. 111–127.
148. H. Yuan, Y. Nishiyama, M. Wada, and S. Kuga, '*Surface acylation of cellulose whiskers by drying aqueous emulsion*', Biomacromolecules, 2006. **7**: pp. 696–700.
149. N. S. Cetin, P. Tingaut, N. Özmen, N. Henry, D. Harper, M. Dadmun, and G. Sèbe, '*Acetylation of cellulose nanowhiskers with vinyl acetate under moderate conditions*', Macromolecular Bioscience, 2009. **9**: pp. 997–1003.
150. S. Berlioz, S. Molina-Boisseau, Y. Nishiyama, and L. Heux, '*Gas-phase surface esterification of cellulose microfibrils and whiskers*', Biomacromolecules, 2009. **10**: pp. 2144–2151.
151. A. de Menezes Jr., G. Siqueira, A. A. S. Curvelo, and A. Dufresne, '*Extrusion and characterisation of functionalised cellulose whiskers reinforced polyethylene nano-composites*', Polymer, 2009. **50**: pp. 4552–4563.
152. C. Gousse, H. Chanzy, M. L. Cerrada, and E. Fleury, '*Surface silylation of cellulose microfibrils: Preparation and rheological properties*', Polymer, 2004. **45**: pp. 1569–1575.

153. M. Andresen, L. S. Johansson, B. S. Tanem, and P. Stenius, '*Properties and characterization of hydrophobized microfibrillated cellulose*', Cellulose, 2006. **13**: pp. 665–677.
154. N. Bordeanu, C. Eyholzer, and T. Zimmermann, '*Surface modified cellulose nanofibers*'. International Patent: WO 2010/066905 A1: 2010.
155. S. Dong, and M. Roman, '*Fluorescently labeled cellulose nanocrystals for bioimaging applications*', Journal of the American Chemical Society, 2007. **129**: pp. 13810–13811.
156. Y. Habibi, A. Goffin, N. Schiltz, E. Duquesne, P. Dubois, and A. Dufresne, '*Bionanocomposites based on poly (ε-caprolactone)-grafted cellulose nanocrystals by ring-opening polymerisation*', Journal of Materials Chemistry, 2008. **18**: pp. 5002–5010.
157. G. Morandi, L. Heath, and W. Thielemans, '*Cellulose nanocrystals grafted with polystyrene chains through surface-initiated atom transfer radical polymerisation (SI-ATRP)*', Langmuir, 2009. **25**: pp. 8280–8286.
158. J. Yi, Q. Xu, X. Zhang, and H. Zhang, '*Chiral-nematic self-ordering of rodlike cellulose nanocrystals grafted with poly(styrene) in both thermotropic and lyotropic states*', Polymer, 2008. **49**: pp. 4406–4412.
159. H. Lönnberg, L. Fogelström, M. A. S. A. Samir, L. Berglund, E. Malmström, and A. Hult, '*Surface grafting of microfibrillated cellulose with poly(ε-caprolactone) – synthesis and characterization*', European Polymer Journal, 2008. **44**(9): pp. 2991–2997.
160. W. Thielemans, M. N. Belgacem, and A. Dufresne, '*Starch nanocrystals with large chain surface modifications*', Langmuir, 2006. **22**: pp. 4804–4810.
161. X. Cao, Y. Habibi, and L. A. Lucia, '*One-pot polymerization, surface grafting, and processing of waterborne polyurethane–cellulose nanocrystal nanocomposites*', Journal of Materials Chemistry, 2009. **19**: pp. 7137–7145.
162. F. Azzam, L. Heux, J. L. Putaux, and B. Jean, '*Preparation by grafting onto, characterisation, and properties of thermally responsive polymer-decorated cellulose nanocrystals*', Biomacromolecules, 2010. **11**: pp. 3652–3659.
163. N. Ljungberg, C. Bonini, F. Bortolussi, C. Boisson, L. Heux, and J. Y. Cavaillé, '*New nanocomposite materials reinforced with cellulose whiskers in atactic polypropylene: Effect of surface and dispersion characteristics*', Biomacromolecules, 2005. **6**: pp. 2732–2739.
164. A. P. Mangalam, J. Simonsen, and A. S. Benight, '*Cellulose/DNA hybrid nanomaterials*', Biomacromolecules, 2009. **10**: pp. 497–504.
165. J. Wang, and K. Matyjaszewski, '*Controlled/living radical polymerisation. Atom transfer radical polymerisation in the presence of transition-metal complexes*', Journal of the American Chemical Society, 1995. **117**: pp. 5614–5615.
166. G. Chen, A. Dufresne, J. Huang, and P. R. Chang, '*A novel thermoformable bionanocomposite based on cellulose nanocrystal-graft-poly(ε-caprolactone)*', Macromolecular Materials and Engineering, 2009. **294**: pp. 59–67.
167. N. Lin, G. Chen, J. Huang, A. Dufresne, and P. R. Chang, '*Effects of polymer-grafted natural nanocrystals on the structure and mechanical properties of poly(lactic acid): A case of cellulose whisker-graft-polycaprolactone*', Journal of Applied Polymer Science, 2009. **113**: pp. 3417–3425.
168. TAPPI Publication, '*Roadmap for the development of international standards for nanocellulose*', 2011. pp. 1–33.

169. W. K. Wan, and L. Millon, '*Poly (vinyl alcohol)–bacterial cellulose nanocomposite*', US Patent: US200900289: 2009.
170. K. Tetsuo, '*Material comprising polylactic acid and cellulose fiber*', International Patent : WO2007136086A1: 2007.
171. J. M. Catchmark, D. M. Mears, J. Siggins, and N. Mishra, '*Composites containing polypeptides attached to polysaccharides and molecules*', International Patent: WO2011047047A2: 2011.
172. N. Kotov, B. S. Shim, and P. Podsiadlo, '*Layer-by-layer assemblies having preferential alignment of deposited axially anisotropic species and methods for preparation and use thereof*', US Patent: US20100098902A1: 2010.
173. C. H. Hong, and S. Han, '*Nylon-4 composite*', US Patent: US 2011 0086948A1: 2011.
174. Y. Oku, '*Flexible substrate and manufacturing method thereof*', US Patent: US20090202843A1: 2009.
175. C. Weder, J. Capadona, and O. Van Den Berg, '*Self-assembled nanofibers templates: Versatile approaches for polymer nanocomposites*', US Patent: US2008242765A1: 2008.
176. S. Iwamoto, A. N. Nakagaito, and H. Yano, '*Nano-fibrillation of pulp fibers for the processing of transparent nanocomposites*', Applied Physics A – Materials Science and Processing, 2007. **89**: pp. 461–466.
177. A. N. Nakagaito, and H. Yano, '*The effect of morphological changes from pulp fiber towards nano-scale fibrillated cellulose on the mechanical properties of high strength plant fiber based composites*', Applied Physics A – Materials Science and Processing, 2004. **78**: pp. 547–552.
178. A. N. Nakagaito, and H. Yano, '*Toughness enhancement of cellulose nanocomposites by alkali treatment of the reinforcing cellulose nanofibers*', Cellulose, 2008. **15**: pp. 323–331.
179. M. Henriksson, and L. A. Berglund, '*Structure and properties of cellulose nanocomposite films containing melamine formaldehyde*', Journal of Applied Polymer Science, 2007. **106**: pp. 2817–2824.
180. D. M. Bruce, R. N. Hobson, J. W. Farrent, and D. G. Hepworth, '*High-performance composites from low-cost plant primary cell walls*', Composites Part A – Applied Science and Manufacturing, 2005. **36**: pp. 1486–1493.
181. J. Zeng, '*Experimental study on tensile properties of cellulose nanocrystal reinforced resin*', Master's thesis, 2009. Oregon State University, USA.
182. M. E. Malainine, M. Mahrouz, and A. Dufresne, '*Thermoplastic nanocomposites based on cellulose microfibrils from Opuntia ficus-indica parenchyma cell*', Composites Science and Technology, 2005. **65**: pp. 1520–1526.
183. F. Dalmas, J. Y. Cavaillé, C. Gauthier, L. Chazeau, and R. Dendievel, '*Viscoelastic behavior and electrical properties of flexible nanofiber filled polymer nanocomposites. Influence of processing conditions*', Composites Science and Technology, 2007. **67**: pp. 829–839.
184. A. Fernandez, M. D. Sanchez, M, Ankerfors, and J. M. Lagaron, '*Effects of ionizing radiation in ethylene–vinyl alcohol copolymers and in composites containing microfibrillated cellulose*', Journal of Applied Polymer Science, 2008. **109**: pp. 126–134.
185. M. O. Seydibeyoglu, and K. Oksman, '*Novel nanocomposites based on polyurethane and microfibrillated cellulose*', Composites Science and Technology, 2008. **68**: pp. 908–914.

186. W. K. Wan, J. L. Hutter, L. E. Millon, and G. Guhados, '*Bacterial cellulose and its nanocomposites for biomedical application*', in *Cellulose nanocomposites – Processing, Characterization and Properties*, K. Oksman, and M. Sain, Editors. 2006, ACS Publication, Washington, DC.
187. L. E. Millon, and W. K. Wan, '*The polyvinyl alcohol–bacterial cellulose system as a new nanocomposite for biomedical applications*', Journal of Biomedical Materials Research Part B: Applied Biomaterials, 2006. **79B**: pp. 245–253
188. T. Zimmermann, E. Pohler, and T. Geiger, '*Cellulose fibrils for polymer reinforcement*', Advanced Engineering Materials, 2004. **6**: pp. 754–761.
189. J. Leitner, B. Hinterstoisser, M. Wastyn, J. Keckes, and W. Gindl, '*Sugar beet cellulose nanofibril-reinforced composites*', Cellulose, 2007. **14**: pp. 419–425.
190. J. Lu, T. Wang, and L. T Drzal, '*Preparation and properties of microfibrillated cellulose polyvinyl alcohol composite materials*', Composites Part A: Applied Science and Manufacturing, 2008. **39**: pp. 738–746.
191. X. Cao, Y. Chen, P. R. Chang, A. D. Muir, and G. Falk, '*Starch based nanocomposites reinforced with flax cellulose nanocrystals*', Express Polymer Letters, 2008. **2**: pp. 502–510.
192. A. P. Mathew, W Thielemans, and A. Dufresne, '*Mechanical properties of nanocomposites from sorbitol plasticized starch and tunicin whiskers*', Journal of Applied Polymer Science, 2008. **109**: pp. 4065–4074.
193. A. P. Kumar, and R. P. Singh, '*Biocomposites of cellulose reinforced starch: Improvement of properties by photoinduced crosslinking*', Bioresource Technology, 2008. **99**: pp. 8803–8809.
194. X. F. Ma, P. R. Chang, and J. G. Yu, '*Properties of biodegradable thermoplastic pea starch/carboxymethyl cellulose and pea starch/microcrystalline cellulose composites*', Carbohydrate Polymers, 2008. **72**: pp. 369–375.
195. J. Kadokawa, M. Murakami, A. Takegawa, and Y. Kaneko, '*Preparation of cellulose-starch composite gel and fibrous material from a mixture of the polysaccharides in ionic liquid*', Carbohydrate Polymers, 2009. **75**: pp. 180–183.
196. C. J. Grande, F. G. Torres, C. M. Gomez, O. P. Troncoso, J. Canet-Ferrer, and J. Martinez-Pastor, '*Morphological characterisation of bacterial cellulose-starch nanocomposites*', Polymers and Polymer Composites, 2008. **16**: pp. 181–185.
197. M. Mondragon, K. Arroyo, and J. Romero-Garcıa, '*Biocomposites of thermoplastic starch with surfactant*', Carbohydrate Polymers, 2008. **74**: pp. 201–208.
198. M. S. Sreekala, K. Goda, and P. V. Devi, '*Sorption characteristics of water, oil and diesel in cellulose nanofiber reinforced corn starch resin/ramie fabric composites*', Composite Interfaces, 2008. **15**: pp. 281–299.
199. A. Chakraborty, M. Sain, M. Kortschot, and S. Cutler, '*Dispersion of wood microfibers in a matrix of thermoplastic starch and starch–polylactic acid blend*', Journal of Biobased Materials and Bioenergy, 2007. **1**: pp. 71–77.
200. A. Lopez-Rubio, J. M. Lagaron, M. Ankerfors, T. Lindstrom, D. Nordqvist, A. Mattozzi, and M. S. Hedenqvist '*Enhanced film forming and film properties of amylopectin using micro-fibrillated cellulose*', Carbohydrate Polymers, 2007. **68**: pp. 718–727.
201. A. J. Svagan, M. A. S. A. Samir, and L. A. Berglund, '*Biomimetic polysaccharide nanocomposites of high cellulose content and high toughness*', Biomacromolecules, 2007. **8**: pp. 2556–2563.

202. A. J. Svagan, M. S. Hedenqvist, and L. Berglund '*Reduced water vapour sorption in cellulose nanocomposites with starch matrix*', Composites Science and Technology, 2009. **69**: pp. 500–506.
203. E. E. Brown, and M. P. G. Laborie, '*Bloengineering bacterial cellulose/poly(ethylene oxide) nanocomposites*', Biomacromolecules, 2007. **8**: pp. 3074–3081.
204. H. Kaczmarek, K. Bajer, P. Galka, and B. Kotnowska, '*Photodegradation studies of novel biodegradable blends based on poly(ethylene oxide) and pectin*', Polymer Degradation and Stability, 2007. **92**: pp. 2058–2069.
205. M. Gallstedt, and M. S. Hedenqvist, '*Packaging-related mechanical and barrier properties of pulp-fiber–chitosan sheets*', Carbohydrate Polymers, 2006. **63**: pp. 46–53.
206. D. Nordqvist, J. Idermark, and M. S. Hedenqvist, '*Enhancement of the wet properties of transparent chitosan–acetic-acid salt films using microfibrillated cellulose*', Biomacromolecules, 2007. **8**: pp. 2398–2403.
207. J. Hosokawa, M. Nishiyama, K. Yoshihara, and T. Kubo '*Biodegradable film derived from chitosan and homogenized cellulose*', Industrial and Engineering Chemistry Research, 1990. **29**: pp. 800–805.
208. J. Hosokawa, M. Nishiyama, K. Yoshihara, T. Kubo, and A. Terabe, '*Reaction between chitosan and cellulose on biodegradable composite film formation*', Industrial and Engineering Chemistry Research, 1991. **30**: pp. 788–792.
209. D. Ciechańska, '*Multifunctional bacterial cellulose/chitosan composite materials for medical applications*', Fibres and Textiles in Eastern Europe, 2004. **12**: pp. 69–72.
210. M. A. S. A. Samir, F. Alloin, J. Y. Sanchez, N. El Kissi, and A. Dufresne, '*Preparation of cellulose whiskers reinforced nanocomposites from an organic medium suspension*', Macromolecules, 2004. **37**: pp. 1386–1393.
211. O. Van den Berg, J. R. Capadona, and C. Weder, '*Preparation of homogeneous dispersions of tunicate cellulose whiskers in organic solvents*', Biomacromolecules, 2007. **8**: pp. 1353–1357.
212. D. Viet, S. Beck-Candanedo, and D. G. Gray, '*Dispersion of cellulose nanocrystals in polar organic solvents*', Cellulose, 2007. **14**: pp. 109–113.
213. J. R. Capadona, O. Van Den Berg, L. A. Capadona, M. Schroeter, S. J. Rowan, D. J. Tyler, and C. Weder, '*A versatile approach for the processing of polymer nano-composites with self-assembled nanofibre templates*', Nature Nanotechnology, 2007. **2**: pp. 765–769.
214. O. J. Rojas, G. A. Montero, and Y. Habibi, '*Electrospun nanocomposites from polystyrene loaded with cellulose nanowhiskers*', Journal of Applied Polymer Science, 2009. **113**: pp. 927–935.
215. A. Iwatake, M.Nogi, and H. Yano, '*Cellulose nanofiber-reinforced polylactic acid*', Composites Science and Technology, 2008. **68**: pp. 2103–2106.
216. A. N. Nakagaito, A. Fujimura, T. Sakai, Y. Hama, and H. Yano, '*Production of microfibrillated cellulose (MFC)-reinforced polylactic acid (PLA) nanocomposites from sheets obtained by a papermaking-like process*', Composites Science and Technology, 2009. **69**: pp. 1293–1297.
217. L. Suryanegara, A. N. Nakagaito, and H. Yano, '*The effect of crystallization of PLA on the thermal and mechanical properties of microfibrillated cellulose-reinforced PLA composites*', Composites Science and Technology, 2009. **69**: pp. 1187–1192.

218. Q. Cheng, S. Q. Wang, T.G. Rials, and S. H. Lee, '*Physical and mechanical properties of polyvinyl alcohol and polypropylene composite materials reinforced with fibril aggregates isolated from regenerated cellulose fibers*', Cellulose, 2007. **14**: pp. 593–602.
219. N. Ljungberg, J.-Y. Cavaillé, and L. Heux, '*Nanocomposites of isotactic polypropylene reinforced with rod-like cellulose whiskers*', Polymer, 2006. **47**: pp. 6285–6292.
220. A. P. Mathew, K. Oksman, and M. Sain, '*Mechanical properties of biodegradable composites from polylactic acid (PLA) and microcrystalline cellulose (MCC)*', Journal of Applied Polymer Science, 2005. **97**: pp. 2014–2025.
221. A. P. Mathew, A. Chakraborty, K. Oksman, and M. Sain, '*The structure and mechanical properties of cellulose nanocomposites prepared by twin screw extrusion*', in *Cellulose Nanocomposites – Processing, Characterization and Properties*, K. Oksman, and M. Sain, Editors. 2006, ACS Publication, Washington, DC.
222. H. Lönnberg, L. Fogelström, E. Malström, Q. Zhou, H. Brumer, T. T. Teeri, S. Samir, L. Berglund, and A Hult, '*Grafting of poly(ε-caprolactone) from microfibrillated cellulose films for biocomposite applications*', in 234th ACS National Meeting. 2007. Boston, MA, USA.
223. L. Pranger, and R. Tannenbaum, '*Biobased nanocomposites prepared by in situ polymerization of furfuryl alcohol with cellulose whiskers or montmorillonite clay*', Macromolecules, 2008. **41**: pp. 8682–8687.
224. J. Churusciel, and L. Slusarski, '*Synthesis of nanosilica by the sol-gel and its activity toward polymer*', Materials Science, 2003. **21**: pp. 463–469.
225. F. Haupert, and B. Wetzel, '*Reinforcement of thermosetting polymers by the incorporation of micro- and nanoparticles*', in *Polymer Composites from Nano to Macro Scale*, 2005, Springer, New York, pp. 45–62.
226. V. Favier, H. Chanzy, and J. Y. Cavaillé, '*Polymer nanocomposites reinforced by cellulose whiskers*', Macromolecules, 1995. **28**: pp. 6365–6367.
227. K. Oksman, A. P. Mathew, D. Bondeson, and I. Kvien, '*Manufacturing process of cellulose whiskers/polylactic acid nanocomposites*', Composites Science and Technology, 2006. **66**: pp. 2776–2784.
228. A. Sorrentino, G. Gorrasi, and V. Vittoria, '*Potential perspectives of bio-nanocomposites for food packaging applications*', Trends in Food Science and Technology, 2007. **18**: pp. 84–95.
229. S. Bandyopadhyay-Ghosh, S. B. Ghosh, and M. Sain, '*Cellulose nanocomposites*', in *Industrial Applications of Natural Fibres: Structure, Properties and Technical Applications*, J. Müssig, Editor, 2010, Wiley, pp. 459–480.
230. A. K. Bledzki, and J. Gassan, '*Composites reinforced with cellulose based fibres*', Progress in Polymer Science, 1999. **24**: pp. 221–274.
231. D. N. Saheb, and J. P. Jog, '*Natural fiber polymer composites: A review*', Advances in Polymer Technology, 1999. **18**: pp. 351–363.
232. S. J. Eichhorn, C. A. Baillie, N. Zafeiropoulos, L. Y. Mwaikambo, M. P. Ansell, A, Dufresne, K. M. Entwistle, P. J. Herrera-Franco, G. C. Escamilla, L. Groom, M. Hughes, C . Hill, T. G. Rials, and P. M. Wild, '*Review: Current international research into cellulosic fibres and composites*', Journal of Materials Science, 2001. **36**: pp. 2107–2131.
233. P. Wambua, J. Ivens, and I. Verpoest, '*Natural fibres: Can they replace glass in fibre reinforced plastics?*', Composites Science and Technology, 2003. **63**: pp. 1259–1264.

234. M. J. John and R. D. Anandjiwala, '*Recent developments in chemical modification and characterization of natural fiber-reinforced composites*', Polymer Composites, 2008. **29**: pp. 187–207.
235. M. J. John, and S. Thomas, '*Biofibres and biocomposites*', Carbohydrate Polymers, 2008. **71**: pp. 343–364.
236. O. Faruk, A. K. Bledzki, H.-P. Fink, and M. Sain, '*Biocomposites reinforced with natural fibers: 2000–2010*', Progress in Polymer Science, 2012. **37**: pp. 1552–1596.
237. N. Ouali, J. Y. Cavaillé, and J. Perez, '*Elastic, viscoelastic and plastic behavior of multiphase polymer blends*', Plastics, Rubber and Composites Processing Applications, 1991. **16**: pp. 55–60.
238. J. C. Halpin, and J. L. Kardos, '*The Halpin–Tsai equations: A review*', Polymer Engineering and Science, 1976. **16**: pp. 344–352.
239. J. C. Halpin, and J. L. Kardos, '*Moduli of crystalline polymers employing composite theory*', Journal of Applied Physics, 1972. **43**: pp. 2235–2241.
240. G. Grimmett, *Percolation*, second edition, 1999, Springer.
241. D. Stauffer, *Introduction to Percolation Theory*. 1985, Taylor & Francis, Philadelphia.
242. J. M. Hammersley, '*Percolation processes – lower bounds for the critical probability*', Annals of Mathematical Statistics, 1957. **28**: pp. 790–795.
243. E. J. Garboczi, K. A. Snyder, J. F. Douglas, and M. F. Thorpe, '*Geometrical percolation threshold of overlapping ellipsoids*', Physical Review E, 1995. **52**: pp. 819–828.
244. D. Dubief, E. Samain, and A. Dufresne, '*Polysaccharide microcrystals reinforced amorphous poly(beta-hydroxyoctanoate) nanocomposite materials*', Macromolecules, 1999. **32**: pp. 5765–5771.
245. A. V. Kyrylyuk, and P. Schoot, '*Continuum percolation of carbon nanotubes in polymeric and colloidal media*', Proceedings of the National Academy of Sciences (PNAS), 2008. **105**: pp. 8221–8227.
246. M. Takayanagi, S. Minami, and S. Uemura, '*Application of equivalent model method to dynamic rheo-optical properties of a crystalline polymer*', Journal of Polymer Science Part C: Polymer Symposia, 1964. **5**: pp. 113–122.
247. Y. Habibi, L. A. Lucia, and O. J. Rojas, '*Cellulose nano-crystals: Chemistry, self-assembly, and applications*', Chemical Review, 2010. **110**: pp. 3479–3500.
248. S. J. Eichhorn, A. Dufresne, M. Aranguren, N. E. Marco vich, J. R. Capadona, S. J. Rowan, C. Weder, W. Thiele mans, M. Roman, S. Renneckar, W. Gindl, S. Veigel, J. Keckes, H. Yano, K. Abe, M. Nogi, A. N. Nakagaito, A. Mangalam, J. Simonsen, A. S. Benight, A. Bismarck, L. A. Berglund, and T. Peijs, '*Review: Current international research into cellulose nanofibres and nanocomposites*', Journal of Materials Science, 2010. **45**: pp. 1–33.
249. J. O. Zoppe, M. S. Peresin, Y. Habibi, R. A. Venditti, and O. J. Rojas, '*Reinforcing poly(ε-caprolactone) nanofibers with cellulose nanocrystals*', ACS Applied Materials Interfaces, 2009. **1**: pp. 1996–2004.
250. W.-I. Park, M. Kang, H. S. Kim, and H.-J. Jin, '*Electrospinning of poly(ethylene oxide) with bacterial cellulose whiskers*', Macromolecular Symposia, 2007. **249-250**: pp. 289–294.
251. W. L. E. Magalhäes, X. Cao, and L. A. Lucia, '*Cellulose nanocrystals/ cellulose core-in-shell nanocomposite assemblies*', Langmuir, 2009. **25**(22): pp. 13250–13257.

252. S. Kamel, '*Nanotechnology and its applications in lignocellulosic composites, a mini review*', Express Polymer Letters, 2007. **1**: pp. 546–575.
253. M. Schroers, A. Kokil, and C. Weder, '*Solid polymer electrolytes based on nanocomposites of ethylene oxide–epichlorohydrin copolymers and cellulose whiskers*', Journal of Applied Polymer Science, 2004. **93**(6): pp. 2883–2888.
254. D. Bondeson, and K. Oksman, '*Polylactic acid/cellulose whisker nanocomposites modified by polyvinyl alcohol*', Composites Science and Technology, 2007. **38**: pp. 2486–2492.
255. L. Petersson, and K. Oksman, '*Preparation and properties of biopolymer-based nanocomposite films using microcrystalline cellulose*', in *Cellulose Nanocomposites – Processing, Characterization and Properties*, K. Oksman, and M. Sain, Editors. 2006, ACS Publication, Washington DC.
256. A. N. Nakagaito and H. Yano, '*Nanocomposites based on cellulose microfibril*', in *Cellulose Nanocomposites – Processing, Characterization and Properties*, K. Oksman, and M. Sain, Editors. 2006, ACS Publication, Washigton DC.
257. J. F. Revol, L. Godbout, X. M. Dong, D. G. Gray, H. Chanzy, and G. Maret, '*Chiral nematic suspensions of cellulose crystallites – phase-separation and magnetic-field orientation*', Liquid Crystals, 1994. **16**: pp. 127–134.
258. Y. Nishiyama, S. Kuga. M. Wada, and T. Okano, '*Cellulose microcrystal film of high uniaxial orientation*', Macromolecules, 1997. **30**: pp. 6395–6397.
259. F. Kimura, T. Kimura, M. Tamura, A. Hirai, M. Ikuno, and F. Horii, '*Magnetic alignment of the chiral nematic phase of a cellulose microfibril suspension*', Langmuir, 2005. **21**: pp. 2034–2037.
260. I. Kvien, and K. Oksman, '*Orientation of cellulose nanowhiskers in polyvinyl alcohol*', Applied Physics A: Materials Science and Processing, 2007. **87**: pp. 641–643.
261. J. Sugiyama, H. Chanzy, and G. Maret, '*Orientation of cellulose microcrystals by strong magnetic fields*', Macromolecules, 1992. **25**: pp. 4232–4234.
262. W. Gindl, and J. Keckes, '*Drawing of self-reinforced cellulose films*', Journal of Applied Polymer Science, 2007. **103**: pp. 2703–2708.
263. T. Ebeling, M. Paillet, R. Borsali, O. Diat, A. Dufresne, J. Y. Cavaillé, and H. Chanzy, '*Shear-induced orientation phenomena in suspensions of cellulose microcrystals, revealed by small angle X-ray scattering*', Langmuir, 1999. **15**: pp. 6123–6126.
264. D. Bordel, J. L. Putaux, and L. Heux, '*Orientation of native cellulose in an electric field*', Langmuir, 2006. **22**: pp. 4899–4901.
265. W. Gindl, and J. Keckes, '*All-cellulose nanocomposite*', Polymer, 2005. **46**: pp. 10221–10225.
266. P. Mathew. L. Petersson, and K. Oksman, '*Dispersion and properties of cellulose nanowhiskers and layered silicates in cellulose acetate butyrate nanocomposites*', Journal of Applied Polymer Science, 2009. **112**: pp. 2001–2009.
267. L. Petersson, I. Kvien, and K. Oksman, '*Structure and thermal properties of poly(lactic acid)/cellulose whiskers nanocomposites materials*', Composites Science and Technology, 2007. **67**: pp. 2535–2544.
268. M. N. Angles and A. Dufresne, '*Plasticized starch/tunicin whiskers nanocomposites. 1. Structural analysis*', Macromolecules, 2000. **33**: pp. 8344–8353.
269. M. A. S. A. Samir, A. M. Mateos, F. Alloin, J. Y. Sanchez, and A. Dufresne, '*Plasticized nanocomposite polymer electrolytes based on poly(oxyethylene) and cellulose whiskers*', Electrochimica Acta, 2004. **49**: pp. 4667–4677.

270. P. Hajji, J. Y. Cavaillé, V. Favier, C. Gauthier, and G. Vigier, '*Tensile behavior of nanocomposites from latex and cellulose whiskers*', Polymer Composites, 1996. **17**: pp. 612–619.
271. A. P. Mathew and A. Dufresne, '*Morphological investigation of nanocomposites from sorbitol plasticized starch and tunicin whiskers*', Biomacromolecules, 2002. **3**: pp. 609–617.
272. M. B. Kellerhals, A. Dufresne, and B. Witholt, '*Transcrystallization in Mcl-PHAs/cellulose whiskers composites*', Macromolecules, 1999. **32**: pp. 7396–7401.
273. C. Aulin, G. Salazar-Alvarez, and T. Lindström, '*High strength, flexible and transparent nanofibrillated cellulose–nanoclay biohybrid films with tunable oxygen and water vapor permeability*', Nanoscale, 2012. **4**: pp. 6622–6628.
274. C. Aulin, M. Gällstedt, and T. Lindström, '*Oxygen and oil barrier properties of microfibrillated cellulose films and coatings*', Cellulose, 2010. **17**(3): pp. 559–574.
275. K. Syverud, and P. Stenius, '*Strength and barrier properties of MFC films*', Cellulose, 2009. **16**: pp. 75–85.
276. U. Pagga, D. B. Beimborn, and M. Yamamoto, '*Biodegradability and compostability of polymers – test methods and criteria for evaluation*', Journal of Environmental Polymer Degradation, 1996. **4**: pp. 173–178.
277. A. Ashori, '*Wood–plastic composites as promising green-composites for automotive industries*', Bioresource Technology, 2008. **99**: pp. 4661–4667.
278. S. V. Joshi, L. T. Drzal, A. K. Mohanty, and S. Arora, '*Are natural fiber composites environmentally superior to glass fiber reinforced composites?*', Composites Part A: Applied Science and Manufacturing, 2004. **35**: pp. 371–376.
279. E. Ozen, A. Kiziltas, E. E. Kiziltas, and D. J. Gardner, '*Natural fiber blends filled engineering thermoplastic composites for automobile industry*', in *Proceedings of SPE, Automotive Composites Conference and Exhibition (ACCE)*. 2012. Troy, MI, USA.
280. M. Kamath, G. Bhat, D. Mueller, and D. Parikh, '*Biodegradable/compostable composites from lingo-cellulosic fibers for automotive applications*', SAE Technical Paper, 2005.
281. A. Kiziltas, E. E. Kiziltas, S. Boran, and D. J. Gardner, '*Micro- and nanocellulose composites for automotive applications*', in *13th Annual SPE Automotive Composites Conference & Exhibition (ACCE)*. 2013. Novi, MI, USA.
282. O. Faruk, M. Sain, R. Farnood, Y. Pan, and H. Xiao, '*Development of lignin and nanocellulose enhanced bio PU foams for automotive parts*', Journal of Polymers and the Environment, 2013: pp. 1–10.
283. Chalmers University of Technology, '*Nerve cells grow on nanocellulose*', press release, 19 March 2012.
284. L. E. Millon, G. Guhados, and W. K. Wan, '*Anisotropic polyvinyl alcohol–bacterial cellulose nanocomposite for biomedical applications*', Journal of Biomedical Materials Research Part B: Applied Biomaterials, 2008. **86B**: pp. 444–452.
285. Y. Z. Wan, Y. Huang, C. D. Yuan, S. Raman, Y. Zhu, H. J. Jiang, F. He, and C. Gao, '*Biomimetic synthesis of hydroxyapatite/bacterial cellulose nanocomposites for biomedical applications*', Materials Science and Engineering C, 2007. **27**: pp. 855–864.
286. J. Huia, J. Yuanyuan, W. Jiao, H. Yuan, Z. Yuan, and J. Shiru, '*Potentiality of bacterial cellulose as the scaffold of tissue engineering of cornea*', in *2nd*

International Conference on Biomedical Engineering and Informatics (BMEI), 2009. China.

287. D. A. Schumann, J. Wippermann, D. O. Klemm, F. Kramer, D. Koth, H. Kosmehl, T. Wahlers, and S. Salehi-Gelani, '*Artificial vascular implants from bacterial cellulose: Preliminary results of small arterial substitutes*', Cellulose, 2009. **16**: pp. 877–885.
288. Y. Z. Wan, L. Hong, S. R. Jia, Y. Huang, Y. Zhu, Y. L. Wang, and H. J. Jiang, '*Synthesis and characterization of hydroxyapatite–bacterial cellulose nanocomposites*', Composites Science and Technology, 2006. **66**: pp. 1825–1832.
289. D. Klemm, D. Schumann, U. Udhardt, and S. Marsch, '*Bacterial synthesized cellulose – artificial blood vessels for microsurgery*', Progress in Polymer Science, 2001. **26**: pp. 1561–1603.
290. S. Bandyopadhyay-Ghosh, and M. Sain, '*Cellulose microfibres: Recent advances and future prospects in medical applications*', GAK Gummi Fasern Kunststoffe, 2009. **9**: pp. 564–568.
291. P. G. Bruce, *Solid State Electrochemistry*. 1995, Cambridge University Press, Cambridge, UK.
292. H. Jain, J. O. Thomas, and M. S. Whittingham, '*Solid electrolytes: Advances in science and technology*', Materials Research Bulletin, 2000. **25**: pp. 11–15.
293. F. Alloin, M. A. S. A. Samir, W. Gorecki, J. Y. Sanchez, and A. Dufresne, '*Nanocomposite polymer electrolytes based on poly(oxyethylene) and cellulose nanocrystals*', Journal of Physical Chemistry B, 2004. **108**: pp. 10845–10852.
294. M. A. S. A. Samir, F. Alloin, J. Y. Sanchez, and A. Dufresne, '*Cross-linked nanocomposite polymer electrolytes reinforced with cellulose whiskers*', Macromolecules, 2004. **37**: pp. 4839–4844.
295. S. Ummartyotin, J. Juntaro, M. Sain, and H. Manuspiya, '*Development of transparent bacterial cellulose nanocomposite film as substrate for flexible organic light emitting diode (OLED) display*', Industrial Crops and Products, 2012. **35**(1): pp. 92–97.
296. R. Jonas, and L. F. Farah, '*Production and application of microbial cellulose*', Polymer Degradation and Stability, 1998. **59**: pp. 101–106.
297. E. Sourty, D. H. Ryan, and R. H. Marchessault, '*Ferrite-loaded membranes of microfibrillar bacterial cellulose prepared by in situ precipitation*', Chemistry of Materials, 1998. **10**: pp. 1755–1757.
298. J.-W. Rhim, and K. W. N. Perry, '*Natural biopolymer-based nanocomposite films for packaging applications*', Critical Reviews in Food Science and Nutrition, 2007. **47**: pp. 411–433.
299. J. F. Revol, L. Godbout, and D. Gray, '*Solidified liquid crystals of cellulose with optically variable properties*', US Patent: US 5,629,055, 1997.
300. J. F. Revol, L. Godbout, and D. G. Gray, '*Solid self-assembled films of cellulose with chiral nematic order and optically variable properties*', Journal of Pulp and Paper Science, 1998. **24**: pp. 146–149.
301. M. A. S. A. Samir, F. Alloin, and A. Dufresne, '*Review of recent research into cellulose whiskers: Their properties and their application in nanocomposite field*', Biomacromolecules, 2005. **6**: pp. 612–626.
302. F. Kramer, D. Klemm, S. Moritz, *et al.* , '*Nanocelluloses: A new family of nature-based materials*', Angewandte Chemie International Edition, 2011. **50**: pp. 5438–5466.

303. 'Research and Markets' Report, '*The global market for nanocellulose to 2017*'. 2013: Dublin, Ireland.
304. N. E. Zafeiropoulos, D. Williams, C. Baillie, and F. Matthews, '*Engineering and characterization of the interface in flax fibre/polypropylene composite materials. Part I: Development and investigation of surface treatments*', Composites Part A: Applied Science and Manufacturing, 2002. **33**: pp. 1083–1093.
305. D. B. Warheit, B. R. Laurence, K. L. Reed, D. H. Roach, G. A. Reynolds, and T. R. Webb. '*Comparative pulmonary toxicity assessment of single-wall carbon nanotubes in rats*', Toxicological Sciences, 2004. **77**: pp. 117–125.
306. Y. Umeda, T. Kasai, M. Saito, H. Kondo, T. Toya, S. Aiso, H. Okuda, T. Nishizawa, and S. Fukushima, '*Two-week toxicity of multi-walled carbon nanotubes by whole-body inhalation exposure in rats*', Journal of Toxicologic Pathology, 2013. **26**: pp. 131–140.
307. A. J. Thorley, and T. D. Tetley, '*New perspectives in nanomedicine*', Pharmacology and Therapeutics, 2013. **140**: pp. 176–185.
308. B. N. Snyder-Talkington, D. Schwegler-Berry, V. Castranova, Y. Qian, and N. L. Guo, '*Multi-walled carbon nanotubes induce human microvascular endothelial cellular effects in an alveolar-capillary co-culture with small airway epithelial cells*', Particle and Fibre Toxicology, 2013. **10**: p. 35.
309. S. Lanone, P. Andujar, A. Kermanizadeh, and J. Boczkowski, '*Determinants of carbon nanotube toxicity*', Advanced Drug Delivery Reviews, 2013. **65**: pp. 2063–2069.
310. K. Donaldson, C. A. Poland, F. A. Murphy, M. Macfarlane, T. Chernova, and A. Schinwald, '*Pulmonary toxicity of carbon nanotubes and asbestos – similarities and differences*', Advanced Drug Delivery Reviews, 2013. **65**: pp. 2078–2086.
311. Y. Zhang, J. Deng, Y. Zhang, F. Guo, C. Li, Z. Zou, W. Xi, J. Tang, Y. Sun, P. Yang, Z. Han, D. Li, and C. Jiang, '*Functionalized single-walled carbon nanotubes cause reversible acute lung injury and induce fibrosis in mice*', Journal of Molecular Medicine, 2013. **91**: pp. 117–128.
312. F. N. Yue, S. M. Luo, and C. D. Zhang, '*Degradation and transformation of engineering carbon nanomaterials in the environment: A review*', Chinese Journal of Applied Ecology, 2013. **24**: pp. 589–596.
313. M. Song, L. Zeng, S. Yuan, J. Yin, H. Wang, and G. Jiang, '*Study of cytotoxic effects of single-walled carbon nanotubes functionalized with different chemical groups on human mcf7 cells*', Chemosphere, 2013. **92**: pp. 576–582.
314. K. Shimizu, A. Uchiyama, M. Yamashita, A. Hirose, T. Nishimura, and N. Oku, '*Biomembrane damage caused by exposure to multi-walled carbon nanotubes*', Journal of Toxicological Sciences, 2013. **38**: pp. 7–12.
315. Y. Shi and X. Li, '*Biomedical applications and adverse health effects of nanomaterials*', Journal of Nanoscience and Nanotechnology, 2012. **12**: pp. 8231–8240.
316. R. F. Service, '*Nanomaterials show signs of toxicity*', Science, 2003. **300**: p. 243.
317. D. F. Rodrigues, D. P. Jaisi, and M. Elimelech, '*Toxicity of functionalized single-walled carbon nanotubes on soil microbial communities: Implications for nutrient cycling in soil*', Environmental Science and Technology, 2013. **47**: pp. 625–633.
318. J. Pelka, H. Gehrke, A. Rechel, M. Kappes, F. Hennrich, C. G. Hartinger, and D. Marko, '*DNA damaging properties of single walled carbon nanotubes in human colon carcinoma cells*', Nanotoxicology, 2013. **7**: pp. 2–20.

319. A. N. Parks, L. M. Portis, P. A. Schierz, K. M. Washburn, M. M. Perron, R. M. Burgess, K. T. Ho, G. T. Chandler, and P. L. Ferguson, '*Bioaccumulation and toxicity of single-walled carbon nanotubes to benthic organisms at the base of the marine food chain*', Environmental Toxicology and Chemistry, 2013. **32**: pp. 1270–1277.
320. P. Newman, A. Minett, R. Ellis-Behnke, and H. Zreiqat, '*Carbon nanotubes: Their potential and pitfalls for bone tissue regeneration and engineering*', Nanomedicine, 2013. **9**: pp. 1139–1158.
321. O. Neskovic, G. Joksic, A. Valenta-Sobot, J. Cveticanin, D. Trpkov, A. Leskovac, and S. Petrovic, '*Genotoxic assessment of carbon nanotubes*', Methods in Molecular Biology, 2013. **991**: pp. 315–323.
322. S. Mullick Chowdhury, G. Lalwani, K. Zhang, J. Y. Yang, K. Neville, and B. Sitharaman, '*Cell specific cytotoxicity and uptake of graphene nanoribbons*', Biomaterials, 2013. **34**: pp. 283–293.
323. C. Muhlfeld, C. A. Poland, R. Duffin, C. Brandenberger, F. A. Murphy, B. Rothen-Rutishauser, P. Gehr, and K. Donaldson, '*Differential effects of long and short carbon nanotubes on the gas-exchange region of the mouse lung*', Nanotoxicology, 2012. **6**: pp. 867–879.
324. Y. Morimoto, M. Hirohashi, N. Kobayashi, A. Ogami, M. Horie, T. Oyabu, T. Myojo, M. Hashiba, Y. Mizuguchi, T. Kambara, B. W. Lee, E. Kuroda, M. Shimada, W. N. Wang, K. Mizuno, K. Yamamoto, K. Fujita, J. Nakanishi, and I. Tanaka, '*Pulmonary toxicity of well-dispersed single-wall carbon nanotubes after inhalation*', Nanotoxicology, 2012. **6**: pp. 766–775.
325. S. L. Montes-Fonseca, E. Orrantia-Borunda, A. Aguilar-Elguezabal, C. Gonzalez Horta, P. Talamas-Rohana, and B. Sanchez-Ramirez, '*Cytotoxicity of functionalized carbon nanotubes in j774a macrophages*', Nanomedicine, 2012. **8**: pp. 853–859.
326. N. A. Monteiro-Riviere, R. J. Nemanich, A. O. Inman, Y. Y. Wang, and J. E. Riviere, '*Multi-walled carbon nanotube interactions with human epidermal keratinocytes*', Toxicology Letters, 2005. **155**: pp. 377–384.
327. P. Moller, P. H. Danielsen, K. Jantzen, M. Roursgaard, and S. Loft, '*Oxidatively damaged DNA in animals exposed to particles*', Critical Reviews in Toxicology, 2013. **43**: pp. 96–118.
328. A. Magrez, S. Kasas, V. Salicio, N. Pasquier, J. W. Seo, M. Celio, S. Catsicas, B. Schwaller, and L. Forro, '*Cellular toxicity of carbon-based nanomaterials*', Nano Letters, 2006. **6**: pp. 1121–1125.
329. Y. Liu, Y. Zhao, B. Sun, and C. Chen, '*Understanding the toxicity of carbon nanotubes*', Accounts of Chemical Research, 2013. **19:** pp. 702–713.
330. C. W. Lam, J. T. James, R. McCluskey, and R. L. Hunter, '*Pulmonary toxicity of single-wall carbon nanotubes in mice 7 and 90 days after intratracheal instillation*', Toxicological Science, 2004. **77**: pp. 126–134.
331. C. W. Lam, J. T. James, R. McCluskey, S. Arepalli, and R. L. Hunter, '*A review of carbon nanotube toxicity and assessment of potential occupational and environmental health risks*', Critical Reviews in Toxicology, 2006. **36**: pp. 189–217.
332. J. S. Kim, K. S. Song, and I. J. Yu, '*Evaluation of in vitro and in vivo genotoxicity of single-walled carbon nanotubes*', Toxicology and Industrial Health, 2013. Published online doi: 10.1177/0748233713483201.

333. S. Ivani, I. Karimi, and S. R. Tabatabaei, '*Biosafety of multiwalled carbon nanotube in mice: A behavioral toxicological approach*', Journal of Toxicological Sciences, 2012. **37**(6): pp. 1191–1205.
334. X. Guo, C. Jagannath, M. G. Espitia, and X. Zhou, '*Uptake of silica and carbon nanotubes by human macrophages/monocytes induces activation of fibroblasts in vitro – potential implication for pathogenesis of inflammation and fibrotic diseases*', International Journal of Immunopathology and Pharmacology, 2012. **25**: pp. 713–719.
335. L. de Gabory, R. Bareille, R. Daculsi, R. Daculsi, B. J. L'Azou, E. Flahaut, and L. Bordenave, '*Carbon nanotubes have a deleterious effect on the nose: The first in vitro data*', Rhinology, 2011. **49**: pp. 445–452.
336. S. Clichici, A. R. Biris, C. Catoi, A. Filip, and F. Tabaran, '*Short-term splenic impact of single-strand DNA functionalized multi-walled carbon nanotubes intraperitoneally injected in rats*', Journal of Applied Toxicology, 2014. **34**: pp. 332–344.
337. M. Bottini, S. Bruckner, K. Nika, N. Bottini, S. Bellucci, A. Magrini, A. Bergamaschi, and T. Mustelin, '*Multi-walled carbon nanotubes induce T lymphocyte apoptosis*', Toxicology Letters, 2006. **160**: pp. 121–126.
338. K. Bhattacharya, F. T. Andon, R. El-Sayed, and B. Fadeel, '*Mechanisms of carbon nanotube-induced toxicity: Focus on pulmonary inflammation*', Advanced Drug Delivery Reviews, 2013. **65:** pp. 2087–2097.
339. B. O'Connor, '*Ensuring the safety of manufactured nanocrystalline cellulose*', in *OECD Conference on Potential Environmental Benefits of Nanotechnology*. 2009. Paris.
340. G. A. Montero and M. Osterberg. and O. J. Rojas, '*Nanocellulose – materials, functions and environmental aspects*', in *OECD Conference on Potential Environmental Benefits of Nanotechnology*. 2009. Paris.
341. S. Moreira, N. B. Silva, J. Almeida-Lima, H. A. O. Rocha, S. R. B. Medeiros, C. Alves Jr, and F. M. Gama, '*BC nanofibres: In vitro study of genotoxicity and cell proliferation*', Toxicology Letters, 2009. **189**: pp. 235–241.
342. J. Vartiainen, T. Pöhler, K. Sirola, L. Pylkkänen, H. Alenius, J. Hokkinen, U. Tapper, P. Lahtinen, A. Kapanen, K. Putkisto, P. Hiekkataipale, P. Eronen, J. Ruokolainen, and A Laukkanen, '*Health and environmental safety aspects of friction grinding and spray drying of microfibrillated cellulose*', Cellulose, 2011. **18**: pp. 775–786.
343. M. H. U. Pitkänen, A. von Wright, A. Sneck, H.-P. Hentze. and J. Sievänen, '*Nanofibrillar cellulose – in vitro study of cytotoxic and genotoxic properties*', in *TAPPI International Conference on Nanotechnology for the Forest Product Industry*, 2010. TAPPI: Otaniemi, Espoo, Finland.
344. D. H. Milanez, R. M. do Amaral, L. I. L de Faria, and J. A. R. Gregolin, '*Assessing nanocellulose developments using science and technology indicators*', Materials Research, 2013. **16**: pp. 635–641.

20

The use of wood fibers as reinforcements in composites

L. M. MATUANA, Michigan State University, USA and
N. M. STARK, USDA Forest Service, Forest Products Laboratory, USA

DOI: 10.1533/9781782421276.5.648

Abstract: The chapter begins by discussing the major characteristics of wood in terms of structure, major chemical constituents and wood fiber production methods, which is followed by the description of major manufacturing processes for wood plastic composites (WPCs). It then reviews the effects of different variables such as the nature of wood flour (particle size, wood species, and wood flour content), filler–matrix adhesion, characteristics of polymer matrix, processing conditions, performance-enhancing additives, and foaming agents on the physical, mechanical, thermal, and durability properties of WPCs. Finally, the chapter discusses future trends in WPCs including new materials, novel manufacturing techniques, and emerging applications.

Key words: wood plastic composites, wood flour, mechanical properties, durability, foaming.

20.1 Introduction: characteristics of wood

Wood is classified as a lignocellulosic material. It is made up of major constituents (cellulose, hemicellulose, and lignin) and minor constituents (ash and extractives) (Baeza and Freer, 2001). The major constituents are structural components with a high molecular weight. Wood is approximately 60–75% cellulose, 20–30% lignin, 1–10% extractives and 0–0.5% ash (Marra, 1992). The chemical composition of wood varies between species.

20.1.1 Cellulose and hemicellulose

Cellulose is the most abundant and the main structural component of wood. The cellulose molecule is largely linear. Three elements, namely carbon, hydrogen, and oxygen, are organized into β-D-glucose units and linked together to form long linear chains that are arranged in one plane. The molecular weight of cellulose varies between 50,000 and 2.5 million depending on the origin of the sample (Fengel and Wegener, 1983). Figure 20.1 shows a combination of three cellulose building blocks. A high proportion of cellulose is crystalline, held together by intermolecular

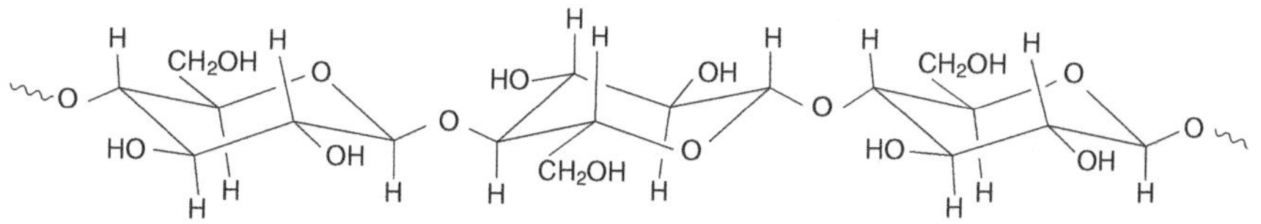

20.1 A combination of three cellulose building blocks (Pettersen, 1984).

hydrogen bonding. The degree of crystallinity is approximately 60–70% for pulp wood (Fengel and Wegener, 1983; Pettersen, 1984).

The hydroxyl groups on cellulose are largely responsible for its reactive nature. Hydrogen bonds exist not only between cellulose hydroxyl groups, but also between cellulose and water. Cellulose is hygroscopic (attractive to water) because it is a polar molecule and easily undergoes hydrogen bonding (Marra, 1992). The absorption of water by cellulose depends on the number of free hydroxyl groups, not those linked with each other. Therefore, water cannot enter crystallites, only amorphous regions are accessible by water. The amount of swelling due to moisture absorption is limited because chain movement is limited. In addition, only hydroxyl groups present in amorphous cellulose are available for chemical reactions (Kazayawoko, 1996).

Hemicellulose is similar to cellulose in that it is arranged in five or six carbon sugars in chains. However, chains are relatively short, therefore are soluble or easily degraded to be soluble (Marra, 1992). The degree of polymerization may be only tens or hundreds of repeating units (Kazayawoko, 1996). Hardwoods contain more hemicelluloses than softwoods.

20.1.2 Lignin

Lignin can be thought of as a matrix that holds cellulose fibers together. It is a brittle, relatively inert material that acts as both a bonding agent and a stiffening agent. Diffusion of lignin into the fiber wall increases stiffness and allows for stress transfer between matrix and fiber. Generally, softwoods have a larger percentage of lignin than hardwoods, accounting for 23–33% in softwoods and 16–25% in hardwoods, respectively (Fengel and Wegener, 1983; Pettersen, 1984).

Lignin is also composed of carbon, hydrogen, and oxygen. The basic precursor is a phenolic material combined in many ways to form a highly branched, three-dimensional network (Fig. 20.2). The result is an isotropic substance. The six member rings of lignin are made up of only carbon atoms forming a stable benzene ring. In addition, there is a low occurrence of hydroxyl sites. Therefore, lignin is not as reactive as cellulose.

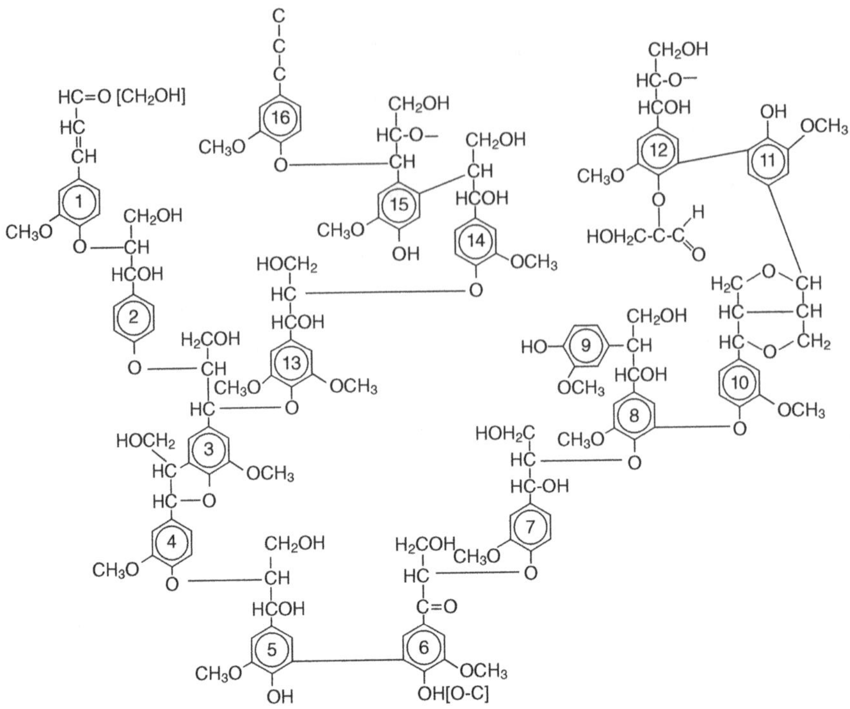

20.2 Structural scheme of spruce (*Picea* spp.) lignin (Pettersen, 1984).

20.1.3 Extractives

Wood extractives are organic substances that can be removed from wood with solvents. Extractives can include organic waxes, oils, fats, tannins, carbohydrates, acids, gums, and resins. Their greatest effects are on hygroscopicity, permeability, and durability of wood. They are deposits, meaning they are not strongly bound in the wood and are free to move (Marra, 1992). There are three types of wood extractives: aliphatic compounds, terpenes and terpenoids, and phenolic compounds. Aliphatic compounds include *n*-alkanes, fatty alcohols, fatty acids, fats (esters) and waxes. Terpenoids include turpentine and resin acids. Phenolic compounds include tannins, flavonoids, lignans, stilbenes, and tropolones (Kazayawoko, 1996). Extractives are typically removed through steam distillation, ether extraction, alcohol extraction, or water extraction. Steam distillation is used to remove terpenes and ether extraction is used to remove fatty alcohols, fatty acids, and waxes. Tannins are alcohol soluble while other carbohydrates and inorganic materials are water soluble (Baeza and Freer, 2001; Fengel and Wegener, 1983).

20.1.4 Wood structure

The familiar concentric ring pattern apparent on the stump of a tree is due to annual growth rings. The lighter material associated with the early growing season is called earlywood. Latewood is the darker ring and is formed later in the growing season. Earlywood is lighter in weight, softer, and weaker than latewood (Wiedenhoeft, 2010).

Wood is composed of several different types of specialty cells. The majority of wood cells are elongated, pointed at the end, and oriented with the stem axis. Cell sizes are different for earlywood and latewood. Earlywood cells typically have relatively large cavities and thin walls. In contrast, latewood cells have smaller cavities and thicker walls.

Cell walls are layered structures arranged concentrically and characterized by differences in chemical composition and orientation (Figure 20.3). Individual cells are glued together by the middle lamella (ML), which is

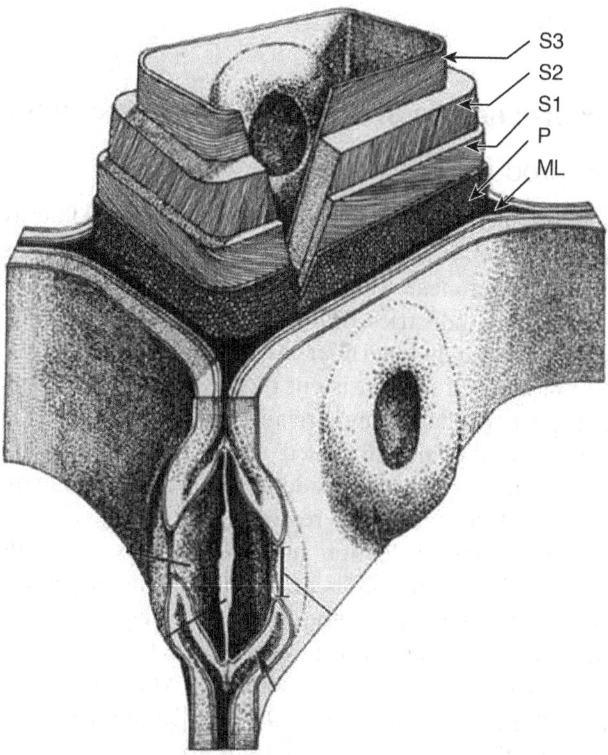

20.3 Drawing of the cell wall. Layers shown include the middle lamella (ML), the primary wall (P), and the secondary wall in three layers (S1, S2, and S3) (adapted from Wiedenhoeft, 2010).

generally free of cellulose. The primary wall (P), secondary wall 1 (S1), secondary wall 2 (S2) and secondary wall 3 (S3), are distinguished by the orientation of cellulose fibrils in the walls. In P, cellulose fibrils are arranged in thin crossing layers, in S1 the fibrils are arranged in a slight slope perpendicular to the fiber direction, and in S2 the fibrils are roughly aligned with the fiber direction. The P and S3 layers have less cellulose present than the S1 and S2 layers. The S3 layer has a high concentration of non-structural substances. The cavity bound by the cell wall is referred to as the lumen. The change in wall thickness between earlywood and latewood is mainly due to changes in S2 thickness, which is also where the majority of cellulose is found. The density of the cell wall is approximately 1.49 g/cm^3 (Marra, 1992).

Softwood cells aligned in the axial direction are primarily tracheids, accounting for over 90% of the volume of wood. Hardwood cells aligned in the axial direction consist primarily of fibers, but also include vessels and parenchyma cells. Softwood tracheids can range from 1 to 10 mm in length while hardwood fibers are on average 0.2 to 1.2 mm in length while (Wiedenhoeft, 2010).

20.2 Fiber processing and composite manufacturing

20.2.1 Fiber production methods

Common wood filler for thermoplastics is wood flour. Wood flour is defined as a finely ground wood. It is derived from various wood planer shavings, chips, sawdust, and other clean waste wood from saw mills and residues of other wood processing industries. The production method often leads to fiber bundles rather than individual fibers. Moisture, particle size distribution, and density are monitored for consistent physical and chemical properties during processing. It is available commercially in a variety of size distributions and species. The most commonly used wood flour species for plastic composites in the United States are made from pine, oak, and maple. Other species are also used depending on regional availability. The main steps in its production are size reduction by several types of grinders (e.g., pulverization by crushing, Wiley mills, attrition mills, hammermill, etc.) and size classification by screening methods. The typical length-to-diameter (L/D) ratio of wood flour ranges from about 3 : 1 to 5 : 1. Wood flour is graded by mesh size. For example, wood flour with 30 and 120 mesh corresponds to 500 and 125 μm, respectively. Detailed information on wood flour production methods is given by Clemons (2010).

Wood flour is available commercially in a variety of size distributions and species. Stark and Berger (1997) reported that commercially produced wood flour has broad particle distribution. They showed that typical

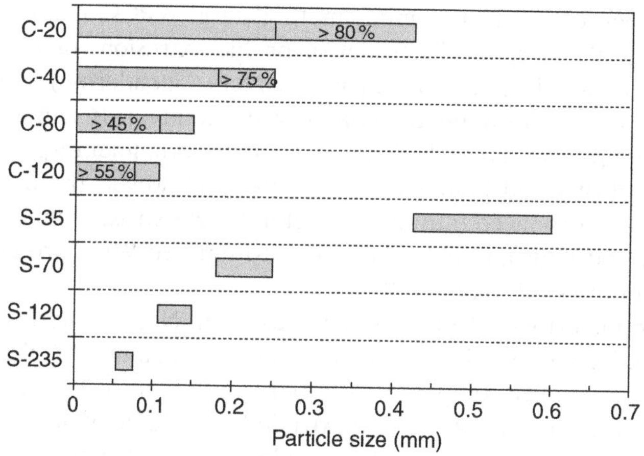

20.4 Particle size range for ponderosa pine (*Pinus ponderosa*) wood flours. S designates screened particle size distributions and C designates commercial particle size distributions (adapted from Stark and Berger, 1997).

commercial grades of wood flour include a mixture of particle sizes, which overlap particularly at larger mesh sizes (smaller particle sizes). Generally, commercial suppliers of wood flour avoid extra screening to narrow the particle size distribution because this process results in a more expensive product. Comparison of ponderosa pine wood flour particles screened to specific sizes (discrete and narrow size distribution) and the particle size of commercially produced wood flour is shown in Fig. 20.4. The discreteness of size ranges for the screened wood flour and the overlap of size ranges for the commercial wood flour are clearly illustrated in this figure.

Generally, wood flour (10–120 mesh) is easier to handle during processing than long wood fibers, which tend to agglomerate and cause dispersion problems during mixing. However, because of their higher aspect ratio, long fibers provide greater reinforcing effects in wood plastic composites (WPCs) than particles. The fiber can be difficult to disperse but the dispersion problems can be offset by using compatibilizers during processing. The use of long fibers in WPCs is described in other chapters in this book.

20.2.2 Composite manufacturing

Presently, major manufacturing processes for wood plastic composites (WPCs) use extrusion. A two-step process consisting of pelletizing the blend prior to extrusion of profiles persists as the most widely used method of manufacturing WPCs (Matuana, 2009). The raw materials are first mixed together, and the composite blend is then formed into a product. The most common types of product-forming methods for WPCs involve forcing molten material through a die (sheet or profile extrusion) or into a cold mold (injection molding), or pressing in calenders (calendering) or between mold halves (thermoforming and compression molding) (Clemons, 2002). Pelletized compounds have a higher degree of uniform filler dispersion and better control of wood flour moisture content. However, the equipment suppliers have introduced into the market a direct extrusion that converts non-dried wood flour and plastic matrix into finished WPC products in a single-step process (Matuana, 2009).

While the majority of WPCs produced today in North America involve extrusion, injection molding and compression molding of WPCs are proportionally smaller compared with extrusion. Most of the composites used in construction applications are extruded to a profile of uniform cross-section (solid or hollow) and any practical length, whereas products having more complex shapes such as those used in the automotive industry or other consumer products are injection or compression molded (Matuana, 2009). Products such as decking, railings, and window profiles readily lend themselves to extrusion through a two-dimensional die. Injection-molded applications such as consumer household goods and furniture parts are gaining importance. Thermoforming or compression molding is the forming method of choice for the automotive industry.

In North American WPC manufacturing, over 80% of wood fiber–plastic composites are made with wood flour (mesh size up to 140) and polyolefin matrices, including polyethylene (PE) and polypropylene (PP) thermoplastics (Zazyczny and Matuana, 2005) although polystyrene (PS) is also used. Currently, poly(vinyl chloride) (PVC) matrix represents only a small portion of thermoplastics used in WPCs but its use will continue to grow due to its superior performance and excellent weatherability compared to olefins (Matuana, 2009). Biodegradable plastics produced from renewable resources are also gaining prominence as matrices for WPCs due to their biomass origin and ability to decompose and maintain carbon dioxide balance (Afrifah and Matuana, 2012). Biodegradable bioplastics include polylactic acid (PLA) and starch acetate, non-biodegradable bioplastics such as bio-derived PE and polyamide (Lampinen, 2009). PLA has received the most attention from researchers as it is widely available commercially.

These plastics are chosen because of their lower processing temperatures (150–220°C) which prevent the deterioration of the added wood materials. When wood is heated, the production of volatiles begins at processing temperatures exceeding 200°C (Saheb and Jog, 1999). While wood starts degrading at 210°C, this is mainly due to extractives and hemicelluloses. Cellulose does not begin to degrade until 270°C (Sears *et al.*, 2001) and lignin at 280°C (Saheb and Jog, 1999). Therefore, it is preferable to set the processing temperature below 220°C to prevent not only the thermal degradation of the wood flour but also that of heat-sensitive polymers such as PLA and PVC (Matuana and Diaz, 2013). In addition to the degradation issue, processing at elevated temperature (above 220°C) results in the production of volatile organic compounds, which can lead to foamed (porous) products with inferior mechanical properties (Guo *et al.*, 2004). Wood flour is also very sensitive to moisture. A moisture level above 1% in wood flour makes the composite foam in the extruder, i.e., causes production of microvoids of irregular and inhomogeneous shapes which affect the performance of the products (Rizvi *et al.*, 2000; Matuana and Mengeloglu, 2002). Consequently, great care must be taken when processing with wood flour. Generally, drying of wood flour prior to processing and keeping compounded feedstock sealed from atmospheric moisture are necessary not only to ease the manufacturing process, but also to enhance the quality and mechanical performance of the products (Matuana, 2009).

Wood flour and plastic matrix constitute the main ingredients in WPC formulations. The amounts of wood flour vary in the formulations and wood flour content of up to 70% is currently used in commercial WPC products. Additionally, WPC producers extensively utilize a broad spectrum of performance-enhancing additives to improve the appearance, reduce the maintenance requirements, and prolong the service life of WPC products. Among these additives, coupling agents and lubricants are specialty chemicals widely used in WPCs because they play crucial roles in the processing and performance of the products.

Coupling agents improve the dispersion and the bond between the wood fiber reinforcements and the plastic matrix, which reduces the rate of water uptake and improves the durability as well as the strength performance (Felix and Gatenholm, 1991; Kazayawoko *et al.*, 1999; Matuana *et al.*, 1998a, 1998b; Li and Matuana, 2003a). They are necessary because the wood surface is largely hydrophilic in contrast to the hydrophobic surface of the plastic polymer (Shah *et al.*, 2005). Since the wood and plastic surfaces are chemically incompatible, coupling agents therefore function as a bridge between these two incompatible surfaces. Several types of coupling agents and compatibilizers have been investigated, including silanes (Matuana *et al.*, 1998a, 1998b; Razi and Raman, 2000; Balasuriya *et al.*, 2002), titanates (Liao *et al.*, 1997; Rodríguez *et al.*, 2003), maleated polyolefins (Felix and Gatenholm, 1991; Kazayawoko *et al.*, 1999; Li and Matuana, 2003a), etc. Their functioning mechanisms are well established (Kazayawoko *et al.*, 1999). Widely used in WPCs, the lubricants improve the surface quality of WPCs by reducing edge tearing, ease the problems of extruding viscous mixtures, and thus increase extrusion production rates (Matuana, 2009). The addition of these two performance-enhancing additives significantly adds to production cost and market price.

A broad spectrum of other performance-enhancing additives includes biocides, antimicrobials/antifungals, UV stabilizers, pigments/light stabilizers, and flame-retardants (Matuana *et al.*, 2001b; Stark and Matuana, 2003, 2006). Added to WPC products, these additives prevent staining and slow color change. Other methods used in the industry to increase color stability include coextruding a cap layer over WPC substrate (Jin and Matuana, 2008, 2010; Matuana *et al.*, 2011; Stark and Matuana, 2009) and fluoro-oxidation of WPC followed by coating (Matuana, 2009). These methods seal the surface of WPC products, thus preventing moisture absorption and surface staining.

20.3 Mechanical performance of wood plastic composites (WPCs)

Property evaluation of WPCs depends strongly on their application. For example, in semi-structural building applications, such as decking, window and door frames, the aesthetics (wood look), the improved thermal and creep performance compared with unfilled plastics, as well as the improved durability to biological agents (e.g., mold, fungi, termites, etc.) and other environmental factors (e.g., freeze–thaw, water, etc.) compared to solid wood are extremely important. On the other hand, the light weight of WPCs (lower specific gravity) compared to plastic filled with inorganic materials makes them attractive for automotive applications. Therefore, properties such as strength, stiffness, impact resistance, density, heat deflection temperature, color stability, resistance to decay, ultraviolet light, moisture, and so on are important considerations in many WPC applications. These properties are evaluated in accordance with the procedures outlined in various ASTM standards.

Most of the available literature on the mechanical performance of wood flour filled thermoplastic matrices can be classified in four categories:

- The nature of wood flour
- The filler–matrix adhesion
- The characteristics of the polymer matrix
- The additives used in WPC formulations.

These topics are reviewed in the following sections.

20.3.1 Nature of wood flour

The nature of wood flour includes variables such as particle size and distribution, wood species, wood flour addition levels, etc., and the effects of these variables are discussed below.

Particle size

One of the factors to consider when using wood flour as filler in wood plastic composites is particle size. This variable has significant influence on some mechanical properties of WPCs. The effect of particle size on properties of injection molded wood flour filled polypropylene composites has been reported (Stark and Berger, 1997; Stark and Rowlands, 2003). Since it is difficult to characterize properties of WPCs based on specific particle sizes using commercial wood flour, due to the size overlap (Fig. 20.4), Stark and coworkers (Stark and Berger, 1997; Stark and Rowlands, 2003) examined the narrow size distributions of wood flour particles, which they obtained by discrete screening of wood flour to designated commercial sizes (Table 20.1). They observed that at wood flour loading of 40% (by weight), particle size did not affect specific gravity (Table 20.2). The higher density of WPCs compared to the unfilled plastic is mainly due to the compression of wood cell walls during processing, which is independent of wood flour particle size. However, the particle size did affect other properties. The melt flow index (MFI) increased as the particle size increased from 0.064 mm (120 mesh) to 0.513 (20 mesh) because the volume of unfilled regions within the polymer increased, resulting in greater flow mobility for the polymer and increased MFI values. Composites with finer particles (70 mesh and higher) had better tensile strength and modulus than their counterparts with coarser particles (35 mesh). Flexural properties followed similar trends. Generally, the larger the particle size, the greater the notched impact energy as a result of the increase in fracture surface area. Conversely, the larger the particle size, the lower the unnotched impact energy as a result of higher stress.

Table 20.1 Mesh sizes and particle sizes of ponderosa pine (*Pinus ponderosa*) wood flour screened to designated sizes by American Wood Fibers, Schofield, Wisconsin, USA

Mesh size range			Screen hole size (mm)		
Maximum	Minimum	Average	Maximum	Minimum	Particle size
40	30	35	0.425	0.600	0.513
80	60	70	0.180	0.250	0.215
140	11	120	0.106	0.150	0.128
270	200	235	0.053	0.075	0.064

Source: adapted from Stark and Rowlands, 2003; Stark and Berger, 1997.

Table 20.2 Influence of wood flour particle size on the physical and mechanical properties of injection-molded polypropylene composites filled with 40% wood flour (ponderosa pine)

Wood flour		Specific gravity (g/cm^3)	Melt flow index (MFI)[1] (g/10 min)	Tensile properties			Flexural properties		Izod impact energy	
Mesh size	Particle size (mm)			Strength (MPa)	Modulus (GPa)	Elongation (%)	Strength (MPa)	Modulus (GPa)	Notched (J/m)	Unnotched (J/m)
35	0.513	1.05	4.13	21.7	3.20	2.27	38.7	2.69	22.3	54
70	0.215	1.05	1.50	25.5	3.61	2.27	42.6	3.15	19.7	79
120	0.128	1.04	2.03	24.9	3.47	2.29	42.9	3.00	18.7	84
235	0.064	1.04	1.20	24.3	3.46	2.11	41.4	2.89	16.0	91

[1] Melt flow index tests were carried out at 190°C.

Source: adapted from Stark and Rowlands, 2003; Stark and Berger, 1997.

Table 20.3 Effect of wood species on the flexural properties of extruded WPCs made with different thermoplastic matrices

Flexural property	Rigid PVC-based comppsites[1]		HDPE-based composites[2]	
	Pine flour	Maple flour	Pine flour	Maple flour
Strength (MPa)[3]	$37.60^{A} \pm 2.10$	$31.55^{B} \pm 0.93$	$24.68^{C} \pm 1.04$	$20.73^{D} \pm 2.26$
Modulus (GPa)[3]	$3.34^{A} \pm 0.24$	$2.54^{B} \pm 0.14$	$2.21^{C} \pm 1.04$	$1.63^{D} \pm 0.33$

[1]The formulations and processing conditions used to produce rigid PVC/wood-flour composites are given in Pilarski and Matuana (2005).
[2]The formulations and processing conditions used to produce rigid HDPE/wood-flour composites are given in Pilarski and Matuana (2006).
[3]Means with the same letter are not significantly different ($p < 0.05$).
Note: The composites contained 50% by weight wood flour of 40 mesh size.
Source: adapted from Pilarski and Matuana, 2005, 2006.

Wood species

The type of wood species constitutes another important factor to consider in WPCs. Generally, wood flours from softwood species are usually preferable to those from hardwood for easy processing and increasing mechanical properties. This trend is clearly seen in the flexural properties of extruded WPCs made with rigid PVC and HDPE matrices summarized in Table 20.3 where composites with softwood flour outperformed their counterparts with hardwood flour. Maldas and coworkers (1989) explained this disparity by the morphological difference between the more flexible softwood and stiffer hardwood as well the chemical compositions of the wood flours from the two wood species (Zazyczny and Matuana, 2005).

Wood flour concentration

The amount of wood flour in WPCs significantly influences both processing and end-use properties. In the melting state, the viscosity of composites increases with wood flour addition level, since the wood flour hinders the movement of polymer chain (Shah and Matuana, 2004; Li and Wolcott, 2005; Guo *et al.*, 2008). Plots of melt flow index as a function of wood flour contents for PLA-based composites are illustrated in Fig. 20.5. Increasing wood flour content reduced the MFI of the composites, suggesting that increasing wood flour content makes the composite melts more viscous. Interactions between wood flour particles and the matrix may account for this decreased trend, since studies have shown that solid wood flour particles embedded in the polymer matrix increase the friction towards the flow of the melt, resulting in higher viscosity (Matuana and Diaz, 2013).

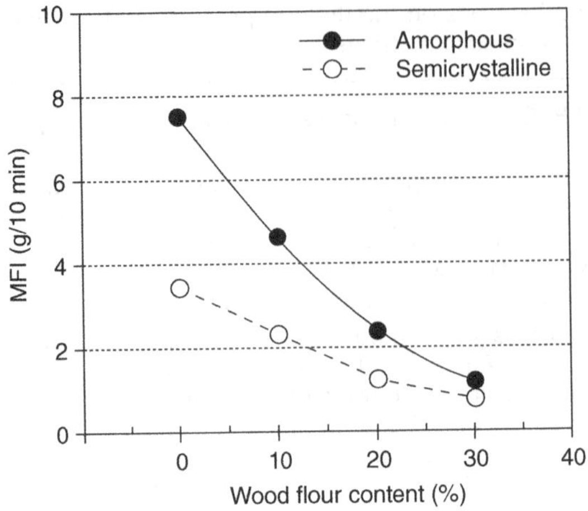

20.5 Effect of wood flour content on the MFI of PLA: semicrystalline (PLA 2002D) and amorphous (PLA 8302D) from NatureWorks®.

Generally, the mechanical properties of composites follow the rule of mixtures. Due to the higher strength but lower modulus of thermoplastic compared to wood, generally the strength property of WPCs decreases as the wood flour content increases in the composites (Kazayawoko *et al.*, 1999; Li and Matuana, 2003a). This is a result of poor interfacial adhesion due to the incompatibility between hydrophilic wood flour and hydrophobic plastics (Table 20.4). In contrast, the stiffness of the composites increases linearly as the amount of wood flour in the composites increases, in accordance with the rule of mixtures, due to the greater stiffness of wood flour compared to plastic. Both the elongation at break and the impact strength of wood flour filled plastic decrease as the wood flour content increases in the composites. The addition of wood flour into plastic matrices reduces the ductile behavior of the matrix by making the composites more brittle. The higher degree of brittleness introduced by the addition of wood flour into the plastic matrix causes this decreased elongation at break and impact strength (Matuana *et al.*, 1997, 1998b). Rigid fillers restrict elongation, limiting almost all the stretching that occurs to the polymeric matrix. This difference is a typical trade-off for improving the stiffness of the matrix by incorporating wood flour into the plastic matrix. In general, wood flour improves the heat deflection of unfilled polymer matrix, which increases with increasing wood flour loading.

It should be pointed out that a recent study that assessed the effect of wood flour content on the impact strength of PLA/wood-flour composites (Afrifah and Matuana, 2012) depicted two distinct behaviors depending on

Table 20.4 Effect of wood flour content (loblolly pine, 40 mesh) on the mechanical and physical properties of injection-molded polypropylene-based WPCs

Weight percent (%) wood flour in the composites	Tensile properties			Flexural properties		Izod impact energy		Heat deflection (°C)
	Strength (MPa)	Modulus (GPa)	Elongation (%)	Strength (MPa)	Modulus (GPa)	Notched (J/m)	Unnotched (J/m)	
0	28.45	1.31	10.44	34.7	1.03	15.0	600	55
20%	24.92	2.14	4.83	40.6	1.71	12.4	120	64
40%	21.43	3.28	1.76	39.3	2.84	13.7	49	78
60%	17.81	4.32	0.92	34.1	3.81	10.4	31	77

Source: adapted from Berger and Stark, 1997.

Table 20.5 Notched Izod impact strength of injection-molded PLA/maple flour composites as a function of wood flour content and particle size

Wood flour content in the composites (% by weight)[1]	Notched Izod impact strength (J/m)[2]			
	20 mesh	40 mesh	60 mesh	100 mesh
0	16.1 ± 1.6^A	16.1 ± 1.6^A	16.1 ± 1.6^A	16.1 ± 1.6^A
5	16.8 ± 1.8^A	15.4 ± 1.3^A	16.4 ± 1.3^A	16.7 ± 1.7^A
15	14.9 ± 2.1^A	15.6 ± 2.0^A	14.5 ± 2.0^A	14.9 ± 1.8^A
25	16.5 ± 1.5^A	–	–	13.5 ± 2.5^B
30	–	15.3 ± 0.9^A	13.6 ± 2.0^B	–
40	15.6 ± 2.0^A	15.4 ± 0.7^A	10.3 ± 2.6^C	6.5 ± 1.9^C

[1]The PLA resin used in the composites was PLA 3001D from NatureWorks, USA, with the following characteristics: density of 1.144 g.cm^{-3}, 8% of D-lactide enantiomer, crystallinity of 15%, and melt flow index of 12.1 g.(10 min)$^{-1}$ at 190°C.
[2]The different superscript letters for each particle size denote that the difference between two wood flour contents is statistically significant ($p < 0.05$).
Source: adapted from Afrifah and Matuana, 2012.

wood flour particle size (Table 20.5). The impact strengths of composites with coarse wood particles (up to 40 mesh) were not significantly affected by the wood flour concentration. Unlike fine particles, it is believed that coarse ones create fewer cracks in the composites due to their lower surface area per unit weight, reducing the propensity of brittle failure which decreases the impact strength. On the other hand, the impact strengths of composites with finer wood particles (60 mesh and up) decreased drastically after adding 15% or more wood flour into the matrix. The observed trend is probably due to wood flour agglomeration in the matrix, which resulted in stress concentration promoting crack propagation with low impact strength (Afrifah and Matuana, 2012).

20.3.2 Filler–matrix adhesion

The polar or hydrophilic nature of wood flour due to the existence of many hydroxyl groups in cellulose, hemicelluloses and lignin is a major concern for its use in WPCs. This is because it results in low compatibility with the non-polar or hydrophobic polymer matrices leading to poor wetting of the wood flour by the polymer, inhomogeneous dispersion of the fillers in the matrix polymer, and poor adhesion between the filler and the polymer. Accordingly, this incompatibility leads to composites with lower strength properties (e.g., tensile and flexural strengths) than the unfilled polymer (Kazayawoko *et al.*, 1999). Effective interfacial adhesion between hydrophilic, polar wood flour and hydrophobic, non-polar plastics is crucial for both the processing and performance of WPC materials. Consequently it is important to enhance the surface properties of the wood flour and/or the polymer matrix to improve interfacial adhesion between components and derive optimum strength properties from the composites.

Since the performance of fiber-reinforced thermoplastic composites strongly depends on solid fiber–matrix adhesion to allow stress transfer between the phases, surface modification of wood flour by coupling (or other compatibilizing) agents is usually performed to attain improved interfacial adhesion and subsequently satisfactory strength properties in the composites (Matuana *et al.*, 1999, 2001a). Substantial research has been carried out on the surface modification of wood fiber using coupling agents to improve the strength properties of WPCs. Among them, the influence of surface treatment of wood flour with maleated polyolefins on the strength properties of polyolefin-based WPCs has received particular attention (Felix and Gatenholm, 1991; Kazayawoko *et al.*, 1999; Li and Matuana, 2003a, 2003b). Many in-depth studies have reported improvement in the strength properties of the composites when maleated polyolefins were used in the formulation and the mechanisms of adhesion between wood flour and polyolefin matrices that caused the improvement. Maleated polyolefins are known to promote the affinity between the wood flour and the plastics (Li and Matuana, 2003b; Carlborn and Matuana, 2006). They induce the bond formation between the wood flour and plastic matrix. When strong chemical bonds are formed, stress transfer from the polymer to the load-bearing filler takes place, resulting in the enhancement of the strength properties of the composites.

The results summarized in Table 20.6 demonstrate the importance of coupling agents on the performance of WPCs. Without coupling agents, the addition of wood flour into the plastic matrix significantly reduced both the tensile and flexural strengths of the composites due to poor interfacial adhesion between wood flour and the matrix. As expected, the addition of coupling agent into the formulations had a positive effect on the strength of the composites due to the enhanced interfacial adhesion between wood flour and plastic matrix.

The effectiveness of maleated polyolefins in improving the adhesion between the wood flour and polymer matrix is attributed to two mechanisms, which have been extensively investigated. Firstly the carboxylic acid portions of the maleated compound react with the hydroxyl groups on the wood surface to form monoester or diester linkages (Carlborn and Matuana, 2006). This leaves the remaining polyolefin portions of the maleated compound as free pendant chains. The second mechanism involves the diffusion of the free pendant polyolefin chain into the polymer matrix which forms a physical bond upon cooling (Kazayawoko *et al.*, 1999). Successfully

Table 20.6 Effect of coupling agent (maleated polypropylene–MAPP) on the mechanical properties of injection-molded polypropylene-based WPCs filled with 40% ponderosa pine wood flour (40 mesh)

Material	Tensile properties			Flexural properties		Izod impact energy	
	Strength (MPa)	Modulus (GPa)	Elongation (%)	Strength (MPa)	Modulus (GPa)	Notched (J/m)	Unnotched (J/m)
Neat polypropylene	28.5	1.53	5.9	38.3	1.19	20.9	656
WPC without MAPP	25.4	3.87	1.9	44.2	3.03	22.2	73
WPC with 3% MAPP	32.3	4.10	[1.9]	53.1	[3.08]	[21.2]	[78]

Note: Brackets indicate values not statistically different from those of uncoupled wood flour composites ($\alpha = 0.05$).
Source: adapted from Stark and Rowlands, 2003.

achieving these two mechanisms results in the formation of an adhesive bridge between the wood fiber and the matrix polymer.

Polyisocyanates, triazines, organosilanes, etc., have also been used as coupling agents. For example, silanes are known to form strong covalent bonds between the wood flour and the polymer matrix (Balasuriya *et al.*, 2002; Matuana *et al.*, 1999). These covalent bonds ensure effective stress transfer from the polymer to the wood flour, resulting in improved strength properties, especially tensile and flexural strengths (Matuana *et al.*, 1998a, 1998b; Razi and Raman, 2000; Balasuriya *et al.*, 2002).

20.3.3 Characteristics of the polymer matrix

The characteristics of polymers, including molecular weight and its distribution, rheology properties, mechanical properties, and degradation temperature, influence the properties and/or manufacturing of WPCs. The melt flow index (MFI) of a polymer, an assessment of average molecular mass and inverse measure of the melt viscosity, is critical to control the processing. Generally, the high MFI (low viscosity) polymer has low strength and modulus due to the low molecular weight. Because polymers with a low MFI often have a high molecular weight (Balasuriya *et al.*, 2001), studies have shown that WPCs with a low MFI polymer matrix have better tensile and impact properties than WPCs with a high MFI polymer matrix (Fig. 20.6) (Balasuriya *et al.*, 2001; Huang and Zhang, 2009).

Generally, WPCs made with PVC matrix outperformed their polyethylene counterparts in flexural strength and stiffness, regardless of wood species, due primarily to the higher stiffness of PVC (Table 20.3).

20.3.4 Additives used in WPCs

It is well established that the ease of processing and end-use performance of WPCs are a function of the additives present in the formulation as well as the characteristics of the plastic matrices and processing conditions used (Zazyczny and Matuana, 2005). As mentioned, a broad spectrum of performance-enhancing additives is employed in WPC formulations in order to enhance the processability and/or the physico-mechanical properties of the composites.

Plasticizers

As shown in Fig. 20.5, the incorporation of wood flour into the plastic resins increases the melt viscosity of the matrix, which worsens the processing, quality, and performance of the products. The addition of plasticizers into WPCs imparts flexibility and plays an important role in heat processing by

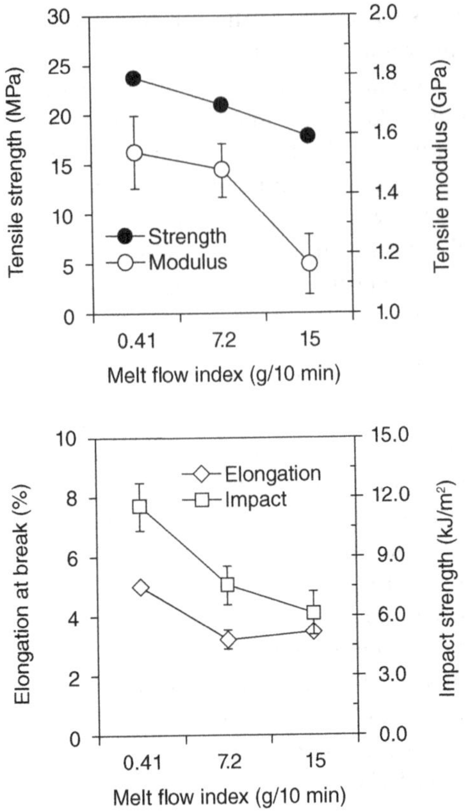

20.6 Effect of polymer MFI on the tensile and notched Izod impact strength properties of HDPE filled with 50% wood flake (40 mesh radiata pine) (adapted from Balasuriya *et al.*, 2001).

reducing the viscosity of the molten material to insure good flow properties (Zazyczny and Matuana, 2005). The available data indicate that both the tensile strength and modulus of WPCs decrease as the plasticizer content increases because of plasticization of the matrix. In contrast, the impact strength of the composites can be significantly improved even with lower plasticizer content in the composites (Matuana *et al.*, 1997).

Impact modifiers

The addition of wood flour into a plastic matrix produces compounds that are more brittle with a much lower capability for absorbing impact energy than unfilled plastics (Matuana *et al.*, 1997, 1998b). Since impact strength is a vital property for many WPC applications, the composites must be toughened for successful applications. The conventional approach to toughening WPC compounds is to incorporate impact modifiers within the matrix as part of the formulation prior to compounding (Zazyczny and Matuana, 2005; Afrifah and Matuana, 2012). Studies have shown that toughness and impact strength of WPCs can be improved by impact modifiers (Mengeloglu *et al.*, 2000). Some reported studies used impact modifiers such as ethylene/propylene/diene terpolymers (EPDM), styrene–ethylene/butylene–styrene and their maleated derivatives for improvement of impact strength and toughness of polypropylene/wood-flour composites (Park and Balatinecz, 1997; Oksman and Clemons, 1998). The addition of these impact modifiers increased impact strength and toughness while it reduced stiffness and strength of the composites.

The effects of impact modifier types and addition levels on the mechanical properties of rigid PVC/wood-flour composites (Table 20.7) revealed a strong dependence of impact resistance of the composites on the type and content of impact modifier (Mengeloglu *et al.*, 2000). All impact modifiers listed in Table 20.7 were effective in enhancing the impact strength of the composites. However, of the impact modifiers studied, methacrylate–butadiene–styrene (MBS) and all-acrylic modifiers (ACR) were observed

Table 20.7 Effect of impact modifier types and concentrations on the mechanical properties of rigid PVC/wood-flour composites

Impact modifier types and contents in the composites[1]	Izod impact strength[2] (J/m)	Tensile strength[2] (MPa)	Tensile modulus[2] (GPa)	Elongation at break[2] (%)
Crosslink modifier (ACR):				
0 phr	20.0	30.3	3.38	0.89
10 phr	38.9	25.7	2.50	1.15
20 phr	73.3	21.9	2.24	1.12
Crosslink modifier (MBS):				
0 phr	20.0	30.3	3.38	0.89
10 phr	38.7	25.7	2.45	1.14
20 phr	70.7	22.7	2.47	1.09
Uncrosslink modifier (CPE):				
0 phr	20.0	30.3	3.38	0.89
10 phr	29.1	27.1	2.71	0.95
20 phr	44.1	26.4	2.63	0.97

[1]50 phr of wood flour (maple, 20 mesh) was used in the composites. PVC (*K* value = 57).

[2]The notched Izod impact strength, tensile strength, modulus, and elongation at break values for neat rigid PVC (without any impact modifier) were 34.98 J/m, 44.0 MPa, 2.47 GPa, and 2.23%, respectively.

Source: adapted from Mengeloglu *et al.*, 2000.

as more effective and efficient in improving the impact resistance of the composites compared to the chlorinated polyethylene modifier. The MBS and ACR are crosslinked, whereas CPE is an uncrosslinked modifier. The addition of impact modifiers reduced the tensile strength and modulus of the composites, whereas the elongation at break of the composites was not a function of impact modifier content and was essentially independent of the modifier type.

Foaming agents

WPCs have been commercialized, but their poor nailability, machinability, and high density (900–1360 kg/m^3), which is almost twice that of solid lumber (350–640 kg/m^3 for various pines), are major drawbacks for their use in applications such as decking, siding, window and door frames, etc. Foaming of WPCs has been proven to be an effective approach for reducing the density of the composites and improving their machinability. Foamed WPCs are accomplished through a conventional continuous extrusion process using chemical foaming agents (CFAs) or water contained in wood cell walls as a blowing agent (Mengeloglu and Matuana, 2001, 2003; Matuana and Mengeloglu, 2002; Matuana and Li, 2001, 2004; Li and Matuana, 2003c; Rizvi *et al.*, 2000). More recently, the use of physical blowing agents such as nitrogen and carbon dioxide has become attractive for microcellular foamed WPCs because of their inertness and also because they produce foamed WPCs with cell sizes in the range of 0.1 to 10 μm and a cell-population density of at least 10^9 cells/cm^3, which are desirable characteristics for the properties of microcellular foamed WPCs (Matuana and Mengeloglu, 2001; Matuana and Faruk, 2010; Matuana and Diaz, 2013). Figure 20.7 illustrates images of microcellular foamed neat PLA and PLA with 20% wood-flour composites.

The effects of CFA types (endothermic versus exothermic) and contents, as well as the influence of processing aids on the density and cell size of extrusion-foamed rigid PVC/wood-flour composites, have been investigated (Mengeloglu and Matuana, 2001). Exothermic foaming agent produced foamed samples with smaller average cell sizes compared to endothermic counterparts, due to the lower solubility and higher diffusivity of N_2 in the PVC matrix compared to CO_2, the main decomposition product of the endothermic foaming agent (Table 20.8). The density reduction can be further improved by the addition of an all-acrylic processing aid in the formulation of rigid PVC/wood-flour composite foams. Property evaluation of PVC/wood-flour composite foams has also been performed (Mengeloglu and Matuana, 2003). Typically, PVC/wood-flour composites foamed with both endothermic and exothermic chemical foaming agents exhibit similar mechanical properties (Table 20.9). The tensile strength and modulus are

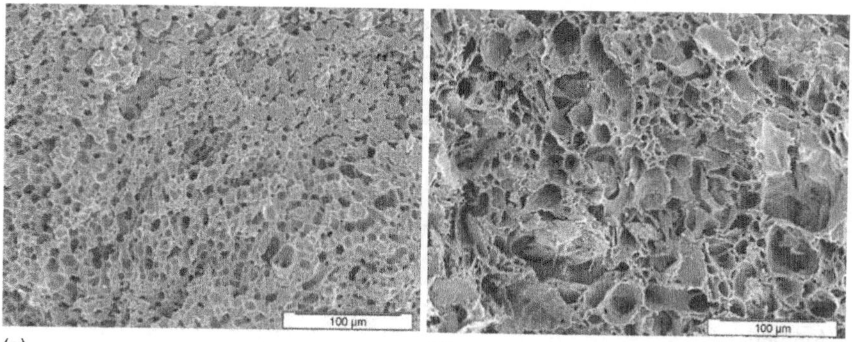

(a) (b)

20.7 Microcellular foamed (a) neat PLA samples (2002D) and (b) PLA with 20% wood-flour composites (pictures originally taken at 350×). Foamed samples were produced in a continuous extrusion process using CO_2 as a blowing agent (Matuana and Diaz, 2013) (adapted with permission from Industrial & Engineering Chemistry Research, ©2013 American Chemical Society).

Table 20.8 Effects of chemical foaming agent (CFA) types and processing aids on the density and average cell size of rigid PVC/30 phr wood-flour composite foamed using 0.5 phr CFA

Type of CFA in the composites[1]	Density (g/cm^3)		Cell size (μm)	
	Without K-400	With K-400[2]	Without K-400	With K-400[2]
Exothermic (modified azodicarbonamide)	0.93	0.58	44	88
Endothermic (baking soda)	0.91	0.57	65	84

[1]Wood flour used in the composites was maple of 100 mesh.
[2]The processing aid utilized was all-acrylic processing aid (6 phr of Paraloid K-400 from Rohm & Haas Co.).
Source: adapted from Mengeloglu and Matuana, 2001.

generally deteriorated by foaming whereas foaming does not compromise the elongation at break of the samples. Foaming reduced the Izod impact resistance of rigid PVC/wood-flour composites but this reduction was not statistically significant for the composites.

20.4 The effect of moisture on composite performance

In WPCs, the hydroxyl groups in wood flour are primarily responsible for the absorption of water, which causes the wood to swell. When WPCs are

Table 20.9 Properties of extrusion-foamed rigid PVC/wood-flour composites

Sample[1]	Density (g/cm^3)	Notched Izod impact strength (J/m)	Tensile properties		
			Ultimate strength (MPa)	Modulus (GPa)	Elongation at break (%)
Unfoamed composite	1.10 ± 0.08	16.8 ± 5.42^A	20.3 ± 2.8^A	2.1 ± 0.6^A	1.3 ± 0.1^A
Composite foamed with BS	0.80 ± 0.08	13.5 ± 4.4^A	11.4 ± 2.9^B	1.1 ± 0.3^B	1.4 ± 0.3^A
Composite foamed with AZRV	0.80 ± 0.04	14.3 ± 3.0^A	14.4 ± 1.8^B	1.5 ± 0.3^B	1.4 ± 0.3^A

[1]Composites contained 50 phr wood flour (maple, 100 mesh) and were foamed using 0.5 phr CFA.
Source: adapted from Mengeloglu and Matuana, 2003.

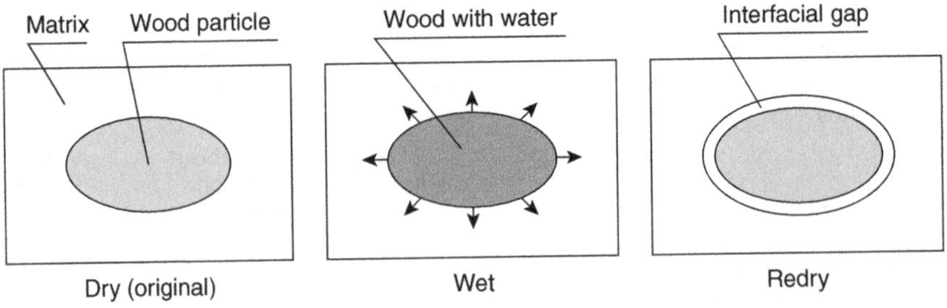

20.8 Schematic of moisture damage mechanism in WPCs (Stark and Gardner, 2008).

exposed to moisture, the swelling wood fiber can cause local yielding of the plastic due to swelling stress, fracture of wood particles due to restrained swelling, and interfacial breakdown (Joseph *et al.*, 1995). Figure 20.8 illustrates this mechanism. Initially, there is adhesion between the wood particle and matrix in a dry WPC. As the wood particle absorbs moisture, it swells. This creates stress in the matrix, leading to the formation of microcracks. It also creates stress in the wood particle, causing damage. After drying the composite, there is no longer adhesion at the matrix and wood–particle interface. Cracks formed in the plastic and the interfacial gap contribute to penetration of water into the composite at a later exposure (Stark and Gardner, 2008).

20.4.1 Moisture exposure

Damage that occurs during moisture exposure degrades mechanical properties. Microcracks in the matrix and damage to wood particles cause a loss in flexural modulus of elasticity (MOE) and strength. Interfacial damage is primarily responsible for a loss in composite strength. The effects can be dramatic. For example, after soaking 40% wood flour filled PP composites in a water bath for 2000 hours the water absorption of the composites was 9%. This corresponded with a 39% decrease in flexural MOE and a 22% decrease in flexural strength (Stark, 2001).

Wood flour content, wood particle size, processing method, and additives influence the amount of water absorbed by the WPC. Table 20.10 shows how increasing wood flour content, removing the original composite surface, and increasing the particle size increase the equilibrium moisture content of WPCs. Because the diffusion of water into WPCs is slow, it can take some time to reach equilibrium moisture content. As the thickness of WPCs increases, the distribution of absorbed water in the material becomes non-uniform with higher moisture content in the outer surface layer than in the core (Wang and Morrell, 2004). Figure 20.9 shows that moisture exposure is cumulative, causing some irreversible damage to WPCs.

Table 20.10 Average moisture content of PP-based composites soaked in distilled water for 238 days

Wood flour content (%)	Wood flour particle size	Coupling agent (%)	Surface treatment	Moisture content (%)
25	coarse	0	none	5.08 (0.03)
25	coarse	0	milled	6.33 (0.21)
25	coarse	3	none	4.42 (0.16)
25	coarse	3	milled	5.75 (0.05)
25	fine	0	none	4.76 (0.17)
25	fine	0	milled	5.73 (0.04)
25	fine	3	none	4.43 (0.06)
25	fine	3	milled	5.19 (0.20)
50	coarse	0	none	13.33 (0.34)
50	coarse	0	milled	14.12 (0.43)
50	coarse	3	none	10.92 (0.27)
50	coarse	3	milled	12.56 (0.16)
50	fine	0	none	11.51 (0.16)
50	fine	0	milled	12.41 (0.07)
50	fine	3	none	10.77 (0.17)
50	fine	3	milled	11.51 (0.10)

Source: adapted from Steckel *et al.*, 2007.

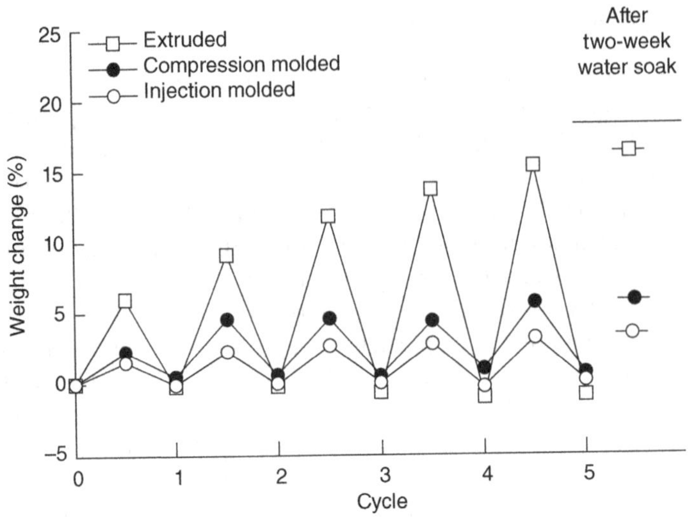

20.9 Moisture sorption of 50% wood flour filled HDPE during five moisture cycles consisting of 2 hours of boiling followed by 22 hours of drying (Clemons and Ibach, 2004).

20.4.2 Freeze–thaw exposure

Moisture exposure also becomes important in cold climates during freeze–thaw cycles. Standard freeze–thaw cyclical testing consists of three parts: (1) water soak until weight gain in a 24-hour period is not more than 1%; (2) freezing for 24 hours at –29°C; and (3) thawing at room temperature for 24 hours (ASTM D6662). Freeze–thaw cycling WPCs results in a loss of mechanical properties. For example, WPCs comprised of 46% pine wood flour filled PVC were exposed to five freeze–thaw cycles. After exposure, the modulus of rupture (MOR) decreased 13% and the MOE decreased 30% (Pilarski and Matuana, 2005).

Coupling agents are commonly used to improve bonding between the hydrophobic matrix and hydrophilic wood fiber. Coupling agents can also reduce the amount of moisture WPCs absorb by reducing gaps at the wood–matrix interface and by reducing the number of hydroxyl groups available for hydrogen bonding with water. However, the decreases in water absorption can be minor compared with other variables, including wood flour content and surface morphology (Table 20.10).

20.5 The effect of temperature on composite performance

Thermal response of polymers can impact WPC mechanical properties. Depending on the climatic exposure, WPCs used in decking applications can experience temperatures ranging from –30°C to +50°C. The primary thermal changes impacting WPC properties include thermal expansion, mechanical creep, and thermal-oxidative degradation. Increased temperatures can also act synergistically with other chemical degradation mechanisms to increase reaction rates (Stark and Gardner, 2008).

20.5.1 Thermal expansion

Thermal expansion in WPCs is anisotropic in extruded solid deck boards. As an example, for PP-based deck boards, the coefficient of thermal expansion was greatest in the thickness direction ($30.03 \times (1/°F) \times 10^{-6}$) followed by the width ($23.61 \times (1/°F) \times 10^{6}$) and is the least in the length ($11.53 \times (1/°F) \times 10^{-6}$) or machine direction of the boards (O'Neill and Gardner, 2004). Both the polymer chains and wood flour align with the flow of extrusion and this behavior is believed to contribute to the anisotropic thermal behavior of extruded deck boards. Thermal expansion behavior of WPC deck boards becomes important during installation because improper gapping can lead to warping. Using a plastic matrix that is not as susceptible to thermal changes, such as nylon, may be an important way to provide thermal stability in the future (Chen and Gardner, 2008). Thermal expansion in WPCs can also be reduced by creating a cellular or foamed microstructure (Finley, 2000). Since wood is more thermally stable with temperature than plastic, WPCs with higher wood contents exhibit lower coefficients of thermal expansion.

20.5.2 Creep

Thermoplastic materials like WPCs will experience changes in mechanical (stiffness) properties as a function of increased temperature. Materials under a mechanical load that might perform adequately at normal service temperatures may experience creep under loads for extended periods of time and at higher temperatures (Brandt and Fridley, 2003). Mechanical creep in WPCs can be reduced by crosslinking the polymer (Bengtsson *et al.*, 2006).

20.5.3 Thermal–oxidative degradation

Thermal–oxidative degradation may also contribute to the degradation of WPC deck boards (Klyosov, 2007). In one case study, WPC deck boards not containing sufficient levels of antioxidant exposed to extreme temperatures in the Arizona desert were experiencing crumbling and subsequent board failure. The failures were attributed to free radical oxidative degradation mechanisms that were exacerbated by high temperatures.

Thermoplastic polymers can experience thermal oxidative effects during processing, especially for PVC–wood composites. Both antioxidants and thermal stabilization aids can be added to WPC formulations to decrease thermal degradation during processing. In addition to protecting the polymer during processing, antioxidants can also help protect the polymer matrix during environmental exposure (Klyosov, 2007).

20.6 The effect of weathering on composite performance

Primary weathering variables include solar radiation, temperature, and water. Secondary variables include seasonal and annual variation, geographical differences, atmospheric gases, and pollution changes. The large number of variables accounts for the high variability in weathering studies, and does not readily allow for comparison between natural weathering studies. Results obtained are specific for the location and timeframe of the test.

Accelerated weathering is a technique used to compare performance by subjecting samples to cycles that are repeatable and reproducible. Primary weathering variables can all be measured during accelerated weathering. During accelerated weathering, test standards are typically followed that prescribe a schedule of radiation (ultraviolet, xenon-arc, etc.) and water spray (number and time of cycles). Weathering variables act independently and synergistically to degrade WPCs.

WPCs exposed to weathering may experience color change, which affects their aesthetic appeal, as well as mechanical property loss, which limits their performance (Matuana and Kamdem, 2002). Changes in mechanical properties after weathering can be due to changes such as composite surface oxidation, matrix crystallinity changes, and interfacial degradation (Stark, 2006; Stark and Matuana, 2004a, 2004b, 2007).

Weathering results in the destruction of the WPC surface (Fig. 20.10). Wood particles exposed at the surface absorb water and swell. In addition, the plastic matrix cracks upon UV exposure. In combination, the result is a flaky, cracked surface. As WPCs weather, the surface chemistry also changes. Oxidation at the surface is one measure of degradation (Stark and Gardner, 2008).

The color of WPCs primarily reflects the color of the wood during weathering, although some whitening is due to stress cracking of the matrix. Water and UV radiation jointly contribute to increasing composite L* (lightness). Exposure to UV radiation degrades lignin, leaving loose fibers at the composite surface. Water spray cycles wash away loose fibers, exposing more material for degradation. The result is a cyclical erosion of the surface (Williams *et al.*, 2001). Additionally, washing the surface can remove some water-soluble extractives that impart color (Stark, 2006). UV radiation and

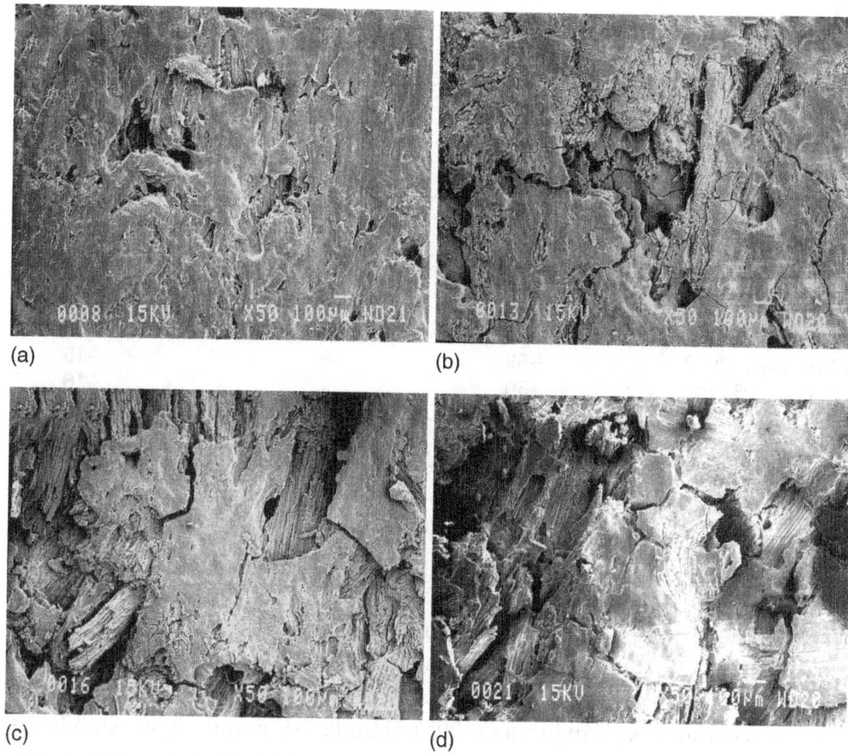

20.10 Micrographs of extruded 50% wood flour filled HDPE composites: (a) before weathering, and after (b) 1000 hours, (c) 2000 hours, and (d) 3000 hours of weathering (adapted from Stark *et al.*, 2004).

water also act synergistically to degrade WPCs in the following ways. Exposing WPCs to UV radiation degrades hydrophobic lignin, leaving hydrophilic cellulose at the surface and increasing surface wettability, causing the surface to become more sensitive to moisture (Kalnins and Feist, 1993). Swelling of the wood fiber also facilitates light penetration into wood and provides sites for further photodegradation (Hon, 2001).

Fiber swelling due to moisture absorption is primarily responsible for the loss in mechanical properties of the WPC after weathering. Cracks form in the polymer matrix due to swelling of the wood fiber and contribute to the loss of composite MOE and strength. The loss in strength is due to moisture penetration into the WPC, which degrades the wood–polymer interface, decreasing the stress transfer efficiency from matrix to the fiber. Synergism between UV radiation and water also contributes to mechanical property losses by eroding the surface and increasing surface wettability as described above, causing exposure to UV radiation with water exposure to be much more damaging than exposure to UV radiation only (Stark, 2006).

The protection of WPCs against weathering generally includes the addition of photostabilizers. Photostabilizers are compounds developed

Table 20.11 Percent change in properties of photostabilized 50% wood flour filled HDPE composites after 3000 hours of accelerated weathering

Formulation	Change in property (%)		
	L^*	Strength	MOE
–	+115	−27	−33
0.5% UVA	+98	−20	−32
1% UVA	+107	−15	−21
1% P	+73	−13	−18
2% P	+61	−5	−18
0.5% UVA, 1% P	+59	−9	−15
1% UVA, 2% P	+50	-2^{NS}	−16

UVA: Hydroxyphenylbenzotriazole, Tinuvin 328, Ciba Specialty Chemicals.
P: Zinc ferrite in carrier wax, Cedar TI-8536, Holland Colors Americas.
NS: Change not significant at $\alpha = 0.05$.
Source: adapted from Stark and Matuana, 2006.

to protect polymers and combat UV degradation. They are generally classified according to the degradation mechanism they hinder. Ultraviolet absorbers (UVAs) and free radical scavengers are important photostabilizers for polyolefins. Commercial UVAs are readily available as benzophenones and benzotriazoles (Gugumus, 1995). Hindered amine light stabilizers (HALS) provide protection as free radical scavengers (Gugumus, 1995). Pigments physically block light, thereby protecting the composite from photodegradation.

Pigments, ultraviolet absorbers, and hindered amine light stablilizers and combinations of photostabilizers have been used with some success in mitigating changes that occur during WPC weathering (Stark and Gardner, 2008). Table 20.11 summarizes the percent change in property that occurs when injection-molded WPCs photostabilized with UVA and/or pigment are exposed to accelerated weathering. In this case UVA reduced lightening by absorbing some UV radiation, resulting in less UV radiation available to bleach the wood component, while P physically blocked UV radiation, which also results in less available UV radiation to the wood component. In addition, P masked some lightening (Stark and Matuana, 2006).

20.7 The effect of biological attack on composite performance

20.7.1 Decay

Wood decays when its moisture content exceeds approximately 20%. In WPCs, it can be assumed that the plastic matrix does not absorb moisture.

Therefore, a 50% wood-filled WPC must reach a moisture content of about 10% for decay to occur (i.e., the wood component reaches a moisture content of 20%). However, the moisture content is typically measured for the bulk of the material. The slow diffusion of water through WPCs results in higher moisture contents at the surface than in the core (Wang and Morrell, 2004). In the field, the moisture content of the bulk WPC material may be significantly lower than the expected point at which decay begins (Stark and Gardner, 2008).

The wood flour in WPCs is responsible for its susceptibility to fungal attack. Because moisture is the critical factor, the variables that influence moisture sorption, including wood flour content and size, also influence decay. Increases in wood flour content increase decay (Mankowski and Morrell, 2000; Verhey *et al.*, 2001), while decreasing wood flour size decreases decay (Verhey and Laks, 2002). Decreasing wood flour content and size both increase encapsulation of the wood flour by the plastic matrix, decreasing decay susceptibility.

WPCs decay only at the surface layer. Stakes cut from a commercial WPC consisting of approximately 56% wood and 44% PE were installed in ground in Hilo, Hawaii and analyzed after 10 years of exposure. Evaluation of the in-ground portion of the stakes using SEM revealed surface pitting due to loss of wood particles. Only 5 mm in from the surface, there was no evidence of microbial attack (Schauwecker *et al.*, 2006). This was attributed to limited moisture movement into the WPC.

The first step in preventing decay is to prevent or limit WPC moisture sorption. Limiting the access of nutrients, i.e. encapsulating the wood in the plastic matrix, also decreases decay. Verhey *et al.* (2001) demonstrated that sanding the surface of a compression-molded WPC, i.e. removing some of the plastic film at the surface and exposing the wood flour, increased the amount of decay at high wood contents (Fig. 20.11). Acetylation and silane treatment of wood fibers decreases moisture sorption, and this has been shown to be beneficial in preventing decay (Hill and Abdul Khalil, 2000). The preservative zinc borate is often used to prevent decay and can be very effective (Ibach *et al.*, 2003; Verhey *et al.*, 2001). For example, mass loss of 50% wood flour filled PP composites was 12.9% after 12 weeks' exposure to brown-rot fungi. Adding zinc borate at 1% reduced the mass loss to less than 1%; adding zinc borate at 3% or 5% resulted in virtually no mass loss (Verhey *et al.*, 2001).

20.7.2 Mold

Moisture exposure of WPCs also relates to mold growth. Dawson-Andoh and coworkers (Dawson-Andoh *et al.*, 2004, 2005; Filson *et al.*, 2009) found more evidence of mold growth when WPCs were directly in contact with moisture

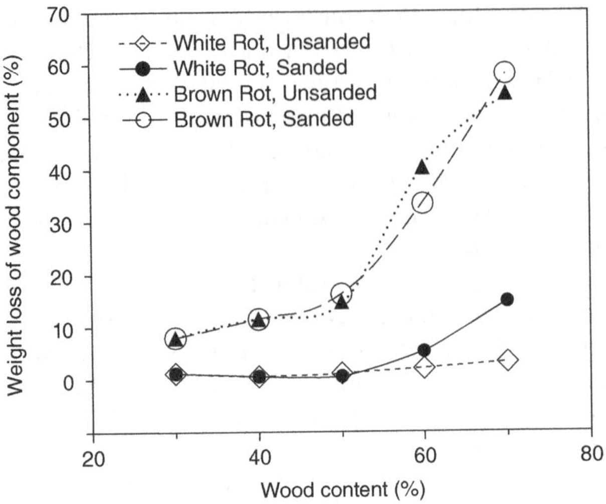

20.11 Percent weight loss in wood content from wood flour/PP as a function of composite wood content after 12-week exposure to either white-rot or brown-rot fungi (adapted from Verhey *et al.*, 2001).

versus exposed to an environment with a high relative humidity. Additionally, pre-conditioning WPCs through either exposure to UV weathering or freeze–thaw cycling was shown to have minimal effect on mold growth (Dawson-Andoh *et al.*, 2004, 2005). Increasing wood flour content increases mold growth (Laks *et al.*, 2005). Lubricants are often used to aid in extrusion. It has been suggested that amines in some lubricants provide sufficient nitrogen for more rapid colonization of mold fungi (Laks *et al.*, 2005).

Because mold and mold spores are always present in the air, it is often recommended to periodically clean WPCs to remove mold that is already growing. Similar to decay prevention, strategies to prevent mold growth often coincide with reducing moisture content and moisture susceptibility. Biocides also can control mold growth. Dylingowski (2003) evaluated an isothiazole and zinc borate moldicide in a model WPC material. Evaluated biocides that have had some success include isothiazole (Dylingowski, 2003), zinc borate (Dylingowski, 2003, Laks *et al.*, 2005), and chlorothalonil (Laks *et al.*, 2005).

20.8 Trends in materials and manufacturing techniques

20.8.1 New materials

Thermoplastics selected for WPCs traditionally have melt temperatures below 200°C (392°F) to maintain processing temperatures below the degradation point of the wood component. Higher processing temperature can result in the release of volatiles, discoloration, odor, and embrittlement of the wood component. Thermoplastics commonly used in WPCs include polyethylene (PE), polypropylene (PP), and polyvinyl chloride (PVC). However, new manufacturing strategies allow for the use of engineering thermoplastics with melting temperatures higher than 230°C (446°F) such as polyethylene terephthalate (PET), polyamide (PA, nylon), and acrylonitrile–butadiene–styrene (ABS) (Gardner *et al.*, 2008; Matuana *et al.*, 2007). The use of engineering thermoplastics for WPCs will grow as new applications requiring superior mechanical and thermal properties are introduced.

A driving force in the development of many new materials is to decrease our use of petroleum. Because the most common plastics used in WPCs, PE and PP, are typically derived from petroleum, there is growing interest in replacing the common WPC thermoplastics with bioplastics. Bioplastics, i.e., plastics derived from renewable biomass, may also be biodegradable. Biodegradable bioplastics include polylactic acid (PLA) and starch acetate; non-biodegradable bioplastics include cellulose acetate and bio-derived PE and polyamide (Lampinen, 2009). PLA has received the most attention from researchers as it is widely available commercially. Bioplastics will be particularly important in automotive and packaging applications.

The wood and/or lignocellulosic material used in WPCs can be derived from a variety of sources. Geographical location often dictates the material choice. In North America, wood is the most common raw material, in Europe natural fibers such as jute, hemp, and kenaf are preferred, while rice hull flour and bamboo fiber are typical in Asia. The wood is incorporated either as fiber bundles with low aspect ratio (wood flour) or as single fibers with higher aspect ratio (wood fiber). Wood flour is processed commercially, often from post-industrial materials such as planer shavings, chips, and sawdust. Wood and lignocellulosic fibers are available from virgin and recycled sources. New lignocellulosic sources include paper mill sludge and biorefinery residues. Because wood and lignocellulosic fibers can lead to superior WPC properties by acting more as a reinforcement than as a filler, in applications requiring additional strength, a trend is to move toward the use of wood and other lignocellulosic fibers.

The adaptation of nanotechnology in WPCs includes the use of nanofibers derived from wood and other lignocellulosics as a reinforcement for plastics. These new composites are termed cellulose nanocomposites. Cellulose nanocomposites are rapidly expected to open new markets in medicine, packaging, electronics, automotive, construction, and other sectors (Oksman *et al.*, 2009). The emergence of non-traditional forest products markets is particularly exciting.

Other materials can be added to affect processing and product performance of WPCs. These additives can improve bonding between the thermoplastic and wood component (for example, coupling agents), product performance (talc, impact modifiers, ultraviolet (UV) light stabilizers, flame retardants), and processability (lubricants). Additives for WPCs are continually evolving, but nanotechnology will drive changes in additive technology as the use of nanomaterials in small amounts in WPCs improves performance. The incorporation of nanomaterials into WPCs is still in its infancy and in the research stage, but includes the use of carbon nanotubes (Jin and Matuana, 2010; Faruk and Matuana, 2008a), nanoclays (Faruk and Matuana, 2008b; Hetzer and DeKee, 2008), and nano-TiO_2 (Stark and Matuana, 2009). It is expected that as the full benefit becomes understood, the use of these and other nanomaterials in WPCs will experience tremendous growth.

20.8.2 New manufacturing techniques

Improvements in and changes to manufacturing methods will help pave the way for the next generation of WPCs. New processes under development include coextrusion, foaming during extrusion, and inline coating technologies.

Coextrusion consists of the extrusion of multiple materials through a single die simultaneously. The most common application in WPCs is coextrusion of an opaque, unfilled plastic cap layer over a WPC core. This process became popular for fencing and is quickly becoming the manufacturing method of choice for decking. The unfilled cap layer enhances durability by improving moisture resistance. In addition, coextrusion allows manufacturers to concentrate expensive additives such as biocides, fungicides, and photostabilizers in the cap layer. Coextrusion is also being evaluated as a method to extrude a clear plastic cap layer over a WPC core (Stark and Matuana, 2009; Matuana *et al.*, 2011; Jin and Matuana, 2008, 2010). This method allows for the more natural appearance of the WPC to show while still providing enhanced moisture resistance and concentrated stabilization. Others are evaluating coextrusion as a method to extrude an all-WPC profile with different additives in different layers as needed (Yao and Wu, 2010).

Growth is also expected in the production of foamed WPCs. Creating a microcellular foamed structure in WPCs not only results in weight reduction, but also improves impact resistance and allows for better surface definition and sharper contours (Faruk *et al.*, 2007). Chemical and physical foaming agents are typically used to foam WPCs. Chemical foaming agents decompose at processing temperatures into gases while physical foaming agents liberate gases as a result of evaporation or desorption at elevated temperatures. A new trend in foaming is to meter and dissolve inert gases in the polymer melt during processing. This is termed extrusion foaming when done during extrusion (Diaz and Matuana, 2009; Matuana and Diaz, 2013), or gas-assist injection molding if done during injection molding. Another option is to use moisture in the wood to foam WPCs (Gardner *et al.*, 2008).

Applying a coating to WPCs post-processing but before they are made available to the consumer is a growing trend to provide increased durability and enhanced aesthetics. Coatings can include latex paints, polyurethanes, or acrylics. However, surface treatments are required to attain adequate adhesion between the WPC and the coating. Treatments that have been found effective include oxygen plasma, flame, chromic acid and benzophenone/UV irradiation (Gupta and Laborie, 2008). Another treatment being commercialized uses fluorooxidation to modify the surface of WPCs for improved coating adhesion. The advantage of this process is that the rapid chemistry allows for inline processing. Fluorooxidation also modifies only the outer few molecular layers, so embossed patterns are unaffected (Williams and Bauman, 2007). New coatings being evaluated include UV-curable coatings and powder coating. Advantages of a UV-curable coating include virtually unlimited color choice and gloss level, long-term resistance to fading, increased scratch, stain, and mar resistance, and prevention of mold and mildew growth (Burton, 2008).

20.9 Current and emerging applications

In the past decade WPCs have become a widely recognized commercial product in construction, automotive, furniture, and other consumer applications. Commercialization of WPCs in North America has been primarily due to penetration into the construction industry. Current WPC applications include decking, railing, window and door lineals, roofing, picnic tables and benches, fencing, landscape timbers, patios, gazebos, pergolas, auto parts, and playground equipment (Smith and Wolcott, 2006). The automotive industry in Europe has been a leader in using WPCs for interior panel parts and is leading the way in developing furniture applications. Manufacturers in Asia are targeting the furniture industry, in addition to interior construction and decorative applications.

Continued research and development will expand the available markets and each application will penetrate the global marketplace. In North America, construction applications that will see increased penetration into the market include siding, fencing, bridge decking, foundation isolation elements, marine structures (chocks, wales, pier decking), laminate flooring, residential furniture (bathroom/kitchen cabinets and patio furniture), utility poles, railroad ties, and exterior and interior molding and millwork (Crespell and Vidal, 2008). Emerging markets are expected as injection molding becomes more common for WPCs, including applications such as cosmetics packaging and toys. Markets in injection-molded automotive application are poised to grow in Europe and include parts such as glove boxes, fixing hooks, sound systems, and fan boxes (Carus *et al.*, 2008).

Growth is also expected as improvements to WPCs allow them to be used as structural members. There are various methods being investigated to improve the strength, stiffness, and creep performance of WPCs. Crosslinking the PE polymer matrix in WPCs using silanes in a reactive extrusion process can improve toughness, reduce creep, and improve durability (Bengtsson *et al.*, 2006, 2007). Using nanoparticle-reinforced plastics as matrices for WPCs is another way of enhancing the structural performance of WPCs (Faruk and Matuana, 2008a, 2008b). Combining WPCs with other non-WPC materials is also a trend that will allow for more structural composites. Currently, some WPC manufacturers in China are extruding WPCs over solid metal or solid wood. This allows for structural members in applications such as pergolas and gazebos to have WPC surfaces that match the non-structural members.

The hygroscopicity of wood can be reduced by replacing some of the hydroxyl groups with alternative chemical groups. Acetylation has been an active area of research in improving the moisture performance of wood and wood composites. Acetic anhydride reacts with hydroxyl groups in the wood cell wall to yield an acetylated fiber. For example, acetylating pine wood fiber reduced its equilibrium moisture content at 90% relative humidity and 27°C from 22% to 8% (Rowell, 1997). This method has been investigated to a more limited extent to provide moisture resistance to WPCs. For example, a 50% wood fiber filled PP composite absorbed 5% moisture after soaking for 34 days while a 50% acetylated wood fiber filled PP composite absorbed only 2.5% (Abdul Khalil *et al.*, 2002).

20.10 References

Abdul Khalil, H.P.S., Ismail Rosfaizal, H. and Ahmad, M.N. (2002) Polypropylene (PP) *Acacia mangium* composites: the effect of acetylation on mechanical and water absorption properties, *Polymer Plast Tech Eng*, 41, 453–468.

Afrifah, K.A. and Matuana, L.M. (2012) Statistical optimization of ternary blends of poly(lactic acid)/ethylene acrylate copolymer/wood-flour composites, *Macromol Mater Eng*, 297, 167–175.

Baeza, J. and Freer, J. (2001) Chemical characterization of wood and its components, Chapter 8 in Hon, D.N.-S. and Shiraishi, N., *Wood and Cellulosic Chemistry*, Marcel Dekker, New York, 275–384.

Balasuriya, P.W., Ye, L. and Mai, Y.-W. (2001) Mechanical properties of wood flake–polyethylene composites. Part I. Effects of processing methods and matrix melt flow behaviour, *Composites: Part A*, 32, 619–629.

Balasuriya, P.W., Ye, L., Mai, Y.-W. and Wu, J. (2002) Mechanical properties of wood flake–polyethylene composites. Part II. Interface modification, *J Appl Polym Sci*, 83, 2505–2521.

Bengtsson, M., Oksman, K. and Stark, N.M. (2006) Profile extrusion and mechanical properties of crosslinked wood–thermoplastic composites, *Polym Compos*, 27, 184–194.

Bengtsson, M., Stark, N.M. and Oksman, K. (2007) Durability and mechanical properties of silane cross-linked wood thermoplastic composites, *Compos Sci Technol*, 67, 2728–2738.

Berger, M.J. and Stark, N.M. (1997) Investigation of species effects in an injection molding grade wood filled polypropylene, in *Proceedings of the Fourth International Conference on Woodfiber–Plastic Composites*, Forest Products Society, Madison, WI, pp. 19–25.

Brandt, C.W. and Fridley, K.J. (2003) Load–duration behavior of wood–plastic composites, *J Mater Civ Eng*, 15, 524–536.

Burton, K. (2008) A cure for wood alternatives: New UV coating technologies can enhance wood plastic composites, fiber cement siding and vinyl substrates used in the construction industry, *Finishing Today*, April 2008, 32–35.

Carlborn, K. and Matuana, L.M. (2006) Functionalization of wood particles through a reactive extrusion process, *J Appl Polym Sci*, 101, 3131–3142.

Carus, M., Gahle, C. and Korte, H. (2008) Market and future trends for wood–polymer composites in Europe: the example of Germany, Chapter 14 in Oksman Niska K. and Sain, M. (eds), *Wood–Polymer Composites*, Woodhead Publishing Ltd, Cambridge, UK.

Chen, J. and Gardner, D.J. (2008) Dynamic mechanical properties of extruded nylon–wood composites, *Polym Compos*, 29, 372–379.

Clemons, C. (2002) Wood–plastic composites in the United States: The interfacing of two industries, *Forest Prod J*, 52, 10–18.

Clemons, C.M. (2010) Wood flour, in Xanthos, M., *Functional Filler for Plastics*, second, updated and enlarged edition, Wiley-VCH, Weinheim, Germany, 269–290.

Clemons, C.M. and Ibach, R.E. (2004) The effects of processing method and moisture history on the laboratory fungal resistance of wood–HDPE composites, *Forest Prod J*, 54, 50–57.

Crespell, P. and Vidal, M. (2008) Market and technology trends and challenges for wood plastic composites in North America, *Proceedings of the 51st International Convention of Society of Wood Science and Technology*, 10–12 November 2008, Concepción, Chile.

Dawson-Andoh, B., Matuana, L.M. and Harrison, J. (2004) Mold susceptibility of rigid PVC/wood-flour composites, *J Vinyl Addit Technol*, 10, 179–186.

Dawson-Andoh, B.E., Matuana, L.M. and Harrison, J. (2005) Susceptibility of high density polyethylene/wood-flour composite to mold discoloration, *J Inst Wood Sci*, 17, 114–119.

Diaz, C.A. and Matuana, L.M. (2009) Continuous extrusion production of microcellular rigid PVC, *J Vinyl Addit Technol*, 15, 211–208.

Dylingowski, P. (2003) Maintaining the aesthetic quality of wood–plastic composite decking with isothiazolone biocide, in *Proceedings, Seventh International Conference on Woodfiber–Plastic Composites*, 19–23 May 2003, Madison, WI, pp. 177–186.

Faruk, O. and Matuana, L.M. (2008a) Reinforcement of rigid PVC/wood-flour composites with multi-walled carbon nanotubes, *J Vinyl Addit Technol*, 14, 60–64.

Faruk, O. and Matuana, L.M. (2008b) Nanoclay reinforced HDPE as a matrix for wood–plastic composites, *Compos Sci Technol*, 68, 2073–2077.

Faruk, O., Bledzki, A.K. and Matuana, L.M. (2007) Microcellular foamed wood–plastic composites by different processes: a review, *Macromol Mater Eng*, 292, 113–127.

Felix, J.M. and Gatenholm, P. (1991) The nature of adhesion in composites of modified cellulose fibers and polypropylene, *J Appl Polym Sci*, 42, 609–620.

Fengel, D. and Wegener, G. (1983) *Wood*, Walter de Gruyter, New York.

Filson, P., Dawson-Andoh, B. and Matuana, L. (2009) Colorimetric and vibrational spectroscopic characterization of weathered surfaces of wood and rigid polyvinyl chloride–wood flour composite lumber, *J Wood Sci Technol*, 43, 669–678.

Finley, M.D. (2000) Foamed thermoplastic polymer and wood fiber profile and member, US Patent 6054207, assigned to Andersen Corporation.

Gardner, D.J., Han, Y. and Song, W. (2008) Wood plastic composites technology trends, *Proceedings of the 51st International Convention of Society of Wood Science and Technology*, 10–12 November 2008, Concepción, Chile.

Gugumus, F. (1995) Light stabilizers, in Gachter, R. and Muller, H., *Plastics Additives Handbook*, Hanser Publishers, New York, 129–262.

Guo, G., Rizvi, G.M., Park, C.B. and Lin, W.S. (2004) Critical processing temperature in the manufacture of fine-celled plastic/wood-fiber composite foams, *J Appl Polym Sci*, 91, 621–629.

Guo, G., Lee, Y.H., Rizvi, G.M. and Park, C.B. (2008) Influence of wood fiber size on extrusion foaming of wood fiber/HDPE composites, *J Appl Polym Sci*, 107, 3505–3511.

Gupta, B.S. and Laborie, M.P. (2008) Surface activation and adhesion properties of wood-fiber reinforced thermoplastic composites, *J Adhesion*, 83, 939–955.

Hetzer, M. and DeKee, D. (2008) Wood/polymer/nanoclay composites, environmentally friendly sustainable technology: a review, *Chem Eng Res Des*, 86, 1083–1093.

Hill, C.A.S. and Abdul Khalil, H.P.S. (2000) The effect of environmental exposure upon the mechanical properties of coir or oil palm fiber reinforced composites, *J Appl Polym Sci*, 77, 1322–1330.

Hon, D.N.S. (2001) Weathering and photochemistry of wood, in Hon, D.N.-S. and Shiraishi, N., *Wood and Cellulosic Chemistry*, Marcel Dekker, New York, 513–543.

Huang, H.X. and Zhang, J.J. (2009) Effects of filler–filler and polymer–filler interactions on rheological and mechanical properties of HDPE–wood composites, *J Appl Polym Sci*, 111, 2806–2812.

Ibach, R.E., Clemons, C.M. and Stark, N.M. (2003) Combined ultraviolet and water exposure as a preconditioning method in laboratory fungal durability testing, in *Proceedings, Seventh International Conference on Woodfiber–Plastic Composites*, 19–23 May 2003, Madison, WI, pp. 61–67.

Jin, S. and Matuana, L.M. (2008) Co-extruded PVC/wood-flour composites with WPC cap layers, *J Vinyl Addit Technol*, 14, 197–203.

Jin, S. and Matuana, L.M. (2010) Wood/plastic composites co-extruded with multi-walled carbon nanotube-filled rigid poly(vinyl chloride) cap layer, *Polym Int*, 59, 648–657.

Joseph, K., Thomas, S. and Pavithran, C. (1995) Effect of ageing on the physical and mechanical properties of sisal-fiber-reinforced polyethylene composites, *Compos Sci Tech*, 53, 99–110.

Kalnins, M.A. and Feist, W.C. (1993) Increase in wettability of wood with weathering, *Forest Prod J*, 43, 55–57.

Kazayawoko, M. (1996) Surface characterisation and mechanisms of adhesion in wood fibre–polypropylene composites, PhD thesis, Department of Forestry, University of Toronto.

Kazayawoko, M., Balatinecz, J.J. and Matuana, L.M. (1999) Surface modification and adhesion mechanisms in wood fiber–polypropylene composites, *J Mater Sci*, 34, 6189–6199.

Klyosov, A.A. (2007) *Wood–Plastic Composites*, John Wiley & Sons, Hoboken, NJ.

Laks, P., Vehring, K., Verhey, S. and Richter, D. (2005) Effect of manufacturing variables on mold susceptibility of wood–plastic composites, in *Proceedings, Eighth International Conference on Woodfiber–Plastic Composites*, 23–25 May 2005, Madison, WI, pp. 265–270.

Lampinen, J. (2009) Biocomposite research as a way to add value and sustainability to composites, *Proceedings of the 4th Wood Fibre Polymer Composites International Symposium*, 30–31 March 2009, Bordeaux, France.

Li, Q. and Matuana, L.M. (2003a) Effectiveness of maleated and acrylic-acid functionalized polyolefin coupling agents for HDPE/wood-flour composites, *J Thermoplast Compos Mater*, 16, 551–564.

Li, Q. and Matuana, L.M. (2003b) Surface of cellulosic materials modified with functionalized polyethylene coupling agents, *J Appl Polym Sci*, 88, 278–286.

Li, Q. and Matuana, L.M. (2003c) Foam extrusion of HDPE/wood-flour composites using chemical foaming agents, *J Appl Polym Sci*, 88, 3139–3150.

Li, T.Q. and Wolcott, M.P. (2005) Rheology of wood plastics melt. Part 1. Capillary rheometry of HDPE filled with maple, *Polym Eng Sci*, 45, 549–559.

Liao, B., Huang, Y. and Cong, G. (1997) Influence of modified wood fibers on the mechanical properties of wood fiber-reinforced polyethylene, *J Appl Polym Sci*, 66, 1561–1568.

Maldas, D., Kokta, B.V. and Daneault, C. (1989) Composites of poly(vinyl chloride)–wood fibers. IV. Effect of the nature of fibers, *J Vinyl Technol*, 11, 90–99.

Mankowski, M. and Morrell, J.J. (2000) Patterns of fungal attack in wood–plastic composites following exposure in a soil block test, *Wood Fiber Sci*, 32, 340–345.

Marra, A.A. (1992) *Technology of Wood Bonding*, Van Nostrand Reinhold, New York.

Matuana, L.M. (2009) Recent developments in wood plastic composites, *J Vinyl Addit Technol*, 15, 136–138.

Matuana, L.M. and Diaz, C.A. (2013) Strategy to produce microcellular foamed poly(lactic acid)/wood-flour composites in a continuous extrusion process, *Ind Eng Chem Res*, 52, 12032–12040.

Matuana, L.M. and Faruk, O. (2010) Effect of gas saturation conditions on the expansion ratio of microcellular PLA/wood-flour composites, *eXPRESS Polym Letters*, 4, 621–631.

Matuana, L.M. and Kamdem, D.P. (2002) Accelerated ultraviolet weathering of PVC/wood-flour composites, *Polym Eng Sci*, 42, 1657–1666.

Matuana, L.M. and Li, Q. (2001) A factorial design applied to the extrusion foaming of polypropylene/wood-flour composites, *Cellular Polym*, 20, 115–130.

Matuana, L.M. and Li, Q. (2004) Statistical modeling and response surface optimization of extruded HDPE/wood-flour composite foams, *J Thermoplast Compos Mater*, 17, 185–199.

Matuana, L.M. and Mengeloglu, F. (2001) Microcellular foaming of impact-modified rigid PVC/wood-flour composites, *J Vinyl Addit Technol*, 7, 67–75.

Matuana, L.M. and Mengeloglu, F. (2002) Manufacture of rigid PVC/wood-flour composite foams using moisture contained in wood as foaming agent, *J Vinyl Addit Technol*, 8, 264–270.

Matuana, L.M., Park, C.B. and Balatinecz, J.J. (1997) The effect of low levels of plasticizer on the rheological and mechanical properties of polyvinyl chloride/newsprint-fiber composites, *J Vinyl Addit Technol*, 3, 265–273.

Matuana, L.M., Balatinecz, J.J. and Park, C.B. (1998a) Effect of surface properties on the adhesion between PVC and wood veneer laminates, *Polym Eng Sci*, 38, 765–773.

Matuana, L.M., Woodhams, R.T., Balatinecz, J.J. and Park, C.B. (1998b) Influence of interfacial interactions on the properties of PVC/cellulosic-fiber composites, *Polym Compos*, 19, 446–455.

Matuana, L.M., Balatinecz, J.J., Park, C.B. and Sodhi, R.N.S. (1999) X-ray photoelectron spectroscopy study of silane-treated newsprint-fibers, *Wood Sci Technol*, 33, 259–270.

Matuana, L.M., Balatinecz, J.J., Sodhi, R.N.S. and Park, C.B. (2001a) Surface characterization of esterified cellulosic fibers by XPS and FTIR spectroscopy, *Wood Sci Technol*, 35, 191–201.

Matuana, L.M., Kamdem, D.P. and Zhang, J. (2001b) Photoageing and stabilization of rigid PVC/wood-fiber composites, *J Appl Polym Sci*, 80, 1943–1950.

Matuana, L.M., Cam, S., Yuhasz, K.B. and Armstrong, Q.J. (2007) Composites of acrylonitrile–butadiene–styrene filled with wood flour, *Polym Polym Compos*, 15, 343–348.

Matuana, L.M., Jin, S. and Stark, N.M. (2011) Ultraviolet weathering of HDPE/wood-flour composites coextruded with a clear HDPE cap layer, *Polym Degrad Stab*, 96, 97–106.

Mengeloglu, F. and Matuana, L.M. (2001) Foaming of rigid PVC/wood-flour composites through a continuous extrusion process, *J Vinyl Addit Technol*, 7, 142–148.

Mengeloglu, F. and Matuana, L.M. (2003) Mechanical properties of extrusion foamed rigid PVC/wood-flour composites, *J Vinyl Addit Technol*, 9, 26–31.

Mengeloglu, F., Matuana, L.M. and King, J.A. (2000) Effects of impact modifiers on the properties of rigid PVC/wood-fiber composites, *J Vinyl Addit Technol*, 6, 153–157.

Oksman, K. and Clemons, C. (1998) Mechanical properties and morphology of impact modified polypropylene–wood flour composites, *J Appl Polym Sci*, 67, 1503–1513.

Oksman, K., Mathew, A.P. and Sain, M. (2009) Novel bionanocomposites: Processing, properties and potential applications, *Plast Rubber Compos*, 38, 396–405.

O'Neill, S. and Gardner, D.J. (2004) Anisotropic thermal expansion of wood–thermoplastic composites, in *Proceedings, Progress in Woodfibre–Plastic Composites*, 10–11 May 2004, Toronto, Canada.

Park, B.D. and Balatinecz, J.J. (1997) Mechanical properties of wood-fiber/toughened isotactic polypropylene composites, *Polym Compos*, 18, 79–89.

Pettersen, R.C. (1984) The chemical composition of wood, Chapter 2 in Rowell, R.M. (ed.), *The Chemistry of Solid Wood*, American Chemical Society, Washington, DC, 76–81.

Pilarski., J.M. and Matuana, L.M. (2005) Durability of wood-flour plastic composites exposed to accelerated freeze–thaw cycling. Part I: Rigid PVC matrix, *J Vinyl Addit Technol*, 11, 1–8.

Pilarski, J.M. and Matuana, L.M. (2006) Durability of wood-flour plastic composites exposed to accelerated freeze–thaw cycling. Part II: High density polyethylene matrix, *J Appl Polym Sci*, 100, 35–39.

Razi, P.S. and Raman, A. (2000) Study on impact fracture properties of wood-polymer composites, *J Compos Mater*, 34, 980–997.

Rizvi, G., Matuana, L.M. and Park, C.B. (2000) Foaming of PS/wood-fiber composites in extrusion using moisture as a blowing agent, *Polym Eng Sci*, 40, 2124–2132.

Rodríguez, C.A., Medina, J.A. and Reinecke, H. (2003) New thermoplastic materials reinforced with cellulose based fibers, *J Appl Polym Sci*, 90, 3466–3472.

Rowell, R.M. (1997) Chemical modification of agro-resources for property enhancement, in Rowell, R.M., Young, R.A. and Rowell, J.K., *Paper and Composites from Agro-Based Resources*, Lewis Publishers, New York, 351–375.

Saheb, D.N. and Jog, J.P. (1999) Natural fiber polymer composites: a review, *Adv Polym Technol*, 18, 351–363.

Schauwecker, C., Morrell, J.J., McDonald, A.G. and Fabiyi, J.S. (2006) Degradation of a wood–plastic composite exposed under tropical conditions, *Forest Prod J*, 56, 123–129.

Sears, K.D., Jacobson, R., Caulfield, D.F. and Underwood, J. (2001) Reinforcement of engineering thermoplastics with high purity wood cellulose fibers, in *Proceedings, 6th International Woodfiber–Plastics Composites Conference*, 15–16 May 2001, Madison, WI, pp. 27–34.

Shah, B.L. and Matuana, L.M. (2004) Online measurement of rheological properties of PVC/wood-flour composites, *J Vinyl Addit Technol*, 10, 121–128.

Shah, B.L., Matuana, L.M. and Heiden, P.A. (2005) Novel coupling agents for PVC/wood-flour composites, *J Vinyl Addit Technol*, 11, 160–165.

Smith, P.M. and Wolcott, M. (2006) Opportunities for wood/natural fiber–plastic composites in residential and industrial applications, *Forest Prod J*, 56, 4–11.

Stark, N. (2001) Influence of moisture absorption on mechanical properties of wood flour–polypropylene composites, *J Thermoplast Compos Mater*, 14, 421–432.

Stark, N.M. (2006) Effect of weathering cycle and manufacturing method on performance of wood flour and high-density polyethylene composites, *J Appl Polym Sci*, 100, 3131–3140.

Stark, N.M. and Berger, M.J. (1997) Effect of particle size on properties of wood flour reinforced polypropylene composites, in *Proceedings of the Fourth International Conference on Woodfiber–Plastic Composites*, Forest Products Society, Madison, WI, pp. 134–143.

Stark, N.M. and Gardner, D.J. (2008) Outdoor durability of wood–polymer composites, in *Wood–Polymer Composites*, CRC Press, Boca Raton, FL.

Stark, N.M. and Matuana, L.M. (2003) Ultraviolet weathering of photostabilized wood-flour-filled high density polyethylene composites, *J Appl Polym Sci*, 90, 2609–2617.

Stark, N.M. and Matuana, L.M. (2004a) Surface chemistry and mechanical property changes of wood-flour/high density polyethylene composites after accelerated weathering, *J Appl Polym Sci*, 94, 2263–2273.

Stark, N.M. and Matuana, L.M. (2004b) Surface chemistry changes of weathered HDPE/wood-flour composites studied by XPS and FTIR spectroscopy, *Polym Degrad Stab*, 86, 1–9.

Stark, N.M. and Matuana, L.M. (2006) Influence of photostabilizers on wood flour–HDPE composites exposed to xenon-arc radiation with and without water spray, *Polym Degrad Stab*, 91, 3048–3056.

Stark, N.M. and Matuana, L.M. (2007) Characterization of weathered wood–plastic composite surfaces using FTIR spectroscopy, contact angle, and XPS, *Polym Degrad Stab*, 92, 1883–1890.

Stark, N.M. and Matuana, L.M. (2009) Co-extrusion of WPCs with a clear cap layer to improve color stability, *Proceedings of the 4th Wood Fibre Polymer Composites International Symposium*, 30–31 March 2009, Bordeaux, France.

Stark, N.M. and Rowlands, R.E. (2003) Effects of wood fiber characteristics on mechanical properties of wood/polypropylene composites, *Wood Fiber Sci*, 35, 167–174.

Stark, N.M., Matuana, L.M. and Clemons, C.M. (2004) Effect of processing method on surface and weathering characteristics of wood-flour/HDPE composites, *J Appl Polym Sci*, 93, 1021–1030.

Steckel, V., Clemons, C.M. and Thoemen, H. (2007) Effects of material parameters on the diffusion and sorption properties of wood-flour/polypropylene composites, *J Appl Polym Sci*, 103, 752–763.

Verhey, S.A. and Laks, P.E. (2002) Wood particle size affects the decay resistance of woodfiber/thermoplastic composites, *Forest Prod J*, 52, 78–81.

Verhey, S., Laks, P. and Richter, D. (2001) Laboratory decay resistance of wood fiber/ thermoplastic composites, *Forest Prod J*, 51, 44–49.

Wang, W. and Morrell, J.J. (2004) Water sorption characteristics of two wood–plastic composites, *Forest Prod J*, 54, 209–212.

Wiedenhoeft, A. (2010) Structure and function of wood, in *Wood Handbook: Wood as an Engineering Material*, United States Department of Agriculture, Forest Service, General Technical Report FPL-GTR-190, Madison, WI.

Williams, K. and Bauman, B. (2007) New technology for enhancing wood–plastic composites, *JCT Coatings Tech*, 4, 52–57.

Williams, R.S., Knaebe, M.T. and Feist, W.C. (2001) Erosion rates of wood during natural weathering. Part II. Earlywood and latewood erosion rates, *Wood Fiber Sci*, 33, 43–49.

Yao, F. and Wu, Q. (2010) Coextruded polyethylene and wood-flour composite: Effect of shell thickness, wood loading, and core quality, *J Appl Polym Sci*, 118, 3594–3601.

Zazyczny, J. and Matuana, L.M. (2005) Fillers and reinforcing agents, Chapter 7 in Wilkes, D.E., Summers, J.W., and Daniels, C.A., *PVC Handbook*, Hanser Publications, Munich, 235–275.

21

The use of *Luffa cylindrica* fibres as reinforcements in composites

D. KOCAK, S. I. MISTIK and M. AKALIN,
Marmara University, Turkey and N. MERDAN,
Istanbul Commerce University, Turkey

DOI: 10.1533/9781782421276.5.689

Abstract: This chapter discusses the use of *Luffa cylindrica fibres*. Areas covered include the properties and surface treatment of fibres, applications and performance as a reinforcement material in composites and nanocomposites incorporating *Luffa cylindrica* fibres.

Key words: *Luffa cylindrica*, natural fibres, lignocellulosic, biocomposites, nanocomposites.

21.1 Introduction

Natural fibres from leaves, seed or bast include coir, sisal, jute, sponge gourd (*Luffa cylindrica*), flax, ramie, abaca, kenaf, cotton, palmyra, bamboo, mesta, henequen, kapok, hemp, sunn, piassava, wood, banana, kusha, sawai grasses, pineapple and more. These materials are all composed of fibrils glued together with natural resinous materials in the plant tissue [1, 2]. These lignocellulosic fibres are biodegradable, ecofriendly and low cost and have high stiffness to density ratios, all properties that are attractive for automotive applications [3].

Luffa fibre is obtained from the fruit of the *Luffa cylindrica* lignocellulosic plant (Fig. 21.1). It is a member of a generic group of eight species known as 'vegetable sponges', which are herbaceous trailing perennials producing fruits with a fibrous vascular system [4]. Another typical source of such fibres is *Luffa aegyptiaca*, a sub-tropical creeping annual plant which is a member of the Cucurbitae family. This is quite common in the mid-south of America, and is cultivated in Africa, Asia and Australia for the cylindrical fruits. Its size varies in relation to the areas in which it is grown, ranging from 15 cm to 1 m or more. In Turkey, *Luffa cylindrical* (luffa) grows well in the areas of Mediterranean climate, with creeping stems reaching 9 m. The diameter of the basic fibre is 20–50 μm.

21.1 Luffa cylindrica fibres.

Table 21.1 Chemical composition of *Luffa cylindrica* fibres [5]

Component	%
Holocellulose	82.4
α-Cellulose	63.0
Hemicellulose	19.4
Lignin	11.2
Extractives	3.2

21.2 Properties and surface treatment of *Luffa cylindrica* fibres

The struts of this natural sponge are characterized by a microcellular architecture with continuous hollow micro-channels (macro-pores with diameter of 10–20 μm) which form vascular bundles. They consist of a cellulosic polymer with various non-cellulosic impurities [5] (Table 21.1).

Luffa fibres are used for many things, including as an exfoliating pad for bathing, as an inner filler for filters, in the leather straps for automobiles, and so on. They are also used for the production of decorative items, for boiler filters and for insulating materials [6].

Luffa fibres have typical lignocellulosic fibre characteristics. Many investigators have explored the capability of additives to enhance the adhesion and thereby improve the properties, such as tensile, impact and flexural strengths, of natural fibre-reinforced composite materials. The methods for surface modification can be physical or chemical. The fibre surface can be modified by bleaching, acetylation and alkali treatments.

Studies have been carried out to improve the hydrophilic characteristic of luffa fibre by means of chemical processes. With this aim, mercerization was applied to luffa fibres by acetylation. In addition, the anion exchange capacity of luffa fibres and their quaternizated specialities were tested. Luffa fibres were coated with different kinds of silane agents in order to improve the characteristics of the composite produced using luffa fibre/polypropylene. The modification of luffa fibres was also performed using ultrasonic and microwave methods. These methods damaged the fibres less than conventional methods and decreased the water, energy and chemicals consumption [7–14].

Plasma treatment at atmospheric pressure was carried out on raw luffa fibres. The orthogonal method was used to optimize the plasma treatment condition. The optimum conditions were obtained at a discharge time of 10 s, a power of 110 W and an electrode distance of 6 mm. Compared with chemical pretreatment, plasma treatment had a significantly shorter duration, used no solvents and improved the cellulose contents. It is important to note that the mechanical strength is an important parameter for evaluating a fibre, and most of these studies are based on removing those impurities which affect the mechanical, chemical, physical and sorption characteristics of luffa fibres [15].

21.3 Applications and performance of *Luffa cylindrica* fibres as reinforcements in composites

Some fibres have been extensively investigated and used in textile fabrics, in composites and for medical purposes, and many other less-known fibres find limited applications, for instance in making ropes, mats, purses and wall hangings. One of the handicaps in finding new uses for many natural fibres is the lack of available scientific data regarding their structure and properties. Natural fibres have little resistance to environmental effects, and their use depends on environmental conditions which are likely to influence ageing and degradation [16].

Natural luffa fibres have been successfully used to advantage in the preparation of moulded composites by compression or resin transfer moulding. Furthermore, unlike many other lignocellulosic materials, the surface of this cellulose has potential for significant surface modification through carbohydrate chemistry. This has led several researchers to use this renewable and abundantly available lignocellulosic material to produce natural fibre-reinforced composites [17].

In this chapter, the use of luffa fibres as reinforcement in resin matrix composite materials is evaluated. The morphology of the fibrous vascular system of the fruit is presented and the advantages of using this natural mat material are highlighted. The results obtained show that the use of luffa has some potential practical advantages over some of the more usual natural fibres used as reinforcement in composite materials. We can highlight a reduction in fibre preparation, because the material occurs naturally as a mat, but the most significant result obtained is related to the change in the failure mode of composites when luffa is used. Improved machanical properties are also obtained when luffa fibre is used in a resin matrix composite between outer layers of E-glass mat [17, 18].

Luffa fibre from Turkey's Aegean region was characterized using Fourier transform infrared spectrophotometry, XPS, scanning electron microscopy, TGA and fibre tensile testing. The cellulose, hemicellulose and lignin content were also obtained. Major findings were as follows: degradation temperature was 381°C, O/C ratio was 0.61, and the distributions of surface functional groups were CAC, CAH (31.9%), CAOH, CAOAC (43.4%), C¼¼O, OACAO (22.5%), and OAC¼¼O (2.2%). Cellulose, hemicellulose and lignin contents were 63.0%, 14.4% and 1.6%, respectively. Tensile strength and tensile modulus of luffa fibres were determined as 385 MPa and 12.2 GPa, respectively. After water ageing, tensile, flexural, ILSSs and elongation at break values of the composite decreased 28%, 24%, 45% and 31%, respectively. Ageing of luffa-reinforced polyester composites under a stream of seawater (5% NaCl) for 170 h at 50°C decreased the mechanical properties of the polymer composite [19].

Preparation and characterization of a vinylester thermoset matrix composite material reinforced by natural luffa fibres after modification treatments were investigated. In this study, extraction treatments with organic solvents, mercerization and a quite new esterification with BTDA dianhydrides were used, and the results show that in all cases the composite materials reinforced by luffa fibres have improvements in mechanical and thermal properties compared to the normal vinylester matrix. As an example, a 50% tensile strength increase was obtained for a composite reinforced by fibres esterified with benzophenone tetracarboxylic dianhydride when compared to normal thermoset matrix [20].

Luffa biocomposites are a novel use of the fibres, but a better understanding of their surface characteristics is necessary to maximize their potential use. In a study, different chemical treatments for luffa fibres were conducted with aqueous solutions of 2% NaOH, and methacrylamide (1–3%), at distinct treatment times. Treatment with NaOH was found to adequately modify the fibre surface in preparation for its use as reinforcement of composite materials, whilst not causing such severe damage to the fibre as the methacrylamide treatment [21].

Luffa fibre was converted into an anion exchange material through etherification using 2-chloroethyldiethylamine hydrochloride (CEDA-HCl) followed by crosslinking using epichlorohydrin. The effects of the different etherification reaction parameters (CEDA concentration, NaOH concentration and temperature) on the extent of modification were assessed by elemental analysis of nitrogen to optimize the reaction. The effect of the etherification on the supramolecular structure was observed by X-ray diffraction. The ability of the crosslinked diethylaminoethylated luffa fibre (DEAE-luffa) after its quaternization with methyl iodide, to adsorb sulfate, nitrate and phosphate anions, was tested. Luffa fibre is characterized by high cellulose content and relatively low lignin content. The lignin in luffa fibre contains both syringyl and guaiacyl types, but the former type predominates. A strong anion exchanger was prepared using luffa through etherification, crosslinking and quaternization. The etherification affected the supramolecular structure of the luffa, as seen from X-ray diffraction, and was not confined to surface modification. Quaternized DEAE luffa showed the ability to adsorb nitrate, sulfate and phosphate anions, and could be used repeatedly without any significant effect on its adsorption capacity [22].

Thermoplastic starch (TPS)/luffa fibre composites were prepared using compression moulding. The luffa fibre contents ranged from 0 wt% to 20 wt%. The tensile strength of the TPS/luffa fibre composite with 10 wt% of luffa fibre increased twofold, to a maximum value of 1.24 MPa, compared to normal TPS. The temperature for maximum weight loss of the TPS/luffa fibre composites was higher than for TPS, and the water absorption of the TPS/luffa fibre composites decreased significantly when the luffa fibre content increased. The results obtained from TGA confirmed that luffa fibre can improve the thermal stability of TPS and that the water absorption of TPS/luffa fibre composites is lower than for TPS, and reduces further when the luffa fibre content is increased [23].

The effects of coupling agents on the mechanical, morphological and water sorption properties of luffa fibre/polypropylene composites were studied. In order to enhance the interfacial interactions between the polypropylene matrix and the luffa fibre, three different types of coupling agents, (3-aminopropyl)-triethoxysilane (AS), 3-(trimethoxysilyl)-1-propanethiol (MS) and maleic anhydride grafted polypropylene (MAPP) were used. Mechanical test results clearly showed that both silane treatment of luffa fibres and reactive treatment of composite with MAPP during compounding increased both the tensile strength and Young's modulus. Composites containing MS-treated luffa fibre showed the most pronounced improvement in mechanical properties compared to the composites containing untreated luffa fibres, due to adhesion and compatibility between the polypropylene and the silane-treated luffa fibre. The improvement in adhesion between polypropylene and luffa fibre treated with coupling agents was also confirmed with the semi-empirical Pukanzy model, and was in agreement with the experimental data and was supported by scanning electron microscopy. AFM studies of the surface of the fibres showed that silane treatment decreased the surface roughness of the fibres. This is also indicative of coverage of the surface of the fibres with a silane layer. Water absorption results showed that silane and MAPP treatment reduced the water absorption capacity compared to untreated composites. Water sorption results can be correlated with mechanical test results, which can be treated as a proof for enhanced interfacial interactions through the use of treatments [10].

Eco-composites were prepared from castor oil polyurethane reinforced with luffa fibres. These fibres were used both without treatment and after mercerization and mercerization/benzylation surface treatments. TG curves of mercerized luffa showed a small increase in thermal stability, associated with the removal of polyoses and soluble lignin. TG curves of benzylated fibres showed a decrease in thermal stability. FTIR spectra and elemental analysis also confirmed benzylation, and X-ray diffractograms showed a decrease in cellulose crystallinity. The eco-composite absorbed more water (14% after four weeks' exposure) than the resin, because of the hydrophilic character of natural fibres, but mercerization and benzylation reduced this value by about 50%. The tensile moduli of the composites reinforced with untreated and mercerized fibres were comparable, but the modulus increased by 33% with the use of mercerized/benzylated luffa fibre [24].

The effects of chemical surface modification of luffa fibres on the mechanical and hygrothermal behaviour of luffa/polyester composites were studied. Unsaturated polyester resin was used as the matrix. Untreated, alkali-treated, combination treated and acetylated luffa fibres were used. Acetylation treatment improved the mechanical properties and decreased the hydrophilic behaviour of the luffa fibres, improving their adhesion to the polyester matrix. The chemical modification of the surface of the luffa fibres also decreased the diffusion coefficient and the maximum amount of water absorbed by these fibres [25].

Table 21.2 compares the physical/chemical and tensile properties of some lignocellulosic fibres used for biocomposites [26].

21.4 Nanocomposites incorporating *Luffa cylindrica* fibres

The incorporation of elongated rod-like crystalline nanocellulose attracted attention in the nanotechnology field with the emergence of nanostructured

Table 21.2 Chemical and mechanical properties of some lignocellulosic fibres

Fibre	Cellulose (%)	Lignin (%)	Young's modulus (GPa)	Ultimate tensile strength (MPa)	Elongation at break (%)
Banana	64	8	30	800	3.1
Jute	65	12.2	20	700	1.6
Ramie	83	0.5	42	750	1.0
Flax	68	4	70	1100	2.0
Pineapple	82	12	80	160	3.1
Sisal	64	10	20	595	4.7
Coir	44	43	5	190	35.6
Luffa cylindrica	62	11.2	11	1100	3

polymer composites. The first reference to the existence of definite crystalline zones in the amorphous structure of cellulose materials was by Nageli and Schwendener in 1870. Many decades later, Rånby and Ribi (1950) were the first to produce a stable suspension of colloidal cellulose crystals by the sulfuric acid hydrolysis of wood and cotton cellulose. The nanocrystals produced were 50–60 nm in length and had a diameter of about 5–10 nm. Since then, many researchers have focused their work on such materials. Pioneering work in cellulose nanocrystals comes mainly from Canada (Marchessault, Gray) and France (Dufresne, Cavaillé) [27–29].

Studies on luffa fibre nanocomposites are quite new. Luffa-based nanocomposites are generally used in medical barrier applications, transparent films and biotechnology. For example, poly(E-prolactone) (PCL) based nanocomposite films reinforced with cellulose nanocrystals isolated from luffa fibre have been prepared by a film casting/evaporation technique in non-aqueous media. The contribution of nanofiller content to the transport properties of the PCL matrix was investigated through water sorption and water permeation processes. In addition, recently developed long-chain isocyanate grafting onto cellulose has been applied to luffa nanocrystals. Its impact on the water transport properties was specifically investigated at nanoparticle loadings higher than the percolation threshold. A peculiar behaviour against water vapour molecules was observed as a function of content, of water activity and of surface-chemical modification. This behaviour was related to the establishment of hydrogen bonding interactions between nanocrystals during the formation of the three-dimensional cellulose, which contributed to a restriction of matrix chain mobility.

The capacity of water molecules to move between barriers and so diffuse through the nanocomposite was investigated by a water permeation process. The tortuous zig-zag diffusion pathways of the PCL matrix were lengthened and water permeability was reduced, while the diffusivity was increased by a classical water plasticizing effect. With surface-modified nanocrystals, a reduction of diffusivity and relative water permeability was found, attributed to water molecule immobilization on the surface nanocrystals. This was evidenced by a reduction of the water plasticization coefficient combined with an increase of the equilibrium water concentration in the nanocomposite. In all cases there is a dependency between the concentration of penetrated water and the diffusion coefficient. Tailoring new eco-friendly nanocomposites by incorporating bio-nanofillers into a non-polar oil-derived biodegradable matrix is of significance for the development of bio-nanocomposites in relation to the growing global need to reduce non-biodegradable material pollution from packaging [30].

Sulfuric acid hydrolysis has been used to prepare cellulose nanocrystals from luffa fibres. The ensuing rod-like nanoparticles have an average length and diameter, determined from microscopic observation, around 242 nm and 5.2 nm, respectively. Polycaprolactone (PCL) was used as matrix to prepare green nanocomposites from dichloromethane, using up to 12 wt% nanocrystals, by a casting–evaporation method. In order to investigate the effect of increasing the compatibility of the filler with the matrix, the surface of cellulose nanocrystals was chemically modified with *n*-octadecyl isocyanate. The effective grafting of long aliphatic chains was observed by infrared spectroscopy and elemental analysis. The degree of substitution was found to be only around 0.05, but sufficient to be detectable when comparing the properties of nanocomposites prepared from unmodified and from chemically modified nanocrystals. A slight increase in the glass transition temperature, the melting point and the degree of crystallinity of PCL was observed when adding cellulose nanocrystals. The degree of crystallinity was further increased when using modified nanoparticles. Both non-linear and linear mechanical tests show an increase of the modulus of the nanocomposites upon addition of luffa nanocrystals. This effect was more marked for modified nanoparticles, probably partly due to the increased crystallinity of the PCL matrix. Moreover, chemical grafting promotes a more homogeneous dispersion of nanocrystals within the PCL matrix, as shown by the significant improvement of the elongation at break compared to unmodified nanoparticles [31].

21.5 Conclusion

Addition of luffa fibres to polymers increases the mechanical properties of the composites. Also, surface-treated luffa fibre reinforcement further improves the mechanical and thermal properties of the polymer composites due to the modified surface of the luffa fibre. Biocomposite structures can be produced by using luffa fibres and degradable polymers. Treated and untreated luffa fibre-reinforced polymer composites can be used in many fields, such as automotive and construction.

21.6 References

1. Chand N., Sood S., Rohatgi P.K., Satyanarayana K.G. (1984) 'Resources, structure, properties and uses of natural fibres of Madhya-Pradesh', *Journal of Scientific and Industrial Research India* 43(9), 489–499.
2. Sen K.K., Reddy S.S. (1994) 'Carbohydrate analysis of some natural fibres', *Research and Industry* 39(4), 258–260.
3. Rijswijk I.V., Brouwer W.D. (2002). 'Benefits of composites made of locally grown natural fibres. In: Mattoso L.H.C., Leão A.L., Frollini E. (eds.), *Proceedings of the Fourth International Symposium on Natural Polymers and Composites*, Publishers: Embrapa Agricultural Instrumentation, São Paulo University (USP), São Paulo State University (UNESP), São Carlos, SP, Brazil, pp. 422–428, ISNaPol-2002.
4. Oswaldo Massambani, respostas tecnicas 01. Agrıcultura E Pecuárıa 2009. http://globoruraltv.globo.com/cgi-in/globorural/montar_texto.pl, 2012.
5. Sharifah H.Z., Ansell M.P., (2004) 'The effect of alkalinization and fibre alignment on the mechanical and thermal properties of kenaf and hemp bast-fibre composites: Part 1 – polyester resin matrix', *Composites Science and Technology*, 64, 1219–1230.
6. Altinisik A., Gur E., Seki Y. (2010) 'A natural sorbent, *Luffa cylindrica* for the removal of a model basic dye', *Journal of Hazardous Materials* 179, 658–664.
7. Boynard C.A., Monteiro S.N., Almedia JRM. (2003) 'Aspects of alkali treatment of sponge gourd (*Luffa cylindrica*) fibres on the flexural properties of polyester matrix composites', *Journal of Applied Polymer Science* 87, 1927–1932.
8. D'Almeida A.L.F.S., Barreto D.W., Calado V., d'Almeida J.R.M. (2006) 'Effects of derivazation on sponge gourd (*Luffa cylindrical*) fibres', *polymers & polymer composites* 14(1), 73–80.
9. Hassan L.M. (2006) 'Quaternization and anion exchange capacity of sponge gourd (*luffa cylindrical*)', *Journal of Applied Polymer Science* 101, 2495–2503.
10. Demir H., Atikler U., Balkose D., Tihminlioglu F. (2006) 'The effects of fibre surface treatments on the tensile and water sorption properties of polypropylene-*Luffa cylindrica* fibre composites', *Composite Part A* 37, 447–456.
11. Kocak D. (2008) 'The influence of ultrasonic energy on chemical treatment of surface properties and the properties of composites made of *luffa cylindrica* fibre–olyester resin', *Journal of Engineering Materials and Technology* 130(4), 1–7.
12. Kocak D. (2008) 'Study of the effects different chemical compound applied on *Luffa cylindrica* fibres with the help of ultrasonic energy', *Journal of Polymer Engineering*, 28(8), 501–515.
13. Merdan N., Sancak E., Kocak D., Yüksek M. (2012) 'Effect of applied different surface modification processes with cellulose enzyme on properties of luffa fibres', *Asian Journal of Chemistry* 24(3), 975–980.

14. Kocak D., Merdan N., Yüksek M., Sancak E. (2013) 'Effects of chemical modifications on mechanical properties of *Luffa cylindrica*', *Asian Journal of Chemistry* 25(12), 637–641.
15. Wang Y., Shen X.Y. (2012) 'Optimum plasma surface treatment of luffa fibres', *Journal of Macromolecular Science Part B: Physics* 51, 662–670.
16. Joseph P.V., Rabello M.S., Mattoso L.H.C., Joseph K., Thomas S. (2002) 'Environmental effects on the degradation behaviour of sisal fibre reinforced polypropylene composites', *Composite Science and Technology* 62(10/11), 1357–1372.
17. Boynard C.A., d'Almeida J.R.M. (2000) 'Morphological characterization and mechanical behaviour of sponge gourd (*Luffa Cylindrica*)–polymer composite materials', *Polymer Plast Technology and Engineering* 39, 489–499.
18. Boynard C.A, d'Almeida J.R.M. (1999) 'Water absorption by sponge gourd (*Luffa cylindrica*)–polyester composite materials', *Journal of Materials Science Letters* 18, 1789–1791.
19. Seki Y., Sever K., Erden S., Sarikanat M., Neser G., Ozes C. (2012) 'Characterization of *Luffa cylindrica* fibres and the effect of water aging on the mechanical properties of ıts composite with polyester', *Journal of Applied Polymer Science* 123(4), 2330–2337.
20. Botaro V.R., Novack K.M., Siqueira E.J. (2012) 'Dynamic mechanical behavior of vinylester matrix composites reinforced by *Luffa cylindrica* modified fibres', *Journal of Applied Polymer Science* 124(3), 1967–1975.
21. Tanobe V.O.A., Sydenstricker T.H.D., Munaro M., Amico S.C. (2005) 'A comprehensive characterization of chemically treated Brazilian sponge-gourds (*Luffa cylindrica*)', *Polymer Testing* 24, 474–482.
22. Hassan, M.L. (2006) 'Quaternization and anion exchange capacity of sponge gourd (*Luffa cylindrica*)', *Journal of Applied Polymer Science* 101(4), 2495–2503.
23. Kaewtatip K., Thongmee J. (2012) 'Studies on the structure and properties of thermoplastic starch/luffa fibre composites', *Materials and Design* 40, 314–318.
24. Melo B., Dos-Santos C., Botaro V., Pasa V. (2008) 'Eco-composites of polyurethane and *Luffa aegyptiaca* modified by mercerization and benzylation', *Polymers & Polymer Composites* 16(4), 249–256.
25. Ghali L., Aloui M., Zidi M., Bendaly H., M'sahli S., Sakli F. (2011) 'Effect of chemical modification of *Luffa cylindrica* fibres on the mechanical and hydrothermal behaviours of polyester/luffa composites', *BioResources* 6(4), 3836–3849.
26. Satyanarayana K.G., Wypych F. (2007) 'Characterization of natural fibres'. In: Fakirov S., Bhattacharyya D., (eds), *Engineering Biopolymers: Homopolymers, Blends and Composites*. Munich: Hanser Publishers, pp. 3–48.
27. Angles M.N., Dufresne A. (2001) 'Plasticized starch/tunicin whiskers nanocomposite materials, mechanical behavior', *Macromolecules* 34, 2921–2931.
28. Samir A.M.A., Alloin F., Dufresne A. (2005) 'Review of recent research into cellulosic whiskers, their properties and their application in nanocomposite field', *Biomacromolecules* 6, 612–626.
29. Lavoine N., Desloges I., Dufresne A., Bras J. (2012) 'Microfibrillated cellulose, ıts barrier properties and applications in cellulosic materials: a review, *Carbohydrate Polymers* 90, 735–764.

30. Follain N., Belbekhouche S., Bras J., Siqueira G. , Marais S., Dufresne A. (2013) 'Water transport properties of bio-nano composites reinforced by *Luffa cylindrica* cellulose nanocrystals', *Journal of Membrane Science* 427, 218–229.
31. Siqueira G., Bras J., Follain N., Belbekhouche S., Marais S., Dufresne A. (2013) 'Thermal and mechanical properties of bio-nanocomposites reinforced by *Luffa cylindrica* cellulose nanocrystals', *Carbohydrate Polymers* 91, 711–717.

22

The use of curaua fibers as reinforcements in composites

S. F. SOUZA and M. FERREIRA, Universidade Federal do ABC, Brazil, M. SAIN, University of Toronto, Canada, and M. Z. FERREIRA, H. F. PUPO, B. M. CHERIAN and A. L. LEÃO, SÃo Paulo State University, Brazil

DOI: 10.1533/9781782421276.5.700

Abstract: Nowadays, there is growing worldwide interest in the use of natural renewable materials to replace synthetic polymers or improve the properties of current materials. Natural fibers seem to offer a great opportunity in this respect, and among them curaua (*Ananas erectifolius* L. B. Smith) fiber has unique properties, such as high strength, low density, a fine structure and high cellulose content. Due to its mechanical properties and recyclability, it is of great interest for developing composite materials. In addition, because of its high cellulose content, curaua fiber is one of the most potentially important sources of raw material for producing nanocellulose for use in the manufacture of nanocomposites.

Key words: curaua fiber, nanofibers, composites, mechanical properties, reinforcement, polymers.

22.1 Introduction

In recent years curaua fiber – from the leaves of *Ananas erectifolius* L. B. Smith – has shown itself to be promising as a reinforcement agent for composites, and thus has been broadly studied where it is cultivated in the Amazon region (Leão *et al.* 1998). It has greater tensile and flexural strength and a finer structure than sisal, coir and jute, and is cheaper than synthetic fibers, and these properties mean that curaua composites can compete with traditional glass fiber composites (Zah *et al.* 2007). Moreover, the fibers are odorless, and so are of interest for use in applications such as the interior of cars, and the Brazilian automobile sector has already invested in this area (Volkswagen 2007).

Recently, the production of nanofibers is another important field where the use of curaua fibers as raw material shows promise, mainly because of the high cellulose content – approximately 70% (Caraschi and Leão, 2000). Different structures and properties of these nanofibers can be achieved depending upon the process applied. The potential of curaua nanofibers or cellulose nanofibers as reinforcement has been studied since 2010; their ability to increase the mechanical properties of polyvinyl alcohol (PVA) was demonstrated by Souza *et al.* (2010).

22.2 Curaua fibers

Curaua (*Ananas erectifolius* L. B. Smith) is a hydrophilous species from the Amazon region, which provides unique lignocellulosic fibers, and has been recognized since pre-Columbian days for its fibers' traditional applications (Leão *et al.* 1998; Tomczak *et al.* 2007). Curaua is found in several neighboring countries, including Guyana, Colombia, Venezuela and Surinam, under different names: kurowa, curaua, caroa, wild pineapple, etc. A genetic study was carried out by Leão *et al.* (2009) which concluded that all the plants have the same genotype. It has long been traded by the local people to make string for their bows, fishing nets, and hammocks, and exchanged for cassava grown by other tribes. In the Brazilian Amazon, curaua fibers are well known in the Amazon River basin, especially in the western region of Pará State, where the first commercial curaua plantations were established (Zah *et al.* 2007).

The curaua plant (Fig. 22.1) has hard, flat, erect leaves of about 1–1.5 m in length, about 40 mm wide and about 5 mm thick (Leão *et al.* 1998; Tomczak *et al.* 2007). In general, each plant yields around 50–60 leaves per year, with an average weight of 150 g each, in a high density stand of 40,000–50,000 plants. This will result in a yield of between 3 and 9 tons of dry fibers per hectare per year, without irrigation where annual precipitation is higher

22.1 Curaua plant before harvest.

than 2000 mm per year. However, with the use of modern techniques for the soil preparation (tillage and leveling using tractors), fertilizer (NPK), pesticides and clear cutting, this production can be doubled (Tomczak *et al.* 2007; Zah *et al.* 2007).

There are four varieties of plants, known as white, purple, red and bright white. The most common are purple (Figure 22.1), with reddish-purple leaves, and white, with light green leaves (Corrêa *et al.* 2010).

In the Pematec–Triangel project, the leaves are harvested manually (Fig. 22.2), and the fibers are extracted from the leaves by means of a rudimentary machine, equipped with rotating knives, which removes around 500 kg/day of mucilage present in the leaves. After that, the fibers are mercerized for 36 h in a water tank and rewashed in order to remove mucilage residue. They are then sun-dried, blended with polypropylene and rolled into bales (Zah *et al.* 2007). Each leaf provides 3–8% of dry fiber (Leão *et al.* 1998; Caraschi and Leão, 2000).

Traditionally, curaua fibrers are used to make hammocks, fishing lines, ropes, etc., but since the recognition of their commercial potential in the 1990s by the Brazilian automotive industries, they have been used to make door panels, dashboards and other interior parts for cars, buses and trucks (Tomczak *et al.* 2007; Zah *et al.* 2007). According to Caraschi and Leão (2000) curaua fibers have further great industrial potential. Fig. 22.3 shows the fibers after mechanical extraction.

The chemical composition of the fibers was reported by Caraschi and Leão (2000) as being 73.6% cellulose, 9.9% hemicellulose and 7.5% lignin.

22.2 Manual harvesting of the curaua leaves (source: Pematec Triangel, www.pematec.com.br/curaua.htm).

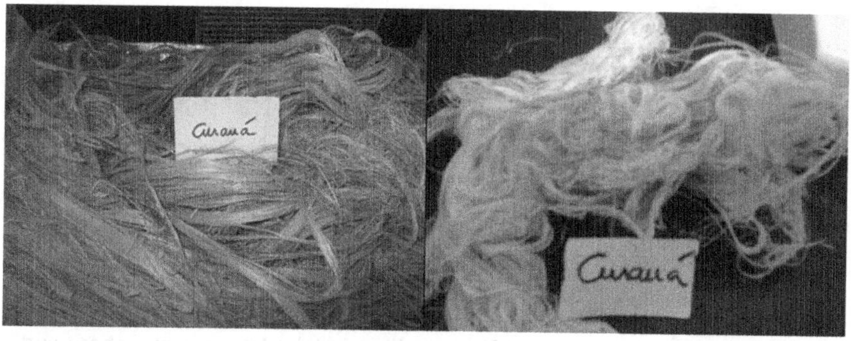

22.3 Curaua fibers extracted mechanically.

Table 22.1 Mechanical properties of curaua fibres

Tensile strength (MPa)	Modulus of elasticity (GPa)	Elongation at break (%)	Source
500–1150	11.8	3.7–4.3	Leão *et al.* (1998)
1250–3000	30–80	4.5–6	Satyanarayana *et al.* (2007)
700–1100	26–46	2–4	Spinacé *et al.* (2009a)
800	10.7	–	Souza *et al.* (2010)

The lignin content is lower than in other kinds of fiber, such as jute (16%), sugar cane bagasse (25%) and coconut (43%) (Trindade *et al.* 2008). Due to the high content of cellulose, the curaua fiber has a degree of crystallinity of some 67–75.6% (Caraschi and Leão, 2000; Tomczak *et al.* 2007, Trindade *et al.* 2008) and higher values of tensile strength (approximately 1000 MPa) and a higher modulus of elasticity (Table 22.1) than other fibers such as sisal, coir and sugar cane bagasse (Spinacé *et al.* 2009a).

The thermo-gravimetric analysis (TG) results of curaua fibers under two different conditions (nitrogen and oxygen) with a heating rate of 20°C min^{-1} are shown in Fig. 22.4 (Tomczak *et al.*, 2007). It can be seen that under a nitrogen atmosphere (Fig. 22.4(a)), all absorbed or combined water present in the fiber was eliminated between room temperature and 150°C. The lignin and cellulose were degraded at 337°C and 520°C, respectively. Under an oxygen atmosphere, the water loss was completed by 72°C followed by decomposition processes of lignin and cellulose at 275°C and 300°C, respectively (Fig. 22.4(b)). Under an oxygen atmosphere, the degradation temperatures for all the fiber compounds – hemicellulose, cellulose and lignin – overlap due to the complex reactions all occur between 220°C and 350°C (Tomczak *et al.* 2007).

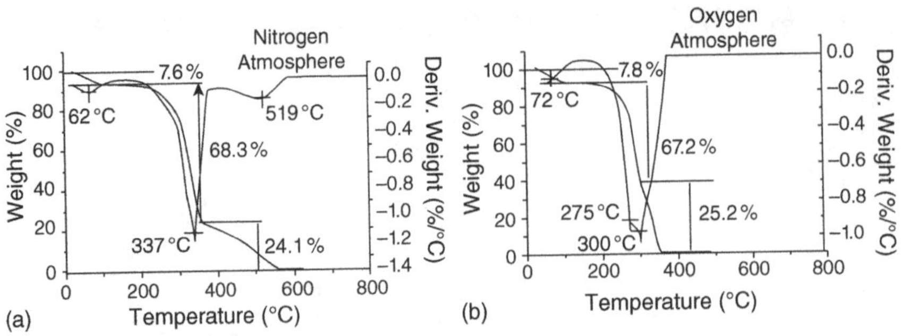

22.4 Thermal analysis of curaua fiber in (a) nitrogen atmosphere and (b) oxygen atmosphere. Reprinted from *Composites*, 38, Tomczak, F., Satyanarayana, K. G. and Sydenstricker, T. H. D., Studies on lignocellulosic fibers of Brazil: Part III – Morphology and properties of Brazilian curaua fibers, 2227–2236, Copyright (2007), with permission from Elsevier.

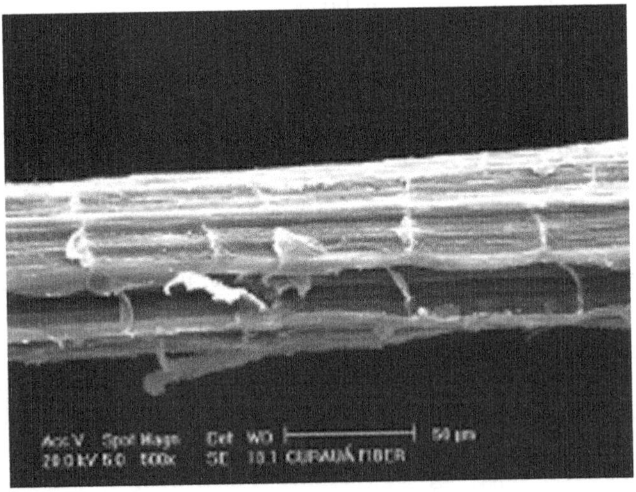

22.5 Eletronic scanning microscopy from curaua fiber. Reprinted from *Composites*, 38, Tomczak, F., Satyanarayana, K. G. and Sydenstricker, T. H. D., Studies on lignocellulosic fibers of Brazil: Part III – Morphology and properties of Brazilian curaua fibers, 2227–2236, Copyright (2007), with permission from Elsevier.

A typical scanning electron microscopy (SEM) image of curaua fiber is shown in Fig. 22.5. The micrograph shows the morphology of the fiber as a compact arrangement of a large number of fibrils linked together by a matrix of lignin and hemicellulose. Some leaf mucilage residue can be seen on the surface, which is rough and can provide good adhesion between the fiber and the matrix of green composites.

22.3 Composites using curaua fibers

The properties of a composite are greater than the properties of its individual components. For example, in a fiber-reinforced composite under tension, it is the fibres which bear the tension rather than the matrix. Natural fibers have a very high modulus (strength/weight), a property which makes them obvious camdidates for use in lightweight composites. Vegetable fiber is one example of a natural composite, where cellulose acts as the reinforcement and lignin as the matrix (Leão *et al.* 1998).

Paiva *et al.* (1999) obtained composites of phenolic thermosetting polymer reinforced with 15% curaua fibers treated with a solution of 10% sodium hydroxide which had greater impact strength than the unreinforced phenolic matrix. Two types of resin were used; one was pure phenolic thermosetting polymer and in the other 40% of the thermosetting weight was substituted by lignin extracted from sugar cane bagasse, using acetone. The composite of phenolic resin/treated curaua fibers had an impact resistance of 65.5 J/m, higher than that of simple phenolic resin (12.8 J/m), and the composite of lignin/phenolic resin/treated curaua fibers had an impact resistance of 51.35 J/m, higher than that of just lignin/phenolic resin (13.3 J/m) (Paiva *et al.*1999).

Spinacé *et al.* (2009b) analyzed curaua fibers as reinforcing agents for recycled polypropylene. The results showed no improvement in the fiber–matrix adhesion when mixed with 20 wt% curaua fibers treated with 5% sodium hydroxide solution compared with untreated fibers, which shows the excellent mechanical properties of the untreated fibers. Another composition was made with 3 wt% MAPP and 20 wt% of short curaua fibers, and the mechanical results showed a 30% increase in tensile strength and a 51% increase in flexural strength when compared with pure recycled polypropylene (Table 22.2). The composition of the composites studied used long and short curaua fibers with content from 10% to 40% long curaua fibers (PP10-I, PP20-I, PP30-I, and PP40-I); 10 wt% of long curaua fibers treated with NaOH (PP10-I-NaOH); 20 wt% short curaua fibers with 3 wt% of MAPP (PP20-s-MA) or without MAPP (PP20-S).

Composites of high-density polyethylene and untreated curaua fiber have higher mechanical strength compared to the pure matrix. Adding curaua fiber, 5–20 wt%, to the composite increases the flexural and tensile strengh, while 25–30 wt% of fiber content decreases the tensile strength compared to 20% (Fig. 22.6(a)), with the flexural strength staying approximately the same (Fig. 22.6(a)). The elastic moduli shown in Fig. 22.6(b), were increased by increasing the load of curaua fibers for flexural and tensile strength (Araújo *et al.* 2010).

Further interesting research has been done by depositing polyaniline on the surface of short curaua fibers. The modified fibers were dispersed by

Table 22.2 Tensile strength (σ), Young's modulus (E), and notched impact strength (IS) for injection molded test samples of post-consumer polypropylene and composites using fibers treated with sodium hydroxide solution

Material	σ (MPa)	E (MPa)	IS (J m^{-1})
pcPP	28.55 (±0.6)	616.0 (±26.0)	20.0 (±0.7)
PP10-I-NaOH-2h[a]	27.60 (±0.4)	753.2 (±18.2)	35.8 (±3.7)
PP10-I-NaOH-4h[a]	27.40 (±0.4)	746.6 (±18.8)	37.8 (±3.4)
PP10-I-NaOH-6h[a]	29.50 (±0.7)	778.1 (±21.7)	40.6 (±3.2)
PP10-I-NaOH-8h[a]	29.80 (±0.5)	752.5 (±15.8)	40.6 (±3.0)
PP10-I-NaOH-12h[a]	30.10 (±1.2)	748.5 (±37.7)	45.3 (±5.5)
PP10-I-NaOH-24h[a]	27.80 (±0.4)	730.0 (±14.1)	33.8 (±5.5)
PP10-I-NaOH-48h[a]	29.30 (±0.4)	746.0 (±28.9)	37.0 (±1.5)

[a]=treatment time of curaua fiber with NaOH solution in hours.
Source: reprinted from *Journal of Applied Polymer Science*, 112, Spinacé, M. A. S., Fermoseli, K. K. G. and De Paoli, M. A. Recycled polypropylene reinforced with curaua fibers by extrusion, 3686–3694, 2009, with permission from the publisher John Wiley & Sons.

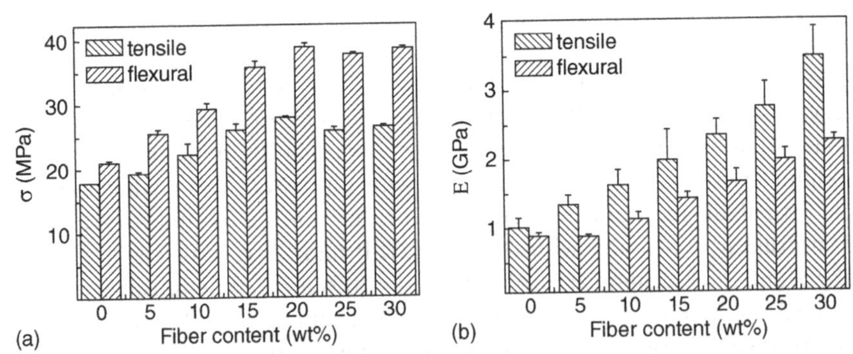

22.6 (a) Tensile strength and (b) elastic modulus of the composites, with increasing fiber content. Reprinted from *Composite Science and Technology*, 70, Araújo, J. R., Mano, B., Teixeira, G.M., Spinacé, M. A. S. and De Paoli, M. A., Biomicrofibrillar composites of high density polyethylene reinforced with curaua fibers: Mechanical, interfacial and morphological properties, 1637–1644, Copyright (2010), with permission from Elsevier.

extrusion in polyamide-6, and the composites were injection molded. The results showed that 5 wt% of curaua fibers the tensile strength of the composite was 52 MPa, 56% higher than for pure polyamide-6. Adding 30 wt% of modified curaua fiber, however, reduced the value to 26 MPa, 23% lower than the values obtained for pure polyamide-6. The best results for both electrical and mechanical properties were obtained with 5 wt% of polyaniline-coated curaua fibers (Araújo *et al.* 2013).

Cement composites have been produced reinforced with 2, 4 and 6 wt% of short curaua fibers. The results with 4 wt% curaua fibers of 50 mm length showed a multiple cracking pattern with a post-crack bending strength of 4.59 ± 0.58 MPa and a toughness of 2.13 ± 0.36 kJ/m^2. For composites reinforced with fibers of 25 mm length, the post-crack bending strength values, when the fiber content increased from 2 to 4%, they increased by 33.44%, and when the fiber content increased from 2 to 6%, increased by 41.48% (D'Almeida *et al.* 2010).

Mothé and Araújo (2004) investigated the thermal properties, the mechanism of thermal degradation and the reaction mechanisms of polyurethane composites with different proportions of curaua fiber (5, 10 and 15 wt%). The results show that the addition of curaua fiber into polyurethane composites increases the Young's modulus and decreases elongation at break.

Silva *et al.* (2009) showed that water absorption in laminated hybrid composites decreases the mechanical properties. The worst results were obtained for flexural modulus when saturated with sea water, and for tensile strength when saturated with distilled water. The results are shown in Table 22.3.

Santos *et al.* (2009) evaluated the use of curaua fibers with polyamide-6. The fibers were treated, and the best results in terms of traction and bending were found for fibers treated with sodium hydroxide when compared to pre-N_2 plasma treatment. The values show that it is possible to prepare composites of polyamide-6 with 20 wt% curaua fibers and obtain similar performance to polyamide-6 reinforced with talc or fiberglass. The behavior of the mechanical properties of composites of polyamide-6 reinforced with curaua fiber and fiberglass was similar when exposed to 2% moisture; the tensile and flexural properties decreased by 25% and 35%, respectively, while the impact strength increased by approximately 60% for both composites (Santos *et al.* 2009).

Table 22.3 Mechanical properties determined in the tensile test

Mechanical property	Condition		
	Dry	Distilled water	Sea water
Tensile strength (MPa)	92.20 (± 6.7%)	72.2 (± 4.0%)	79.20 (± 2.2%)
Young's modulus (GPa)	2.34 (± 4.4%)	2.0 (± 6.6%)	2.29 (± 3.0%)
Elongation (%)	3.70 (± 7.2%)	3.5 (± 8.5%)	3.50 (± 3.9%)

Source: Silva, R. V., Aquino, E. M. F., Rodrigues, L. P. S. and Barros, A. R. F. *Journal of Reinforced Plastics and Composites*, 28, pp. 1857–1868, © 2009 by Sage, reprinted by permission of Sage.

Table 22.4 HDT, tensile strength, tensile modulus, and impact strength of composites as a function of curaua fiber content

Sample	HDT (°C)	Tensile strength (MPa)	Tensile modulus (MPa)	Impact strength (J/m)
EPSr	77.6 (±0.8)	26.5 (±0.3)	3712 (±756)	148.9 (±27.7)
PS	70.2 (±1.1)	35.9 (±1.5)	3284 (±342)	257.9 (±22.3)
EPFC20	86.7 (±1.8)	43.5 (±0.7)	5738 (±339)	166.6 (±28.1)
PFC20	76.9 (±1.3)	48.1 (±1.6)	5799 (±422)	182.5 (±21.5)

Source: reprinted from *Journal of Applied Polymer Science*, 128, Borsoi, C., Scienza, L. C. and Zattera, A. J. Characterization of composites based on recycled expanded polystyrene reinforced with curaua fibers, 653–659, 2013, with permission from the publisher John Wiley & Sons.

Borsoi *et al.* (2013) evaluated composites produced from virgin and recycled matrices (recycled expanded polystyrene) with 20 wt% of curaua fiber. The addition of curaua fibers in these composites resulted in an improvement in the tensile strength, modulus of elasticity, stiffness, thermal stability and melt viscosity. The results showed that tensile strength increased for both matrices when curaua fibers were added. These results are illustrated in Table 22.4.

Lopes *et al.* (2011) showed in their research that the addition of 30 vol% curaua fibers in polyester composites resulted in endothermic peaks around 100°C and an increase in the absolute enthalpy values (Fig. 22.7). The work also showed that the introduction of curaua fiber into the polyester matrix improved the composite hydration rate.

Ferreira *et al.* (2010) investigated the toughness by Charpy impact tests of polyester composites reinforced with 30 vol% alkali-treated continuous and aligned curaua fibers. The incorporation of non-treated curaua fibers in the composites gave better results than the treated fibers. The values found for the composites with 30 vol% non-treated curaua fibers were approximately 150 J/m, while for the treated it fiber was only 35 J/m.

An investigation was carried out by Monteiro *et al.* (2013) in which the authors investigated the properties of a polyester matrix with 30 wt% of continuous and aligned curaua fibers. The curaua composites with higher fiber content showed an increase in notch toughness, measured by Charpy/Izod impact tests. The results indicated an increase of notch toughness greater than 10%. By adding 20 wt% of curaua fiber, composite resistance rose to approximately 100 MPa.

Ornaghi *et al.* (2012) observed that the storage modulus decreased with the increase of temperature when curaua fiber content in polyester resin was increased in hot compression molding. This occurred mainly in the

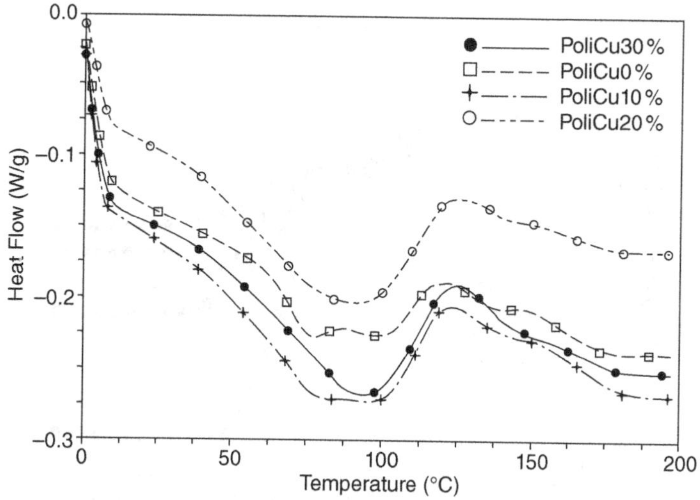

22.7 DSC curve for the pure polyester, 0% fiber as well as 10, 20 and 30% volume fraction of curaua fibers. Reprinted from EPD Congress 2011, The Minerals, Metals and Materials Society – TMS 2011 Annual Meeting & Exhibition, Lopes, F. P. D., Ferreira, A. S., Rodriguez, R. J. S., Castillo, T. E. and Monteiro, S. N., Characterization of thermal behavior of polyester composites reinforced with curaua fibers by differential scanning calorimetry, Copyright (2011), with permission from the publisher John Wiley & Sons.

glass-transition region and is correlated with the stiffness of curaua fiber (Fig. 22.8). This same study showed that the addition of curaua fiber to the composite increased the loss modulus peak (Fig. 22.9).

Rossa *et al.* (2013) showed that increasing curaua fiber content in a poly(hydroxybutyrate-*co*-valerate) matrix increased the tensile strength. The results showed an increase of 56% with 30 wt% fiber content. The modulus of elasticity was 193% higher, when compared with the neat polymer. The flexural strength increased by 33%, and the flexural modulus of elasticity increased by 117%, for the samples with 20% curaua fibers.

Gomes *et al.* (2004) used curaua fibers treated with different concentrations of sodium hydroxide solutions for 1 h and 2 h for composites of corn starch-based biodegradable resin. The average values showed that the tensile strength and fracture strain of the composites using treated fibers increased 10% and 105%, respectively, when compared with the composite using untreated fibers. The fracture strain of fibers treated for 2 h with 15% sodium hydroxide solution was double that of untreated fibers.

Trindade *et al.* (2008) described treatments of curaua fibers with ionized air, and prepared composites with randomly oriented curaua fibers. The results showed that the fibers had a progressively increasing impact on

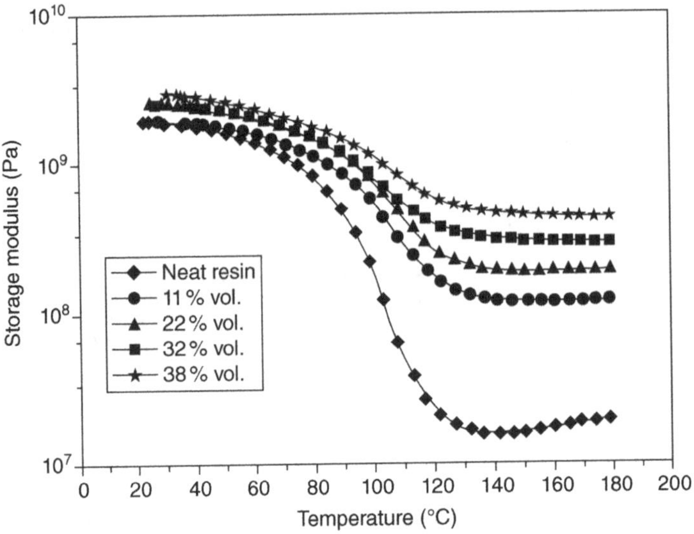

22.8 Variation in the storage modulus with temperature for all of the samples. Reprinted from *Journal of Applied Polymer Science*, 125, Ornaghi Jr, H. L., Silva, H. S. P., Zattera, A. J. and Amico, S. C. Dynamic mechanical properties of curaua composites, 110–116, Copyright (2012), with permission from the publisher John Wiley & Sons.

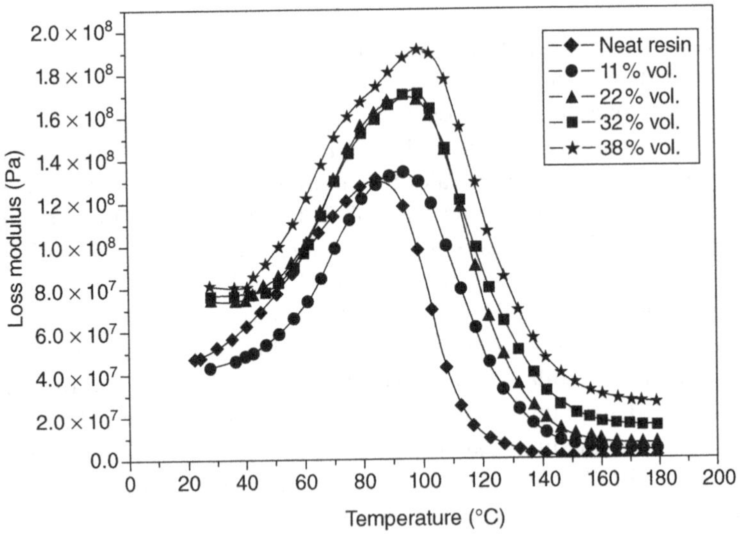

22.9 Variation in the loss modulus with temperature for all samples. Reprinted from *Journal of Applied Polymer Science*, 125, Ornaghi Jr, H. L., Silva, H. S. P., Zattera, A. J. and Amico, S. C. Dynamic mechanical properties of curaua composites, 110–116, Copyright (2012), with permission from the publisher John Wiley & Sons.

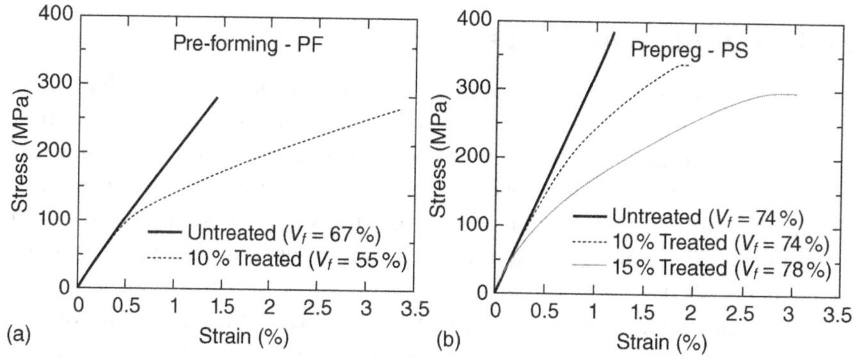

22.10 Stress–strain diagrams of: (a) pre-forming (PF) and (b) prepreg composites (PS). Reprinted from *JSME Jsme International Journal Series A – solid Mechanics and Material Engineering*, 47, Gomes, A., Goda, K. and Ohgi, J., Effects of alkali treatment to reinforcement on tensile properties of curaua fiber green composites, 541–546, Copyright (2007), with permission from the publisher John Wiley & Sons.

strength for up to 7h of treatment, and decreased the composite water absorption capacity.

Composites based on a blend of polycaprolactone (PCL) and cornstarch mixed with curaua fibers treated with sodium hydroxide were prepared using a prepreg sheet method and a preforming method (Gomes *et al.* 2007). A prepreg sheet is a kind of laminated composite with a quasi-isotropic property. The fibers were arranged, strips were placed on a metallic plate and stretched and the resin was dispersed into them. Later, the composite was pressed slightly at 120°C. The preforming involves hot-pressing preforms of resin-pasted fiber slivers. The fiber strips are wound and stretched around a metallic plate, while the resin was dispersed into the strips with a small brush. These preforms of fibers embedded in resin were dried at approximately room temperature for 24h and then a fabricated preform is inserted into the metallic mold and pressed at 150°C for 1h.

Analyzing the material from Fig. 22.10, it is possible to see that the area under the stress–strain curve for the composite made of 10% sodium hydroxide-treated fiber is 3.3 times larger when compared with untreated fibers. For the composite prepared using the prepreg sheet method with 10% sodium hydroxide-treated fibers, the area was 1.8 times larger than for composites with untreated fibers. At the same time, the composites with curaua fibers treated with 15% sodium hydroxide using the prepreg sheet process achieved a greater area than the composites prepared with fibers treated with 10% sodium hydroxide (Gomes *et al.* 2007).

22.4 Curaua nanofibers

Curaua nanofibers have been obtained by different processes. Each process used can give a large variety of final products with distinct properties that can be used for different applications. The properties can be tailor made; it is possible to obtain fragile crystals of cellulose, known as nanowhiskers, or flexible and long nanofibers or even networks of nanofibers called nanofibrillated cellulose. These materials can be used in different matrices (Souza *et al.* 2010).

Curaua has been used as a renewable source to obtain nanocellulose or cellulose nanofibers, because of the high quantity of cellulose available (approximately 70%) (Caraschi and Leão, 2000; Spinacé *et al.* 2009a).

Chemical processes are used to acquire nanofibers and nanowhiskers, and mechanical equipment is used to obtain nanofibrillated or microfibrillated cellulose. Enzymatic processes are also used to isolate the cellulose or break the cellulose chain.

Nanowhisker is also known as cellulose whisker, cellulose nanocrystal or crystal, and other terms are sometimes used for the same product, such as microfibril, microcrystals or microcrystal likes, although its dimensions are at nanoscale. The term *whisker* is used to designate crystalline nanoparticles that are shaped like rods or wire. They are produced by the breakdown of cellulose fibers at the amorphous region, isolating the crystalline regions through chemical processes with high concentrations of acids (Siqueira *et al.* 2010; Brinchi *et al.* 2013). Nanofibrils or nanofibrillated cellulose have crystalline and amorphous regions in their structures.

Figure 22.11 shows two different approaches to producing nanocrystals. The first method uses a pulping and bleaching process to isolate the cellulose, and the second one uses steam to remove the hemicellulose and some lignin, and a fractionation process to isolate the cellulose (Brinchi *et al.* 2013). To isolate just the crystals, acid hydrolysis with high acid concentrations is used, followed by centrifugation to remove all the acids involved and reach a neutral pH. Sonication is the process used to disperse and break down the cellulose chain present in the suspension and to obtain, finally, the whisker or nanocrystalline cellulose (NCC) (Brinchi *et al.* 2013).

It is common to carry out pretreatments before the hydrolysis solution to remove all the non-cellulosic components such as extractives, hemicellulose, lignin and others, thus increasing the cellulose content (Corrêa *et al.* 2010; Taipina *et al.* 2012). A common treatment used is reacting the natural fibers with sodium hydroxide, a process known as mercerization. The mercerization process, depending upon the concentration, can change the structure and morphology of fibers, converting cellulose I into cellulose II, changing the strength and visible brightness of the fibers.

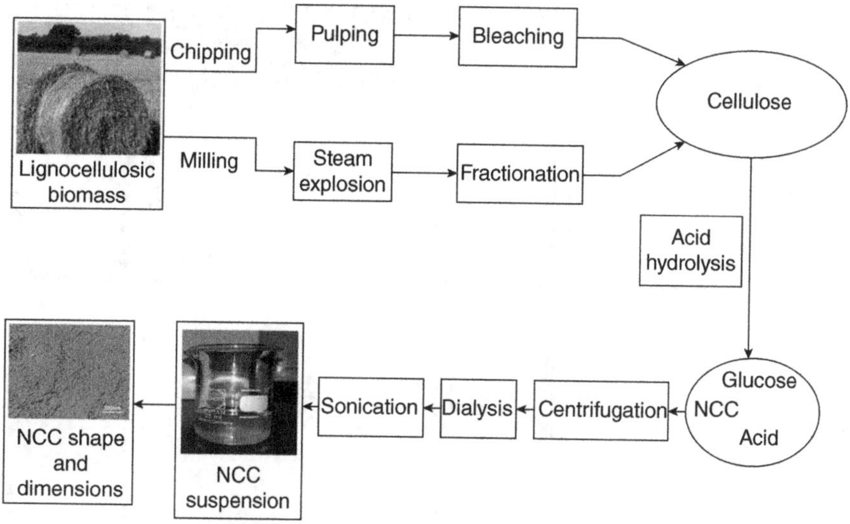

22.11 Scheme of main steps needed to prepare NCC from lignocellulosic biomass. Reprinted from *Carbohydrate Polymers*, 94, Brinchi, L., Cotana, F., Fortunati, E. and Kenny, J. M., Production of nanocrystalline cellulose from lignocellulosic biomass: Technology and applications, 154–169, Copyright (2013), with permission from Elsevier.

This process is used with the objective of increasing the surface area and exposing the fiber for further treatment (Ass *et al.* 2006).

Corrêa *et al.* (2010) prepared nanowhiskers of curaua fibers to analyze the influence of sodium hydroxide solution followed by acid hydrolysis using three different approaches (sulfuric acid solution, a mixture of sulfuric acid and hydrogen chloride and just hydrogen chloride). They showed (Fig. 22.12) that curaua nanowhisker presented a rod-like shape, with an average diameter of 6–10 nm and length of 80–170 nm, with an aspect ratio of around 13–17 (Corrêa *et al.* 2010). Using 17.5% sodium hydroxide solution resulted in more crystalline nanofibers which were thermally less stable, with an inferior degree of polymerization compared to a solution having a concentration of just 5%. The influence of different acids was evaluated, showing that increasing the percentage of hydrogen chloride in the solution undergoing hydrolysis promoted more thermal stability but also more agglomeration (Corrêa *et al.* 2010).

Another interesting methodology for obtaining nanofibers is microfibrillated cellulose (MFC), which produces a cellulose material with a specific morphology where it is possible to identify a network structure of nanofibers. This material is obtained using special (homogenizer) equipment which, using high pressure or grinding, opens up the cellulose fibril to produce the specific network structure.

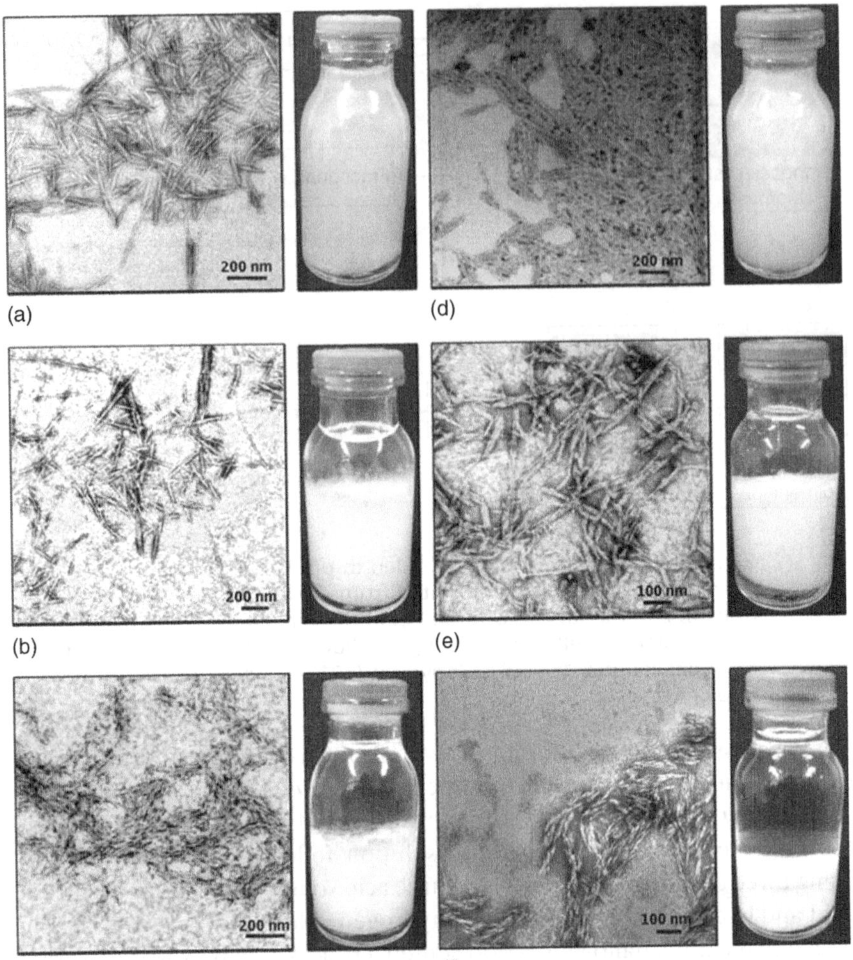

22.12 TEM micrographs of curaua nanofibers and their respective suspensions after 2 months of extraction: (a) nNa5s; (b) nNa5m; (c) nNa5c; (d) nNa17.5s; (e) nNa17.5m; (f) nNa17.5c. With kind permission from Springer Science, Business Media and the original publisher (*Cellulose*, 17, 2010, 1183–1192, Cellulose nanofibers from curaua fibers, Corrêa, A. C., Teixeira, E. M., Pessan, L. A. and Mattoso, L. H. C., Figure 3, original copyright notice).

Souza *et al.* (2010) obtained MFC from curaua fibers using the homogenizer equipment. Initially they isolated the cellulose by employing a pulping process using sodium hydroxide, and later bleached it twice with a solution of 2% $NaClO_2$. To achieve the nanoscale, a cryocrushing procedure was used, utilizing a mortar and pestle to crush the frozen bleached fibers, which were finally fibrillated 80 times by the homogenizer. Using environmental

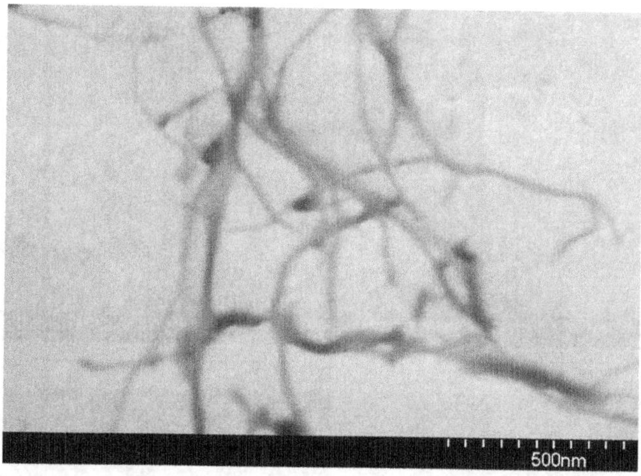

22.13 Scanning Electron Microscope (SEM) image of nanofibrillated curaua. Reprinted from 'Nanocellulose from curaua fibers and their nanocomposites', Souza, S. F., Leão, A. L., Cai, J. H., Crystal, W., Sain, M. and Cherian, B. M., *Molecular Crystal and Liquid Crystals*, 522, 42–52, 2010, reprinted by permission of the publisher, Taylor & Francis Ltd.

scanning electronic microscopy it was shown (Fig. 22.13) that over 75% of the nanofibers obtained have diameters smaller than 50 nm.

Nowadays, with the objective of using fewer chemical reagents and less energy, researchers have been using biotechnology resources, as with the case of producing cellulose using bacteria, *Gluconacetobacter xylinum*, or by just breaking the chain of cellulose to produce glucose for later fermentation to produce second-generation ethanol. One interesting study aimed to break the cellulose chain using enzymes, endoglucanase and a concoction of hemicellulases and pectinases, to obtain cellulosic nanoparticles (Campos *et al.* 2013). A pulping process with sodium hydroxide was used to obtain the cellulose, and after bleaching with sodium chlorite in acetic acid solution, the enzymes were applied and then sonicated; 75% of the nanofibers produced by this process had diameters smaller than 50 nm and lengths in microns (Fig. 22.14), but some damage was observed in the XRD analysis, with a decrease of crystallinity (Campos *et al.* 2013). The morphology presented an interconnected web-like structure similar to that obtained in the mechanical process by Souza *et al.* (2010).

22.5 Nanocomposites with curaua fibers

The potential use of nanofibers as reinforcement material represents a new field. Compared with conventional applications of microfibers, a higher

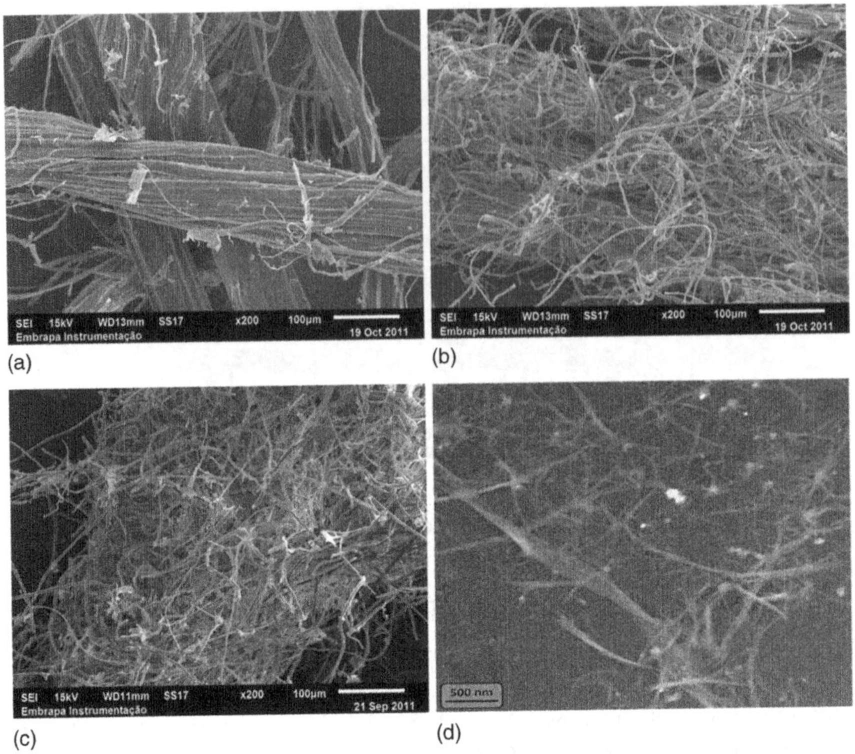

22.14 Curaua samples: SEM micrographs of (a) raw material, (b) bleached fiber, (c) after enzymatic treatment and (d) STEM micrograph after sonication of (c) samples. With kind permission from Springer Science, Business Media and the original publisher (*Cellulose*, 20, 2013, 1491–1500, Obtaining nanofibers from curaua and sugarcane bagasse fibers using enzymatic hydrolysis followed by sonication, Campos, A., Corrêa, A. C., Cannella, D., Teixeira, E. M., Marconcini, J. M., Dufresne, A., Mattoso, L. H. C., Cassland, P. and Sanadi, A. R., Figure 1, original copyright notice).

strength can be obtained due to the greater surface area that can interact with the matrix. In addition they can offer a lower density and transparency, and use abundant, renewable raw materials. All these aspects make nanofibers or nanocellulose in general very attractive for the development of new composites (Seydibeyoglu and Oksman 2008; Habibi *et al.* 2010). The final treatments of the nanofibers – the purification process or chemical modifications – are very important since, depending on the treatments, the cellulose chain can change its structure, surface and composition, affecting its interaction with the matrix.

There is a large range of composites that could be reinforced with nanocellulose. Souza *et al.* (2010) tested MFC as reinforcing component for PVA, combining the properties of strength and transparency, both important properties for this polymer, for example when used for the fabrication of

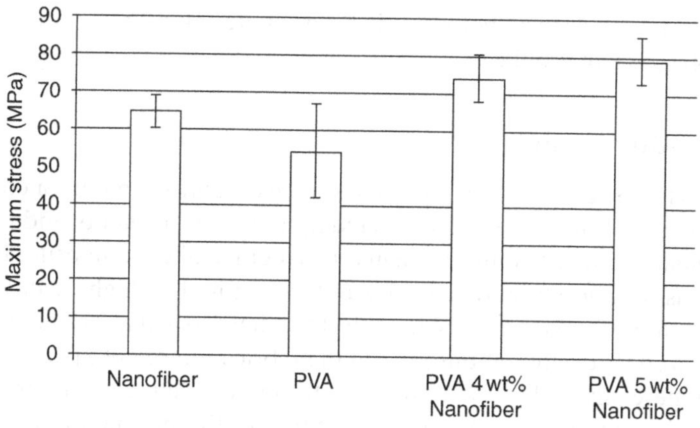

22.15 The results of maximum stress after the incorporation of MFC obtained from curaua fibers. Reprinted from 'Nanocellulose from curaua fibers and their nanocomposites', Souza, S. F., Leão, A. L., Cai, J. H., Crystal, W., Sain, M. and Cherian, B. M., *Molecular Crystal and Liquid Crystals*, 522, 42–52, 2010, reprinted by permission of the publisher, Taylor & Francis Ltd.

contact lenses. The maximum stress is shown in Fig. 22.15, where the increases of 36% and 44% for maximum stress followed the addition of 4% and 5% MFC into PVA.

Even so, not all composites reinforced with nanofibers are totally successful. Today there are still problems with the use of hydrophilic fibres in hydrophobic matrices and the consequent need to remove water from the suspension. One good example is the trial production of ethylene vinyl acetate composites, where nanowhiskers extracted from curaua fibers were utilized for the development of nanocomposites. Corrêa (2010) used the extrusion process to incorporate curaua nanowhiskers (1–7 wt%) in ethylene vinyl acetate. Their properties were compared with composites reinforced with mercerized fibers (2 mm) with a content of 1–30 wt%. It was possible to recognize poor dispersion of nanowhiskers due to the clusters created during the water removal in suspension and their tendency to agglomerate. This nanocomposite gave just a small increase in elastic modulus with a decrease in the strain at break. The best mechanical properties were obtained from mercerized curaua, with an increase at the elastic modulus better than that obtained with nanowhisker.

Some studies have been undertaken with the objective of applying cellulose nanofibers in the biomedical field. This area requires a great amount of testing before any use, since it is important to understand the behavior of this material inside the human body. The genotoxic effects of cellulosic nanowhiskers obtained from curaua fibers were tested *in vitro* using plant and animal cells by Lima *et al.* (2012) who identified that certain aggregations of nanofibers induce a reduction in toxicity. Therefore, further studies are necessary before using this material as nanofiber and not aggregated as microfiber.

22.6 Conclusion

Curaua fibers represent an important source of reinforcement for composite materials in different matrices, including polymers, cements and others. It is possible to identify improvement due to curaua fiber composites in elastic modulus, tensile strength and flexural strength. The high-quality curaua fiber is in use in the automobile industry, where its value has already been recognized. The value of this fiber in the fabrication of auto parts has made it commercially viable for companies to invest in larger-scale production. Moreover, curaua fiber has great potential in the production of cellulose nanofibers because of its high cellulose content. These nanofibers can be produced using different methods, giving varying structures, including nanowhiskers, nanofibers and/or nanofibrillated cellulose. In conclusion, curaua nanofibers can be utilized as reinforcements for several different matrices, improving the qualities of the composite, increasing their mechanical properties and even keeping them transparent for optical applications.

22.7 References

Araújo, J. R., Mano, B., Teixeira, G. M., Spinacé, M. A. S. and De Paoli, M. A. (2010) Biomicrofibrillar composites of high density polyethylene reinforced with curaua fibers: Mechanical, interfacial and morphological properties, *Compos Sci Technol*, 70, 1637–1644.

Araújo, J. R., Adamo, C. B., Costa Silva, M. V. and De Paoli, M. A. (2013) Antistatic-reinforced biocomposites of polyamide-6 and polyaniline-coated curaua A fibers prepared on a pilot plant scale, *Polym Composite*, 34, 1081–1090.

Ass, B. A. P., Belgacem, M. N. and Frollini, E. (2006) Mercerized linters cellulose: Characterization and acetylation in *N,N*-dimethylacetamide/lithium chloride, *Carbohyd Polym*, 63, 19–29.

Borsoi, C., Scienza, L. C. and Zattera, A. J. (2013) Characterization of composites based on recycled expanded polystyrene reinforced with curaua fibers, *J Appl Polym Sci*, 128, 653–659.

Brinchi, L., Cotana, F., Fortunati, E. and Kenny, J. M. (2013) Production of nanocrystalline cellulose from lignocellulosic biomass: Technology and applications, *Carbohyd Polym*, 94, 154–169.

Campos, A., Corrêa, A. C., Cannella, D., Teixeira, E. M., Marconcini, J. M., Dufresne, A., Mattoso, L. H. C., Cassland, P. and Sanadi, A. R. (2013) Obtaining nanofibers from curaua and sugarcane bagasse fibers using enzymatic hydrolysis followed by sonication, *Cellulose*, 20, 1491–1500.

Caraschi, J. C. and Leão, A. L. (2000) Characterization of curaua fiber, *Mol Cryst Liq Cryst*, 353, 149–152.

Corrêa, A. C. (2010) Preparation of cellulose nanofibers from curaua fibers to develop EVA/cellulose nanocomposites. Doctorate thesis, São Carlos, Brazil: Federal University of São Carlos.

Corrêa, A. C., Teixeira, E. M., Pessan, L. A. and Mattoso, L. H. C. (2010) Cellulose nanofibers from curaua fibers, *Cellulose*, 17, 1183–1192.

D'Almeida, A., Toledo Filho, R. and Melo Filho, J. (2010) Cement composites reinforced by short curaua fibers, *Materia*, 15, 151–156.

Ferreira, A. S., Lopes, F. P. D., Monteiro, S. N. and Satyanarayana, K. G. (2010) Charpy impact resistance of alkali treated curaua reinforced polyester composites, *Materia*, 15, 131–137.

Gomes, A., Goda, K. and Ohgi, J. (2004) Effects of alkali treatment to reinforcement on tensile properties of curaua fiber green composites, *JSME Int J A – Solid M*, 47, 541–546.

Gomes, A., Matsuo, T., Goda, K. and Ohgi, J. (2007) Development and effect of alkali treatment on tensile properties of curaua fiber green composites, *Composites: Part A*, 38, 1811–1820.

Habibi, Y., Lucia, A. L. and Rojas, O. J. (2010) Cellulose nanocrystals: Chemistry, self-assembly, and applications, *Chem Ver*, 110, 3479–3500.

Leão, A. L., Rowell, R. and Tavares, N. (1998) Applications of natural fibers in automotive industry in Brazil – thermoforming process, *Sci Technol Adv Mat*, 755–760.

Leão, A. L., Machado, I. S., Souza, S. F. and Soriano, L. (2009) Production of curaua fibers for industrial applications: Characterization and micropropagation, *Acta Hortic*, 822, 227–238.

Lima, R., Feitosa, L. O., Maruyama, C. R., Barga, M. A, Yamawaki, P. C., Vieira, I. J., Teixeira, E. M., Corrêa, A. C., Mattoso, L. H. C. and Fraceto, L. F. (2012) Evaluation of the genotoxicity of cellulose nanofibers, *Int J Nanomed*, 7, 3555–3565.

Lopes, F. P. D., Ferreira, A. S., Rodriguez, R. J. S., Castillo, T. E. and Monteiro, S. N. (2011) Characterization of thermal behavior of polyester composites reinforced with curaua fibers by differential scanning calorimetry. *TMS 2011 Annual Meeting and Exhibition*, EPD Congress 2011, The Minerals, Metals and Materials Society.

Monteiro, S. N., Lopes, F. P. D., Nascimento, D. C. O., Ferreira, A. S. and Satyanarayana, K. G. (2013) Processing and properties of continuous and aligned curaua fibers incorporated polyester composites, *J Mater Sci Technol*, 2, 2–9.

Mothé, C. G. and Araújo, C. R. (2004) Caracterização térmica e mecânica de compósitos de poliuretano com fibras de curauá, *Polimeros*, 14, 274–278.

Ornaghi Jr, H. L., Silva, H. S. P., Zattera, A. J. and Amico, S. C. (2012) Dynamic mechanical properties of curaua composites, *J Appl Polym Sci*, 125, 110–116.

Paiva, J. M. F., Wanderson, G. T. and Frollini, E. (1999) Compósitos de Matriz termofixa fenólica reforçada com fibras vegetais, *Polimeros*, 9, 170–176.

Rossa, L. V., Scienza, L. C. and Zattera A. J. (2013) Effect of curaua fiber content on the properties of poly(hydroxybutyrate-*co*-valerate) composites, *Polym Composite*, 34, 450–456.

Santos, P. A., Spinacé, M. A. S., Fermoselli, K. K. G. and De Paoli, M. A. (2009) Efeito da forma de processamento e do tratamento da fibra de curauá nas propriedades de compósitos com poliamida-6, *Polimeros*, 19, 31–39.

Satyanarayana, K. G., Guimarães, J. L. and Wypych, F. (2007) Studies on lignocellulosic fibers of Brazil, *Composites*, 38, 1694–1709.

Seydibeyoglu, M. O. and Oksman, K. (2008) Novel nanocomposites based on polyurethane and micro fibrillated cellulose, *Compos Sci Technol*, 68, 908–914.

Silva, R. V., Aquino, E. M. F., Rodrigues, L. P. S. and Barros, A. R. F. (2009) Curaua/glass hybrid composite: The Effect of water aging on the mechanical properties. *J Reinf Plast Comp*, 28, 1857–1868.

Siqueira, G., Bras, J. and Dufresne, A. (2010) Luffa cylindrical as a lignocellulosic source of fiber, microfibrillated cellulose, and cellulose microcrystals, *Bioresources*, 5(2), 725–740.

Souza, S. F., Leão, A. L., Cai, J. H., Crystal, W., Sain, M. and Cherian, B. M. (2010) Nanocellulose from curaua fibers and their nanocomposites, *Mol Cryst Liq Cryst*, 522, 42–52.

Spinacé, M. A. S., Lambert, C. S., Fermoselli, K. K. G. and De Paoli, M. A. (2009a) Characterization of lignocellulosic curaua fibers, *Carbohyd Polym*, 77, 47–53.

Spinacé, M. A. S., Fermoseli, K. K. G. and De Paoli, M. A. (2009b) Recycled polypropylene reinforced with curaua fibers by extrusion, *J Appl Polym Sci*, 112, 3686–3694.

Taipina, M. O., Ferrarezi, M. M. F. and Gonçalves, M. C. (2012) Morphological evolution of curaua fibers under acid hydrolysis, *Cellulose*, 19, 1199–1207.

Tomczak, F., Satyanarayana, K. G. and Sydenstricker, T. H. D. (2007) Studies on lignocellulosic fibers of Brazil: Part III – Morphology and properties of Brazilian curaua fibers, *Composites*, 38, 2227–2236.

Trindade, W. G., Paiva, J. M. F., Leão, A. L. and Frollini, E. (2008) Ionized-air-treated curaua fibers as reinforcement for phenolic matrices, *Macromol Mater Eng*, 293, 521–528.

Volkswagen (2007) The Volkswagen Fox 'grows' its roof, in Zah, R., Hischier, R., Leão, A. L. and Braun, I., Curaua fibers in the automotive industry and the sustainability assessment, *J. Clean Prod*, 15, 1032–1040.

Zah, R., Hischier, R., Leao, A. L. and Braun, I. (2007) Curaua fibers in the automotive industry and the sustainability assessment, *J Clean Prod*, 15, 1032–1040.

Index

CPSIA information can be obtained at www.ICGtesting.com
Printed in the USA
LVOW04*2020140914

404040LV00001B/24/P

9 781782 421221